The Baltic State

Since the end of the Cold War there has been an increased interest in the Baltics. *The Baltic States* brings together three titles, *Estonia, Latvia* and *Lithuania*, to provide a comprehensive and analytical guide integrating history, political science, economic development and contemporary events into one account. Since gaining their independence each country has developed at its own pace with its own agenda and facing its own obstacles.

The authors examine the tensions accompanying a post-communist return to Europe after the long years of separation, and how each country has responded to the demands of becoming a modern European state. Estonia was the first of the former Soviet republics to enter membership negotiations with the European Union in 1988 and is a potential candidate for the next round of EU expansion in 2004. Lithuania and Latvia have also expressed their desire for future membership of NATO and the EU.

This book will be of particular use for courses on Eastern Europe and the Baltics. Its structure and detailed country studies, will also interest students and researchers in comparative politics and international relations.

David J. Smith is a lecturer in Contemporary History and International Relations and a member of the Baltic Research Unit at the Department of European Studies, University of Bradford, UK. **Artis Pabriks** is a lecturer at the University of Latvia and Vidzeme University College, Latvia and a frequent political analyst for Latvia's mass media. **Aldis Purs** is a lecturer at Vidzeme University College, Latvia, having previously completed a Research Fellowship at the Woodrow Wilson Center for International Studies, USA. **Thomas Lane** is a lecturer in European History and a member of the Baltic Research Unit at the Department of European Studies, University of Bradford, UK.

The Baltic States

Estonia, Latvia and Lithuania

David J. Smith, Artis Pabriks,
Aldis Purs and Thomas Lane

ROUTLEDGE
Taylor & Francis Group

London and New York

First published 2002
by Routledge
11 New Fetter Lane, London EC4P 4EE

Simultaneously published in the USA and Canada
by Routledge
29 West 35th Street, New York, NY 10001

Routledge is an imprint of the Taylor & Francis Group

© 2002 David J. Smith, Artis Pabriks, Aldis Purs and
Thomas Lane

Printed and bound in Great Britain by
TJ International Ltd, Padstow, Cornwall

British Library Cataloguing in Publication Data
A catalogue record for this book is available from the British
Library

Library of Congress Cataloging in Publication Data
A catalogue record for this book has been requested

ISBN 0–415–28580–1

Estonia
independence and
european integration

David J. Smith

STOP OCCUPA

NIA NEVER JOINED

Estonia

Postcommunist States and Nations

Books in the series

Belarus: A denationalized nation
David R. Marples

Armenia: At the crossroads
Joseph R. Masih and Robert O. Krikorian

Poland: The conquest of history
George Sanford

Kyrgyzstan: Central Asia's island of democracy?
John Anderson

Ukraine: Movement without change, change without movement
Marta Dyczok

The Czech Republic: A nation of velvet
Rick Fawn

Uzbekistan: Transition to authoritarianism on the silk road
Neil J. Melvin

Romania: The unfinished revolution
Steven D. Roper

Lithuania: Stepping westward
Thomas Lane

Latvia: The challenges of change
Artis Pabriks and Aldis Purs

Estonia: Independence and European integration
David J. Smith

Bulgaria: The uneven transition
Vesselin Dimitrov

FOR THOMAS AND EMILY

TABLE OF CONTENTS

Chronology ix

Preface xi

Reflections on the Estonian 'New National Awakening' xix

Glossary of Abbreviations xxvii

Map of Estonia xxx

Map of the Estonian-Russian Border Area xxxi

1 'One Day There Will Be an Estonian State' 1

2 The Long Second World War: Estonia Under Occupation 1940–91 33

3 Old Wine in New Bottles: The Politics of Independence 65

4 'The Little Country that Could': Estonia's Economic Return to Europe 113

5 'The Devil and the Deep Blue Sea': Foreign Policy Between East and West 147

Bibliography 177

Index 189

CHRONOLOGY

March 1917	Formation of Estonian Provincial Assembly (Maapäev)
24 February 1918	Proclamation of the Estonian Republic
November 1918	Invasion by Soviet Russia
2 February 1920	Tartu peace treaty with Soviet Russia
12 March 1934	End of parliamentary democracy in Estonia
23 August 1939	Molotov-Ribbentrop Pact
28 September 1939	Estonia signs mutual assistance pact with USSR
16 June 1940	Estonia undergoes fully military occupation by USSR
July 1940	Puppet assembly proclaims Soviet power in Estonia
August 1940	Estonia annexed by USSR
June 1941	Soviet authorities deport 10,000 Estonian citizens
July–November 1941	German army occupies Estonia
Autumn 1944	Soviet army reconquers Estonia
1944–1953	Guerrilla war against Soviet occupier
1949	Authorities deport a further 60,000 from Estonia
1950	Johannes Käbin becomes CPE first secretary
1978	Karl Vaino becomes CPE first secretary
1980	Youth riots prompt 'letter of forty'
1987	'Phosphate Spring'
23 August 1987	Hirvepark demonstration. MRP-AEG formed
September 1987	IME plan unveiled
April 1988	Popular Front of Estonia formed
September 1988	Vaino Väljas becomes CPE first secretary
February 1989	Start of Citizens' Committee movement
23 August 1989	'Baltic Chain' to mark 50th anniversary of MRP
February 1990	Congress of Estonia elected
March 1990	Elections to ESSR Supreme Soviet
30 March 1990	ESSR Supreme Soviet declares start of transitional period leading to the restoration of independence
April 1990	Edgar Savisaar becomes prime minister
January 1991	Russia under Yeltsin recognises sovereignty of Estonia
20 August 1991	*De facto* restoration of the Republic of Estonia
6 September 1991	USSR recognises Estonian independence
17 September 1991	Estonia enters United Nations

15 October 1991	Estonia enters CSCE
November 1991	Law on Citizenship
January 1992	Tiit Vähi becomes prime minister
May 1992	Estonia signs trade and economic co-operation agreement with EU
June 1992	New constitution approved in referendum
June 1992	Introduction of the Kroon
September 1992	First post-communist parliamentary elections
September 1992	Lennart Meri elected president
October 1992	New Isamaa coalition government under Mart Laar
May 1993	Estonia admitted to the Council of Europe
July 1993	Pullapää Incident
June–October 1993	The 'Aliens Crisis'
February 1994	Estonia joins NATO Partnership for Peace programme
July 1994	Free trade agreement with the EU
31 August 1994	Russian troops withdraw from Estonia
September 1994	Mart Laar steps down following no-confidence vote
November 1994	Andres Tarand becomes prime minister
November 1994	Estonia becomes an associate partner of the WEU
February 1995	Second post-communist parliamentary elections
March 1995	New KMÜ-Centre coalition government under Tiit Vähi
June 1995	Estonia signs association agreement with the EU.
October 1995	Vähi government resigns following Savisaar scandal
November 1995	New KMÜ-Reform coalition government under Vähi
November 1995	Estonia submits application for full EU membership
October 1996	Lennart Meri re-elected president
November 1996	Reform Party withdraws from government
December 1996	Vähi forms new KMÜ minority government
February 1997	Vähi resigns as prime minister
April 1997	New KMÜ minority government under Mart Siimann
July 1997	Estonia obtains favourable avis from EU Commission
December 1997	EU admits Estonia to membership negotiations
January 1998	United States–Baltic Charter of Partnership
March 1998	Official opening of EU membership negotiations
March 1999	Third post-communist parliamentary elections
April 1999	Mart Laar becomes prime minister

During a visit to Tallinn in 1992, Dr Otto Von Habsburg made a speech in which he described the Estonians as 'the best of Europeans.'[1] Over the course of the past decade, the leaders of the restored Estonian Republic have done their utmost to live up to this epithet. But what does it mean to be 'European' at the start of the 21st century. When asked to write a work on post-communist transition in Estonia, I predictably chose to focus upon the dual—and often conflicting—imperatives of restoring national independence whilst simultaneously securing integration with European and Euro-Atlantic transnational organisations. This was the challenge outlined in the remarkably prescient conclusion to Rein Taagepera's work *Estonia: Return to Independence* published in 1993.

When I sat down to write in June 1997, Estonia was already firmly established as the pacesetter for economic and political reform in the Former Soviet Union (FSU). It received due confirmation of this status a month later, becoming the first former Soviet republic to be recommended for admittance to 'fast track' negotiations on membership of the European Union (EU). Since its admittance to these negotiations in 1998, this 'small' country—slightly larger than the Netherlands but with a population of only 1.5 million—has been hailed as the 'model pupil' amongst the current EU applicant states.[2]

For many Estonians, the label 'former Soviet republic' applied to their country has become an understandable source of irritation. In terms of international law the term is indeed a misnomer, for throughout 1940–91, Estonia was deemed a *de jure* independent republic under illegal occupation by the USSR. In this sense, it can be more properly characterised as one of the states of Central and Eastern Europe (CEE). As was the case with Poland, Hungary and Czechoslovakia, Soviet-imposed communism never acquired any genuine legitimacy amongst an Estonian population which had previously experienced 'an alternative non-Soviet and self-determined national existence' during 22 years of independent statehood from 1918–40.[3] This interwar experience is crucial to understanding contemporary state-building and the sustained efforts to 'return to Europe' since 1991. Von Habsburg's comments are entirely consistent with the discourses of current Estonian political leaders, who portray the Soviet period as a hiatus which artificially divided the nation from its western neighbours. In cultural terms, they insist, the Estonians never ceased to be members of the 'family of democratic European nations' after 1940.[4]

The current work begins by examining the formation of the Estonian state following the 'national awakening' of the late 19th–early 20th centuries. Whereas today, independence is often hailed as the fulfilment of the nation's destiny, the poet Juhan Liiv's claim that "one day there will be an Estonian state" sounded fanciful when he pronounced it back in 1910. Indeed, the very concept of an Estonian 'nation'—whilst clearly rooted in a pre-existing ethnic consciousness and a mythologised past— was at that time still fairly recent, having entered political discourse only from the 1850s onwards.

After exploring the origins and development of this national consciousness, Chapter One goes on to consider the period of Estonian Statehood from 1918–40. Whilst it is true that the interwar Estonian Republic did not prove fertile ground for liberal democracy, the same can be said for most other European states during these years. More notable were the achievements in terms of building a viable economy and durable sense of national identity which survived the 'long Second World War' of 1940–91.

After giving a brief account of the sovietisation process in Estonia, Chapter Two considers how the interwar republic—and the circumstances surrounding its demise—came to serve as the ideal for the national movements which sprang up to challenge Soviet power during the late 1980s. Anthony D. Smith writes that the mobilising power of nationalism is "immeasurably increased by the living presence of traditions embodying memories, symbols and values from earlier epochs in the life of a population, community or area".[5] This was as true for the period 1987–91 as it had been for the last decades of Tsarist rule. Yet if previous experiences of statehood provided the essential content for the late-Soviet 'new national awakening', the factors facilitating the resurgence of Estonian nationalism must ultimately be sought in the structure of the Soviet state and the actions of its leaders after 1985 (see the brief introduction to Chapters One and Two, entitled 'Reflections on the Estonian New National Awakening').

By the same token, the restoration of statehood according to the principle of legal continuity could not efface the factual legacies— political, economic, social, cultural, environmental and demographic— left behind by fifty years of sovietisation. In this sense, the use of the term 'post-Soviet' with regard to Estonia is not only valid, but makes the subsequent progress in domestic reform and European integration appear all the more remarkable.

Chapters Three to Five chart the main developments from 1991–99, focusing on constitutional reform and the politics of independence; marketisation of the economy; and foreign affairs respectively. Whilst these chapters seek to uncover those factors which have served to differentiate Estonia's development from that of other FSU states since 1991, they also explore the contradictions and tensions inherent in the concept of 'returning to Europe'. The central goal of the late-Soviet independence movement was to restore sovereign nation-statehood, an aim which had been supported—at least in principle—by western governments throughout the years of the Cold War. Seen in these terms, 'return to Europe' becomes essentially synonymous with ending Soviet occupation and "getting rid of all those elements that had deprived the ... [Estonians] ... of true and undistorted self-expression" throughout 1940–91.[6] Yet, the Europe to which the Estonia 'returned' following the collapse of communism was very different from the one it had left behind fifty years earlier. As Taagepera noted back in 1993, the challenge facing the Estonian was that of "independence in an interdependent world", a world in which economic globalisation and the "growth to prominence of a wide range of ... transnational institutions confront the model of a world divided into sovereign states".[7]

To be sure, interdependence was also an inescapable fact of life during 1918–39, a period when, like today, the Estonian Republic sought to anchor its independence within the broader economic and political framework of the 'New Europe'. Nevertheless, the conditions which exist in the post-Cold War Europe of the 1990s mean that national governments enjoy significantly less room for maneouvre than they did between the wars. In their domestic and foreign policies, successive Estonian administrations have prioritised the goal of integration with the 'Euro-Atlantic Space', the network of transnational organisations which are seen to have laid the foundations for prosperity and stability in western Europe after 1945. Entry to these organisations has been deemed integral to banishing the 'unilateral dependence' on Russia and the FSU bequeathed by sovietisation. Yet existing and prospective membership of bodies such as the Conference (later Organisation) for Security and Cooperation in Europe (CSCE/OSCE), the Council of Europe (CE), NATO and—most especially—the EU has also entailed the acceptance of significant external constraints on state-building. These have not always been welcomed in a country which has so recently dispensed with suzerainty by Moscow, and do much to explain the apparent lack of enthusiasm for EU membership on the part of the Estonian public as a whole.

This divergence between the goals of the political leadership on the one hand and public opinion on the other is but one facet of the problematic relationship between state and society bequeathed by sovietisation. Although Estonia has made rapid progress in building democratic state institutions, these have yet to be consolidated through the emergence of a functioning civil society along western lines. Such considerations account for the title of Chapter Three. Vihalemm, Lauristin and Tallo have spoken of 'new content' being poured into 'old molds'. I prefer to use the analogy of 'old wine in new bottles', a term borrowed and adapted from a thought-provoking article by Jeff Richards.[8]

In addition to the challenges posed by globalisation and the growing influence of transnational institutions, many contemporary European states are also having to contend with the reassertion of sub-state regional and national identities. The question of the state's relationship to 'non-titular' national groups living within its borders has assumed a particular relevance in Estonia, where the legal restoration of the interwar republic meant that automatic citizenship rights were not extended to the large, mainly Russian-speaking population of Soviet citizens which had settled in the country during 1940–91. Policy towards these Soviet-era settlers is also a recurrent theme of this work, partly because it has been central to my own research over the past few years, but mainly because for most western commentators, it has constituted the major point of controversy regarding Estonia's post-communist development. Indeed, it is this issue which most graphically illustrates the underlying tension between the desire for membership of Euro-Atlantic institutions and the project of restoring a sovereign nation-state.

Estonia's remarkable ability to contain ethnic unrest over the past decade has belied initial predictions by outside commentators, many of whom were all too ready to draw ill-judged and often tendentious parallels with the situation in former Yugoslavia. By the same token, early pessimism with regard to Estonia's economic prospects was laid to rest by the reforms launched during 1992–93. Radical measures to stabilise, liberalise and privatise the economy created the conditions for sustained growth over the following five years, a period when Estonia earned the epithet of 'the little country that could.' Consequently, when the European Commission formulated its avis in July 1997, Estonia's market economy was deemed better-equipped than those of its Baltic neighbours in terms of dealing with the competitive pressures of EU membership. Exposure to the world economy carries costs as well as benefits, as became clear when the shock-waves from successive financial crises in East Asia and Russia hit Estonia during 1997–98. The effects of the Russian 'meltdown', one should add, also underlined the continued importance of FSU markets to certain sectors of the Estonian economy since independence. Neither crisis, however, has seriously compromised Estonia's hard-earned international reputation for sound economic management. The ability to initiate and sustain 'shock therapy' can be attributed at least partly to the political marginalisation of the Russian-speaking settler community, the group which had most to lose from the transition to a market economy. At the same time, it has been possible to speak of an overarching consensus regarding the direction of economic development, at least during the initial stages of reform. When questioned, most inhabitants of Estonia still rate the new market economy more highly than they do that of the old Soviet Union or the new Russia. Even so, the growing income disparities caused by marketisation have inevitably proved a source of social tension, all the more so now that the 'extraordinary politics' of the early 1990s has largely run its course.

The biggest question mark over Estonia's future development relates not to internal affairs but to the external situation, most notably the country's problematic relations with Russia. Domestic policy towards the Russian-speaking settler population, for instance, rests largely on the contention that the 'Eastern Neighbour' poses an active threat to Estonian independence. This contention has been reinforced by ongoing political conflicts in Russia, where the fate of 'compatriots' abroad has been consistently invoked as justification for an active sphere-of-influence policy within the former Soviet space. For Estonian leaders, historically-rooted fear and suspicion of Russia is offset by a desire for 'normalisation' of inter-state relations, most notably in the sphere of trade. In this regard, the Estonians' dilemma is little different to that faced by the EU and—more especially—NATO as they contemplate enlargement to the East. On the one hand, these organisations have to address the legitimate security concerns and aspirations of CEE applicant countries. At the same time, they have been understandably reluctant to antagonise Moscow lest this create a new East-West line of division in the post-Cold War Europe.

The Estonians have needed little reminding that their long-term security will rest in large part upon a successful resolution of the 'Russian question'. Even so, the lack of any clear and unequivocal enlargement policy on the part of organisations such as NATO has made the latter a 'moving target' for the CEE applicant countries.[9] The same was true of the period 1918–21, when western leaders continued to link the Estonians' fate to the overall evolution of Russia. This perennial problem reflects the country's location at the crossroads of East and West, a position which former foreign minister Jüri Luik has termed "between the devil and the deep blue sea". I have adopted this as the title for Chapter Five, which outlines the main developments in foreign and security policy since 1991 whilst drawing comparisons with the situation between the wars. This approach constitutes a departure from the otherwise largely chronological structure of the work, and some readers may consider the parallels to be overstated. Nonetheless, the frequent references to the inter-war period in recent foreign policy discourse underline the extent to which these years have influenced the thinking of Estonia's leaders since 1990.

I am deeply grateful to the many people who have provided me with help, advice and inspiration during the preparation of this work. In particular, I wish to thank my colleagues John Hiden and Thomas Lane of the Baltic Research Unit for their continued support and their valuable comments on the initial draft. Thanks also to Jeffrey Harrop for his comments on the economics chapter, and to Einar Ligema for his patient efforts in teaching me to read Estonian. In the course of my research I was extremely fortunate to receive guidance from the late Graham Smith, whose work will always remain a tremendous source of inspiration. Trips to Estonia would have been considerably more difficult and less enjoyable had it not been for the generous support provided by the Eesti Instituut and its representatives Krista Kaer and Tiina Lats. I an also indebted to Rein Ruutsoo for many stimulating conversations during my visits to Tallinn. In recent years, I have benefited greatly from my work with Marko Lehti of the University of Turku, Finland. He too has provided many useful comments on the first draft of this book. My association with Turku has proved especially fruitful. In this connection, I also wish to thank Esa Sundbäck, formerly head of the Baltic Sea Region Studies Programme, and Professor Kalervo Hovi of the Department of General History. As always, my biggest thanks go to Sanna and to all my family, who have given me the things that matter most.

1   Norman Davies, *Europe: A History* (London, 1997), p.944.
2   *Helsingin Sanomat*, 13 November 1999.
3   V. Stanley Vardys, "Modernisation and Baltic Nationalism," *Problems of Communism*, September–October 1975, p.36.
4   Lennart Meri, "Iseseisvus Laiemas Kontekstis," in Lennart Meri, *Presidendikõned* (Tartu, 1996), p.227; Toomas H. Ilves, statement at the opening of Estonia's negotiations with the European Union, Brussels, 31 March 1998. *http://www.vm.ee/eng/index.html/*.

5   Anthony D. Smith, *National Identity* (London, 1991), p.20

6   Pertti Joenniemi, "The Baltic Countries as Deviant Cases; Small States in Search of Foreign Policies," in *New Actors on the International Arena: The Foreign Policies of the Baltic Countries*, eds. Pertti Joenniemi and Peeter Vares (Tampere, 1993), p.188–189.

7   Rein Taagepera, *Estonia: Return to Independence* (Boulder, 1993), pp.215–232; Joseph A. Camilleri and Jim Falk, *The End of Sovereignty? The Politics of a Shrinking and Fragmenting World* (Aldershot, 1992), p.7.

8   Peeter Vihalemm, Marju Lauristin and Ivar Tallo, "Development of Political Culture in Estonia," in *Return to the Western World: Cultural and Political Perspectives on the Estonian Post-Communist Transition*, eds. Marju Lauristin and Peeter Vihalemm with Karl Erik Rosengren and Lennart Weibull (Tartu, 1997), p.204; Jeff Richards, "Old Wine in New Bottles: The Resurgence of Nationalism in the New Europe," in *Ethnicity and Nationalism in Russia, the CIS and the Baltic States*, eds. Christopher Williams and Thanasis D. Sfikas (Aldershot, 1999), pp.11–23.

9   This analogy has been used in a slightly different context by Judy Batt and Kataryna Wolczuk within the context of their current ESRC-funded research project on "Fuzzy Statehood and European Integration" (CREES, University of Birmingham, 1998).

The current period of Estonia's existence as a sovereign state dates from 20 August 1991. Amidst the dramatic events then unfolding in Moscow, the first—and only—freely-elected parliament in Soviet Estonia gathered in extraordinary session to affirm the republic's *de facto* independence from the USSR. The subsequent collapse of the Moscow *putsch* opened the way to full diplomatic recognition by the international community, a process crowned by Estonia's admission to the United Nations during October. Yet, in legal terms at least, this new actor on the international arena does not belong to the ranks of the Soviet 'successor states'. Indeed, in the minds of most of its political leaders and the majority of its population, the Estonian state was already well into its 73rd year by the time of the Soviet collapse. The present-day Republic of Estonia (*Eesti Vabariik*) traces its existence back to 24 February 1918. On this day—still officially commemorated as the anniversary of independence—representatives of the Estonian Provincial Assembly (*Maapäev*) declared Estonia an 'independent and democratic republic' within its 'historical and ethnographic borders'.[1] Independence was later cemented by the Tartu treaty of 2 February 1920, when Russia renounced in perpetuity any claim on the territory of Estonia. On this basis, Estonia proceeded to obtain full diplomatic recognition by the western powers and entry to the League of Nations over the following three years.

The fortunes of the inter-war Estonian Republic rose and fell with those of the 'New Europe' of which it formed an integral part. In the tense atmosphere of September 1939, the Soviet Union proved to be no respecter of the international agreements entered into with the Baltic states during the previous two decades. On the pain of invasion by the forces of the Red Army massed on its eastern border, Estonia was required to consent to an ultimatum demanding the establishment of Soviet military bases on its territory. Latvia and Lithuania gave way to similar ultimatums during the month which followed. In June 1940, Estonia and its neighbours underwent full occupation by Soviet troops, a development which served as the prelude to the forcible incorporation of the three countries into the USSR. With the exception of states such as Nazi Germany and Sweden, the international community condemned this illegal annexation and never gave legal recognition to Soviet rule over the Baltic states. This policy of non-recognition gave rise to the principle of legal continuity, which held that *de jure*, Estonia remained an independent state under illegal occupation throughout the period 1940–91.

Thus, when the Estonian parliament (Supreme Council) gathered at the height of the Moscow coup, it did not formally declare independence from the USSR, for, in the eyes of a majority of its members, Estonia had never legally formed part of the Soviet state. On 30 March 1990, the Supreme Council had declared an end to Soviet power in Estonia and the start of a transition period during which independence was to be restored through negotiations with Moscow. The coup was deemed to have rendered such negotiations impossible, whilst posing a threat to the democratic processes enacted in Estonia. The Supreme Council resolution of 20 August 1991, carried by 73 votes to 0, affirmed the continuity of the 1918 Republic as a subject of international law.[2] On this basis, longer-established states simply restored pre-war diplomatic links rather than recognising a new, post-Soviet 'Third Republic' as a successor to the Estonian Soviet Socialist Republic (ESSR) founded in 1940.

The principle of legal continuity carried not just immense symbolic but also great practical significance for the maintenance of the Estonian national idea after 1940. The memory of the interwar republic was kept alive by a large and well organised community of émigrés—including a large part of the surviving political and intellectual élite—who fled the country prior to its reoccupation by the USSR in 1944. Acting head of state Jüri Uluots and his successors presided over a government-in-exile which met in Oslo right up until 1992. Pre-war diplomatic legations continued to function in the United States, where émigrés founded the Free Estonia Committee during the 1950s. The activities of this and other organisations reminded the international community that Estonia's incorporation into the Soviet state had been anything but voluntary.[3] If further confirmation were needed, it was provided by the 'Forest Brothers' (*Metsavennad*), a guerrilla movement which waged war against occupying Soviet forces from 1944–53. This struggle was conducted in the expectation of an imminent East-West conflict which would lead to the liberation of the Baltic states by the United States and its European allies. The growing futility of these hopes, combined with savage repression by Soviet security forces led to the gradual decline of the movement from the late 1940s onwards. The spirit of resistance, however, lived on in Estonia through the activities of the anti-Soviet dissident movement, which adopted legal continuity as its guiding principle. Many of these long-standing dissidents were to play a key role in the independence movements which emerged during 1987–91. Seen in this light, the experience of the Estonians after 1940 becomes a narrative of national struggle and eventual emancipation from the Soviet 'prison of nations'.[4]

That the Stalinist regime set out to eradicate all traces of the independent Estonian Republic is hardly open to question. Through executions, mass deportations and the thorough-going sovietisation of local society, it effectively decimated those classes and institutions most closely associated with the development of Estonian national identity during 1918–40. As a result of the losses incurred through war, executions, deportation, and flight of refugees during 1939–1944, Estonia is reckoned to have lost 18% of its population. The resultant gap in the local population was filled by an influx of migrant workers from western Russia and Ukraine, 145,000 of whom arrived during the period 1945–49 alone.[5] The task of these new settlers was to build up a heavy industrial base and lock Estonia firmly into the centralized structures of the all-union economy. The mass immigration which began during this period was also seen as desirable from the ideological standpoint of 'internationalising' local society and culture. Soviet theorists envisaged an inexorable process of rapprochment between the different nationalities leading to their eventual merger into a new 'Soviet People'.[6]

Yet, whereas the economic policies of the USSR were largely consistent with this aspiration, the political structure of the Soviet state betrayed its persistent failure to get to grips with the 'national question'. Considerations of political expediency dictated the creation of a federalist model with a complex, four-tiered division of the Soviet state into nationally-defined territorial units.[7] Following its annexation, Estonia more or less retained its existing territorial borders and was transformed into one of 16 'union republics' (SSRs), each of which bore the name of an ethnically-defined national group. These were, as Brubaker has noted, nothing other than nation-states-in-waiting.[8]

This model of ethnoterritorial federalism was meant to be a transitional stage on the road to the new transcendent Soviet identity. Instead, it became a system which 'nourished cultural uniqueness yet denied its expression'.[9] The existence of union republics provided a framework for the continued cultural development (in Soviet parlance *rassvet*) of the various nationalities within the Soviet state. The ESSR possessed its own autonomous institutions, including a republican government and parliament which were in principle entitled to initiate secession from the USSR. Prior to the late 1980s, this apparent sovereignty remained no more than a constitutional fiction, since the hegemonic political control exercised by the all-union Communist Party (CPSU) meant that real decision-making power resided in Moscow. Prior to the onset of Gorbachev's *perestroika*, the CPSU central leadership and its repressive organs moved swiftly to quash any manifestation of political nationalism, not merely by dissidents, but also on the part of local communist elites in the individual republics.

In their recent work on Lithuania, V. Stanley Vardys and Judith Sedaitis convincingly bring out 'the essential role of coercion critical to ... [the Soviet] ... system.'[10] Yet the policies pursued by the Soviet regime in the individual republics, whilst clearly anti-*nationalist*, were far from being anti-*national*. In the case of Estonia, the nature of Soviet federalism perpetuated the existing link between the ethno-culturally defined Estonian nation and the territory of the Estonian republic. Misiunas and Taagepera, writing in 1980, warn us not to underestimate the boost to national identity supplied by the mere existence of the ESSR. The cultural autonomy granted to the republic included the possibility to maintain a publishing industry and a system of secondary and higher education in the Estonian language. The generations of Estonians who grew up under Soviet rule were thus able to receive schooling in their native tongue and, on this basis, find employment in their native republic. In this way, the Estonians further extended their 'historical depth' as a modern nation in the years after 1940.[11] Most importantly, the Soviet period saw the emergence of new strata of administrative cadres and humanistic intelligentsia who were to become the standard bearers of the Estonian national idea during the late 1980s.

Under these circumstances, the economic policies pursued by the Soviet regime only served to exacerbate national differences rather than bringing about their disappearance. In a situation where almost three quarters of Estonian industry was directly administered by all-union ministries in Moscow, the republican government lacked control over its own resources. Industrial projects in the ESSR were thus conceived to serve the needs of the all-union economy as a whole, without any reference to local needs. The Soviet model of extensive economic development caused massive environmental degradation and brought a continued influx of migrant workers to the ESSR during the 1960s and 1970s. The fact that Soviet censuses and internal passports classified citizens according to their ethnic origin as well as place of residence further institutionalised the boundaries between the Estonian 'titular nation' and 'non-Estonians' not readily identifiable with the territory of the ESSR. Soviet census figures tell their own story. In the four decades after the war, immigration engineered a dramatic demographic shift, with the ethnic Estonian share of the ESSR's population declining from 94% in 1945 to 61% in 1989. By the 1970s, it was widely believed that the titular Estonian nation would soon be reduced to a minority within its own administratively-recognised ethnic homeland.

The primarily Russian-speaking settler population was under no compulsion to learn the Estonian language or integrate into the local cultural milieu. Most settlers worked in all-union enterprises in which little or no Estonian was spoken, whilst their children attended Russian-language schools built specifically to serve their needs. Whereas little or no Estonian language was taught in these schools after 1959, an increasing accent was placed upon the study of Russian in Estonian-language schools. As a result, some 80–90% of Estonians were conversant in Russian by the late 1980s, whilst only 12% of the non-Estonian population could claim fluency in the local language.[12] In the light of these facts, the growing tide of immigration raised the spectre of ultimate russification. Should Estonians become a minority of the total population, it would appear a logical step on the part of the central authorities to deprive the Estonian Republic of its SSR status—with all of its concomitant benefits in terms of cultural autonomy and employment—and amalgamate the territory to the neighbouring RSFSR. A precedent for this already existed in the form of the heavily russified Karelian-Finnish SSR, which in 1956 was stripped of its union republic status and transformed into an autonomous republic of the Russian Federation.[13]

At the level of everyday life, settlers aroused resentment through their insistence on the use of Russian in day-to-day transactions and their privileged access to housing. For the most part, these ethnic tensions remained latent, although they occasionally spilled over into disturb-ances by disaffected youth. Until the mid 1980s, organised political nationalism remained the preserve of the small but growing dissident movement, whose principal activity consisted of circulating *samizdat'* publications and advertising the plight of the Estonians to international agencies such as the United Nations. Unofficial nationalist groups were subject to often brutal repression by the all-pervasive security apparatus of the Soviet state. By the turn of the 1980s, it was unclear how much authority they commanded amongst a population which took an increasingly cynical view of any form of political activity. According to one subsequent independence activist, the experiences of the late Brezhnev period had only served to demonstrate the futility of individual protest in the face of the monolithic power of the CPSU and its repressive organs.[14]

As soon as active repression of non-party groups ceased to be a central element of state policy, the crisis of Soviet power in Estonia could not be long delayed. By the time of Gorbachev's arrival as General Secretary of the CPSU in 1985, declining living standards in the hitherto relatively prosperous ESSR had added to the existing stock of grievances. Once Gorbachev's *Glasnost'* restored freedoms of expression and association, the structure of Soviet society virtually ensured that nationality would become the dominant basis for interest articulation. The restoration of these political freedoms in Estonia thus facilitated the emergence of mass nationalist movements demanding first autonomy and later full independence. The simultaneous democratisation of the political process allowed these movements to capture power at the republican level, where they now enjoyed the institutional resources to mount an effective challenge to Soviet rule. Whilst this pattern was apparent to a greater or lesser degree in all of the Soviet republics, nationalist movements in the Baltic republics possessed a number of distinguishing features. The first of these relates to the unparalleled degree of popular mobilisation amongst the Estonians, Latvians and Lithuanians. A second, related factor was the legal continuity principle which provided the frame of reference for the process of disengagement from Moscow. Rather than asking how they might leave the USSR, the Baltic national movements demonstrated why the USSR should leave them.

The pivotal role of the Baltic Popular Fronts in bringing about the collapse of the Soviet state fully exposed Stalin's folly of incorporating the independent Baltic republics rather than keeping them as nominally independent satellite states. In seeking to explain the scale of popular mobilisation in Estonia during 1988, authors have cited the residual traditions of civil society and individual initiative stemming from the pre-war experience of political democracy and a market economy.[15] After the initial Stalinist terror had run its course, the Soviet regime was never able entirely to eradicate these traditions. Nor, for that matter, did it necessarily seek to do so. Whilst the popular designation 'Soviet West' (*Sovetskii Zapad*) or even 'Our Abroad' (*Nasha Zagranitsia*) encompassed all three Baltic republics, it was especially applicable to Estonia. As the Soviet republic with the smallest population and the highest level of GDP per capita, the ESSR was accorded a status of 'experimental republic' as early as the 1950s. That is to say, it became a testing ground for new initiatives in economics and culture ahead of their eventual introduction to the Soviet Union as a whole.

A further distinguishing factor was Estonia's proximity—both geographic and cultural—to neighbouring Finland, which had the closest relationship to the USSR of any western state. The independent Estonian Republic had already enjoyed an intimate relationship with Finland during the interwar period. From the 1960s onwards, the diplomacy of Finnish president Urho Kekkonen facilitated a resumption of contacts. A regular ferry service between Helsinki and Tallinn brought a steady flow of Finnish tourists, lending a cosmopolitan flavour to the Estonian capital. During the same period, access to Finnish television opened up an 'electronic window on the West' for the residents of northern Estonia. This meant that 'when independence came, they knew exactly what a western economy should look like.'[16] Of more immediate significance was the fact that, despite enjoying the highest living standards of any Soviet nationality, the Estonians merely began to compare their position unfavourably to that of the independent western nations. In so doing, they were given further cause to ponder the divergent fates of Estonia and Finland during World War Two. Many older residents of Tallinn could still remember a time when their living standards had matched those of the Nordic countries. Now, in gazing across at one of the world's most prosperous economies, younger Estonians also began to rue what might have been had their country not succumbed to Soviet aggression in 1939–40. All of this helped further to strengthen the place of the interwar republic in the popular imagination. A consideration of this period thus serves as an essential prelude to discussions not only of the collapse of Communism in Estonia, but also of the processes of state and nation-building after 1991.

1   Mati Graf, *Eesti Rahvusriik* (Tallinn, 1993), p.240.
2   20 August Klubi ja Riigikogu Kantselei, eds. *Kaks Otsustavat Päeva Toompeal (19–20 August 1991)*, (Tallinn, 1996), p.64.
3   On the government in exile and emigre organisations, see Raimo Raig, "*Virolaisek Viron Ulkopuolella*," in *Viro: Historia, Kansa, Kulttuuri*, ed. Seppo Zetterberg (Helsinki, SKS, 1995),

pp.364–365.

4   For good examples of this perspective, see Mart Laar, Urmas Ots and Sirje Endre, *Teine Eesti* (Tallinn, 1996); Mart Laar, *War in the Woods: Estonia's Struggle for Survival 1994–1956* (Washington, 1993).

5   V. Stanley Vardys, "Modernisation and Baltic Nationalism," *Problems of Communism*, September–October 1975, p.36; Tönu Parming, "Population Changes in Estonia, 1935–1970," *Population Studies*, Vol. 26, March 1972, p.60.

6   Graham Smith, "The Soviet State and Nationalities Policy," in *The Nationalities Question in the Post-Soviet States*, ed. Graham Smith (London, Longman, 1996), pp.2–22.

7   Rogers Brubaker, *Nationalism Reframed. Nationhood and the National Question in the New Europe* (Cambridge, 1996), pp.30–32.

8   Ibid.

9   Ronald Suny, "Incomplete Revolution: National Movements and the Collapse of the Soviet Empire," *New Left Review*, No. 189, 1992, p.113.

10 V. Stanley Vardys and Judith Sedaitis, *Lithuania: The Rebel Nation* (Boulder, 1997), p.71.

11 Romuald Misiunas and Rein Taagepera, *The Baltic States: Years of Dependence 1940–1990* (London, 1993), p.273.

12 Toomas Hendrik Ilves, "Reaction: The Intermovement in Estonia," in *Towards Independence: The Baltic Independence Movements*, ed. Jan Arveds Trapans (Boulder, 1991), p.77.

13 Tönu Parming, "Population Processes and the Nationality Issue in the Soviet Baltic." *Soviet Studies*, Vol. 32, No. 3, July 1980, p.405.

14 Author interview with Rein Ruutsoo (Tallinn, 24 April 1996).

15 "The Popular Movements and the Soviet Union: Discussion." in *Towards Independence: The Baltic Independence Movements*, ed. Jan Arveds Trapans (Boulder, 1991), pp.43–53.

16 Riina Kionka, " The Estonians," in *The Nationalities Question in the Soviet Union*, ed. Graham Smith (London, 1991), p.42; quote from Ian Watson, *The Baltic & Russia Through the Back Door* (Riga, 1994), p.83.

GLOSSARY OF ABBREVIATIONS

BOE	Bank of Estonia
CAP	Common Agricultural Policy
CE	Council of Europe
CEE	Central and Eastern Europe
CIS	Confederation of Independent States
CPD	Congress of People's Deputies
CPE	Communist Party of Estonia
CPL	Communist Party of Lithuania
CPSU	Communist Party of the Soviet Union
CSCE	Conference for Security and Cooperation in Europe
EEK	Estonian Kroons
EERE	Eesti Erastamisettevõte: Estonian Privatisation Enterprise
EFTA	European Free Trade Area
EME	Eesti Maarahva Erakond: Estonia Rural People's Party
EMU	Economic and Monetary Union
EPA	Estonian Privatisation Agency
ERSP	Eesti Rahvusliku Sõltumatuse Partei: Estonian National Independence Party
ESDP	Estonian Social Democratic Party
ESSR	Estonian Soviet Socialist Republic
ETRL	Eesti Töötava Rahva Liit: The Estonian Working People's League
EU	European Union
EVL	Eesti Vabadussõjalaste Liit: League of Veterans of the War of Independence
FDI	Foreign Direct Investment
FSU	Former Soviet Union
IME	Isemajandav Eesti: Self-Managing Estonia
IMF	International Monetary Fund
KMÜ	Koonderakonna ja Maarahva Ühendus: Coalition and Country People's Union
MEP	Member of the European Parliament
MKE	Meie Kodu on Eestimaa: Our Home is Estonia
MRP	Molotov-Ribbentrop Pact
MRP-AEG	Molotovi-Ribbentropi Avalikustamise Eesti Grupp: Estonian Group for the Publication of the Secret Protocols of the Molotov-Ribbentrop Pact
NATO	North Atlantic Treaty Organisation
NKVD	Narodnii Komitet Vnutrennykh Del: People's Commissariat for Internal Affairs
OECD	Organisation for Economic Cooperation and Development
ONPE	Ob'edinennaia Narodnaia Partiia Estonii: United National Party of Estonia
OSCE	Organisation for Security and Cooperation in Europe
OSTK	Ob'edinennyi Soyuz Trudovykh Kollektivov: United Council of Work Collectives
RDM	Russian Democratic Movement

RSFSR	Russian Soviet Federal Socialist Republic
SMEs	Small and Medium-sized Enterprises
TCB	Tartu Commercial Bank
UN	United Nations
UWC	Union of Work Collectives
VEB	Vneshekonombank: Foreign Economic Bank of the USSR
WTO	World Trade Organisation

FINLAND

Gulf of Finland

Baltic
Sea
Paldiski
⊛ TALLINN
Narva
Kohtla-
Järve

Hiiumaa

Haapsalu

Vöhma

Lake
Peipus

Saaremaa

Pärnu
Tartu
Lake
Pskov

Vorts-
Järv

Irbe Vain

Valga

RUSSIA

Gulf of
Riga

LATVIA

| 0 | 50 | 100 km |
| 0 | 50 | 100 km |

Map of Estonia

Map of the Estonian-Russian border. The shaded area marks the territory transferred from Estonia to the RSFSR in 1945.

─┤├─┤├─┤├─	1920 Border
─ ─ ─ ─ ─	current physical border

Chapter 1

'ONE DAY THERE WILL BE AN ESTONIAN STATE'

In 1918, as in 1991, an Estonian state emerged from the internal collapse of a multinational empire. Unlike the recent peaceful transition to independence, however, the Estonian national movement of 1918–20 was required to assert its claims to self-determination by force of arms. In what is popularly referred to as the War of Liberation (*Vabadussota*), the hastily-improvised armed forces of the republic successfully repelled an invasion from Soviet Russia which was aided and abetted by pro-Bolshevik elements amongst the local population. Peace was finally secured by the 1920 Tartu treaty, which current Estonian president Lennart Meri has famously termed 'the birth certificate of the Estonian Republic'.[1]

Two years earlier, the fledgling republic had had to contend with the designs of Imperial Germany, whose forces advanced northwards into Estonia following the Russian military collapse in the autumn of 1917. The independence declaration of 24 February 1918 was thus prompted first and foremost by the imminent arrival of German troops, which took the capital, Tallinn, the following day. Under the March 1918 Brest-Litovsk treaty with Germany, Russia's Bolshevik government agreed to relinquish the Baltic provinces of the former Tsarist empire. In the wake of this agreement, occupying German forces began to prepare for the annexation of these territories to an enlarged Reich. In so doing, they enjoyed the support of the greater part of the local German nobility, which had constituted the ruling caste in the Baltic provinces of Estland and Livland for almost seven centuries.

These plans were cut short by the collapse of Imperial Germany in November 1918. As German forces withdrew from Estonia, Bolshevik Russia renounced the Treaty of Brest-Litovsk and sought to return the former Baltic provinces to the Soviet ambit. The invasion was supported by Bolshevik elements amongst the population of Estonia, which established an Estonian Workers' Commune in the border town of Narva. The latter proved short-lived, for by February 1919 the republic had been all but cleared of Soviet forces following a successful counter offensive by the new Estonian national army. Having failed to extend its control to the entire territory of Estonia, the by now beleaguered government of Soviet Russia began peace overtures during the autumn of the same year.

Considerations of political expediency prompted the Bolsheviks to sign a treaty which, it was hoped, would break the anti-Soviet front in the Baltic area and open up a 'window on Europe' in the form of economic ties.[2] The western allies, on the other hand, were somewhat slower to acknowledge the existence of an independent Estonian Republic. Britain's *de facto* recognition of Estonian independence in May 1918 was conceived primarily as a counter to German ambitions in the area during the final stages of World War One. Although Britain later provided naval support to the Estonians at a crucial juncture in the war with Soviet Russia, the Allied Council withheld full *de jure* recognition until 1921. Even then, leading figures within the British Foreign Office remained pessimistic regarding Estonia's long term prospects as an independent state.[3] This pessimism, one should add, was not confined to foreign observers. Whilst the attainment of independent statehood is today hailed as the fulfilment of the nation's destiny, there was nothing inevitable about the creation of the Estonian Republic. Although national self-determination was firmly on the agenda by 1917, most nationally-minded activists continued to tie the Estonians' fate to that of Russia, advocating territorial autonomy within a democratic Russian Federation. This remained the predominant view until the autumn of 1917, when the internal collapse of Russia prompted a quest for other alternatives. Even then, many Estonian leaders continued to advocate membership of a wider federation of Baltic and Scandinavian states (see Chapter Five). In this sense, the proclamation of the Estonian Republic should be viewed primarily as a means of internationalising the issue of national self-determination rather than an end in itself. At the same time, it would be wrong to see independence as a purely incidental outcome of the collapse of empires. As was to be the case in 1987–1991, developments on the wider international stage simply created opportunities. These were skilfully exploited by a determined national movement which succeeded in mobilising a majority of the local population behind it.

THE RISE OF ESTONIAN NATIONAL CONSCIOUSNESS

The Estonian national movement was a product of the 'National Awakening' (*Ärkämisaeg*) of 1860–1917. This period witnessed the emergence of a new nationally-minded community of intellectuals, freehold farmers and urban professionals which began to challenge the political hegemony of the ruling Baltic German elite. Unlike their Polish and Lithuanian counterparts, Estonian nationalists could not draw upon any past experiences of statehood by way of a mobilising ideal. National identity was constructed on the basis of the distinctive Finno-Ugric language and folk culture derived from tribes which moved westwards from the Ural mountains to settle the shores of the eastern Baltic some 5,000 years ago. These migrants subsequently pushed northwards into what is now Finland, a development which explains the close similarity between the modern Finnish and Estonian languages. By 1200, the territory of present-day Estonia was home to a population estimated at 150,000. This was grouped into a loose federation of townships without any single leader.[4] In the early 13th century, this decentralised political structure succumbed to a joint assault by the Danes and the Germanic 'Order of Sword Brethren' who conquered and christianised the indigenous tribes. The Danish hold on northern Estonia (*Estland*) lasted until 1346, when the King of Denmark sold his lands to the Teutonic Order following an uprising by the local population. Its legacy is reflected in the name of the present-day capital Tallinn, whose name derives from the Estonian *taan'i linn* ('Danish town' or 'Danish fortress'). These formerly Danish-controlled territories were subsequently incorporated into the province of Old Livonia (1346–1561), which encompassed most of present-day Estonia and Latvia.[5]

The German conquest and Livonia's status as part of the Holy Roman Empire are today celebrated for having brought the Estonians within the ambit of European culture.[6] The Germanic legacy is immediately apparent in the strikingly-preserved medieval architecture of Tallinn, which became one of the most important ports of the Hanseatic League. Today, the 'Hansa Spirit' is also frequently invoked in connection with emerging projects of regional cooperation amongst the Baltic sea states.[7] From the mid-19th century up to 1940, however, the coming of the Germans was usually portrayed as marking the start of a long period of feudal oppression for the indigenous 'Estonian' population.

The presence of the kingdom of Lithuania to the south barred the way to large-scale colonisation by German settlers and the possible assimilation of the autochthonous inhabitants. The local peasantry of the German-ruled Baltic provinces thus retained its distinctive Finno-Ugric language and culture, but referred to itself simply as *Maarahvas* or 'Country People'. Not until the 1850s did the term Estonian (*Eesti*) emerge, and even then its use was initially restricted to a small educated elite.[8] During the 14–16th centuries, the native population was gradually enserfed by the German nobility, which established a social order based on rigorous caste distinctions. This mentality persisted up until the final end of German hegemony in 1918. Indeed, C.A. Macartney was later to remark that until this date, the Estonians had been accustomed to a pre-modern social order in which ideas of a modern unitary state could scarcely take root.[9] Whilst Macartney's comments clearly constitute an exaggeration, they highlight the gulf between Estonia and Finland, where serfdom was unknown and basic state institutions had emerged already during the 19th century.

The German presence nevertheless held important implications for the eventual development of Estonian national identity. One notable feature was the introduction of Lutheran Protestantism, which was instrumental in the early development of a written Estonian language. The Lutheran insistence on making religion accessible to the population as a whole saw the publication of an Estonian catechism and the start of church services in the local language from the 1530s onwards.[10] It was only in the 1700s that Christianity finally began to supplant the deep-rooted attachment to old pagan customs amongst the peasantry. The activities of the Moravian Brethren during this period encouraged literacy and emphasised the virtues of piety, responsibility and initiative, thereby contributing to what Andrus Park describes as 'a tradition of pragmatic, peaceful and somewhat impassive individualism' within Estonian culture.[11]

The Reformation accentuated already existing social and political divisions amongst the ruling German elite of Old Livonia. After years of decline, the province finally collapsed during the Livonian Wars of 1558–1561, when Estland passed to Sweden and southern Livonia (Livland) to Poland-Lithuania. During the 1620s Livland too went over to Swedish control, a development which brought the southern Estonians back into the Lutheran fold. This period has remained in the collective memory as the 'Happy Swedish time' (*Vana Hea Rootsi Aeg*). Feudalism was alien to the Swedish crown, which undertook measures to improve the lot of peasants working state land. A network of rural schools was also established and the University of Dorpat (present-day Tartu) founded in 1632. The myth of good Swedish rule lacks foundation in the sense that the position of the German nobility remained largely undisturbed, although many landowners seemingly began to fear the loss of their privileges during Karl XI's Reduction of noble estates in the late 17th century.

Sweden's aspiration to dominion of the Baltic ultimately proved unsustainable, however. During the Great Northern War of 1700–1721, control of Estland and Livland passed to Petrine Russia, which swiftly reiterated noble privileges through the *Ostzeiskii Zakon* of 1721.[12] On the basis of these laws, the region remained economically and politically distinct from the rest of the Russian empire right up until its demise in 1917. The Baltic Germans became a trusty bulwark of Tsarist autocracy, enjoying a high degree of influence at court. This position left them well placed to resist the projects of Russification advanced by government officials and Slavophile publicists during the latter half of the 19th century.

Until the late 1700s, Estland and Livland remained virtually isolated from the rest of the empire in terms of their administration. It was during this period that the burdens of serfdom grew to their most onerous, a situation which provoked widespread peasant disturbances. By degrees, however, the ideals of the Enlightenment began to take root. A growing movement in favour of peasant rights and rationalisation of agriculture owed as much to economic necessity as it did to humanitarian concerns. The ideas of Baltic German writers such as Merkel and Petri found support from a Tsarist regime anxious to guarantee social stability during the period of the Napoleonic wars. These reformist impulses meant that the peasantry of Estland and Livland were emancipated in 1816–1819, some forty years before similar reforms were extended to the empire as a whole.[13]

The Baltic German presence also acted as a conduit for Herderian Romanticism, which evoked a new interest in the hitherto despised culture of the peasant masses. An Estonian language lectureship was established at the newly-reopened University of Dorpat in 1803, and Estiphile Germans later established the Estonian Learned Society as a medium for the scholarly investigation of local folklore (1838).[14] In what was an entirely unintended development, this cultural renaissance interacted with ongoing advances in education and agricultural production to bring about the emergence of an Estonian national consciousness.

ÄRKÄMISAEG

Having obtained personal freedom in 1816–1819, the *Maarahvas* was able to form self-governing peasant communities at the local level. During the same period, new legislation provided for the further development of elementary education in the Estonian language.[15] In many respects, the initial reform of agriculture merely served to exacerbate tensions in the countryside. The laws of 1816–1819 were devised by the noble estates in order to thwart more radical proposals by a Tsarist government wishing to set norms for the performance of labour service and distribution of land to peasant households. Instead, the Baltic German variant stated explicitly that land belonged to the nobility, which was left free to lease land to the peasantry on its own terms. On this basis, landowners were able arbitrarily to increase the level of corvée labour, expropriating the holdings of peasants who refused to comply. Many landowners took advantage of this provision to rationalise agricultural production by annexing peasant allotments directly to their estates. This new state of affairs gave rise to a further wave of peasant unrest, culminating in the so-called 'Mahtra War' of 1858.[16]

By this time, however, the old feudal practices were being supplanted by more modern farming methods. In 1860, the corvée principle applied to barely a fifth of peasant farmsteads, whilst further reforms initiated in 1848 meant that peasant farmers were now allowed to buy land or lease it in return for money rent. Farm purchases began on a significant scale during the 1870s and 1880s, and by the turn of century, Estonians already owned a third of the land in Estland and northern Livland.[17] Nevertheless, owners and leaseholders still constituted less than one in three of the rural population at this time. The presence of a large, disaffected pool of landless labourers brought continued outbreaks of unrest, culminating in generalised disturbances during the revolutionary year 1905–6.

Estonians with a foothold in the propertied order, however, began to demonstrate a growing self-confidence. Administrative reforms enacted in 1866 reduced the sway of the nobility over the legal system and the institutions of peasant self-government by providing for the election of rural township assemblies. Fired by parallel developments in the cultural sphere, the new class of independent small farmers and administrators began to reject Germanisation as a prerequisite for social advancement. A widening of Estonian-language education and the abolition of rigid caste distinctions became the primary goals of the early national movement, which still imagined *Eesti* as a cultural community rather a territorial or adminstrative entity.[18] By now, the ranks of the Estonian Learned Society included the sons of peasant estate workers whom their lords had seen fit to educate to university level. Inspired by the appearance of the *Kalevala* in neighbouring

Finland, the physician Friedrich Robert Fählmann (1798–1850) set about preparing a similar epos for the Estonians. This task was later realised by Friedrich Reinhold Kreutzwald (1803–1882), the compiler of *Kalevipoeg* (Son of Kalev) published during 1857–61.[19] By drawing together the traditional folk poetry of the peasantry into a single work, Kreutzwald constructed the mythical past which was a further necessary fundament for the imagined community of the Estonian nation.

The spread of national consciousness was greatly facilitated by the fact that 90% of the rural population were able to read by 1850.[20] This provided a basis for the emergence of a weekly periodical press, which already numbered eight titles by the early 1880s. The first of these was *Perno Postimees* ('The Pärnu Postman'). This was established in 1857 by Johannes Voldemar Jannsen (1819–1890), who later went on to found *Eesti Postimees* in 1864. Drawing on German influences, Jannsen organised the first Estonian song festival in 1869, an event which was attended by 10,000 people.[21] The festival marked the start of an enduring national tradition, one which was revived (albeit in heavily Sovietised form) by the Communist regime during the 1950s. The importance of these occasions in terms of Estonian national consciousness became truly apparent during the drive for independence from the USSR, which has been famously termed the 'Singing Revolution'.

As was to be case during the 1980s, however, the 19th century Estonian nationalists were not all singing from the same hymn sheet. For Jannsen, Estonian cultural identity was clearly rooted within the western traditions inherited from the Baltic Germans. As such, he supported incremental reform of the existing institutional and economic order and was prepared—where possible—to make common cause with the ruling classes to preserve the autonomous character of the Baltic provinces within the empire. Jannsen's views corresponded to those of Jakob Hurt, whose collection of Estonian folklore during the 1880s served further to propagate the national idea amongst the rural masses.[22] This moderate approach was challenged by the 'Saint Petersburg Patriots', an eastern-oriented faction headed by the radical Karl Robert Jakobson and his ally Johann Köler, a Professor of Fine Art in the imperial capital. According to Jakobson, German tutelage had constituted an 'era of darkness' for the Estonians, who had been stripped of their independent, pagan-based national spirit.[23] The future interests of the Estonians, moreover, were deemed to lie with Russia as a whole. Jakobson argued that Russian *Zemstvo* institutions should be extended to the Baltic provinces in order to break the power of the

German nobility and give Estonians a greater voice in the political process. In this regard, Köler and Jakobson made common cause with Russian Slavophiles such as Samarin, who had been advocating the abolition of Baltic German privileges since the 1840s.[24]

In the 1880s, the Köler-Jakobsen line appeared the more representative of a population which had traditionally looked to the Tsar as a protector against the oppressive rule of local landlords. This loyalty had aided the cause of a religious conversion campaign launched in the 1840s, when around 17% of the northern Livland peasantry adopted the Orthodox faith in the mistaken belief that this might confer easier access to land.[25] The case for administrative Russification was strengthened by the Russian land reform of 1861, which—unlike the early laws in the Baltic provinces—provided for the mandatory sale of land to peasants at state-regulated prices. Extension of this law to Estland and Livland remained the key demand for the Estonian radicals, who had not yet espoused the revolutionary idea of expropriating and redistributing noble lands.[26]

Great hopes were aroused by the accession of the Slavophile-minded Alexander III to the Russian throne in 1881. Breaking with precedent, the new Tsar refused to reconfirm the rights and privileges of the Baltic German nobility. At this time, the attentions of the central authorities were again drawn to the Baltic provinces by a wave of rural unrest, and the Tsar was confronted with a wave of petitions for reform from the increasingly organised Estonian population. In 1881, Jakobson headed an Estonian delegation to St Petersburg. Amongst other things, it presented Alexander with radical proposals for the abolition of the existing provincial boundaries and the incorporation of all ethnic Estonians into a single administrative district. The Estonians' faith in the Tsarist regime proved ill-founded. Plans were drawn up for a far-reaching reform of Baltic institutions, but were shelved in the mid 1890s as a hard-pressed government reverted to its traditional policy of appeasing conservative German interests in the provinces.[27]

If the administrative structures of the Baltic provinces were never wholly remodelled along Russian lines before 1917, the last years of the 19th century were nevertheless marked by intensified efforts to promote cultural Russification. From the 1860s, Russian governments had voiced increasing concern over the possible Germanisation of the local peasantry, whilst remaining confident that educated and socially mobile Estonians would voluntarily embrace Russian language and culture if opportunities for them to do so were expanded.[28] A gradual increase in the weight of Russian language instruction in the Baltic provinces

intensified during the last two decades of the century, when many
elementary schools in the rural townships began an almost total shift
towards instruction in the Russian language. The ending of German-
language instruction at the University of Dorpat in 1894 prompted
many Baltic Germans to seek higher education and careers abroad,
contributing to a gradual demographic decline of this group in the years
before 1914. These Russifying measures seriously backfired. By undermining the educational privileges of the ruling elite, they offered the
Estonians increased access to education and social advancement. At the
same time, it became abundantly clear that government reforms had
only served to deepen Estonian national consciousness. Opposition to
Russified elementary schools, for instance, became one of the focal
points for popular mobilisation against the Tsarist regime during the
1905 revolution.[29]

The fortunes of the Estonian national movement had gone into
temporary decline in the 1880s, when it lost several of its most
influential leaders through death or infirmity. Ideological splits within
the movement also impeded the campaign for secondary education in
the Estonian language, at a time when the pressures of Russification
increased the need for solidarity. The mood of pessimism was quickly
dispelled during the following decade, as a new generation emerged to
restore dynamic leadership. By this time, to quote Toivo Raun,
'Estonian political and social thought showed much diversity and
reflected the changes wrought by socio-economic modernisation'.[30] The
industrialisation of the Baltic provinces had gathered pace following the
arrival of the railways in the 1870s. The new opportunities which it
created brought the landless peasantry flocking to the towns, where it
formed the basis for a new ethnic Estonian working class. Similarly, the
abolition of archaic guild practices and the expansion of state
bureaucracy expanded the middle class in a society where one in five
inhabitants lived in towns by the eve of World War One. The growing
wealth of the Estonian bourgeoisie enabled it to obtain control of
municipal government in six of the ten major towns in Estland and
Livland during 1901–1914.

1905 AND ITS AFTERMATH

The 1905 revolution marked a crucial stage in the evolution of the
national movement. An All-Estonian Congress convened in Tartu was
united in its calls for an end to Russification and the establishment of

territorial autonomy for the Estonians within a single administrative district. Yet the Congress quickly split along proto-party lines which reflected the growing class distinctions amongst the Estonian population. Representatives of the bourgeoisie and landowning farmers gravitated towards the Estonian Progressive People's Party under Jaan Tõnisson. As editor of *Postimees* from 1896, Tõnisson had spearheaded the revival of the national movement. Stressing the sanctity of private property, his party continued to propound the Hurt-Jannsenist line of constructive co-operation with the Baltic Germans.[31] In so doing, it quickly parted company with a more amorphous radical faction headed by Tõnisson's former classmate and lifelong rival Konstantin Päts. Born into an Orthodox family in Pärnu, Päts had briefly attended a religious seminary in Riga before graduating in law from the University of Tartu. As editor of the Tallinn-based *Teataja* from 1901–1905, he continued to link the future of the Estonians to the overall evolution of the Russian state, reiterating Jakobson's demands for the extension of *Zemstva* to the Baltic provinces. Unlike Jakobson, though, Päts could have no truck with Tsardom, regarding national self determination as a means to improving the lot of the working class and landless peasantry whose interests he championed during the early stages of his political career.[32]

Still further to the left were the revolutionary Estonian Social Democrats, who became the largest and best organised of the proto-parties formed in 1905. This fact reflects the growing militancy of the labouring classes in Estland, which engaged in widespread attacks on manor houses. In the same year, workers and peasants also began to establish revolutionary councils in the rural townships and main urban centres. This unrest met with brutal retribution in the form of executions and mass deportations to the Russian interior. In other respects, the central government proved more accommodating towards its Baltic minorities during 1906–13, allowing the establishment of private Estonian-language secondary schools before reverting to the previous policy of Russification on the eve of World War One.

For the Estonian nation as a whole, the last decade of Tsarist rule was a time of continued cultural and economic advancement, although hopes for sweeping political reform quickly faded. Estonian delegates to the four Russian Dumas gravitated to the opposition Kadet or Social Democrat blocs, where they joined other minority representatives in calling for greater national autonomy. More restrictive electoral laws introduced in 1907 tilted the balance of Duma representation in favour of the Baltic Germans, whose support was solicited by the government to an even greater degree after the 1905 revolution. Yet the events of this year had underlined the vulnerability of the local élite, positioned uncomfortably between the growing Slavophile nationalism of Russian conservatives and the ethnic Estonian national awakening. Increasingly, the Baltic Germans began to think of themselves in national terms rather than as representatives of a a social estate.[33] This heightened sense of national self-awareness complicated efforts by more liberal elements amongst the German population to reach accommodation with the Estonians. It also foreshadowed a shift in loyalty 'from Tsar to Kaiser' on the part of many Baltic Germans after 1914.[34]

WAR AND REVOLUTION

It is difficult to say how quickly Estonian aspirations for national self-determination would have been realised without the catalyst of war and revolution. According to a later author, only a 'prophet or a mental patient' could have predicted independence in 1910, when the poet Juhan Liiv declared that 'one day there will be an Estonian state'. Liiv—a schizophrenic—was both.[35] Anti-German sentiment caused the Estonians to rally behind the regime at the outbreak of World War One. The renewed political demands advanced by Tõnisson's Northern Baltic Committee a year later were for autonomy within a reformed Russian state. The evolution of the left-wing federalist orientation during the war years reflected the influence of the radical cultural movement 'Young Estonia' (*Noor Eesti*), founded by Gustav Suits in 1906. *Noor Eesti* propounded a radical synthesis of moderate nationalism and social democracy, urging its followers to 'remain Estonians but become Europeans'.[36] Former adherents of the movement later assumed key positions in the parties of the non-Marxist Left which were to play a crucial role in state-building during 1918–20.

The Estonians' path to statehood and the New Europe began with the Russian Revolution of February 1917. Responding to pressure from Tõnisson's moderate nationalists and a 40,000 strong demonstration by Estonians in St Petersburg, the new Russian provisional government quickly agreed to unify Estland and Northern Livland administratively into a single district. In line with the long-standing demand for implementation of the *Zemstvo* system, the German-controlled provincial estates were abolished and replaced by an elected provincial assembly (*Maapäev*) which would exercise power pending the establishment of an all-Russian constituent assembly. Elections to the *Maapäev* hastened the evolution of the party system and resulted in a majority for the forces of the centre and the right. The Agrarian League (*Maaliit*, 13 deputies) was the largest of the nine fractions within the 62-member assembly, followed by Otto Strandmann's Labour Party (11). Tõnisson's Democratic Party emerged with seven seats, while Konstantin Päts—elected as a nonparty candidate—affiliated himself to *Maaliit*.[37] Politically, Päts had moved gradually towards the right throughout 1905–1915. Even so, this final stage in the transition from socialist radical to conservative agrarian carries a whiff of the political opportunism which Päts was to display so amply in his later career.

The deliberations of the *Maapäev* began at a time when prospects for national self-determination appeared increasingly bleak. When German forces seized the islands of Saaremaa and Hiumaa in late August, Tõnisson argued that the Baltic and Finnish peoples should sever their links with Russia and unite with Scandinavia within a new 'Baltoscandian Federation'.[38] The idea of separation from Russia gained ground after the Bolshevik takeover in Petrograd. At this point, Estonian Bolsheviks Viktor Kingisepp and Jaan Anvelt assumed power in the province in the name of the Estonian Military Revolutionary Committee. The Bolsheviks had obtained only five seats in the *Maapäev* elections, yet were buoyed by the presence of large numbers of mutinous Russian troops in Estonia. On this basis they had gradually assumed control of local Soviets established after the February revolution. The new regime promptly disbanded the *Maapäev*, promising full elections to an Estonian constituent assembly. Its subsequent failure to deliver on these promises, however, is a mark of the Bolshevik distaste for the principle of democratic accountability. Initially, at least, the Bolsheviks enjoyed considerable support amongst the local population, garnering 40% of the vote in elections to the all-Russian constituent assembly during November 1917. Although the socialist and non-socialist vote was almost evenly divided in this poll, opposition parties were denied any voice in government, whilst the authorities also took steps to curb the non-Bolshevik press. Above all, the Bolsheviks squandered popular goodwill through their refusal to redistribute land to the peasantry. Noble estates were nationalised, but were promptly transformed into state-run collective farms. Consequently, the first round of elections to the Estonian constituent assembly in January 1918 brought significant gains for the opposition Labour Party. Citing an alleged bourgeois conspiracy to overthrow Soviet power, the authorities cancelled the remainder of the poll.[39]

THE FORMATION OF THE ESTONIAN STATE 1918–20

A campaign of non-compliance by the local bureaucracy ensured that the Bolshevik grip on power was less than complete by the time German forces reached Tallinn in February 1918. The representatives of the *Maapäev* had gone underground the previous autumn, entrusting power to a Committee of Elders consisting of Päts, Konstantin Konik and the Labour Party's Jüri Vilms. On 24 February the Committee emerged to declare independence and form a provisional government, before German occupation again rendered normal political activity impossible. In the interim, representatives had been dispatched to the major European capitals to promote the Estonian cause amongst the western powers. In spite of these efforts, the restored provisional government was left largely to its own devices when German occupation gave way to an invasion by Bolshevik Russia in November 1918, although external military support was provided by the intervention of British naval forces and the arrival of some 3,700 volunteers from neighbouring Finland during December and January.[40] Only with considerable difficulty was the government able to marshal its own fledgling armed forces under the leadership of former Tsarist officer Johannes Laidoner. The key to mobilising the local peasantry behind the struggle lay not through lofty appeals to the national idea, but in promises of land for those who fought.[41]

In the midst of the war, elections to an Estonian constituent assembly gave a clear majority to the parties of the democratic left. The assembly moved quickly to solve the agrarian question by nationalising estate land and providing for its redistribution in the form of smallholdings (av. 24 ha.). On this basis, some 32,000 new farms were created during the early 1920s, with precedence given to war veterans and the landless. Tenants on the former estates were also given security in the form of permanent leases.[42] This radical approach to the land issue had the practical effect of undermining residual pro-Bolshevisk sentiment amongst the rural population at a crucial juncture. Yet the main thrust of the reform was directed against the Baltic Germans, who still held 58% of the land in 1918. The fact that at least some local Germans enlisted to fight in the war of independence against Soviet Russia could not dispel deeply-felt popular resentment towards the former overlords, a sentiment which was sharpened by the recent experience of German occupation. It is notable that the deliberations of the constituent assembly coincided with the 'Landeswehr war', in which the Estonians helped to defeat renegade German forces which had overthrown the government in neighbouring Latvia. In the minds of many Estonians, this constituted the defining moment of the independence struggle. The land reform soured Estonia's relations with Germany during the early years of independence and was viewed—erroneously—in other western capitals as being little different from the Bolshevik approach to land reform. Former landlords were eventually granted partial compensation in 1926, although this amounted to less than 10% of the true value of the land.[43]

In other respects, western states such as Weimar Germany and, most especially, Switzerland served as an inspiration to the founders of the Estonian Republic. The liberal-left majority within the constituent assembly resolutely rejected calls for a powerful executive. Instead, supreme power was vested in a 100-member parliament (*Riigikogu*) elected by proportional representation at three-yearly intervals. The head of government (*Riigivanem*—State Elder) had no right of veto over parliament, which could only be dissolved by popular referendum. Notwithstanding the agrarian reform, the constitution of December 1920 was also notable for its tolerance towards ethnic minorities, which were accorded the right to receive education in their native tongue along with promises of full cultural autonomy. Estonian state-builders regarded this new constitutional order as a means of staking a place in the 'New Europe' which was expected to arise from the ruin of war. In certain respects—most notably its provisions for minorities—Estonia actually exceeded the minimum requirements later established by the League of Nations. This shows that far from simply being a testing ground for imported western models, Estonia made its own indigenous contribution to the reconstruction of Europe. However, ultra-liberal provisions such as the right of the electorate to initiate laws through referenda meant that the political system ultimately proved 'too democratic for its own good' during the interwar period.[44]

DEMOCRATIC POLITICS IN ESTONIA 1920–34

The task of consolidating democracy was complicated by the extreme fragmentation of the party system. The five sets of parliamentary elections during 1920–1932 were contested by as many as twenty six different groupings, whilst the number of parties actually gaining representation was anywhere between six (1932) and fourteen (1923). This was not a recipe for stable government: from 1920 to 1934 Estonia had seventeen cabinets lasting an average of nine months.[45] As Hiden and Salmon remind us, such an experience was not unusual amongst the inter-war European democracies, where stable government was the exception rather than the rule.[46] In the Estonian case, parliamentary politics was rendered intelligible by the presence of left, centre and right-wing blocs, with the latter two always able to command an overall majority in successive parliaments. Although Päts' *Maaliit* had opposed radical land reform within the constituent assembly, the agrarian right ultimately proved to be the main beneficiary of the 1919 legislation. Newly-propertied elements amongst the rural population began to turn to *Maaliit* (later renamed the *Põllumeeste Kogud* or Farmers' Party) and the new Smallholders' Party in defence of their interests during the 1920s. From 1926, the democratic right constituted the largest bloc within parliament. The Farmers' Party consequently became the mainstay of government in what remained a predominantly agrarian state, supplying the State Elder in 10 out of 17 of the coalitions. Päts himself was *Riigivanem* a total of five times during the years of liberal democracy, whilst Jaan Tõnisson (now of the centrist National Party) occupied the post twice. Continuity of key personnel within successive administrations constituted a further anchor of stability amidst the carousel of short-lived coalition government.[47]

The first few years of the Estonian Republic were economically fraught, as the government set about the task of constructing an independent economic base. Industrial workers were especially hard-hit, since the effective closure of the Russian market from 1922 sounded the death knell for many of the enterprises built during the late Tsarist era. These economic difficulties brought an upsurge in industrial action and a growth in support for the Estonian Communist Party. Although formally proscribed following the independence war, the Communists contested local and national elections under an assumed name, gaining a third of the municipal vote in Tallinn and Tartu and ten parliamentary seats during 1923. Over the next year, the Comintern helped to orchestrate a wave of communist agitation across Estonia. This culminated in an attempt to overthrow the Estonian Republic on 1 December 1924, when a small but well-armed band of insurrection-aries failed in their attempt to seize government buildings and key installations in the capital. More importantly, the coup did not attract support from local workers or members of the armed forces.[48] The attempted uprising was crushed decisively by the military authorities under acting Commander-in-Chief Laidoner, who invoked emergency powers vested in him by the government. Many of the participants were summarily executed whilst others were either imprisoned or managed to flee to the USSR.

The putsch had an important influence on subsequent political developments. The Communist Party ceased to be a major force after 1924, whilst the hand of the nationalist right was strengthened by the resurrection of the civilian Defence League (*Kaitseliit*) constituted during the independence war. In the short-term, the crisis also encouraged the democratic parties to set aside their differences and tackle pressing economic and social questions in a new spirit of co-operation. In particular, the events of December 1924 lent further impetus to ongoing discussions of cultural autonomy for Estonia's minorities. Whilst the Baltic Germans had shown their loyalty to the existing state order during the putsch, there were worrying signs of pro-communist feeling amongst the impoverished Russian peasantry in the eastern border districts of Pechora and Prinarova. Early in 1925, the *Riigikogu* approved new legislation allowing minority groups with more than 3,000 members to constitute themselves as public corporations. They were then entitled to elect cultural councils enjoying full administrative and supervisory powers over minority schools and other cultural institutions. The councils were to be funded mainly by central and local government, although they also had the power to raise taxes from amongst the relevant ethnic group. The Estonian Republic was the only state in inter-war Europe to adopt such a system of personal autonomy for its minorities. The 1925 law—hailed by contemporaries as an exemplary solution to the minorities problem—reflected the thinking of the Austrian Marxists Renner and Bauer, who had argued that the issue of minority rights in multiethnic states could not be settled according to the territorial principle alone.[49] Unlike today, inter-war Estonia was amongst the most ethnically homogeneous of the successor states. Minorities made up around 12% of the population, the largest groups being Russians (9%) Germans (1.5%) Swedes (0.7%) and Jews (0.4%). These statistics, however, mask the importance of ethnic minorities to the economic life of the new republic. The position of the Germans within commerce and the professions remained vastly disproportionate to their small numerical size during the 1920s. The German minority also displayed a high degree of political organisation. Having played a major role in drafting the Law on Cultural Autonomy, the Germans were able to implement the scheme within a year, closely followed by the Jewish minority in 1926. Estonia's Russians, on the other hand, proved too politically disunited to institute cultural autonomy. This was despite tireless campaigning by the academic and politician Mikhail Kurchinskii, whose efforts could stand as a shining example to Russian political leaders in contemporary Estonia.[50]

This inability to overcome narrow sectional interests in pursuit of wider goals was not simply confined to national minority parties. After a brief period of 'wall-to-wall' coalition under the centrist Jüri Jaakson in 1925, the main Estonian parties quickly reverted to the fractionalism which they had displayed during the early years of independence.[51] The fragmented party system did not impede the successful functioning of liberal democracy during the prosperous years of the late 1920s, when a thriving agricultural export sector allowed Estonia to carve out a niche within the European economic system (see Chapter Four). The economic slowdown after 1928, however, further intensified the calls for a strengthening of the executive which had begun following the 1924 putsch. Päts in particular became one of the leading advocates of constitutional reform during this period[52], yet successive proposals by the Farmers' and National Parties foundered due to resolute opposition from the Social Democrats. A series of party mergers along with measures to streamline the governmental administration during the ealy 1930s had little effect in terms of instilling greater political stability. As such, the democratic system was singularly ill-equipped to cope when Estonia encountered the full force of economic depression after 1931.

THE CONSTITUTIONAL CRISIS 1932–34

The effects of the depression were felt most deeply by the urban population. Although the worldwide slump in agricultural prices had a predictable effect on agricultural incomes, the status of the propertied rural classes remained, on the whole, secure. The farmers enjoyed self-sufficiency in food. Agrarian-led governments provided them with subsidies, while state-owned banks refrained from pressing unduly for loan repayments. Trade contraction, coupled with the government's recourse to devaluation and fiscal austerity, posed a greater threat to the position of smaller entrepreneurs, and representatives of the professions and state bureaucracy, groups which encountered the highest levels of unemployment during the depression.[53]

The bottoming-out of the slump coincided with a growing impression of political paralysis, as the established parties proved unable to overcome their differences and act as effective 'crisis mediators'. This contrived to diminish their already low standing in the eyes of the population. According to Parming, it is these factors—rather than the large number of political parties or constitutional arrangements *per se*— which explains the subsequent eclipse of liberal democracy.[54] Fearing for its recently-acquired social status, the ethnic Estonian bourgeoisie progressively abandoned the established parties after 1931. Instead, they rallied to a campaign for strong executive government spearheaded by the non-parliamentary radical right. 1929 had seen the formation of a Central League of Veterans of the War of Independence (*Eesti Vabadussõjalaste Keskliit*—later abbreviated to *Eesti Vabadussõjalaste Liit* (EVL), popularly known as the '*Vapsid*'). Driven by the oratorical skills of the young lawyer Artur Sirk, EVL evolved into a paramilitary-style organisation sharing certain common features with fascist movements elsewhere in Europe.[55] In the course of 1932–33 the *Vapsid* were able to seize the initiative in debates over constitutional reform, by now the dominant issue in Estonian politics.

During the two years in question, power was exercised by no fewer than six different coalition governments, five of which were agrarian-led. In the face of external agitation by the Veterans' League, the *Riigikogu* twice attempted to secure approval for a new constitution through popular referendum. Opposition from the *Vapsid* and the Social Democrats, however, ensured that neither draft was approved by the electorate. The *Vapsid*, meanwhile, comfortably obtained the 25,000 signatures necessary to place their own constitutional draft to a referendum. This provided for a popularly-elected President with the power to appoint cabinets and, 'in case of urgent necessity', to rule by decree. This proposal was endorsed by referendum in October 1933, a development which paved the way for the establishment of authoritarian rule less than six months later.

The Veterans' constitutional draft had received the formal backing of Konstantin Päts already in June 1933. Having been forced to resign as *Riigivanem* in April of that year, the Agrarian leader later returned as head of a new coalition government in October. By now, however, he faced growing dissent within his own party. When the new constitution came into force in January 1934, Päts assumed the role of acting head of state—with extensive executive powers—pending new presidential and parliamentary elections in April. During the previous autumn, Päts, along with former Commander-in-Chief Johannes Laidoner, had been approached by the Veterans' League with a view to becoming the movement's presidential candidate. Negotiations were held, but the

EVL congress eventually opted for the more pliant General Andres Larka. The movement apparently suspected that Päts or Laidoner would have simply used their powers to curb the EVL if elected.[56] In the event, it was Laidoner who was selected as the Agrarian Union (and Centrist) candidate, whereas Päts had to rely on the support of the original Farmers' Party.

In the local elections of January 1934, the *Vapsid* made sweeping gains amongst the urban vote, securing an absolute majority in Tallinn and several other major towns. Although there was every indication that EVL would gain power at the national level, the movement persisted with a campaign of anti-government agitation which had begun following its referendum victory the previous autumn.[57] This furnished Päts with a suitable pretext for mounting a pre-emptive strike against his opponents. Matters came to a head on 12 March. Larka was by now way ahead in the race for the presidency, whereas the acting head of state had not yet secured the minimum 10,000 signatures necessary to stand as a candidate. Having sounded out the military on its attitude to the crisis, Päts invoked his power to rule by decree, proclaiming a six-month state of emergency. Laidoner was re-appointed Commander-in Chief with a remit covering all aspects of state security. The Veterans' League was promptly disbanded, and 400 of its leading members arrested. Päts justified his actions by claiming that the *Vapsid* were planning a coup against the constitutional order. On this basis, he was able to obtain the consent of all the *Riigikogu* parties, including the Social Democrats, for the temporary state of emergency.[58]

In reality, there is no convincing evidence to support claims of an impending insurrection by the non-parliamentary right.[59] Instead, it seems that Päts invented the threat of a *Vapsid* putsch as a pretext for mounting his own coup d'état, thereby attaining the authoritarian power which would not have become available to him through the ballot box. Päts' true intentions became abundantly clear during the course of 1934. Having obtained parliamentary consent for emergency rule, he quickly postponed the upcoming elections and closed the current session of the *Riigikogu*. In August, the Agrarian and former *Riigivanem* Karl Einbund was appointed to the combined posts of prime minister and interior minister. Henceforth, effective power was exercised by a triumvirate of Päts, Laidoner and Einbund. Upon appointment, Einbund declared that the role of Estonia's 'authoritarian' government was to carry the country through its present deep crisis rather than to create an authoritarian order. He thus called for internal calm and an end to party political squabbles.[60] In fact, a return to political stability seemed largely assured by the end of Summer 1934, yet Päts still decreed an extension to the state of emergency when its original term expired in September. During the same month, the *Riigkogu* was reconvened in extraordinary session, but, 'counting on the people's support', Einbund promptly dismissed it—this time for good—after deputies refused to accept the government's nominee to the post of chairman of the assembly.[61]

AUTHORITARIAN ESTONIA 1934–40

Thus began the period referred to in Estonia as the 'Era of Silence' (*vaikiv ajastu*).[62] The closure of parliament was followed by the suspension and eventual prohibition of all political parties save Päts' newly-created Fatherland League (*Isamaaliit*), a highly-centralised governmental party dedicated to building 'national unity'.[63] The same role was assigned to the State Propaganda Service, which unleashed a campaign of censorship against the non-government press. Jaan Tõnisson, traditional guardian of the liberal 'spirit of Tartu' (*Tartu Vaim*), was at the centre of democratic opposition to the dictatorship during its early stages. Consequently, the government sequestered the assets of *Postimees* in July 1935, replacing Tõnisson with the head of *Isamaaliit*, Tartu University Professor Jüri Uluots.[64] The government also clamped down on opposition within the city's university by abolishing the existing student union and appointing the rector over the heads of faculty members. Opposition to the new regime was also apparent amongst the ranks of the labour movement, where strike activity reached unprecedented levels during 1935. Consequently, the government abolished the existing Central Council of Labour Unions and replaced it with an appointed body. This formed part of a fascist-style corporatisation of the economy which built upon the earlier extension of state control during the depression. 1935 also saw the final settlement of accounts with the Veterans' League. In a series of court-martial trials of *Vapsid* leaders during the Summer and autumn, the government was able to make its charges stick in fewer than forty cases. Those convicted received sentences ranging from a few months to six years. Artur Sirk, however, had already escaped from prison and fled to Finland in 1934. From exile he orchestrated plans for an armed uprising which were uncovered by Päts' security police in December 1935. On this occasion, the ringleaders received hefty prison terms. Sirk, meanwhile, moved to Luxemburg, where he fell to his death from a hotel window in mysterious circumstances two years later.[65]

In seeking to legitimate his regime, Päts sought to convey the impression that it had prevented more extreme forces from coming to power. Ironically, he now denounced the 1934 constitution as being conducive to dictatorial rule. Early promises of further constitutional reform were delivered following a 1936 referendum in which the electorate approved government proposals for a new constituent assembly. This body convened early in 1937 following elections in which only government-nominated representatives were allowed to stand. It duly gave its approval to a new form of government. The 1938 constitution merits close inspection. The focus of a campaign by restorationist nationalists during the early 1990s, it also forms the basis for subsequent claims that Päts was seeking to restore democracy during the last years of his rule.[66] On the face of it, the powers of the head of state were curtailed under the new system, which provided for a bi-cameral parliament alongside the executive branch. The lower chamber (*Riigivolikogu*) was to be popularly elected, whilst the Council of State (*Riiginõukogu*) consisted of presidental appointees and candidates nominated by various state and non-governmental organisations. The head of state, or president, would now be appointed by an electoral college consisting of parliamentary delegates and representatives of local authorities.

In reality, the restoration of parliamentary institutions did little to alter the substance of authoritarian rule. Whilst opposition candidates were allowed to contest elections to the lower house, they did not enjoy the right to organise along party lines. Also, the president still enjoyed the power to appoint the government, veto legislation and, in certain cases, rule by decree. In this sense, the constitutional reform is best seen as a token gesture designed to confer legitimacy to the Päts dictatorship. Ruusmann is thus correct to distinguish between the provisional nature of the 'era of silence' and the more consolidated authoritarian regime which emerged after 1938.[67]

It is, therefore, difficult to sustain the argument that Päts 'assumed absolute power in defence of democracy', for it seems probable that had the non-parliamentary right prevailed in 1934–35, its policies would have been little different.[68] The light sentences issued to EVL leaders and the subsequent formation of *Isamaaliit* further reinforce the contention that Päts shared similar aims to the League, but was determined to fulfil them on his terms alone. What united Päts and the

EVL leadership was a disdain for politics based on class conflicts and a hankering for stability and national unity.[69]

Unlike the predominantly urban-based EVL, however, *Isamaaliit* drew the bulk of its membership from the former Agrarian Party. In what remained a largely agrarian state, the party's leadership had come to equate the 'national interest' with the interests of an idealised peasant community.[70] In seeking to explain the events of 1934, it is revealing to contrast this image of the nation with the conception held by elites in neighbouring Finland. When Finnish democracy came under attack from the Lapua movement during 1931–32, the government initiated a clamp down on the extra-parliamentary extreme right. Although leading state representatives from President Svinhufvud down had acquiesced in Lapua's initial campaign against communism, none of the established parties went so far as to question the existing political order. Part of the explanation lies in the fact that even in the case of conservative politicians such as Svinhufvud, Finnish national identity was rooted in the defence of constitutional principles and long-established state institutions.[71] In Estonia, on the other hand, the nation was still envisioned first and foremost in cultural rather than political terms. Although Päts had played a major role in the formation of the Estonian state, he evidently did not consider parliament to be the legitimate representative of the Estonian people. His disdain for party squabbles and his talk of the need for a 'master in the house' betray the paternalistic outlook of a man who essentially saw himself as the embodiment of the national interest.[72]

Päts 'conservative peasant authoritarianism'[73] was close to fascism in many important respects, but did not seek to emulate Nazi totalitarianism. The best thing which can be said for it is that it was more benign than other authoritarian regimes in inter-war Europe and, for the overwhelming majority of Estonia's residents, infinitely preferable to what followed after 1940. There was no systematic use of terror against political opponents, while many imprisoned communist and Veterans' League representatives were released under an amnesty in 1938. The Päts regime also proved remarkably successful in sustaining economic recovery following the depression when, notwithstanding its agrarian roots, it began diversification into industry. In this respect, it seemingly secured the approval of a significant section of the population, although the results of the 1938 *Riigivolikogu* elections hardly constituted a resounding victory for the ruling party.[74]

Discontent was perhaps most widespread amongst the ethnic minorities, which were subjected to a more centralising and assimilative line in nationalities policy after 1934. According to Ruutsoo, Estonia never became a 'nation-state' (*Rahvusriik*) in the narrow sense applicable to other states of inter-war east-central Europe.[75] In its attempts to divert public attention away from politics, the State Propaganda Office devoted much of its energies to what were ostensibly benign cultural initiatives. Examples included campaigns to promote the use of the national flag, the Estonianisation of surnames and even the external redecoration of houses (*kodukaunistamine*)! The institutions of minority cultural autonomy were now brought under the auspices of the Ministry of Education, but were never formally proscribed. The fact that Mikhail Kurchinskii intensified his efforts to achieve cultural autonomy for the Russians after 1934 suggests that the organised minorities remained best-placed to defend their interests under the new regime.[76] Even so, Päts and his government consciously sought to strengthen the position of the titular Estonian nation in the economic life of the republic, whilst new laws on the definition of nationality were designed to limit the number of residents classifiable as representatives of an ethnic minority.[77] Sadly, these 'nationalising' policies merely reinforced the growing allure of Nazism amongst Estonia's German population during the 1930s.[78]

THE END OF INDEPENDENCE 1939–40

The demise of the inter-war Estonian Republic, however, was occasioned not by internal developments, but by external factors. The eclipse of democracy in Estonia and the subsequent destruction of independent statehood must be seen within the context of a generalised deterioration in the European security environment on the back of the Great Depression (see Chapter Five). After drifting further into international isolation during the 1930s, the Estonian Republic fell victim to the cynical spheres of influence agreement concluded by Nazi Germany and the USSR on 23 August 1939. A secret protocol to this Molotov-Ribbentrop Pact stated that 'in the event of a territorial and political rearrangement in the areas belonging to the Baltic states (Finland, Estonia, Latvia and Lithuania), the northern boundary of Lithuania shall represent the boundary of the spheres of influence of Germany and the USSR.'[79]

In the immediate term, the agreement paved the way for the partition of Poland. Following Germany's invasion from the West, Soviet troops moved in to occupy eastern Poland on 17 September. A week later,

Estonian foreign minister Karl Selter travelled to Moscow, where he was presented with demands for a 'mutual assistance pact' between his country and the USSR. Faced with a massive Soviet military build-up on the eastern border, Päts and his government subsequently gave their assent to an agreement which allowed the USSR to establish military bases and station 25,000 troops on Estonian soil for the duration of the European war. Article 5 of the pact stipulated that the agreement would in no way impair Estonia's sovereign rights or its economic and political structure.[80] Latvia and Lithuania were pressurised into signing similar pacts during October.

In seeking to explain the events of 1939, western authors have frequently portrayed Estonia and its Baltic neighbours as 'helpless victims of Soviet expansionism ... [which] ... had no alternative but to submit.'[81] However, realist interpretations alone cannot adequately explain the contrast between the Estonian response and that of neighbouring Finland, which chose to resist Soviet demands in the face of similarly overwhelming odds. Conventional wisdom holds that Finland was better prepared militarily than its Baltic neighbours in 1939. In fact, a realistic appreciation of the country's military weakness meant that the Finnish General Staff consistently urged accommodation with the USSR, only to see its advice disregarded by the political leadership.[82] By way of justification for its unflinching resistance to Soviet demands, the Finnish government could point to the support which it enjoyed both in a democratically-elected parliament and amongst public opinion. Notwithstanding the residual divisions caused by the Civil War of 1918, dominant discourses on Finnish national identity had been informed by an often virulent strain of Russophobia between the wars. This was not so much the case in Estonia, where anti-Germanism had been at least as powerful a factor, if not more so, in the construction of the national self-image. This perhaps explains why Päts and his circle seemed more willing to give the Soviets the benefit of the doubt in the uncertain climate of September 1939 (see below).

By way of further justification for his system of rule, Päts had claimed that the restoration of multi-party democracy would have undermined national unity and threatened independence in the dangerous international climate of the late 1930s.[83] In fact, as Taagepera points out, the absence of any real public debate in Estonia simply made it easier for the government to yield in 1939.[84] Unlike its Finnish counterpart, the Päts regime did not enjoy the legitimacy conferred by democratic election when contemplating the possibility of war against the Soviet neighbour, and clearly did not feel confident in committing the nation to this course.

In opting for capitulation, the Estonian leaders were apparently convinced of the invincibility of the Soviet armed forces, an impression not borne out by the population's subsequent encounters with the poorly-equipped and ill-disciplined army which entered Estonia in October 1939.[85] The Finns, on the other hand, went to the other extreme, over-exaggerating the weakness of the Red Army and gambling that the USSR would not resort to the military option. These calculations, of course, proved sadly mistaken. Even so, Stalin only reluctantly ordered an invasion of Finland, continuing diplomatic overtures until the eleventh hour. Similarly, there is anecdotal evidence to suggest that the Soviet dictator feared the possible international repercussions of an all-out attack on Estonia in September 1939.[86] Indeed, had this not been the case, it seems doubtful that he would have been so scrupulous in attempting to give a legal veneer to the subsequent forcible annexation of the Baltics. It is thus intriguing to speculate whether a more robust response on the part of the Estonian leadership would have given the Soviets pause for thought.

In late August 1939, Päts had pledged to 'struggle until the last drop of blood' in defence of national sovereignty.[87] Yet this rhetoric is wholly inconsistent with his refusal to order immediate mobilisation as soon as the substance of the Molotov-Ribbentrop Pact became known. In fact, given the absence of any formal guarantee of outside support, the government was fearful of the possible repercussions of undertaking such a step. Thus, when the Soviet Union presented its demands a month later, Estonia possessed a poorly-equipped standing army of only 8,000 men. As Taagepera notes, the almost complete absence of military preparations in the international climate of the late 1930s pushes one 'to the rationale of lack of will and foresight rather than inherent defencelessness.'[88] To be sure, Estonia could never have hoped to match the Soviet invader in numerical terms (160,000 troops were massed on its borders in September 1939). However, the Finns held out for nearly three months in the face of equally overwhelming odds.

In spite of the obvious futility and possibly catastrophic consequences of resistance, General Laidoner initially advocated the rejection of Moscow's ultimatum. This, he argued, would leave world opinion in no doubt of Estonia's determination to defend its independence at any cost.[89] As the government deliberated upon its response, the Soviet air force undertook low sorties over Estonia's major towns. Soviet control of territorial waters was by now also firmly established. This demonstrative show of force sapped morale, as did the continued fruitless attempts to secure outside support. On 24 September, an Estonian military delegation travelled to Königsberg to sound out the Nazi response to a Soviet attack on Estonia. The response was predictable: Germany would provide no aid in the event of an Estonian-Soviet war, nor would it allow the transit of materials from western Europe through the Baltic.[90]

Since simultaneous overtures to Finland and Latvia also came to nothing, Päts and his advisors—Laidoner included—ultimately decided that it was better to face an uncertain future under Soviet auspices than to risk the decimation of a large part of the Estonian nation.[91] In taking this decision, Päts mistakenly calculated that a Nazi-Soviet war would break out within twelve months, thereby opening up the prospect of German assistance. Later authors have also criticised Päts' alleged belief that the USSR had grown to respect international agreements and would, therefore, honour its promise not interfere in the internal political and economic arrangements of the Estonian state.[92]

Accounts of the panic and paralysis gripping the government in the last months of 1939 mean that one should at the very least question Päts' capacity for decision-making at this crucial juncture. Debates relating to role of Estonian leadership in the crisis have been sharpened by recent findings of historian Magnus Ilmajärv, who has produced evidence that Päts was paid by Moscow to lobby for Soviet economic and political interests during 1924–34.[93] Such revelations compound the controversial claims by Eino Saaremaa that the future *Riigivanem* acted as an agent both for the Tsarist *Okhrana* and the Germans earlier in his political career. In his account of 1939–40, the émigré Saaremaa—an open *Vapsid* sympathiser—goes so far as to hold Päts and Laidoner directly responsible for the spiritual collapse of the nation and the loss of independence.[94] Whilst this is plainly an exaggeration, Päts' subsequent moves to circumscribe civil rights and prevent the flight of persons and capital strengthen the picture of a self-serving and unscrupulous politician whose primary object was to cling to power at all costs, be this as head of a Soviet or subsequent Nazi puppet state.

THE SOVIET ANNEXATION

The policy of the administration during 1939–40 was to avoid any move which might antagonise the USSR. It therefore failed to lodge a protest when the Soviet air forced launched attacks from its Estonian bases during the war with Finland, despite the fact that this constituted a clear violation of the mutual assistance pact. The state-controlled media, meanwhile, continued to profess optimism, suggesting that the agreement with the USSR had enabled Estonia to avoid the fate of its northern neighbour. Contemporary accounts suggest that this optimism was not shared by the population as a whole.[95]

In an ominous portent of things to come, Hitler initiated the evacuation of the vast majority of the Baltic German population to the Reich during October-December 1939. This inaptly-named 'Resettlement to the Reich' *(Nachumsiedlung)* represented a further piece of cynicism on the part of the Nazi leadership, which was primarily interested in procuring 'ethnically reliable' labour for those territories recently-annexed from Poland.[96] Whilst some Estonian leaders welcomed the end of 700 years of German settlement, others could see that it marked the beginning of the end for the Estonian Republic. It is indeed ironic that numerous Estonians now rushed to claim 'German' affiliation or ancestry, thereby availing themselves of the opportunity to flee the country. At the same time, hundreds of men smuggled themselves across the Gulf of Finland to enlist in the armed struggle against Soviet aggression.[97]

The fact that the Estonian government deposited 11 tons of its Gold reserve in Britain, Sweden and Switzerland during the first half of 1940 suggests that by this time, it understood only too well what was in the offing. As soon as the attentions of Germany and the western powers were diverted by war in France and the low countries during the early summer, the USSR duly moved to consolidate its grip over the Baltic states. On 16 June, Estonia was issued with a further ultimatum demanding the establishment of a government 'capable and willing to warrant honest execution of the Soviet-Estonian mutual assistance pact'. The pretext on this occasion was the alleged pro-allied sympathies of the Estonian government and its Baltic neighbours, which had belatedly attempted to forge closer mutual ties during the year. With at least 25,000 Soviet troops already in place, resistance was futile, although Päts again sounded out Germany before bowing to the inevitable.[98] The following day, some 90,000 additional troops entered the country. The military occupation of Estonia was now complete.

Having pointedly denied any intention of annexing the three Baltic states, Stalin dispatched emissaries to each of the countries' capitals. Andrei Zhdanov, arriving in Tallinn on 19 June, dictated the composition of a new 'people's government' containing socialists and left-leaning intellectuals but no known communists. Headed by the poet Johannes Vares, this new cabinet was installed amidst 'spontaneous' popular demonstrations which were in fact carefully orchestrated by occupying military forces.[99] Vares and his colleagues similarly promised to uphold Estonia's sovereignty, ruling out the establishment of a Soviet-style regime. Yet the Popular Front-style government was no more than a puppet of the Soviet legation headed by Zhdanov. The true intentions of the occupying regime with regard to the Baltic countries were revealed by Molotov in July 1940 during a conversation with Vincas Krévé-Mickevicius, Deputy Prime Minister of an identical 'people's government' installed in Lithuania. In response to Mickevicius' complaint that Moscow's plenipotentiaries were overstepping their powers, Molotov claimed that Russia had aspired to possess the Baltic region since the time of Ivan the Terrible. In the modern world, he said, small states have no future. The Baltic countries were to be incorporated into the USSR.[100]

The Soviet Union went to great lengths to dress up this forcible incorporation as a popular revolution. This myth was assiduously cultivated by a Soviet historiography which branded the Päts regime 'fascist' and Soviet power the true expression of the will of the Estonian people. This was used to underpin notions of Russia's 'historic claim' to the territory of the three Baltic states.[101] The events of June–August 1940 were more accurately characterised by U.S. Under Secretary Sumner Welles as 'devious processes whereunder the political independence and territorial integrity of the three small Baltic republics ... [were] deliberately annihilated by one of their more powerful neighbours'.[102] In early July, the by now ailing President Päts consented to a flagrant violation of the existing consitution by ordering new parliamentary elections in Estonia within ten days. Through a combination of intimidation of independent opposition candidates and further violations of existing electoral laws, these elections were, to use Taagepera's phrase, 'de-choiced' entirely.[103] In all but one of the eighty electoral districts, ballots carried the name of a single candidate nominated by the Estonian Working People's League (Eesti Töötava Rahva Liit—ETRL), a front organisation for the newly-legalised Communist Party of Estonia. No other political party was allowed to put forward candidates. As election day approached, the new government daily *Rahva Hääl* warned that only 'enemies of the people' would stay at home on polling day. Most did not in fact turn out to vote. Nevertheless, in true Stalinist fashion, official results registered 92.8% approval for the ETRL on the basis of an 84.1%

turnout. Subsequent evidence of widespread ballot-rigging by the Central Electoral Committee and its local representatives testifies to the obvious absurdity of Soviet claims.[104]

At its first meeting on 21–23 July, the new puppet assembly drafted an application for membership of the USSR. A few days later, President Päts was forced to tender his resignation and was then promptly deported to Russia along with other representatives of the political and military leadership.[105] On 6 August, the Supreme Soviet of the USSR acceded to the membership request of the Estonian 'parliament', as *Pravda* hailed 'the birth of the 16th Soviet Republic'. In all of this—as Mart Laar has noted—'no-one asked the Estonian population for its opinion.'[106] As we shall see in the next chapter, the western allies also wrote off the Estonians during World War Two, when the need to maintain the anti-Nazi alliance with the USSR ruled out the possibility of any active support for the restoration of Baltic independence. The Estonians thus found that their interests had again been subsumed within the broader 'Russian question', much as they were in 1918–20.[107] Nevertheless, the crucial commitment to legal continuity, coupled with the inter-war experience of nation-building, gave grounds to believe that, one day, an Estonian state would be restored.

1   *The Baltic Independent*, 20 July 1995.
2   Eesti NSV Teaduste Akademia, *Eesti NSV Ajalugu III*, (Tallinn, 1971), p.177.
3   John Hiden and Patrick Salmon, *The Baltic Nations and Europe* (London, 1992), pp.43–44.
4   Rein Taagepera, *Estonia: Return to Independence* (Boulder, 1993), p.14; Toivo Raun, *Estonia and the Estonians* (Stanford, 1991), pp.11–12.
5   On the conquest of the eastern Baltic lands, see Eric Christiansen, *The Northern Crusades* (London, 1997).
6   Toomas H. Ilves, "Estonia on its Way to an Integrated Europe," speech given at the University of Latvia, 10 March 1998. *http://www.vm.ee/eng/index.html*; Lennart Meri, "Läänemeri on Meie Elu Telg," in Lennart Meri, *Presidendikõned* (Tartu, 1996), p.279; Lennart Meri, "Meie Piir on Euroopa Väärtuste Piir," in Meri, p.326.
7   See, for instance, the speech given by President Meri in Hamburg on 25 February 1994. Lennart Meri, "Mis Käärib Praegu Venemaa Avarustes?," in Meri, p.381.
8   Raun, pp.56–57; Marko Lehti, *A Baltic League as a Construct of the New Europe* (Frankfurt, 1999), pp.61–67.
9   C.A. Macartney, *National States and National Minorities* (New York, 1968), p.406.
10 Taagepera, p.22.
11 Andrus Park, "Ethnicity and Independence: The Case of Estonia in Comparative Perspective", *Europe-Asia Studies*, Vol.46, No.1, 1994, pp.69–70; Raun, p.53.
12 Raun, p.29.
13 Ibid, pp.42–46.
14 Taagepera, pp.26–31; Raun, p.56.

15 Edward C. Thaden, "Reform and Russification in the Western Borderlands 1796–1855," in *Russification in the Baltic Provinces and Finland 1855–1914*, ed. Edward C. Thaden (Princeton, 1981), p.19.

16 Edward C. Thaden, "Administrative Russification in the Baltic Provinces 1855–1881," in Thaden ed., p.34.

17 Ray Abrahams and Juhan Kahk. *Barons and Farmers: Continuity and Trasnformation in Rural Estonia (1816–1994)*, (Gteborg, 1994), pp.21–26.

18 Lehti, pp.63–64.

19 Taagepera, p.30.

20 Raun, pp.54–55.

21 Taagepera, p.32.

22 Raun, p.64.

23 Lehti, p.63; James D. White, "Nationalism and Socialism in Historical Perspective," in *The Baltic States: The Self-Determination of Estonia, Latvia and Lithuania*, ed. Graham Smith (Basingstoke, 1994), pp.22–23.

24 White, loc cit.

25 Raun, p.45

26 Abrahams and Kahk, pp.24–27.

27 Edward C. Thaden, "The Abortive Experiment: Cultural Russification in the Baltic Provinces 1881–1914," in Thaden ed., pp.54–75.

28 Ibid, p.58.

29 Ibid, p.72.

30 Raun, p.63.

31 Ibid, p.82.

32 White, p.29; Eino Saaremaa, *Eestlaste Ajalugu 1820–1945* (Tallinn, 1997), p.83.

33 Hiden and Salmon, p.22–23.

34 Ibid; Georg Von Rauch, *The Baltic States: The Years of Independence 1917–1940* (London, 1995), pp.16–17.

35 Mati Graf, *Eesti Rahvusriik* (Tallinn, 1993), p.37; Kai Laitinen, "Kirjallisuus Vuoteen 1940," in *Viro: Historia, Kansa, Kulttuuri*, ed. Seppo Zetterberg (Helsinki, 1995) p.239.

36 Ibid, p.240.

37 Saaremaa, p.89; Raun, p.100.

38 Von Rauch, p.31.

39 Raun, pp.101–104.

40 Hiden and Salmon, pp.32–33.

41 Tönu Parming and Elmar Järvesoo, "Introduction," in *A Case Study of a Soviet Republic: The Estonian SSR*, ed. Tönu Parming and Elmar Järvesoo (Boulder, 1978), p.6.

42 Anu Mai Kõll, *Peasants on the World Market: Agricultural Experience of Independent Estonia 1919–1939* (Stockholm, 1994), pp.43–45.

43 Von Rauch, p.90.

44 Anatol Lieven, *The Baltic Revolution: Estonia, Latvia, Lithuania and the Path to Independence* (New Haven and London, 1993), p.64.

45 Tönu Parming, *The Collapse of Liberal Democracy and the Rise of Authoritarianism in Estonia*, (London, 1975), pp.14–16.

46 Hiden and Salmon, p.50.

47 Ibid; Parming, pp.13–17.

48 Parming, pp.10–13; Von Rauch, pp.111–117.

49 Macartney, p.408; Von Rauch 1993, pp.135–145.

50 David J. Smith, "Retracing Estonia's Russians: Mikhail Kurchinskii and Interwar Cultural Autonomy," *Nationalities Papers*, Vol.27, No.3, 1999, pp.455–474.

51 Parming, p.13.

52 Andres Kasekamp, "The Nature of Authoritarianism in Interwar Estonia," *International Politics* (formerly *Coexistence*), No.33, March 1996, p.57.

53 Parming, pp.36–37.

54 Ibid, p.17.
55 On the EVL, see: Andres Kasekamp, "The Estonian Veterans' League: A Fascist Movement?" *Journal of Baltic Studies*, Vol.24, 1993, pp.263–68; Andres Kasekamp, "Radical Right-Wing Movements in the North-East Baltic," *Journal of Contemporary History*, Vol.34, No.4, 1999, pp.587–600.
56 Ants Ruusmann, *Eesti Vabariik 1920–1940* (Tallinn, 1997), p.102.
57 Kasekamp 1996, pp.58–59.
58 Ibid; Raun, p.119.
59 Raun, p.119; Kasekamp 1996, p.59; Von Rauch, p.151.
60 Ruusmann, pp.111–112.
61 Ibid, p.113.
62 The term derives from comments by Einbund, who announced that, henceforth, the Assembly would assume 'a silent existence.' Kasekamp 1996, p.59.
63 Parming, p.57.
64 Ruusmann, p.115; Von Rauch, p.159; Parming, p.58.
65 Parming, pp.56–57; Hiden and Salmon, pp.51–52.
66 Taagepera, p.56.
67 Ruusmann, p.134.
68 Von Rauch, p.174; Parming 1975, p.57.
69 Nicholas Hope, "Interwar Statehood: Symbol and Reality," in *The Baltic States: The Self-Determination of Estonia, Latvia and Lithuania*, ed. Graham Smith (Basingstoke, 1994), p.63.
70 Kõll, p.112.
71 David Kirby, *Finland in the Twentieth Century* (London, 1979), pp.87–88; see also Kasekamp 1999, p.599.
72 Hope, loc cit; 62–63.
73 Kõll, p.111.
74 Parming, pp.63–64; David Kirby, *The Baltic World 1772–1993: Europe's Northern Periphery in an Age of Change*, (London, 1995), p.327; Raun, p.121.
75 Rein Ruutsoo, "Rahvusvähemused Eesti Vabariigis," in *Vähemusrahvuste Kultuurielu Eesti Vabariigis* (Tallinn, 1993), pp.13–14.
76 David Smith, pp.468–469.
77 Iur'evskoe Russkoe Sobranie, *Materialy, Kasayushchiesia Kul'turnoi Avtonomii Russkogo Men'shinstva* (Tartu, 1936), Eesti Ajalooarhiiv f.2097, n.1, su.16.
78 Hiden and Salmon, pp.57–58.
79 Cited in John Alexander Swettenham, *The Tragedy of the Baltic States*, (London, 1952), p.24. A supplementary protocol to the agreement, signed on 28 September 1939, assigned Lithuania to the Soviet sphere. For a full discussion of Estonia's international relations between the wars, see Chapter Five.
80 Ibid, p.30.
81 Raun, p.141.
82 Anthony Upton, *Finland 1939–40* (London, 1974), pp.21–43.
83 Kasekamp 1996, pp.61–62.
84 Taagepera, p.73.
85 W.H. Galliene (British Ambassador to Estonia), Despatch to Viscount Halifax, 5 December 1939, FO 419/33189, *Public Record Office*, London.
86 Saaremaa, pp.154–155.
87 Ibid, p.150.
88 Taagepera, p.73.
89 Jaakko Korjus, *Viron Kunniaksi* (Hämeenlinna, 1998), p.20.
90 Ibid, pp.20–21.
91 Hiden & Salmon, p. 111.
92 Raun, p.143; *Baltic Independent*, 11–17 August 1995; Mart Raud, *Kaks Suurt: Jaan Tõnisson, Konstantin Päts ja Nende Ajastu* (Toronto, 1953), pp.317–318.
93 Jukka Rislakki, "Virolaisen Historiantutkijan Väite: Valtionpäämies Päts sai rahaa Moskovasta," *Helsingin Sanomat*, 8 September 1999.

94   Saaremaa, pp.151–152.
95   Galliene, 5 December 1939; W.H. Galliene, Despatch to Viscount Halifax, 9 October 1939, FO 419/3398, *Public Record Office*, London.
96   Hiden and Salmon, p.115.
97   Korjus, pp.48–49.
98   Raun, p.146.
99   Taagepera, p.61.
100 Cited in V. Stanley Vardys and Judith Sedaitis, *Lithuania. The Rebel Nation* (Boulder, 1997), pp.51–52.
101 Eesti NSV Teaduste Akademia, pp.479–502; Kristian Gerner and Stefan Hedlund, *The Baltic States and the End of the Soviet Empire* (London, 1993), pp.57–60.
102 Swettenham, p.50.
103 Taagepera, p.62.
104 Romuald Misiunas and Rein Taagepera, *The Baltic States. Years of Dependence 1940–1990* (London, 1993), p.28.
105 Päts died in a psychiatric hospital near Kalinin in 1956. His remains were returned to Tallinn for reburial during 1990.
106 Mart Laar, *War in the Woods: Estonia's Struggle for Survival 1944–1956* (Washington, 1993), p.7.
107 Hiden and Salmon, pp.121–125.

Chapter 2

THE LONG SECOND WORLD WAR: ESTONIA UNDER OCCUPATION 1940–91

Molotov's pronouncement that small states had no future would return to haunt a later generation of Soviet leaders. This was despite the best efforts of the new regime, which set out to eradicate all traces of the inter-war Estonian Republic following its assumption of power. The first phase of Soviet rule lasted only until August–October 1941, when Estonia fell to the invading forces of Nazi Germany. In his work dealing with the rise and fall of Communism in the states of the former eastern bloc, Patrick Brogan refers to the period 1939–89 'the Fifty Years War.'[1] This idea assumes a particular relevance in the case of the Baltic peoples. For the previously sovereign nations of central and eastern Europe subjected to German occupation during 1939–45, the defeat of Nazism at least brought the physical restoration of independent statehood, however nominal. For the Estonians and their neighbours, on the other hand, 'liberation' by Soviet forces was simply the prelude to forcible reincorporation into the USSR. The *de facto* restoration of the Estonian Republic therefore had to wait until 1991. In the intervening years, the Baltic question was left in abeyance, pending settlement of the 'unfinished business' of World War Two.

THE FIRST YEAR OF SOVIET OCCUPATION

In 1940–41, the Soviet regime devoted most of its energies to transforming the urban economy. Banks and large-scale industry were nationalised almost overnight, whereas smaller private enterprises were gradually forced out of business by high rents and taxation.[2] Wage and monetary policy—including the introduction of the Soviet rouble at an artificially low exchange rate and the confiscation of all savings accounts over 1,000 roubles (approx £50 at 1940 prices)—combined with rising prices to bring about a plunge in living standards. New regulations simultaneously increased the length of the working day and restricted the free movement of labour.

In the countryside, all land in excess of 75 acres in private possession was confiscated and placed into a state reserve along with Church and local authority land. The reserve was used to supply the poorer elements of the rural population with land up to a maximum of 30 acres.

Notwithstanding official rhetoric regarding the creation of self-supporting agricultural households, the new farms were too small to function as independent units. In effect, the land reform was simply an intermediate stage in the transition to collectivised agriculture on the Soviet model. To this end, around a hundred collective farms were created in Estonia during 1940–41, while medium-sized farmers and so-called 'kulaks' were subjected to punitive taxation and compulsory deliveries to the state at artificially low prices.[3]

Sovietisation was only partially complete by the time German troops overran the country. Even so, in twelve months the authorities are estimated to have executed 2,000 Estonian citizens and deported 18–19,000 more to labour camps or exile in the Soviet interior.[4] Of these, over half (10,517) were removed in the space of a few days prior to the German attack on the USSR.[5] This action by the forces of the Soviet NKVD (Narodnii Komitet Vnutrennykh Del—the forerunner to the KGB) had been planned already in late 1940, central guidelines establishing 14 categories of individuals earmarked for deportation. The latter included former state officials and army officers, industrialists, large landowners and clergymen. Many of the above managed to slip through the net in June 1941, since the figure of 10,000 represented less than half the total number slated for deportation. Available evidence, moreover, suggests that victims were often selected at random.[6]

To losses through deportation can be added the 33,000 Estonians who were forcibly conscripted into the Soviet army during 1941. Although at least some deportees and draftees survived to return to Estonia after Stalin's death, a considerable proportion perished either through forced labour or in armed conflict. Total population losses for the year have been estimated at 54,000–60,000, almost 6% of the total population.[7]

These initial experiences of Soviet rule suggested that even token resistance in 1939–40 would have been preferable to capitulation. As it was, in response to the wave of deportations and forced conscription, many men took to the forests, where they formed the nucleus of a resistance movement. As the occupying forces began their withdrawal in the face of the German advance, guerrilla units numbering several thousand engaged Soviet 'destruction battalions' charged with carrying out a scorched earth policy. In some areas of southern Estonia, pro-independence administrations were already in place by the time German troops arrived. Having somehow managed to escape deportation,

Jüri Uluots (Prime Minister from October 1939–June 1940) set up a co-ordinating council in Tartu, yet stopped short of declaring a provisional government.

THE NAZI OCCUPATION 1941–44

The experiences of the previous year led many Estonians to greet the Germans as liberators, an illusion which was swiftly dispelled during the early months of Nazi occupation. On the subject of Estonian independence, the views of Alfred Rosenberg (Hitler's Reich Minister for the Occupied Eastern Territories and a Baltic German born and raised in Tallinn) coincided entirely with those of Molotov.[8] The incoming German military and civil administration disarmed the local resistance movement and flatly rejected Uluots' proposals for an Estonian government and army. Nor did the occupying regime take any significant steps to reverse the economic changes perpetrated by the Soviets. The long-term intention of the Nazi leadership was to annex the territory of Estonia directly to the Reich. Under the *Generalplan Ost* devised in 1942, half of the population was slated for resettlement to the East over the coming 25 years. This was to pave the way for settlement by German colonists who would gradually assimilate the remaining indigenous inhabitants.[9]

In the more immediate term, the Germans' main interest lay in harnessing the Estonian manpower and resources to the war effort. The new regime did not restore private property confiscated by the Soviets, and proceeded to exploit ruthlessly the economic base of the republic. As elsewhere in Nazi-occupied eastern Europe, Hitler's 'New Order' found at least some active adherents amongst the local population. In an attempt to consolidate the occupying regime, the ruling *Reichskommissariat* set up a puppet native administration headed by the previously-exiled former Veterans' League leader Hjalmar Mäe. This engaged in score-settling with Communists and other former opponents from the independence era. A newly-established Estonian Home Guard (*Omakaitse*) also abetted SS *Einsatzgruppe A* in the extermination of the thousand or so Estonian Jews who remained in the country following the Soviet withdrawal.[10]

A further 6,000 Estonian citizens were executed during the course of the Nazi occupation.[11] Many more prisoners from elsewhere in German-occupied Europe perished in slave-labour camps established on Estonian territory. Whilst the Mäe administration must be held at least partly responsible for these atrocities, the vast majority of Estonians were understandably alienated both by the Nazi regime and its Estonian stooges. Resistance to the exactions of the occupier was expressed through a campaign of non-compliance, co-ordinated by underground political circles and a clandestine press. In many cases, representatives of the official bureaucracy also proved less than scrupulous in their implementation of German directives. Thus, efforts to mobilise local manpower to work in the Reich yielded meagre results, as did an initial call for volunteers to an Estonian Legion of the Waffen-SS founded in late 1942.[12] In response, the Germans resorted to forced mobilisation the following Spring, but draft-dodging remained widespread. A further 6,000 Estonian men of fighting age escaped to Finland, where many enlisted to fight Soviet forces on the Karelian front.[13]

THE SOVIET RECONQUEST 1944–53

As the tide of war turned inexorably against the Germans, Soviet troops stood poised to re-enter Estonia by the spring of 1944. At this point, underground resistance circles came together to form a National Committee of the Republic of Estonia. In the first instance, the initiative came from representatives of the pre-war opposition parties, but the latter were quickly joined by Uluots and other supporters of the former Päts regime, who had hitherto constituted a separate wing of the resistance movement.[14] The Committee aimed to organise resistance to the Soviet invader and to prepare for the establishment of a provisional government following the expected German withdrawal. Largely in response to a radio broadcast by Uluots, some 38,000 Estonians had answered a call for general mobilisation in February. These were later joined by Estonian units of the Finnish army, returning under an amnesty granted by the hard-pressed occupying regime. Although poorly equipped, these reinforcements helped to stabilise the front until July, when a renewed Soviet onslaught conquered mainland Estonia within two months.[15] In the midst of the German evacuation, Uluots appointed a provisional government. It sat for just two weeks before its capture by the Red Army. The acting Head of State, meanwhile, was amongst the 70,000 Estonians who fled westwards to Germany and Sweden in full awareness of the fate which awaited them should they choose to stay in the country.[16]

Soviet terror duly resumed in the autumn of 1944, when an estimated 30,000 people were deported from 'liberated' areas of the Baltic States. Most at risk were those who had served in the German or Finnish armed forces, although no-one who had lived through the Nazi occupation was above suspicion. Many former combatants had in any case already fled the country or taken refuge in the forests, where they continued the armed struggle against units of the NKVD. Partisan groups were joined by escapees from the Soviet draft and other individuals who simply 'could no longer tolerate the insecurity of civilian life.'[17]

In the course of its nine-year existence, the 'Forest Brotherhood' may have involved as many as 30,000 people. With hindsight, it has been criticised for its emphasis on military struggle at the expense of longer-term political organisation. From the perspective of the late 1940s, however, it was still not unreasonable to expect military assistance from the West. Through the intermediary of the diplomatic legation in Helsinki and political exiles in Sweden, the Estonian National Committee had opened channels to the western allies during 1944. In so doing it invested undue hopes in the 1941 Atlantic Charter pledging the restoration of sovereign rights to peoples under Nazi occupation.[18] Compared to 1918–20, however, the international situation proved far less favourable to Estonian aspirations. For the duration of the war, Britain and the USA were determined not to let the fate of the Baltic countries become a sticking point in dealings with their Soviet ally. In practice, they had little option but to accept de facto Soviet control of the region, which was effectively conceded at the Teheran conference of 1943 and definitively at Yalta in 1945.

As the Cold War gathered momentum, the local guerrilla movements could draw some encouragement from the start of British intelligence operations in the Baltic. Yet these networks provided little in the way of material support, and were in any case quickly infiltrated by the NKVD. In the absence of more effective outside assistance, the Forest Brethren faced insurmountable odds. The movement obtained fresh recruits during the collectivisation of agriculture in 1949, when an estimated 60,000 people were deported from Estonia within the space of a single month.[19] Yet collectivisation and deportation also severely disrupted the movement's support network in the countryside, whilst the remnants of an already beleaguered rural population increasingly began to resent food requisitions by the partisans. Although some fugitives remained at large in the forests until the 1970s, armed resistance had effectively come to an end by the time of Stalin's death in 1953. Soviet rule, it seemed, was back to stay.

The massive population losses sustained during the war were compounded by a frontier revision which transferred the eastern districts of Petseri (Pechora) and Prinarova to the Russian Republic of the USSR in January 1945. This deprived Estonia of much of its inter-war Russian minority, leaving a population which was 93% ethnically Estonian immediately after the war. Of the 7,000 members of the Communist Party of Estonia (CPE) at this time, however, only 27% were native Estonians.[20] At the time of the original Soviet occupation, the CPE had numbered only 133 in Estonia. Many of its original leaders—including Jaan Anvelt—had met their end in the USSR during the pre-war Stalinist purges, whilst other native members of the 1940–41 Soviet government had perished during the war or its immediate aftermath. From 1946–50, the CPE was headed by the native Estonian Nikolai Karotamm, yet the bulk of the post-1944 party and administrative elite consisted either of ethnic Russians or Russified Estonians born and raised in the pre-war Soviet Union. The preponderance of these groups was partly born of necessity, yet the Russian-Estonians were also deemed more politically reliable than party members who had lived in Estonia during the inter-war years.

The extent of Stalin's paranoia in this regard became clear when Karotamm was dismissed on the grounds of his 'bourgeois nationalist' tendencies. He was replaced by the Russian-Estonian Johannes Käbin (then known by the Russified variant Ivan Kebin), who was to occupy the post until 1978. Karotamm's closest circle of associates was also purged, and by 1952 the republican government and the CPE Politburo and Secretariat were run entirely by Russians or Estonian-Russians.[21] The republican Supreme Soviet, on the other hand, contained a native Estonian majority as early as 1955. Up until the late 1980s, however, this institution remained little more than a rubber stamp for policies which were, for the most part, decided in Moscow.

'A NEW PRAGMATISM': 1956–1968

Developments nevertheless entered a qualitatively new phase follow-ing the death of Stalin in 1953. During Khrushchev's tenure as CPSU First Secretary (1956–1964), the doctrine of 'different roads to socialism' appeared to offer the prospect of a meaningful national existence within the parameters of Soviet communism. Khrushchev's message was duly heeded by the CPE leadership, which sought to achieve a *modus vivendi* with the local population. In 1957, teaching of Estonian history and geography was reintroduced into schools, albeit within a Soviet ideological framework. Käbin himself underwent a process of re-Estonianisation during the years which followed. The Russified variant of his surname was dropped and he began to use the Estonian language more frequently in official communications.

The Hungarian uprising of 1956 led briefly to an upsurge of dissident activity in Estonia, yet the subsequent military intervention by Soviet forces in Hungary quickly dispelled any lingering hopes that the West might intercede to restore the sovereignty of Moscow's satellites in east-central Europe.[22] Instead, Estonians who came of age in the late 1950s—early 1960s increasingly displayed what has been termed a 'new pragmatism' towards the Soviet system. Communist Party membership had reached 40,000 by the mid 1960s, half of which was drawn from the titular nation.[23]

For the most part, newer party members adopted a 'National Communist' ideology, attributing economic and cultural progress in the ESSR to the efforts of the Estonians themselves rather than to 'fraternal' aid in the form of subsidies from Moscow.[24] In this respect, they diverged from the top party leadership, which would never go so far as to advocate a clean break with the USSR. As was the case in Latvia in 1959, Moscow moved quickly to quash any overt manifestations of nationalism on the part of republican leaders. Käbin therefore sought to act as a buffer between Moscow-based centralisers on the one hand, and advocates of Estonian national self-determination on the other. It was a role which he was to fulfil with varying degrees of success throughout his long period in office.

The early 1960s were particularly notable for a renaissance of cultural activity compared to the suffocating strictures of the Stalinist period. Estonian artists and writers benefitted from the limited restoration of contacts with the exile community, which also provided valuable material support to relatives left behind in 1944.[25] In the economic sphere, ESSR Prime Minister Aleksei Müürisepp strongly supported Khrushchev's *Sovnarkhoz* system of territorially-based economic management. Under the new arrangements, control of 80% of Estonia's industry passed to a Regional Economic Council, whereas previously three quarters of enterprises had been in the hands of sectoral ministries in Moscow. Consequently, local elites were able to stem the flow of Russian immigration to the republic in the late 1950s. During this same period, Käbin and his associates secured the status of 'experimental republic' for the ESSR.[26]

Brezhnev's accession to the post of CPSU First Secretary in 1964 heralded the end of the *Sovnarkhoz* experiment and a reversion to centralised economic control over the union republics. In cultural affairs, however, the curtailment of autonomy was not so immediately apparent. Estonia also retained its status as a laboratory for economic experiments, playing a pioneering role in the Kosygin managerial reforms introduced in 1965.[27] Agriculture benefitted from increased investment and innovation, including the introduction of local self management on state farms and the more extensive use of private plots. According to one author, Estonian GDP per capita rose by 90% during 1958–1968, compared to a Soviet average of 67%.[28] Whilst one obviously hesitates to take these statistics at face value, it is clear that in Soviet terms at least, Estonia enjoyed high living standards during this period.

In the course of the decade, a pro-reform movement emerged within the *Komsomol* of Tartu University, spearheaded by a generation which was too young to remember the full horrors of Stalinism. Within its ranks were later leaders of the 1980s 'new national awakening' such as Marju Lauristin, Rein Ruutsoo, Siim Kallas and Trivimi Velliste.[29] Thus, in spite of its best efforts, the Party leadership could not insulate Estonia from the wave of unrest which swept Europe during 1968. The subsequent intervention by Warsaw Pact forces in Czechoslovakia marked an important watershed, signalling an end to hopes of establishing an autonomous brand of 'socialism with a human face' in Estonia. As elsewhere in the USSR, the 'Brezhnev Doctrine' led to a crackdown against western-inspired 'ideological diversions': liberal editors of cultural journals were purged, and there was a return to discipline at the University of Tartu. Some former *Komsomol* activists focused on advancing their party careers; others renounced politics in favour of scientific research or cultural life. A third group, however, began to gravitate towards the dissident movement which underwent a major revival from 1968 onwards.[30]

BREZHNEVITE STAGNATION

Emergent nationalist tendencies continued to be contained through the ruthless repression of dissident groups.[31] At the same time, however, the Brezhnev regime resorted to 'corporatist politics' in an attempt to maintain the cohesion of the Soviet state. A policy of 'trust in cadres' led to further indigenisation of the CPE membership during the 1970s. Across the USSR, local party bosses were given stability in office and a greater degree of personal autonomy over their fiefdoms, in return for which they were expected to maintain social stability.[32]

This system proved workable during the 1970s, when most Estonians 'created an adaptive way of life ... [allowing] ... them to pursue their own personal interests without overt conflict with the official political system.'[33] Yet internal migration from politics went hand in hand with growing cynicism and materialism. Misiunas and Taagepera claim that by 1980 'consumerism was ... weakening the ideological underpinnings both of the regime and of dissent. It was not clear ... which one would be more affected'.[34] This remark highlights the wider debate amongst western scholars regarding the future of the Soviet Union. For the vast majority, any suggestion of imminent collapse seemed preposterous. As late as 1987, Alexander Motyl predicted that the practice of co-opting potential leaders amongst the national minorities would enable the regime to nip any manifestations of ethnic discontent in the bud.[35]

In reality, 'national communists' within the Estonian hierarchy were becoming increasingly conscious of their lack of real political power by the start of the 1980s. In 1978, the aging Johannes Käbin was kicked upstairs to the largely decorative post of Chairman of the ESSR Supreme Soviet. Native Estonian Vaino Väljas (Ideology Secretary to the CPE Central Committee) was widely tipped for the leadership, but was passed over in favour of Karl Vaino, a Russian-Estonian who had never mastered the local language. Väljas, meanwhile, was sent abroad to serve as Soviet ambassador to Venezuela and Nicaragua.

Vaino's appointment could hardly have been more inopportune, for it came at a time when the national question was beginning to challenge the 'regime stasis' of the Brezhnev era.[36] In this regard, the major focus of discontent related to mass immigration, which had resumed following the abolition of territorially-based economic management in 1964. The settler population was concentrated mainly in Tallinn (50% Russian by 1980) and the north-eastern cities of Narva and Sillamäe, where barely 5% of inhabitants were of Estonian nationality. Immigrants were typically industrial workers and technical personnel, many of whom stayed no more than a couple of years before moving on. This transient element was especially liable to arouse resentment, being the least inclined to learn the Estonian language and the most likely to obtain access to new housing. Inadequate living quarters have been cited as one factor behind a continued decline in the birth rate amongst the Estonians, who had one of the lowest levels of fertility in Europe already during the inter-war period.[37]

Despite the high turnover of immigrant labour, 43% of non-Estonians residing in the ESSR in 1989 had been born there. Of the remainder, over half had lived there for more than 25 years.[38] Those who did make the effort to learn Estonian generally found acceptance amongst the titular nationality, which regarded them as a group of 'local Russians' distinct from the mass of the settler population.[39] Yet with Russian assuming an ever-greater role in education and scientific research, many began to fear for the long-term future of the Estonian language within a modern industrialised society. In 1978 the CPSU Central Committee decreed a further increase in Russian language teaching in non-Russian schools right down to kindergarten level. The same principle was applied to the universities, where all doctoral dissertations were now to be written in Russian.

The CPE leadership under Vaino adopted these new proposals without reservation, even though the majority of Estonians perceived them as assimilative in intention and effect.[40] Dissident tracts had steadily assumed a more ethno-nationalist tone during the 1970s, when settlers were branded as a 'civil garrison' and 'an ominous tumour in the body of the Estonian nation.'[41] In 1980, educational policy provided the explicit focus for large-scale youth disturbances in Tallinn. The official designation of the rioters as mere 'hooligans' prompted 40 representatives of the scientific and cultural establishment—including Marju Lauristin, Andres Tarand and Rein Ruutsoo—to publish the so-called 'Letter of Forty', in which they attributed the disturbances to Soviet nationalities policy and its attendant social problems.[42] Although this initial appeal for reasoned debate fell on deaf ears, it marked the first open manifestation of a more pragmatic strand of nationalism committed to undermining the system from within.[43] Its opportunity came following Gorbachev's accession to the post of CPSU General Secretary in 1985.

THE COLLAPSE OF SOVIET POWER 1985–91

Gorbachev recognised that reform of the Soviet system was both necessary and long overdue. Despite some suggestions to the contrary, however, he never contemplated relinquishing Moscow's grip on the Baltic states.[44] His aim was rather to strengthen the USSR through a process of managed change from above, with fundamental restructuring (*perestroika*) of the command economy being seen as the key to accelerating growth.

In an effort to circumvent inevitable opposition from the bureaucracy—the main beneficiaries of Brezhnevite stagnation—*perestroika* was accompanied by a limited restoration of civic freedoms. Individual citizens were given the possibility to criticise the shortcomings of the Soviet system and to form popular associations independent of the CPSU. The introduction of multi-candidate elections within a continuing framework of single-party rule was similarly designed to frighten the ruling *apparat* into compliance with reformist directives issued by the centre. The subsequent downfall of the Soviet system stemmed from Gorbachev's inability to control the forces which he had unleashed. In embarking upon *perestroika*, the last Soviet leader patently underestimated the gravity of the 'national question'. Consequently, he proved incapable of dealing with emergent nationalist tendencies and became perpetually torn between a desire to advance the reform drive and the need to preserve the integrity of the Soviet state.

Accusations of 'naive optimism' levelled at Gorbachev over the national question seem particularly apt with regard to his initial policies towards the Baltic Republics.[45] By virtue of their more recent traditions of market economy and civic participation, the ESSR and its neighbours were again promoted as a showcase for Soviet reform. In 1985–87, Estonia in particular became a testing ground for innovations in the spheres of services and light industry. A year later, it became home to the 'Popular Front in Support of Perestroika'—a prototype for the unofficial mass movements which quickly emerged throughout the Union Republics of the USSR.

The Estonians, however, drew their own conclusions from this acknowledgement of their special status within the USSR. In launching *perestroika*, Gorbachev called for a return to authentic Leninist principles, blaming the current shortcomings of the Soviet system on the flawed legacy of Stalinism. In the non-Russian republics, however, Stalinism was viewed firmly through a nationalist prism. This was especially so in the case of the Baltic peoples, whose very presence within the Union had come about as a direct result of Stalinist policies. Uncovering the 'blank spots' of Soviet history in these republics would thus almost inevitably call into question the current borders of the Soviet state. In the words of Marju Lauristin, a co-founder of the Popular Front of Estonia: 'once it could be said publicly that Estonia and the other two Baltic states [were] illegally annexed territories, countries under foreign, Soviet occupation, something entirely unexpected occurred: the renaissance of a nation, the rebirth of society'.[46]

THE 'NEW NATIONAL AWAKENING': JANUARY 1987–MARCH 1990

Something of Gorbachev's insensitivity towards the national question can be gleaned from comments he made during his visit to Estonia in

February 1987—the first, incidentally, by any Soviet leader. The man who six months earlier had referred to the Soviet Union as 'Russia' during a speech in Ukraine now informed an audience in Tallinn of the need to strengthen 'internationalist' education in the individual republics. Locally, of course, this slogan was perceived as a by-word for Russification.[47]

To add insult to injury, Gorbachev implied that the ESSR was an economic burden to the central government, receiving 3 billion rubles annually from union budget and providing only 2.5 billion in return. In response, local commentators claimed that Estonia exported goods to other union republics at below world prices whereas most of its imports were priced above them.[48] The final straw for an increasingly critical Estonian intelligentsia came on 25 February, when all-union ministries announced plans to commence large-scale mining of phosphate deposits in central and north-eastern Estonia. This scheme would not only have wrought further destruction on the local environment, but would also have brought a further 30,000 settlers and their families to the republic. It became the object of a wave of demonstrations and media criticism which have been described as 'environmental in form ... [but] nationalist in content'. Once again, the Tartu University *Komsomol* was at the centre of the protests. The subsequent decision to shelve the project was the first victory scored by popular mobilisation. The ostrich-like stance assumed by the republican government and party leadership during the crisis, meanwhile, further diminished the credibility of the ruling elite in the eyes of public opinion.

The new permissive climate of *glasnost'* soon paved the way for more radical expressions of nationalism by independence-minded activists. In the early 1980s, Brezhnev and his immediate successor had devoted particular energy to silencing these out-and-out dissidents, one of whom—physicist Jüri Kukk—met an unexplained death in a Soviet prison camp during 1981. Under Gorbachev, however, political prisoners were released from labour camps. Amongst the most prominent were Lagle Parek and Tiit Madisson, who on 23 August 1987 called a demonstration in Tallinn's Hirvepark to mark the 48th anniversary of the Molotov-Ribbentrop Pact. The meeting, which attracted at least 2,000 participants, resulted in the formation of the Estonian Group for the Publication of the Secret Protocols of the Molotov-Ribbentrop Pact (Molotovi-Ribbentropi Avalikustamise Eesti Grupp—MRP-AEG).

In January 1988, this group became the first organisation to call publicly for the immediate and unconditional restoration of Estonian independence, basing its demands upon the argument that Estonia was a *de jure* independent state under illegal foreign occupation. In the interim, a group of activists, including the young historian Mart Laar, founded the Estonian Heritage Society devoted to restoring monuments and other cultural artefacts from the inter-war period.

For the former dissident wing of the nationalist movements, Hirvepark is considered the real onset of *glasnost'* in Estonia, setting the scene for the development of all subsequent political movements.[49] One month after the demonstration, a group consisting of officials, journalists and academics— Edgar Savisaar, Siim Kallas, Mikk Titma and Tiit Made—published a more modest proposal for economic autonomy within the USSR. This 'Self-Managing Estonia' (*'Isemajandav Eesti'* or *IME*) programme was nonetheless far more ambitious in scope than the economic reform plans recently unveiled by the central government. If realised, it would effectively have transformed Estonia into a 'Soviet Hong Kong' with its own budget and tax system; local control over resources (including labour) and fiscal and monetary instruments; self-financing enterprises and prices determined by market forces. There were also proposals to encourage large-scale foreign investment and to conduct trade with other Soviet republics and outside states along market lines using a new convertible Estonian currency.[50]

IME—the acronym means 'miracle' in Estonian—predictably met with rejection from the CPE Central Committee, whose counter-arguments merely served to hammer home the extent of Estonia's dependence on Moscow in the minds of the public. As preparations began for the 19th All-Union Communist Party conference in June 1988, Vaino and his colleagues received a vote of no-confidence from the Estonian intelligentsia. The plenum of Estonian Cultural Unions in April 1988 was the occasion for lively discussions on the problems facing the republic. It was here that writer, director and future president Lennart Meri provided the frame of reference for subsequent wide-ranging debates on national identity, posing the questions 'Who are we? Where have we come from? Where are we going?'.[51] Cultural representatives called upon the forthcoming party conference to institute sovereignty for the republics, organise multi-candidate elections and guarantee the cultural rights of all Soviet nationalities. Some advocated the introduction of separate citizenship for residents of the ESSR and new laws making Estonian the sole official state language.[52] It was on the basis of this meeting that Edgar Savisaar proposed the creation of the Popular Front of Estonia during a debate broadcast live on TV.

Many of the initial founders of the Popular Front were also members of the Communist Party. This movement of the republican intelligentsia attempted to fill the gap between the ultra-conservative CPE leadership and what Savisaar described as the 'unrealistic restorationists' of *MRP-AEG*.[53] Like the Hurt-Jannsen faction a century before, the Popular Front initially sought to advance national interests by working within existing institutions. Savisaar and his colleagues feared that radical demands for the immediate and unconditional restoration of independence might provoke a clampdown by the central authorities which would prejudice any chance of reform.[54]

In reality, the repressive will of the state was waning fast. On 2 February 1988, a demonstration in Tartu commemorating the anniversary of the 1920 peace treaty had been forcibly dispersed by riot police. This, however, proved to be the last occasion on which such methods were used. Three weeks later, 10,000 people gathered in Tallinn to mark the 70th anniversary of Estonian independence. As the authorities again sanctioned discussion of the events surrounding the birth of the Estonian Republic, the Estonian Heritage Society publically displayed the still-prohibited flag of inter-war Estonia in Tartu during April. The re-appearance of the blue, black and white tricolor gave an enormous boost to national consciousness. In June the flag was widely in evidence during a four-day festival of music in Tallinn, which drew a crowd of 60,000. It was during this event that artist Heinz Valk first coined the expression 'singing revolution'.[55]

The revolutionary implications of these events were not lost on Karl Vaino, who attempted to stifle meaningful change in time-honoured bureaucratic fashion. With the approval of the CPE Central Committee, Vaino simply appointed the ESSR delegation to the 19th Party Conference rather than holding the multi-candidate elections envisaged by Gorbachev's reforms. The refusal to include any Popular Front representatives prompted calls by the opposition for a meeting between conference delegates and the people. At this point, Vaino appealed to Moscow for a military clampdown to restore order, yet Gorbachev merely acknowledged popular discontent by dismissing the aged CPE First Secretary and assigning him to a post in his native Russia. Vaino Väljas, recalled from Latin America, was installed as head of the Estonian party. The following day, 100,000 people gathered at a meeting attended by five of the 32 delegates to the all-union conference. By affirming the growing legitimacy of the Popular Front in the eyes of the population, this gathering seemingly vindicated the cautious, gradualist approach at the expense of the radical solutions offered by MRP-AEG.[56]

If Gorbachev had hoped that Vaino's removal would stem the growing tide of nationalism, he was very much mistaken. The new CPE First Secretary quickly showed his national communist credentials, supporting Popular Front demands in his speech to the party conference. When the reformist Indrek Toome replaced Bruno Saul as prime minister in September 1988, the ESSR government espoused the hitherto maligned concept of economic autonomy. Two months later, it announced further plans to make Estonian the official state language of the republic. It was at this point that the republican Supreme Soviet—headed by former CPE agriculture secretary Arnold Rüütel since 1983—finally came into own as medium for advancing the nationalist agenda. The new CPE leadership proved able to push reformist legislation through a body in which three quarters of the delegates were ethnic Estonians.[57]

These measures were undertaken in response to the renaissance of civil society during the second half of 1988. By the autumn, the Popular Front had already attracted a membership of 100,000, roughly equal to that of the CPE.[58] A rally convened by the Front on 11 September 1988 attracted 300,000 participants, almost a quarter of the total population. On the eve of the meeting, Väljas stated that the demands of the people and the Popular Front were now the demands of the CPE.[59] In fact, the Popular Front leadership moved one step ahead of the party during the autumn, when it called for the transformation of the USSR into a confederation of independent states. More radical members, meanwhile, were already openly calling for complete secession from the USSR using the rights enshrined in article 72 of the Soviet constitution.

Events in Estonia and the other Baltic republics were now moving far beyond the bounds envisaged by Gorbachev. In response, the Soviet leader proposed constitutional amendments which would have made it all but impossible for individual republics to leave the USSR. Whilst article 72 was left untouched, an alteration to a later clause left 'decisions on questions of the composition of the USSR' in the hands of the Congress of People's Deputies, a new all-union parliament which was to meet in Moscow.

When the ESSR Supreme Soviet gathered on 16 November, it was told that 861,000 signatures had been collected for a petition opposing the proposed constitutional changes. Unlike his counterparts in Lithuania and Latvia, Väljas oversaw the immediate adoption of a counter declaration 'about the sovereignty of the Estonian SSR'. This stated that amendments to the Soviet constitution would only come into force in the Estonian republic upon approval by the ESSR Supreme Soviet. The declaration earned a stiff rebuke from Moscow and led to a state of constitutional impasse between the two sides. Through its actions, the Soviet leadership thereby dealt 'a crushing blow ... to any belief in *perestroika* as a way of realising the ambitions being voiced by the Popular Front'.[60]

These attempts by the national communist establishment to harness the nationalist wave were also attacked by former dissidents. Following the publication of the secret protocols of the Molotov-Ribbentrop Pact by the Party Daily *Rahva Hääl* during August 1988, MRP-AEG disbanded and reconstituted itself as the Estonian National Independence Party (*Eesti Rahvusliku Sõltumatuse Partei—ERSP*). As far as the radicals were concerned, the halfway house of autonomy within the USSR merely threatened to undermine the drive to restore full independence. Only the latter, they argued, could provide adequate guarantees of national survival in the long-term.[61]

When the republican leadership marked the 71st anniversary of Estonian independence by replacing the flag of Soviet Estonia with the inter-war tricolor, radical nationalists condemned this move as 'an insult to the national flag by the occupying forces'.[62] On the same day— 24 February 1989—ERSP announced the formation of 'Citizens' Committees', whose task was to register all citizens of the inter-war republic and their descendants. Once this process was complete, elections would be held to an unofficial parliament, the Congress of Estonia. Citing the legal continuity argument, the Committee movement argued that only citizens of the inter-war republic were legally entitled to take decisions regarding Estonia's future. Participation in elections to the Supreme Soviet—scheduled for March 1990—was, of course, open to all residents of the ESSR, including the large Russian-speaking population which had settled after 1940. The radicals thus feared—with some justification—that the Popular Front might fail to obtain the two-thirds majority which was required in order to carry a vote for independence under the Soviet constitution.

RUSSIAN REACTIONS

The hoisting of the inter-war Estonian tricolor over government buildings also sparked the first significant protest by those elements of

Estonia's Russian-speaking population which were committed to upholding the status quo. On 14 March 1989 an estimated 30–50,000 people assembled in Tallinn to demand the restoration of the flag of Soviet Estonia.[63] The demonstration also denounced the new language law which required all officials and service personnel to become conversant in Estonian within four years. Although the legislation offered continued guarantees for the official use of the Russian language alongside Estonian, many settlers attacked this new linguistic parity as 'discriminatory'.

The March 1989 demonstration was jointly organised by the Internationalist Movement of the Estonian SSR (Intermovement or *Interdvizhenie*) and the United Council of Work Collectives (*Ob'edinennyi Soyuz Trudovykh Kollektivov—OSTK*). The leadership of these sister organisations mainly consisted of members of the Russian-speaking party-state apparatus resolutely opposed to the concept of republican autonomy. Also well-represented were managers of all-union enterprises which relied on Moscow for subsidised raw materials and a constant flow of transient labour.[64]

Interdvizhenie had been formed in June 1988 following the removal of the old-guard CPE leadership, yet its appeals to the population had proved singularly unsuccessful in mobilising popular support.[65] The focus of activity was thus shifted to the work collective, where managers were able to exert direct influence over their workers. Here, initial attempts to form a single organisation of work collectives uniting both Estonian and non-Estonian enterprises foundered. Estonian representatives, having expressed their support for the Popular Front, broke away to form their own body, the Union of Work Collectives (UWC).[66] The UWC-OSTK split was symptomatic of a growing political division along ethnic lines. At the founding congress of the Popular Front in October 1988, all but 5% of the delegates were ethnic Estonians, a fact generally attributed to the lack of a sizeable humanistic intelligentsia amongst the non-Estonian population.[67]

At the same time, it would be wrong to speak of an ethnically polarised society. The all-union Soviet media, for instance, frequently claimed that the 'Russian-speaking population' in the Baltic was united in its desire for intervention to overthrow 'irresponsible' nationalist governments. Yet a survey conducted in April 1989 actually found that less than a third of non-Estonians supported *Interdvizhenie* and *OSTK*.[68] When elections were held to a new all-union parliament, the Congress of People's

Deputies (CPD) in March 1989, *Interdvizhenie* candidates fared poorly, gaining only five seats out of the thirty-six allocated to the ESSR. The overwhelming majority of votes went to pro-autonomy candidates backed by the Popular Front. The fact that many Russians actually voted for Estonian candidates suggests that the reform communist leadership had retained the confidence of the non-Estonian population.

This in turn belies the picture of an 'irresponsible' nationalist movement painted by Moscow. Although the Supreme Soviet declaration of November 1988 had referred to Estonians as the 'indigenous people', it nonetheless urged 'all those who have tied their fate to Estonia' to participate in the building of a 'democratic and socialist society'. The declaration explicitly ruled out discrimination on the grounds of nationality. The same was true of the January 1989 language law, which was preceded by widespread consultation and discussions with Russian-speaking work collectives.[69]

FROM AUTONOMY TO INDEPENDENCE

By May 1989, a quarter of non-Estonians living in the ESSR supported the Popular Front platform of a 'sovereign republic in a Soviet confederation'. Only 5%, however, favoured an independent Estonia outside the USSR. Amongst the ethnic Estonian population, support for the latter goal was already 56%, with 39% expressing support for the confederation option.[70] Nine months later, prominent nationalist Endel Lippmaa stated that 'over the past two years it has become clear that it is beyond anyone's power to improve the Soviet Union, be it as a group or alone ... [There is] thus ... a universal understanding that Estonia should become an independent state as it once used to be'.[71]

Lippmaa's article was written shortly before elections to the unofficial Congress of Estonia. By the time this 499-member body convened on 11 March (one week ahead of elections to the ESSR Supreme Soviet), all Estonian political forces, including the nationalist wing of the CPE, had rejected autonomy in favour of outright independence. At this time, Lippmaa's 'universal understanding' could not be taken to include non-Estonian residents of the republic. In the Summer of 1989, the ESSR government had attempted to introduce a residence requirement for voters which would have debarred more recent immigrants to the republic. This provision was later rescinded after *OSTK* organised a series of strikes with the tacit approval of Moscow. Even so, the proportion of non-Estonians expressing support for pro-independence parties had risen to 29% by December 1989. More important was the fact that the non-Estonian population had not been persuaded to resist the independence drive by violent means. This testifies to the Estonian nationalists' success in transforming the conflict with Moscow from an ethnic confrontation into a question of territorial sovereignty.[72]

In spite of a virtual blackout of its activities in the official media, the Committee movement was a startling success, registering an estimated 700,000 pre-war citizens and their descendants for the unofficial elections of February 1990. It is notable that registration was also open to post-war immigrants and their descendants who wished to apply for citizenship of a restored Estonian republic. Around 30,000 settlers availed themselves of this opportunity during 1989.

The scale of the Committee movement reflected the groundswell of popular opinion in favour of outright independence. It also posed a clear challenge to the Popular Front and its strategy of gradual emancipation through existing Soviet structures. As its delegates left Tallinn to take up their seats in the Congress of Peoples Deputies in May 1989, the Popular Front again re-assessed its goals. Sovereignty within a new Soviet confederation remained the immediate aim. This, however, was to be merely an intermediate stage on the road to full independence.[73] In October, the Front made a more detailed and unequivocal declaration of its intent to break with the USSR. The steps to achieving this would be: democratic elections, followed by a declaration of Estonia's status as an occupied territory and the abolition of the Estonian SSR; a referendum on independence; and a temporary treaty of confederation with the USSR which would set a date for the establishment of an independent Republic of Estonia.[74]

Behind the radicalisation of nationalist demands lay a growing recognition that Moscow would never sanction genuine autonomy for Estonia within the Soviet system. Any remaining hopes in this regard were quickly dispelled once the CPD convened in Moscow. The new Soviet parliament quickly agreed in principle to grant economic autonomy to the Baltic republics. Actual legislation, however, was delayed until December 1989, as Gorbachev strove to water down the proposals put forward by Baltic delegates. The law which eventually emerged was a 'messy compromise' which pleased no-one: crucial issues such as the ownership of land and natural resources were fudged, whilst the supply system and price control mechanism was to stay in the hands of Moscow.[75]

The intervening months witnessed further acrimonious disputes regarding the Molotov-Ribbentrop Pact. In response to prompting by Baltic delegates, Gorbachev ordered the formation of a 26-member CPD commission to investigate the pact. Chaired by Gorbachev's close ally Aleksandr Yakovlev, this body included 11 representatives from the Baltic republics. Fourteen members of the commission subsequently attested to the authenticity of the secret protocols to the 1939 agreement, which were further deemed to have infringed the rights of the Baltic peoples. Unsurprisingly, however, Yakovlev and Gorbachev refused to accept the link between the Nazi-Soviet pact and the subsequent annexation of the Baltic states, reiterating the fiction that the June 1940 incorporation into the USSR had been based on consent.

The Kremlin was delivered with a potent rejoinder to such claims on the 23 August 1989 (the 50th anniversary of MRP), when tens of thousands of Estonians joined a 2 million-strong human chain stretching from Tallinn to Vilnius. With the eyes of the world upon them, the Baltic Popular Fronts used the occasion to call for the peaceful restoration of statehood by parliamentary means. The 'Baltic Chain' coincided with the start of dramatic changes in east-central Europe, which lent further impetus to the Estonian independence drive during the autumn. At the same time, Estonian nationalist leaders were anxious lest they once again find themselves excluded from the reshaping of Europe. Of particular concern was the prospect that western states might choose to recognise the Baltic republics as autonomous territories within a revamped Soviet federation (see Chapter Five). Gorbachev's furious response to the human chain, meanwhile, had made it clear that patience with the Estonians was wearing thin in Moscow. A resolution of the CPSU Central Committee slammed Baltic 'extremists' for their separatist line and alleged discrimination against non-titular nationalities residing in the republics. Its thinly-veiled threat of a military crackdown foreshadowed Gorbachev's moves to create a strong executive presidency empowered to declare a national emergency and rule by decree.

The ruling party was nevertheless forced to recognise that the tide of change could not be halted at the borders of the USSR. A new nationalities policy unveiled in September 1989 incorporated early Baltic demands for a return to 'Leninist' principles of federation. Similarly, the grudging consent given to economic autonomy at the end of the year could be seen as a belated and largely fruitless attempt to satisfy Baltic demands for sovereignty. Whereas these concessions might have been gratefully received a year earlier, they could not by this time stifle the rising demand for outright independence.

The new conditions of (still informal) political pluralism led Communist parties in the Baltic republics to sever their links with Moscow—rather as their counterparts in eastern Europe had done—in an attempt to avoid eclipse in the upcoming republican elections. On 20 December 1989, the Communist Party of Lithuania (CPL) voted to separate from the CPSU, thereby smashing the main pillar of Moscow's political control over the republic. Two weeks before, the Lithuanian Supreme Soviet had voted overwhelmingly to abolish the CPL's constitutional monopoly on power and legalise a multi-party system. Although Gorbachev initially condemned the move, he too was soon forced to bow to the inevitable. In February 1990, he moved to delete the CPSU's leading role from the Soviet constitution, arguing that the party had to become a democratically recognised force in society.

In Estonia, the leading role of the party was not formally abolished until after the elections to the Supreme Soviet. By this time, however, the reformist wing of the CPE had already declared itself in favour of a socialist Estonia independent of the USSR. This grouping sought to disassociate itself from the party, contesting the elections under the name *Vaba Eesti* (Free Estonia). The pro-independence communists subsequently voted to separate from the CPSU, leaving a residual, mainly Russian-speaking group which continued to profess loyalty to Moscow.

Earlier, on 12 November 1989, the ESSR Supreme Soviet had ruled that the vote to join the USSR in July 1940 had been illegal. Remarkably, this declaration denouncing 'Soviet aggression, military occupation and annexation of Estonia' was echoed by the Congress of Peoples Deputies in Moscow. On 24 December 1989 the Soviet parliament condemned the Molotov-Ribbentrop Pact for its violation of the sovereignty and independence of other nations. It also declared the secret protocols of the pact to be null and void. By virtually admitting to the forcible annexation of Estonia, the CPD declaration strengthened the hand of radical nationalists arguing for the legal restoration of the inter-war republic. For the Committee movement, there could be no talk of secession from the USSR, for Estonia could not leave a union which it had never joined. Instead, the radicals called upon Moscow to withdraw its troops as the first step towards ending the illegal occupation of the country. This, they pointed out, could be done without invoking article 72 of the Soviet constitution. In this way, the restoration of Baltic independence would not set a legal precedent for those republics which belonged to the USSR before 1940.

By degrees, the previously 'unrealistic' restorationist position began to attract influential 'establishment' nationalists such as Lippmaa, one of the leaders of the Union of Work Collectives. On 2 February 1990, UWC organised an informal assembly to coincide with the 70th anniversary of the Estonian-Soviet peace settlement. Some 3,000 Estonian political representatives from the local, republican and all-union levels called upon Moscow to restore Estonian independence on the basis of the 1920 Tartu treaty.[76]

Gorbachev, however, still refused to entertain the idea that Estonia had been forcibly annexed by the USSR. The Soviet leader did mention the possibility, at least in principle, of a 'divorce' between Estonia and the USSR. Any such settlement, however, would have to take full account of the far-reaching changes wrought by Sovietisation, including compensation for Soviet investment losses and military bases left behind.

Proposals for the restoration of independence on the basis of legal continuity also aroused concern amongst post-war settlers, who would lack any automatic rights to citizenship within a future Estonian Republic. The prospect of a violent backlash by this group was the main factor militating against a restorationist approach. Popular Front leaders such as Savisaar thus continued to advocate the creation of a post-Soviet 'Third Republic' giving citizenship to *all* residents of the existing ESSR.[77]

This approach was resolutely opposed by the Congress of Estonia, which on 11 March 1990 declared itself the representative body of the citizens of Estonia. The ideological differences between the Popular Front and the Congress, however, were not always clear to ordinary people. Nor should the two organisations be regarded as mutually exclusive. The Committees and the Front presented joint lists of candidates during local elections in December 1989, trouncing the CPE in many districts. Numerous members of the Popular Front also stood for election to the Congress, which consequently incorporated a broad spectrum of nationalist opinion.

Unlike in Latvia, Congress leaders did not call for a boycott of elections to the official parliament in March 1990. Although the more fundamentalist *ERSP* declined to participate, more moderate Congress parties (Liberal Democrats, Christian Democrats) presented candidates. The elections, held on 18 March, left anti-independence groups (OSTK,

CPSU and candidates elected by the Soviet armed forces) with 27 seats in the 105-seat legislature, 8 short of the number required to exercise a veto.[78] The Popular Front returned 43 deputies, *Vaba Eesti* 25, and the Congress parties 10. In all, 44 deputies to the new parliament were also members of the Congress of Estonia. At the beginning of April, Savisaar narrowly squeezed home as prime minister at the head of a loose Popular Front-*Vaba Eesti* coalition government.

THE TRANSITIONAL IMPASSE: MARCH 1990–AUGUST 1991

Fears that pro-independence parties might not obtain the necessary majority had thus proved groundless. The Supreme Soviet not only enjoyed a popular mandate to declare independence, it was also the only body which was in practice capable of conducting negotiations with Moscow. At the same time, however, the government had to take account of the radical caucus within parliament. Nor could it disregard the immense moral authority which the Congress commanded amongst the Estonian population.

Prior to the elections, Popular Front leaders had promised to hold full consultations with the Congress over the transition to independence. The need to accommodate the wide spectrum of nationalist opinion is reflected in the declaration adopted by the Supreme Soviet at its first session on 30 March 1990. Mindful of recent events in neighbouring Lithuania and threats by *OSTK* extremists, the new parliament was pragmatic enough to avoid a declaration of outright independence. Instead, it stated that Estonia was a *de jure* independent state under illegal Soviet occupation. The 'constitutional institutions of the Republic of Estonia' were to be restored following a transition period, during which inter-state negotiations would be held with the USSR.

The Supreme Soviet invested the Congress of Estonia with the symbolic title 'restorer of Estonian independence'; in return, the Congress agreed to 'delegate' power to the Supreme Soviet during the transitional period. Once this had occurred, the Estonian population threw its weight behind the official parliament, whilst the influence of the Congress declined correspondingly. The standing of the Congress parties was further undermined by their attempts to forge an unholy alliance with the Communists against Savisaar in November 1990.

In May 1990 the new parliament renamed itself the Supreme Council and reinstated the title (*Eesti Vabariik*) along with the state symbols of the inter-war republic. These symbolic gestures, however, were not matched by real progress towards the restoration of sovereign statehood. The 30 March declaration on eventual independence had narrowly preceded new Soviet legislation making it all but impossible for individual republics to secede from the union. Gorbachev, recently appointed (not elected) President of the USSR, denounced the Estonian declaration as 'illegal and invalid', offering instead the prospect of 'special status' within a renewed Soviet confederation. This attempt to undermine the solidarity of the three Baltic republics met with a refusal from the Estonian government, yet Moscow similarly refused to consent to demands for full negotiations on independence. Relations with the centre thus became stuck in an impasse during the months which followed, at a time when internal squabbles over market reform and the shape of future citizenship legislation threatened to undermine the unity of the independence movement. This came against the background of a deepening economic crisis which sorely tested the patience of the local population.

Estonia's fate was henceforth largely to be determined by developments at the centre. Deprived of any effective instrument of political control over the Baltics, Moscow turned to more nefarious methods in an attempt to keep the breakaway republics within the union. The economic autonomy granted in December 1989 was effectively rescinded, as union ministries increased the pressure on Estonia through a series of obstructive measures. They found support from the heads of Estonia's all-union enterprises, who established a new industrial conglomerate—*Integral*—operating independently of the republican government. The flag of Soviet Estonia continued to flutter on town halls in the largely Russian-populated north–east, where a number of localities refused to recognise laws passed by the Estonian Supreme Council.[79] In July, pro-Soviet representatives in the area came together to form an Inter-Regional Council of Peoples Deputies which had close links to conservative circles in Moscow.

On 30 March 1990—and again on 20 August 1991—Russian-speaking deputies to the Supreme Council refused to participate in the vote on Estonian independence. Many Russian political representatives were not opposed to independence *per se*, but rather to the principle of legal continuity which now stood at the heart of the Estonian national movement. Viktor Andreev, Deputy Speaker of the Supreme Council from 1990–92, speaks for more moderate Russians when he claims that: '[for many] the decision to restore independence aroused suspicions. This was not because they were opposed to the restoration of the

Estonian Republic as a subject of international law, but out of fear of the impetus—already apparent—of the nationalist wave.' However, Andreev goes on to add that 'the predominant expectation ... was that a prudent and civilised Estonian nation would not take revenge against thousands of innocent people for the crimes committed by a despotic regime. For this reason, if you look at the voting minutes of the then Supreme Council, you will find no voters against independence.'[80]

This expectation owed much to the cautious and pragmatic line of the Savisaar government. In its dealings with the non-Estonian population, the Popular Front-led administration repeatedly advocated a 'zero option' variant of citizenship, without, however, making any explicit commitment to this end. In this way, it hoped to placate local Russians whilst avoiding open conflict with the Congress of Estonia.[81] This moderate approach was in sharp contrast to the extreme language and behaviour employed by hard-line opponents of independence and their allies in Moscow. On 15 May 1990, Estonia came periliously close to bloodshed, as a 5,000-strong crowd attempted to storm the Supreme Council and restore the flag of Soviet Estonia. In response to an appeal broadcast by radio, an estimated 15,000 people flocked to defend the parliament. The crowd demonstrated remarkable restraint, and the pro-Soviet demonstrators eventually dispersed peacefully.

The fact that an identical attack was mounted in Latvia on the same day suggests coordination from Moscow. Such displays of force, however, further diminished the already low standing of reactionary forces amongst local Russians. Similarly, Gorbachev's prediction that the Baltic republics would find themselves in an 'economic swamp' following independence were barely credible given the precipitous decline of the Soviet economy during 1990.[82] In the light of this, many local Russians came round to the view that their economic interests might after all be better served by an independent Estonia able to function as a 'bridge' between East and West.

The misgivings of the non-indigenous population with regard to independence were further reduced when the newly-sovereign Russian Republic gave its backing to the Baltic independence movements in June 1990. In January of the following year, Boris Yeltsin visited Tallinn and signed a treaty which recognised the sovereignty of Estonia. Despite scepticism from some quarters, this document did much to assuage the fears of local Russians concerning citizenship and the future of economic links between Estonia and Russia.[83] The emergence of a sovereign Russian republic also dealt a decisive blow to the authority of the Soviet government and its attempts to preserve a centralised Union.[84] In August 1990, Gorbachev was forced to accede to the drafting of a new union treaty placing relations between the republics on a confederal basis. Estonia was one of five republics which refused to sign this so-called '9+1' treaty. By the end of the year, even the conservative Russian-language daily *Molodezh Estonii* was forced to admit that this course corresponded to the wishes of an absolute majority of both Estonians and non-Estonians.[85]

Yet the proposed treaty was anathema to hardliners within the Soviet armed forces and the *Soyuz* group of the USSR Supreme Soviet, the new parliament elected by the CPD in March 1990. From the autumn of the same year, Gorbachev increasingly came under the sway of these groups. By appointing hardliners such as Valentin Pavlov and Boris Pugo to his government, the Soviet president promoted the very men who were to move against him less than a year later. The reactionary forces grouped around the presidency began to prepare a counterstrike against the Baltic republics. In view of the importance attached to keeping good relations with the West, this was timed to coincide with the start of the Gulf War.

The day after Yeltsin's visit to Estonia on 12 January 1991, Soviet troops in Lithuania killed fifteen unarmed demonstrators during an attack on the television centre in Vilnius. A week later, Soviet special forces stormed the Latvian interior ministry in Riga, killing a further six people. Estonia was alone amongst the Baltic republics in avoiding bloodshed during 1991. Yeltsin's visit came against the background of a military build-up, ostensibly to round up Estonians who had refused to serve in the Soviet armed forces. In the same week, *OSTK* called for a mass demonstration against the 'anti-national' course of the Savisaar government.[86] Events thus seemed to be going to the way of Latvia and Lithuania, where pro-Soviet groups organised 'National Salvation Committees' ahead of the assaults by the military. By appealing for moderation on the part of the non-Estonian population and members of the Soviet armed forces, the Russian leader may have helped to forestall an attempted coup in Tallinn. Yet the determinant role in avoiding bloodshed probably belonged to the Savisaar government, which held frequent consultations with Soviet military commanders in late 1990–early 1991, thereby helping to alleviate tensions.

Gorbachev's exact role in the events of January 1991 remains unclear. Although in public he sought firmly to distance himself from the attacks, it seems almost certain that they enjoyed the half-hearted approval of a man who on 10 January had threatened to introduce direct presidential rule in Latvia and Lithuania. Ultimately, no-one was willing to take responsibility for the botched attacks, which brought opprobrium from the West in spite of the latter's preoccupation with events in the Gulf. Gorbachev's international reputation was therefore tarnished at a time when Yeltsin was establishing his credentials as the possible head of a new, more decentralised union. With hindsight, the Baltic republics thus appear to have gained from the bloody events of January 1991. In Estonia, the crisis strengthened the resolve of the population and brought a rapprochment between the Supreme Council and the Congress of Estonia. In Moscow, the crisis deprived Gorbachev of any remaining credibility he enjoyed amongst liberals and hardliners alike. After this, the final reckoning of the Soviet system could not be long delayed.

Late in 1990, Gorbachev had asserted that republican leaders unwilling to sign the new union treaty did not represent the will of their peoples. Questions relating to the treaty were thus to be decided through a union-wide referendum to be held in the spring of 1991. When the Estonian parliament refused to participate, Gorbachev claimed that it was afraid of the outcome. In reponse, the ruling Popular Front made good its previous commitment to hold a separate referendum on Estonian independence. Controversially, this was made open to all residents, including post-war settlers. The Congress parties (*ERSP* excepted) eventually agreed not to call for a boycott of the poll, which went ahead in March 1991. The question posed was 'Do you want restoration of the independence of the Republic of Estonia?' Misgivings regarding the participation of settlers again proved groundless, for the referendum obtained a 78% majority in favour, based on a turnout of 950,000.[87] On this basis, it is reckoned that anywhere between 25–40% of the non-Estonian population voted for independence.[88] When Gorbachev's own referendum was held a month later, the pro-Moscow rump of the CPE was permitted to conduct a non-legally binding 'consultation' on the new union treaty. Amidst widespread allegations of vote-rigging and participation by Soviet military personnel, some 250,000—predominantly non-Estonian—residents turned out to vote. Of these, 95% were in favour of the new union treaty.[89]

THE AUGUST COUP AND THE RESTORATION OF INDEPENDENCE

Further consultations between the Estonian and Soviet governments over the proposed transition to independence yielded few concrete results. In spite of this, many could now see that the Soviet Union's days were numbered. Representatives of certain all-union enterprises, for instance, began negotiations with the Savisaar government on their future status within an independent Estonia. Similarly, Russian political representatives in the north-east also dropped their previous stance of non-cooperation in favour of demands for economic and political autonomy.[90] In the face of bitter opposition from radical nationalist critics, the Savisaar government assented to these demands in principle during April 1991, thereby accentuating fissures within the independence movement.[91]

The state of deadlock was broken by the botched coup of August 1991, which drew a declaration of support from the OSTK leadership. As was the case with the Emergency Committee established in Moscow, however, OSTK's action failed to achieve any real measure of support amongst Estonia's Russian-speaking population. In the still uncertain atmosphere of 19–20 August, a far more potent threat came from the military units dispatched to bring the Baltic republics to heel. By way of a footnote to these events, one should perhaps mention the role of future Chechen president Dzokhar Dudayev—at that time commander of the Soviet military airbase at Raadi, near Tartu. Dudayev is still fondly remembered in the city for his refusal to give landing rights to Soviet airborne units at the height of the coup. More radically-nationalist circles in Estonia have consistently expressed sympathy for the Chechen cause since 1991, a fact which has further complicated the already problematic relations between Tallinn and Moscow (see Chapter Five).[92]

As armoured columns moved towards Tallinn in August 1991, moderate and radical nationalists temporarily buried their differences in order to formulate a common response to the crisis. After intense debate, the Congress of Estonia finally gave its backing to a Supreme Council declaration which re-affirmed the legal continuity of the Estonian Republic as a subject of international law and called for the restoration of pre-1940 diplomatic links on this basis.[93] On the back of Yeltsin's resistance to the failed coup, Russia became one of the first states to recognise Estonian independence on 24 August. In the light of this decision, the Supreme Soviet of the USSR had little option but to follow suit in early September, opening the way for a flood of recognition from western states over the following month. Fifty-one years after its disappearance, the Estonian Republic had reclaimed its place on the map of Europe.

1 Patrick Brogan, *Eastern Europe 1939–1989: The Fifty Years War* (London, 1990).
2 Toivo Raun, *Estonia and the Estonians* (Stanford, 1991), pp.151–153.
3 John Alexander Swettenham, *The Tragedy of the Baltic States* (London, 1952), pp.71–102.
4 Tiit Made, *Eesti Tee* (Stockholm, 1989), p.19; Tönu Parming, "Population Changes in Estonia, 1935–1970," *Population Studies*, Vol.26, March 1972, p.54.
5 Made, p.18.
6 Romuald Misiunas and Rein Taagepera, *The Baltic States: Years of Dependence 1940–1990* (London, 1993), p.41; Rein Taagepera, *Estonia: Return to Independence* (Boulder, 1993), p.67.
7 Parming, p.54; Made, p.19.
8 John Hiden and Patrick Salmon, *The Baltic Nations and Europe* (London, 1992), p.116.
9 Misiunas and Taagepera, p.49.
10 Dov Levin, "*Estonia*," in Encyclopedia of the Holocaust, Vol.2, ed. Israel Gutman (New York, 1990).
11 Parming, p.55.
12 Raun, p.158.
13 Parming, p.55.
14 Raun, p.163.
15 Ibid, p.159.
16 Some of these refugees perished, either at sea en route for Sweden or during further fighting in Germany and central Europe. Whilst some 35,000 Baltic refugees were allowed to settle in Sweden after the war, 167 men who had fought with the Germans were subsequently extradited to the USSR. See: Stig Hadenius, *La Politique de La Suède au XXe Siècle* (Stockholm, 1989), p.63.
17 Misiunas and Taagepara, p.84.
18 Raun, p.163; Misiunas and Taagepera, p.85; Mart Laar, *War in the Woods: Estonia's Struggle for Survival 1944–1956* (Washington, 1993), pp.53–76.
19 Misiunas & Taagepera, p.99.
20 Kaija Virta, "Älymystö Haaveli Virossakin 1960-luvulla Sosialismin Ihmiskasvoista: Allikin Perheessä on Koettu Kommunismin Kaikki Käänteet," *Helsingin Sanomat*, 14 July 1997.
21 Seppo Zetterberg, "Historian Jännevälit," in *Viro: Historia, Kansa, Kulttuuri*, ed. Seppo Zetterberg (Helsinki, 1995), p.134.
22 Mart Laar, Urmas Ots and Sirje Endre, *Teine Eesti* (Tallinn, 1996), pp.58–59; Laar, p.194.
23 Edgar Kaskla, "Five Nationalisms: Estonian Nationalism in Comparative Perspective," *Journal of Baltic Studies*, Vol.23, No.2, Summer 1992, p.168; Jaan Pennar, "Soviet Nationality Policy and the Estonian Communist Elite", in *A Case Study of a Soviet Republic: The Estonian SSR*, eds. Tönu Parming and Elmar Järvesoo (Boulder, 1978), p.117.
24 V. Stanley Vardys, "Modernisation and Baltic Nationalism," *Problems of Communism*, September–October 1975, pp.45–46.
25 Tönu Parming and Elmar Jarvesoo, "Introduction," in *A Case Study of a Soviet Republic: The Estonian SSR*, eds. Tönu Parming and Elmar Järvesoo (Boulder, 1978), pp.6–7; Misiunas and Taagepera, pp.179–183.
26 Mare Kukk, "Political Opposition in Soviet Estonia 1940–1987," *Journal of Baltic Studies*, Vol.24, No.4, Winter 1993, p.370.
27 Riina Kionka and Raivo Vetik, "Estonia and the Estonians," in *The Nationalities Question in the Post-Soviet States*, ed. Graham Smith (London, 1996), p.133.
28 Zetterberg, p.138.
29 Laar, Ots and Endre, p.62–63; Virta, loc cit.
30 Virta, loc cit; Andres Kng, *A Dream of Freedom* (Cardiff, 1991), p.81.
31 Laar, pp.224–225.
32 Graham Smith, "The Resurgence of Nationalism," in *The Baltic States: The Self-Determination of Estonia, Latvia and Lithuania*, ed. Graham Smith (Basingstoke, 1994), p.122; Graham Smith, "The State, Nationalism and the Nationalities Question in the Soviet Republics," in *Perestroika: The Historical Perspective*, ed. C. Merridale and C. Ward (London, 1991), p.204.
33 Marju Lauristin and Peeter Vihalemm, "Recent Historical Developments in Estonia: Three Stages of Transition (1987–1997)," in *Return to the Western World: Cultural and Political Perspectives*

on the Estonian Post-Communist Transition, eds. Marju Lauristin and Peeter Vihalemm with Karl Erik Rosengren and Lennart Weibull (Tartu, 1997), p.76.

34 Misiunas and Taagepera, p.210.

35 Alexander Motyl, *Will the Non-Russians Rebel? State, Ethnicity and Stability in the USSR* (Ithaca, 1987). Cited in Kristian Gerner and Stefan Hedlund, *The Baltic States and the End of the Soviet Empire* (London, 1993), p.27.

36 Smith, 1994, p.126.

37 Rein Taagepera, "Estonia's Road to Independence," *Problems of Communism*, Vol.38, December 1989, p.12.

38 Cynthia Kaplan, "Estonia: A Plural Society on the Road to Independence," in *Nations and Politics in the Soviet Successor States*, eds. Ian Bremmer and Ray Taras (Cambridge, 1993), p.209.

39 Rasma Karklins, *Ethnic Relations in the USSR* (Boston, 1986), pp.54–55.

40 Tönu Parming, "Population Processes and the Nationality Issue in the Soviet Baltic," *Soviet Studies*, Vol.32, No.3, July 1980, p.405.

41 From the Baltic dissident journal *Lituanus* (vol.22, no.1, 1976), pp.65–71. Cited in ibid, p.403.

42 Sirje Kiin, Rein Ruutsoo and Andres Tarand, *40 Kirja Lugu* (Tallinn, 1990), pp.3–7.

43 Author interview with Rein Ruutsoo (Tallinn, 16 April 1996).

44 See, for instance, the arguments advanced in Taagepera, 1989; also Walter C. Clemens, *Baltic Independence and Russian Empire* (London, 1991), p.3; Brogan, p.251.

45 Alexander Motyl, cited in Smith, 1991, p.12.

46 Marju Lauristin, "Estonia: A Popular Front Looks to the West," in *Towards Independence: The Baltic Popular Movements*, ed. Jan Arveds Trapans (Boulder, 1991), p.46.

47 Gerner and Hedlund, p.75.

48 Ibid.

49 Laar, Ots and Endre, p.710.

50 Brian Van Arkadie and Mats Karlsson, *Economic Survey of the Baltic States* (London, 1992), pp.104–5.

51 Lauristin and Vihalemm, pp.85–86.

52 Raun, p.224.

53 Edgar Savisaar, quoted in ibid, p.225.

54 Taagepera, 1993, pp.127–133.

55 Gerner and Hedlund, p.81.

56 Taagepera, 1993, p.136.

57 Kionka and Vetik, p.138.

58 Party membership stood at 110,000 at this time. The CPSU and the Popular Front had an overlapping membership of some 25,000. Rein Taagepera, "A Note on the March 1989 Elections in Estonia," *Soviet Studies*, Vol.42, No.2, April 1990, p.333.

59 Gerner and Hedlund, p.82.

60 Ibid, p.100.

61 Laar, Ots and Endre, p.715.

62 Mikk Titma, *Estoniia: Chto U Nas Proiskhodit?* (Tallinn, 1989), p.85.

63 *Estonian Independent*, 22 March 1989.

64 *Estonian Independent*, 15 March 1989.

65 Oleg Morozov, "Memoirs of a Leader of the Intermovement of Estonia," *The Monthly Survey of Baltic and Post-Soviet Politics*, September 1993, p.76.

66 Ibid.

67 Misiunas and Taagepera, p.317.

68 Toomas H. Ilves, "Reaction: The Intermovement in Estonia," in Trapans ed., p.80.

69 Author interviews with Rein Ruutsoo (Tallinn, 16 April 1996) and Sergei Sovetnikov (Narva, 20 February 1995).

70 Ilves, p.80.

71 Endel Lippmaa, "How to Regain Estonia's Statehood," *Homeland*, No.6–7, 21 February 1990.

72 Gerner and Hedlund, p.136.

73 Stephen White, *Gorbachev and After* (New York, 1992), p.161.

74 Taagepera, 1993, pp.170–171.

75 Michael Bradshaw, Philip Hanson and Denis Shaw, "Economic Restructuring," in Smith ed., p.169.

76 Riina Kionka, "Three Paths to Estonia's Future," *Radio Liberty Report on the USSR*, Vol.2, No.8, 23 February 1990, pp.32–33.

77 Laar, Ots and Endre, p.712; David Smith, *Legal Continuity and Post-Soviet Reality: Ethnic Relations in Estonia 1991–95*, unpublished Ph.D. Dissertation, University of Bradford, 1997, pp.90–96.

78 *Homeland*, 28 March 1990; Riina Kionka, "Elections to the Estonian Supreme Council," *Radio Liberty Report on the USSR*, Vol.2, No.14, 6 April 1990, pp.22–24.

79 Riina Kionka, "'*Integral*' and Estonian Independence," *Radio Free Europe Report on the USSR*, Vol.2, No.30, 27 July 1990, pp.20–21.

80 Viktor Andrejev, "*Kommentaarid*," in *Kaks Otsustavat Päeva Toompeal (19–20 August 1991)*, eds. 20 August Klubi ja Riigikogu Kantselei (Tallinn, 1996), p.83.

81 David Smith, loc cit.

82 Quoted in: J. Jekabson, "Economic Independence is Not Enough for Lithuania, Estonia and Latvia," in *Without Force or Lies: Voices from the Revolution of Central Europe in 1989–90*, eds. W.M. Brinton and A. Rinzler (San Fransisco, 1990), p.353.

83 "Nadezhnoe Plecho Rossii," *Sovetskaia Estoniia*, 15 January 1991; "Vstrecha prodolzhalis' 10 minut," *Sovetskaia Estoniia*, 17 January 1991.

84 Gail Lapidus, "From Democratisation to Disintegration: The Impact of Perestroika on the National Question," in *From Union to Commonwealth: Nationalism and Separatism in the Soviet Republics*, eds. Gail Lapidus and Viktor Zaslavsky (Cambridge, 1992), p.59.

85 "*Voprosi Oprosa 'Kakoi by vy Khoteli Videt' Estoniyu?*" *Molodezh' Estonii*, 16 January 1991.

86 "*Obrashchenie Deputatov Verkhovnogo Soveta Estonskoi Respubliki k Izbirateliam, Vsem Zhiteliam Estonii*," *Molodezh Estonii*, 15 January 1991.

87 *Estonian Independent*, 21 March 1991.

88 Rein Taagepera, "Ethnic Relations in Estonia, 1991," *Journal of Baltic Studies*, Vol.23, No.29, Summer 1992, p.126.

89 *Estonian Independent*, 21 March 1991.

90 Iaroslav Tolstikov, ""My Mozhem Stat' Soyuznikami Estonskogo Pravitel'stva", Govoriat Rukovoditeli Predpriiatii Soyuznogo Podchineniia," *Sovetskaia Estoniia*, 25 January 1991; Iaroslav Tolstikov, "Dogovorit'cia Vse-Taki Mozhno!" *Sovetskaia Estoniia*, 1 February 1991; Oleg Morozov, "Memoirs of a Leader of the Intermovement of Estonia—II," *The Monthly Survey of Baltic and Post-Soviet Politics*, January 1994, pp. 51–60.

91 *Estonian Independent*, 25 April 1991.

92 It is important to note that no Estonian government has ever advocated formal recognition of Chechenya's independence from Russia. However, a declaration of support for independence passed by the *Riigikogu* in February 1995 caused outrage in Moscow. Accusations of Estonian support for the Chechen rebels has formed an important part of Russian propaganda against the Baltic states since 1992.

93 Mart Laar, "Kommentarid," in *Kaks Otsustavat Päeva Toompeal (19.–20 August 1991)*, eds. 20 August Klubi ja Riigikogu Kantselei (Tallinn, 1996), pp.104–107.

Chapter 3

OLD WINE IN NEW BOTTLES: THE POLITICS OF INDEPENDENCE

In addition to affirming the restoration of the Estonian Republic, the Supreme Council declaration of 20 August 1991 provided for the establishment of a constituent assembly, membership of which was to be divided equally between the Council and the Congress of Estonia. Once the assembly had drafted a constitution, this was to be submitted to referendum and fresh parliamentary elections called during 1992. Marju Lauristin, one of the architects of the declaration, has described it as a compromise 'third way' which guaranteed the legal continuity of statehood, yet allowed for radical renewal of the constitutional order according to the democratic principles of the late 20th century. The two alternatives—restoration of the 1938 constitution or the declaration of a new 'Third Republic'—would both have deepened existing divisions within society. They would also have complicated Estonia's integration with the West—the former by reinstating an authoritarian state order, the latter by preserving existing Soviet structures for an indeterminate period.[1]

Lauristin's emphasis on westward integration is entirely representative of an Estonian political elite which had placed the slogan 'Return to Europe' at the heart of its campaign for independence over the previous three years.[2] Having already entered the Conference for Security and Cooperation in Europe (CSCE) in the autumn of 1991, Estonia promptly applied for Council of Europe (CE) membership and began negotiations with the European Union (EU) on a trade and cooperation agreement. These first steps towards association and eventual membership of the EU were paralleled by attempts to forge closer ties with the NATO alliance over the next year. However, as noted in the preface to this work, existing and prospective membership of these organisations has imposed conditions which amount to external constraints on the state-building process. Indeed, one cannot adequately explain political developments in Estonia over the past decade without constant reference to this external dimension, such has been the blurring of boundaries between domestic and foreign policy.

At its Copenhagen meeting in June 1993, the European Council of Ministers declared that associated countries of Central and Eastern

Europe would be allowed to join the EU as soon as they were able to assume the obligations of full membership. As regards their political system, candidate countries were required to demonstrate stability of institutions guaranteeing democracy, the rule of law, compliance with human rights norms and respect for and protection of minorities.[3]

Estonia's admission to negotiations on full membership during 1997 testifies to its success in meeting these criteria as defined by the EU and associated organisations such as the Council of Europe. The 1992 constitution has given rise to a functioning set of democratic institutions in the form of a freely-elected legislature (*Riigikogu*), an executive branch with powers clearly defined and limited by law and an independent judiciary. The three sets of parliamentary elections to date have been deemed free and fair, and transfer of power occurred smoothly in 1992, 1995 and—most recently—1999. Nor can one point to any concerted challenge to the existing constitutional order during the period in question. Press freedom appears firmly established along with basic freedoms of speech. Indeed, the media has arguably played a key role in Estonia's overall democratic development through its frequent exposure of political corruption.

Citizens of the Estonian Republic have been free to form political parties, whilst freedom to found non-political organisations is open to all residents regardless of citizenship. In 1998, Estonia became the first of the former Soviet republics to abolish the death penalty. It has also worked hard to remedy the continued shortcomings in its judicial system highlighted by the European Commission.

However, the emergence of a truly stable democratic order hinges not simply upon state institutions, but also upon the emergence of a supportive civic culture amongst the population as a whole. In order to instill widespread public trust in the system, politicians must believe in their duty to serve the people. Individual citizens must in turn believe in their own capacity to influence government decisions and their duty to participate in political life. Recent literature on postcommunist transition stresses the crucial role of political parties and non-governmental associations in promoting democratic political culture and structuring civic participation.[4]

Whilst comparatively few questions have been posed with regard to formal institutional structures in Estonia, there remains considerable debate surrounding this 'bottom-up' societal perspective so integral to democratic consolidation. Lauristin and Vihalemm, for instance, maintain that Estonia is already well on the way to becoming a 'normal' European country: by 1997, it had already entered a 'post-transitional' stage of development characterised by the existence of stable political parties and a firmly rooted democratic political culture.[5] Much of the available evidence, however, suggests that eight years on from independence, political society still remains 'fundamentally weak', as President Meri himself admitted in his February 1999 independence day speech.[6] Formally speaking, the share of the population participating in non-governmental organisations is roughly comparable to that of most EU countries. Yet, as Rein Ruutsoo points out, 'organisational density is not necessarily an expression of civil ethos,' for civil associations in Estonia have a lower economic capacity and more limited social functions than their western counterparts.[7] The fact that the democratisation of political culture has lagged behind institutional reform is hardly surprising given the legacy of Soviet Communism, a system which engendered deep mistrust of the state whilst simultaneously fostering a culture of dependency upon it.[8] As we shall see, popular cynicism towards the political process has been heightened by the behaviour of a ruling elite which is increasingly perceived as inefficient, self-serving and corrupt.

The problematic consolidation of democracy has not proved to be a great impediment in terms of Estonia's progress towards EU negotiations. Compared to the subsequent nitty-gritty of implementing the *acquis communautaire*, the 'Copenhagen criteria' were rather vague. Notions of democratic stability are in any case less easily quantifiable than economic indicators and hence more open to politicisation on the part of the Commission and existing member states. In this regard, therefore, Estonia's 'Return to Europe' has arguably been about 'image-building' above all else.[9]

One major factor differentiating Estonia from neighbouring Latvia is the way in which it has addressed the somewhat thorny criteria relating to 'respect for and protection of minorities.' As noted in the preface to this work, the refusal to grant automatic citizenship rights to the large Russian-speaking settler population is the issue which has aroused the greatest controversy amongst outside observers since 1991. The EU and the Council of Europe have been ill-placed to question the juridicial bases of Estonia's citizenship and minority legislation, for these organisations are still far from achieving consensus on the actual definition of 'minority' and the future nature of citizenship within a united Europe. Nevertheless, a situation where almost a third of Estonia's residents lacked full citizenship rights aroused understandable fears of ethnic unrest which might prejudice the very existence of the state. Although these fears have proved greatly exaggerated, western transnational organisations and their individual member states have consistently urged the speedy integration of non-citizens into the polity. Mindful of the possible repercussions in terms of westward integration, successive Estonian governments have generally taken heed of these recommendations, yet they have been forced to balance them with their own domestic agenda of 're-nationalising' the state through the restoration of Estonian as the sole official language.

1991–92: REBUILDING THE STATE

The need to establish a new constitutional order was only one of a number of pressing tasks confronting the authorities after August 1991. Unlike the former Soviet satellite countries of central and eastern Europe, newly-independent Estonia possessed none of the formal attributes of sovereign statehood.[10] Alongside the almost complete absence of autonomous financial institutions (see Chapter Four), one can list the absence of a functioning external border with Russia and—most notably—the continued presence of thousands of former Soviet troops. In tracing political developments up to the final withdrawal of these forces on 31 August 1994, it is therefore important to bear in mind that state-building took place under conditions in which formal sovereignty was less than assured.

Following the collapse of the August coup, the government dismissed all those in positions of influence who had openly supported the Moscow plotters. The pro-Soviet United Council of Work Collectives (OSTK) was dissolved, and its leaders relieved of their positions as directors of former all-union enterprises. Armed worker's detachments established within these factories were dissolved, and certain of their members imprisoned. The CPSU and the KGB were also banned on the grounds that they were agencies of a foreign state. In spite of a belated pledge of allegiance to the Estonian Republic, city council leaders in Narva and Sillamäe were also dismissed on account of their previous refusal to observe Estonian legislation during 1990–91. Unlike in Latvia and Lithuania, however, the Estonian government allowed former council members to stand as candidates in fresh elections. The old leaderships were duly re-elected in October 1991, albeit on a turnout of only 35% in Narva. Having retained a power base in the Russian-speaking north-east, these former opponents of independence were well-placed to mount a challenge to the new state order over the next two years.

The government's pragmatic stance towards the north-east was symptomatic of its continued efforts to build bridges with the non-Estonian population. In September 1991, Savisaar sponsored the launch of the Russian Democratic Movement (RDM), a new pro-independence, pro-integrationist party. RDM helped to fill the 'vacant political niche' left by the disappearance of the Communist Party, bringing to prominence a new generation of liberally-minded Russian intellectuals. By co-opting former opponents of independence such as Vladimir Lebedev, it also helped to prevent a possible drift towards more extreme groups during the crucial months which followed. Savisaar's support for the RDM was far from purely altruistic, however. By wooing the Russian-speaking population, the prime minister was also seeking to revive his waning political fortunes. Savisaar had never enjoyed unquestioned leadership of the diverse nationalist groupings within the Popular Front. Divisions began to crystallise around the March 1990 elections, which saw the birth of a number of small party groupings. Amongst the first to emerge were Marju Lauristin's Estonian Social Democratic Party (ESDP), Ivar Raig's Rural Centre Party and a Liberal Democratic grouping which included more moderate members of the Congress of Estonia. In all, twelve distinct factions emerged within the 1990–92 Supreme Council. As head of the Centre caucus in parliament, Savisaar was elected as prime minister by a margin of only one vote.

Amidst the unprecedented national unity of the 'movement society', fellow politicians already suspected Savisaar of having pretensions to authoritarian leadership.[11] As long as independence remained the overarching goal, the Popular Front leader was allowed to preside over a loose coalition government containing three reform communists. The restoration of statehood, however, deprived the Front of its original *raison d'être*, and fragmentation soon ensued. In September, Savisaar consolidated the residual core of the movement by forming the People's Centre Party, which combined pragmatic nationalism with a gradualist approach to economic reform. According to Savisaar's successor, Tiit Vähi, this approach had worked well as long as Estonia needed to become independent. The achievement of independence, however, called for radical policies to achieve a definitive break with the socialist past.[12] As winter approached and acute shortages of fuel and food took hold, Vähi and fellow reform communist Jaak Tamm (Minister for Industry) persistently criticised Savisaar's 'outdated' approach to economic management. Following Tamm's resignation in December 1991, the two men founded the Coalition Party (*Koonderakond*), pledging to build a 'socially-oriented' market economy along Nordic lines. The Coalition was in fact conceived as a vehicle for advancing the interests of the former state managerial class previously represented by *Vaba Eesti* and the Union of Work Collectives. As elsewhere in the postcommunist world, this group underwent an apparently effortless transition from Communist *nomenklatura* to free marketeer.[13]

In January 1992, Savisaar asked the Supreme Council to grant him emergency powers to tackle the economic crisis, deeming the proposal a vote of confidence in the government. Two thirds of ethnic Estonian deputies voted against the motion, which only scraped home with the support of the Russian-speaking caucus. This fact led to a further rash of defections from the government camp, prompting Savisaar to submit his resignation. His demise was hastened by attacks from the newly-revitalised radical nationalist forces within the Congress of Estonia, which, in return for cooperation with the Supreme Council on 20 August 1991, had now obtained a role in state-building wholly disproportionate to their representation within the official parliament. Like the Supreme Council, the Congress was home to a number of small proto-parties formed during 1988–90. Of these, the best established was the National Independence Party *ERSP*, which was the only party to boast nationwide organisation by the time of the 1992 elections. In September 1991, a group of smaller Congress parties (The Christian Democrats, Liberal Democrats, Conservative People's Party and Republican Coalition Party) came together to form a coalition known as *Isamaa* ('Fatherland' or, as it prefers to be known, '*Pro Patria*'). The platform of this new grouping was based around the principles of individual freedom, accountable government, low taxes and the nation-state.[14]

All of these radical nationalist parties were informed to various degrees by the 'restorationist principle' of state-building.[15] Having installed legal continuity as the cornerstone of the independence drive, they now sought to tap inherited memories of the inter-war republic. *Isamaa's* campaign slogan 'Cleaning House' ("Plats puhtaks!") signalled its intention to instigate a thorough-going de-Sovietisation of society. In the year which followed the restoration of independence this message carried a broader appeal than the somewhat vaguer principles enunciated by Savisaar. According to some authors, time and Soviet rule had encouraged an 'unanalysed nostalgia' for the inter-war period on the part of many Estonians.[16] Yet the widespread appeal of parties such as *Isamaa* arguably rested on more concrete considerations: namely, their promise to restore property confiscated by the Soviet regime after June 1940. Here the radicals made considerable political capital, criticising the Savisaar government's failure to enact a law on property restitution approved by parliament back in June 1991 (see Chapter Four). Yet the growing influence of the Congress parties was most striking in the acrimonious debates over citizenship and the constitution. Having ceded ground to their radical opponents in August 1991, pragmatic nationalists were now ill-placed to reassert their primacy in these areas.

THE CONSTITUENT ASSEMBLY

Riina Kionka has rightly characterised the state-building debates of 1991–92 as a thinly-veiled political struggle rather than a contest of high ideals.[17] Whilst the name *Isamaa* immediately suggests continuity from the inter-war Päts dictatorship, such comparisons obscure more than they reveal. As quickly became apparent, the 'golden age' label did not extend to all aspects of the inter-war period. Rather, nationalist forces have simply drawn on these images when it has been politically expedient to do so. When the constituent assembly began its deliberations in the autumn of 1991, pressure for a strong executive came in the first instance not from radical nationalists but from former communists such as Arnold Rüütel, who as head of the Supreme Council enjoyed a position akin to that of president under the existing constitutional arrangements. Rüütel's high approval rating amongst ordinary Estonians rested on his role in overseeing the transition to independence and his demeanour of silver-haired elder statesman. Politically, he was closest to the Coalition Party in late 1991, yet was careful to distance himself from party politics.

The fact that Rüütel opposed Estonian membership of NATO and dismissed the need for a national army made him all the more unpalatable to radical nationalists, who were perhaps all too mindful of the 'national betrayal' perpetrated by a strong president back in 1939–40. Constitutent assembly delegates drawn from the Congress thus advocated a political system weighted towards the legislature, yet incorporating sufficent checks and balances to avoid the pitfalls encountered during 1920–1934. This position quickly found majority support within the assembly, which endeavoured to create a synthesis between the 1920 constitution and the political system of the post-war Federal Republic of Germany.[18] The draft constitution of December 1991 gave a limited role to the Head of State, whose ostensibly ceremonial duties include representing Estonia in international relations and acting as supreme commander of the defence forces. The president nominates the prime minister, yet approval of a new government rests firmly with parliament. Perhaps the most significant power attaching to the presidential office is the right to return legislation to parliament for revision. If parliament refuses to comply, the president must either promulgate the law or refer it to the National Court for final adjudication on its constitutional validity.

DEBATES OVER CITIZENSHIP

This basic division of powers established by the first draft remained unchanged following four months of debate and revision, when Estonia's lawmakers sought advice from the Council of Europe before presenting a final version to the electorate. The finished document contained a more precise definition of the rights and responsibilities of non-citizen residents, a change which reflected western concern over the Estonian citizenship law passed in February 1992. In line with the principle of legal continuity, the latter granted automatic citizenship rights only to pre-war citizens and their descendants. Soviet-era settlers and their descendants wishing to obtain citizenship were required to undergo a process of naturalisation, the terms for which were modelled on a citizenship law introduced by the Päts regime in 1938.

Applicants for naturalisation were required to: take an oath of loyalty to the Estonian Republic; possess a basic—albeit unspecified—knowledge of the Estonian language and to have resided permanently in Estonia for two years after 30 March 1990 (officially deemed the end of the Soviet occupation). They then had to undergo a further waiting period of one year following submission of the citizenship application. Naturalisation was made open to all permanent residents of Estonia, save those who had served in the armed forces or security services of the Soviet state. The law also included a clause on 'citizenship for special services' whereby certain individuals (e.g. members of the 1990–92 Supreme Council and local authority representatives) can be granted citizenship without undergoing the normal naturalisation procedures.

The Supreme Council's approval of this legislation in November 1991 dealt a further blow to the dwindling band of pragmatic nationalists headed by Savisaar. Just prior to the August coup, the Popular Front had eschewed naturalisation in favour of an 'option' (*optsioon*) variant which would have given all residents of the former ESSR a choice between taking Estonian citizenship or retaining their Soviet citizenship. Those who applied for Estonian citizenship would then be granted it without further conditions.[19] Advocates of this approach maintained that it would ensure loyalty to an independent Estonia on the part of all residents. The alternative, they argued, was a large population of non-citizens susceptible to outside influence from Moscow.[20] Finally, an inclusive approach to citizenship would serve as 'proof of deference to western liberalism', thereby facilitating trade, access to credit and the quest for integration into international organisations.[21] Once independence was restored, pragmatic nationalists had to take account of the legal continuity idea. Initial draft legislation proposed by the citizenship commission of the Supreme Council in September 1991 thus gave automatic citizenship only to citizens of the first republic and their descendants, and imposed naturalisation procedures far stricter than those established under the 1938 law.[22] However, the draft also contained a series of waivers which would have allowed virtually the entire settler population to become citizens without fulfilling residence and language requirements.

Such suggestions were of course wholly inimical to the radical nationalist vision of a restored Estonian nation-state. Opinion formers close to the Congress of Estonia accused the citizenship commission of 'national treason' and undue deference to external pressure from Russia.[23] Citizenship legislation, they argued, was strictly an internal affair of each sovereign state. In any case, the 1938 citizenship law was entirely consistent with existing citizenship legislation throughout Europe and the world. Citizenship entails responsibilities as well as rights: should post-war 'colonists' be granted citizenship without first being required to prove their loyalty to the state, they might simply use their considerable political influence to slow down economic reform and press for close political ties with Moscow.[24] According to the radicals' definition of citizenship, loyalty to the state was synonymous with knowledge of the Estonian language. In the absence of any formal requirement to integrate with the state-bearing nation, settlers would merely insist upon the introduction of Russian as a second official state language, thereby perpetuating the situation arrived at during the Soviet occupation. The question of whether naturalisation as a citizen would prompt local Russians to drop such demands is a moot point. It appears that the radicals' arguments in favour of the 1938 law simply provided a convenient rationalisation for their immediate agenda, which was to exclude the Russian-speaking 'fifth column' from political influence. In this respect they were successful, for the naturalisation criteria introduced in 1992 ensured that settlers and their descendents would be unable to participate in the first post-Soviet parliamentary elections.

The public outcry occasioned by the initial draft citizenship law led many Popular Front deputies within the Supreme Council to defect to the restorationist camp. This group included Marju Lauristin, the principal architect of the draft. Lauristin's own comments on the affair are revealing. When the law was thrown out by the Council, she declared that 'I now support going back to the naturalisation law of 1938. We have no other choice, given the deep conflicts which exist over the issue. We have to go back to the only law that is there, that has *legitimacy*'.[25] With independent statehood restored, radical nationalists were able to tap the rich vein of popular resentment against Soviet-era settlers. The depiction of this group as a possible 'fifth column' appeared plausible at a time when the continued presence of Soviet troops evoked uncomfortable memories of June 1940. In addition to these serving units, Estonia was also home to around 30,000 Russian military 'pensioners', some of whom were reserve officers as young as thirty. Citizenship, one should add, traditionally entails an obligation to bear arms in defence of the state. When the Supreme Council passed legislation providing for the formation of an Estonian army in September 1991, many questioned whether local Russians would answer the call in the event of an attack from the East.

'THE ETHNIC DEMOCRACY' THESIS

The nature of the political system established after independence has prompted certain commentators to label Estonia an 'ethnic democracy'. This latter term is used to denote a multi-ethnic state in which the 'core nation' possesses a superior institutional status beyond its numerical proportion within the state; certain civil and political rights are open to all; and certain collective rights are extended to ethnic minorities.[26] The Estonian constitution of 1992 contains provisions making Estonian the sole official state language and guaranteeing citizenship according to *jus sanguinis*.[27] In all but exceptional cases, positions in state and local government are to be filled by Estonian citizens, whilst knowledge of the state language has progressively been deemed a requirement for public sector employment. At the same time, all individuals are deemed equal before the law. Citizens of foreign states and stateless persons resident in Estonia enjoy the same fundamental freedoms as Estonian citizens, and full economic and social rights unless otherwise determined by law. In addition, all permanent residents have the right to vote in local elections, regardless of citizenship status.

The constitution also extends collective rights to ethnic minorities in accordance with the Law on Cultural Autonomy, subsequently restored to existence in 1993. The term 'ethnic minority', however, is deemed to apply only to ethnically non-Estonian residents with citizenship. This position is entirely consistent with the Council of Europe convention for the protection of national minorities, which leaves it up to individual states to define the term 'minority'.[28] Defined in this way, Estonia's ethnic minorities correspond to inter-war proportions and are not sufficiently numerous to challenge the political hegemony of the core Estonian nation and/or the predominance of Estonian language and culture within the state. Post-war settlers and their descendants have a distinct legal status similar to that of, say, Turkish *Gastarbeiter* in Germany. The Estonian authorities insist that settlers were citizens of an illegally-occupying foreign state (the USSR). Once Russia assumed the role of legal successor to the USSR in December 1991, they were simply deemed to be citizens of the Russian Federation living and working in Estonia. As *de jure* immigrants, they enjoy access to social and economic rights and fundamental freedoms, but not full political rights. Prior to legislative amendments introduced in 1998, the *jus sanguinis* concept of nationhood dictated that children born to non-citizens living in Estonia did not obtain Estonian citizenship at birth. However, in its expectation that immigrants should integrate linguistically with the titular state-bearing nation, Estonia is no different from other 'nationalising' states such as France which operate the *jus solis* principle of nationhood. Estonian officials also point out that naturalisation provisions under the 1992 citizenship law are generous compared to those in other European states. The same can be said of the political rights accorded to non-citizens.

Comparisons between German *Gastarbeiter* and Russian-speaking non-citizens in Estonia, however, ignore the particular context arising from the collapse of the USSR. Whereas in existing EU states, immigrants typically constitute less than 10% of the total population, in Estonia this figure is more than 30%. More importantly, Russians who settled in Estonia during 1944–1990 could not conceive of the fact that they were moving to a different country. As Soviet citizens, they had the same rights—however limited—as indigenous inhabitants of the Estonian SSR, as well as additional privileges in the form of access to employment and education in their native tongue.

Western criticism of Estonian nationalities policy thus relates not to the juridical aspects of state-building, but rather to its potential sociological and political implications. The 'ethnic democracy' thesis is derived from literature on conflict regulation in ethnically-divided plural societies. Seen from this perspective, Estonia's nationalities policy rests on a classic strategy of divide and rule. To quote Smith, Aasland and Mole: 'in combining some elements of civil and political democracy with explicit ethnic dominance, ethnic democracy attempts to preserve ethno-political stability based on the contradictions and tensions inherent in such a system'.[29]

To date, Estonia's success in avoiding major unrest has owed much to the socio-economic, cultural and political divisions within the 'Russian-speaking population' which were so apparent during 1988–91. The fact that some 80,000 ethnic Russians (members of the inter-war minority and their descendants) were able to obtain automatic citizenship in 1992 introduced a further element of division. Amongst the non-citizen Russian population, a lack of clearly-established institutional bases for nationally-based mobilisation has been compounded by an apparent lack of interest in politics. In this respect, lack of citizenship has not proved such a burning issue as is often supposed by outside commentators. In the early stages of independence at least, many non-citizens have seemingly been prepared to forego full political rights provided that social and economic rights are not infringed, an attitude which betrays the legacy of Soviet political culture.[30]

Theorists of ethnic conflict regulation argue that systems of 'hegemonic control' by a core nation are unlikely to remain stable in the long-term, since they are invariably characterised by injustice.[31] Yet the system put in place in 1991–92 was mooted not as a long-term solution to ethnic issues, but as the start of a developmental process. The restored Estonian state established mechanisms for the absorption of non-citizens into the polity and the creation of an integrated multi-ethnic society in which minority groups can claim far-reaching collective rights alongside those conferred by individual citizenship. The state of affairs arrived at in 1992 was thus by no means immutable. This was the view taken by western governments, which consider a speedy resolution of the citizenship issue as the key to removing potential sources of instability and conflict. The same stance has been taken by the CSCE/OSCE and its commissioner for national minorities Max Van der Stoël, who insists that, *de facto*, Estonia is accountable for non-citizens residing on its territory even if, *de jure*, these cannot be classed as 'stateless persons'.[32]

Yet the framing of citizenship legislation could not dispel the deep divisions between pragmatic and restitutionist nationalists over the 'Russian question'. Although the national radicals achieved their bottom line of barring settlers from participation in the 1992 elections, a durable consensus over the future direction of nation-building has remained elusive. Whereas more moderately nationalist forces have continued to regard the gradual integration of settlers as both a desirable and viable course of action, more uncompromising elements within the Congress clearly regarded 'ethnic democracy' as an end in itself rather than a temporary state of affairs. In March 1992, for instance, the ruling body of the Congress attacked the Supreme Council for having legalised the presence of 'colonists' in Estonia. Elements of *ERSP*, meanwhile, repeated calls for a formal programme of decolonisation under the auspices of the United Nations. For those who continued to harbour this aim, the ambiguities inherent in the constitution offered ample scope for creating a climate of uncertainty amongst the non-citizen population. This still unresolved tension within domestic politics was to prove especially uncomfortable for the right-wing coalition government which took office following the elections of September 1992.

VÄHI'S CARETAKER GOVERNMENT

The task of steering Estonia through to these elections fell to Tiit Vähi, who was appointed in February 1992 to head a caretaker government composed of 'specialists, not politicians'. To avoid further political wrangling, ministers agreed to suspend their party membership whilst in office and forego participation in the forthcoming elections. Vähi pledged to build up a democratic Estonia with a functioning market economy. Dismissing the need for an economic state of emergency, he placed the accent upon privatisation and currency reform. The subsequent introduction of a national currency—the Kroon—along with the signing of an IMF memorandum and the creation of an Estonian Privatisation Enterprise, later formed the basis for Vähi's questionable claims to be the architect of economic reform in Estonia (see Chapter Four).[33] During his brief premiership, Vähi also stood accused of facilitating corruption and 'spontaneous privatisation' by the state managerial elite. His timely departure from government in October 1992, however, allowed him to bask in the glow of monetary reform whilst avoiding the ensuing popular backlash against economic 'shock therapy'.

The introduction of the Kroon in June 1992 was swiftly followed by a referendum on the draft constitution. After a further bout of wrangling, it was decided that only citizens of the inter-war republic and their descendents should be allowed to participate. The electorate approved the constitution by a comfortable majority, yet narrowly rejected a proposal—included at the behest of the Centre Party—that non-citizens who applied for naturalisation before the elections should be allowed to vote. The subsequent law on the implementation of the constitution nevertheless stipulated that the first *Riigikogu* would sit for only two and a half years rather than the four stipulated under the constitution. In this way, non-citizens who underwent naturalisation promptly would not have to wait too long to exercise their vote.[34]

Fears that disenfranchised Russians might take to the streets to vent their anger proved greatly exaggerated, yet the very fact that they were voiced underlines the fragility of the Estonian state during the first year of independence. In the north-east, Narva Council Chairman Vladimir Chuikin called for the restoration of the USSR and hinted at organising a referendum on territorial autonomy for the region. In addition to threats from the extreme left, the government also had to contend with the spectre of extreme restitutionist nationalism harking back to the authoritarian 1930s. This challenge from the right was spearheaded by émigrés who professed loyalty to the government-in-exile rather than the elected government of the Estonian Republic. It sought to exploit the politicisation of Estonia's fledgling defence forces, whose development had reflected the ideological splits within the nationalist movement. In Spring 1990, when the Popular Front-led government initiated the formation of a Home Guard (*Kodukaitse*) and new border guard detachments, forces grouped around the Congress simultaneously restored the inter-war Defence League (*Kaitseliit*). An émigré group later broke with the *Kaitseliit* command and formed the so-called Defence Initiative Centre, which in turn established its own volunteer Läänemaa Infantry Company (LIC) based at Pullapää in western Estonia.

These developments coincided with attempts to build up a functioning national army during the spring and summer of 1992. Vähi's appointment of former Soviet army officer Ulo Uluots to the newly-created post of defence minister led to friction with LIC and other radically nationalist elements of *Kaitseliit*, which engaged in shooting matches with Russian troops. As relations with Russia progressively deteriorated, a new émigré-sponsored pressure group—'Restitution'—argued for a temporary transfer of power to the government in exile which would serve to reaffirm the principle of legal continuity and pave the way for new elections under the terms of the 1938 constitution. Support for the exile government within the defence forces prompted an unsuccessful attempt to organise a coup during June. This followed earlier attempts to undermine the Vähi government following its accession in February.[35] The restitutionist challenge evoked a new display of solidarity on the part of established political forces. In June, 13 parties from the Centre to the *ESRP* issuing a joint appeal to vote in favour of the constitution, arguing—quite plausibly—that its adoption in no way prejudiced the legal continuity of Estonian statehood.[36]

THE PARTY SYSTEM AND THE 1992 ELECTIONS

The fact that a majority of the electorate heeded this call suggests that Estonians had absorbed the lessons of history with regard to the authoritarian 1930s. At the same time, the experience of 1920–34 offered a salutary reminder of the problems inherent in trying to build a functioning pluralist democracy virtually from scratch. According to Rosimannus, the fractious parliamentary politics of the 1920s—'more democratic façade than democratic substance'—had ultimately alienated the people from state institutions, with fatal consequences. The explanation lay in a mismatch between a political system borrowed from the West and a political culture which lacked any deep-rooted tradition of democratic politics.[37]

As was the case in the 1920s, most politicians regarded a multi-party system as a logical and necessary facet of western-inspired institutional development. The September 1992 elections were contested by over 20 political parties, many of which were modelled on long-established parties in western Europe. Attempts to draw close parallels with western 'sister parties' were largely misleading, however, for in the absence of well-defined categories of socio-economic class, late-Soviet political mobilisation had occurred primarily along national lines. As the façade of national unity began to crack during 1990, parties were built on the basis of the various small, issue-based groupings within the popular movements. More often than not, these 'parties' began life as little more than cliques based around one or more prominent individuals.

On the premise that elections would henceforth be multi-party affairs, the new electoral law of 1992 replaced the single transferable vote system of 1989–90 with a proportional representation list system. Scope for expression of the 'personality factor' was retained in so far as voters still cast their ballot for individual candidates rather than parties or electoral coalitions.[38] Independent candidates are allowed to stand, and individuals who fulfil a given quota (the total number of votes cast divided by the number of seats to be filled) within a particular district are automatically elected to parliament. However, generally high quotas meant that only 17% of *Riigikogu* seats were allocated this way in September 1992.[39]

The remaining seats are allocated according to the total number of votes cast for each party or electoral coalition. Provided a party obtains more than 5% of the vote at the nationwide level, its total vote within the constituency is divided by the local electoral quota to determine the number of seats obtained in the given district. In the first instance, these seats will go to local candidates in descending order of their personal vote (provided this exceeds 10% of the district quota). Those seats which remain unfilled are transferred to a national pool where they are allocated in proportion to the total national vote for each eligible party. Under this final 'compensation' mechanism, candidates are selected according to their ranking on the national party list rather than their personal vote.[40]

The 5% threshold was intended as a safeguard against an unduly fragmented parliament. Hopes that smaller parties would merge to form larger groupings were only partially realised in the first two sets of post-Soviet elections, for the rules governing electoral coalitions allowed different parties to run under a single list whilst still maintaining their distinct identities. The loose nature of these alliances provided no guarantee of continued cooperation within parliament, where the coalition partners were free to part company and form their own separate caucuses.

Although the institutional bases of a multi-party system were in place by the time Estonians went to the polls in September 1992, the fledgling political parties still faced an uphill struggle in order to establish themselves as legitimate agents of representation. Surveys taken before and after revealed that only 13% of the population identified with a particular party, compared to the EU average of 56%.[41] If the mushrooming of parties and electoral platforms posed difficulties for political analysts, it proved yet more confusing to an Estonian electorate whose interest in politics was already waning following the exertions of the independence campaign.

Prior to independence, the major issues of the day had been clear and unambiguous, requiring little in the way of rational deliberation. The appeal of the independence movements rested on promises of the bright future which beckoned once statehood was restored. Growing economic hardship during 1991–92, it seems, could not dispel this sense of optimism, so total was the rejection of Soviet communism. Yet in as much as the new ideological discourses espoused by politicians remained largely untested, public understanding of them was necessarily limited. Best placed to prosper were parties such as *Isamaa* which promised a rapid break with the Soviet past. For over half of voters casting ballots in September 1992, however, the personality of the candidate counted for more than his/her party affiliation.

At 66%, participation in the elections was high by western European standards, but already well down on the levels witnessed in Estonia during 1989–91. Due to the exclusion of settlers and their descendants, the electorate was now 90% ethnically Estonian, as opposed to 65% two years two years earlier. It was thus hardly surprising that the new 101-member *Riigikogu* was made up entirely of Estonian representatives. Of the 15 parties and coalitions which contested the elections, a total of seven crossed the 5% barrier.

After a campaign heavily coloured by nationalist issues, the new parliament is most easily rendered intelligible with reference to the pre-existing cleavages within the independence movement. Of the Congress groupings, by far the largest number of seats (29) fell to *Isamaa*, which drew most of its support from urban areas. The *ERSP* (10 seats) vote was more evenly distributed nationwide, with the bulk of its support coming from lower income groups. The second most popular group amongst urban voters was the Moderates (*Mõõdukad*) coalition consisting of Lauristin's ESDP and Raig's Rural Centre Party (12 seats). These parties belonged to that section of the Popular Front which had defected to a more exclusivist stance on citizenship during late 1991. In keeping with this new orientation, the Moderates' electoral programme included provision for future cooperation with *Isamaa* and *ERSP*. Together, these three groupings commanded a slim 3 seat majority and entered government in October 1992.

The rump Popular Front coalition, meanwhile, was left with only 15 seats. Savisaar's high visibility and residual popularity probably helped to avert a more disastrous slump in support for a party perceived as overly-compliant towards Russian-speaking settlers. Nationalist attacks on the Centre and its allies had come not just from the right, but also from the old reform communist wing of the independence movement, now grouped within the 'Secure Home' *(Kindel Kodu)* coalition. Secure Home's emphasis on 'security', 'home' and 'family' found widespread support amongst the elderly and the impoverished rural population, giving the coalition 17 seats in the new parliament.[42] The inaptly-named 'Left Opportunity'—drawn from the rump CPE—failed to cross the 5% threshold, making Estonia the first post-socialist state in which the official successor to the Communist Party lacked any representation in parliament.[43]

Another notable feature of the elections was the success encountered by two 'anti-system' groups.[44] Out of the Restitution movement came Estonian Citizen (8 seats) led by charismatic Estonian-American Vietnam veteran Jüri Toomepuu. Thanks to a well-funded and organised electoral campaign, Toomepuu garnered the highest personal vote of any candidate in September 1992. Slightly more quaint were the Royalists, a group of media celebrities and young intellectuals headed by Kalle Kulbok. Kulbok himself is a committed monarchist who in 1994 courted Britain's Prince Edward as a possible candidate for King of Estonia. Many Royalists, however, simply sought to lend colour to what they considered a thoroughly dull election. This assessment of the campaign was seemingly shared by younger voters, whose high support for the Royalists gave the party 8% of the vote and 8 seats in the new *Riigikogu.*

Despite the avowed preference for personalities over parties expressed by the electorate, 60 of the 101 *Riigikogu* seats were allocated through the 'compensation' mechanism which left the final choice of candidate to the party. This fact was later seized upon by the demagogue Toomepuu, who contrasted his 14,000 personal votes with the mere handful (often less than 100) obtained by other *Riigikogu* deputies.

THE PRESIDENTIAL ELECTIONS

The July 1992 law on the implementation of the constitution had recognised the salience of the 'personality factor' by providing for a one-off popular election to the presidency, held simultaneously to the parliamentary poll.[45] Under these regulations, any candidate obtaining more than 50% of the popular vote would be automatically elected, otherwise the matter would be referred to parliament.

The candidates for the post (and their party sponsors) were as follows: Arnold Rüütel (*Kindel Kodu*) and Lennart Meri (*Isamaa*) emerged as the front runners, whilst émigré Professor Rein Taagepera (Centre) and former dissident Lagle Parek (*ERSP*) declared that they were competing against Rüütel rather than one another. In spite of the heightened attention given to Rüütel's Communist past, the 'Silver Fox' was not far short of an absolute majority, obtaining 42% of the vote. Meri—who also found himself under the microscope regarding alleged former links with the KGB—gained 29%, Taagepera 23% and Parek 4%. When the two leading candidates were referred to the *Riigikogu*, the right-of-centre majority predictably opted for Meri over Rüütel. Despite lacking the popular legitimacy conferred by direct election, Meri soon grew to command the respect of the electorate. In this regard, he was to fare better than parliament or government during 1992–95.

'JUST DO IT': THE ISAMAA GOVERNMENT 1992–95

Meri's first act as president was to invite *Isamaa*'s Mart Laar to form a government. Laar duly concluded a coalition agreement with *ERSP* and the Moderates, giving the former the portfolios of interior, defence and transport and the latter reform, social affairs, agriculture and the environment. The rest of the 14 principal cabinet posts went to representatives of his own coalition. The new government took office on 21 October 1992. Headed by a 32 year-old prime minister, it was young and relatively untainted by associations with the former regime. Non-party technocratic experts were strongly represented, as were émigré Estonians.

Buoyed by massive popular expectations for change, Laar pledged decisive economic reform regardless of the short-term pain which this entailed, a philosophy which he summed up using the Nike advertising slogan 'Just Do It'.[46] The new government duly delivered on its promise of 'shock therapy' over the next two years (see Chapter Four). Like his idol Margaret Thatcher, however, Laar alienated many through his domineering style of leadership. Pretensions to political 'house cleaning' were also diminished by a string of ministerial improprieties which dispelled the government's initial reputation for honesty.

The other major priorities set by Laar in his inaugural speech to the *Riigikogu* were withdrawal of Russian troops; membership of the

Council of Europe; an association agreement with the EU; reorganisation of the defence forces; and cooperation with NATO. Emphasising that 'cleaning house' was a creative not a destructive concept, the new prime minister pledged to protect human rights and give residence guarantees to non-citizens whilst working to promote their integration into Estonian society. At the same time, he promised to aid the 'repatriation' of settlers wishing to leave Estonia, thereby hinting at the intra-coalition tensions which were to dog the government throughout its existence.[47]

In spite of Laar's calls for cooperation and constructive dialogue between all political forces, his honeymoon period proved short-lived. After less than three months in office, support for the government had slumped in the face of fierce attacks by the opposition.[48] *Isamaa*'s radical economic policies attracted plaudits from the international community, yet made life especially uncomfortable for Social Democrat ministers faced with popular demands for increased pensions and wages.

The government also had to contend with continued unrest within the defence forces, which cast a shadow over the new democratic order during the first year of its existence. In a move timed to coincide with the handover of power during September–October 1992, the Läänemaa Light Infantry Company and four other units announced that they no longer recognised the authority of the *Kaitseliit* central command. These units subsequently made a play for leadership of the Defence League with the support of the Defence Initiative Council and Estonian Citizen's Jüri Toomepuu. The appointment of Estonian-Swedish émigré Hein Rebas to the post of defence minister failed to placate the more more extreme nationalist elements within the defence forces. Rebas and *ERSP* colleague Lagle Parek (minister for the interior) seemed powerless to prevent armed clashes between the still largely Russian-staffed civilian police and the more cavalier elements of *Kaitseliit* during 1992–93. As the government struggled to assert civilian leadership, its political opponents sought to exploit the divisions within the armed forces for their own ends.

Rebas was especially criticised for failing to prevent the entry of an extra 250 Russian troops into Estonia during February 1993. The opportunistic Edgar Savisaar used the affair to launch a generalised attack on émigrés within the government, but still had no qualms about joining forces with Toomepuu to accuse the government of 'wholesale surrender of Estonian national interests'.[49] Savisaar and other opposition politicians also took the side of the Light Infantry Company in its dispute with the government, which decided to disband the unit at the end of

July. LIC withdrew to its base in Pullapää, but its leaders had to be arrested by military police after they refused to hand over their weapons. The action split the ruling coalition, with one wing of *ERSP* expressing support for the renegade unit.[50] The affair proved to be the final downfall not only of Rebas, but also of Parek, who tendered her resignation after an LIC fugitive shot and wounded a police officer during November.

The appointment of *Isamaa*'s Jüri Luik—one of the most impressive of Estonia's post-Soviet politicians—as defence minister stabilised the situation somewhat, yet a commission of inquiry set up to investigate the situation in the armed forces was overshadowed by a further scandal relating to the purchase of $49 million of arms from Israel. The deal was concluded by the government without consultation either with the *Riigikogu* or the commander-in-chief of the armed forces, Estonian-American Aleksander Einseln. The government justified its decision in terms of the need to develop a western orientation in defence policy. Most of the weapons, however, were later found to be obsolete. In the wake of the affair, Einseln—formerly a Colonel in the US army—submitted his resignation, which the president refused to accept. Throughout the remainder of his period in office, Einseln's public attacks on corruption and incompetence within the armed forces brought him into frequent conflict with his political masters, who accused him of disregarding the principle of civilian control of the military.[51]

'FIREFIGHTING': THE ALIENS CRISIS

The 'Pullapää incident' of July 1993 was made doubly worrying by the fact that it coincided with the first significant manifestation of unrest amongst Estonia's large non-citizen population. Predictably, this was centred around the cities of the north-east. At the national level, the Russian Democratic Movement had quickly descended into political in-fighting following its exclusion from the 1992 elections. In January 1993, more moderate elements formed the Representative Assembly, an unofficial movement dedicated to the defence of Russian interests within an independent Estonia. A hard-line faction, meanwhile, founded the more overtly nationalist Russian Assembly (*Russkii Sobor*). As squabbling intensified amongst the political elite, most ordinary Russians remained preoccupied with ekeing out a living under difficult economic circumstances, continuing to reside in Estonia on the basis of their now defunct Soviet internal passports.

Measures to formalise the legal status of non-citizens were both necessary and long overdue by the middle of 1993. When it came, however, this legislation seemed calculated to cause the maximum anxiety. The 'Law on Aliens' adopted by the *Riigikogu* on 8 June gave civilians residing in Estonia on Soviet or Russian passports one year in which to apply for new residence and work permits. Failure to do so would confer illegal immigrant status and the prospect of deportation from the country. The law neglected to draw any distinction between an immigrant who had arrived in the country the previous day, and a former Soviet citizen who had been born in Estonia or lived there for twenty or more years. Its psychologically unsettling effect was heightened by the fact that only temporary five-year permits were to be issued in the first instance. In order to qualify for a permit, applicants were required to possess a 'lawful source of income', a category only vaguely defined under the law. Such ambiguities caused obvious alarm amongst the large body of unemployed non-citizens in the north-east. Council leaders in Narva and Sillamäe—who as non-citizens were by this time barred from standing for re-election in the October 1993 local elections—now used the law as a pretext for organising a referendum on 'national-territorial autonomy within the Republic of Estonia'.

Notwithstanding the parallels drawn by some western commentators, the 'Aliens Crisis' never threatened to descend into the kind of violent ethnic conflict previously witnessed in post-Soviet Moldova. The term 'crisis' is perhaps best used to describe the position of the Laar government, which was by now precariously balanced between the competing dictates of western integration and domestic nationalism. By early 1993, the coalition government had made it clear that social stability, foreign investment and membership of the Council of Europe counted for more than the restoration of an Estonian nation-state. In line with recommendations made by western organisations, the government pushed through amendments to citizenship legislation, fixing the linguistic requirement for naturalisation at a level corresponding to a basic working knowledge of the Estonian language (level C under the 1989 language law). According to many experts, the low number of applications for naturalisation during 1992–93 could be explained at least partly by the lack of any firm guidelines regarding linguistic requirements. This had left scope for arbitrariness on the part of the individual departments dealing with applications.[52] A further amendment specified that henceforth citizenship would be passed automatically via the maternal as well as the paternal line.

These expressions of goodwill facilitated Estonia's entry to the Council of Europe in May 1993, a full two years ahead of neighbouring Latvia. Yet the acceptance of external constraints over nationalities policy was fast becoming a major bone of contention within domestic politics. Addressing the nation on this historic occasion, President Meri felt obliged to reassure his countrymen that integrating into Europe was not the same as 'dissolving' into Europe. Estonia's return to Europe, he claimed, had aroused suspicion amongst Estonian intellectuals who claimed that the country would lose its distinct identity, culture and language.[53] Discontent over the citizenship law amendments extended to the ranks of the ruling coalition, where many *ERSP* and *Isamaa* deputies were deeply unhappy at measures which stood to increase the number of ethnic Russians with citizenship. Radicals were further piqued by President Meri's decision to veto a draft privatisation law which barred former Soviet functionaries from receiving privatisation vouchers.[54]

In spite of these amendments, Estonia was still able to gain admittance to the European 'club of democracies' with its controversial citizenship legislation more or less intact. Once the hurdle of CE membership had been negotiated, more radical elements within the ruling coalition clearly felt at liberty to pursue a more assertive line in nationalities policy. When a new Law on Local Elections came before parliament in May 1993, it contained a clause allowing non-citizens not only to vote but also to stand for office. This provision was reputedly included by the government in line with assurances given to the CE prior to Estonia's accession. For many deputies, however, allowing non-citizens to stand for office represented one concession too many. To the acute discomfort of the government, the *Riigikogu* voted to delete this provision from the final law.[55]

Worse embarrassment was to follow over the Law on Aliens a month later. The shortcomings of this legislation have been attributed to 'sloppy drafting', yet in fact, the ambiguity inherent in certain provisions appears to have been entirely calculated. By leaving the wording of the law open to differing interpretations, it was hoped to satisfy western organisations whilst simultaneously leaving the way open to a more stringent interpretation of the law by the local authorities charged with its implementation.[56] There seems little doubt that for more radical nationalists, the aliens law was conceived as a means of intensifying the pressure upon non-citizens to 'repatriate' themselves to Russia. In the case of Soviet military pensioners and their families, this aim was quite explicit, for the law stated that residence permits would not be made available to non-citizens who had served in the armed forces of a foreign state. The same applied to family members who had entered Estonia in connection with the service or retirement of military personnel.

In an echo of the wildly emotive rhetoric then emanating from Moscow, opposition deputies from the Centre Party claimed that policies towards aliens had taken the direction of 'ethnic cleansing' and 'decolonisation' following Estonia's acceptance to the Council of Europe.[57] Other leading figures, including some Social Democratic MPs, made a more measured yet nevertheless highly critical attack on the law in an open letter published at the end of June. Reservations were also expressed regarding the new Law on Education, which provided for a complete switch to teaching in Estonian in all Russian-language gymnasiums (upper secondary schools) and higher education establishments by the year 2000. In view of the chronic shortage of suitably qualified Estonian native language teachers, this target was rightly seen as wildly over-ambitious.[58] The same was true of the one-year deadline for submission of residence permit applications. This would clearly have proved overwhelming for a state bureaucracy which one *ERSP* parliamentary deputy later described as 'inflexible ... apathetic and underpaid' as well as lacking in impartiality.[59]

Such instances of the gap between legislation and its implementation have led certain western commentators to question the effectiveness of the rule of law in Estonia.[60] The political furore sparked by the aliens law proved highly embarrassing at a time when Estonia was still under close international scrutiny over the Law on Local Elections. In response to an appeal by CSCE Commissioner Max Van der Stoel, President Meri refused to promulgate the Law on Aliens, insisting that it be scrutinised by experts from the CSCE and the Council of Europe. Laar duly promised that the government would abide by any recommendations put forward.

Clearly encouraged by the international reaction, the leadership of Narva city council announced its decision to hold the referendum on territorial autonomy. Although this step is best regarded as a cynical piece of opportunism, the fact that the the stand-off between Tallinn and Narva was resolved entirely peacefully is a testament to the political culture of restraint in Estonia. This assertion holds true not just for the Estonian government, but for local leaders in Narva, where, 'for all their headline-grabbing intransigence, ... officials [took] seriously their ability to keep the city under control.'[61] Ignoring radical nationalist appeals to disband the Narva council, the government referred the matter to the Chancellor of Law, who ruled that although the referendum could have no legal force, there was no impediment to its actually being held. On the same day, parliament approved a revised draft of the aliens law which took on board many of the recommendations made by the CE and the CSCE. The only point on which the government refused to budge was its insistence that no permits be made available to military pensioners and their families.

In the hope of extracting further concessions, Sillamäe town council announced its own referendum on territorial autonomy to be held simultaneously with the Narva poll on 17 August. In the event, popular endorsement of the autonomy proposals proved less than wholehearted. Whilst 97% of local residents who voted came out in favour, officially-recorded turnouts were only 54.8% in Narva and 61.4% in Sillamäe, and this in a poll marred by widespread allegations of vote-rigging.[62] Subsequent fears that Chuikin and his allies would seek to organise alternative local elections also proved unfounded, for the old leaders became marginalised within their own constituencies ahead of the October poll.[63] Concerned at the implications for foreign investment and economic development, an influential body of local opinion came out in favour of ending the stand-off with Tallinn. The Estonian authorities were thus able to come up with a sufficient number of 'loyal' candidates who either already possessed citizenship or could be co-opted by means of accelerated naturalisation.[64] Locals turned out in force to cast their votes, electing councils which subsequently proved far more amenable to cooperation with central government.

Contrary to the belief of some commentators, however, the peaceful resolution of the 'Aliens Crisis' did not mark a major turning point in Estonian nationalities policy. The legal status of the non-citizen population had not changed. Nor had there been any resolution of the cardinal issues at the heart of the 'Russian question'. In this regard, Georg Sootla is correct to describe the international mediation of July 1993 as 'fire-fighting' which could not liquidate the sources of tension. 'Unless the main sources of argument are solved clearly and unambiguously on an international level', he continued, 'it is difficult to avoid the ... dilemma between international mediation and internal politics.'[65] This indeed proved to be the case throughout the remainder of *Isamaa*'s term of office.

In November, the government finally bowed to western pressure by agreeing to grant residence permits to military pensioners. The only category now excluded were those demobilised from the Soviet army after August 1991. This was the final straw for *ERSP* party chairman Ants Erm, who broke away to found the National Progress Party. Calls for Erm's resignation had begun following his public display of support for the Light Infantry Company during the Summer. Following the loss of the defence portfolio to *Isamaa* in August, *ERSP* nominee Arvo Siir and his supporters defected to form their own Estonian Future Party. Both parties espoused hard-line nationalism, declaring their opposition to EU membership for Estonia. Their deputies in the *Riigikogu* were joined in a new independents' faction by *Isamaa* defectors such as the former dissident Enn Tarto.

THE END OF THE LAAR GOVERNMENT

The local elections of October 1993 represented a further significant setback to *Isamaa*, which obtained only 5 of the 64 seats on Tallinn city council. On this occasion, non-citizens made the most of their opportunity to vote, something which cannot be said for the ethnically Estonian electorate. Two ethnic Russian lists captured a third of the seats in Tallinn, thereby heightening nationalist fears of what might happen in future general elections should large numbers of settlers obtain citizenship and the vote. Nationwide, the overall victor was the Coalition Party, which swept to power in the capital on a platform of continued economic reform cushioned by greater social guarantees. Tiit Vähi, elected party chairman six months earlier, returned to public office as head of Tallinn city council.

In the wake of the elections, Laar stated that Estonia could either 'clench its teeth' and continue along the path of economic reform and European integration or else find itself one day a member of the Confederation of Independent States (CIS) along with many other former Soviet republics.[66] This categorical 'either/or' discourse by the ruling parties was increasingly challenged by the Coalition. Backed by powerful industrial interests, Vähi insisted that an independent Estonia could and should maintain close ties with the CIS whilst continuing to integrate with the West.[67]

Popular discontent at the social consequences of economic shock therapy—including burgeoning crime rates—was undoubtedly one factor behind *Isamaa*'s eclipse in the elections. Yet perhaps equally determinant were the charges of corruption, incompetence and arrogance levelled at the government over the previous twelve months.[68] This arrogance was exemplified by attempts to pin the defeat on the apathy shown by a large section of the Estonian electorate. Entirely characteristic was the an outburst by Lagle Parek, who claimed 'the elections showed that the Estonian people are not yet ready for independence'.[69] More fitting was Lennart Meri's comment that the restored Estonian Republic had proved rather more successful than the current Estonian government.[70] Parek herself was to resign in ignominious circumstances just over a month later. Further defections and splits occurred during November, when six Liberal Democrat deputies broke with *Isamaa* to form their own parliamentary fraction. For the moment, the Liberals continued to vote with the government, allowing it to survive a vote of no-confidence and carry on into the new year.

In March 1994, polls revealed that support for *Isamaa* was running at barely 5%. The most popular party by far was still the Coalition, which enjoyed the support of 23% of respondents.[71] Later in the year, Laar could point to the resumption of economic growth and the achievement of a free trade treaty with the EU as a vindication of economic 'shock therapy'. By this time, however, the problems besetting the government seemingly had more to do with the prime minister's personality than his policies.[72] Following a failed attempt to depose Laar during June, deputies from the Conservative People's and Republican Coalition Parties quit the *Isamaa* coalition, joining earlier defectors within a new nine-member fraction known as the Rightists (*Parempoolsed*). They were promptly followed by the Liberals, who stated that they could no longer tolerate the prime minister's arbitrary decision making.

Nor could the government derive much political benefit from the final departure of former Soviet troops—arguably the most significant event of the period 1992–95 (see Chapter Five). When President Meri signed the final withdrawal agreements in Moscow at the end of July, he did so without obtaining approval from the government, thereby contravening the basic law on the conduct of foreign affairs. Yet more controversial was a clause in the agreement which provided for the granting of residence permits to Soviet military pensioners excluded under existing legislation. Although the government was at pains to stress that residency would be denied to anyone who constituted a danger to state security, the opposition—left as well as right—made the most of this further opportunity to play the 'national card'.[73]

Calls for an end to Laar's 'scandal-ridden leadership' intensified following revelations that the 2.3 billion Soviet roubles withdrawn from circulation in June 1992 had been sold to Chechnya without any consultation with parliament. Part of the proceeds were alleged to have gone to companies closely associated with Laar. At this point, the Moderates became the latest party to quit the ruling coalition, depriving the government of its parliamentary majority and thereby allowing the opposition to carry a vote of no-confidence in the Laar government on 26 September 1994.

TARAND'S 'CHRISTMAS CABINET'

With an eye to maintaining the momentum of economic reform, President Meri subsequently chose Central Bank chairman Siim Kallas as his nominee for prime minister. Kallas was popular with the electorate, yet his 'strong man' reputation did not find favour with a parliament which had just dispensed with Laar. A compromise candidate emerged in the form of environment minister Andres Tarand (Moderates) who received the backing of all the original 1992 coalition partners, including the Rightists and the Liberals.

The aforementioned splits within the nationalist right were followed by a further realignment of political forces during the autumn. Significant new additions to the party scene were Rüütel's Estonian Country People's Party (*Eesti Maarahva Erakond—EME*) and Kallas' Reform Party, which incorporated the Liberals and a number of defectors from the Moderates at its initial congress in November 1994. In an attempt to instill parties with a genuine mass base, legislation passed in May 1994 had set a minimum membership requirement of one thousand. This provision, though, was only enforced in 1998. The second parliamentary elections of March 1995 were thus contested by a total of 33 parties grouped within 10 electoral unions. Of these, no more than three could boast a thousand members, whilst many barely met the existing requirement of two hundred.

As the various political groupings jockeyed for position, Tarand was left to exercise a caretaker role until February 1995. Only five changes were made to the cabinet bequeathed by Laar, the most notable being the appointment of the Rightist Kaido Kama as minister of the interior. At Meri's request, the impressive Jüri Luik was maintained as foreign minister in order to ensure continuity in foreign policy.[74] It was in this latter sphere where Tarand had the greatest impact. Building on the recent withdrawal of Russian troops, the prime minister unveiled new proposals towards an eventual settlement of Estonia's border dispute with the Russian Federation (see Chapter Five). In domestic affairs, the course of economic policy remained unchanged, whilst Kama's tenure of office was most notable for a crackdown on organised crime and corruption in the state bureaucracy. As investigations temporarily halted the work of the department of citizenship and migration, the government was predictably accused by the Russian-language media of seeking to slow up the naturalisation of non-citizens ahead of the elections.

With the non-Estonian share of the electorate now standing at around 18%, Russian political leaders too had begun to form their own parties during late 1994. The fact that three parties were necessary rather than one is further testimony to the deep ideological divisions amongst the Russian-speaking population. In this regard, the internationalist and pro-integrationist outlook of the United National Party of Estonia (*Ob'edinennaia Narodnaia Partiia Estonii—ONPE*) stood in sharp contrast to the more 'patriotic' Russian nationalist orientation of the Russian Party of Estonia and the Estonian-Russian People's Party. Considerations of political expediency, however, prompted the three parties to unite within the electoral union 'Our Home is Estonia' *(Meie Kodu on Eestimaa—MKE)*, which campaigned on a platform of equal rights for all residents.[75]

With *MKE* tipped to return as many as 15–20 deputies, the by now demoralised and fragmented nationalist right expressed alarm that a future 'ex-communist' government might become prey to demands for a revision of nationalities policy. Such concerns appear exaggerated in the light of the cross-party support given to a new, more stringent citizenship law introduced in January 1995. Ostensibly designed to bring Estonia closer into line with European standards, this increased the requirement for naturalisation to five years' permanent residence and introduced further tests requiring applicants to demonstrate a detailed knowledge of the constitution and political system.[76] In the light of Moscow's recent initiatives regarding ethnic Russians in the 'near abroad', a further clause stated that no Estonian citizen could simultaneously hold the citizenship of another state.

THE FEBRUARY 1995 ELECTIONS

In the event, talk of a renewed challenge by Russian political forces proved premature. *MKE* cleared the 5% barrier in February 1995, yet returned only six deputies to the *Riigikogu*. Those Russian-speaking citizens who actually bothered to vote were seemingly just as likely to opt for the Centre, the Coalition or the former Communist list *Õiglus* (Justice) as they were for the Russian parties. Such voting behaviour was typical in an election largely dominated by economic issues. The declining salience of the 'national question' was exemplified by the poor showing of the extreme nationalist *Parem Eesti/Eesti Kodanik* ('Better Estonia'/Estonian Citizen) coalition, whose campaign slogan 'Estonia for the Estonians' made little impact with the electorate. This alliance of right-wing fringe organisations went the same way as the Central League of Nationalists, the Future Party, the Royalist-Green 'Fourth Force' and the Blue Party, all of which failed to attain the 5% threshold. Similarly, *Õiglus* again failed to enter parliament despite its strong showing in the Russophone north-east. If the absence of 'anti-system' groups suggested a growth in political stability compared to 1992, the new parliament still contained 14 different parties grouped into seven fractions. The second *Riigikogu* was thus to be characterised by the same pattern of unstable coalition government, party splits and unconventional political alliances as the first.

The main question in the minds of most commentators was whether the elections would lead to a repudiation of the reform policies pursued during 1992–95. As expected, the ruling parties fared badly. The rump *Isamaa* and *ERSP* united into a single electoral union—presaging an eventual merger in December 1995—but lost half of their previous vote and two thirds of their seats. Together they were left with just 8 deputies in the new parliament. The Moderates (6 seats) were saved by the high popularity rating of prime minister Tarand, whilst the Rightists scraped into parliament by the skin of their teeth, obtaining 5% of the vote and 5 seats. Much of the 1992 *Isamaa* vote flowed to the Reform Party, whose radical pledge to abolish corporation tax found considerable support amongst new private business interests. A well-funded electoral campaign yielded 16% of the vote and 19 seats, making Kallas' party the second largest in parliament.

A majority of seats, however, went to parties exhibiting strongly populist campaign rhetoric. Overall victor in the elections was *KMÜ* (*Koonderakonna ja Maarahva Uhendus*), an electoral union consisting of the Coalition Party, three agrarian parties (the Country People's Party, the Rural Union, plus the smaller Farmers' Assembly) and the Association of Pensioners and Families (41 seats). On the one hand, the parties' joint manifesto pledged to pursue existing policies such as integration with the EU and NATO and the maintenance of a balanced budget. At the same time it promised to improve relations with Russia and increase social security payments through a more effective use of existing resources. *KMÜ*'s biggest gains came in the countryside, where it wooed farmers with promises of preferential tax rates and access to low interest loans. In this regard, the list profited greatly from the presence of Arnold Rüütel, who polled 17,000 personal votes in the southern Estonia.[77] Still more populist in tone was the Centre Party (16 seats), which combined promises of reduced unemployment, higher pensions and protection for agriculture with pledges to protect the interests of middle-class small entrepreneurs.

The fragmented nature of the new parliament left considerable room for political horse trading. Called upon to nominate a new prime minister, President Meri shunned *KMÜ* leader Vähi in favour of Kallas. Any new government, however, would necessarily have had to include *KMÜ*, and the Reformists refused to enter into a coalition agreement with the agrarian parties. Vähi was thus required, somewhat reluctantly, to broker a deal with the Centre Party in order to obtain a parliamentary majority. Typically, Savisaar exacted a high price in return for his cooperation, securing for himself the key posts of deputy prime minister and minister of the interior. In addition, the Centre also obtained economic affairs, transport, culture and education, and social affairs.[78] Eight of the 15 ministerial posts (including finance, justice, foreign affairs and the new European integration portfolio), however, went to the Coalition, leaving the agrarian parties with agriculture and the environment.

BUSINESS AS USUAL: THE COALITION PARTY IN POWER 1995–99

According to some commentators, the 1995 election did indeed represent the kind of 'turn to the left' witnessed in other postcommunist states of central and eastern Europe. Yet predictions of a retreat from reform were not borne out by the policies of the new government. According to Laar, Vähi had little choice but to adhere to the same line as his predecessors, so decisive had been the political and economic changes of the last three years.[79] There is at least some truth in this assertion. The institutional framework established in 1992, for instance, afforded precious little room for maneouvre in economic policy, whilst the scale and pace of private sector development had made it difficult to turn the clock back.

Having said this, Laar predictably exaggerates the extent to which the new government actually questioned the pre-existing course of economic reform, democratisation and European integration. According to another commentator, Enn Soosaar, opposition to this course was largely confined to the agrarian parties, the Centre and at least some members of *MKE*. Together, these parties had obtained only a third of seats in the *Riigikogu*. This in turn demonstrated that a clear majority of the Estonian electorate wanted to 'continue moving forwards.'[80]

The balance of forces within the new parliament did indeed ensure that economic reform and European integration stayed firmly on track over the next four years. The contradictions at the heart of the victorious *KMÜ* coalition, on the other hand, were to prove a persistent source of governmental instability. In drawing a distinction between the Coalition and the United Agrarians, Soosaar's analysis highlights the real essence of *KMÜ*, which was little more than a tactical alliance between parties with differing ideological standpoints. Perhaps the most fundamental source of disagreement related to the question of agricultural import tariffs. Whereas the Coalition under Vähi was committed to maintaining the ultra-liberal trade policy bequeathed by its predecessor, the rural parties contained an ardent protectionist lobby which was to grow progressively more vociferous over time.

With the right-wing opposition cheerfully predicting the speedy collapse of this bizarre political coupling, Vähi was required to draw upon his considerable reserves of political acumen. The prime minister successfully fended off calls for a revision of economic policy by citing commitments under the IMF memorandum, which was not set to expire until May 1997. Opponents of radical economic reform were forced to stand by helplessly as the government failed to fulfil its electoral promises of increased pensions and social security benefits.[81]

Whereas the agrarian parties seemed content to bide their time over the tariff issue, the ever-ambitious Edgar Savisaar was not so easily brought to heel. The new interior minister quickly made his presence felt by announcing plans for a crackdown on organised crime. Those worried by Savisaar's authoritarian tendencies had their fears amply confirmed in September 1995, when a police investigation into a private security firm revealed that Savisaar had arranged for the phones of leading politicians to be tapped during negotiations on a new government earlier in the year. The revelations raised serious issues regarding democracy, not least in view of Savisaar's reluctance to resign. President Meri ultimately had to prevail upon Vähi to submit his own resignation along with that of the entire government.[82]

In a clear statement of his political preferences, Meri helped to engineer the entry of the Reform Party to a new coalition government under Vähi's leadership. The Reformists took over the portfolios previously held by the Centre plus foreign affairs, which was occupied by Kallas himself. The new government took office in October 1995, with Vähi confidently predicting that it would carry out reforms more quickly and efficiently. The Reform Party can hardly have concurred with this assessment, for it soon began to criticise what it regarded as a lack of progress towards EU membership following Estonia's achievement of a full association agreement in July 1995.

That the coalition partners proved to be uneasy bedfellows when it came to economic policy is hardly surprising. The Reformists' sponsorship of a corporate tax exemption bill for companies reinvesting profits was opposed by Vähi, who feared that the loss of revenue might prejudice the achievement of a balanced budget. In addition, the prime minister faced increased pressure from more 'nationally'-minded quarters. Frustrated at its marginalisation within government, the Rural Union began to flex its muscles, threatening to withdraw from the ruling coalition if its demands for protective tariffs were not met. In August 1996, Vähi's own party was rocked by the maverick Endel Lippmaa's resignation as Minister for European Affairs. In withdrawing from the post, Lippmaa cited the government's apparent readiness to make concessions over the eastern border, along with its decision to extend the vetting of residency applications by Russian military pensioners.

Two months previously, Lippmaa's name had been mentioned in connection with the arrest of radical nationalist Tiit Madisson, who had published leaflets claiming that an underground Estonian 'Liberation Army' was ready to overthrow the government. The two-year imprisonment of this hero of the independence movement became something of a cause célèbre, for an unrepentant Madisson continued to call for direct action against the 'Communist'-ruled state. Although Madisson's name was invoked by certain circles of the nationalist right, another prominent ex-dissident, Enn Tarto, underlined the need to 'get over the Pullapää syndrome' by solving issues democratically.[83] The issue of civil-military relations had again came to the fore in December 1995, when President Meri finally accepted the resignation of Defence Forces Chief Aleksander Einseln. Einseln's departure followed another highly-publicised spat with Defence Minister Andres Oovel over the issue of corruption in the armed forces. Einseln, typically, refused to bow out gracefully, suggesting that he had in fact been dismissed at the behest of Moscow.[84]

By the autumn of 1996, tensions within the government coalition were beginning to take their toll. Vähi was forced to take an increasingly equivocal stance on major policy issues such as agriculture, enticing farmers with the prospect of tax cuts whilst also hinting at the possibility of tariffs on non-EU food imports. With the Reform Party ahead in the opinion polls, Vähi could not countenance early parliamentary elections. In the event, matters came to a head after the ruling parties failed to cooperate in October's local elections. Having emerged as the largest party on Tallinn Council, the Reformists joined a right-of-centre administration under Mart Laar, insisting that it could not work with Coalition Party councillors tainted by corruption. Laar's tenure proved short-lived, however, for the Coalition promptly struck a bargain with the Centre and the Russian-speaking list which propelled Edgar Savisaar to leadership of the city government. This marked a further stage in a remarkable political comeback by the former Popular Front leader, who had bowed out of politics for half a year following his disgrace in September 1995. His subsequent re-relection as Centre Party chairman during the Spring provoked a split, as former chair Andra Veidemann led a breakaway group to form the Development Party.[85]

The Centre-Coalition deal in Tallinn served as the prelude to a cooperation agreement at the national level. When Vähi refused to cancel the pact, the Reformists withdrew from the government on 23 November. According to Kallas, Vähi was seeking unilaterally to expand the government 'in a direction ... [which] ... endangers Estonia's independence, development and moral atmosphere. ... We will be moving closer to Russia, ignoring western integration.'[86] Vähi's decision to sack Kallas on account of these remarks was endorsed by the President, who was dismayed by the alarmist tenor of the Reformist leader's statement. However, Kallas was probably correct to question the wisdom of including the Centrists in government at a time when EU deliberations on enlargement were reaching a crucial stage.

In the event, Vähi's subsequent negotiations with the Centre Party came to nought, since the Coalition leader opted to continue governing without formal majority support within the *Riigikogu*. The courting of Savisaar may have been no more than a political ploy designed to secure the approval of the right-wing opposition for his new minority government. Vähi's return to power was facilitated by the appointment of Toomas Hendrik Ilves—previously Estonian Ambassador to Washington and not affiliated to any political party—as foreign minister. With hindsight, Ilves can be seen to have played a key role in smoothing Estonia's path to negotiations on EU membership.[87]

Continuity at the prime-ministerial level was matched by Lennart Meri's re-election as president in September 1996. Meri's subsequent receipt of the 'European of the Year' award in 1998 testifies to the high esteem in which he is held abroad. Once again, the election of Meri's main rival Arnold Rüütel at this juncture may have proved somewhat less propitious in terms of enhancing Estonia's profile internationally. Domestically, though, Meri faced a rough ride, securing election only at the fifth time of asking. After three inconclusive votes in the *Riigikogu*, the question was referred to an electoral college. Here, Meri finally triumphed over Rüütel in a second-round run-off on 20 September. Meri had evoked the ire of many parliamentarians during his first term, when he frequently stood accused of exceeding his constitutional mandate. It appears that the rural parties in particular were baying for blood over the incumbent president's role in the 1994 troop withdrawal. During the campaign, the nationalist right also renewed its allegations of Meri's former links with the KGB. Such allegations, however, seem to have done little to dent Meri's high popularity ratings amongst the people as a whole. Few indigenous Estonian politicians could claim that they were in no way compromised by dealings with the Soviet regime. Perhaps more significant was the fact that Meri still appeared untainted by the sleaze which has pervaded the younger post-Soviet political generation.

Nevertheless, a suitably chastened president promised greater cooperation and consultation with parliament in the wake of his re-election. In a similar vein, the newly-reinstated Tiit Vähi spoke of the need to restore confidence amongst the electorate, where support for the Coalition Party was running at less than 5%.[88] This task proved beyond the prime minister, whose third government was to last all of three months. At the end of 1996, Vähi unveiled his 'ten commandments' designed to foster economic growth and bring about the creation of a 'socially-oriented' market economy. Such pronouncements, however, had a hollow ring at a time when the government had yet to tackle thorny issues such as pension reform.

The passage of the 1997 budget, moreover, marked the start of a heated debate over priorities, most notably the relative weight which should be accorded to social and defence spending. The implicit rejection of Estonia's bid for NATO membership during the autumn (see Chapter Five) raised further questions regarding the state of the country's armed forces. Consequently, the defence ministry unveiled a new four-year development programme which would raise spending to 2% of GDP (the NATO minimum).[89] NATO's equivocal policy on enlargement, however, increased scepticism amongst certain members of the ruling coalition, who argued that if membership were not a realistic prospect, the money could be more usefully employed elsewhere.

VÄHI'S DOWNFALL

Vähi's luck finally ran out in January 1997, when he was confronted with evidence of 'unethical' privatisation during his tenure as head of Tallinn City Council from 1993–1995. Following his humiliating removal from the latter post the previous autumn, *Isamaa*'s Mart Laar revealed that under Vähi, some 200 flats in the capital had been sold to prominent individuals for as little as 5% of their market value. Amongst those to benefit was Vähi's daughter. Although Vähi protested his innocence, he resigned after the right-wing opposition came within a whisker of securing a vote of no-confidence in early February.

Opposition parties such as *Isamaa* and the Reformists could derive little satisfaction from Vähi's departure, since many of their members had themselves become mired in scandal. Siim Kallas faced, for instance, now faced allegations relating to the disappearance of $10 million from the North Estonian Bank prior to its subsequent merger with the Union Bank of Estonia. Kallas later relinquished his parliamentary immunity in order to fight the charges, and was cleared in April 1999. Yet the perceived corruption and ineptitude of the Estonian political class was steadily becoming a source of disillusion amongst the population as a whole, a fact which does much to explain the much-reduced turnout of 57% in the elections of March 1999. Thus, although the right-wing opposition clearly derived some political capital from its continued attacks on government ministers during 1997–99, no party grew to command widespread public respect. Consequently, opinion polls suggested that under the existing rules, an early election would bring no change to the fragmented parliamentary landscape. Throughout this period, the irrepressible Lennart Meri frequently warned of the threat to political stability posed by discredited politicians and bickering between small parties.[90]

THE SIIMANN GOVERNMENT

The need for political stability was rendered all the more pressing by the impending verdict on Estonia's application for EU membership. This fact was emphasised by Coalition Party deputy chairman Mart Siimann, who was nominated to form a new government in February 1997.[91] Siimann could count on continued support from the Centre, but, like his predecessor, he chose not to bring Savisaar's party into government. Instead, he opted for a minority KMÜ administration which retained all but two of the ministers from the previous cabinet. By pledging to maintain the existing course of economic policy and emphasising his determination to speed up EU integration, Siimann initially obtained backing from the Reformists and the Development Party.

Shortly after his accession, the new prime minister assumed personal control of the integration process and identified the 50 areas in most urgent need of harmonisation. This was in response to a visit by EU Commissioner for External Affairs Hans Van der Broek, who expressed concern that only a small proportion of Estonia's laws had been brought into line with EU standards. Van der Broek's fulsome praise for the economic policies pursued since independence, however, hinted at the subsequent outcome of the enlargement deliberations during July–December 1997.

ETHNOPOLITICS AND THE EU

Whilst Estonia had unquestionably been the pace-setter in economic reform during the early 1990s, the other two Baltic EU candidates understandably disputed the notion of a major lag in development between themselves and their northern neighbour by 1997. Their argument was vindicated at least partly by Latvia's subsequent accession to the World Trade Organisation ahead of Estonia. The enlargement decision thus cannot be explained purely on the basis of economic factors. In comparison with Latvia, Estonia was further distinguished by its faster progress in terms of drafting citizenship legislation and regulating the legal status of non-citizens. Following the tensions of 1993, developments during the subsequent four years had pointed to a growth in ethno-political stability. Confirmation of this came in January 1997, when the Council of Europe decided to end its monitoring of the situation in Estonia.

Even so, the European Commission *avis* of July 1997 rightly highlighted the need to speed up the political and social integration of non-citizens. Once again, the Siimann government worked hard to address international concerns in this area, shuffling Andra Veidemann from European Integration to a newly-created ministerial post devoted entirely to ethnic affairs. In response to pressure from the EU and the OSCE, the government proposed legislative amendments granting automatic citizenship to children of non-citizen parentage born in Estonia after February 1992. After a debate lasting almost a year, the *Riigikogu* eventually assented to these proposals in December 1998, at which point leading Estonian politicians urged the two organisations to refrain from any further demands.

Notwithstanding these changes, however, non-citizens have expressed growing dissatisfaction with the procedures for naturalisation established under the 1995 Law on Citizenship.[92] Whilst the younger generation of Russian-speakers has expressed a growing desire to learn Estonian, 1997 saw a decline in the number of naturalisations, allegedly on account of applicants' inability to pass the required examination in the Estonian language.[93] In response, the government unveiled a new 'national integration strategy' designed to channel more resources into Estonian language teaching. Yet it emphatically rejected suggested amendments—tabled by the Russian parliamentary fraction—granting automatic citizenship to all pre-1990 permanent residents, on the grounds that this would create too many citizens unable to speak the state language.

The period 1995–99 thus brought no end to the underlying tension between international integration and 'nationalising' impulses at the state centre. The extension of automatic citizenship rights, for instance, has been paralleled by amendments to the language law requiring parliamentary deputies and local government officials to demonstrate a working knowledge of Estonian. The *Riigikogu* also instructed the government to formulate new language requirements for employees working in the service sector. These new provisions have proved controversial, attracting further criticism from the OSCE.[94] The president, for his part, denounced them as unconstitutional, claiming that they handed undue powers to the government. Meri twice refused to promulgate the amendments, which were referred to the Supreme Court prior to their final adoption in February 1999.

Furthermore, leading nationalities experts in Estonia have questioned the need for linguistic regulation in the private business sector, claiming that priority should instead be given to the improvement of Estonian language teaching in Russian schools.[95] Given the continued shortage of suitably qualified personnel, this seems likely to be a long process. The government integration strategy has sought to address the problem by appointing more teachers of Estonian as a Foreign Language at the primary level. Yet plans for a complete switch to Estonian-language teaching in upper secondary schools—originally scheduled for the year 2000—have now had to be postponed until 2007. Pending the necessary improvements in this sector, language will inevitably remain a source of social closure for graduates of Russian-language schools.

The EU's approach to the 'national question' is perhaps one reason why local Russians express higher levels of support for membership than their Estonian co-habitants. In August 1997, for instance, Swedish-Finnish MEP Jörn Donner elicited a predictably furious response in the Estonian-language media when he suggested that the country should adopt Russian as a second official language. Yet support for the EU amongst local Russians is also conditioned by their 'western' outlook and their negative view of economic and political developments in neighbouring Russia. Ethnic Russians born and raised in the 'Soviet West' felt little affinity with the Russian republic of the USSR prior to 1991. The new Russian Federation's subsequent pretensions as an 'external national homeland' for the Baltic Russians have been further undermined by its policy of economic sanctions, which hurts the very population which it purports to protect. Few Estonian Russians would dispute that Moscow's allegations of mass human rights abuses are inspired by geopolitical objectives rather than any real concern for the cultural and material needs of its so-called 'compatriots' in the near abroad. Yet as long as Russia aspires to bring the Baltic states within a sphere of influence, Estonian legislators are even less likely to countenance any fundamental modification to existing nationalities policy.

ECONOMIC WOES AND POLITICAL DISPUTES 1998–99

Over the past three years, geopolitical insecurities with regard to Russia have been counterbalanced by growing 'geoeconomic' pressures for an accommodation with the 'eastern neighbour' (see Chapter Four). Prior to his final departure from politics, ex-prime minister Vähi aroused a storm of controversy after he was granted a 90-minute audience with Yevgenii Primakov during a 'private' visit to Russia. Vähi, who at that time was still Chairman of the ruling Coalition Party, stood accused of trying to influence government policy on the national question in return for an amelioration of Russia's trade policy towards Estonia.[96] The subsequent financial crisis in Russia during 1998 further accentuated the difficulties faced by those Estonian industries still reliant upon eastern markets, especially agricultural producers and food manufacturers. This in turn served to increase the vociferousness of the rural lobby, a development which cast doubt over Siimann's ability to fulfil earlier pledge that he would maintain the existing course of economic policy.

As early as December 1996, the Coalition Party had bowed to pressure from its agrarian partners by abandoning a clause banning import tariffs from the government's programme. Following Vähi's resignation, the Country People's Party began to assert itself within the new minority coalition after Andres Varik replaced the much-criticised Ilmar Mandmets as minister for agriculture. With the IMF agreement set to expire, the government approved a new package of subsidies to grain and dairy farmers, supplemented by a proposed 10% tariff on non-EU imports of pork.

The right-wing opposition parties vigorously opposed the measures, claiming that they would harm Estonia's reputation with the IMF, complicate EU accession and ongoing negotiations with the WTO, and reduce the export competitiveness of an agricultural sector which, they claimed, was simply unwilling to embrace reform. Proponents of the bill retorted—quite justifiably—that agricultural support measures are commonplace in EU member countries, and that a tariff law was actually a prerequisite for Estonian membership of the WTO. Although there was general agreement on the need for a technical law, the opposition insisted that the existence of such legislation did not necessarily entail the introduction of tariffs. After half a year of fierce debate, the *Riigikogu* gave its consent to a watered-down bill permitting the government to levy tariffs on specified foodstuffs for a six-month period. Yet even this law was referred to the constitutional court, which ruled that the government had exceeded its powers in claiming the right to impose tariffs without parliamentary approval. As such, the law on tariffs had still not been implemented by the time of the March 1999 elections.

These successful blocking tactics were the work of the 'United Opposition' (Reform, *Isamaa-ERSP*, Moderates, Right-Wingers) formed in November 1997. Although Siimann could still rely upon the cooperation of the Centre Party in parliament, the strains of running a minority coalition began to tell over the following year, when fallout from the Russian crisis slowed economic growth and further narrowed the margin for compromise. If the rural parties were generally satisfied with farm subsidies as an alternative to tariffs, the ruling coalition remained beset by tensions. Interior minister Robert Lepikson was sacked for his outspoken attacks on cabinet colleagues, while Toomas

Hendrik Ilves was forced to depart after his newly-created United People's Party refused to toe the government line in parliament.

Increasingly, Siimann's grip on power seemed to rest upon day-to-day political expedients rather than any coherent policy programme. After a failed attempt to bring the Reform Party into the ruling coalition, the prime minister decided upon an early election. Yet with support for the Coalition Party running at less than 5%, its only hope of retaining office lay in forming an electoral alliance similar to that which had brought it to power three years earlier. Bigger, more popular parties such as the Centre and the Reformists insisted that they would only agree to early elections if such alliances were banned, thereby forcing Siimann to withdraw his request in June 1998. As an increasingly beleagured government limped on towards the end of its term, the KMÜ coalition finally came asunder during the autumn, when the support of the Country People's Party ensured the passage of a Centre Party motion banning electoral alliances. The amendment stipulated that parties were not allowed to cooperate unless they agreed either to merge formally or to run under the banner of a single political party.

The new electoral law was hailed as a necessary step towards ending the fragmented parliaments and unstable coalition government so characteristic of the period 1992–99. The enforcement of the 1,000-member rule ahead of the elections reduced the number of officially registered parties to 16 compared to 30 in 1995. The subsequent ban on alliances brought a further round of mergers in the three months which followed. Even so, seven parties managed to clear the 5% barrier in March 1999, making the new parliament little different from its predecessor. The degree of governmental stability will therefore depend upon the cohesion of the three party right-of-centre coalition (*Isamaa*, Moderates and Reform Party) which came to power in April 1999 with a slim 3-seat majority.

The fortunes of the by now doomed KMÜ coalition were not helped by Lennart Meri's pre-election broadcast complaining of 'stagnation' in public life.[97] Of the original partners, the Rural Union and the Pensioners and Families Party merged with the Coalition, which gained only 7 seats in the new parliament. The Country People's Party, on the other hand, chose to run singly, hoping to capitalise on rural discontent. The bulk of the vote from socially-disadvantaged groups flowed to the Centre Party which ran a high-profile campaign centred around proposals for a progressive income tax. Despite a further warning from Meri concerning 'politicians using authoritarian and undemocratic means'—seen by many as a thinly-veiled attack on Edgar Savisaar—the Centre emerged as the largest single party with 28 seats.[98] However, even with the support of the Country People's Party (7 seats), the United National Party (6 seats) and—possibly—the Coalition, it was still three short of an overall majority.

When called upon to nominate a new prime minister, the president predictably opted not for Savisaar, but for *Isamaa* chairman Mart Laar, whose party (18 seats) was joint-second largest group in parliament along with the Reformists (18) and the Moderates (17). Faced with renewed economic difficulties after years of growth, the electorate—albeit by a narrow margin—ultimately retained its faith in established policies. In this regard, Laar could at least point to his resolve in tackling earlier, far more serious crises, and the economic track record of a party 'that does what it promises'.[99]

FUTURE PROSPECTS

Whether the *Isamaa* leader will be able to fulfil his party's campaign commitment to EU membership by 2003 still remains to be seen. Estonia has already made considerable progress in its EU membership negotiations, where discussions have largely focused upon the considerable challenges—both technical and financial—of meeting the *acquis communautaire*. In the final analysis, however, it seems that Estonia's progress will ultimately be determined by the degree of political will on the part of existing member states vis-à-vis enlargement (see Chapter Five).

The Estonian delegation insists that a long delay would lead to widespread public disillusionment and a consequent slowdown in the reform process.[100] Such statements highlight the challenge faced by the political leadership in terms of convincing the public of the merits of EU membership. In a statement which typifies the attitude of the ruling elite since independence, Toomas Hendrik Ilves claims that EU integration is 'a natural part of our development rather than a process forced upon us from outside'. In his eyes, the real essence of Estonia's 'Return to Europe' has been 'the building of a society that is prosperous, democratic and stable.'[101] Yet although a majority of Estonia's population continues to subscribe to this latter aim, this does not necessarily mean that entry to the EU is viewed as a natural part of Estonia's development. As Ilves admits, EU accession was viewed with indifference in most circles of Estonian society during the early 1990s.

Interest picked up around 1995, and yet even after the positive *avis* obtained from the European Commission, support for EU entry stood at only 40% at the end of 1997, with 35% still undecided. A year later, active support was down to 26%, with almost 60% falling into the undecided category.[102]

A widespread view has been to regard EU membership as the lesser of two evils, the only other alternative being membership of the CIS. Yet the growing body of Eurosceptics suggests that EU entry would simply mean the replacement of previous control from Moscow by control from Brussels. The consequent loss of national sovereignty would entail the sacrifice of Estonia's liberal market economy. It would also complicate trade with non-EU states at a time when the majority of Estonia's production does not correspond to European standards. Aside from the debates on the 'national question' alluded to above, other negative perceptions relate to fears of a further growth in bureaucracy and a possible erosion of national identity through global homogenisation of culture and unrestricted immigration.[103] According to many Eurosceptics, the political elite has been so anxious to enter the EU that it has not actually paused to consider whether membership would actually be beneficial.[104] Ilves freely admits that the principal problem in terms of public opinion concerns the lack of adequate information regarding the potential benefits and drawbacks of EU entry. Genuine public debate over these questions is indeed only just beginning. Given that Euroscepticism has not yet taken on the appearance of a well-organised political movement, it is difficult to say whether it is representative of a broad section of the population or not. In this regard, it is worth noting that the share of survey respondents actively opposed to EU membership remained steady at around 14% during 1997–98.[105]

If the European Union has already provided considerable aid in terms of formal institution-building in Estonia, it has done less to address the continued weaknesses of civil society. Indeed, the often 'mechanical' adoption of EU norms may have contributed to the already problematic state-society relationship and the apparent loss of faith in representative democracy evident during 1995–98.[106] The task of selling the merits of EU membership to a sceptical public is, of course, not simply confined to the applicant countries of Central and Eastern Europe. It applies equally to governments of existing member states as they negotiate the deepening of economic and political union. In so doing, Europe's leaders are increasingly being made aware of the need to render existing institutions more accountable to the people. The EU itself has yet to seriously address the affective dimension of building a 'New Europe'. If formal European citizenship is already a reality, EU institutions still lack the legitimacy conferred by a popular sense of European identity. For the latter to emerge will require nothing less than a fundamental reappraisal of the established nation-state model. This was a problem which preoccupied Estonian thinkers on cultural autonomy back in the 1920s. Indeed, with its often overlooked heritage of managing multiculturalism, Estonia could have much to contribute to the future evolution of an expanded EU.[107] For the moment, 'Euroscepticism', ethnic tensions, suspicion of bureaucracy and declining faith in representation remain generalised problems across a continent still undergoing rapid and fundamental change following the end of the Cold War. In this sense, at least, Estonia is well on its way to becoming a 'normal' European country.

1   Marju Lauristin, "Kommentarid," in *Kaks Otsustavat Päeva Toompeal (19–20 August 1991)* (Tallinn, 1996), p.81.

2   Lennart Meri, "Õiguste ja Kohutuste Tasakaal," speech on the occasion of Estonia's admittance to the Council of Europe, 13 May 1993, in Lennart Meri, *Presidendikõned* (Tartu, 1996), p.335.

3   Cited in: Joan Löfgren, Helena Mannonen, Olli-Pekka Jalonen and Jouko Huru, *The Integration of the Baltic States into the EU*, Project Report for the Finnish Ministry of Foreign Affairs (Tampere, 1997), p.85.

4   Karen Dawisha and Bruce Parrott, *Russia and the New States of Eurasia: The Politics of Upheaval* (New York, 1994), pp.123–126.

5   Marju Lauristin and Peeter Vihalemm. "Recent Historical Developments in Estonia: Three Stages of Transition (1987–1997)," in *Return to the Western World: Cultural and Political Perspectives on the Estonian Post-Communist Transition*, eds. Marju Lauristin and Peeter Vihalemm with Karl Erik Rosengren and Lennart Weibull (Tartu, 1997), pp.77–80.

6   David Arter, *Parties and Democracy in the Post-Soviet Republics*
    The Case of Estonia (Aldershot, 1996), p.11; "Speech of Lennart Meri on the 81st Anniversary of the Republic of Estonia, February 24th 1999," *The Monthly Survey of Baltic and Post-Soviet Politics Extra. Estonia: Elections 7 March 1999*, pp.20–27. See also: Rain Rosimannus, "Political Parties: Identity and Identification," *Nationalities Papers*, Vol.23, No.1, 1995; Kristi Raik, "Democratisation and Integration into the European Union: The Case of Estonia." Unpublished paper presented at the seminar *Linking the European East and West: The Baltic Sea Area within the Framework of European Integration*, University of Turku, Finland, 3 June 1999; *Baltic Times*, 11–17 March 1999.

7   Rein Ruutsoo, "Estonian Post-Communist Transition, Civil Society and Social Sciences in the Context of EU Enlargement," in Chancellery of the Riigikogu, *Society, Parliament and Legislation* (Tallinn, 1999), pp.71–73.

8   Shirley A. Woods, "Ethnicity and Nationalism in Contemporary Estonia," in *Ethnicity and Nationalism in Russia, the CIS and the Baltic States*, eds. Christopher Williams and Thanasis D. Sfikas (Aldershot, 1999), pp.283–284.

9   Vahur Made, "Estonia and International Organisations," unpublished paper presented at the seminar *Linking the European East and West: The Baltic Sea Area within the Framework of European Integration*, University of Turku, Finland, 3 June 1999; Löfgren, Mannonen, Jalonen and Huru, p.85.

10 See, for instance, the comments in Mart Laar, "Eesti Vabariigi Sisepoliitika ja Riigikorralduse Probleemid," in *Kaks Algust: Eesti Vabariik 1920. ja 1990. Aastad*, ed. Jüri Ant (Tallinn, 1998), p.151.

11 *Baltic Independent*, 7–13 February 1992.

12 Mart Laar, Urmas Ots and Sirje Endre, *Teine Eesti* (Tallinn, 1996), p.661–662. Vähi was minister for transport in the Savisaar government during 1990–92.

13 Arter, pp.181–182; "'Kindel Kodu' Tsentrist Veidi Paremal," *Rahva Hääl*, 12 August 1992.

14 Igor Rotov, "Parempoolsed Uhenisid," *Rahva Hääl*, 28 September 1991.

15 David Smith, "The Restorationist Principle in Post-Communist Estonia," in Williams and Sfikas eds., pp.287–323; Andrus Park, "Ethnicity and Independence: The Case of Estonia in Comparative Perspective," *Europe-Asia Studies*, Vol.46, No.1, 1994, pp.69–87.

16 Anatol Lieven, The Baltic Revolution: Estonia, Latvia, Lithuania and the Path to Independence (London, 1993), p.55; Graham Smith, "Introduction: The Baltic Nations and National Self-Determination," in *The Baltic States: The Self-Determination of Estonia, Latvia and Lithuania*, ed. Graham Smith (Basingstoke, 1994), p.5.

17 Riina Kionka, "Estonian Political Struggle Centers on Voting Rights," *RFE/RL Research Report*, Vol.1, No.24, 12 June 1992, p.17.

18 Riina Kionka, "Drafting New Constitutions: Estonia," *RFE/RL Research Report*, Vol.1, No.27, 3 July 1992; Lauristin and Vihalemm, p.101.

19 Igor Rotov, "Kodakondsusest: Optsioon või Naturalisatsioon?" *Rahva Hääl*, 8 September 1991.

20 Ibid

21 Ibid; Riina Kionka, "Who Should Become a Citizen of Estonia?" *RFE/RL Research Report on the USSR*, Vol.3, No.39, 27 September 1991, p.25.

22 Kionka, 27 September 1991, pp.23–26.

23 Ibid, p.24.

24 For a good summary of the main arguments, see: Vesa Saarikoski, "Russian Minorities in the Baltic States," in Joenniemi and Vares eds.

25 Quoted in Lieven, p.277.

26 Graham Smith, Aadne Aasland and Richard Mole, "Statehood, Ethnic Relations and Citizenship," in Graham Smith ed., pp.189–190; See also Lauristin and Vihalemm, p.101.

27 The principle which holds that a person's nationality at birth is the same as that of his/her natural parents. As opposed to *jus soli*, which holds that a person's nationality at birth is determined by the territory in which he/she was born.

28 Löfgren, Mannonen, Jalonen and Huru, p.90. In order to address the controversy elicited by the 1992 citizenship law, the Vhi government took the precaution of submitting it to scrutiny by CE experts.

29 Smith, Aasland and Mole, pp.189–90.

30 David Smith, "Russia, Estonia and the Search for a Stable Ethno-Politics," *Journal of Baltic Studies*, Vol.29, No.1 (Spring 1998), pp.3–18.

31 John McGarry and Brendan O'Leary, "Introduction: The Macro-Political Regulation of Ethnic Conflict," in *The Politics of Ethnic Conflict Regulation*, eds John McGarry and Brendan O'Leary (London, 1993), p.23.

32 Riina Kionka, "Estonia: A Difficult Transition," *RFE/RL Research Report*, Vol.2, No.1, 1 January 1993, p.90.

33 "Coalition Party in Competing Visions of an Estonian Future," *Nationalities Papers*, Vol.23, No.1, 1995, pp.195–198.

34 Estonian Institute, "The Estonian Parliament—The '*Riigikogu*'," *Estonia in Facts*, (Tallinn, Estonian Institute, 1993).

35 Laar, p.151; For a full background to these events, see: Georg Sootla, "Political Background and Possible Consequences of the Summer Crises in Estonia," *The Monthly Survey of Baltic and Post-Soviet Politics*, July 1993, pp.54–57.

36 Kionka, 3 July 1992, loc cit.

37 Rosimannus, p.40.

38 Arter, p.195.

39 Ibid.
40 Vahur Kalmre, ed., *Postimehe Valimisteatmik* (Tartu, 1995), pp.127–130.
41 Rosimanus, p.30.
42 *Rahva Hääl*, 12 August 1992.
43 Dawisha and Parrott, p.144.
44 Arter, pp.195–213.
45 According to Article 79 of the 1992 constitution, the President is to elected either by a two-thirds majority vote in the *Riigikogu* or, in the event of three inclusive votes, by an electoral college consisting of the members of the *Riigikogu* and representatives of local government councils.
46 Mart Laar, "Estonia's Success Story," *Journal of Democracy*, Vol.7, No.1, 1 January 1996, p.97.
47 "Mart Laari Kõne Riigikogus 19. Oktoobril 1992," *Rahva Hääl*, 20 October 1992.
48 *The Baltic Independent*, 11 February 1993.
49 *The Baltic Independent*, 6 May 1993.
50 Sootla, loc cit; *The Baltic Independent* 16 September 1993.
51 *Economist Intelligence Unit Country Report*, 1st Quarter 1996, p.7; *The Baltic Independent*, 17–23 April 1997.
52 *The Baltic Independent*, 12–18 February 1993.
53 Lennart Meri, 13 May 1993, pp.336–337.
54 *The Baltic Independent*, 13 May 1993.
55 Kalle Muuli, "Kohalikesse Volikogudesse Pääsevad 17 Oktoobril Üksnes Eesti Kodanikud," *Postimees*, 20 May 1993.
56 For a detailed discussion of the 'Aliens Crisis' see: David Smith, "Legal Continuity and Post-Soviet Reality: Ethnic Relations in Estonia 1991–95," unpublished Ph.D. dissertation, University of Bradford, 1997, pp.205–245.
57 Hannes Rumm, "Keskfraktsioon Kardab Eestis Etnilist Puhardust," *Postimees*, 11 June 1993.
58 Estonian Ministry of Culture and Education, *"Report on Estonian-language Learning Initiatives,"* Tallinn, 11 March 1994, p.2.
59 Author interview with Jaak Roosaare, Tallinn, 10 February 1995.
60 Löfgren, Mannonen, Jalonen and Huru, p.89.
61 Commission for Security and Cooperation in Europe, *Russians in Estonia: Problems and Prospects* (Washington, 1992).
62 Sootla, p.25.
63 CSCE Mission to Estonia Narva Office, Internal Memo, 22 September 1993.
64 CSCE Mission Jõhvi office, Internal Memo, 28 September 1993.
65 Sootla, p.52.
66 Quoted in John Hiden, "The Baltic Republics," in *The Annual Register: A Record of World Events 1993*, ed
    Alan Day (Harlow, 1994), p.146.
67 Jorma Rotko, "Viron Vähi Moitti Edeltäjiensa Idänpolitikkaa," *Helsingin Sanomat*, 22 April 1995.
68 *The Baltic Independent*, 28 October 1993.
69 I. Nikifirov, "Iz Tallinna v Revel'," *Nezavisimaia Gazeta*, 21 October 1993.
70 Hiden, 1994, p.146.
71 *The Baltic Independent*, 14 April 1994.
72 *Eesti Ringvaade*, 16–22 May 1994.
73 *RFE/RL Daily Report*, No.141, 27 July 1994.
74 Luik replaced Trivimi Velliste (*ERSP*) as foreign minister in January 1994.
75 "Vene Valimisliit Loodab Valimiste Järel Pusima Jääda," *Hommikuleht*, 8 February 1995; United National Party of Estonia, *Programma Ob'edinennoi Narodnoi Partii Estonii (Proekt)*, draft programme circulated at the party's founding congress in Tallinn, 9 October 1994.
76 *Law on Citizenship. Adopted by the Riigikogu on 19 January 1995*, unofficial translation, 24 January 1995.
77 *Eesti Ringvaade*, 26 February–4 March 1995; Kalmre, ed., pp.68–72.
78 *The Baltic Independent*, 30 March 1995; *Eesti Ringvaade*, 14–20 April 1995.
79 Laar, p.100.

80   Enn Soosaar, "Eesti Rahva Enamus Tahab Edasi Minna," *Rahva Hääl*, 7 March 1995.

81   *The Baltic Independent*, 28 July–3 August 1995.

82   John Hiden, "The Baltic Republics," in *The Annual Register: A Record of World Events 1995*, ed. Alan Day (London, 1996), p.141.

83   *The Baltic Independent*, 28 August–3 September 1997.

84   *Economist Intelligence Unit Country Report*, 1st Quarter 1996, p.7.

85   Peeter Kaldre, "Savisaare Sinusoid," *Luup*, no.24, 25 November 1996.

86   *The Baltic Independent*, 28 November–4 December 1997.

87   "Toomas Ilves, Estonia's American-European," *The Economist*, 31 October 1998, p.52.

88   *Economist Intelligence Unit Country Report*, 1st Quarter 1997, p.10.

89   Ibid, p.13.

90   *The Baltic Times*, 5–11 March 1998; "Speech of Lennart Meri...," loc cit.

91   A fifty year-old former psychologist, Siimann worked as a censorship official during the 1980s before becoming director-general of state-run *Eesti TV*.

92   Kristi Malmberg and Rein Sikk, "Muulased Tahavad Lastele Eesti Kodakondsust," *Eesti Päevaleht*, 4 April 1997; Tartu University Market Research Team, *The Attitude of Town Residents of North-Eastern Estonia towards Estonian Reforms and Social Policy: A Comparative Study of 1993, 1994 and 1995* (Tartu, 1995), p.8.

93   Jorma Rotko, "Viron Venäläisten Asenne Virolaisuuteen Riipuu Iästä," *Helsingin Sanomat*, 6 January 1997; *Economist Intelligence Unit Country Report*, 1st Quarter 1998, p.13.

94   *Monthly Survey of Baltic and Post-Soviet Politics*, 7 March 1999, pp.2–5.

95   *The Baltic Times*, 8–14 January 1998.

96   *The Baltic Independent*, 17–23 April 1997.

97   *The Baltic Times* 10–16 December 1998.

98   "Speech of Lennart Meri...," p.23; *The Baltic Times*, 11–17 March 1999.

99   *The Baltic Times*, 11–17 February 1999.

100 *The Baltic Times*, 19–25 November 1998.

101 Toomas H. Ilves, Minister for Foreign Affairs of Estonia, "Estonia on its Way to an Integrated Europe," speech at the University of Latvia, 10 March 1998, *http://www.vm.ee/eng/index.html*.

102 Ibid; Andrus Saar, "Euro-Barometer and Estonian Experiences: European Union Integration and Enlargement—Attitudes in Estonia," in Chancellery of the Riigikogu, *Society, Parliament and Legislation*. Tallinn, Department of Economic and Social Information of the Chancellery of the Riigikogu, 1999, p.53.

103 Ibid, pp.53–56; Mait Talts and Aksel Kirch, "Eesti ja Euroopa Liit: Poolt ja Vastu Eesti Ajakirjanduses," in *Eesti Euroopa Liidu Lävepakul*, ed. Rein Ruutsoo and Aksel Kirch (Tallinn, 1998), pp.100–113; Author interview with Marina Kaas, Tallinn, 23 April 1996.

104 Saar, p.53; *The Baltic Times*, 13–19 August 1998.

105 *The Baltic Times*, 14–20 January 1999.

106 Raik, loc cit.

107 Lennart Meri, "Eesti Roll Uues Euroopas," in Meri, 1996, p.240; David Smith, "Retracing Estonia's Russians: Mikhail Kurchinskii and Interwar Cultural Autonomy," *Nationalities Papers*, Vol.27, No.3, 1999.

Chapter 4

'THE LITTLE COUNTRY THAT COULD': ESTONIA'S ECONOMIC RETURN TO EUROPE

The centrally-planned economy of the USSR was expressly designed to integrate individual union republics into a unitary Soviet state and counteract any attempts by local elites to assert political sovereignty vis-à-vis Moscow. Consequently, those successor states seeking a rapid and complete break with the Soviet system have faced a dual challenge: alongside the already daunting task of moving from plan to market, they have been required to transform what were in effect largely provincial economies into ones befitting independent countries operating in a globalised market environment. Most commentators would agree that Estonia—'the shining star from the Baltics'—has been the most successful in meeting the economic challenges of independent statehood in an interdependent world.[1] Post-Soviet Estonia has become—to quote *Newsweek* magazine—'the little country that could', confounding the earlier predictions not only of the Soviet government but also of many western specialists.[2]

Notwithstanding the frequent changes of government described in the previous chapter, economic policy has notability for its consistent adherence to the set of parameters devised in conjunction with the International Monetary Fund (IMF) during 1992. These were expressly designed to meet the needs of a small, open economy, whilst simultaneously erecting reliable bulwarks against international economic fluctuations and political pressures for a retreat from reform. Thus far, such pressures have been muted, reflecting an enduring consensus based on the rejection of Communism and the desire for an 'economic return to Europe'.[3] Yet recent evidence suggests that this period of 'extraordinary politics' has now run its course. Continued plaudits from the International Monetary Fund and other quarters cannot disguise the negative social consequences of market reform such as crime, unemployment, income disparities and a further steep decline in the already low rate of fertility. These issues all require urgent action if Estonia is to sustain its economic progress in the medium to long term.

THE SOVIET LEGACY

Estonia's progress over the past decade appears all the more remarkable when set against the forecasts which accompanied the restoration of independence. In the West, many commentators seemed all too ready to accept Gorbachev's assessment that the Baltic states would find themselves in an 'economic swamp' should they try to go it alone outside the USSR. The small size of the domestic market meant that the restored Estonian Republic would again be highly reliant on trade, yet in this respect post-war Sovietisation had bequeathed a dependency on Russia which was higher than average for the successor states as a whole.[4] Compared to its Baltic neighbours, Estonia was better placed by virtue of a power generating capacity based on oil shale deposits in the north-east of the republic. These catered for half of the country's total energy requirements at the time of the Soviet collapse. Even so, Estonia remained entirely dependent upon Russia for imports of petroleum and natural gas. In the light of Russia's progressive shift to trading at world prices with non-CIS republics during 1992, it was therefore predicted that Estonia would have considerable difficulties in paying for its energy imports should it attempt to pursue total political independence from Moscow and marketisation simultaneously.[5]

Such predictions failed to reckon with the Estonians' own determination to restore the kind of independent economic existence enjoyed between the wars. This determination goes a long way towards explaining the perspectives and policies advanced during 1988–91. Following the half-hearted approval given to Baltic proposals for economic autonomy in November 1989, Estonia quickly became the pacesetter for reform in the USSR, introducing key elements of a market economy already prior to the restoration of full independence. Price liberalisation took gradual effect during 1990–91, when the Savisaar government slashed subsidies on alcohol, foodstuffs, energy, transport and telecommunications. These measures proved partially successful in limiting export sales and increasing imports, thereby helping to relieve the supply shortages which had begun to plague the economy.[6] The same period also saw important progress towards nationalising and modernising the tax system. With Moscow's consent, the government began to collect all taxes locally and send a single negotiated lump sum to the all-union budget. The Soviet system of turnover and corporate tax was replaced by uniform rate VAT at 10% and progressive taxes on corporate and personal income. The successful application of these rules in practice is attested by the government's ability to maintain a consolidated budget surplus in 1991 and its drafting of a balanced budget for 1992.[7]

Such fiscal prudence is most easily explicable in terms of a desire to assert political sovereignty vis-à-vis Moscow. As long as Estonia remained part of the rouble zone, the short term interests of the population would arguably have been better served by fiscal irresponsibility, for while the benefits of such a policy would have been reaped locally, the costs of the resultant financial instability would have been dispersed across the entire territory of the former USSR.[8] Yet, if the striving for economic autonomy was ultimately a means to the political end of securing full independence from the USSR, experiences during 1990 merely confirmed that further meaningful economic reform would be impossible without the prior attainment of full political sovereignty.[9] Moscow's continued adherence to the principles of a unitary command economy set clear limits to the possibilities for autonomous economic development envisaged under the original *IME* plan. The refusal to transfer 21 all-union enterprises to Estonian jurisdiction, for instance, impeded implementation of property reform. The Estonian Central Bank *(Eesti Pank*, or Bank of Estonia (BOE)), established in January 1990, lacked recognition from the Soviet government and was thus required to vie for precedence with the all-union *Gosbank*. In what became known as the 'war on banking', the USSR's Foreign Economic Bank *(Vneshekonombank*—VEB) closed all of its accounts in Estonia and withdrew the funds to Moscow. Although the government passed new laws making the conduct of foreign trade independent from the centre, these were difficult to implement as long as the Soviet authorities retained control over Estonia's frontiers with the West.[10] The same was true of plans to restore the inter-war currency, the Kroon. Given that this would almost certainly have to be minted abroad, many predicted that it would simply be confiscated as soon as it arrived in Estonia.[11]

A far more serious obstacle to creating a national currency was the prospect that Moscow would charge world prices for energy exports and erect tariff barriers for states remaining outside the new Union Treaty. Under these circumstances Estonian economic experts felt that external aid would be required in order to guarantee the Kroon's convertibility. The international situation prior to August 1991, however, meant that western states could not be counted on to provide a stabilisation fund.[12] If the restoration of independence removed the remaining political obstacles to interaction with the West, the impasse of the past eighteen months had seemingly dented Savisaar's resolve to pursue a definitive break with the FSU.[13] In the short term, the collapse of the Soviet economy accentuated Estonia's mounting economic difficulties. Declining production and disruption to former supply networks brought acute shortages of fuel and other raw materials as winter approached. An Estonian-Russian agreement providing for barter trade at world prices was signed in December 1991 but did not come into effect for almost three months. In the intervening period, industrial and agricultural production in Estonia declined by 36–46%, leading to shortages and panic buying.[14] Inflation, already running at 300% in 1991, soared to 940% over the following year as further domestic price liberalisation occurred and Russia and other former republics began to charge world prices for their exports. The resultant terms of trade shock was estimated as being equivalent to a 16–20% drop in GDP. Consequently, industrial production fell by nearly 40% during the first nine months of 1992, whilst the number of unemployed climbed to 7%.[15]

In raising prices whilst simultaneously cutting exports, Russia made no secret of its intention to use energy policy as a lever for realising its political aims with regard to Estonia. Edgar Savisaar, it seemed, had been banking on continued goodwill from the eastern neighbour. When this failed to materialise, the prime minister was unable to come up with radical measures to combat the economic crisis. Instead, he responded by attempting to impose greater centralised control. Had Savisaar succeeded in imposing a state of emergency in January 1992, the subsequent course of Estonia's economic development might have been very different. In the event, his successor Tiit Vähi proved more willing to grasp the nettle of reform. In order to combat the energy crisis, alternative supplies were sought from individual Russian regions and from foreign states. These efforts were later aided by credits obtained from the World Bank and the European Bank for Reconstruction and Development. Perhaps most significantly, the government sought to reduce energy usage by raising prices sharply in March 1992.[16]

A more robust approach was also evident in the sphere of currency reform. The inflationary chaos afflicting the entire rouble zone, coupled with a severe shortage of cash roubles, made the case for a national currency more pressing. Whereas Savisaar had expressed a preference for the kind of gradual strategy subsequently adopted by most FSU states (allowing substitute 'coupons' to circulate in parallel to the rouble prior to the full introduction of a national currency), the new administration opted for a straight transition to the Kroon in June 1992. The shift towards radical reformism owed much to the new government's reliance on technocratic experts—including young émigré Estonians such as Ardo Hansson—working in tandem with the IMF. The ability of these 'change teams' to exert a decisive influence over policy rested in turn upon a relative lack of structural and political constraints compared to other FSU states. Estonia was more fortunate than its Baltic neighbours in the sense that it inherited a smaller share of Soviet-era heavy industry and former all-union enterprises.[17] Many of the latter were heavily staffed by Russian-speaking settlers who had no opportunity to express discontent via the ballot box in the September 1992 elections. Conversely, the outright rejection of Soviet communism amongst the Estonian population brought a heightened receptiveness to radical reform. Although the price increases of 1990–91 had proved politically unpopular, the Estonian trade union movement was prepared to sanction successive decreases in real wages in the run-up to independence. In the immediate term, at least, outright opponents of economic 'shock therapy' such as Savisaar were also hampered by the absence of a well-developed party system acting as a conduit for popular discontent.[18]

MONETARY REFORM AND MACROECONOMIC STABILISATION

Savisaar's reluctance to sever ties with Russia was shared, at least initially, by the European Community and the IMF.[19] In spite of the rouble zone's dwindling credibility as an optimal currency area, the IMF in particular was wary of provoking further dislocation of trade within the former Soviet space. By the Summer of 1992, however, the Russian government was beginning to regard the uncontrolled monetary policies pursued by some former republics as a major obstacle to achieving macroeconomic stabilisation in Russia proper. For this reason, the substitution of the rouble by the Kroon was not hampered by any major disputes with Moscow. The Estonian insistence on a rapid transition to a new currency was vindicated by the events of the next eighteen months, when virtually all of the former republics decided to opt for their own currencies, albeit with varying degrees of success. Indeed, by early 1993, the IMF was threatening to withhold its support from successor states unwilling to take this step.

Representatives of the IMF also doubted whether Estonia would be able to accumulate the reserves necessary to sustain the Kroon. The key to overcoming these reservations lay in the 11-ton bullion reserves of the inter-war Estonian Republic, which had been deposited in Britain, Switzerland and Sweden back in 1940. Following his appointment as Bank of Estonia chairman in September 1991, Siim Kallas was instrumental in negotiating the return of these reserves.[20] Once Estonia recovered the gold, the IMF changed heart, thereby paving the way for the immediate reintroduction of the Kroon. The main impact of monetary reform lay in its accompanying policy measures. Under the 1992 constitution, the government is obliged to maintain a balanced budget, a provision which can only be modified with a two-thirds majority vote in parliament. With a view to meeting this requirement, the currency reform was complemented by a package of fiscal measures raising VAT from 10% to 18% and establishing a flat rate corporation tax of 35% (an 8% increase on the previous average of 27%). The top rate of personal income tax was simultaneously increased from 33% to 50%. These rates were incorporated into a balanced budget for the second half of 1992. An earlier law on the security of the Estonian Kroon, passed in May 1992, required all cash and bank deposits in the new currency to be fully backed by gold and foreign currency reserves. The law provided for a currency board system whereby additional cash and reserve deposits can be issued only in exchange for new foreign exchange receipts. This means that the Bank of Estonia can, in principle, redeem every Kroon without running out of reserves. In addition to the $120 million of gold and foreign currency reserves already held by the bank at the time of reform, the Kroon was underwritten by loans of $40 million from the IMF and a further $30 million credit line from the World Bank. Finally, the currency board system pegged the Kroon to the German Mark at a rate of 8 EEK-1 DM. This rate can only be changed with the approval of parliament, providing a further legal guarantee of exchange rate stability. The credibility of this system became immediately apparent. When turbulence gripped international currency markets in September 1992, the Kroon maintained its parity to the German mark whereas the rouble plummeted.[21]

BANKING REFORM

Currency reform was further underpinned by the creation of an effective two-tier banking system. The Bank of Estonia was made fully independent, and is not permitted to lend to the government or to commercial banks. The bank thus has no power to determine interest rates, although it can and does set obligatory reserve ratios for commercial banks. By adopting a tight regulatory system borrowed from the Nordic countries, Estonia has managed to avoid the experience of neighbouring Latvia, where initially lax supervision contributed to the disastrous collapse of the Baltija Bank in 1995.[22] The determination to enforce these regulations became clear in November 1992, when the Bank of Estonia moved quickly to forestall a crisis in the mushrooming commercial banking sector. In this, as in so many other spheres, Estonia had taken the lead during the late Soviet period. The Tartu Commercial Bank (TCB—formed in 1988) was the first of twelve private banks founded prior to independence. As inflation progressively eroded the real value of minimum capital requirements over the following year, the number grew to 43. Foreign currency trading and continued access to cheap central bank credits made banking a profitable business prior to June 1992, although the simultaneous decline of industry meant that many of the new banks were in fact effectively insolvent. Monetary reform, coupled with the earlier freezing of banks' VEB accounts in Moscow sparked a growing liquidity crisis, yet the banks seemingly remained confident that the BOE would intervene to bail them out. In fact, the central bank showed its teeth by placing a moratorium on the three largest commercial banks: TCB, Union Baltic, and PEAP (Northern Estonian Shareholders Bank).[23] Minimum reserve requirements were promptly raised, and a further seven licences suspended early the following year. Mergers and bankruptcies reduced the number of banks to 23 by the Spring of 1993, at which point the BOE temporarily suspended the issue of new licences.[24]

This decisive intervention helped to bring about an improvement in lending policy and a corresponding decrease in the number of bad debts held by banks. The introduction of a functioning bankruptcy law in September 1992 represented another important factor in this evolution. The subsequent decision to liquidate the TCB debunked the conventional wisdom that bankrupting major banks would be politically impossible.[25] In its principled opposition to an inflationary bail-out which might prejudice the stability of the currency board system, the Bank of Estonia was at one with the government of Mart Laar.[26] In keeping with its neo-liberal philosophy, the new administration adhered strictly to the balanced budget principle. On taking office, it also replaced the existing progressive income tax system with a flat rate of 26%, in order to encourage 'free and productive activity' of the individual.[27] Perhaps the most important impact of the Laar government lay in further liberalising measures designed to facilitate trade and inward investment (see below).

INFLATION

The macroeconomic benefits of strict monetary policy quickly became apparent. Economic growth resumed as early as 1994 and registered successive increases of 4% over the following two years.[28] The annual rate of inflation fell to 35% in 1993. After rising to 47% the following year, it has since undergone successive annual decreases, standing at 8% in 1998.[29] A further considerable decline will therefore be necessary before Estonia meets the EMU reference value for EU countries, although meeting EMU criteria is not a precondition for accession to the Union. The persistence of inflation within the Estonian economy can be partly explained by the adjustment of prices to international levels, most notably the progressive elimination of state subsidies to transport, housing and energy. Statistics from 1996 pointed to the continued presence of inflationary pressures in this regard, suggesting that Estonian rates will remain above western European levels for some time to come.[30] Another factor relates to the significant undervaluation of the Kroon at the time of its introduction. Individuals' savings and bank deposits were converted at a rate of ten roubles to one kroon in June 1992. This figure was based on the market rate of the rouble as determined by interbank auctions, where foreign currency was scarce and roubles abundant. The initial pegging to the Deutschemark is thus reckoned to have undervalued the Kroon by as much as 600%. This step was intended to promote exports whilst protecting domestic producers from import competition. The consequent increase in import prices, however, fuelled inflation and eroded any initial advantages enjoyed by exporters in this respect. Flexible labour markets and the weak bargaining power of trade unions have led to an institutional framework which is not conducive to strong wage pressures. The average salary nevertheless increased by 78% during 1995–98, far outstripping the corresponding growth in consumer prices, although the same period also witnessed a progressive slowdown in the rate of wage growth.[31] Finally as regards inflation, one could also say that Estonia has been a victim of its own success, with a high rate of economic activity partly spilling over into price increases. The credible economic policies put in place during 1992 have meant that the inflow of foreign currency from exports, loans and foreign investments exceeded all expectations, leading to fears of economic overheating during 1997.

THE LIBERALISATION OF TRADE

The swift introduction of a stable and convertible national currency has been one major factor behind the speed of Estonia's 'economic return to Europe'. Another has been trade policies which are both liberal and predictable.[32] Unlike Latvia and Lithuania, Estonia immediately abolished virtually all remaining import and export restrictions once the Laar government took office. Consequently, it has proved more successful than its Baltic neighbours in diversifying its trade and breaking unilateral dependence on the FSU. During 1992–93, all of the Baltic states concluded bilateral free trade agreements with the EFTA states. Of the three, Estonia was particularly well-placed to benefit by virtue of its proximity to Finland, which replaced Russia as its leading trade partner as early as 1993. Trade transparency has also paid off in dealings with the European Union. When EU-Baltic agreements on free trade in manufactured goods were signed in July 1994, Estonia was exempted from the transitional period applied to Latvia and Lithuania. As such, the agreement became effective already in January 1995. By 1997, the majority of Estonia's foreign trade was with the countries of the European Union, Finland remaining the most important partner. For the other two Baltic states, on the other hand, the most important partner at this time was still Russia.[33]

This rapid geographical shift was underpinned by a genuine growth in trade and exports. Total turnover increased by two and a half times during 1993–1994, when exports to the West registered annual increases of 53% and 95%. The corresponding export growth figures for Lithuania were only 29% and 21%, whereas Latvia registered a meagre 5% in both years.[34] More encouragingly, over half of Estonian exports consisted of processed goods, as opposed to one tenth in Latvia and a quarter in Lithuania.[35] Much of the initial export trade, however, had only limited growth potential, relying on the undervalued Kroon and low labour and input costs. One third of processed exports—especially in the engineering sector—was low value-added, being directly linked to imports.[36] The main export growth sectors have been resource or labour-intensive industries such as timber, pulp, furniture, textiles, clothing and food processing. Particularly ill-equipped to compete were energy-intensive heavy industrial sectors (chemicals and engineering). Generally speaking, the combined effects of an open trade regime, a small domestic market and an uncompetitive product base have meant that the structural shift away from heavy industry—and the corresponding rise in services—has been far more dramatic than in other postcommunist states of central and eastern Europe.[37]

The manufacturing sector nevertheless underwent something of a renaissance during 1997. After two years of sluggish growth, industrial production jumped by 13%, whilst the growth of value-added was faster than that of output as a whole.[38] Moving from the first 'easy' export stage towards a more sustainable manufacturing base will inevitably take time, since enterprises need to invest in new technology in order to modernise production and improve productivity. The fact that as much as two-thirds of current imports consists of the capital goods necessary to bring about this transformation gives room for optimism regarding future development.[39] In the interim, interest in trade with the FSU has remained strong, especially for sectors (e.g. agriculture) incapable of quick reorientation to western markets. After a considerable contraction during 1992, the value of Estonian exports to Russia and the CIS increased by two and a half times over the following year.[40] Three years later, FSU countries (including Latvia and Lithuania) continued to take 40% of Estonian exports, whilst Russia remained the second largest trade partner in 1997.[41] The obvious potential for growth in eastern trade has been thwarted by the poor state of Estonian-Russian relations. Moscow has consistently refused to grant Most Favoured Nation status, and levied a 60% tariff on imports from Estonia from 1994 (see Chapter Five). The situation has been eased somewhat by the signing of free trade agreements with Latvia and Ukraine, allowing Estonian firms to export to Russia via these countries.[42] Ukraine and Latvia have since become important trade partners in their own right. Nevertheless, one business leader noted that the lack of access to Russian markets was forcing many Estonian enterprises to run at undercapacity during 1996.[43] Often, the enterprises hardest hit by this policy are staffed by the very ethnic Russians which Moscow purports to protect.[44]

The problems alluded to above led the European Commission to express concern at Estonia's narrow export base when formulating its *avis* in 1997.[45] Low productivity and high domestic demand meant that growth in imports consistently outstripped that of exports during 1994–97, when the balance of trade deficit increased from EEK 4.6 billion (15% of GDP) to EEK 21 billion (32% of GDP).[46] The overall current account deficit has been mitigated by a strong surplus on services, 70% of which derives from spending by foreign tourists. In this, as in so many other respects, proximity to Finland has been integral to Estonia's economic development. Of the 2.5 million people who visited Estonia in 1996, 67% were Finnish, most of whom came on day trips from Helsinki ($3\frac{1}{2}$ hours away by ferry or 90 minutes by hydrofoil).[47] Certain commentators see a risk that Finnish custom may decline as rising prices continue to undermine Tallinn's attractiveness as a shopping centre. Such claims appear exaggerated at the time of writing, since the number of foreign tourists has continued to rise following the introduction of visa-free travel with the Nordic countries in May 1997. A greater long-term threat stems from growing resentment at what are perceived as Finnish 'neo-colonialist' attitudes towards Estonia. According to author Maimu Berg, too many Finnish visitors treat the country as a 'giant supermarket ... where people speak some kind of Finnish dialect'.[48] Relations between the two states have been strained by the 'shipping war' which broke out when dockers in Finland refused to unload Estonian ships on the grounds that the wages of seamen working on them are barely a third of those earned by their Finnish counterparts. This action was not based on any agreement with the corresponding Estonian trade union, prompting allegations that its real motive was to drive out Estonian competition on routes used by Finnish carriers.[49] On the back of this dispute, many Estonian parliamentarians have also expressed concern at growing penetration of the economy by Finnish capital.[50]

TRANSIT TRADE

Another key source of service income has derived from the lucrative East-West transit trade. In 1990, the Baltic republics handled almost a quarter of total freight shipments going through Soviet ports. In 1993, they continued to handle a similar proportion of Russia's foreign trade shipments, whilst the amount of cargo passing through Estonian ports almost doubled during the next five years. Of the 21 million tons handled in 1998, almost three quarters was transit. Although Estonia lacks the pipeline capacity of neighbouring Latvia, its railways and ports handle as much as 20% of Russia's diesel and black oil exports to the West. One 'silver lining' of the Soviet collapse has been the opportunity to charge world prices, payable in hard currency, for transit and loading fees.[51] Transit has fast become one of Estonia's most profitable industries, accounting for 7.4% of employment by 1997.[52] In comparison to Finland, Estonian firms can offer lower prices, knowledge of the Russian language and more active ties with the East. Similarly, many German firms have opted to use Estonian ports in preference to the sometimes problematic overland route via Poland.[53]

The major long-term threat to this trade lies in Russia's plans to upgrade its own port capacity. In June 1997, President Yeltsin approved a scheme to renovate and enlarge the port of St. Petersburg and to build three new ports on the Gulf of Finland. One of these—Ust-Luuga—is already under construction just east of the Estonian border. If implemented, this scheme would more than double the port capacity which Russia 'lost' with the collapse of the USSR.[54] Initial projections by the Russian government envisaged a 75% drop in Russian transit trade through the Baltic states by 2010, with an annual saving of around $1 billion in transit fees. Whilst Russia certainly has a strong economic incentive to develop its own capacity, the Estonian harbours still enjoy considerable advantages in terms of their depth and relatively ice-free location. For this reason, one suspects, the new scheme should perhaps be viewed primarily as a further attempt to increase political leverage over the Baltic states. In this regard, Yeltsin's comment that 'the Baltic countries should think hard about their policy towards Russia' following his approval of the plan is instructive.[55] Such bluster failed to reckon with the Russian 'financial meltdown' of 1998, which led to the suspension of work on the new Ust-Luuga freight terminal. Estonian ports, meanwhile, were hard-pressed to cope with the upsurge in Russian exports brought about by the devalued rouble and an increase in oil shipments to the West. From a western perspective, Estonia's more favourable tax and customs regime, added to lower levels of corruption and organised crime, also leave it well placed to deal with future Russian competition. Planned improvements to the railway network, in any case vital if Estonia is to continue to compete with Finland, should also help it keep pace.[56] The Russian government's ability to engage in 'state-spawned economic geopoliticking' over energy supplies has been further undermined by the continued rise of powerful private business interests with competing agendas towards the Baltic states. Rather than investing in new Russian capacity, the firm LUKoil put its money into a new oil terminal at the Estonian port of Muuga during 1996. This decision built on strong ties between LUKoil and Packterminal, Estonia's largest and most profitable oil shipment firm.[57] Whilst such ties could secure the long-term future of the transit business, they continue to arouse suspicion, for it is felt that Russian economic penetration might quickly translate into undue political influence—witness the controversy caused by Tiit Vähi's private visit to Moscow in 1997.

FOREIGN DIRECT INVESTMENT

Estonia's ability to finance a current account deficit since 1994 has rested on a veritable influx of foreign direct investment (FDI), which totalled 24.3 billion EEK ($2.025 billion) by the end of 1998.[58] Total FDI in Estonia has fallen far short of the amount deemed necessary to achieve rapid economic parity with the West, but, per capita, it has nonetheless been on a par with more advanced transition economies in central Europe.[59] This has been enough to ensure an overall balance of payments surplus throughout 1994–98. The biggest share of investment has come from Sweden (32%) and Finland (27%) which together account for almost two thirds of the total. In 1998, the share of these countries rose to 80%. By sector, industry accounted for the largest share of investments (30%) followed by retail and wholesale trade (23%) and finance (22%).[60] FDI has thus proved sufficient to cover much of the bill for the import of capital goods, the area where Estonia's trade deficit is at its widest. Since industrial investment has usually been geared towards export, it should contribute to a narrowing of the trade deficit in the long-term.

On the minus side, the high volume of FDI contributed to fears of possible economic overheating during 1997, when the rate of economic growth climbed to a staggering 10.6%. The current account deficit more than doubled during 1995–97, whilst in 1996 the overall balance of payments surplus was down 40% on the previous year. Consequently many experts warned of the risk of a far-eastern style currency crisis leading to irresistible pressures for devaluation.[61] The problem was exacerbated by the situation in the booming financial sector, where access to cheap foreign loans meant that growth of banks' loan portfolios far outstripped that of deposits.[62] The government and the Bank of Estonia duly heeded the advice of the IMF and its own experts by applying the brakes to the economy. The BOE sought to curb credit growth by increasing minimum capital requirements and adequacy ratios for commercial banks, while the government opted to run a budget surplus in 1998, channelling excess revenues into a stabilisation fund designed to insulate the economy against future shocks. This fund amounted to 3.8 billion EEK by March 1999, most of which is held in German banks.[63]

This timely intervention enabled the Estonian currency board to weather the international turbulence caused by the far eastern and Russian financial crises during 1997–98. Credit growth slowed from 70% in 1997 to only 15% in the first half of 1998, and the current account deficit narrowed from 12% to 8.4% of GDP over the year. Foreign currency reserves were briefly dented by the Russian crisis, but had recovered to 1997 levels by December.[64] The same period brought a corresponding decline in the rate of economic growth, which fell to 4% in 1998. Growth of industrial sales, buoyant during the previous year, barely reached 1%. The Russian crisis proved disastrous for those export industries—especially agricultural—oriented towards eastern markets, again exposing Estonia's continued vulnerability to economic development in the CIS. The tightening measures announced by the BOE in autumn 1997 also precipitated a further crisis in the banking sector, which sustained major losses over the next 12 months. Part of the problem lay in commercial banks' over-exposure to fluctuations on the Tallinn stock exchange (opened in May 1996 with capitalisation largely based on bank stocks), which continued to fall following an initial crash in October 1997. Initially lax regulation in this area meant that banks were not required to ringfence stock trading from their other activities. This allowed them to use their stock portfolio as collateral for loans as the market soared during early 1997.[65] Subsequent stock losses led to the bankruptcy of *Maapank* (Rural Bank) in mid-1998. Later in the year, the BOE was forced to declare a moratorium on three other banks, two of which were heavily engaged in Russian markets.[66] The lessons of the crisis appear to have been heeded, for the government and central bank have since taken steps to set up a single regulatory body covering the financial sector as a whole. A deposit guarantee fund has also been created to protect investors against any future bankruptcies.[67]

PRIVATISATION

The sum total of measures taken during 1997–98 have, it seems, proved sufficient to maintain Estonia's credibility in the eyes of the international community. Experts continue to predict moderate economic growth and a narrowing of the trade deficit during 1999–2001, whilst foreign investment reached record levels during 1998. Estonia's success in attracting FDI is testimony not only to a stable macroeconomic environment but also to a strategy for industrial privatisation which has tended to favour strategic—and especially western—investors. The Estonian privatisation programme began back in 1990, and has since been hailed as the most successful in the former USSR.[68] Privatisation of small and medium sized enterprises—already well underway in 1992— was all but complete by the end of 1994. After a shaky start, large-scale privatisation took off during the same year. As a complex process of institutional change, privatisation belongs to the 'consolidation' as opposed to the 'launch' phase of economic transition. Compared to initial macroeconomic stabilisation and liberalisation it has generally proved to be a protracted and politically controversial process. Those in favour of rapid denationalisation of state property have tended to favour voucher schemes giving priority to existing management and employees. Others advocate the more piecemeal approach of direct sale to strategic investors, arguing that efficient management should take precedence over simple transfer of ownership. These debates are informed not simply by economic objectives, but also by political calculations and notions of social fairness.[69]

The complexities of the process are reflected in Estonia's comprehensive privatisation law of July 1993, which established a hybrid approach based on: national capital vouchers and public offering of shares; sales to a core investor on the basis of tender; and public or restricted auction. The new law both amended and consolidated a series of earlier legal acts introduced by the Savisaar and Vähi governments. Initial legislation passed in 1990 provided for the privatisation of small and medium-sized enterprises in the sphere of services, trade and catering. Participation was made open to anyone who had resided in Estonia for over 10 years, with first right of purchase going to the employees of the enterprise.[70] Initial progress was slow and dogged by controversy. High inflation and the lack of any effective mechanism for the valuation of assets brought accusations that small enterprises were simply being acquired by their existing managements at knock-down prices.[71] Monetary reform lent further impetus to privatisation of shops and small enterprises, although Vähi's tenure as prime minister did little to dispel accusations of 'insider privatisation'.

A 1992 Law on Privatisation of major industrial enterprises eschewed mass voucher privatisation in favour of cash sales to a core investor, a method which was mooted as the most efficient means of engineering a rapid return to economic viability. Estonia was the first of the former Soviet republics to develop a privatisation agency (The Estonian Privatisation Enterprise (EERE) to oversee enterprise restructuring ahead of sale. Headed by Estonian-German émigré Andres Bergman, EERE was modelled on the German *Treuhand*, whose officials acted as key advisors. In its early stages, however, large-scale privatisation was attacked for being insufficiently regulated, prone to corruption and detrimental to national interests. Right of purchase was made open to all permanent residents of Estonia, who were allowed to pay in instalments over ten years. Existing management and workers nevertheless still enjoyed certain privileges under this scheme. Many were able to engage in 'asset stripping' by forming new private companies which acquired existing state enterprises cheaply before selling them on at market prices. In certain cases, local buyers were able to act as front men for buyers outside Estonia.[72] Particular concern related to the involvement of Russian business interests in the process, whilst EERE also stood accused of bias towards German firms when targeting overseas investors.[73]

Radical nationalist attacks on the privatisation process focused upon the need to restore rightful ownership of land and property confiscated by the Soviet regime after 1940. Espousal of the legal continuity principle by all nationalist forces during the drive for independence served to strengthen popular demands for restitution of nationalised property to pre-war private owners and their descendants. During 1990–1991, the ruling Popular Front attempted to downplay the issue, opting instead for the distribution of assets to all residents save the most recently-arrived settlers. Yet the Savisaar government was forced to acknowledge the strength of popular feeling over the restitution issue, as reflected in the December 1990 Supreme Council resolution restoring inheritance rights. This was followed up by the June 1991 Law on the Bases of Property Reform, which provided for physical restitution of property to former owners or their compensation through the issue of vouchers. The value of state property distributed in this way is estimated to have come to 29 billion EEK at 1993 prices.[74] The law also provided for the distribution of national capital vouchers (total 8.5 billion EEK at 1993 prices) to all residents based on the number of years worked in Estonia. National capital vouchers were mainly earmarked for the privatisation of housing, yet were made freely tradeable from May 1994 and can be used to buy land or shares in state enterprises and investment funds.

Notwithstanding their earlier emphasis on restitution, the *Isamaa* parties did not deviate from the cash sale approach to industrial privatisation once in power. In this area, the position of voucher holders remained secondary to that of large investors. The major change discernible in this area was a determination to curb the inflow of Russian capital by giving priority to western buyers. Minister for Privatisation Liia Hanni claimed that hitherto it had been virtually impossible to distinguish Russian investment from others, whilst the

Ministry of Justice alleged that some of the enterprises sold had included assets confiscated by the Soviet regime after 1940.[75] These political conflicts plunged the privatisation process into disarray. Bergmann's EERE had frequent clashes with the Finance Ministry Department for State Property, which, it is said, devoted as much energy to renationalising improperly privatised enterprises as it did to disbursing state assets.[76] In November 1992 Bergmann was dismissed by the government only 10 days after EERE had put out tenders for 38 enterprises.[77] However, when Estonia's Supreme Court attempted to reverse the two largest sales to date on the grounds of legal irregularity, the government blocked this move lest it should undermine confidence in the entire process.[78]

The new privatisation law of July 1993 merged EERE and the Department of State Property into a single Estonian Privatisation Agency (EPA) operating under the jurisdiction of the finance ministry. The formation of EPA answered demands for an acceleration of the privatisation process. From the start of 1993 to the middle of 1996, 433 enterprises were sold for a total of $277 million.[79] Legislation on collateralised lending and the establishment of a land registry during 1993 made it easier for Estonian companies to raise finance, and in April of the same year, parliament approved a law allowing foreign investors to buy land in connection with the purchase of an enterprise.[80] However, land reform was impeded by the lengthy and highly complex nature of the restitution process. Ownership disputes meant that the claims submitted by inter-war citizens and their descendents amounted to two and half times the total land area of Estonia. In early 1994, some commentators estimated that it would take 200 years to settle outstanding claims and fully register all restituted land.[81] Spurred on by external pressure from the European Union, subsequent governments have managed to accelerate the process. The total area of land registered in 1997, for instance, was double that of the previous year. A recent survey suggests that the process of restitution was already nearing its end by 1998, whilst amendments to the land law brought about a sharp rise in the pre-emptive privatisation of land during 1997–98.[82]

Progress over land reform has been matched in the sphere of housing, where almost 90% of residences had passed into private hands by the start of 1998.[83] An initial housing privatisation deadline of 1 December 1994 proved to be wildly over-optimistic, since only 1,500 apartments had been privatised by March of that year. High maintenance costs and the dilapidated state of the housing stock seem to have been the main factors behind the low take-up rate, although the possibility of a restitution claim by former owners was clearly an issue in the case of older properties.[84] The subsequent take-off of privatisation came too late for the *Isamaa* government, whose strategy became the object of growing opposition attacks during 1994. The slow pace of restitution was in itself a source of discontent, doubly so in the sense that it impeded the sale of land and privatisation of housing. Over industrial privatisation, the ruling coalition was accused by its populist opponents of 'selling' Estonia following the granting of land purchase rights to foreigners in 1993. Privatisation statistics for 1992–94 revealed that only 20% of enterprises had been sold directly to foreigners.[85] Nevertheless, opinion polls revealed low levels of public confidence in the EPA, which was alleged to have placed speed of sale ahead of restitution and the long-term interests of the economy.[86]

The sale of Tallinn department store *Kaubamaja* early in 1995 marked the start of a growing trend towards public offering of shares based on tradeable national capital vouchers. At the same time, however, subsequent Coalition Party-led governments have actually taken steps to strengthen the position of foreign investors within the privatisation process.[87] The number of privatisations decreased dramatically during 1996–99, mainly because the majority of enterprises had already been sold. The final stage of the programme has therefore focused upon the time-consuming and politically sensitive process of privatising the major utilities. The first major privatisation in this area came in May 1996, when Maersk Air of Denmark obtained a majority stake in the national carrier Estonian Air. Plans for further privatisations in the sphere of energy and communications have necessitated new legislation governing the operation of private sector monopolies, whilst there has been fierce debate as to whether certain industries might be too stategically important to remove from state ownership.[88] A number of MPs, for instance, attempted to block the recent sale of all but 27% of *Eesti Telekom* (worth 3.061 billion EEK), when a 49% share of the company passed to Nordic firms Sonera and Telia.[89] The two remaining priorities are rail freight and the energy sector, both of which have experienced long delays after being slated for privatisation in 1995–96. At the time of writing, the government is still engaged in a four-year negotiation with American firm NRG, which has promised long-term investments totalling 9 billion EEK in return for a 49% share in two power stations located in north-east Estonia.[90]

The other major facet of private sector development has involved the foundation of entirely new privately-owned businesses. Here too, Estonia has made encouraging progress. Already during the *perestroika* period, Estonia became home to a thriving co-operative sector. Almost 2,500 such ventures employing 50,000 workers were in existence by 1991.[91] Five years later there were already 64,000 registered companies in Estonia. Of these, 87% were indigenously-owned private businesses. The vast majority were small and medium-sized enterprises (SMEs), most of which employed less than 10 people. The total share of the workforce employed by SMEs grew from 26% in 1992 to 40% in 1994. During the same period, as many as two thirds increased their number of employees, thereby underlining the crucial role of the new private sector in counteracting unemployment brought about by the decline of heavy industry.[92] Studies of economic transition assign an important role to SMEs in terms of the shift towards a more flexible, export-oriented manufacturing base. A survey conducted in 1994 found that 61% of Estonia's SMEs were engaged in export, as opposed to 36% in Lithuania and only 26% in Latvia. In 1996, the number of enterprises engaged in manufacturing stood at 9,300. Although this figure represented a thirty-fold increase over the preceding decade, it accounted for less than one in six companies. This export trade, moreover, was again heavily based on subcontracting and low labour costs.[93]

A major constraint on the continued development of the SME sector has related to what is perceived as an unduly high burden of corporate taxation, a fact that does much to account for the formation and subsequent appeal of the Reform Party. The employer's obligation to pay the entire national insurance contribution of 33% has also been cited as a disincentive to hiring new staff, although avoidance of NI payments was apparently rife in 1996. Low levels of domestic savings and high inflation leading to high interest rates also discouraged small businessmen from making long-term investments in manufacturing capacity during the first five years of independence. Rather, the tendency has been to invest in trade-related activities offering high turnover and profit margins. These problems have fuelled demands for a more active policy on the part of the government. A considerable portion of funds given to Estonia under the EU PHARE programme have been earmarked for the development of export-oriented SMEs through the creation of enterprise centres offering help and advice.

WINNERS AND LOSERS

Much of the literature on postcommunist transition in central and eastern Europe has focused upon the exceptional challenge of effecting simultaneous economic and political liberalisation. Here, the early 'conventional wisdom' was to regard democratic regimes as disadvantaged when it comes to undertaking macroeconomic stabilisation, since 'the early effects of market-oriented economic reforms increase insecurity, inequality and apparent foreign influence, just as major sectors of the population begin to find themselves empowered politically. Under such conditions, democracy in the political realm easily works against economic reform.'[94] Estonia's ability to sustain radical economic 'shock therapy' challenges this conventional wisdom, albeit only partially in the sense that political empowerment did not extend to the Russian-speaking settler population. Even so, the lack of any major electoral backlash against reform seems remarkable when one considers that during 1991–93, average purchasing power declined to 32.5% of its previous level and unemployment—hitherto virtually unknown—soared to around 10%.[95] The explanation lies in the period of 'exceptional politics' which followed independence, when outright rejection of Soviet-imposed communism and mobilisation around national goals on the part of the mainly Estonian electorate translated into a readiness to accept radical reform.[96] If the initial reforms of 1992 were largely the preserve of technocratic experts, the *Isamaa* government could point to the legitimacy conferred by free elections. In this sense one can argue that far from being an impediment to radical change, democracy served as a necessary precondition for sustaining policies of macroeconomic stabilisation.[97]

Marja Nissinen notes that 'while a negative consensus (re: the Soviet regime) may suffice for *launching* a reform, later stages depend on the building of a positive consensus on the ultimate goals of transition.'[98] Surveys conducted during the stabilisation period of 1991–93 found that although no Estonian respondent advocated a return to Communist rule, most still rated the former Soviet economy more favourably than they did the existing market system. By April 1995, the two systems were rated equally, whereas by November 1996, the new market system scored far higher than the Soviet past. These findings diverged significantly from those obtained in Latvia and Lithuania, where the old system still scored 50% higher during 1996. Perhaps the most significant aspect of the 1993–96 surveys were the consistently high levels of optimism (83–88%) regarding Estonia's future economic development. This suggests that a 'shock therapy' approach bringing about speedy macroeconomic stabilisation and a resumption of growth proved more successful in mitigating societal tensions than the gradualist strategy applied in, say, Lithuania.[99] In this regard, it is instructive to note that by the end of 1998, average monthly wages ($284) and pensions ($90) in Estonia were the highest amongst the three Baltic states.[100]

THE ETHNIC DIMENSION

In the wake of independence, Estonia's Russians were widely tipped as the principal 'losers' from market reform, with some commentators predicting that this would further exacerbate already existing ethnic tensions.[101] The decline of Soviet-era heavy industry since 1991 has indeed led to higher than average rates of unemployment amongst the Russian-speaking population.[102] Other authors claim that the settler population was greatly disadvantaged when it came to privatisation: the shift towards property restitution during 1991 clearly catered for the needs of the Estonian majority, whilst the quick adoption of citizenship legislation meant that the former Russian-speaking economic elite had less time to engage in 'spontaneous privatisation' than it did in Latvia.[103] Yet, in practice, ethnic and socio-economic cleavages have not reinforced one another to the extent which was anticipated in the aftermath of independence. Attitudes of Estonians and non-Estonians towards economic transition have proved remarkably similar. Although Russian-speaking respondents have generally expressed greater nostalgia for the Soviet economy, they are equally optimistic with regard to future development under the new market system. This is perhaps symptomatic of a widely-held belief that the collapse of the USSR was inevitable.[104] In 1994, non-citizens were asked to state whether, in the interests of economic utility and state security, Estonia ought to join the European Union or the CIS. Almost 50% of non-citizen respondents in Tallinn and the north-east opted for the EU on both counts, and less than 10% for the CIS. Around 30% were in favour of joining both, yet it is notable that 20% of Estonian respondents shared this view.[105] Most implacably opposed to the independence drive during 1988–91 were workers from former all-union—often military—enterprises who had settled in the ESSR during the 1970s and 1980s. Unlikely to speak Estonian or feel any affinity with the culture of the ethnic majority, these recent settlers have been most prone to re-emigrate to Russia and the CIS following the collapse of the USSR. In this sense, emigration probably did act as a 'safety valve' for ethnic tensions during 1991–93, when an estimated 56,000 people left Estonia, many from recently-built suburbs of Tallinn such as Lasnamäe.[106]

The subsequent decline in the level of emigration, however, has belied earlier claims that independence would provoke a mass exodus. The optimism which many local Russians have expressed with regard to their future economic prospects in Estonia rests partly on a 'negative consensus' with regard to economic and political developments in Russia, yet it also highlights the absence of any overt 'winners-losers' split along ethnic lines, at least during the initial stages of market reform. Research conducted in 1993–94 found that on average, ethnic Russians still earned more than Estonians. Wages in urban areas have remained significantly higher than those in the countryside, and Russians still retain a significant presence in industries (mining, energy, transport, certain manufacturing branches) paying significantly higher than the national average.[107] Russians have also participated actively in the foundation of new private businesses. By 1995, the Estonian-Russian Chamber of Entrepreneurs represented fifteen commercial organisations and as many as ninety different firms with partners in Russia, Germany and the Nordic countries. Russian entrepreneurs have worked hard to circumvent poor inter-state relations with Russia by forging closer economic ties with neighbouring regions. Their capital has been central to the financing and development of the Representative Assembly and the United National Party.[108]

Similarly, figures showing a higher-than-average unemployment rate amongst non-Estonians mask significant disparities between the main areas of Russian settlement. Almost half of Estonia's Russian-speaking population lives in and around the capital, which possesses the largest concentration of new private businesses and the biggest share of tourism and foreign investment. Real unemployment in Tallinn was running at just over 1% in 1996, with some sectors suffering from a shortage of skilled labour.[109] In this regard, the contrast with the north-east cities of Narva and Sillamae could hardly be more striking. Official statistics placed unemployment in the region at 13–15% in 1998, yet many suggested that it was in fact nearer 20%. Back in 1993, real unemployment was estimated at anywhere between 30 and 35%.[110] This state of affairs clearly contributed to ethnic tensions during the 'aliens crisis' of that year, yet its seems that most local residents blamed the local authorities for their plight rather than the Estonian government. Under later administrations, Narva and Sillamae have done much to shed the 'red' image which had previously deterred inward investment. Swedish textile firm Boras Wäfveri acquired a 75% stake in Narva's *Kreenholm* textile mill in 1995, whilst NRG is currently poised to invest significantly in local power plants. Planned improvements to transport infrastructure and border crossings mean that the region would be well-placed to benefit from any improvement in Estonian-Russian relations, an aim which is cherished by local elites in the neighbouring Leningrad *Oblast'* of the Russian Federation. Many residents in the impoverished Russian border town of Ivangorod look with envy towards their compatriots across the Narva river, and a petiton has even been generated in favour of adhesion to the Estonian Republic. Even so, the economic situation in Narva remains bleak by Estonian standards. Younger residents who go to study in Tallinn or Russia rarely return after graduation, whilst more and more are leaving to start new lives in the West. These trends are deeply worrying for local leaders in Narva, who foresee a 'lumpenized city' populated by the elderly.[111] Forthcoming privatisations should guarantee the long-term future of the power industry—the largest regional employer—yet the short-term restructuring which this will entail is likely to prove painful, bringing yet more unemployment. Continued economic deprivation, meanwhile, contrived to push the future of the north-east back to the top of the political agenda by early 1999.

During the first five years of independence, survey data suggested that a majority of Russians were able to get by at work using their native language.[112] For those not conversant in Estonian, the position has become progressively more difficult as 'nationalising' tendencies have extended to cover more and more occupations. This is especially so for school leavers, since the education system is still ill-equipped to integrate non-native speakers adequately into an Estonian language environment. Indeed, a government report published early in 1999 identified inadequate language knowledge as one of the main reasons for the higher levels of unemployment found amongst ethnic Russians.[113] It may be that in the short to medium term, more and more younger, better educated Russians might be impelled to seek careers outside Estonia. Whilst some more nationally-minded politicians appear to regard this prospect with equanimity, others argue that the country cannot afford a 'brain drain' on a significant scale. This debate highlights wider concerns with regard to Estonia's demographic development. Amongst the titular Estonian nation, birth rates have been in decline ever since the 1920s. In 1998, the total number of births in Estonia was 10,000 lower than in 1988, a fact which is attributed to the heightened insecurity engendered by economic transition.[114] With life expectancy also in decline, the total population fell by almost 100,000 during 1991–96. If this trend is not arrested, an ageing population and poor levels of health will ultimately place a massive burden on state expenditure and threaten the fiscal balance.[115] In 1997, the Siimann government therefore proposed a new pension system based around three pillars: an earnings-related scheme funded by the state; obligatory national insurance payments by the employee; and an additional voluntary scheme aided by income tax relief. This was due to come into operation by 2000-01.[116]

AGRICULTURE

In the continued absence of large-scale Russian representation at the parliamentary level, political debates over the economy have increasingly followed urban-rural rather than ethnically-based cleavages. From the perspective of the rural periphery, talk of an Estonian economic miracle seems fanciful to say the least. Wages paid to workers on former state and collective farms remained the lowest of any sector in 1998, whilst unemployment in the mainly agricultural counties of the south-east stood at 11–13%. If the future of the countryside is today a major issue in almost all European states, it is especially so in a country such as Estonia which has undergone the transition to a fully urbanised society only within the last half a century. Indeed, even after the sustained industrialisation of the Soviet period, agriculture still accounted for 18% of employment and 20–30% of GDP in 1990. As is also the case in the Nordic countries, the land remains a key element of Estonian social, cultural and political identity.[117] The formation of an independent smallholder class was integral to nation-building between the wars, when an estimated 140,000 family farms were created. Considered individually, many of these holdings were too small to be economically viable, yet Estonian farmers successfully adopted the Danish co-operative model. On this basis, Estonia developed a thriving export trade, rivalling the Danes in the British market for butter and bacon. Rural living standards during these years are generally considered to have been on a par with neighbouring Scandinavia.

This state of affairs was rudely interrupted by the 'social and economic disaster' of Stalinist collectivisation, which was undertaken primarily in order to 'obliterate the ... foundations on which the prewar independent republic had rested.'[118] Later Soviet leaders, however, implicitly acknowledged inter-war traditions by granting considerable powers of self-management to state and collective farms in Estonia. From the mid-1960s, yields, productivity and rural incomes—although meagre by Scandinavian and west European standards—were the highest of any Soviet republic, and Estonian produce acquired a reputation for quality across the USSR. The period since 1991 has marked a return to the small-scale agriculture of the inter-war period. As Kirby notes, however, 'the restoration of the private farm ... is ... as much to do with cultural and national values as it is with productivity.'[119] Notwithstanding the important place of family farming in Estonian national identity, the position of small producers across western Europe has grown steadily more precarious in the years since World War Two. For this reason, the Estonians have been warned not to pin too many hopes on this sector as the main form of agricultural development.[120] Yet simultaneous attempts to place large-scale agriculture and auxiliary processing industries on a new economic footing have been fraught with difficulty. Ten years on from the start of reform, Estonia's farming sector appears stranded between two worlds at a time when EU agricultural policy itself stands at a crossroads.

The first significant step towards the re-establishment of private agriculture was the Law on Peasant Farming approved by the ESSR Supreme Soviet in December 1989. This recognised the principle of private farming but not that of private land ownership. By allowing residents of farmhouses to lease up to 50 ha of land from a state or collective farm, it was intended both to restore former family farms and to create new ones. The aim was to engineer a gradual shift towards new, more effective modes of production without challenging the legal status quo. More extensive reorganisation began during 1991–93, when a flurry of new legislation restored the rights of pre-1940 owners and re-established private land ownership as the norm.[121] Consequently, the number of new family farms (average size 20 ha) grew from 7,000 to 19,767 during 1992–96. In the same period, the 360 former state and collective farms were reorganised into just under 1,000 smaller production enterprises in the form of limited companies, joint stock companies and co-operatives.[122] The Law on Agricultural Reform adopted in March 1992 established procedures for the dismemberment of former state and collective farms, obliging each to establish a reform commission consisting of farm members and representatives of local and state government. Once restitution claims on land and other farm assets had been settled, the commissions were to distribute remaining non-land assets amongst employees in the form of labour shares. The latter could then either be transferred to a new successor enterprise or used to acquire property to be withdrawn from the existing collective. The overall distribution and re-organisation of farm assets were to be determined according to a strategic plan drawn up by the reform commission.

The reform process sparked heated debates regarding the proper function of land. Should it be viewed simply as private property, or rather as a productive asset to be used for the benefit of society as a whole? Within the Estonian Farmers' Union, these divergent ideological viewpoints led to conflict between commercially-oriented 'production' farmers and the so-called subsistence or 'hobby' farmers which appeared in increasing numbers after 1991.[123] More often than not, however, those who opposed restitution did so out of self-interest rather than any deeply-held ideological conviction. For instance, former employees of state and collective farms who had created new family holdings under the 1989 law now became subject to claims on their land by pre-war owners and their descendants. The Soviet system had also created powerful vested interests in the form of collective and state farm management, who saw the continuation of large-scale agriculture as the key to maintaining the status which they had enjoyed prior to independence. Certain of these so-called 'Red Barons' engaged in 'spontaneous privatisation' of farm assets whilst mounting a fierce rearguard action against the resurrection of family farming.[124]

Reform of larger-scale units has also been retarded by the prolonged and highly complex nature of the restitution process. Restitution claims are thought to have affected 50% of all agricultural land in Estonia, of which almost half has been subject to competing claims. In this situation it became extremely difficult to evaluate or implement the plans for the transformation of state and collective farms. In 1996, large-scale units—many barely reformed—accounted for 60% of agricultural land use, yet barely half of total production. Continued uncertainties relating to ownership meant that 90% of *all* farms were still operating on short term leases from the state, giving farmers neither the incentive nor the collateral to raise loans for much needed investment.[125] The need for a quick resolution of the land question was made all the more urgent by a dramatic deterioration in market conditions, which brought about a 44% decline in gross agricultural output during 1989–1994. The restoration of independence marked the end of subsidised imports of feed, fuel, fertilisers and machinery from the FSU. A seventeenfold increase in input costs during the three years after 1991 was not matched by a corresponding rise in producer output prices. At the same time, retail food prices grew by a factor of 29, leading to a sharp decrease in consumer demand.[126]

If one examines the development of agricultural policies since independence, a number of ironies become apparent. In the run-up to the 1992 elections, the nationalist parties of the *Isamaa* coalition drew heavily upon images of Estonia's pre-Soviet agrarian past. Yet the policies pursued by these parties in power were largely inimical to the regeneration of the countryside. In pursuing an economic policy based on undiluted neo-liberalism, *Isamaa* eschewed any recourse either to import tariffs or agricultural subsidies. This policy disregarded the fact that such agricultural support measures have been commonplace in the capitalist economies of the European Union which Estonia today aspires to join. In this sense, to quote a later Minister for European Affairs, Estonia can be accused of being 'more pious than the Pope'.[127] Estonian farmers justifiably complain that they are denied the opportunity to compete on an equal footing with western European producers who have flooded the domestic market with subsidised imports since 1992. Whilst agricultural exports registered an impressive 300% increase during 1992–94, this amounted to less than half of the corresponding increase in imports. By 1995, the country had become a net importer of food and agricultural products.[128] The free trade agreement signed with the European Union in July 1994 entailed a 60% reduction in EU tariffs levied on Estonian imports, whilst quotas on meat and dairy products were set at a level which exceeded existing volumes of Estonian exports to EU countries. In spite of this, non-tariff barriers such as high veterinary and sanitary standards have contrived to impede the growth of exports towards EU and other OECD economies of the West.[129] To date, the countries of the FSU—including Latvia and Lithuania—have continued to take almost 70% of agricultural exports. Yet here too, market growth is restricted due to the punitive tariffs imposed by Russia.

It is this situation which explains the persistent calls by the agricultural lobby for subsidies and the imposition of protective tariffs. The acrimonious debates over these issues during 1997–99 coincided with a deepening of the rural crisis occasioned by the collapse of *Maapank*, the disastrous 1998 harvest and the devaluation of the

Russian rouble. Yet the future of the countryside now hinges largely on Estonia's current negotiations with the European Union. The imposition of tariffs on EU imports has been ruled out on the grounds that it would breach the existing association agreement and seriously hinder the accession process. This same process also dictates that any tariffs on non-EU imports be introduced according to the EU general foreign trade regime. The EU itself is seeking to shift the Common Agricultural Policy (CAP) away from price support towards rural development subsidies. Already prior to accession, Estonia could receive up to 350 million EEK annually under the EU Rural Life and Agriculture Programme for future members, doubling the existing subsidies paid to farmers. These allocations would be used to improve farm productivity, develop rural infrastructure and bring production into line with European veterinary and sanitary requirements.[130] Once full membership is achieved, analysts suggest that the agricultural sector would receive 700 million kroons annually in direct subsidies. Yet, even if this additional support materialises, many farmers still feel that they would be unable to meet the stringent production guidelines imposed by the EU.[131]

THE WAY AHEAD

The preceding point brings one to the wider question of whether the Union and its individual member states will provide sufficient financial aid to enable Estonia to meet the *acquis communautaire*.[132] Given the overall importance of trade to the economy, the Estonian delegation to the accession talks has been anxious to deflect demands for any transitional period following entry to the European Union. Yet despite their own sustained efforts to fulfil the criteria for full membership, the Estonians still face formidable political obstacles in the form of those existing member states seeking to retard eastward enlargement. Transitional periods may thus prove inevitable in some areas, and indeed have already been applied in the case of environmental protection and energy policy.[133] In the short term, Estonia's accession to the World Trade Organisation during 1999 should not only facilitate negotiations on EU entry, but also improve prospects for expanded access to Russian markets.[134] At the start of the 1990s, it was widely felt that Estonia's long-term economic viability would depend upon good relations with Russia. Western specialists felt that the country's future niche lay in east-west transit trade and providing a base for overseas companies seeking to tap into Russian markets.[135] Similar arguments were advanced back in the the early 1920s, yet the absence of normal relations with Russia ultimately proved no impediment to the development of a viable economic base between the wars. Similarly today, foreign companies are beginning to invest in Estonia with a view to exporting their production not to Russia, but to the wider Baltic states' market or the EU itself.[136] Some predict that the Russian financial crisis of 1998 will accelerate integration with the EU by forcing unprofitable industries to adapt themselves more quickly to western markets. Also, the projected NRG investment in *Eesti Energia* would do more than modernise environmentally-unfriendly oil shale-fired power stations. It would also place considerable American capital on the Estonian-Russian border and improve access to the 'Baltic Ring'—the Nordic-sponsored scheme to create a common energy market in the Baltic sea region.[137]

Toomas Hendrik Ilves, re-appointed as foreign minister in April 1999, has expressed a hope that Estonia will in future become 'just another boring Nordic country'. Such talk seems premature when one considers that Estonian GDP per capita was only 28% of the European Union average in 1998.[138] Aside from highlighting the gulf separating Estonia from existing member states, average indicators of economic prosperity also mask significant income disparities within society.[139] The challenge over the medium to long-term will therefore be to narrow these external and internal disparities without prejudicing export competitiveness or the budgetary constraints and exchange rate stability requirements imposed by future membership of the EU. If recent elections did not entail any rejection of the policies in place since 1992, the success of the Centre Party was one indication that the positive consensus surrounding economic development was beginning to wear thin. By the same token, the eclipse of the 1995 coalition partners rested partly on their failure to honour promises to distribute the fruits of economic growth more equitably. In this respect, the new *Isamaa*-led coalition will thus have to do more than simply revive an economy flagging after the 1998 economic down-turn.[140] Yet, as a *Baltic Times* editorial noted in 1998, even countries like Sweden face significant social problems.[141] One might add that EU member states such as the United Kingdom are home to ethnic tensions which are more deeply-rooted and intractable than those currently existing in Estonia. In this sense, talk of economic consolidation and adjustment to the European norm becomes misleading, for, as the same paper noted after the elections, 'with normality come the everyday problems and hard decisions faced by all democratic, market-oriented countries.'[142]

1 Hansen, John and Piritta Sorsa, "Estonia: A Shining Star from the Baltics," in *Trade in the New Independent States*, eds. Constantine Michalopoulos and David Tarr (Washington, 1994), pp.115–132.
2 Headline cited in Mart Laar, "Estonia's Success Story," *Journal of Democracy*, Vol.7, No.1, 1 January 1996, p.97.
3 Ardo H. Hansson, "Transforming an Economy while Building a Nation: The Case of Estonia," revised version of a paper presented at the 24th National Convention of the American Association for the Advancement of Slavic Studies, Phoenix, 19–22 November 1992; Ole Norgaard with Dan Hindsgaul, Lars Johannsen and Helle Willumsen, *The Baltic States after Independence* (Cheltenham, 1996), p.145; Marju Lauristin and Peeter Vihalemm. "Recent Historical Developments in Estonia: Three Stages of Transition (1987–1997)," in *Return to the Western World: Cultural and Political Perspectives on the Estonian Post-Communist Transition*, eds. Marju Lauristin and Peeter Vihalemm with Karl Erik Rosengren and Lennart Weibull (Tartu, 1997), pp.113–116.
4 In 1989, 90–95% of Estonia's trade was with the rest of the USSR. Philip Hanson, "Estonia: Radical Economic Reform and the Russian Enclaves." Unpublished draft chapter, 5 February 1994, p.2.
5 Ibid; John M. Kramer, "'Energy Shock' from Russia Jolts Baltic States," *RFE/RL Research Report*, Vol.2, No.17, 23 April 1993. pp.41–49. Norgaard with Hindsgaul, Johannsen and Willumsen, p.125.
6 Roman Frydman, Andrzej Rapaczynski, John S. Earle et al, *The Privatisation Process in Russia, Ukraine and the Baltic States* (Budapest, 1993), p.131.
7 Hansson, p.8.
8 Ibid, pp.8–9.
9 Kristian Gerner and Stefan Hedlund, *The Baltic States and the End of the Soviet Empire* (London, 1993), p.79.
10 Ardo Hansson, "Reforming the Banking System in Estonia," in *Banking Reform in Central Europe and the Former Soviet Union*, ed. Jacek Rostowski (Budapest, 1995), p.146; Brian Van Arkadie and Mats Karlsson, *Economic Survey of the Baltic States* (London, 1992), p.108.
11 Riina Kionka, "How will Estonia Cope after the Union Treaty?" *RFE/RL Report on the USSR*, Vol.3, No.30, 26 July 1991.
12 Ibid.
13 Riina Kionka, "Plea for Special Powers Topples Estonian Government," *RFE/RL Research Report*, Vol.1, No.7, 14 February 1992, p.33.
14 Kramer, loc cit; Frydman, Rapaczynski, Earle et al, pp.132–135;
15 Hansen and Sorsa, p.116.
16 Kramer, loc cit.
17 Norgaard with Hindsgaul, Johannsen and Willumsen, p.125.
18 Ibid; Van Arkadie and Karlsson, p.109.
19 S. Lainela and P. Sutela, "Introducing New Currencies in the Baltic Countries," in *The Transition to a Market Economy. Transformation and Reform in the Baltic States*, ed. Tarmo Haavisto (Cheltenham, 1997), pp.67–68.
20 Following its annexation of Estonia, the USSR demanded the return of the gold reserves from the countries concerned. The British government of Harold Wilson recognised Soviet claims to ownership in 1969, but agreed to pay compensation should Estonia recover its independence. Sweden, on the other hand, handed over the gold to Moscow at the first time of asking. It too agreed to pay compensation to Estonia in 1992. See: Siim Kallas, "1992. Aasta Rahareform ja Selle Mõju Eesti Arengule," in *Kaks Algust: Eesti Vabariik—1920. ja 1990. Aastad*, ed. Jüri Ant

(Tallinn, 1998), pp.170–171; Riina Kionka, "A Break with the Past," *RFE/RL Research Report*, Vol.1, No.1, 3 January 1992, p.66.

21 Hansson, 1992, pp.9–11.

22 Norgaard with Hindsgaul, Johannsen and Willumsen, p.132; *The Baltic Independent*, 29 April 1993.

23 Hansson, 1993, pp.146–154.

24 Niels Mygind, "A Comparative Analysis of the Economic Transition in the Baltic Countries— Barriers, Strategies, Perspectives," in Haavisto ed., p.31; *The Baltic Independent*, 29 April 1993.

25 Hansson, 1993, p.162.

26 *Baltic Independent*, 14 January 1993.

27 Laar, 1996, p.99.

28 Bank of Estonia, "Indicators of the Estonian Economy 1993–1998," at *http://www.ee/epbe/ datasheet/macroeconomics/table12b.html*.

29 Statistical Office of Estonia/Bank of Estonia, "Indicators for Estonian Economy 1994–1998," in *Yearbook of the Estonian Economy 1998*, Estonian Ministry of Economic Affairs, *http:// www.mineco.ee/eng/sisu.html*, section 1.2, p.3.

30 Toivo Kruus, "Estonia and EMU Prospect," *Review of Economies in Transition*, No.7, (Helsinki, November 1997), pp.16–18.

31 Ibid; p.18; Estonian Ministry of Economic Affairs Yearbook 1998, Section 2.3, pp.4–5; *Economist Intelligence Unit Country Report*, 1st Quarter 1999, p.18.

32 Hansson, 1992, p.20; Hansen and Sorsa, p.132.

33 Michael Wyzan, "Economies Show Solid Performance Despite Many Obstacles," *Transition*, 4 April 1997, p.13.

34 Piritta Sorsa, "Regional Integration in the Baltics and the Global Context," in *Regional Integration and Transition Economies: The Case of the Baltic Sea Rim*, OECD Centre for Cooperation with the Economies in Transition (Geneva, 1996), p.32.

35 Hansen and Sorsa, p.128.

36 Sorsa, p.32.

37 Wyzan, p.12.

38 Bank of Estonia, loc cit; Estonian Ministry of Economic Affairs, *Yearbook of the Estonian Economy 1998*, http://www.mineco.ee/eng/sisu.html, section 1.2, p.4.

39 Commission of the European Union, *Agenda 2000. Commission Opinion on Estonia's Application for Membership of the European Union*, http://europa.eu.int/comm/dg1a/agenda2000/en/opnions/ estonia/b21.htm.

40 Hansen and Sorsa, p.125.

41 Wyzan, p.13.

42 Author interview with Evgenii Stepanov (Tallinn, 9 March 1995).

43 Author interview with Marina Kaas (Tallinn, 23 April 1995).

44 Yaroslav Tolstikov, "Ekonomika i Politika: Paldiski—Berlin—Stokgol'm—Oslo, no ne Moskva ili Sankt-Peterburg," *Estoniia*, 13 October 1994; Author interview with Vladimir Chuikin (Narva, 26 April 1996).

45 Commission of the European Union, p.6.

46 Statistical Office of Estonia/Bank of Estonia, loc cit.

47 Commission of the European Union, p.3.

48 Jukka Rislakki, 'Virossa Kasvaa Ärtymys Suomea Kohtaan', *Helsingin Sanomat*, 11 April 1999.

49 *The Baltic Times*, 1–13 January 1999.

50 *The Baltic Times*, 4–10 February and 18–24 February 1999.

51 Hanson (1993), op cit, p.5; *The Baltic Times* 11–17 February 1998; Kramer, loc cit; Mikhail A. Alekseev and Vladimir Vagin, "Russian Regions in Expanding Europe: The Pskov Connection," *Europe-Asia Studies*, Vol.51, No.1, 1999, pp.55–56.

52 *The Baltic Times*, 11–17 March 1999.

53 "'Estma': po chasti transita Estoniia sposobna konkurirovat' c Finliandiey," *Russkii Ekspress*, No.20, 1995.

54 Alekseev and Vagin, p.53.

55 Ibid.

56 *IEWS Russian Regional Report*, Vol.2, No.35, 16 October 1997, *The Baltic Times*, 20–26 August 1998 and 11–17 March 1999.

57 Alekseev & Vagin, pp.58–59.

58 Estonian Ministry of Economic Affairs Yearbook, Section 5, p.1.

59 David Arter, *Parties and Democracy in the Post-Soviet Republics. The Case of Estonia* (Aldershot, 1996), p.99.

60 Estonian Ministry of Economic Affairs, Section 5, p.3.

61 *Economist Intelligence Unit Country Report*, 1st Quarter 1998, p.13.

62 The loans portfolio of commercial banks registered a 70% increase during 1996, whereas deposits grew by only 31%. *Eesti Ringvaade*, 19–25 January 1997.

63 *The Baltic Independent*, 7–13 August 1997; Estonian Ministry of Economic Affairs, Section 1, p.2.

64 *Economist Intelligence Unit Country Report*, 4th Quarter 1998, p.9; Estonian Ministry of Economic Affairs, Section 1, p.2.

65 *The Baltic Independent*, 13–19 November 1997; *Economist Intelligence Unit Country Report*, 4th Quarter 1998, p.8.

66 *The Baltic Times*, 15–21 October 1998.

67 *Eesti Ringvaade*, 5–11 April 1998 & June 28–July 04 1998.

68 Norgaard with Hindsgaul, Johannsen and Willumsen, p.135 and p.170.

69 Marja Nissinen, *Latvia's Transition to a Market Economy: Political Determinants of Economic Reform Policy* (Basingstoke, 1999), pp.79–80.

70 Erik Andre Andersen, "The Legal Status of Russians in Estonian Privatisation Legislation," *Europe-Asia Studies*, Vol.49, No.2, 1997, p.308.

71 Lieven, p.340; Frydman, Rapaczynski, Earle et al, p.170.

72 Lieven, p.340.

73 *The Baltic Independent*, 10 July and 28 August 1992.

74 Andersen, p.305.

75 Lieven, p.344; Andersen, p.309.

76 *The Baltic Independent*, 25 February 1993.

77 *The Baltic Independent*, 14 January 1993.

78 Lieven, p.343.

79 *Economist Intelligence Unit Country Report*, 3rd Quarter 1996, p.13.

80 Hansson, 1993, p.145; *Baltic Independent*, 29 April 1993.

81 Norgaard with Hindsgaul, Johannsen and Willumsen, p.138.

82 *Economist Intelligence Unit Country Report*, 3rd Quarter 1996, p.13; *Eesti Ringvaade*, 24–30 May 1998; Estonian Ministry of Economic Affairs Yearbook, Section 15, p.1.

83 Estonian Ministry of Economic Affairs Yearbook, Section 11, p.1.

84 *Baltic Independent*, 14 April 1994; Andersen, p.310.

85 Although the real figure was probably nearer 40%, since some Estonian buyers had foreign financial backing. *The Baltic Independent*, 5 May 1994.

86 Poll cited in Norgaard with Hindsgaul, Johannsen and Willumsen, p.136.

87 *Economist Intelligence Unit Country Report*, 3rd Quarter 1996, p.13.

88 Ibid, p.15; *Economist Intelligence Unit Country Report*, 1st Quarter 1997, pp.14–15.

89 *The Baltic Times*, 4–10 February and 18–24 February 1999.

90 *The Baltic Times*, 30 July–5 August 1998.

91 Michael Bradshaw, Philip Hanson and Denis Shaw, "Economic Restructuring," in *The Baltic States: The Self-Determination of Estonia, Latvia and Lithuania*, ed. Graham Smith (Basingstoke, 1994), p.178.

92 It should be noted, however, that one third of the 48,000 registered enterprises in 1993 existed only on paper. All statistics taken from Urve Venesaar and David Smallbone, "The Development of Manufacturing SMEs in Estonia," paper presented at the conference *Estonia: A Development Agenda*, Royal Institute of Geography, London, 26 March 1996.

93 Ibid.

94 Nissinen, p.3.

95 Norgaard with Hindsgaul, Johannsen and Willumsen, p.143.

96 Nissinen, p.42; Lauristin & Vihalemm, p.78; Lainela and Sutela, p.67.

97 Laar, p.97.

98 Nissinen, p.42.

99 Lauristin and Vihalemm, pp.123–126.

100 Economist Intelligence Unit Country Report, 1st Quarter 1999, p.18.

101 Raivo Vetik, "Ethnic Conflict and Accommodation in Post-Communist Estonia," *Journal of Peace Research*, Vol.30, No.3, 1993, p.276.

102 L. Hansson, "Attitudes Towards Unemployment and Self-Employment: Reality and New Opportunities," in *Social Strata and Occupational Groups in the Baltic States*, eds. Blom, Harri Melin and Nikula (Tampere, 1995), p.24; Estonian Ministry of Economic Affairs Yearbook, Section 2, p.3.

103 Andersen, pp.303–316; Norgaard with Hindsgaul, Johannsen and Willumsen, p.133 and p.146.

104 "Russkaia Real'nost' v Estonii," *Digest*, No.7, May 1995; Richard Rose and William Maley, *Nationalities in the Baltic: A Survey Study* (Strathclyde, 1994), p.27.

105 Tartu Ülikooli Turu-Uurimisrühm, *Kirde-Eesti Linnaelanike Suhtumine Eesti Reformidesse ja Sotsiaalpoliitikasse: 1993-ja 1994. Aasta Võrdlev Analüüs* (Tartu, September 1994), p.17

106 Anu Toots, "Immigration in Estonia: Politics and Policy," *Monthly Survey of Baltic and Post-Soviet Politics*, April 1994, p.78. Aksel Kirch, Marika Kirch and Tarmo Tuisk, "Russians in the Baltic States: To Be or Not to Be?", *Journal of Baltic Studies*, Vol.24, No.1, Summer 1993, p.181.

107 Harri Melin, "Ethnicity and Social Class in the Baltic Countries," working paper for the project "Social Change in the Baltic and Nordic Countries," University of Tampere Department of Sociology and Social Psychology, 1995, pp.10–11; Estonian Ministry of Economic Affairs Yearbook 1998, Section 2, p.5.

108 K. Petti, "Majandusliku Käitumise Soodumused Turu-Uuringute Põhjal," paper presented at the conference "*Sotsiaalsed Probleemid Eestis Rahvusgruppide Lôikes: 1988–1993.aastate Sotsioloogiste Uringute Pôhjal*", Tallinn, 16 June 1994. Author interviews with Ants Paju (Tallinn, 14 June 1994) and Evegenii Stepanov (9 March 1995).

109 *Eesti Ringvaade*, 13–19 May 1996.

110 Estonian Ministry of Economic Affairs Yearbook 1998, Section 2, p.2; Hanson, 1993, p.21.

111 Author interview with Nikolai Zolin (Narva, 26 April 1996).

112 Tartu Ülikooli Turu-Uurimisrhm, p.33.

113 Estonian Ministry of Economic Affairs Yearbook 1998, Section 2, p.3.

114 *Baltic Times*, 6–12 August 1998.

115 Kruus, p.24.

116 *Economist Intelligence Unit Country Report*, 3rd Quarter 1997, p.13.

117 David Kirby, *The Baltic World 1772–1993: Europe's Northern Periphery in an Age of Change* (London, 1995), p.401.

118 Ibid, pp.411–12; Ray Abrahams and Juhan Kahk, *Barons and Farmers: Continuity and Transformation in Rural Estonia (1816–1994)* (Gteborg, 1994), pp.50–61.

119 Ibid, p.417.

120 Ray Abrahams, "Continuity and Change in the Estonian Agrarian Sector," paper presented at the conference *Estonia: A Development Agenda*, Royal Institute of Geography, London, 23 March 1996, p.17.

121 OECD Centre for Co-operation with the Economies in Transition, *Review of Agricultural Policies: Estonia* (Paris, 1996), pp.65–68.

122 Ibid, p.25.

123 Abrahams, pp.5–16.

124 Abrahams and Kahk, pp.112–119.

125 OECD Centre for Co-operation with the Economies in Transition, p.18 and pp.23–25.

126 Ibid, pp.17–19.

127 Abrahams, p.19.

128 OECD Centre for Co-operation with the Economies in Transition, pp.17–19.

129 Ibid, pp.19–20.

130 *Eesti Ringvaade*, April 19 1998.

131 *The Baltic Times*, 13–19 August 1998.

132 *The Baltic Times*, 26 March–1 April 1998.

133 *The Baltic Times*, 18–24 March 1999.

134 Sorsa, p.36.

135 Bradshaw, Hanson and Shaw, pp.158–180.

136 Estonian Ministry of Economic Affairs Yearbook 1998, Section 5, p.4.

137 David Smith, "Nordic, Baltic and Arctic Organisations," in *The Annual Register 1998*, ed. Alan J. Day (Keesing's Worldwide, 1999), pp.445–447; *Baltic Times*, 25 February–3 March 1999.

138 Government of the Republic of Estonia, *Roadmap to Reform: Estonia's Future Plans in the Field of European Integration* (Tallinn, September 1997), p.7.

139 Estonian Ministry of Economic Affairs Yearbook 1998, Section 2.3, pp.4–6; *Economist Intelligence Unit Country Report*, 1st Quarter 1999, p.18.

140 *The Baltic Times*, 11–17 March 1999.

141 *The Baltic Times*, 20–26 August 1998.

142 *The Baltic Times*, 11–17 March 1999.

Chapter 5

'THE DEVIL AND THE DEEP BLUE SEA': FOREIGN POLICY
BETWEEN EAST AND WEST

Estonia's inclusion in the first group of applicant states to begin negotiations on entry to the European Union constitutes the most significant achievement to date of a post-Soviet foreign policy largely based around the aspiration for a 'return to Europe'.[1] Over the past decade, narratives on external relations have consistently expressed a belief that Estonia's interests will be best served by membership of the network of European and transatlantic international organisations which laid the foundations for prosperity and stability in western Europe after 1945.[2] Entry to these organisations is also seen as a safeguard against a recurrence of the 'worst case scenarios' of history, for, as Estonia emerged from fifty years of Soviet occupation, memories of the fateful year 1939–40 inevitably loomed large.[3] The complex of unresolved issues arising from occupation has led to an acute sense of security threat from the 'eastern neighbour', Russia. Drawing on the experience of the inter-war period, foreign policy makers have expressed anxiety that the country might again find itself part of a 'grey area' between a predatory East and an indifferent West.[4]

Entirely typical is the following statement by Jüri Luik, who in March 1994 expressed the belief that 'security in our corner of the world is integral to European security in a wider sense. This belief is based upon the geopolitical facts of life. We find ourselves located on the front line ... of the growing crisis in the East. At the same time, we are at the frontier of democratic and free-market thinking prevalent amongst our closest neighbours, with whom we share a coastline. Some would characterise our position as being between the Devil and the Deep Blue Sea.'[5]

Luik's assessment of Estonia's position seems particularly apt. To the East, the turbulence which continues to afflict Russia and other former Soviet republics has led to a growth in geopolitical insecurity across the FSU as a whole. Influential circles in Moscow insist that the Baltic states' aspiration to integrate politically, economically and, above all, militarily with the West constitutes a threat to Russia's still vaguely-defined 'national interest'. From the Estonian perspective, these objections to NATO and EU membership are merely reflective of an enduring tradition of Russian imperialism which seeks to lock the Baltic states into a perpetual state of unilateral dependence.[6] Statements by leading foreign policy actors have therefore stressed the inalienable right of every sovereign state to determine its own security arrangements, whilst simultaneously expressing a belief in the indivisibility of European security as a whole. The 'New Europe' of the post-Cold War era is portrayed not as a geographical expression, but as a 'philosophical, ethical, political and economic programme' based around the key principles of democracy, individual rights, rule of law and a market economy. The eastwards enlargement of the 'Euro-Atlantic space' should thus be effected not on the basis of geopolitical criteria, but according to the applicant countries' progress in implementing the programme outlined above.[7]

By this definition, membership of organisations such as NATO and the EU remains open to any FSU country—including Russia—willing to adhere to these principles. In practice, the predominant discourses on Estonian national identity have drawn firm boundaries between the western 'us' and the eastern 'other'. Russia is deemed to be excluded from integration with the Euro-Atlantic space on the grounds that it is a state which is unable to behave in a 'European-like' manner. Conversely, the Baltic states are portrayed as the litmus test of Russia's real intentions towards Europe—only by respecting the rights of the Baltic nations to sovereign statehood will Russia prove its European credentials.

In November 1994, Luik expressed the guiding principles of Estonian foreign policy using the formula 'security equals normalisation [with the East] plus integration [with the West]'.[8] By 'normalisation', he understood 'mutual respect for sovereignty, mutual respect for national security interests, mutual refraining from verbal and other confrontation, mutual respect for international norms of behaviour, most importantly in the area of human rights'.[9] These goals have remained elusive. As Toomas Hendrik Ilves pointed out in 1998, normal inter-state relations 'cannot develop without a required legal base—i.e. a minimal foundation of bilateral agreements' such as a border treaty and a most favoured nation trade regime.[10] Early in 1999, this legal base was still absent. During the three years after August 1991, when Estonia remained occupied by Russian-controlled units of the former Soviet army, Estonian-Russian relations descended into barely disguised hostility. The final withdrawal of Russian troops spurred the Estonian side towards a more active strategy of 'pacification with the East', yet the subsequent five years have seen no liquidation of the underlying sources of disagreement.[11] For the Estonians, it seems that western integration—bringing with it concrete guarantees of state security—remains the essential prerequisite for full normalisation with the East. For, to quote Ilves again, such normalisation will be difficult as long as Estonia is unable to relate to Russia as 'a normal Western state, free and confident of its independence and not as a former colony or *oblast*' burdened by complexes and doubts'.[12]

Westward integration is complicated by the fact that the main European and transatlantic international organisations are themselves struggling to redefine their identity in the aftermath of the Cold War. Notwithstanding current debates on the enlargement of the EU and the future shape of its institutions, swift accession to this organisation remains a realistic prospect. The same cannot be said for membership of the NATO alliance, whose 'identity crisis' has deepened following the conflict in Kosovo. Seen by many Estonians as the 'ultimate guarantor' of independence, NATO has appeared torn between a desire to support the aspirations of CEE applicant countries and a fear of provoking further political instability in Russia. This explains the organisation's manifest reluctance to admit Estonia and its Baltic neighbours without tacit approval from Moscow.

In effect, NATO is arguing that normalisation of Estonian-Russian relations must necessarily precede integration into the alliance. Moscow, it seems, understands this only too well, and has sought to exploit western prevarication in order to increase its own leverage in the eastern Baltic region. Russia's continued refusal to sign a border treaty with Estonia, along with its allegations of mass human rights abuses against the country's 'Russian-speaking population' have proved less successful in terms of blocking Estonia's progress towards the EU. However, the latter organisation too has been ready to take heed of Russian concerns, and one cannot entirely rule out the possibility that the citizenship question might yet complicate Estonia's accession negotiations. This final chapter traces the evolution of Estonia's foreign policy against the background of the eastern question and the changing position of the West. Given the salience of historical parallels, it seems apt to begin with a brief overview of the country's international relations between the two World Wars.

ESTONIAN FOREIGN POLICY AND THE QUEST FOR A 'NEW EUROPE' 1918–40

In a speech to the Council of Europe in November 1995, the then foreign minister Siim Kallas claimed that 'viewing Europe today, I am struck by the parallels between the end of World War One and the end of the Cold War. After the Treaty of Versailles, western nations were secure in their false peace and chose isolationism. ... These ... naive hopes and deceptive realities soon gave way to international financial turmoil, a rejection of democratic traditions and war. I see a similar complacency emerging in the West. This complacency is characterised by a reluctance to make difficult decisions regarding the future of European integration'.[13] From the standpoint of mid-1999, Kallas' analysis perhaps appears unduly pessimistic. The significant progress towards the EU during the intervening period means that Estonia's international position—and, indeed, the state of European international relations as a whole—appears more secure than it did eight years into the first period of independence. Even so, Luik's depiction of a state lying between the 'Devil and the Deep Blue Sea' remains valid, and is an analogy with which his inter-war predecessors would doubtless have readily identified.

The Versailles settlement of 1919 signally failed to realise aspirations for a 'New Europe' in the wake of World War One. The League of Nations, supposedly the embodiment of this new order, was fatally weakened at the outset by the United States' reversion to an isolationist stance and by the exclusion of Germany and Russia. Under these circumstances, Estonia's strategic position at the crossroads of East and West spelled threat rather than opportunity. The new state was left sandwiched between two revisionist great powers, both of which had harboured designs on the eastern littoral of the Baltic during 1918–20. These experiences of invasion and occupation by Germany and Soviet Russia meant that voluntary acceptance of bilateral security guarantees from either power was out of the question.

The Estonians naturally looked first and foremost to the western allies for support, yet here too they had to contend with Franco-British divisions regarding the implementation of the peace settlement. From the standpoint of London and Paris, Estonian aspirations for national self-determination were strictly secondary to the struggle against Germany and the wider question of the future of Russia during 1918–20. In the year following the armistice with Germany, at a time when the Bolshevik grip on Russia still hung in the balance, the allies were not prepared to alienate a future White Russian government by giving full *de jure* recognition to the Baltic states. This became clear at the Versailles peace conference, which came as a bitter disappointment to the Estonian delegation headed by Jaan Poska.

The end of allied intervention in the Russian Civil War and the resultant French-led proposals for an eastern barrier between Germany and Soviet Russia during late 1919 improved the prospects for *de jure* recognition. The allies' waning interest in the white Russian cause meant that the Estonians enjoyed an increasingly free hand in their dealings with the eastern neighbour. The end of the allied blockade against Soviet Russia removed the last obstacle to the regulation of Estonian-Russian relations, achieved under the Tartu treaty of February 1920. Even so, the subsequent outbreak of war between Poland and Soviet Russia led Britain and France to delay *de jure* recognition of Estonia until January 1921, since neither power was prepared to extend security guarantees in the event of a renewed Bolshevik threat to the country's integrity.

The long wait for international recognition gave added impetus to the project for a federation of Baltic and Scandinavian states, first mooted by Tõnisson in September 1917. During the following year Ants Piip and Karel Pusta, based respectively in London and Paris as representatives of the Estonian foreign delegation, unveiled proposals for a Baltic 'League of Free Peoples' which would safeguard the small Baltic and Scandinavian nations against the threat of German and Russian revanchism. In its original incarnation, the Baltic League was intended as far more than a traditional military alliance between sovereign states, including as it did proposals for economic integration and common political institutions.[14] At this time, it seems that Estonian policy-makers found it hard to conceive of anything other than 'limited sovereignty' as part of some larger entity.[15] In this regard, the Baltic League was only one of several ideas—including proposals for a Finnish-Estonian dual state—which were floated. Konstantin Päts, for instance, was a leading advocate of union with Finland around this time. Interestingly, he was to resurrect this proposal on the eve of his deportation in 1940, when he issued guidelines to be used in the event of a future peace conference.[16]

As Estonian statehood was consolidated during the early 1920s, conceptions of the Baltic League idea changed accordingly. Discussions around this theme reached what was, by the European standards of the day, an 'exceptional level of intensity'.[17] The period 1919–1926 witnessed some forty joint conferences comprising varying constellations of the following states: Estonia, Latvia, Lithuania, Finland and Poland.

In spite of this, the proposals for a Baltic League—either as a Federation or more narrowly-defined politico-military alliance—never came to fruition. The Scandinavian states rejected overtures from the

Baltic states at an early stage, opting instead for the isolationism which had seen them through World War One but was to serve them less well twenty years later. From 1920 onwards the Baltic League idea became confined to the eastern border states, yet discussions quickly foundered on the differing security conceptions and mutual rivalries of the nations concerned. The intractable Lithuanian-Polish dispute over Vilnius, for instance, ruled out the possibility of a united Baltic-Polish bloc after 1920. In this regard, Polish designs on the Latvian province of Latgale also had a part to play. Moreover, the profound anti-Germanism of the Estonians and Latvians found little echo in Finland, which from 1922 began to eschew Baltic cooperation in favour of a more Scandinavian orientation. Lithuania's territorial disputes with Poland and Germany (over Memel) meant that, from the point of view of Tallinn or Riga, a Baltic triple alliance seemed as likely to endanger peace as it was to guarantee it. The only remaining option became an Estonian-Latvian alliance, duly concluded in 1923. Even this agreement contained few military provisions, whilst the aim of a customs union between the two countries proved elusive. Estonia's neighbours also proved susceptible to the strategy of divide and rule pursued by the USSR. Then, as now, Estonian attempts to obtain guarantees of security through integration were invariably portrayed as a hostile act by Moscow. Soviet foreign policy thus sought to play on border state divisions by proposing bilateral trade and non-aggression treaties with the individual countries concerned.

Estonia's state of international isolation, however, was only rendered threatening by the failure to achieve a durable settlement of European affairs as a whole. For the Estonians, proposed regional arrangements in the eastern Baltic were never seen as an end in themselves, merely one factor in an overarching European security architecture. Piip and Pusta's 1918 proposals, for instance, corresponded closely to western thinking on the 'New Europe', which envisaged a series of regional leagues under the overall auspices of the League of Nations. Having gained membership of the latter organisation, Estonia participated actively in its work to promote reconstruction and alleviate sources of conflict between states.[18] In this regard, much was expected of the 1922 Genoa conference, when the statesmen of Europe gathered to discuss the economic reconstruction of Europe. The prospect of a normalisation of relations between the West and Soviet Russia was eagerly anticipated in Estonia, which stood to profit handsomely from a potential boom in transit trade. Instead, the Soviet government was able to exploit western European divisions by signing the treaty of Rapallo with Germany.

With hindsight, Rapallo did not constitute such a negative development, since it provided a stable basis for German-Soviet relations over the coming decade. Yet the agreement could not fail to arouse distrust in Estonia, since it increased the possibility of great power cooperation directed against Baltic interests. Throughout the 1920s, foreign policy makers in Estonia continued to cherish the aim of creating an eastern alliance under Polish auspices. What united Berlin and Moscow above all was their mutual desire to prevent such a development. In the face of this joint German-Soviet opposition, the viability of any proposed border states alliance depended heavily upon continued backing from the western powers. The *de jure* recognition of the Soviet regime by Britain and France during 1924 signalled their declining interest in maintaining an anti-Bolshevik cordon sanitaire. Support for an eastern barrier directed against Germany also declined following the signing of the 1925 Locarno pact, an agreement which hastened the decline of discussions on an eastern Baltic alliance.

Locarno was nevertheless hailed as the start of a new era in European affairs, in which all outstanding disputes might be settled according to the principle of mutual guarantees adopted by the western powers. The Estonian-Latvian 'Baltic protocol' of 1925 was one of number of initiatives designed to integrate the eastern Baltic region into these new European political arrangements.[19] In spite of such efforts, the Locarno system was never extended to cover eastern Europe. Germany refused to subscribe to a system of guarantees covering its eastern borders, whilst the Soviet Union, suspicious of the new spirit of cooperation amongst the western capitalist powers, intensified its efforts to create a separate Soviet-dominated eastern-alliance system as a counterweight to Locarno. The new sense of optimism in European affairs made the Estonians somewhat less wary of Soviet overtures. Thus, in 1929 Estonia signed the 'Litvinov Protocol'—the Soviet response to the 1927 Briand-Kellogg peace pact—and concluded a new trade agreement with the USSR.

To the West, however, the 'Spirit of Locarno' could not weather the onset of the Great Depression. Growing political instability in Germany lay at the root of a gradual Franco-Soviet rapprochement during the early 1930s, as Paris sought a stronger eastern alliance partner and Moscow assured the West that it had abandoned the goal of world revolution in favour of peaceful coexistence and collective security.[20] As part of this charm offensive, the USSR signed a series of non-aggression pacts with its western neighbours, including a 1932 Estonian-Soviet treaty guaranteeing the 1920 border between the two states. These ultimately worthless assurances are best regarded as an attempt to bring the Baltic states within a Soviet sphere of influence, thereby pre-empting any future initiative by a Germany fast coming under Nazi dominance.

Following Hitler's assumption of power in 1933, Estonia and its neighbours were duly transformed into a political arena where two totalitarian regimes vied for supremacy.[21] In January 1934, Hitler scored a diplomatic coup in the form of a non-aggression treaty between Germany and Poland, thereby destroying one of the cornerstones of the Soviet security system in eastern Europe. Henceforth, Nazi expansionist aims towards the Baltic region were shared more or less openly by a Soviet leadership anxious to obtain a territorial glacis for the defence of Leningrad. In April of the same year, Moscow welcomed Barthou's initiative for a Locarno-style 'Eastern Pact' guaranteed by France. Franco-Soviet proposals—rejected by Germany and Poland—aroused understandable suspicion in the Baltic states, for, unlike the non-aggression treaties of 1932, they provided for 'mutual aid', including provision for Soviet forces to enter the territory of neighbouring countries.

In response to Soviet overtures, Estonia and its neighbours reiterated their strict neutrality and desire to remain clear of any power bloc. However, this stance was not underpinned by any effective deterrent against an outside aggressor. The tripartite Baltic Entente signed in September 1934 provided for coordination of foreign policy, but contained no formal military clauses and was undermined by petty nationalist frictions between the three states. Under these circumstances, professions of neutrality were barely credible, and were certainly not regarded as such by the USSR.

Ruling circles in Estonia were certainly more anti-Soviet than anti-German during the late 1930s. After Britain effectively conceded German naval hegemony in the Baltic during 1935, many political and military leaders believed that the only hope lay in German intervention to uphold the independence and sovereignty of the Baltic states. Hitler cynically exploited such sentiments, thereby increasing Soviet mistrust towards the Estonian government.[22] By the time Britain and France finally decided to draw a line against German expansionism in March 1939, they were indeed forced to recognise that they could not offer direct military guarantee to the states of the eastern Baltic. Subsequent attempts to negotiate a security pact with the USSR foundered on Moscow's insistence that it be allowed to establish military bases in Finland and the Baltic states as a defence against the anticipated attack by German forces. The western allies' reticence on this point stemmed partly from a principled commitment to maintaining the integrity of the Baltic states, yet most of all from a realisation that acceding to Soviet demands would deliver Estonia and its neighbours into the arms of Hitler. As the negotiations dragged on, Stalin ultimately decided that he could realise his aims more quickly by coming to an agreement with Nazi Germany. Thus was the stage set for the Molotov-Ribbentrop Pact and the subsequent annexation of Estonia by the USSR.

RESTORING INDEPENDENCE. THE EXTERNAL DIMENSION

Throughout the ensuing occupation, external propagation of the Estonian cause fell to émigré political organisations and the domestic dissident movement. From the early 1970s, Baltic dissidents issued a series of joint appeals to the United Nations and other international fora demanding the restoration of independent statehood.[23] Perhaps the best known of these, made on the fortieth anniversary of the Molotov-Ribbentrop Pact in 1979, prompted the European Parliament to call for the Baltic question to be examined within the UN Committee for Decolonisation.[24] According to one commentator, the ability of these groups in keeping the legal continuity principle alive for half a century made this period the most fruitful so far in terms of Estonian interaction with international organisations.[25]

In Estonia proper, governments of the ESSR contained a 'foreign minister' from the 1960s, but it was not until the Savisaar government took office in April 1990 that Lennart Meri was able to breathe life into this hitherto largely decorative post. The charismatic writer and director (born 1929) possessed a strong pedigree for the role. The son of a diplomat of the inter-war Estonian Republic, he had spent part of his childhood in Paris and Berlin before being deported to Siberia in 1941.

With Savisaar at the helm, Estonia was now in a position to formulate, if not yet to implement, an independent foreign policy. The primary goal in terms of external relations during the transitional period of 1990–91 was to achieve the factual restoration of statehood on the basis of negotiations with foreign states, including the USSR.[26] This goal had been espoused as early as May 1989, when the leaders of the three Baltic Popular Fronts demanded independence for their republics within a 'neutral and demilitarised Baltoscandinavia'.[27]

The use of this term was of course strongly evocative of the inter-war period. In the context of 1989, it was also consistent with the Popular Front approach of working with Moscow to secure independence, since demands for a nuclear free zone in northern Europe had been a long-standing feature of Soviet policy initiatives towards the Nordic region. In seeking to align themselves with a Nordic Europe located between the two superpower blocs, the Estonians and their neighbours could depict independence as something not inherently threatening to the interests of the USSR.[28]

Statements by Meri during 1990–91 repeated the argument that an independent Estonia forming part of Northern Europe would be in the best interests of Moscow. The need to establish stable and good neighbourly relations with the 'Eastern Neighbour', he maintained, was the leading principle of an Estonian foreign policy rooted in geopolitical necessity.[29] On this basis, a leading member of the Popular Front claimed that the Estonians had learnt the virtues of the post-war Finnish approach when dealing with the Russians.[30]

However, Meri's statements around this time also emphasised that the sole legal basis for relations with Moscow remained the Tartu Treaty of 1920 under which Soviet Russia had renounced in perpetuity all claims on Estonian territory.[31] This demonstrates the growing salience of the legal continuity principle within the politics of the independence movement during 1989–90. The success of the radical Committee movement around this time rested partly on a growing popular realisation of the limited possibilities for evolutionary change within the Soviet system. Gorbachev's clear reluctance to sanction even significant autonomy for the Baltic republics made it all the more important to internationalise the issue in the hope of attracting support from the western powers. By insisting that their country had remained an independent republic under illegal Soviet occupation during the past 50 years, Estonia's leaders sought to place themselves on a par with the former Soviet satellites of eastern Europe, whilst reminding the West that the Baltic states alone had suffered the injustice of being wiped from the map after World War Two.

As the former wartime allies began to discuss the reunification of Germany, the Estonians were therefore at pains to remind them of the 'unfinished business' of World War Two in the form of the Baltic question. Baltic independence, it was argued, was the final prerequisite for realising Gorbachev's vision of a 'Common European Home' following the democratisation of eastern Europe and German reunification. The

Soviet leader was therefore urged to apply international law to both great powers and small.[32] Stressing that there 'cannot be a free Europe without a free Baltic', Meri also sought to underline the indivisibility of European security, noting that the Molotov-Ribbentrop Pact had been the precursor of the invasion of France as well as the annexation of the three Baltic states.[33]

The most obvious forum to internationalise the question of Estonian independence became the CSCE, 'the only pan-European forum embracing all the sovereign nations of Europe'.[34] From its inception in the 1970s, the CSCE process had provided a focus for dissident groups within Estonia and also for the émigré World Council of Estonians, which had used meetings of the organisation to draw attention to the plight of its homeland.[35] The strongest backing for Estonian attempts to acquire full membership or observer status within the CSCE during 1990–91 came from the Nordic countries. The latter proved unwilling to accede to Baltic requests for observer status within the Nordic Council, arguing that this would dilute the specifically 'Nordic' character of the organisation,[36] yet the Balts were invited to attend the October 1990 CSCE foreign ministers' meeting as guests of the Nordic delegation. Ultimately, however, the CSCE deferred to strong objections from Moscow regarding Baltic participation at the organisation's Paris summit a month later. The reluctance of the western states to undermine Gorbachev's position during this period heightened fears that the long-standing support for legal continuity would be compromised and the 'Baltic question' categorised as an internal affair of the USSR.[37]

1991–1994: FOREIGN POLICY AND DESOVIETISATION

At first sight, the sudden and relatively peaceful collapse of Soviet power in August 1991 stood in marked contrast to the turmoil which accompanied the dissolution of the Tsarist empire after World War One. This fact meant that the task of obtaining full diplomatic recognition was more straightforward than it had been in 1918. In reality, the formal restoration of the Estonian Republic to the international arena could not efface the complex of outstanding problems left by the Soviet occupation.

In particular, sovereign status remained far from assured as long as former units of the Soviet army remained stationed on Estonian territory. Once the Russian Federation had assumed the mantle of legal successor to the USSR, these units came under Russian command at the beginning of 1992. In the period before 31 August 1994, securing withdrawal of the remaining Russian troops constituted the overriding priority of Estonian foreign policy. As Joenniemi notes, however, it is difficult to speak of a 'foreign policy' during this period, since troop withdrawal and relations with Russia were inextricably bound up with domestic debates relating to the status of the large Russian-speaking settler population.[38]

At the heart of this 'Russian question'[39] lay Russia's refusal to recognise the principle of legal continuity underpinning Estonia's restored statehood. The period of fruitful Estonian-Russian cooperation during 1990–91—exemplified by Russia's prompt recognition of Estonian independence on 24 August 1991—rested on the mutual interest of the two parties in undermining the power of the Soviet central government. This imperative led both governments to dodge awkward questions relating to the future status of Estonia's Russian-speaking population. In so far as its thinking extended to other issues during this period, the new Russian leadership was preoccupied above all by the question of relations with the West, where it advocated an 'Atlanticist' line of partnership and cooperation.[40]

Initially, at least, Boris Yeltsin was primarily a 'nation-builder' rather than an 'empire saver'. His ideology has been described as a civic-based 'new Russian nationalism' based around the borders of the Russian Republic at the time of its declaration of sovereignty in June 1990.[41] A policy agenda driven by Russian domestic concerns made relations with the non-Russian Soviet successor states a question of second order, to be resolved according to the vague dictum of 'good neighbourly relations'.[42] Whilst seeking to maintain long-standing economic ties between the former Soviet republics, the Yeltsin administration saw no place for interference in the internal political affairs of its neighbours. In an address of 30 September 1991, Yeltsin pledged to defend the rights and interests of Russians abroad. This, however, was to be done solely according to international law. In line with the civic vision of Russian national identity, it was assumed that Russians living outside Russia would be granted the citizenship of their host state.[43]

When the Confederation of Independent States (CIS) was founded in December 1991, Estonia declined to participate, maintaining the stance of outright and unconditional independence to which it had adhered throughout Gorbachev's previous attempts to negotiate a new union treaty. Whilst this attitude apparently came as no surprise to the

Russian Ministry of Foreign Affairs, it nevertheless expected that Estonia and Russia would continue to maintain close economic and political relations.[44] In line with this expectation, Russia rejected the legal continuity argument, insisting that the current Estonian Republic came into existence in August 1991 as a Soviet successor state. The Estonian-Russian treaty of January 1991 is thus deemed to be the only valid basis for interstate relations. Continuing the line of the old Soviet central government, the Yeltsin administration holds that the takeover of Estonia in June 1940 did not amount to illegal occupation, but was in fact based on consent. At this point, the pre-war Estonian republic ceased to exist and the Tartu treaty of 1920 became invalid.[45]

This stance was guaranteed to clash headlong with the legal continuity principle which formed the point of departure for Estonian state-building. The July 1992 constitution, for instance, stipulates that the land borders of the Estonian Republic correspond to those established under the Tartu treaty of 1920.[46] The territorial changes effected immediately after World War Two, when some 2,000 square kilometres of pre-war Estonian territory were transferred to the RSFSR, are thus deemed illegal. On this basis, the Russian government accused Estonia of harbouring a territorial claim against it. This, it was feared, might spark an avalanche of claims for frontier revision within an already unstable Russian Federation. On the question of Estonian citizenship policy, Russia also claims that the January 1991 treaty provides for a 'zero option' variant, an interpretation which is disputed by the Estonian side.

Estonian nationalist perceptions of a Russian-speaking 'fifth column' were reinforced by the rise of discourses on Russia's 'near abroad' following the collapse of the USSR. By the end of 1991, the 'Atlanticist' orientation of Yeltsin's foreign policy found itself under increasing attack by opposition forces within the Russian parliament. These ranged from 'statist' advocates of a more assertive Russian policy within the former Soviet space to the so-called 'National Bolsheviks', for whom 'Russia' was synonymous with the borders of the old USSR.[47] These groups enjoyed much support within the armed forces, where leading elements of the officer corps were alarmed at the disintegrationist tendencies still at work within the FSU. In particular, it was feared that ports and military bases in the former Baltic republics might fall into the hands of a 'hostile' western alliance.

The rise of this domestic opposition impelled the Yeltsin regime to assume a 'big brother' role towards the former Soviet republics.[48] The mantle of legal successor to the Soviet Union in turn made it harder to reject calls for Russia to assert itself as the leading power within the post-Soviet space.[49] The doctrine of the 'near abroad', elaborated during the early months of 1992, held that Russia had a set of interests and obligations towards the former Soviet republics distinct from those governing relations with the states of the 'far abroad'. Henceforth, the dictum of 'good neighbourly relations' became essentially synonymous with the establishment of a Russian sphere of influence within the FSU.

In April 1992, Estonian and Russian delegates gathered for an initial round of negotiations on troop withdrawal and other outstanding issues. Citing 'technical difficulties' connected with the rehousing of troops, Russian representatives refused to commit themselves to any firm date for withdrawal beyond vague references to 1999.[50] Appeals to the West for financial aid to cover the withdrawal were accompanied by attempts to gain Estonian consent for 'military cooperation' which would allow Russia to maintain its bases in the country. In response, the Estonian government merely reiterated its demand for a complete withdrawal of troops by the end of 1992.

The opposing positions of the two parties at the negotiations fanned the flames of nationalism in both countries. An agreement between the Vähi and Gaidar governments on Most Favoured Nation trading status for Estonia fell victim to the growing political instability in Russia, where the parliament refused to ratify the treaty.

Relations were not helped by Lennart Meri's call for a new 'cordon sanitaire' against Russia where, he claimed, hunger and cannibalism were developing on a massive scale.[51] Meri's handling of the foreign affairs portfolio had already attracted criticism from former communist circles in Estonia, whose representatives fell victim to 'house cleaning' within the ministry. For many, the ill-advised remarks about Russia were the final straw, forcing him to submit his resignation in March 1992.

The subsequent appointment of Estonian-Swedish emigre Jaan Manitski as caretaker foreign minister was nevertheless portrayed as a defeat for the communists and a victory for the advocates of a westward-leaning foreign policy. Potentially more serious in terms of external relations was the challenge mounted by extreme restitutionist nationalists during the Spring and Summer of 1992. In May, the Congress of Estonia approved a resolution demanding that negotiations with Russia be suspended pending Moscow's recognition of the need for decolonisation of Estonia.[52]

This resolution had little more than symbolic value, for, if Russia was not about to acquiesce in demands for decolonisation, the same can be said of the United States and its allies. Support for legal continuity notwithstanding, the latter had always made clear their unwillingness to reverse the isation of the area by physical means. The subsequent challenge to the new constitutional order mounted by restitution prompted the mainstream political parties to close ranks. In answer to radical demands on negotiations with Russia, it was insisted that this process would be guided by *realpolitik* rather than legal principles.[53]

In the face of Russian truculence over troop withdrawal, Estonia joined its Baltic neighbours in lobbying western states and every international forum at its disposal in order to drum up support. The Balts' efforts were rewarded at the Helsinki summit of the CSCE in July 1992, which called for 'an early, orderly and complete' withdrawal of Russian troops from the territories of the Baltic states.[54] A month earlier, the United States Senate had threatened to suspend humanitarian aid to Russia if no progress over the withdrawal of troops was noted within a year.[55] The senate decision exposed the central dilemma of the Yeltsin administration. Whilst it had to answer domestic criticism, it could not go so far as to prejudice the Atlanticist plank of foreign policy and the receipt of western credits which were urgently needed to revitalise the economy. The withdrawal of troops from Estonia therefore continued apace, regardless of statements made by leading foreign policy makers in Moscow.

In response to setbacks in its Baltic policy, Russia began to employ a new strategy, which Estonians frequently describe as the 'Karaganov doctrine'. In the autumn of 1992, Sergei Karaganov of Moscow's European Institute suggested that ethnic Russians residing in the 'near abroad' should be employed as a tool of Russian foreign policy towards the region. According to Karaganov, Moscow should seek international approval for its aims by posing as the defender of human and minority rights throughout the former Soviet Union. In particular, he argued, propaganda against the 'political leverage points' of Estonia and Latvia should assume a central role in this strategy.[56] Shortly after Karaganov published his article, Russian foreign policy statements indeed began to insist that the withdrawal of troops from Estonia would be contingent upon an end to alleged 'systematic discrimination' against the Russian-speaking population residing there. Moscow's efforts to gain international sanction for this position within the United Nations and other fora presaged later attempts to attain a peacekeeping role within the territory of the former Soviet Union.[57]

Just as it had previously enlisted the support of the CSCE, so Estonia was able successfully to counter Russian attempts within the UN to link troop withdrawal to the nationalities issue.[58] As Joenniemi notes, Estonian interventions within these fora were loaded with a strong moral connotation and argued in categorical terms. This style, he maintains, was not in keeping with general expectations concerning the behaviour of small states, which normally act in a 'modest, pragmatic, flexible and realist' manner.[59] It was, however, precisely such modest, pragmatic behaviour which had proved to be the undoing of the Estonian Republic back in 1939–40. If, as Taagepera suggests, Estonia's inter-war leaders had been socio-psychologically ill-equipped to deal with the demands of a great power neighbour, the generation which took power in 1992 was of an altogether different mindset. The need to erect firm boundaries against the 'eastern other' became a stock-in-trade of foreign policy statements in the two years before September 1994, when *ERSP*'s Trivimi Velliste (October 1992–January 1994) and *Isamaa*'s Jüri Luik occupied the post of foreign minister. In its refusal to bow to the dictates of a great power, Estonia's early postcommunist foreign policy resembled that of inter-war Finland, a state whose strong sense of national identity had been constructed in opposition to Russifying pressures during the last years of the Tsarist empire.

The refusal by most western states to recognise Soviet military occupation during the Cold War, one need hardly add, made it easier for the Estonians to claim the moral high ground over the troop withdrawal issue. As already noted, however, the desire for an uncompromising stance against Russia was tempered by a realisation of the limits to western goodwill. Whilst failing in their cardinal aim of halting troop withdrawal, Russia's allegations of mass human rights abuses in Estonia fuelled fears of possible ethnic unrest, briefly casting doubt over Estonian entry to Council of Europe and the signature of a trade treaty with the EC during 1992.

Unable to compete with the might of the Russian propaganda machine, Estonia responded by submitting controversial legislation to international expertise and inviting fact-finding delegations—including an OSCE Mission—to assess the human rights situation. None of these groups found evidence to support allegations of systematic ethnic discrimination. Estonia's success in shaping the international agenda over troop withdrawal therefore owed much to weaknesses and inconsistencies in Russia's own arguments. The crude and emotive nature of Russian attacks—which used terms like 'ethnic cleansing' and 'apartheid'—made it easy for Estonian foreign policy makers to portray their country as an outpost of western values threatened by a state whose foreign policy precepts were alien to the principles governing the 'New Europe'. Foreign policy statements around this time were thus replete with references to Russia's 'state-sanctioned concept of the Near Abroad'. The now president Lennart Meri condemned what he termed Russia's 'Monroe doctrine' and spoke of Estonia's eastern border as the border of western values.[60]

Estonia's subsequent entry to the Council of Europe in May 1993 could thus be reasonably portrayed as something of a diplomatic triumph over Russia. Yet, as indicted earlier in this work, the concessions over the nationalities issue which represented the price of CE membership led to discontent amongst radical nationalists. One ERSP minister, for instance, later spoke of his 'dual attitude' towards membership of international organisations. In view of Estonia's geopolitical situation, he argued, any international support was welcome. However, should such organisations 'ignore the fact of [Soviet] occupation' and attempt to treat post-war settlers as a conventional ethnic minority, this would amount to unwelcome interference in Estonia's internal affairs.[61]

Radicals were particularly incensed by the willingness of western leaders to bow to Russian demands concerning the status of Soviet military pensioners residing in Estonia. As the United States strove to secure the simultaneous withdrawal of Russian troops from all central and east European states, Russia made continued residence guarantees for ex-servicemen a central condition of any agreement in the case of Estonia and Latvia. Estonian concessions over the issue in November 1993 caused shock waves in domestic politics, but did not remove the sticking point of servicemen demobilised after 1991. In May 1994, Latvia assented to a US-brokered agreement providing social guarantees to military pensioners living in the Republic in return for a Russian withdrawal by 31 August. Perhaps more significantly from Moscow's point of view, it allowed the Russian military to rent the former Soviet early warning station at Skrunda until 1999.[62] Estonia, on the other hand, rejected any further concessions over the issue of pensioners. Nor would it accede to Russian demands for continued use of the submarine training facility at Paldiski.

Russia intensified the pressure by granting Most Favoured Nation trading status to Latvia, whilst simultaneously doubling duty on Estonian imports to a staggering 60%. During the same month, it began unilateral demarcation of the Estonian-Russian border along the lines agreed by the January 1991 treaty, thereby signaling in the clearest possible terms that it would not countenance discussion of the 1920 Tartu treaty as a basis for inter-state relations.

The resultant deadlock was broken through a combination of carrots and sticks from the United States. With Congress threatening to withhold $839 million in aid to Russia if it did not withdraw by the August deadline, President Clinton visited Riga, where he made a speech calling for tolerance towards Soviet-era settlers in the Baltic states.[63] As late as July, Yeltsin was still refusing to commit himself publicly to a withdrawal from Estonia by 31 August, yet an offer of a further U.S. aid to finance troop resettlement of troops apparently changed his mind.

After a flurry of diplomatic activity, Lennart Meri's visit to the Kremlin on 26 July proved decisive. Russia agreed to withdraw all of its forces, save a small detachment of experts who would remain until 31 August 1995 to dismantle the nuclear reactors at the Paldiski naval base. In return, the Russian embassy in Tallinn was to provide the Estonian authorities with a full list of the 10,000 military pensioners residing in Estonia. The latter were granted the right to obtain residence permits subject to the findings of a special government commission, which would consider each application on a case-by-case basis.

The subsequent departure of Russian troops on 31 August was hailed by many Estonians as the final end of World War Two. The low-key nature of the celebrations, however, reflected the deep political controversy surrounding the withdrawal agreement. The nature of the bargain struck by Meri presented the *Riigikogu* with a dilemma. If it ratified the agreement, it would be sanctioning an unconstitutional action by the president. If, on the other hand, it chose to withhold ratification, this might undermine Estonia's credibility in the negotiation of future international agreements. The protracted debate over the treaty did little to improve Estonian-Russian relations in the year which followed. The same was true of a statement by Russian foreign minister Andrei Kozyrev, who promised that the troop withdrawal would only lead his government to protest more vigorously against the alleged mistreatment of ethnic Russians in Estonia.

THE 'EUROPEANISATION' OF ESTONIAN FOREIGN POLICY

The final departure of Russian troops nevertheless marked an important turning point in the evolution of Estonian foreign policy. In an address to the *Riigikogu* on 12 February 1998, Foreign Minister Toomas Hendrik Ilves claimed that the three years after September 1994 were characterised by 'the consolidation of our independence ... and the establishing of a foreign policy that coincides with Estonia's interests in Europe and in the transatlantic space.'[64]

Achievement of these goals entailed a qualitative change in approach. Ole Waever argues that during the early years of independence, 'politicians of the Baltic states used ... moral arguments and presented themselves as abused victims of Soviet communism in dire need of western assistance and protection.' He goes on to state that by 1997 the central and east European nations had begun to 'market themselves as normal, democratic market economies, who can easily ... be fitted into the West European circle. Among the political elites in the Baltic states, there are signs of a similar cognizance.'[65]

In Estonia, this cognizance brought a striving for what Ilves terms the 'Europeanisation' of Estonian foreign policy, a term he defines as 'adherence to the behaviour standards of European foreign policy.'[66] This new approach has been most evident in the sphere of relations with Russia, where talk of 'positive engagement' has been transformed from mere rhetoric into concrete policy initiatives. The change in dominant discourses towards the 'eastern neighbour' was occasioned partly by the change of government in 1995 and the rise of voices citing the 'geoeconomic' necessity of trading links with the CIS. Most important, however, were the ever clearer signals from western organisations that Estonian membership would be contingent upon the settlement of outstanding disputes with Russia.[67] Building on the recent proposal for a 'European Stability Pact' unveiled by French premier Eduard Balladur, prime minister Andres Tarand announced in December 1994 that Estonia would formally relinquish its claim to the 'lost territories' in return for Russian recognition of the Tartu Treaty as the only valid basis for relations between the two countries.[68]

Russia, however, would only agree to recognise the 'historical significance' of the Tartu treaty, repeating the view that the latter became invalid after 1940. Given that Tiit Vähi had consistently criticised the unwillingness of *Isamaa* to engage with Russia, the start of Coalition Party-led government in April 1995 might have been expected to bring about a breakthrough in relations.[69] The new *Riigikogu*, however, proved no less nationalist than its predecessor when it came to the question of ratifying the agreements on troop withdrawal. On this point, the Coalition's Endel Lippmaa—appointed

Minister for European Affairs in August 1995—had been amongst the leading opponents of the agreements. Failure to ratify the withdrawal accords patently worked to Estonia's diplomatic disadvantage as the government launched the country's bid for full EU membership (24 November 1995). The unseemly dispute was finally settled in December 1995 when the *Riigikogu* ratified the withdrawal treaty and the agreement on social guarantees as two separate bills. Codicils were added to the effect that the 1940 annexation had been illegal and that residence guarantees only applied to those ex-servicemen already receiving a pension at the time of the agreement.

A few days before the accords were ratified, the EU Madrid summit had instructed the European Commission to report on the fitness of CEE applicant countries by mid-1997 with a view to starting full membership negotiations thereafter. In the light of the impending decision on enlargement, the need to reach agreement with Russia over the eastern border became all the more urgent. The biggest volte-face in Estonia's eastern policy came during a further round of border negotiations in November 1996. Meeting with his Russian counterpart Yevgenii Primakov, acting foreign minister Siim Kallas declared that Estonia would abandon its prior insistence on the Tartu treaty as the only valid basis for an agreement.

The speedy resolution of technical issues relating to physical demarcation of the border meant that a treaty lay ready for signing. Russia, however, has continued to withhold its consent, citing continued mistreatment of Estonia's Russian-speaking population. Hopes for a quick settlement were soon dispelled by Primakov, who in January 1997 reiterated claims of massive human rights abuses and threats of a total trade boycott against Estonia.[70] The timing of this renewed propaganda offensive was instructive, coming just ahead of the Council of Europe's decision to end its monitoring of Estonia. As noted in Chapter Three, Russia's internationalisation of the issue has led the European Union to insist on amendments to citizenship legislation as a precondition for Estonian membership. Thus far, the EU has stopped short of espousing Moscow's demand that citizenship be granted unconditionally to all Soviet-era settlers. In the light of recent experience, many Estonian political leaders feel that further concessions will inevitably be met by an intensification of Russian demands. According to this interpretation, consenting to the 'zero option' would simply lead to calls for a system of dual Estonian-Russian citizenship for Moscow's 'compatriots' in the Baltic.[71]

To date, however, Estonian concessions have gone a long way towards maintaining western goodwill and keeping the country's EU ambitions on track. On the decision to agree to a border treaty with Russia, Ilves has noted that this step was 'not merely necessary for the demarcation of Estonia, ... it was also the first step showing that Estonia is a responsible nation with a European-like behaviour. For different reasons there was hitherto a rather well-spread notion that Estonia is a nation that unreasonably avoids cooperation and is unable to react in a European manner.'[72] The recent policy of 'positive engagement' has indeed ensured that it is Russia, rather than Estonia, which appears the uncooperative partner in the relationship. To quote a further statement by Ilves, '1997 proved that Estonia can achieve its foreign policy goals regardless of whether it has or does not have a border agreement with the Russian Federation.'[73]

ESTONIA AND NATO

Whilst this last statement holds true in the case of Estonia's continued progress towards the European Union, it is less valid with regard to the goal of NATO membership. From the autumn of 1996, securing inclusion in the first round of EU expansion supplanted accession to NATO as the immediate priority of Estonian foreign policy.[74] This followed a statement by US defence secretary William Perry confirming that the three Baltic states would not be included in the first wave of NATO enlargement to be announced in 1997.[75]

In the light of Estonian experiences during the previous six years, Perry's statement can hardly have come as a great surprise. Following the country's admission the North Atlantic Cooperation Council in December 1991, many Estonian leaders entertained hopes of a rapid accession to the alliance. These were dampened in March 1992, when NATO Secretary General Manfred Wörner made it plain that full membership would not be on the agenda in the near future. Nor, he added, was NATO prepared to extend any guarantee of military support to non-member states.[76]

According to Perry, the Baltic states were ruled out of the first wave of NATO enlargement on the grounds that their armed forces were insufficiently prepared for membership. A further string of accidents and scandals within the Estonian military—including the tragic death of 14 soldiers on a training exercise in September 1997—lends credence to such claims.[77] However, the small size of the armed forces suggests that problems such as inadequate officer training could be quickly addressed with the active help of western partners. As such, this 'interoperability' argument is best regarded as a stalling tactic masking the more unpalatable reality. Namely, the fact that NATO expansion is contingent upon geopolitically-based argumentation. The limited enlargement announced in 1997, coming shortly after the signing of the NATO-Russia Founding Act, confirmed the lack of a political will to expand the alliance beyond the boundaries of the former USSR. The reluctance on the part of most member states to sanction such a move is a product of the alarm which the Baltic states' membership of NATO would cause in Russia. In the realm of security, Estonia has therefore found itself lumped into a single bloc with the other two Baltic states, although the absence of a large Russian-speaking population has perhaps placed Lithuania in a better position than its northern neighbours.[78]

In the absence of any clear prospects for full NATO membership, the Estonians have had to content themselves with the various other security initiatives on offer from the West. Estonia was amongst the first CEE states to sign up for NATO's Partnership for Peace initiative unveiled late in 1993, and has since taken an active role in the programme. Shortly afterwards, in May 1994, it obtained associate membership of WEU, the defence arm of the European Union. The common predicament of the three Baltic states has enhanced the importance of joint initiatives such as the Baltic peacekeeping battalion *Baltbat* and the joint naval squadron *Balttron*. The latter have been portrayed as a means of proving that the Balts are 'producers' rather than simply 'consumers' of security.

Baltbat—which has already seen service in Bosnia and the Lebanon—operates under the overall coordination of the Danish peacekeeping battalion. Denmark's support for Baltic membership of NATO, along with Finland's training of the Estonian military, underline the crucial importance of Baltic independence to the Nordic states. At the same time, Nordic governments have emphatically rejected suggestions that they should extend formal security guarantees to their eastern neighbours. The Estonian side has been equally emphatic in its rejection of any scheme which might be perceived as an alternative to full NATO membership. The primacy of this goal was again reiterated during the signing of the US-Baltic joint charter in January 1998, when Washington pledged its interest in the continued independence of the Baltic states without extending any automatic guarantees of their security.

Officially at least, the Estonians have expressed satisfaction with NATO's rhetoric of the 'open door' and the inadmissibility of creating new dividing lines in Europe. Whilst arguing forcefully but unsuccessfully for a second round of enlargement in 1999, Estonia has had little choice but to jump through the hoops which the alliance holds up. However, NATO's ambiguous stance towards the Baltic runs the risk of undermining the wider impetus towards westward integration. The insistence on increased defence spending, for instance, has served to increase internal political divisions whilst diverting resources away from EU integration. Russia, meanwhile, has sought to capitalise on these divisions by offering unilateral security guarantees to the three Baltic states.

THE EUROPEAN UNION AND THE BALTIC STATES

If the decision on NATO membership remains largely out of the Estonians' hands, the principles governing EU enlargement have offered them far more scope to determine their own fate. In 1995 the Union made it clear that applicant states would not be divided into groups according to their geopolitical location, but would be considered individually on the basis of their progress in achieving economic and political reform.[79] In its prior dealings with Estonia, Latvia and Lithuania, the EU had clearly treated the three countries as a single bloc, encouraging them to engage in projects of mutual cooperation as a necessary step towards the integration of the region into the EU.[80]

Such cooperation has been apparent in the establishment of an inter-parliamentary Baltic Assembly (1991) and the Baltic Council of Ministers (1994). Although closely modelled on Nordic regional institutions, these bodies have commanded a lesser degree of authority over individual governments. Progress was clearly discernible during 1994, when the three states reached agreement on free trade in industrial goods. Yet negotiations on agricultural free trade proved more complicated, and it was not until 1997 that the agreement was extended to cover this sphere. In the interim, Estonian-Latvian relations were marred by a dispute over the two countries' maritime border, in which the Latvian government threatened to send gunboats to protect what it claimed were its fishing grounds in the Gulf of Riga. This 'pilchard war' foreshadowed the more recent 'pork war' in which Estonia sought to impose quotas on cheap imports of Latvian pork. Such disputes are all too reminiscent of the nationalist squabbles which undermined projects for a Baltic League back in the 1920s.

Whereas NATO's stance on enlargement has tended to reinforce the solidarity of the three Baltic states, the EU policy of considering each applicant on its individual merits has worked in the opposite direction. In this regard, Estonia's faster progress in implementing economic reform increasingly served to differentiate it from its neighbours in the eyes of the West. When the three states signed free trade agreements in manufactured goods with the EU in July 1994, Estonia's liberal foreign trade regime meant that it was exempt from the transition period applied to Latvia and Lithuania. The enlargement policy announced in 1995 only increased the temptation to strike out alone. In a situation where Estonia lacked the necessary resources to harmonise its own legislation with EU whilst simultaneously ensuring coordination with its neighbours, it seemed entirely logical to favour European integration at the expense of intra-Baltic ties. Thus, shortly after the association agreement came into force in August 1995, former Prime Minister Mart Laar became the first of many prominent Estonians to suggest that Baltic co-operation constituted a drag on the country's development, and should be abandoned in favour of a more Nordic orientation.[81]

Officially at least, the three countries subsequently agreed to abide by the 'gentlemanly' formula 'all for one and one for all' in their common pursuit of western integration. Yet accusations of Estonian 'beggar-thy-neighbourliness' resurfaced following the 1997 enlargement decision, particularly when Lennart Meri hinted at the possible imposition of a visa requirement for Latvians and Lithuanians should Estonia enter the EU ahead of its neighbours.

In the wake of the EU Commission *avis* on the Baltic countries, Latvia and Lithuania understandably disputed the notion of a continued lag in economic development between themselves and their northern neighbour. This argument was at least vindicated when Latvia became the first Baltic state to join the World Trade Organisation in 1998, a development which shook the Estonians' belief in their own economic pre-eminence.[82]

Debates surrounding the relative merits of admitting the Baltic states singly or as a group are fast being overtaken by events, for recent evidence suggests that Latvia and Lithuania will be admitted to 'fast track' membership negotiations by the end of 1999. Depending on their progress in implementing the acquis, it is therefore entirely possible that these countries will enter the EU at the same time as Estonia. In this regard, the biggest challenge faced by *all* applicant countries will be to overcome the political obstacles to rapid eastward enlargement erected by existing member states. Whilst the Balts can count on solid backing from Nordic members of the EU, Gerhard Schröder's election as German Chancellor in 1998 led to predictions that the Social Democrat government would seek to divert resources away from EU enlargement in order to fund domestic social programmes. Germany and Austria have also sought to impose a transition period in order to prevent the free movement of labour from CEE countries following their accession. In addition, France has considerable vested interests with regard to the Common Agricultural Policy, whilst Spain, Greece, Portugal and Ireland fear that enlargement will dilute the considerable income transfers which they receive under existing EU budgetary arrangements.[83]

Treating the Baltic applicants as a bloc might slow down Estonia's accession, but it would at least serve to undermine Russia's traditional policy of divide and rule towards these countries. In security terms, full membership would have the virtue of making Russia's behaviour towards the Baltic states indistinguishable from its behaviour towards the EU as a whole, thereby ending Moscow's current pretensions to establish a sphere of influence over the area. The persistent use of the 'ethnic discrimination' card against Estonia and Latvia suggests that Russia's leaders understand the implications of EU enlargement only too well.

The prevailing western attitude to the national question in these countries can be summed up in the following quote by a western diplomat, who in 1997 claimed that: 'ethnic Russians, if they're not exactly being abused, certainly don't get fair treatment. If Russia decides to make an issue of it, that won't help stability in the region—to put it mildly'.[84] In point of fact, the status of the 'Russian-speaking population' became the central issue in Estonian-Russian relations as early as 1992. Moscow's persistent internationalisation of the question has in turn been an important factor behind western initiatives designed to push Estonia towards a more accommodating nationalities policy. Yet the successful long-term regulation of latent ethnic tensions will necessarily entail a delicate balancing act between the rights of the Russian minority and the interests of an Estonian majority still labouring under the burden of past injustices. Estonia's ability to contain ethnic unrest during the first eight years of independence constitutes a remarkable achievement, all the more so when set against the carnage which followed the break-up of Yugoslavia. In the light of this, Russia's accusations of 'ethnic cleansing' and 'apartheid' have not only been thoroughly unhelpful, but also have also lacked any basis in reality. The fact that Russian leaders refused to accept the use of the term 'ethnic cleansing' with regard to Serb actions in Bosnia demonstrates beyond all doubt that Moscow's rhetoric towards Estonia and Latvia is inspired by geopolitical objectives rather than any real concern for the fate of ethnic Russians living in these countries.[85]

Thus far, western states have tended to regard the presence of a large non-citizen population as an understandable state of affairs to be rectified in the due course of time. However, one could easily foresee a situation whereby this issue is used as a pretext for slowing down EU enlargement towards the Baltic states. If however, in the current climate, the EU were to espouse Russia's demands for 'zero option' citizenship in Estonia, this might conceivably undermine rather than enhance ethno-political stability. In the immediate term, such a step would almost certainly lead to a significant political backlash against EU membership, which is probably exactly what Russia hopes to achieve.

The dilemmas posed by Estonia's quest for membership of NATO and the EU thus bring into focus much bigger questions concerning East-West relations and Russia's relationship to the 'Euro-Atlantic space' following the end of the Cold War and the collapse of Communism. In turn, these questions feed into perennial debates over the nature of Russian identity. Namely, is Russia part of 'Europe', or should it be classed as a distinct 'Eurasian' state? If Russia does indeed fall outside, then to what extent should it be allowed to dictate to neighbouring peoples who feel themselves to be 'European'? Back in the 1920s, Count Coudenhove-Kalergi asserted that 'the Russian question stands at the very centre of the question of Europe'.[86] This assertion remains equally valid in the late 1990s, when, to quote Toomas Hendrik Ilves, Russia has become the 'keystone of European security and stability'.[87] In so far as Estonia's fate remains inextricably bound up with these wider issues, it will continue to constitute the 'litmus test' not only of Russia's attitudes towards Europe, but of the 'New Europe' as a whole.

1   Statement by Toomas Hendrik Ilves, Minister for Foreign Affairs of Estonia, at the opening of Estonia's negotiations with the European Union, Brussels, 31 March 1998, *http://www.vm.ee/eng/index.html*.

2   Ibid; Lennart Meri, "Iseseisvus Laiemas Kontekstis," speech at the meeting of the CSCE, Copenhagen, 5 June 1990. In Lennart Meri, *Presidendi Kõned* (Tartu, 1996) p.227.

3 T.H. Ilves, "Estonia on its Way to an Integrated Europe," speech at the University of Latvia, 10 March 1998, *http://www.vm.ee/eng/index.html*.

4 Andris Ozolins, "The Policies of the Baltic Countries vis-à-vis the CSCE, NATO and WEU," in *The Foreign Policies of the Baltic Countries: Basic Issues*, eds. Pertti Joenniemi and Juris Prikulis (Riga, 1994), p.62.

5 J. Luik, "Security in the Baltic Sea Region—A Guarantee of Stability in Europe," address at the Swedish Institute of International Affairs, Stockholm, 2 March 1994, *http://www.vm.ee/eng/index.html*.

6 Ozolins, p.50.

7 Lennart Meri, "Euroopa on Programm," speech at the conference "The Baltic States in the Architecture of the New Europe", Tallinn, 2 March 1991, in Meri, 1996, p.255; Lennart Meri, "Euroopa kui Eesmärk," speech at the Council of Europe, Strasbourg, 26 November 1991, in Meri, 1996, p.268.

8 *Baltic Independent*, 28 October–3 November 1994.

9 Address by Jüri Luik to the United Nations, 28 September 1994, *http://www.vm.ee/eng/index.html*.

10 Address by T.H. Ilves in the name of the Government to the *Riigikogu*, 12 February 1998, *http://www.vm.ee/eng/index.html*.

11 Mikhail A. Alexseev and Vladimir Vagin, "Russian Regions in Expanding Europe: The Pskov Connection," *Europe-Asia Studies*, Vol.51, No.1, 1999, p.54.

12 Address by Toomas Hendrik Ilves in the name of the government to the *Riigikogu*, 5 December 1996, *http://www.vm.ee/eng/index.html*.

13 Remarks by Siim Kallas at the 97th Session of the Committee of Ministers, Council of Europe, Strasbourg, 9 November 1995, *http://www.vm.ee/eng/index.html*.

14 On the Baltic League, see Marko Lehti, *A Baltic League as a Construct of the New Europe* (Frankfurt, 1999).

15 Marko Lehti, "The Baltic League and the Idea of Limited Sovereignty," *Jahrbucher fur Geschichte Europas*, No.45, 1997, pp.450–465.

16 On the 1940 proposal see Jaakko Korjus, *Viron Kunniaksi* (Hämeenlinna, 1998), pp.62–64; On the earlier period, see Lehti, 1999, pp.103–180.

17 Lehti, 1999, p.11.

18 Vahur Made, "Eesti ja Rahvasteliit 1918–1925," in *Kaks Algust: Eesti Vabariik 1920. ja 1990. Aastad*, ed Jüri Ant (Tallinn, 1998), pp.46–72.

19 Lehti, 1999, pp.459–491.

20 Eero Medijainen, *Maailm Prowintsionu Peeglis* (Tartu, 1998), pp.61–64; Georg Von Rauch, *The Baltic States: The Years of Independence 1917–1940*, (London, 1995), pp.171–175.

21 Von Rauch, p.175.

22 Medijainen, p.135; Von Rauch, pp.200–203; John Hiden and Patrick Salmon, *The Baltic Nations and Europe* (London, 1992), pp.102–103; Toivo Raun, *Estonia and the Estonians* (Stanford, 1991), p.142.

23 On the dissident movement, see: Andres Küng, *A Dream of Freedom* (Cardiff, 1991); Mare Kukk, "Political Opposition in Soviet Estonia 1940–1987," *Journal of Baltic Studies*, Vol.24, No.4, Winter 1993, pp.369–384.

24 Juris Prikulis, "The European Policies of the Baltic Countries," in Joenniemi and Prikulis eds., p.92.

25 Made, loc cit.

26 L. Meri, "*Mis Tahes Poliitika on Alati Dialoog*", interview to *Päevaleht*, 19 April 1990, in Meri 1996, p.219.

27 Stephen White, *Gorbachev and After* (New York, 1992), p.161.

28 David Smith, "Bridging East and West? The Baltic Sea Region as a Coordinate in the Construction of Post-Soviet Identities", paper presented at the seminar *Linking the European East and West: The Baltic Sea Area within the Framework of European Integration*, University of Turku, Finland, 3 June 1999.

29 Meri, 19 April 1990, p.222; Meri, 5 June 1990, pp.227–228.

30 Rein Ruutsoo, "The Perception of Historical Identity and the Restoration of Estonian National Independence," *Nationalities Papers*, Vol.23, No.1, 1995, p.177.

31 Meri, 19 April 1990, p.219.

32 Endel Lippmaa, "How to Regain Estonia's Statehood," *Homeland*, No.6–7, 21 February 1990, p.3.

33 Lennart Meri, "Ei Saa Olla Vaba Euroopat Vaba Baltikumilta," speech on Norrmalmstorg, Stockholm, 18 June 1990, in Meri, 1996, pp.232–234; Lennart Meri, 'Eesti Roll Uues Euroopas', speech at the Royal Institute of International Affairs, London, 6 November 1990, in Meri, 1996, pp.239–242.

34 Ozolins, p.55; Meri, 5 June 1990, pp.228–229.

35 Ozolins, pp.53–54.

36 Lena Jonson, "Russia in the Nordic Region in a Period of Change," in *The Baltic Sea Area: A Region in the Making*, eds. Mare Kukk, Sverre Jervell and Pertti Joenniemi (Oslo/Stockholm, 1992), pp.87–89; John Fitzmaurice, *The Baltic: A Regional Future?* (Basingstoke, 1992), p.150.

37 Endel Lippmaa, "How to Regain Estonia's Statehood," *Homeland*, no.6–7 (267–268), 21 February 1990, p.3; H. Tiido, "Estonia and Eastern Europe," *Homeland*, no.6–7 (267–268), 21 February 1990, p.2.

38 Pertti Joenniemi, "The Baltic Countries as Deviant Cases; Small States in Search of Foreign Policies", in *New Actors on the International Arena: The Foreign Policies of the Baltic Countries*, eds. Pertti Joenniemi and Peeter Vares (Tampere, 1993), p.188.

39 N. Dormas, "'Russkii Vopros' po Estonski," *Den' za Dnem*, 17 February 1995.

40 Jonson, pp.176–177; Neil Melvin, *Russians Beyond Russia: The Politics of National Identity* (London, 1995), p.10.

41 Roman Szporluk, "Dilemmas of Russian Nationalism," *Problems of Communism*, July/August 1989, pp.15–35. Cited in Ian Bremmer, "Reassessing Soviet Nationalities Theory," in *Nations and Politics in the Soviet Successor States*, eds. Ian Bremmer and Ray Taras (Cambridge, 1993), p.18; See also Jonson, p.175.

42 Karen Dawisha and Bruce Parrott, *Russia and the New States of Eurasia: The Politics of Upheaval* (New York, 1994), p.199; Author Interview with Consul Vasilii Svirin, Head of the Russian Federation Delegation to Inter-State Negotiations with Estonia (Moscow, 21 May 1995).

43 Melvin, p.11.

44 Svirin, loc cit.

45 Ibid.

46 Edgar Mattisen, *Estoniia-Rossiia: Istoriia Granitsy i ee Problemy* (Tallinn, 1995), p.188.

47 Jonson, p.178; Melvin, p.12; Dawisha and Parrott, p.200.

48 Jonson, p.176.

49 O. Zhuryari-Ossipova, "Russian Factor in Estonian Foreign Policy: Reaction to the Limitation of Sovereignty," in Joenniemi and Vares eds., pp.123–4.

50 *Eesti Ringvaade*, 29 April 1992 and 6 May 1992.

51 Anatol Lieven, *The Baltic Revolution: Estonia, Latvia and Lithuania and the Path to Independence* (London, 1993), p.288.

52 *Eesti Ringvaade*, 9 May and 25 May 1992.

53 Riin Kionka, "Drafting New Constitutions: Estonia," *RFE/RL Research Report*, Vol.1, No.27, 3 July 1992, p.60.

54 Ozolins, pp.57–58.

55 *Eesti Ringvaade*, 29 June–5 July 1992.

56 See especially Trivimi Velliste, "The 'Near Abroad' in the Baltic Republics: the View from Estonia," in *Nordic-Baltic Security: An International Perspective*, eds. A.O. Brundtland and D.M. Snider (Washington, CSIS, 1994), pp.57–62. Velliste (*ERSP*) was Foreign Minister from October 1992 to January 1994. Karaganov's article was published in *Diplomaticheskii Vestnik* in September 1992.

57 Neil Melvin, "The Baltic Russians in Latvia and Estonia: From Germans to Austrians?" unpublished draft paper presented to study group at Chatham House, 31 January 1994 as part of Royal Institute of International Affairs Russian and CIS Programme.

58 Rita P. Peters, "Small State Diplomacy in International Organizations: The Baltic States and the United Nations," unpublished paper presented at the First Conference on Baltic Studies in Europe, University of Latvia, Riga, 17 June 1995.

59 Joenniemi, p.189.

60 *The Baltic Independent*, 14–20 May 1993; Lennart Meri, *"Meie Piir on Euroopa Väärtuste Piir,"* speech at the "Estonia" Concert Hall, Tallinn on 75th anniversary of Estonian independence, 24 February 1993, in Meri, pp.325–331.

61 Author interview with Peeter Olesk (*ERSP*), Minister of Citizenship and Migration of Estonia October 1993–June 1994, Tallinn, 17 June 1994.

62 *Baltic Independent*, 6–12 May 1994.

63 William Safire, "Bill 'n' Boris Fix it in the Baltics," *The Guardian*, 31 August 1994.

64 Ilves, 12 February 1998, loc cit.

65 Ole Waever, "The Baltic Sea: A Region after Post-Modernity?" in *Neo Nationalism or Regionality? The Restructuring of Political Space around the Baltic Rim*, ed. Pertti Joenniemi (Stockholm, 1997), p.16.

66 Toomas Hendrik Ilves, "Eesti Poliitika Euroopastumine," *Luup*, No.6, 17 March 1997.

67 *Baltic Independent*, 3–6 June 1994.

68 "Viro ei enää Vaadi Petserinmaata Venäjältä," *Turun Sanomat*, 3 February 1995.

69 Jorma Rotko, "Viron Vahi Moitti Edeltajiensa Idanpolitikkaa," *Helsingin Sanomat*, 22 April 1995.

70 Kari Koponen, "Venäjä Väläytti Viron Vastaisten Pakotteiden Mahdollisuutta," *Helsingin Sanomat*, 10 January 1997.

71 Author interview with Priit Järve, Chairman of Estonia's Presidential Round Table of Nationalities 1996–1998, Bradford, 19 February 1997.

72 Ilves, 17 March 1997, loc cit.

73 Ilves, 12 February 1998, loc cit.

74 Ibid.

75 *Baltic News Service*, 11 October and 9 November 1996.

76 Ozolins, p.62.

77 On problems in the military, see: *Economist Intelligence Unit Country Report*, 1st Quarter 1997, p.13; *The Baltic Times*, 25 September–1 October 1997, 2–8 October 1997 & May 27–2 June 1999.

78 Ron Asmus and Bob Nurick, *NATO Enlargement: An Alliance Strategy for the Baltic States*, Report of the RAND Corporation, April 1996, pp.2–3.

79 "Baltic States to Sign Europe Agreements," *The Baltic Review*, Spring/Summer 1995, p.5; Atis Leijins, "The Quest for Baltic Unity: Chimera or Reality?" in *Small States in a Turbulent Environment: The Baltic Perspective*, eds. Atis Leijins and Zaneta Ozolina (Riga, 1997), p.173.

80 Ozolins, pp.106–107.

81 *The Baltic Independent*, 8–14 September 1995.

82 Aap Neljas, "Eesti Konarlik Tee WTOsse," *Postimees*, 21 October 1998.

83 *The Baltic Times*, 8–14 October 1998 and 12–18 November 1998.

84 Jon Henley, "Estonia freezes out its Russians," *The Guardian*, 11 January 1997.

85 Peters, loc cit.

86 Mikhail Kurchinskii, *Soedinennie Shtati Evropy. Ekonomicheskie i Politicheskie Perspektivy Etoi Idei'*, (Tartu, 1930), p.27.

87 Statement by Toomas H. Ilves at the meeting of the foreign ministers of the Baltic states and the USA in Vilnius, 13 July 1997, to discuss NATO enlargement, the U.S—Baltic Charter and regional cooperation, *http://www.vm.ee/eng/index.html*; Toomas H. Ilves, "Implications of EU and NATO Enlargement Policies for the Baltic States," remarks at the Conference on the Northern Dimension of the CFSP, Helsinki, 8 November 1997, *http://www.vm.ee/eng/index.html*.

BIBLIOGRAPHY

Abrahams, Ray. "Continuity and Change in the Estonian Agrarian Sector." Paper presented at the conference *Estonia: A Development Agenda*, Royal Institute of Geography, London, 26 March 1996.

Abrahams, Ray and Juhan Kukk. *Barons and Farmers: Continuity and Transformation in Rural Estonia (1816–1994)*. Göteborg, Europaprogrammet, 1994.

Alexseev, Mikhail A and Vladimir Vagin. "Russian Regions in Expanding Europe: The Pskov Connection," *Europe-Asia Studies*, Vol.51, No.1, 1999.

Andersen, Erik Andre. "The Legal Status of Russians in Estonian Privatisation Legislation," *Europe-Asia Studies*, Vol.49, No.2, 1997.

Andrejev, Viktor. "*Kommentaarid.*" In *Kaks Otsustavat Päeva Toompeal (19–20 August 1991)*, edited by 20 August Klubi ja Riigikogu Kantselei. Tallinn, Eesti Entsüklopeediakirjastus, 1996.

Arter, David. *Parties and Democracy in the Post-Soviet Republics. The Case of Estonia*. Aldershot, Dartmouth, 1996.

Asmus, Ron and Bob Nurick. *NATO Enlargement: An Alliance Strategy for the Baltic States*, RAND Corporation Report, April 1996.

"Baltic States to Sign Europe Agreements," *The Baltic Review*, Spring/Summer 1995.

Bank of Estonia. "Indicators of the Estonian Economy 1993–1998," at *http://www.ee/epbe/datasheet/macroeconomics/table12b.html*.

Bradshaw, Michael, Philip Hanson and Denis Shaw. "Economic Restructuring." In *The Baltic States: The Self-Determination of Estonia, Latvia and Lithuania*, edited by Graham Smith. Basingstoke, Macmillan, 1994.

Bremmer, Ian. "Reassessing Soviet Nationalities Theory." In *Nations and Politics in the Soviet Successor States*, edited by Ian Bremmer and Ray Taras. Cambridge, Cambridge University Press, 1993.

Brogan, Patrick. *Eastern Europe 1939–1989: The Fifty Years War*. London, Bloomsbury, 1990.

Brubaker, Rogers. *Nationalism Reframed. Nationhood and the National Question in the New Europe*. Cambridge, Cambridge University Press, 1996.

Camilleri, Joseph A. and Jim Falk. *The End of Sovereignty? The Politics of a Shrinking and Fragmenting World*. Aldershot, Edward Elgar, 1992.

Christiansen, Eric. *The Northern Crusades*. London, Penguin, 1997.

Clemens, Walter C. *Baltic Independence and Russian Empire*. London, Macmillan, 1991.

"Coalition Party in Competing Visions of an Estonian Future," *Nationalities Papers*, Vol.23, No.1, 1995.

Commission of the European Union, *Agenda 2000. Commission Opinion on Estonia's Application for Membership of the European Union*, http://europa.eu.int/comm/dg1a/agenda2000/en/opinions/estonia/.

Commission for Security and Cooperation in Europe. *Russians in Estonia: Problems and Prospects*. Washington, CSCE, 1992.

Davies, Norman. *Europe: A History*. London, Pimlico, 1997.

Dawisha, Karen and Bruce Parrott. *Russia and the New States of Eurasia: The Politics of Upheaval*. New York, Cambridge University Press, 1994.

Dormas, N. "'Russkii Vopros' po Estonski," *Den' za Dnem*, 17 February 1995.

Eesti NSV Teaduste Akademia. *Eesti NSV Ajalugu III*. Tallinn, Eesti Raamat, 1971.

"'Estma': Po Chasti Transita Estoniia Sposobna Konkurirovat' c Finliandiey," *Russkii Ekspress*, No.20, 1995.

Estonian Institute. "The Estonian Parliament—The *'Riigikogu'*." *Estonia in Facts*. Tallinn, Estonian Institute, 1993.

Estonian Ministry of Culture and Education. "Report on Estonian-language Learning Initiatives." Tallinn, 11 March 1994.

Fitzmaurice, John. *The Baltic: A Regional Future?* Basingstoke, Macmillan, 1992.

Frydman, Roman, Andrzej Rapaczynski, John S. Earle et al. *The Privatisation Process in Russia, Ukraine and the Baltic States*. Budapest, Central European University Press, 1993.

Gerner, Kristian and Stefan Hedlund. *The Baltic States and the End of the Soviet Empire*. London, Routledge, 1993.

Government of the Republic of Estonia. "Roadmap to Reform: Estonia's Future Plans in the Field of European Integration." Tallinn, September 1997.

Graf, Mati. *Eesti Rahvusriik*. Tallinn, Tallinna Raamatutrükikoda, 1993.

Hadenius, Stig. *La Politique de La Suède au XXe Siècle*. Stockholm, Institut Suedois, 1989.

Hansen, John and Piritta Sorsa. "Estonia: A Shining Star from the Baltics." In *Trade in the New Independent States*, edited by Constantine Michalopoulos and David Tarr. Washington, World Bank/UNDP, 1994.

Hanson, Philip. "Estonia: Radical Economic Reform and the Russian Enclaves." Unpublished draft chapter, 5 February 1994.

Hansson, Ardo H. "Transforming an Economy while Building a Nation: The Case of Estonia." Revised version of a paper presented at the 24th National Convention of the American Association for the Advancement of Slavic Studies, Phoenix, 19–22 November 1992.

Hansson, Ardo. "Reforming the Banking System in Estonia." In *Banking Reform in Central Europe and the Former Soviet Union*, edited by Jacek Rostowski. Budapest, Central European University Press, 1995.

Hansson, L. "Attitudes Towards Unemployment and Self-Employment: Reality and New Opportunities," in *Social Strata and Occupational Groups in the Baltic States*, eds. Raimo Blom, Harri Melin and Jouko Nikula. Tampere, 1995.

Henley, Jon. "Estonia freezes out its Russians," *The Guardian*, 11 January 1997.

Hiden, John. "The Baltic Republics." In *The Annual Register: A Record of World Events 1993*, edited by Alan Day. Harlow, Longman, 1994.

Hiden, John. "The Baltic Republics." In *The Annual Register: A Record of World Events 1995*, edited by Alan J. Day. London, Catermill International, 1996.

Hiden, John and Patrick Salmon. *The Baltic Nations and Europe*. London, Longman, 1992.

Hope, Nicholas. "Interwar Statehood: Symbol and Reality." In *The Baltic States: The Self-Determination of Estonia, Latvia and Lithuania*, edited by Graham Smith. Basingstoke, Macmillan, 1994.

Ilves, Toomas H. "Estonian Supreme Soviet Declares Continued *de jure* Existence of Estonian Republic as an Occupied State," *Radio Liberty Report on the USSR*, Vol.2, No.15, 13 April 1990.

Ilves, Toomas H. "Reaction: The Intermovement in Estonia." In *Towards Independence: The Baltic Independence Movements*, edited by Jan Arveds Trapans. Boulder, Westview Press, 1991.

Ilves, Toomas H. Address made to the *Riigikogu* in the name of the government, 5 December 1996. *http://www.vm.ee/eng/index.html/*.

Ilves, Toomas H. "Eesti Poliitika Euroopastumine," *Luup*, No.6, 17 March 1997.

Ilves, Toomas H. Statement at the meeting of the foreign ministers of the Baltic states and the USA in Vilnius, 13 July 1997, to discuss NATO enlargement, the U.S—Baltic Charter and regional cooperation. *http://www.vm.ee/eng/index.html*.

Ilves, Toomas H., "Implications of EU and NATO Enlargement Policies for the Baltic states," speech given at the Conference on the Northern Dimension of the CFSP, Helsinki, 8 November 1997. *http://www.vm.ee/eng/index.html*

Ilves, Toomas H. Address made to the *Riigikogu* in the name of the government, 12 February 1998, *http://www.vm.ee/eng/index.html*.

Ilves, Toomas H. Statement at the opening of Estonia's negotiations with the European Union, Brussels, 31 March 1998. *http://www.vm.ee/eng/index.html*

Ilves, Toomas H. "Estonia on its Way to an Integrated Europe," speech given at the University of Latvia, 10 March 1998. *http://www.vm.ee/eng/index.html*.

Iur'evskoe Russkoe Sobranie. "Materialy, Kasayushchiesia Kul'turnoi Avtonomii Russkogo Men'shinstva." Tartu, 1936. Eesti Ajalooarhiiv f.2097, n.1, su.16.

Jekabson, John. "Economic Independence is Not Enough for Lithuania, Estonia and Latvia." In *Without Force or Lies: Voices from the Revolution of Central Europe in 1989–90*, edited by William M. Brinton and Alan Rinzler. San Fransisco, Mercury House, 1990.

Joenniemi, Pertti. "The Baltic Countries as Deviant Cases; Small States in Search of Foreign Policies." In *New Actors on the International Arena: The Foreign Policies of the Baltic Countries*, edited by Pertti Joenniemi and Peeter Vares. Tampere, TAPRI, 1993.

Jonson, Lena. "Russia in the Nordic Region in a Period of Change." In *The Baltic Sea Area: A Region in the Making*, edited by Mare Kukk, Sverre Jervell and Pertti Joenniemi. Oslo/Stockholm, Europaprogrammet, 1992.

Kaldre, Peeter. "Savisaare Sinusoid," *Luup*, No.24, 25 November 1996.

Kallas, Siim. Remarks at the 97th Session of the Committee of Ministers, Council of Europe, Strasbourg, 9 November 1995. *http://www.vm.ee/eng/index.html*.

Kallas, Siim. "1992. Aasta Rahareform ja Selle Mõju Eesti Arengule." In *Kaks Algust: Eesti Vabariik –1920. ja 1990. Aastad*, edited by Jüri Ant. Tallinn, Eesti Riigiarhiiv, 1998.

Kalmre, Vahur, ed. *Postimehe Valimisteatmik*. Tartu, Postimehe Kirjastus, 1995.

Kaplan, Cynthia. "Estonia: A Plural Society on the Road to Independence." In *Nations and Politics in the Soviet Successor States*, edited by Ian Bremmer and Ray Taras. Cambridge, Cambridge University Press, 1993.

Karklins, Rasma. *Ethnic Relations in the USSR*. Boston, Allen and Unwin 1986.

Kasekamp, Andres. "The Estonian Veterans' League: A Fascist Movement?" *Journal of Baltic Studies*, Vol.24, 1993.

Kasekamp, Andres. "The Nature of Authoritarianism in Interwar Estonia," *International Politics* (formerly *Coexistence*), No.33, March 1996.

Kasekamp, Andres. "Radical Right-Wing Movements in the North-East Baltic," *Journal of Contemporary History*, Vol.34, No.4, 1999.

Kaskla, Edgar. "Five Nationalisms: Estonian Nationalism in Comparative Perspective," *Journal of Baltic Studies*, Vol.23, No.2, Summer 1992.

Kiin, Sirje, Rein Ruutsoo and Andres Tarand. *40 Kirja Lugu*. Tallinn, Orion, 1990.

"'Kindel Kodu' Tsentrist Veidi Paremal," *Rahva Hääl*, 12 August 1992.

Kionka, Riina. "Three Paths to Estonia's Future," *Radio Liberty Report on the USSR*, Vol.2, No.8, 23 February 1990, pp.32–33.

Kionka, Riina. "Elections to the Estonian Supreme Council," *Radio Liberty Report on the USSR*, Vol.2, No.14, 6 April 1990.

Kionka, Riina. "*Integral*' and Estonian Independence," *Radio Free Europe Report on the USSR*, Vol.2, No.30, 27 July 1990.

Kionka, Riina. "The Estonians." In *The Nationalities Question in the Soviet Union*, edited by Graham Smith. London, Longman, 1991.

Kionka, Riina. "How will Estonia Cope after the Union Treaty?" *RFE/RL Report on the USSR*, Vol.3, No.30, 26 July 1991.

Kionka, Riina. "Who Should Become a Citizen of Estonia?" *RFE/RL Research Report on the USSR*, Vol.3, No.39, 27 September 1991.

Kionka, Riina. "Debate about New Constitution Sparks Old Rivalries in Estonia," *RFE/RL Report on the USSR*, Vol.3, No.50, 13 December 1991.

Kionka, Riina. "A Break with the Past," *RFE/RL Research Report*, Vol.1, No.1, 3 January 1992.

Kionka, Riina. "Plea for Special Powers Topples Estonian Government," *RFE/RL Research Report*, Vol.1, No.7, 14 February 1992.

Kionka, Riina. "Estonian Political Struggle Centers on Voting Rights," *RFE/RL Research Report*, Vol.1, No.24, 12 June 1992.

Kionka, Riina. "Drafting New Constitutions: Estonia," *RFE/RL Research Report*, Vol.1, No.27, 3 July 1992.

Kionka, Riina. "Estonia: A Difficult Transition," *RFE/RL Research Report*, Vol.2, No.1, 1 January 1993.

Kionka, Riina and Raivo Vetik. "Estonia and the Estonians." In *The Nationalities Question in the Post-Soviet States*, edited by Graham Smith. London, Longman, 1996.

Kirby, David. *Finland in the Twentieth Century*. London, Hurst and Company, 1979.

Kirby, David. *The Baltic World 1772–1993: Europe's Northern Periphery in an Age of Change*. London, Longman, 1995.

Kirch, Aksel, Marika Kirch and Tarmo Tuisk, "Russians in the Baltic States: To Be or Not to Be?" *Journal of Baltic Studies*, Vol.24, No.1, Summer 1993.

Kõll, Anu Mai. *Peasants on the World Market: Agricultural Experience of Independent Estonia 1919–1939*. Acta Universitatis Stockholmiensis, Studia Baltica Stockholmiensa No.14, Stockholm, University of Stockholm 1994.

Koponen, Kari. "Venäjä Väläytti Viron Vastaisten Pakotteiden Mahdollisuutta," *Helsingin Sanomat*, 10 January 1997.

Korjus, Jaakko. *Viron Kunniaksi*. Hämeenlinna, Karisto, 1998.

Kramer, John M. "'Energy Shock' from Russia Jolts Baltic States," *RFE/RL Research Report*, Vol.2, No.17, 23 April 1993.

Kukk, Mare. "Political Opposition in Soviet Estonia 1940–1987," *Journal of Baltic Studies*, Vol.24, No.4, Winter 1993.

Kruus, Toivo. "Estonia and EMU Prospect." *Review of Economies in Transition*, No.7. Helsinki, Bank of Finland Unit for Eastern European Economies, November 1997.

Kurchinskii, Mikhail. *Soedinennie Shtati Evropy. Ekonomicheskie i Politicheskie Perspektivy Etoi Idei'*. Tartu, University of Tartu, 1930.

Küng, Andres. *A Dream of Freedom*. Cardiff, Boreas, 1991.

Laar, Mart. *War in the Woods: Estonia's Struggle for Survival 1944–1956*. Washington, Compass Press, 1993.

Laar, Mart. "Estonia's Success Story," *Journal of Democracy*, Vol.7, No.1, 1 January 1996.

Laar, Mart. "Kommentarid." In *Kaks Otsustavat Päeva Toompeal (19–20 August 1991)*, edited by 20 August Klubi ja Riigikogu Kantselei. Tallinn, Eesti Entsüklopeediakirjastus, 1996.

Laar, Mart. "Eesti Vabariigi Sisepoliitika ja Riigikorralduse Probleemid," in *Kaks Algust: Eesti Vabariik 1920. ja 1990. Aastad*, edited by Jüri Ant. Tallinn, Eesti Riigiarhiiv, 1998.

Laar, Mart, Urmas Ots and Sirje Endre. *Teine Eesti*. Tallinn, SE & JS, 1996.

Lainela, S. and P. Sutela, "Introducing New Currencies in the Baltic Countries." In *The Transition to a Market Economy. Transformation and Reform in the Baltic States*, edited by Tarmo Haavisto. Cheltenham, Edward Elgar, 1997.

Laitinen, Kai. "Kirjallisuus Vuoteen 1940." In *Viro: Historia, Kansa, Kulttuuri*, edited by Seppo Zetterberg. Helsinki, SKS, 1995.

Lapidus, Gail. "From Democratisation to Disintegration: The Impact of Perestroika on the National Question." In *From Union to Commonwealth: Nationalism and Separatism in the Soviet Republics*, edited by Gail Lapidus and Viktor Zaslavsky. Cambridge, Cambridge University Press, 1992.

Lauristin, Marju. "Estonia: A Popular Front Looks to the West," in *Towards Independence: The Baltic Popular Movements*, edited by Jan Arveds Trapans. Boulder, Westview Press, 1991.

Lauristin, Marju. "Kommentaarid." In *Kaks Otsustavat Päeva Toompeal (19–20 August 1991)*, edited by 20 August Klubi ja Riigikogu Kantselei. Tallinn, Eesti Entsüklopeediakirjastus, 1996.

Lauristin, Marju and Peeter Vihalemm. "Recent Historical Developments in Estonia: Three Stages of Transition (1987–1997)." In *Return to the Western World: Cultural and Political Perspectives on the Estonian Post-Communist Transition*, edited by Marju Lauristin and Peeter Vihalemm with Karl Erik Rosengren and Lennart Weibull (Tartu, Tartu University Press, 1997).

"Law on Citizenship. Adopted by the Riigikogu on 19 January 1995," unofficial translation, 24 January 1995.

Lehti, Marko. "The Baltic League and the Idea of Limited Sovereignty," *Jahrbucher für Geschichte Europas*, No.45, 1997.

Lehti, Marko. *A Baltic League as a Construct of the New Europe*. Frankfurt, Peter Lang, 1999.

Leijins, Atis. "The Quest for Baltic Unity: Chimera or Reality?" In *Small States in a Turbulent Environment: The Baltic Perspective*, edited by Atis Leijins and Zaneta Ozolina. Riga, Latvian Institute of International Affairs, 1997.

Levin, Dov. "Estonia." In *Encyclopedia of the Holocaust, Vol.2*, edited by Israel Gutman. New York, Macmillan, 1990.

Lieven, Anatol. *The Baltic Revolution: Estonia, Latvia and Lithuania and the Path to Independence*. London, Yale University Press, 1993.

Lippmaa, Endel. "How to Regain Estonia's Statehood," *Homeland*, No.6–7, 21 February 1990.

Löfgren, Joan, Helena Mannonen, Olli-Pekka Jalonen and Jouko Huru. "The Integration of the Baltic States into the EU," project report for the Finnish Ministry of Foreign Affairs. Tampere, TAPRI, 1997.

Luik, Jüri. "Security in the Baltic Sea Region—A Guarantee of Stability in Europe," speech at the Swedish Institute of International Affairs, Stockholm, 2 March 1994. *http://www.vm.ee/eng/index.html*.

Luik, Jüri. Address to the United Nations, 28 September 1994. *http://www.vm.ee/eng/index.html*.

Macartney, C.A. *National States and National Minorities*. New York, Russell and Russell, 1968.

Made, Tiit. *Eesti Tee*. Stockholm, Kirjastus Välis-Eesti & EMP, 1989.

Made, Vahur. "Eesti ja Rahvasteliit 1918–1925." In *Kaks Algust: Eesti Vabariik 1920. ja 1990. Aastad*, edited by Jüri Ant. Tallinn, Eesti Riigiarhiiv, 1998.

Made, Vahur. "Estonia and International Organisations," unpublished paper presented at the seminar "Linking the European East and West: The Baltic Sea Area within the Framework of European Integration," University of Turku, Finland, 3 June 1999.

Malmberg, Kristi and Rein Sikk. "Muulased Tahavad Lastele Eesti Kodakond-sust," *Eesti Päevaleht*, 4 April 1997.

Mart Laari Kõne Riigikogus 19. Oktoobril 1992, *Rahva Hääl*, 20 October 1992.

Mattisen, Edgar. *Estoniia-Rossiia: Istoriia Granitsy i ee Problemy.* Tallinn, Ilo, 1995.

McGarry, John and Brendan O'Leary. "Introduction: The Macro-Political Regulation of Ethnic Conflict." In *The Politics of Ethnic Conflict Regulation*, edited by John McGarry and Brendan O'Leary. London, Routledge, 1993.

Medijainen, Eero. *Maailm Prowintsionu Peeglis.* Tartu, Kleio, 1998.

Melin, Harri. "Ethnicity and Social Class in the Baltic Countries," working paper for the project "Social Change in the Baltic and Nordic Countries," University of Tampere Department of Sociology and Social Psychology, 1995.

Melvin, Neil. "The Baltic Russians in Latvia and Estonia: From Germans to Austrians?" Unpublished draft paper presented to study group at Chatham House, 31 January 1994 as part of Royal Institute of International Affairs' Russian and CIS Programme.

Melvin, Neil. *Russians Beyond Russia: The Politics of National Identity.* London, Pinter, 1995.

Meri, Lennart. "Mis Tahes Poliitika on Alati Dialoog." In Lennart Meri, *Presidendikõned.* Tartu, Ilmamaa, 1996.

Meri, Lennart. "Iseseisvus Laiemas Kontekstis." In Lennart Meri, *Presidendikõned.* Tartu, Ilmamaa, 1996.

Meri, Lennart. "Ei Saa Olla Vaba Euroopat Vaba Baltikumilta." In Lennart Meri, *Presidendikõned.* Tartu, Ilmamaa, 1996.

Meri, Lennart. "Eesti Roll Uues Euroopas." In Lennart Meri, *Presidendikõned.* Tartu, Ilmamaa, 1996.

Meri, Lennart. "Euroopa on Programm." In Lennart Meri, *Presidendikõned.* Tartu, Ilmamaa, 1996.

Meri, Lennart. "Euroopa kui Eesmärk." In Lennart Meri, *Presidendikõned.* Tartu, Ilmamaa, 1996.

Meri, Lennart. "Meie Piir on Euroopa Väärtuste Piir." In Lennart Meri, *Presidendikõned.* Tartu, Ilmamaa, 1996.

Meri, Lennart. "'iguste ja Kohutuste Tasakaal." In Lennart Meri, *Presidendikõned.* Tartu, Ilmamaa, 1996.

Meri, Lennart. "Mis Käärib Praegu Venemaa Avarustes?" In Lennart Meri, *Presidendikõned.* Tartu, Ilmamaa, 1996.

Misiunas, Romuald and Rein Taagepera, *The Baltic States: Years of Dependence 1940–1990.* London, Hurst and Company, 1993.

Morozov, Oleg. "Memoirs of a Leader of the Intermovement of Estonia," *The Monthly Survey of Baltic and Post-Soviet Politics*, September 1993.

Morozov, Oleg. "Memoirs of a Leader of the Intermovement of Estonia—II," *The Monthly Survey of Baltic and Post-Soviet Politics*, January 1994.

Muuli, Kalle. "Kohalikesse Volikogudesse Pääsevad 17 Oktoobril Üksnes Eesti Kodanikud," *Postimees*, 20 May 1993.

Mygind, Niels. "A Comparative Analysis of the Economic Transition in the Baltic Countries—Barriers, Strategies, Perspectives." In *The Transition to a Market Economy. Transformation and Reform in the Baltic States*, edited by Tarmo Haavisto. Cheltenham, Edward Elgar, 1997.

"Nadezhnoe Plecho Rossii," *Sovetskaia Estoniia*, 15 January 1991.

Neljas, Aap. "Eesti Konarlik Tee WTOsse," *Postimees*, 21 October 1998.

Nissinen, Marja. *Latvia's Transition to a Market Economy: Political Determinants of Economic Reform Policy*. Basingstoke, Macmillan, 1999.

Norgaard, Ole with Dan Hindsgaul, Lars Johannsen and Helle Willumsen. *The Baltic States after Independence*. Cheltenham, Edward Elgar, 1996.

"Obrashchenie Deputatov Verkhovnogo Soveta Estonskoi Respubliki k Izbirateliam, Vsem Zhiteliam Estonii," *Molodezh Estonii*, 15 January 1991.

OECD Centre for Co-operation with the Economies in Transition. *Review of Agricultural Policies: Estonia*. Paris, OECD, 1996.

Ozolins, Andris. "The Policies of the Baltic countries vis-à-vis the CSCE, NATO and WEU." In *The Foreign Policies of the Baltic Countries: Basic Issues*, edited by Pertti Joenniemi and Juris Prikulis. Riga, Centre of Baltic-Nordic History and Political Studies, 1994.

Park, Andrus. "Ethnicity and Independence: The Case of Estonia in Comparative Perspective," *Europe-Asia Studies*, Vol.46, No.1, 1994.

Parming, Tõnu, "Population Changes in Estonia, 1935–1970," *Population Studies*, Vol.26, March 1972.

Parming, Tõnu. *The Collapse of Liberal Democracy and the Rise of Authoritarianism in Estonia*. London, SAGE, 1975.

Parming, Tõnu. "Population Processes and the Nationality Issue in the Soviet Baltic," *Soviet Studies*, Vol.32, No.3, July 1980.

Parming, Tõnu and Elmar Järvesoo. "Introduction." In *A Case Study of a Soviet Republic: The Estonian SSR*, edited by Tõnu Parming and Elmar Järvesoo. Boulder, Westview Press, 1978.

Pennar, Jaan. "Soviet Nationality Policy and the Estonian Communist Elite." In *A Case Study of a Soviet Republic: The Estonian SSR*, edited by Tõnu Parming and Elmar Järvesoo. Boulder, Westview Press, 1978.

Peters, Rita P. "Small State Diplomacy in International Organizations: The Baltic States and the United Nations," paper presented at the First Conference on Baltic Studies in Europe, University of Latvia, Riga, 17 June 1995.

Petti, Kalev. "Majandusliku Käitumise Soodumused Turu-Uuringute Põhjal," paper presented at the conference "Sotsiaalsed Probleemid Eestis Rahvusgruppide Lõikes: 1988–1993.aastate Sotsioloogiste Uringute Põhjal," Tallinn, 16 June 1994.

Prikulis, Juris. "The European Policies of the Baltic Countries," in *The Foreign Policies of the Baltic Countries: Basic Issues*, edited by Pertti Joenniemi and Juris Prikulis. Riga, Centre of Baltic-Nordic History and Political Studies, 1994.

Raig, Raimo. "*Virolaiset Viron Ulkopuolella*." In *Viro: Historia, Kansa, Kulttuuri*, edited by Seppo Zetterberg. Helsinki, SKS, 1995.

Raik, Kristi. "Democratisation and Integration into the European Union: The Case of Estonia," unpublished paper presented at the seminar Linking the European East and West: The Baltic Sea Area within the Framework of European Integration, University of Turku, Finland, 3 June 1999.

Raud, Mart. *Kaks Suurt: Jaan Tõnisson, Konstantin Päts ja Nende Ajastu*. Toronto, Orto, 1953.

Raun, Toivo. *Estonia and the Estonians*. Stanford, Hoover Institution Press, 1991.

Richards, Jeff. "Old Wine in New Bottles: The Resurgence of Nationalism in the New Europe." In *Ethnicity and Nationalism in Russia, the CIS and the Baltic States*, edited by Christopher Williams and Thanasis D. Sfikas. Aldershot, Ashgate, 1999.

Rislakki, Jukka. "Virossa Kasvaa Ärtymys Suomea Kohtaan," *Helsingin Sanomat*, 11 April 1999.

Rislakki, Jukka. "Virolaisen Historiantutkijan Väite: Valtionpäämies Päts Sai Rahaa Moskovasta," *Helsingin Sanomat*, 8 September 1999.

Rose, Richard and William Maley. *Nationalities in the Baltic: A Survey Study*. Strathclyde, Centre for the Study of Public Policy, 1994.

Rosimannus, Rain. "Political Parties: Identity and Identification," *Nationalities Papers*, Vol.23, No.1, 1995.

Rotko, Jorma. "Viron Vähi Moitti Edeltajiensa Idanpolitikkaa," *Helsingin Sanomat*, 22 April 1995.

Rotko, Jorma. "Viron Venäläisten Asenne Virolaisuuteen Riipuu Iästä," *Helsingin Sanomat*, 6 January 1997.

Rotov, Igor. "Kodakondsusest: Optsioon või Naturalisatsioon?" *Rahva Hääl*, 8 September 1991.

Rotov, Igor. "Parempoolsed Uhenisid," *Rahva Hääl*, 28 September 1991.

Rumm, Hannes. "*Keskfraktsioon Kardab Eestis Etnilist Puhardust*," *Postimees*, 11 June 1993.

"Russkaia Real'nost' v Estonii," *Digest*, No.7, May 1995.

Ruusmann, Ants. *Eesti Vabariik 1920–1940*. Tallinn, TPÜ Kirjastus, 1997.

Ruutsoo, Rein. "Rahvusvähemused Eesti Vabariigis." In *Vähemusrahvuste Kultuurielu Eesti Vabariigis*. Tallinn, Olion, 1993.

Ruutsoo, Rein. "The Perception of Historical Identity and the Restoration of Estonian National Independence," *Nationalities Papers*, Vol.23, No.1, 1995.

Ruutsoo, Rein. "Estonian Post-Communist Transition, Civil Society and Social Sciences in the Context of EU Enlargement." In Chancellery of the Riigikogu, *Society, Parliament and Legislation*. Tallinn, Department of Economic and Social Information of the Chancellery of the Riigikogu, 1999.

Saar, Andrus. "Euro-Barometer and Estonian Experiences: European Union Integration and Enlargement—Attitudes in Estonia." In Chancellery of the Riigikogu, *Society, Parliament and Legislation*. Tallinn, Department of Economic and Social Information of the Chancellery of the Riigikogu, 1999.

Saaremaa, EiNo.*Eestlaste Ajalugu 1820–1945*. Tallinn, Elva, 1997.

Saarikoski, Vesa. "Russian Minorities in the Baltic States." In *New Actors on the International Arena: The Foreign Policies of the Baltic Countries*, edited by Pertti Joenniemi and Peeter Vares. Tampere, TAPRI, 1993.

Safire, William. "Bill 'n' Boris Fix It in the Baltics," *The Guardian*, 31 August 1994.

Smith, Anthony D. *National Identity*. London, Penguin, 1991.

Smith, David. "Legal Continuity and Post-Soviet Reality: Ethnic Relations in Estonia 1991–95," unpublished Ph.D. dissertation, University of Bradford, 1997.

Smith, David. "Russia, Estonia and the Search for a Stable Ethno-Politics," *Journal of Baltic Studies*, Vol.29, No.1, 1998.

Smith, David. "The Restorationist Principle in Post-Communist Estonia," in *Ethnicity and Nationalism in Russia, the CIS and the Baltic States*, edited by Christopher Williams and Thanasis D. Sfikas. Aldershot, Ashgate, 1999.

Smith, David. "Bridging East and West? The Baltic Sea Region as a Coordinate in the Construction of Post-Soviet Identities", unpublished paper presented at the seminar "Linking the European East and West: The Baltic Sea Area within the Framework of European Integration," University of Turku, Finland, 3 June 1999.

Smith, David. "Nordic, Baltic and Arctic Organisations." In *The Annual Register 1998*, edited by Alan J. Day. Rockville, Keesing's Worldwide, 1999.

Smith, David. "Retracing Estonia's Russians: Mikhail Kurchinskii and Interwar Cultural Autonomy," *Nationalities Papers*, Vol.27, No.3, 1999.

Smith, Graham. "The State, Nationalism and the Nationalities Question in the Soviet Republics." In *Perestroika: The Historical Perspective*, edited by C. Merridale and C. Ward. London, Arnold, 1991.

Smith, Graham. "Introduction." In *The Baltic States: The Self-Determination of Estonia, Latvia and Lithuania*, edited by Graham Smith. Basingstoke, Macmillan, 1994.

Smith, Graham. "The Resurgence of Nationalism." In *The Baltic States: The Self-Determination of Estonia, Latvia and Lithuania*, edited by Graham Smith. Basingstoke, Macmillan, 1994.

Smith, Graham, Aadne Aasland and Richard Mole. "Statehood, Ethnic Relations and Citizenship." In *The Baltic States: The Self-Determination of Estonia, Latvia and Lithuania*, edited by Graham Smith. Basingstoke, Macmillan, 1994.

Smith, Graham. "The Soviet State and Nationalities Policy." In *The Nationalities Question in the Post-Soviet States*, edited by Graham Smith. London, Longman, 1996.

Soosaar, Enn. "Eesti Rahva Enamus Tahab Edasi Minna," *Rahva Hääl*, 7 March 1995.

Sootla, Georg. "Political Background and Possible Consequences of the Summer Crises in Estonia," *The Monthly Survey of Baltic and Post-Soviet Politics*, July 1993.

Sorsa, Piritta. "Regional Integration in the Baltics and the Global Context." In OECD Centre for Cooperation with the Economies in Transition, *Regional Integration and Transition Economies: The Case of the Baltic Sea Rim*. Geneva, OECD, 1996.

"Speech of Lennart Meri on the 81st Anniversary of the Republic of Estonia, February 24th 1999," *The Monthly Survey of Baltic and Post-Soviet Politics Extra. Estonia: Elections 7 March 1999*.

Statistical Office of Estonia/Bank of Estonia. "Indicators for Estonian Economy 1994–1998." In Economics Ministry of the Republic of Estonia, *Yearbook of the Estonian Economy 1998. http://www.mineco.ee/eng/sisu.html*.

Suny, Ronald. "Incomplete Revolution: National Movements and the Collapse of the Soviet Empire," *New Left Review*, No.189, 1992.

Swettenham, John Alexander. *The Tragedy of the Baltic States*. London, Hollis and Carter, 1952.

Taagepera, Rein. "Estonia's Road to Independence," *Problems of Communism*, Vol.38, December 1989.

Taagepera, Rein. "A Note on the March 1989 Elections in Estonia," *Soviet Studies*, Vol.42, No.2, April 1990.

Taagepera, Rein. "Ethnic Relations in Estonia, 1991," *Journal of Baltic Studies*, Vol.23, No.29, Summer 1992.

Taagepera, Rein. *Estonia: Return to Independence*. Boulder, Westview Press, 1993.

Talts, Mait and Aksel Kirch. "Eesti ja Euroopa Liit: Poolt ja Vastu Eesti Ajakirjanduses." In *Eesti Euroopa Liidu Lävepakul*, edited by Rein Ruutsoo and Aksel Kirch. Tallinn, Teaduste Akadeemia Kirjastus, 1998.

Tartu Ülikooli Turu-Uurimisrühm. *Kirde-Eesti Linnaelanike Suhtumine Eesti Reformidesse ja Sotsiaalpoliitikasse: 1993-ja 1994. Aasta Võrdlev Analüüs*. Tartu, University of Tartu, September 1994.

Tartu University Market Research Team. *The Attitude of Town Residents of North-Eastern Estonia towards Estonian Reforms and Social Policy: A Comparative Study of 1993, 1994 and 1995*. Tartu, University of Tartu, 1995.

Thaden, Edward C. "Reform and Russification in the Western Borderlands 1796–1855." In *Russification in the Baltic Provinces and Finland 1855–1914*, edited by Edward Thaden. Princeton, Princeton University Press, 1981.

Thaden, Edward C. "Administrative Russification in the Baltic Provinces 1855–1881." In *Russification in the Baltic Provinces and Finland 1855–1914*, edited by Edward C. Thaden. Princeton, Princeton University Press, 1981.

Thaden, Edward C. "The Abortive Experiment: Cultural Russification in the Baltic Provinces 1881–1914." In *Russification in the Baltic Provinces and Finland 1855–1914*, edited by Edward Thaden. Princeton, Princeton University Press, 1981.

"The Popular Movements and the Soviet Union: Discussion." In *Towards Independence: The Baltic Independence Movements*, edited by Jan Arveds Trapans. Boulder, Westview Press, 1991.

Tolstikov, Iaroslav. "'My Mozhem Stat' Soyuznikami Estonskogo Pravitel'stva', Govoriat Rukovoditeli Predpriiatii Soyuznogo Podchineniia," *Sovetskaia Estoniia*, 25 January 1991.

Tolstikov, Iaroslav. "Dogovorit'cia Vse-Taki Mozhno!" *Sovetskaia Estoniia*, 1 February 1991

Tolstikov, Iaroslav. "Ekonomika i Politika: Paldiski—Berlin—Stokgol'm—Oslo, no ne Moskva ili Sankt-Peterburg," *Estoniia*, 13 October 1994.

"Toomas Ilves, Estonia's American-European," *The Economist*, 31 October 1998.

Toots, Anu. "Immigration in Estonia: Politics and Policy," *Monthly Survey of Baltic and Post-Soviet Politics*, April 1994.

20 August Klubi ja Riigikogu Kantselei, eds. *Kaks Otsustavat Päeva Toompeal (19.–20 August 1991)*. Tallinn, Eesti Entsüklopeediakirjastus, 1996.

United National Party of Estonia. "Programma Ob'edinennoi Narodnoi Partii Estonii (Proekt)," draft programme circulated at the party's founding congress in Tallinn, 9 October 1994.

Upton, Anthony. *Finland 1939–40*. London, Davis-Poynter, 1974.

Van Arkadie, Brian and Mats Karlsson. *Economic Survey of the Baltic States*. London, Pinter, 1992.

Vardys, V. Stanley. "Modernisation and Baltic Nationalism," *Problems of Communism*, September–October 1975.

Vardys, V. Stanley and Judith Sedaitis. *Lithuania. The Rebel Nation*. Boulder, Westview Press, 1997.

Velliste, Trivimi. "The 'Near Abroad' in the Baltic Republics: The View from Estonia." In *Nordic-Baltic Security: An International Perspective*, edited by A.O. Brundtland and D.M. Snider. Washington, CSIS, 1994.

"Vene Valimisliit Loodab Valimiste Järel Pusima Jääda," *Hommikuleht*, 8 February 1995.

Venesaar, Urve and David Smallbone. "The Development of Manufacturing SMEs in Estonia." Paper presented at the conference *Estonia: A Development Agenda*, Royal Institute of Geography, London, 26 March 1996.

Vetik, Raivo. "Ethnic Conflict and Accommodation in Post-Communist Estonia," *Journal of Peace Research*, Vol.30, No.3, 1993.

Vihalemm, Peeter, Marju Lauristin and Ivar Tallo. "Development of Political Culture in Estonia." In *Return to the Western World: Cultural and Political Perspectives on the Estonian Post-Communist Transition*, edited by Marju Lauristin and Peeter Vihalemm with Karl Erik Rosengren and Lennart Weibull. Tartu, Tartu University Press, 1997.

"Viro ei enää Vaadi Petserinmaata Venäjältä," *Turun Sanomat*, 3 February 1995.

Virta, Kaija. "Älymystö Haaveli Virossakin 1960-luvulla Sosialismin Ihmiskasvoista: Allikin Perheessä on Koettu Kommunismin Kaikki Käänteet," *Helsingin Sanomat*, 14 July 1997.

Von Rauch, Georg. *The Baltic States: The Years of Independence 1917–1940*. London, Hurst and Company, 1995.

"Voprosi Oprosa 'Kakoi by vy Khoteli Videt' Estoniyu?" *Molodezh' Estonii*, 16 January 1991.

"Vstrecha prodolzhalis' 10 minut," *Sovetskaia Estoniia*, 17 January 1991.

Waever, Ole. "The Baltic Sea: A Region after Post-Modernity?" In *Neo Nationalism or Regionality? The Restructuring of Political Space around the Baltic Rim*, edited by Pertti Joenniemi. Stockholm, NordRefo, 1997.

Watson, Ian. *The Baltics & Russia Through the Back Door*. Riga, Rick Steves' Europe Through the Back Door, Inc., 1994.

White, James D. "Nationalism and Socialism in Historical Perspective." In *The Baltic States: The Self-Determination of Estonia, Latvia and Lithuania*, edited by Graham Smith. Basingstoke, Macmillan, 1994.

White, Stephen. *Gorbachev and After*. New York, Cambridge University Press, 1992.

Woods, Shirley A. "Ethnicity and Nationalism in Contemporary Estonia." In *Ethnicity and Nationalism in Russia, the CIS and the Baltic States*, edited by Christopher Williams and Thanasis D. Sfikas. Aldershot, Ashgate, 1999.

Zetterberg, Seppo. "Historian Jännevälit," in *Viro: Historia, Kansa, Kulttuuri*, edited by Seppo Zetterberg. Helsinki, SKS, 1995.

Zhuryari-Ossipova, Olga. "Russian Factor in Estonian Foreign Policy: Reaction to the Limitation of Sovereignty." In *New Actors on the International Arena: The Foreign Policies of the Baltic Countries*, edited by Pertti Joenniemi and Peeter Vares. Tampere, TAPRI, 1993.

INTERVIEWS CONDUCTED BY THE AUTHOR

Chuikin, Vladimir. Marketing Director of Narova Furniture Factory, Narva. Head of Narva City Council December 1989–October 1993 (Narva, 26 April 1996).

Järve, Priit. Chairman of Estonia's Presidential Round Table of Nationalities 1996–1998 (Bradford, 19 February 1997).

Kaas, Marina. Head of the Estonian Small Business Association (Tallinn, 23 April 1996).

Olesk Peeter. Minister of Citizenship and Migration of Estonia October 1993-June 1994 (Tallinn, 17 June 1994).

Paju, Ants, Chairman of Estonia's Presidential Round Table of Nationalities 1993–1996 (Tallinn, 14 June 1994).

Roosaare, Jaak. *ERSP* Member of Riigikogu 1992–1995 (Tallinn, 10 February 1995).

Sovetnikov, Sergei. Member of Estonian Supreme Council 1990–92 (Narva, 20 February 1995).

Stepanov, Evgenii. Advisor to the Estonian-Russian Chamber of Entrepreneurs (Tallinn, 9 March 1995).

Svirin, Vasilii. Head of the Russian Federation Delegation to Inter-State Negotiations with Estonia (Moscow, 21 May 1995).

Zolin, Nikolai. Member of Estonian Supreme Council 1990–92. Deputy Chairman of Narva City Council 1993–96 (Narva, 26 April 1996).

Index

Agriculture: 1–2, 4, 5–6, 8, 13,
    33–34, 40, 83, 94–95, 96, 98,
    103, 104, 122, 136–140, 169
Alexander III: 8
'Aliens' Crisis': 85–90, 134
Andreev, Viktor: 56
Anvelt, Jaan: 11, 38
Ärkämisaeg: 2, 5–9
Armed Forces: 1, 13, 15–16, 25–26,
    71–72, 74, 78–79, 84–85, 97,
    99–100, 167–169
Austria: 17, 171
Authoritarianism: 18–23, 65, 69,
    78–79, 96, 105

Balladur, Eduard: 165
Baltic Germans (see also National
    minorities: Germans): 1–11,
    13–14
Baltic League: 151–152, 169
Baltic States: xx, xxiv–xxv, 23,
    27–28, 42–43, 52, 103,
    114, 121, 124, 133, 150,
    152, 154, 156–157, 161,
    164–165, 167–172
    Baltbat: 168
    Balttron: 168
    Baltic Assembly: 169
    Baltic Council of Ministers: 169
    Baltic Entente: 154
Banking: 17, 33, 92, 100, 115, 116,
    117–119, 120, 125–126,
Bauer, Otto: 17
Berg, Maimu: 123
Bergmann, Andres: 127
Birth rate: 41, 113, 135–136
Border Disputes: 68, 92, 148, 159,
    164, 166, 169

Brezhnev, Leonid: xxiv, 39–40, 41,
    42, 44
Briand-Kellogg Pact: 153

Central and Eastern Europe: xi,
    xv–xvi, 125, 153, 156, 163,
    165, 166, 168, 171
Chechnya: 60, 92
Chuikin, Vladimir: 78, 89
Citizenship: 45, 51, 54, 56, 57, 66,
    67, 71–77, 81, 86–90, 93,
    101–102, 108, 133, 166, 172
Civil Society: xiv, xxiv, 47, 66–67,
    107
Clinton, William: 164
Comintern: 15
Commonwealth of Independent States
    (CIS): 90, 107, 122, 126, 133,
    158, 165,
Communism: xi, xxv, 7, 15–16, 22,
    28, 33, 38, 47–48, 50, 67, 81,
    83, 97, 113, 117, 132, 165,
    172
    1924 Putsch: 15–16
    Communist Party of Estonia:
        15–16, 28, 38–42, 45–47,
        49–50, 53, 59, 82, 94
    Communist Party of the Soviet
        Union: xxii, xxiv, 38–39,
        42–43, 52, 53, 55, 69
    'Estonian Working People's
        League': 28
Conference (Organisation) for
    Security and Cooperation in
    Europe: xiii, 65, 77,
    88–89, 102, 157, 161–162
Congress of People's Deputies: 47,
    49–50, 51–53, 58

Constituent Assembly (1919–20): 12–15
Constituent Assembly (1937): 21
Constituent Assembly (1991–92): 65, 71–72
Constitutions (Republic of Estonia): (1920): 14; (1934): 18, 21; (1938): 21–23, 79; (1992): 65, 71–72, 74–79, 82–83, 91, 93, 118, 159, 164
Corruption: 67, 77–78, 85, 90–93, 96–98, 100, 124, 127
Coudenhove-Kalergi: 172
Council of Europe: xiv, 65–67, 72, 75, 83–84, 86–89, 101, 149, 162–163, 166
Currency Reform: 45, 77–78, 115–121, 125
Czechoslovakia: xi, 40

Defence: (see Armed Forces)
Defence Initiative Centre: 78, 84
Denmark: 3, 130, 136, 168
Deportations: xxi, 10, 29, 34–37, 155
Dissident Movement: xx, xxii–xxiii, 39–40, 42, 44–45, 48, 83, 90, 97, 155, 157
'Letter of Forty': 42
Donner, Jörn: 103
Dudayev, Dzokhar: 60

Education: xxii–xxiii, 4–6, 9, 14, 16, 23, 38, 42, 44, 75–76, 88, 95, 102–103, 135
Einbund (Eenpalu), Karl: 19–20
Einseln, Aleksander: 85, 97
Elections:
(1917–40): 11–15, 18–19, 21, 24, 28–29
(1988–91): 43, 48–51, 69
(1992–): 54–55, 65–66, 68, 70, 74, 75, 77–83, 85, 86, 87, 88–95, 98–100, 104–106, 117, 142

Émigré Estonians: xx, 26, 78–79, 82, 84–85, 116, 127, 160
Energy: 114–116, 120, 124, 140, 141
Environment: 44, 83, 92, 95, 140
Erm, Ants: 90
Estland: 1, 3, 5, 6, 8–11
Estonian Soviet Socialist Republic (ESSR): xx–xxv, 29, 39, 41, 43, 44, 46–51, 54, 56–57, 73, 133, 155
ESSR Supreme Soviet (Supreme Council): xx, 38, 41, 47, 48, 50, 53, 54–55, 56, 57, 59, 60, 72, 128 137
Estonian Workers' Commune: 1
'Ethnic Democracy': 74–77
European Bank for Reconstruction and Development: 116
European Free Trade Association (EFTA): 121
European Union: xi, xiii–xv, 65–67, 75, 80, 84, 90, 94, 97, 99–102, 104, 106, 108, 117, 121, 122, 129, 133, 139–142, 147–149, 150, 162, 165, 167, 169–172
Aid: 130
Association Agreement: 140
Commission Avis (1997): xiv, 101, 107, 122, 165, 170
Common Agricultural Policy: 140, 171
Free Trade Agreement: 91, 170
Membership Negotiations: xi, 106, 140, 170–172
Trade and Cooperation Agreement: 65, 162

Fählmann, Friedrich Robert: 7
Finland: xxv, 3–4, 7, 12–13, 20, 22–27, 36–37, 121–123, 125, 151–152, 155–156, 161–162, 168
Foreign Investment: 45, 54, 86, 89, 119, 125–130, 134–135, 140–141

'Forest Brothers': xx, 34–35, 37
Former Soviet Union (FSU): xi, xii, xiii, 115–117, 122, 126, 147–148, 160–161
France: 27, 75, 150–151, 153–155, 157, 171

Gaidar, Yegor: 60
Gastarbeiter: 75
Genoa Conference (1922): 152
Germany: 1–2, 3, 11–14, 23, 26–27, 29, 33–36, 72, 75, 123, 127, 134, 150, 151, 153–156, 171
  Holy Roman Empire: 3
  Second Empire (1871–1918): 1–2, 11–14, 150
    Occupation of Estonia (1918): 1–2, 12–14, 150
    and Baltic Germans: 1, 11
  Weimar Republic: 14, 150–151, 153
  Third Reich: xix, 23, 26–27, 29, 33–36, 154–155
  and Baltic Germans: 23, 27
  Occupation of Estonia (1941–1944): 33–36
  Federal Republic: 72, 75, 123, 127, 134, 156, 171
Gorbachev, Mikhail: xxii, xxiv, 42–44, 46, 47, 51–54, 56–59, 114, 156–157, 158
Government in Exile: xx, 78
Great Britain: 2, 13, 37, 117, 142, 150–151, 153–155
Great Depression: 17–18, 20, 22–23, 153
Greece: 171

Hanni, Liia: 128
Hanseatic League: 3
Hansson, Ardo: 116
Hirvepark: 44–45
Hitler, Adolf: 27, 35, 154–155
Hungary: xi, 39
Hurt, Jakob: 7, 46

Ilves, Toomas Hendrik: 98, 105–106, 107, 141, 148–149, 164, 167, 172
IME Plan: 45, 115
Immigration: xxi–xxii, 39, 41, 107
Industry: 9, 15, 22, 33, 39, 41, 43, 56, 69, 90, 116–117, 119, 121–122, 125–128, 130–131, 133, 135–136, 169
Inflation: 116, 119–120, 130
'Integral': 56
International Monetary Fund (IMF): 77, 96, 104, 113, 117–118, 125
Ireland: 171
Israel: 85

Jaakson, Jüri: 17
Jakobson, Karl Robert: 7–8, 10
Jannsen, Johannes Voldemar: 7, 46

Käbin, Johannes: 38–39, 41
Kaitseliit: 16, 78–79, 84
Kalevala: 6
Kalevipoeg: 7
Kallas, Siim: 40, 45, 92, 94–95, 97–98, 100, 117, 149–150, 166
Kama, Kaido: 92–93
Karaganov, Sergei: 161
Karotamm, Nikolai: 38
Kekkonen, Urho: xxv.
KGB: 34, 68, 83, 99
Khrushchev, Nikita: 38–39
Kingisepp, Viktor: 12
Kodukaitse: 78
Köler, Johan: 7–8
Komsomol: 40–44
Kosovo: 149
Kozyrev, Andrei: 164
Kreutzwald, Friedrich Reinhold: 7
Kukk, Jüri: 44
Kulbok, Kalle: 82
Kurchinskii, Mikhail: 17, 23

Läänemaa Light Infantry Company:
  78, 84–85, 90
Laar, Mart: 29, 44, 83–84, 86, 88,
  90–92, 95, 98, 100, 106, 119,
  121, 170
Laidoner, Johannes: 13, 15, 18–19,
  25–26
Language Law: 45, 47, 49–50, 86,
  102–103
Larka, Andres: 19
Latvia: xix, xxiv, 3, 14, 23, 24, 26,
  39, 47, 54, 57–59, 67–68, 87,
  101, 118–119, 121–123,
  131–133, 139, 151–153, 161,
  163–164, 169–170, 171
Lauristin, Marju: xiv, 40, 42–43,
  65–67, 69, 74, 81
League of Nations: xix, 14, 150,
  152
Lebedev, Vladimir: 69
Legal Continuity of Independence
  (1940–1991): xi, xx,
  xxiv–xxv, 29, 48, 54, 56, 60,
  65, 70, 72–73, 79, 128,
  155–159, 161
Leninism: 43, 52
Lepikson, Robert: 104
Liiv, Juhan: xii, 11
Lippmaa, Endel: 50, 54, 97, 165
Lithuania: xix, xxii, xxiv, 3–4, 23–24,
  28, 47, 53, 55, 58–59, 68,
  121–122, 131–133, 151–152,
  169–170
Litvinov Protocol (1929): 153
Livland: 4–6, 8–9, 11
Livonia: 3–4
Local Government: 19, 54, 56,
  68–69, 75, 86–90, 98, 100,
  134–135, 137–138
Locarno, Treaties of (1925): 153–154
Luik, Jüri: xvi, 85, 92, 147, 150, 162

Maapäev: xix, 11–13

Maapank: 126, 139
Maarahvas: 3, 5
Made, Tiit: 45
Madisson, Tiit: 44, 97
Mäe, Hjälmar: 33–34
'Mahtra War': 6
Mandmets, Ilmar: 104
Manitski, Jaan: 160
Media: xxv, 7, 10, 12, 20, 27–29, 36,
  40, 44–45, 48–49, 51, 58, 66,
  82, 93, 103
Meri, Lennart: 1, 45, 67, 83, 87, 88,
  91, 92, 95–100, 105,
  155–157, 160, 163, 164, 170
Moldova: 86
Molotov, Vyacheslav: 28, 33
Molotov-Ribbentrop Pact: 23, 25,
  44, 48, 52–53, 147, 155, 157
Müürisepp, Aleksei: 39

Narva: 1, 41, 56, 68, 78, 86, 88, 89,
  135
National Minorities 16, 22–23, 27,
  35, 37–38, 67–68, 76
  Germans (see also Baltic Germans):
    16, 23, 27, 35
  Jews: 16, 35
  Russians (see also 'Russian-
    speaking Population):
    16–17, 23, 37–38, 76
  Swedes: 16
Nationalism: xii, xxii–xxv, 3, 7, 11,
  16, 21, 38, 40, 42–45, 47–55,
  57, 60, 69–74, 77–79, 81–82,
  84–90, 92–94, 97, 99, 128,
  139, 154, 158–160, 163, 165,
  169
NATO: xiii, xv, 65, 71, 84,
  94, 99–100, 147–149,
  167–170
  NACC: 167
  Partnership for Peace: 168
Netherlands xi
NKVD: 34, 37

Nordic Council: 157, 169
Nordic Countries: 2, 12, 70, 118,
    123, 130, 134, 136–137, 141,
    151, 156–157, 168–171
North Estonia Shareholders' Bank:
    100, 119
Norway: xx

Omakaitse: 35
Oovel, Andres: 97

Paldiski: 164
Parek, Lagle: 44, 83–85, 90
Päts, Konstantin: 10, 12–13, 15,
    17–29, 36, 71–72, 151
Pensions: 84, 95–96, 99, 133, 136,
    166
Perestroika: xxii, 42–43, 48–131
Perry, William: 167
Piip, Ants: 151–152
Poland: xi, 2, 4, 23, 27, 123, 151–154
Portugal: 171
Political Parties, Coalitions and
    Movements (1905–1917):
  Democratic Party: 12
  Estonian Progressive People's
    Party: 10
  Estonian Social Democratic Party:
    10
  Noor Eesti: 11
Political Parties, Coalitions and
    Movements (1917–1940):
  Agrarian League (Farmers' Party):
    12, 15, 22
  Estonian Social Democratic Party:
    17–19
  Isamaaliit: 20–22
  Labour Party: 12–13
  League of Veterans of the War of
    Independence: 18–22, 26,
    35
  National Party: 15
Political Parties, Coalitions and
    Movements (1987-)

Blue Party: 94
Centre Party: 70, 78–79, 82–83,
    88, 94, 106, 141
Christian Democratic Party: 54, 70
Coalition Party: 70–71, 90–91,
    94–96, 98–100,
    103–106, 130, 165
Committee Movement / Congress
    of Estonia: 48, 50–51,
    54–55, 57, 59–60, 65,
    69–71, 73, 77–78, 81,
    156, 160
Conservative People's Party: 70,
    91
Eesti Kodanik, 82, 84, 94
Estonian Development Party: 98,
    101
Estonian Future Party: 90, 94
Estonian Country People's Party:
    92, 94–96, 99, 104–106
Estonian National Independence
    Party (ERSP): 48, 54, 59,
    70, 77, 79–80, 83–84,
    87–88, 90, 94, 104, 161,
    163
Estonian-Russian People's Party:
    93
Estonian Social Democratic Party:
Farmers' Assembly: 94–96, 99
Fourth Force: 94
Intermovement: 49
Isamaa: 70–71, 81, 83, 84–85, 87,
    89–91, 94, 100, 104–106,
    128, 130, 132, 139, 141,
    162, 165
Kindel Kodu: 82–83
KMÜ: 94–96, 101, 105
Left Opportunity: 82
Liberal Democrats: 54, 69–70, 91,
    92
Moderates: 81, 83, 92, 94,
    104–106
MRP-AEG: 44, 46–48
National Progress Party: 90

Õiglus: 94
Our Home is Estonia: 93, 96
Parem Eesti: 94
People's Party: 105
Popular Front of Estonia: xxiv, 43,
    45–52, 54–55, 57, 59, 69,
    73–74, 78, 81–82, 98, 128,
    155–156
Reform Party: 92, 94–95, 97–98,
    100–101, 104–106, 130
Republican Coalition Party: 70, 91
Representative Assembly: 85, 134
Restitution: 79, 82
Royalist Party: 82, 94
Rural Centre Party: 69, 81
Rural Union: 94–97, 99, 105
Russian Assembly: 85
Russian Democratic Movement:
    69, 85
Russian Party of Estonia: 93
Union of Work Collectives: 49, 54,
    70
United Council of Work Collectives
    (OSTK): 49–50, 54–55, 58,
    60, 68
United National Party of Estonia:
    93, 106, 134
Vaba Eesti: 53, 55, 70
Press: (see Media)
Primakov, Yevgenii: 103, 166
Privatisation: 77–78, 87, 100,
    126–130, 133, 135, 138
    Estonian Privatisation Enterprise:
        77, 127, 128, 129
    Estonian Privatisation Agency: 129,
        130
Property Restitution: 71, 128–130,
    133, 138
'Pullapää Incident' (1993): 78, 85, 97
Pusta, Karel: 151–152

Raig, Ivar: 69, 81
Rapallo, Treaty of (1922): 153
Rebas, Hein: 84, 85

Referenda: 14, 18, 19, 21, 51, 59, 65,
    78, 86, 88–89
Religion: 4, 8, 10, 33
    Lutheranism: 4
    Orthodoxy: 8, 10
Renner, Karl: 17
Riigikogu: 14, 16, 18–19, 66, 78,
    80–83, 85–87, 90, 93–94, 96,
    98–99, 102, 104, 164–166
Rosenberg, Alfred: 35
Russia: 1–2, 5, 6–13, 37, 57–58, 60,
    68, 73–75, 79, 83, 91–93,
    97–99, 103, 114, 116–117,
    121–124, 126, 128, 131–135,
    139–141, 147–152, 157–166
    Tsarist Empire: 2, 5, 6–11, 157
    1905 Revolution: 9–11
    under Provisional Government
        (1917): 2, 11–12
    Bolshevik Russia (1917–1922): xix,
        12–13, 15, 140–141,
        150–152
    Civil War: 150–151
    Invasion of Estonia (1918): xix,
        1–2, 13
    RSFSR: xxviii, xxxi, 38, 57–58, 60
    Russian Federation (1992-): xv, 68,
        73, 75, 79, 97–98, 117, 128,
        147–149, 157–161, 165,
        171–172
    and 'Near Abroad': xv, 159–160,
        163
    and NATO enlargement: 168
    and Russians in Estonia: xv, 75,
        88, 93, 103, 133–135,
        158–159, 161–166,
        171–172
    and withdrawal of former Soviet
        troops from Estonia: 68,
        74, 79, 83, 84, 91–92, 99,
        148, 157–158, 160–166
    Estonian trade with: xiv,
        103–104, 114, 116,
        121–124, 126, 139–141,
        163–164

'Russian-speaking Population' (see
    also National Minorities:
    Russians): xiv–xv, xxiii, 16,
    41–42, 48–51, 54, 56–60,
    67–69, 72–78, 81, 84–91,
    93–94, 97–98, 101–103, 117,
    128, 132, 133–136, 141, 149,
    158–159, 161–163, 166,
    171–172
    Soviet Military Pensioners: 72, 74,
        87–91, 97, 163–164,
Russification: xxiii, 5, 8–10, 44
Rüütel, Arnold: 47, 71, 83, 92, 95, 99
Ruutsoo, Rein: 23, 40, 42, 67

Saul, Bruno: 47
Savisaar, Edgar: 45–46, 54–58, 60,
    69–71, 73, 82, 84, 95–96, 98,
    100, 106, 114–115,
    116–117, 127–128, 155
Scandinavia: (see Nordic Countries)
Schröder, Gerhard: 171
Selter, Karl: 24
Serbia: 172
Serfdom: 4–5
Siimann, Mart: 100–101, 103–105,
    135–136
Siir, Arvo: 90
Sillamäe: 41, 56, 68, 86, 89
'Singing Revolution': 7, 46
Sirk, Artur: 18, 20–21
Soosaar, Enn: 96
Sovnarkhozy: 39
Spain: 171
Stalin, Josef: xxiv, 25, 27, 34, 37, 38,
    155
Stalinism: xxi, xxiv, 28, 38–40, 43, 136
Strandmann, Otto: 12
Suits, Gustav: 11
Supreme Council (See ESSR Supreme
    Soviet)
Svinhufvud, Per: 22
Sweden: xix, 4–5, 27, 36, 37, 117,
    125, 135, 141
Switzerland: 14, 27, 117

Taagepera, Rein: xi, xiii, xxii, 24–25,
    28, 41, 85, 162
Tallinn: xi, xxv, 1, 3, 10, 13, 15, 19,
    27, 35, 41, 42, 44, 46, 49,
    51–52, 57–58, 60, 88–90, 98,
    100, 123, 126, 130, 133,
    134–135, 152, 164
    City Council: 90, 98, 100
Tamm, Jaak: 69–70
Tarand, Andres: 42, 92, 94, 165
Tariff Policy: 96–98, 104, 115, 122,
    139, 140
Tarto, Enn: 90, 97
Tartu:
    Commercial Bank: 119
    Treaty of (1920): xix, 1, 46, 54,
        151, 156, 159, 164, 165,
        166
    University of: 4, 5, 9–10, 20, 40, 44
Taxation: 16, 33–34, 45, 70, 94–95,
    97–98, 105, 114, 118–119,
    124, 131, 136
Thatcher, Margaret: 83
Titma, Mikk: 45
Tõnisson, Jaan: 10–12, 15, 20, 151
Toome, Indrek: 47
Toomepuu, Jüri: 82, 84
Trade: 17, 45, 65, 73, 91, 96, 103,
    107, 114–117, 119, 121–125,
    126, 131, 136, 139–140, 148,
    152–153, 160, 162–163,
    165–166, 169–170
Trade Unions: 117, 119, 123, 138

Ukraine: xxi, 44, 122
Uluots, Jüri: xx, 20, 35, 36
Uluots, Ulo: 78
Unemployment: 17, 86, 95, 113, 116,
    131–136
United Kingdom: (see Great Britain)
United Nations: xix, xxiii, 77, 155,
    161–162
United States of America: xx, 37, 85,
    130, 141, 150, 161, 163, 164,
    167

USSR: xi, xix–xxi, xxiv, 16, 23–29,
33, 38, 40–42, 44–45, 47–48,
50–56, 59, 74–75, 78,
113–115, 124, 134, 137,
152–157
Constitution: 47–48, 53
Occupation of Estonia (1939–
1940): xi, xix, xxiv, 23–26
Annexation of Estonia (1940): xix,
26–29, 53–54, 74
Non-Aggression Treaty with
Estonia (1932): 154
Supreme Soviet of: 58, 60

Vähi, Tiit: 69, 77–79, 90, 95–100,
103, 116, 124, 127, 160, 165
Vaino, Karl: 41–42, 45–47
Väljas, Vaino: 41, 46
Valk, Heinz: 46
Van der Broek, Hans: 101
Van der Stoël, Max: 77–78
Varik, Andres, 104
Veidemann, Andra: 98, 101
Velliste, Trivimi: 40, 162

Versailles, Treaty of (1919): 150
Vneshkombank: 115, 119
Von Habsburg, Otto: xi

Wages: 84, 117, 120, 123, 133–134,
136
West European Union: 168
World Bank: 116, 118
World Trade Organisation: 101, 104,
140, 170
World War I: 2, 9–11, 150, 152, 157
World War II: xii, xxv, 29, 33, 137,
156, 159, 164
Wörner, Manfred: 167

Yakovlev, Aleksandr: 52
Yalta Conference: 37
Yel'tsin, Boris: 57–60, 124, 158–159,
161, 164
Yugoslavia: xiv, 171

Zemstva: 7
Zhdanov, Andrei: 27–28

Latvia

the challenges of change

Artis Pabriks and Aldis Purs

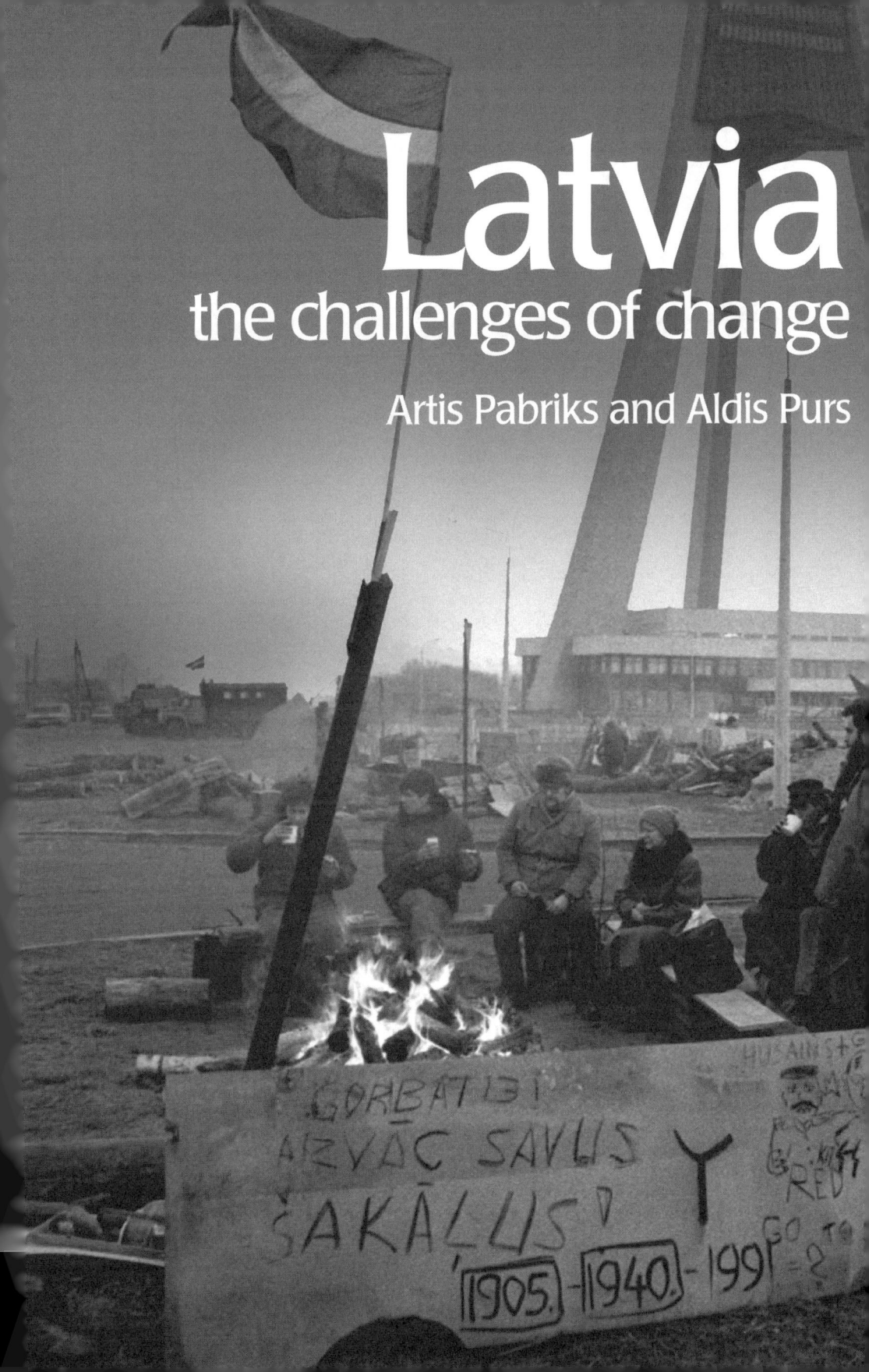

GORBATJI
AIZVĀC SAVUS
ŠAKĀĻUS!

HUSAIN +

GO TO
RED

GO TO

1905. - 1940. - 1991 = ?

Latvia

Postcommunist States and Nations

Books in the series

Belarus: A denationalized nation
David R. Marples

Armenia: At the crossroads
Joseph R. Masih and Robert O. Krikorian

Poland: The conquest of history
George Sanford

Kyrgyzstan: Central Asia's island of democracy?
John Anderson

Ukraine: Movement without change, change without movement
Marta Dyczok

The Czech Republic: A nation of velvet
Rick Fawn

Uzbekistan: Transition to authoritarianism on the silk road
Neil J. Melvin

Romania: The unfinished revolution
Steven D. Roper

Lithuania: Stepping westward
Thomas Lane

Latvia: The challenges of change
Artis Pabriks and Aldis Purs

Estonia: Independence and European integration
David J. Smith

Bulgaria: The uneven transition
Vesselin Dimitrov

TABLE OF CONTENTS

Figures and Tables   vii

Chronology   ix

Preface   xi

Map of Latvia   xiii

1 A Historical Introduction to Modern Latvia   1

2 Latvia's Politics 1987–1991: The Thorny Road Towards
   Independence   45

3 Latvia's Democracy Examined: 1991–1999   67

4 Latvia's Economy since 1991   89

5 The Foreign Policy of Latvia   119

6 Conclusions   149

   Bibliography   155

   Index   163

Figure 3.1   The three most favoured citizenship models   76

Table 1.1   Per cent of the vote to political parties in Latvia's
            four *Saeimas*   18
Table 1.2   Political arrests during the first year of the Soviet
            occupation   27
Table 1.3   Political arrests during the Stalinist era   33
Table 1.4   Political arrests during the rise and fall of the
            National Communists   36
Table 1.5   Political arrests from 1963 to 1986   40
Table 2.1   Decrease in number of ethnic Latvians among
            engineers and policemen   48
Table 2.2   Communist Party size and ethnicity 1960–1990   49
Table 2.3   Support for independence according to ethnicity   64
Table 3.1   Distribution of seats to the fifth *Saeima*, 1993 71
Table 3.2   The attitude toward Latvia joining Russia   80
Table 3.3   Number of party lists and their success in the
            parliamentary elections   83
Table 3.4   Parties according to their orientation   83
Table 3.5   Latvian Prime Ministers and their party affiliation 83
Table 4.1   Jurisdiction of industry in Latvian SSR   91
Table 4.2   A comparison of Latvian and Swedish health
            statistics   114

CHRONOLOGY

1817, 1819	Serf emancipation in Baltic provinces
1850s	Beginning of Latvian national awakening
1861	Serf emancipation in Latgale
1868	Riga Latvian Association founded
1873	First Latvian song festival
1880–1900	Beginning of industrialisation
1890s	Social democratic thought enters Latvia
1905–1907	Revolution and punitive expeditions
1914–1920	World War One and revolution
November 18, 1918	Latvian independence declared
August 11, 1920	Peace treaty with Soviet Russia
September 1921	Latvia admitted to the League of Nations
1920–1922	Latvia adopts constitution and agrarian reform
May 15, 1934	Ulmanis's *coup d'état*
August 23, 1939	Non-aggression and friendship pact between Germany and USSR
October 5, 1939	Treaty stations Soviet troops in Latvia
Winter 1939	'Repatriation' of Baltic Germans
1939–1945	World War Two
June 1940	Soviet ultimatum to Latvia
June 1940– June 1941	Sovietisation of Latvia
June 16, 1940	Soviet troops enter Latvia
August 5, 1940	Latvia 'admitted' into the USSR
June 14, 1941	First mass deportations
June 21, 1941	Germany invades the USSR
July 5, 1941	All of Latvia under German control
July 1941– December 1941	Holocaust of Latvia's Jews
1944–1956	Partisan war against Soviet power
March 24–30, 1949	Mass deportations of 'Kulaks'
1953–1959	National Communist reforms
October 14, 1986	Article against hydroelectric dam
June 14, 1987	First mass calendar demonstration
October 8–9, 1988	First congress of Latvian Popular Front
August 23, 1989	Baltic Chain

May 4, 1990	Supreme council votes to renew Latvian independence
January 1991	Crackdown in Latvia and Lithuania
August 21, 1991	Latvia renews independence
August 24, 1991	Russian federation recognises Latvia
September 1991	Latvia admitted to United Nations
March 5, 1993	Latvian currency, Lats, re-introduced
June 5–6, 1993	Elections to *Saeima* (fifth for Republic of Latvia)
July 6, 1993	Fifth *Saeima* convened
July 7, 1993 and July 8, 1996	Guntis Ulmanis elected State President
October 31, 1993	Baltic council founded
September 30 –October 1, 1995	Elections to sixth *Saeima*
January 16, 1998	Charter of Partnership signed between US and Baltic States
October 3, 1998	Elections to seventh *Saeima* and citizenship referendum
June 17, 1999	Vaira Vike-Freiberga elected State President
August 13, 1999	Secretariat of Baltic assembly founded

PREFACE

In the second half of the 1980s, poster artists helped to define the beginnings of *glasnost* in the USSR. In the Soviet Baltic Republics, one particularly incisive Latvian poster summed up the danger of the moment and the determination for greater autonomy. On a deep green background stood a rooster and a hatchet: the rooster was Latvia, the hatchet was Moscow, the caption: 'Dialogue'.

More than a decade later, Latvia is an independent state. The growth of democracy and the burgeoning of political parties and opinions has demonstrated that, for all of the problems of the new state, a lack of dialogue is not one of them.

This book on modern Latvia is also a dialogue, now between the book and its readers, but first of all between its two authors: Artis Pabriks, a political scientist, and Aldis Purs, a historian.

Pabriks was born and raised in Soviet Latvia; he was an active participant in the independence movement and was one of the first to benefit from the new opportunities available to Latvians. He studied in Denmark and the United States and returned to Latvia to play a role in the setting up of Vidzeme University College, a new provincial university. Pabriks remains politically active—a liberal republican— standing for civil society, the rule of law, democracy and minority rights. Purs is the child of Latvian émigrés to the West. He grew up surrounded by the intense conservatism and nationalism of the émigré community. Purs is the turn-of-the-century academic: cynical and irreverent of most things.

The authors have their differences, but they rarely disagree. They have worked together at Vidzeme University College, collaborated on academic panels, and now they have written the present volume together. Pabriks largely authored the chapters on politics and international relations, and Purs largely those on history and economics, and the conclusion. But at every stage they have worked through each other's writing and each has throughout the endeavor benefited from the support and advice of the other.

We hope that the reader finds the conversation stimulating.

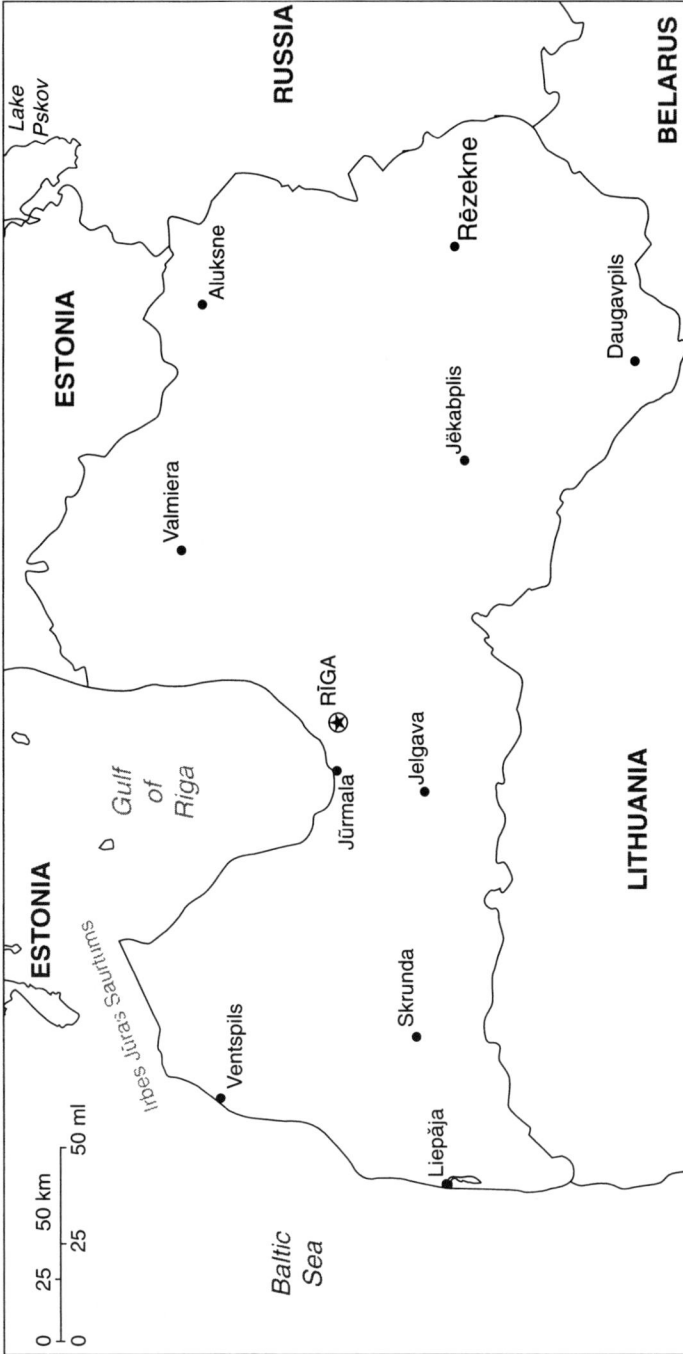

Map of Latvia

Chapter 1

A HISTORICAL INTRODUCTION TO MODERN LATVIA

A HISTORICAL INTRODUCTION TO MODERN LATVIA

Modern Latvia encompasses 64,589 square kilometres on the eastern shore of the Baltic Sea. The land itself is not immediately imposing, but has a gentle, pastoral beauty and a surprising variety of landscapes. A coastal zone of long beaches, a few natural harbours and navigable rivers yields to a patchwork of fields, forests, lakes, marshes and low hills. The landscape also reveals the history of Latvia before the nineteenth century. The ancestors of the Latvians built fortifications on the hilltops for centuries prior to the arrival of German Teutonic Knights and missionaries. Archaeological digs continue to uncover the contours of ancient Baltic society. In 1201, under the Germans, Riga was built. Latvia's landscape is dotted with relics of the following seven hundred years of different ruling powers' hold on the eastern Baltic littoral. Most castles, churches and palaces tell of the local control of Baltic Germans, but others attest to the periods of Polish and Swedish rule. Many ruined castles bear witness to Russia's expansion into the Baltic area in the eighteenth century. By the end of that century, all of modern Latvia was within the realm of the Russian Empire. Baltic Germans continued to hold power locally, but the nineteenth century witnessed the rise of a new force. The peasant nation that constituted the bulk of the population began to see its ethnic differentness as a unique identity. This identity consisted of social and economic demands as well as the political mantra of modern nationalism: that the political unit should be synonymous with the ethnic. Latvia's history stretches for millennia, but the modern history of the Latvians begins essentially in the nineteenth century.

THE NINETEENTH CENTURY, 1800–1917

Latvians had always been ethnically unique with a distinct culture and language, but only in the nineteenth century did this ethnic uniqueness become the basis for social and state organisation. The emergence of these identities coincided with pan-Germanic philosophical currents of the late eighteenth century, influenced by the French Revolution. Johann Gottfried von Herder, who taught and preached in Riga from 1764 to 1769, inspired modern nationalism in Latvia. Herder's stay in Riga was largely responsible for his interest in peasants. He participated in peasant traditions such as Midsummer's Eve and collected and published Latvian folk songs in his *Stimmern der Volker in Liedern*.[1] Herder was effectively "the first intellectual who acknowledged the Latvian people as a *nation worthy of an individual identity*".[2] Garlieb Merkel, a Baltic German, echoed Herder's belief. Unlike Herder, however, Merkel was blatant in his condemnation of Baltic German lords. He warned that Latvian serfs could be a revolutionary force. The growing frequency and severity of serf rebellions, the words of humanists like Merkel and Herder, and the fear of events in France led Tsar Alexander I to contemplate a change in peasant–lord relations.

A general emancipation of Baltic serfs was in its early stages when Alexander I was sidetracked by the Napoleonic Wars. Although Alexander I's reign became more conservative following the wars, the plans for serf emancipation in the Baltic Provinces proceeded in the 1810s. In 1817 and 1819 Baltic serfs were emancipated.[3] This newly granted freedom had few privileges; ex-serfs were not given land, nor could they purchase it. They were denied freedom of movement and were still obligated to the social control of their former lords. The emancipation, however, came more than forty years before the emancipation of Russian serfs, guaranteeing that the Baltic Provinces followed a radically different course of development during the nineteenth century. The south-eastern corner of modern Latvia, however, was not administratively a part of the Baltic Provinces. This corner, Latgale, would develop differently from most of contemporary Latvia.

The restrictions accompanying emancipation guaranteed continued Baltic German control and impoverished Latvian peasants. Prompted by poverty and despair, some Latvian peasants converted to Orthodoxy en masse, hoping to have better treatment under the religion of the Tsar. Legal changes in the 1840s and 1850s allowed Latvians the right to buy land and move more freely. Gradual Latvian purchases of land and the ability to move to towns, coupled with a demographic explosion, fuelled the growth of Latvian nationalism in the second half of the nineteenth century. Latvian smallholders and townsmen looked for a movement that would protect their interests against the enormously privileged Baltic German nobility.[4]

As Latvians purchased land, they moved into junior positions in local administration and the judiciary. Many became schoolteachers. Previously, the most gifted children rose above their station in life through education that involved cultural assimilation. Latvian peasant identity was shed, and people became Baltic Germans. Baltic German-ness was associated more with education and position in society than blood. In the mid-nineteenth century, the need to assimilate waned. Sons of Latvian peasants (at the time almost solely sons) entered a Baltic German world of education which included humanist streams of thought that praised the spirit of Latvians and accorded them equal value to other nations. By mid-century, about thirty Latvians were enrolled at the University of Dorpat (Tartu), the only institution of higher learning in the Baltic Provinces and a bastion of the Baltic German world in the Tsarist Empire. This group consciously identified themselves as Latvians, chose not to assimilate and began the Latvian national awakening. Inspired by the Young Italy and Young Germany movements, they called themselves the Young Latvians (*Jaunlatviesi*).

The Young Latvians did not completely divorce themselves from the traditional Baltic German cultural milieu. Most married Baltic German women, wrote in German and were equally comfortable in Russian. Their ability to manoeuvre for limited Latvian rights and to gain more popular acceptance for the idea of a Latvian nation depended upon what Benedict Anderson called "their bilingual literacy, or rather literacy and bilingualism."[5] They worked as intermediaries between the Baltic Germans, the Russian Court and the Latvian peasants in a society that increasingly needed an active and literate citizenry. Anderson's observations on why the beginnings of nationalism are invariably referred to as "awakenings" also fit the Latvian example. Awakening from sleep connected the new polity with a distant past. Anderson's words echo the Young Latvians:

> the vanguard of most European popular nationalist movements were literate people often unaccustomed to using these vernaculars, this anomaly needed explanation. None seemed better than 'sleep,' for it permitted those intelligentsias and bourgeoisies who were becoming conscious of themselves as Czechs, Hungarians, or Finns to figure their study of Czech, Magyar, or Finnish languages, folklore, and musics as 'rediscovering' something deep-down always known.[6]

This first generation "rediscovered" Latvian language, tradition and folklore. It created for the Latvian nation all of the accoutrements of a modern, civilised nation, from a national epic to Latvian translations of the Classics to scientific language. Their published output found an eager market in the new Latvian smallholders and those moving to the city.

The Latvian peasantry was an eager market because literacy rates in the Baltic Provinces were abnormally high for the Russian Empire.[7] From the mid-nineteenth century, literacy rates increased markedly every decade and there was a boom in school construction. At least 80 per cent of the Latvian population could read—a rate that dwarfed Russian rates. Increasing literacy, sophistication and wealth led to the formation of Latvian self-help organisations. The Latvian Association of Riga, founded in 1868, was the first and most important. Provincial branches co-ordinated a campaign promoting education and culture across the Baltic Provinces. In 1873, the Latvian Association organised the first Latvian Song Festival in Riga, a visual and audible demonstration of Latvian cultural merit. The Latvian Association of Riga and the Young Latvian Krisjanis Valdemars epitomised the Latvian nation from 1850 to 1885. They struggled for a cultural awakening, but had a gradual programme for social and political change. They did not advocate social rebellion, but defended the values of the middle class from private property to their leading role within the national awakening. Krisjanis Valdemars, whose formula for the national revival was education + technology + capital, found no contradiction in working within Tsarist ministries while becoming the material godfather of the movement. The Young Latvians were loyal to the Tsar; the extent of their political programme was indigenisation or the replacement of the Baltic Germans' privileged position in Baltic society with Young Latvians.

Economic and demographic changes surpassed the modest plans of the Young Latvians in the 1880s and 1890s. In the second half of the nineteenth century, the Empire began a vigorous drive for industrialisation. Riga was a natural starting point. The city's port was a natural conduit for trade and the Baltic German elite was eager to convert earlier land-holding wealth to wealth based on merchant ties and industrial strength. The liberalisation of restrictions on peasant movement also gave Riga and the towns of the Baltic Provinces the work force for the new industrial and mercantile concerns. If, in 1861, Riga was not much different from its Hanseatic former self (that is, controlled by guilds and predominantly export oriented), by the turn of the century Riga was a city transformed. Its economy became increasingly diversified; its engineering, lumber, chemicals and manufacturing industries all vied with the more traditional textile industry for the floods of workers arriving every decade. Riga's population more than doubled from 1867 to 1897; its workforce quadrupled. The Baltic

Germans were able to maintain their control over the city despite the revolutionary changes of early industrialisation. Their percentage of the population decreased markedly, but their wealth and power remained unchallenged. Even in 1900, Baltic Germans owned more than 80 per cent of Riga's largest businesses.[8] Riga was not alone; the towns of Mitau, Vendau and Libau all became industrial centres (if small ones) in the last decades of the nineteenth century.

As already mentioned, the Young Latvians did not see Imperial power as a threat to Latvian identity. They identified Baltic German hegemony as the obstacle to their national aspirations. After their years at the University of Dorpat, many of the Young Latvians assembled in 1863 as the editorial staff of the *Petersburg Newspaper* (*Peterburgas avize*). Although the paper was clearly critical of Baltic German hegemony, it also contained articles on self-improvement, the latest farming techniques, and debates about economics. Baltic German pressure closed the paper in 1866. The experience pushed many of the Young Latvians closer to the Slavophiles and the idea that Russification could curb the power of the Baltic Germans.

From 1880 to 1900, the Tsarist authorities began a vigorous campaign of Russification in the Baltic Provinces and Finland. Ideological Russifiers were not content with the simple removal of German influence. They had as little respect for the Latvian peasantry as the Baltic German aristocracy did, and hoped to transform Latvian peasants into good Russians. Younger Latvians resented the "Young Latvians" compromise with Tsarism as a compromise with autocracy and looked beyond the ideas of nationalism. Finally, because the Russifiers targeted the school system, Baltic Germans withheld funding for Russified schools. Tsarist financial support often did not make up the financial shortfall, and Latvian schoolchildren were faced with sub-par education at the very moment that the demand for better, mass education increased. Ultimately, Russification was a terrible failure. Neither Baltic German political, economic, social nor cultural hegemony was removed from the Baltic Provinces. The "Young Latvians" were compromised in the eyes of a new, more radical generation of Latvians. The Provinces were neither culturally Russified nor administratively standardised with the rest of the Empire.[9]

Aggravating the tensions between Baltic Germans, Young Latvians and the Tsarist administration was the emergence of a working class and a radical new generation of Latvian intellectuals that claimed to speak for the workers. In the mid-1880s, a movement named the New

Current (*Jaunstravnieki*) began to coalesce around the daily newspaper, *Daily Page* (*Dienas Lapa*). The newspaper introduced basic social democratic theory through reviews of literature and within news stories. The anger of the New Current was not simply directed against the Baltic Germans, but against the Riga Latvian Association as well. If in the 1850s and 1860s the Young Latvians confronted a Baltic countryside controlled by the Baltic German aristocracy, the New Current saw a countryside in the 1890s with divided loyalties and a differentiation among peasants. The countryside was divided between Latvians who had acquired land (418,000) and those who were still landless agricultural labourers (591,000). Likewise, if Riga had about 6,000 factory workers in the 1860s, by the 1890s there were over 24,000. The Riga Latvian Association seemed to represent merchants and property owners, not the growing working class. The New Current thrived on the support of the landless and the factory workers.[10]

The New Current did not support political nationalism; its adherents argued that workers across the Russian Empire had a common cause. It would be incorrect, however, to view the New Current as a refutation of the national awakening. Instead, as Andrejs Plakans has brilliantly argued, the New Current challenged the Riga Latvian Association's right to speak in the name of the nation. In other words, the New Current offered an alternative concept of Latvianness. As Plakans argued: "By staying inside the Latvian-language world to sharply attack Latvian nationalism, the *jauna strava* demonstrated that the Latvian intelligentsia could sustain deep cleavages without diminishing the Latvian presence in the Baltic littoral."[11]

Workers' unrest in 1897 convinced Tsarist authorities that the message of the New Current was becoming dangerous. The authorities banned the newspaper and arrested eighty-seven members of the movement. Despite the arrests, increased police surveillance and general crackdown against the New Current, social democratic thought was not distinguished from the Baltic Provinces. In 1904, an underground Social Democratic Workers' Party was founded as Latvia's first political party. The growing importance of socialist thought in the Baltic provinces should not eclipse equally important events of the turn of the century.

The *fin de siècle* was an exciting time in Latvia and Riga. In the countryside, rudimentary education was available to all. Literacy rates remained the highest in the Empire and educational opportunities, although constricted by Russification, continued to grow. Private schools, from teachers' colleges to commercial schools, opened and gave a new generation of boys and girls greater educational options. Likewise, local self-help organisations, such as co-operatives, insurance plans, choirs and cultural associations blossomed. The rail and road system continued to improve and the Baltic Provinces seemed to become a much smaller place. The countryside was a complex, variegated world. In less than a hundred years, Latvian peasants had gone from a lifetime defined by serfdom to a dizzying array of variables: smallholding, landless, school teacher, civil servant or migration to the towns.

If the countryside was rapidly changing, the towns and particularly the city of Riga changed beyond description.[12] The demographics of the town skyrocketed; landless Latvians flocked to Riga (and the other towns) by the tens of thousands and overwhelmed the old mechanisms for assimilation into Baltic German culture and social control.[13] These thousands created a Latvian Riga, something that had not existed before. They consumed the written Latvian word with a passion, creating an avid market for everything from poetry and translations of classics to pseudo-scientific literature and pulp fiction. This thirst extended to a demand for Latvian theatre, art and popular entertainment.

The demographic change altered city politics. Within Tsarist autocracy there had been municipal elections, but the elected body's authority was severely limited and the electorate was minuscule owing to property requirements. City government was the arena of the Baltic Germans. By the turn of the century, there were substantial numbers of Latvians that met the property requirements, including phenomenally wealthy ones. They challenged Baltic German hegemony in city politics.[14] To further confuse the political picture, a substantial number of Russians, as well as Jews, Poles and Lithuanians, lived in the city. The Baltic German elite manoeuvred one group against another and remained in control of the city throughout the Imperial period, but their hold grew tenuous.[15] The affluence and political division of the city combined to create the lasting monument to the turn of the century in Riga, its architecture. Political enfranchisement rested on property ownership, which, coupled with the demographic explosion, led to a building boom. The Baltic German cultural hold on the city included the attachment to the *Jugendstil* prevalent in Western Europe at the time, and as a result, Riga has block after city block of the best examples of *Jugendstil* architecture in all of Europe.[16] Much of contemporary Riga was built from 1890 to 1914.

The cities, and again particularly Riga, were industrial centres. Year after year, the number and size of factories increased. If, in 1884, there were 47 factories that employed from 51 to 200 workers and four that employed more than 200, by 1910 there were 151 employing from 51 to 200 and 77 employing more than 200. The factory workers were employed in all types of industries, and increasingly in towns outside Riga. In the decade from 1900 to 1910, industry grew faster in smaller towns than in Riga proper. For example, for every 100 workers in Riga in 1900, there were 141 workers in 1910. In the provincial towns, however, this change was much more marked. For every 100 workers in 1900, in 1910 there were 168 in Libau (Liepaja), 187 in Mitau (Jelgava), 192 in Ventau (Ventspils), 197 in Dvinsk (Daugavpils), 261 in Sloka, and 302 in Limbazi.[17] The statistics, however, belie the complexity of the industrial picture. There was a tremendous difference between the types of industries and the experience of its workers. Some were privileged, skilled workers with relatively good wages who subscribed to a variety of newspapers, enjoyed Riga's cultural offerings and lived considerably better than their fathers and forefathers, while others toiled long hours for substandard wages and lived in abject poverty. Generalisations about the working class, workers' reactions or loyalties are as simplistic as generalisations about Latvian peasants or Latvians themselves.

Riga was one of the most modern cities of the Russian Empire. Its prosperity and bright economic prospects attracted considerable foreign capital that brought the best of Western European technology to the city. Riga was awash in electric light, telegraphs (even telephones), streetcars and the amenities of *fin de siècle* Western Europe. Riga's rapid industrialisation and modernisation, however, left a substantial part of the population behind in its rush to prosperity. These people, unskilled workers, migrant farm labourers and peripheral religious or ethnic communities, contrasted sharply with the wealth of the upper and middle classes. Riga and much of Latvia approached Western European standards of cultural achievement, literacy, economic productivity and social welfare, yet its political institutions lagged desperately behind.[18] The limited rights of city government paled in comparison to the autocratic rights of the Tsar or the rights of the Tsar's representative, the governor-general of the Baltic Provinces. The lack of political rights frustrated all ranks of society: the workers could not legally organise, the middle class Latvians could not challenge the rights of the Baltic Germans, Russification of the school system could not be opposed, religious liberties were threatened by the official Orthodoxy of the Empire, and so on. Revolution meant an opportunity to redress grievances by violent or at least exceptional means.

The Revolution of 1905 suggests that generalisations about Latvians seem justified. Soviet scholars generalised when claiming that working class radicalism and landless peasant fury proved a Latvian affinity with Bolshevism, whereas nationalist historians claimed the Revolution was a national rising.[19] Even Ronald Suny's more sophisticated attempt to describe both 1905 and 1917 as instances where "class and ethnic identities overlapped and reinforced one another, but the form of expression was socialist than predominantly nationalist" falls short of the mark.[20] Although all explanations draw upon legitimate evidence, they only present a partial picture of the Revolution. The events of 1905 in the Baltic Provinces showed the rough convergence between Latvian nationalists and socialists on some issues, but highlighted their differences in other areas.

In January 1905, Tsarist police fired on demonstrators on the banks of the Daugava River, killing dozens. The Baltic Provinces erupted in revolution.[21] Workers' councils effectively controlled urban centres and peasants rose in rebellion. Armed peasant formations successfully captured small towns, burned dozens of Baltic German manors (45 in Kurzeme and 85 in Vidzeme) and formed autonomous local govern-ments. There seemed to be common ground between all participants, as lists of demands from various congresses of teachers, local governments and workers read similarly. The October Manifesto, however, exposed the divisions within the Latvian camp. Backed into a corner, Tsar Nicholas II agreed in the October Manifesto to consider a constitutional monarchy and a degree of civil rights. Progressives throughout the Empire and nationalists (the followers of the Young Latvian ideal) in the Baltic Provinces saw this as the important victory. They believed that constitutional monarchy would rein in autocracy and that the gradual extension of the franchise would lead to the ousting of the Baltic Germans from political power. The nationalists at this point wanted to return to normalcy and work within the new system.

The socialists, however, saw the October Manifesto as the last gasp of a dying beast. Autocracy was on the verge of defeat and therefore the Revolution should continue with its more radical demands. The socialists worried that the Tsar was "buying off" the middle class in order to dilute the strength of the Revolution and crush the workers. The question of land and property rights, however, dogged the socialists through out 1905. The majority of peasants were landless and the majority of workers were unskilled. Both were receptive to much of the Social Democratic programme. It seems likely, however, that they did not want state or collective ownership of property. They wanted to seize the considerable remaining land holdings of the Baltic German aristocracy and divide it among themselves. Once divided, land would become private property. Likewise, the nationalists who had built their programme on the economic self-help basis of the Young Latvian movement would not consider an assault on property rights. Therefore on the surface, the Tsar's ploy worked. The October Manifesto succeeded in dividing the opponents of autocracy so that they could be conquered one by one. On a deeper level, however, the October Manifesto exposed the crucial difference between nationalists and socialists. The nationalists used extreme measures as a last option and preferred an evolutionary movement towards the Western European ideals of constitutional government based on property and national rights. The socialists saw Revolution as means to an end and desired social and economic transformation equal to political change.

If in 1905 the revolutionaries carried the initiative, in 1906 autocracy did the same as it ruthlessly crushed the revolution.[22] Three thousand Tsarist troops arrived in December and aligned with the Baltic Germans' self-defence militias (the *zelbstschutz* that were formed as revolutionaries burned the manors) to form punitive expeditions. These expeditions defeated peasant groups across the Baltic Provinces, at times using artillery. The armed units burned farmsteads, schools and public buildings and delivered summary justice at rapidly held field courts martial. Thousands were forced into exile and despite a short-lived guerrilla movement that took to the forests, the Revolution was crushed militarily by mid-1906.

The political end of the Revolution of 1905 was much more ambiguous. The October Manifesto provided for an elected, consultative body—the *duma*. The elections to the first *Duma* brought good returns for Latvian nationalists, who, although working within the system, pushed it to the extreme. The Tsar disliked the radical nature of the first *duma* and dissolved it. The progressive candidates, including Latvian nationalist deputies such as the future first State President of Latvia, Janis Cakste, withdrew to Finland and authored the Vyborg Declaration. The Declaration called on peasants and workers to continue to resist Tsarist autocracy and to withhold taxes, but the momentum of the Revolution was lost. Tsarist autocracy diluted each successive *duma* with new franchise restrictions, and the number of

Latvian deputies declined progressively. Despite these setbacks, there were considerable political gains. An electoral and parliamentary system held the promise for reform from within. Skilled Latvians moved into elective positions, into lower levels of ministries and administrative bodies and into the army and police force in greater numbers. The professionalisation of ministries lessened the caprice of the Royal court and encouraged skilled, qualified nationalists to work within the government. Many future Latvian leaders, after participating in the Revolution of 1905, spent the years until 1914 in government service. Russification was more or less abandoned. The censorship of the press was loosened. The legal procedure for founding associations (and their ability to work) was simplified. Peasant land banks and easier conditions for purchasing land led to increased farm purchasing and the growth of the co-operative movement. Agricultural and industrial production increased. After a long industrial slump (a hidden, often overlooked contributor to worker unrest in 1905) that lasted from 1903 to 1908 ended, new factories were built. More jobs and higher wages followed. Building construction continued at the earlier feverish pace. Importantly, after 1905 the Latgalian dialect was legalised in print and considerable steps were taken to bring the Latgalian population in step with the Latvian national movement. These progressive changes, however, highlighted the redundancy of autocracy. Despite all of the new freedoms and development, a Russian Tsarist bureaucrat or a Baltic German lord could still arbitrarily overrule or obstruct the emerging civil society. Frustrations with the system became more apparent as more and more Latvians worked within that very same system.

Many Latvians chose to work definitively against the system. The above-mentioned avenues of advancement seemed open to so very few. Latvian socialists from prison and exile, workers left without a political voice, and landless peasants who bore the brunt of the punitive expeditions and had little hope of purchasing land looked at the return of Tsarist autocracy with bitterness. To these radicals and other disgruntled victims of the Revolution, the emerging middle class and prosperous farmers had been co-opted into the system. This seemed all the more apparent because it fell into standard Marxist theory. The course of action for the future lay with more radical steps and a break with the nationalists and bourgeoisie. Not surprisingly, Latvian socialists turned overwhelmingly to the extremism of Bolshevism. As the Tsarist Empire stumbled towards World War One, Latvian

Bolsheviks were just as active as the nationalists were. If nationalists worked within the system, Latvian social democrats (largely Bolsheviks) revived strikes and labour unrest, and maintained an underground party despite police repression.

World War One was a disaster for Russia.[23] In the summer of 1914, the Latvian members of the *duma* spoke dutifully of their unquestioned support for the war effort and their belief in a speedy victory. The first great Russian offensive showed considerable progress, but the German counter-offensive turned quickly to a rout. By the spring of 1915, the Russian armies were defeated and in full retreat. The Tsarist army called for the evacuation of Kurzeme and Zemgale, throwing tens of thousands of Latvians into exile. The front stabilised along the Daugava River, effectively halfway through Latvian territory, and remained there until the second half of 1917. The majority of the population of the German-occupied area was evacuated, but the Tsarist system had no provisions for war refugees. Autocracy feared the impact of thousands of hungry refugees in the hungry cities of Moscow and Petrograd, and encouraged them to settle in the Far East and the Caucasus. Riga was within artillery range of the German forces and was partially evacuated. Most of the industrial machinery and its workforce were transported to central Russia.

Facing a refugee crisis and military defeat, the Tsarist autocracy recognised two Latvian organisations in the hope of turning the tide. Latvian Refugee Associations were allowed to operate across the empire to look after Latvian war refugees. Similarly, Latvian regiments were formed within the Tsarist army. These, it was argued, would more passionately defend their homeland. In 1916, the two organisations seemed the basis for nationalist hopes. The Refugee Associations would develop the skills of governing and legitimacy in the eyes of the refugees. They would be the nucleus of a future Latvian state. The Latvian regiments, more popularly known as Latvian Rifles, would be the nucleus of a Latvian army. Individual Latvian nationalists began to talk about Latvian autonomy in a post-war Russia.

The hopes of Latvian nationalists seemed to rise as the popularity of Latvian Bolsheviks fell. Lenin, after all had, proclaimed defeatism steadfastly since the beginning of the war. Defeat for all countries, in Lenin's eyes, would weaken the structures of the state and give the proletariat a chance to rise. Strategically, Lenin proved to be prophetic, but until late 1916 it was an unpopular proposition in Latvia. Defeatism sounded like suicide. By early 1917, the status quo was suicide as well.

The Latvian Rifles' Christmas offensive was initially successful, but incompetence in the Tsarist army negated the early victory. The battle seemed an analogy for the war—a pointless loss of lives after heroism and bravery. To the soldiers it became obvious that the Tsarist army could not win the war, but that they had no qualms about sacrificing the Latvian nation in a losing effort. The war had to be stopped. The February Revolution accelerated the radicalisation of the Baltic Provinces.

In Petrograd and throughout most of the Empire, dual power characterised 1917.[24] Dual power was the curious situation in which Tsarist autocracy was not replaced by one government in power, but by two conflicting institutions—the provisional government and the Soviet of workers, peasants and soldiers. Conditions were impossibly complex and further complicated by efforts to continue the war and prepare for a Constituent Assembly that would determine the future of Russia. By October, political momentum and power shifted to the Bolsheviks in Russia's great cities. Dual power ended when the Bolsheviks claimed sole power in a coup, the October Revolution. The Russian Civil War soon followed complete with foreign intervention and the foundation of the Soviet state. Latvia emerged from the chaos as an independent state, one of what Richard Pipes referred to as the nations that got away—Poland, Finland, Estonia, Latvia and Lithuania. Ronald Suny instead referred to their independence as a consequence of "political circumstances, rather than from the plans or intentions of the dominant political movement".[25]

Dual power was a short-lived phenomenon in the Baltic Provinces.[26] The Provisional Government appointed a Latvian mayor to Riga, hoping to assuage national concerns; but it was too little, too late. By May 1917, the Latvian Rifles, once the darlings of the nationalists, went over to the Bolshevik side. With the Rifles as military support, the Riga Soviet pushed ahead of the representatives of the Provisional Government. Similarly, radical soviets in the countryside followed the Riga example. Riga became known as "Riga the Red" throughout the Russian empire, emboldening Bolsheviks everywhere and drawing contempt and disgust from conservatives and progressives. The red city did not last; a German counter-offensive captured Riga. German occupation gradually extended through almost all of the Baltic Provinces (complete occupation occurred during the Brest-Litovsk treaty negotiations). Latvia was a nation occupied, and Latvians a nation divided. Latvian Bolsheviks chose to defend the October

Revolution while nationalists bided their time. Latvians at home lived under German military occupation while tens of thousands of refugees were scattered across a Russia at civil war.

German occupation was concerned with provisioning the army and with accommodating Baltic German requests to be annexed into the German Empire. Many Baltic Germans found their position within the Tsarist Empire under threat. They considered themselves to have been loyal subjects of the Tsar and bringers of culture to Latvian peasants. The malice of Latvian peasants in 1905 alarmed them. Anti-German sentiments in the Royal court unsettled their hopes in the Imperial Court. Furthermore, the strength of Imperial Germany in 1917, compared to the chaos and anarchy of 1917 Russia, made the choice of alignment with Germany rather obvious.[27] The early spring of 1918 seemed to be the high point of German rule in the Baltic. By September and October, however, their fortunes had turned. German troubles on the Western front pushed Imperial Germany to accommodate Latvians (and other nationalities) to preserve their victories in the East. This was the setting for the birth of the Latvian state. To preserve the October Revolution, the Bolsheviks had abandoned the Baltic Provinces (and much more) to the Germans in the Treaty of Brest-Litovsk in the spring of 1918. In the fall of 1918, however, a power vacuum existed from the Baltic Sea to the Black Sea. Latvian nationalists, like other nationalists throughout Eastern and Central Europe, stepped to the fore. Demands for autonomy within an Empire fell away. The unimaginable dream of independence seemed possible. Simultaneously, the Bolsheviks, including Latvian Bolsheviks, saw an opportunity to return to the lands surrendered at Brest-Litovsk and push westward to see a European-wide revolution.

THE REPUBLIC OF LATVIA, 1918–1940

The Latvian National Council, a collection of self-appointed representatives from various Latvian organisations, declared the independence of Latvia on November 18, 1918. Karlis Ulmanis, the leader of the Farmers' Union (after the Social Democrats, Latvia's most popular party), became Latvia's first Minister President and the talented Zigfrids Anna Meierovics the Minister of Foreign Affairs. Initially, it seemed as if all of the new state's supporters were gathered in the Theatre Hall where independence was proclaimed. The Latvian National Army was composed of a handful of patriotic students and former Latvian Tsarist junior officers. The ruling administration in the Baltic Provinces was still the German army, but with the November 11th armistice signed, most German soldiers assumed they would soon return to Germany. The Bolsheviks counted on a German evacuation or a German revolution and invaded Latvia in December 1918. Declaring a Soviet Republic of Latvia, the Red Army accompanied by Latvian Bolsheviks met little resistance as they occupied Riga in January 1919. Latvian nationalists, uneasily aligned with the remaining German forces, retreated to the far west of Latvia. In April, the Germans staged a coup that cast aside the provisional Latvian national government and set up a conservative Baltic German government headed by the puppet pastor, Andrievs Niedra.[28] Ulmanis and his ministers took refuge on a British ship in Liepaja harbour. The nationalists seemed a spent force.

Within a year the Latvian national government staged a miraculous comeback, winning all Latvian territory and receiving *de facto* international recognition.[29] Within two more years, Latvia was a full member of the League of Nations, convened its Constituent Assembly and began the long, arduous path of post-war reconstruction. How was this possible? To begin with, the Soviet government lost popular support quickly. Latvian Bolsheviks would not sanction private ownership of land, instead proposing state farms. Furthermore, their heavy-handed treatment of all but their most ardent supporters sealed their fate. In the Latvian Bolsheviks' last weeks, many enemies (judged solely by class) were executed and others sent to hard labour. The German government, of course, did not offer a credible alternative. The allied powers tolerated their presence in the Baltic littoral only in the hopes of staving off Bolshevik incursions into the power vacuum of Eastern Europe. Once the Latvian government proved its legitimacy and the Latvian national army proved its relative merit on the battlefields, the Allies demanded the evacuation of German troops from Latvia. The Latvian national government won this legitimacy, and troops for the army, by their guarantee of radical agrarian reform and a democratically elected Constituent Assembly. The government guaranteed veterans of the war for independence first access to land in the reform, a particularly effective incentive to join the Latvian national army. This democratic promise meant that Latvia's peasants, workers and intellectuals could create their own society. This programme, regardless of its imperfections, far surpassed the promises held out by Soviet or German rule of Latvia.

The state that emerged from the ashes of extended war and revolution was inevitably scarred by the experience of the previous six years. The lasting effects of the war and its devastation cannot be underestimated. To begin with, the industrial nature of Riga and the other industrial centres of the Baltic Provinces was erased by the evacuation of 1915. In 1918, at the declaration of independence, Riga had no operating factories. The Soviets had no intention of returning the factory machinery. As a result, cities shrivelled. In 1914, more than a million people lived in Latvia's cities, almost 40 per cent of the population. In 1920, this figure was less than 400,000. Along with the collapse of industry, trade stopped. Riga's port was silent for nearly five and a half years, and with little industry in post-war Latvia and the Soviet Union curtailing foreign trade, commerce never fully returned to Latvia. Post-war Latvia was forced to address massive economic restructuring from an industrial society to a smallholding one.

The countryside, however, was equally scarred. Latvia was one of the battlefields of World War One and few civil parishes (*pagasti*) did not show the scars of war. Almost half (251 of 499 civil parishes) saw military action within their boundaries between 1915 and 1920. The civil parishes, districts and provinces near the Daugava River, which was the front between Germany and Russia for more than two years, showed the most damage. Within the province of Zemgale, 66 per cent of all civil parishes witnessed the war, while 74 per cent of civil parishes within Latgale suffered from the physical effects of battle. Similarly, 55 per cent of all of Latvia's civil parishes were covered with trenches. Riga, which suffered several long, protracted sieges, had 95 per cent of its district's civil parishes covered with trenches and barbed wire. These physical scars of battle handicapped post-war reconstruction. Peasants returning to fallow fields had to spend countless hours cutting barbed wire, filling trenches and bomb craters and rebuilding homes, barns and other buildings. One-tenth of all of Latvia's buildings were destroyed in the war. Seven of the unfortunate civil parishes near Riga lost more than 80 per cent of their pre-war buildings. Rebuilding agricultural communities complicated the difficulty of adjusting to a new agrarian order following the land reforms of the Constituent Assembly.

The greatest catastrophe of war and revolution was the tremendous loss of life. Latvia was depopulated. Our most accurate comparisons are with the first and last census of Imperial Russia, that of 1897. For every one hundred people in 1897, in 1920 there were ninety-six in Latgale, eighty-four in Kurzeme, eighty-two in Vidzeme, and seventy in Zemgale (the four provinces of Latvia). The majority of lives lost were young men killed in battle, but rarely is a sexual imbalance properly understood. In 1897, there were 37,515 more women than men in the territory that would be Latvia. After the war, this figure had increased to 152,277. The demographic pyramid of the state suggested a difficult future for the new state. There were few children under the age of ten (there had been nearly 535,000 before the war, but there were now fewer than 250,000), suggesting long-term problems in societal function. Most troublesome for the short-term existence of the new state was the lack of able-bodied adults needed for reconstruction. Extrapolating from pre-war statistical information, by 1920 there should have been about 2,500,000 people in Latvia, with 780,000 between the ages of twenty and forty (the working prime). Instead, in 1920 there were only 383,000 people between the ages of twenty and forty to rebuild the state.[30] All of Latvia's accomplishments need to be valued in relation to the tragic demographic backdrop against which they were achieved.

The Constituent Assembly enacted radical agrarian reform and authored a Constitution for the new state. The two pieces of legislation defined the strengths and cleavages of independent Latvia. Land reform broke the Baltic Germans' hold on rural life, but questions remained about the efficiency of the smallholding agriculture the reform created. The reform did, however, bring social peace to the countryside. Most landless peasants received land and almost overnight went from a "landless proletariat" supporting revolution to smallholders worried about building subsidies, agricultural credits and market prices. The Baltic Germans, after fruitless protests to the League of Nations, abandoned the countryside, but maintained their pre-war domination of Latvia's industrial and mercantile sectors.[31]

The countryside divided along a new cleavage: beneficiaries of the agrarian reform versus the small established Latvian, farmers from before the war (as well as a significant minority of the landless). The political weight of the new smallholders, coupled with their dire conditions on a war-devastated land, often without seed, livestock, supplies or even homes, led the government to grant tax waivers and subsidies. Furthermore, popular aspirations for improved education and health care were also brought to fruition in a legislative flurry. The established farmers, however, often almost as devastated by the long years of war, resisted. They saw the state's provision of greater services to people with tax waivers as unfair.

The Constitution defined many other divisions in independent Latvia. Power was invested in the people of Latvia, not in the Latvian people. The nation was a geographic unit, not an ethnic one. The shock to Latvian nationalists came when the Constituent Assembly and subsequent parliaments (*Saeimas*) returned substantial numbers of ethnic minority deputies (Table 1). Latvia's Social Democrats, always numerically strongest in Parliament, could not be counted on to protect ethnic Latvian interests. Latvian nationalists believed that at the moment the Latvian state was achieved, its development was robbed of the proper doses of ethnic Latvian content thanks to a "too democratic" Constitution and the uncooperative nature of the electorate at the polling booth. Latvia's nationalists faced the dilemma of which was a higher principle, democracy or nationalism. Minority parties, however, were equally unhappy that specific minority rights were not enshrined in the constitution. Minority education, press, association and community regulations were within the preserve of common law and under the constant threat of removal by a simple majority in the *Saeima*. Nationalists and minorities alike disliked the Constitution.

The Constitution also left several loose political ends.[32] The Assembly deferred the Constitution's second part, the rights of individuals, indefinitely (finally accepted in 1998!). Furthermore, the Constitution created a State President and Minister President. The Minister President was the equivalent of a Prime Minister, and the State President, elected by the *Saeima,* was a largely ceremonial post. Latvia's conservatives called consistently for increasing the powers of the executive post, and for electing the incumbent popularly. The left and the minority parties, however, resisted, fearing that the Farmers' Union and Karlis Ulmanis would capture the post, and use it to dominate the political landscape.

TABLE 1 PERCENTAGE OF THE VOTE TO POLITICAL PARTIES IN LATVIA'S FOUR *SAEIMAS*

Party	1st SAEIMA	2nd SAEIMA	3rd SAEIMA	4th SAEIMA
Social Democrats	30	32	26	20
Other Socialists	8	5	10	8
Left Centre	13	11	9	11
Farmers' Union	16	16	16	14
Other Agrarian	9	12	13	20
Right Wing	9	9	8	8
German	6	5	6	6
Russia	3	5	6	6
Jewish	5 (+1)	4 (+1)	4 (+1)	5
Polish	1	2	2	2

Source: Modified from Bilmanis, Alfreds, *A History of Latvia* (Princeton: Princeton University Press, 1951), pp. 342–343.

Finally, electoral law allowed parties with 1 per cent of the vote to claim a seat in the *Saeima*. As a result, Latvia's political life saw dozens of parties, the majority of which were little more than special interest groups. The necessity of including many parties in coalitions exposed politics to corruption and parliamentary instability.

Latvian inter-war politics is often caricatured with the blatant bribery involved in creating cabinets, the inherent weakness of all governments, and the ransom of national interests by minority deputies in return for coalition support. That Latvia had fourteen cabinets in twelve years of parliamentary rule has been a source of criticism, but the changes need closer scrutiny. The first cabinet was inherited from the Constituent Assembly. The death of the last cabinet in the coup of May 1934 resulted from unnatural circumstances, rather than by parliamentary means. Furthermore, it became political habit to dissolve cabinet on the verge of new elections so that each new parliament saw a new cabinet. The number of cabinets therefore seems less shocking. More significantly, despite cabinet changes, ministers' portfolios remained fairly constant. Politicians earned reputations as experts in certain fields and repeatedly held the same ministerial portfolios. Beyond the minister, the civil servants and memberships of parliamentary commissions remained remarkably consistent. Directors of departments within ministries stayed at their posts regardless of cabinet changes. Throughout all twenty-two years of Latvia's existence, for example, there was only one Director of the Department of Local Government.[33] Latvia's parliamentary era suggests instability, but the system worked well enough to pass legislation consistently. Except for the corporatist Civil Code of 1937, almost every defining aspect of political, economic and social life in Latvia was passed by the *Saeima*, and was not a product of the Ulmanis dictatorship of the 1930s.

If the 1920s are casually dismissed as a time of political instability, weak coalition governments, bribery and corruption, then the rich complexity of independent Latvia is lost. Latvia witnessed a renaissance in literary and artistic creativity, theatrical productions, scholarship, economic activity, improvements in health, education and welfare, and popular culture. The best example of how thorough and varied this renaissance was is the jubilant book published for the tenth anniversary of Latvian independence, *The First Ten Years of the Republic of Latvia*.[34] The book celebrated the achievements of the new state, but was confident enough to point to shortcomings as well. The leading representative of each national minority, for example, wrote a summary of their development within the Latvian state. All identified continuing shortcomings in minority legislation, but they bragged of their community newspapers, journals, theatres, economic enterprises and political organisations. The book also included essays on the development of social welfare, the sports movement, the youth movement, the women's movement, the Red Cross, associations, and societies, the labour movement, the scout movement, religious life, and more. The hundreds of advertisements and photographs put a further human face on what the Latvian state meant to its inhabitants. Directors and workers pose proudly with their factory's bicycles or barrels of beer and butter. Nurses and doctors show off their hospital's latest equipment. The book is very much like a proud family album, but a family album of a nation. More telling is the book's publication date, 1928. A year later Latvia would slide into the Depression and calls for abandoning democracy would grow louder.

An authoritarian coup ended the democratic experiment on the night of May 15, 1934. The roots of this coup go back to the nationalists' disillusionment with the form of the Republic following the Constituent Assembly. The agrarian right and Karlis Ulmanis were the leading lights of almost all cabinet governments, but they felt increasingly hampered by the rules of parliamentary democracy and the growth of competitive parties. Parliament confined action; the Social Democrats, with two exceptions, remained in opposition for ideological reasons, effectively removing thirty seats out of the pool for coalition governments. Minority parties bargained their increased political value to gain protection for minority rights, something that nationalists were increasingly reluctant to sanction. A curious paradox emerged. The conservative, agrarian and nationalist concerns of Karlis Ulmanis and his Farmers' Union could not be realised through parliament, despite their control of most cabinets. Elections to the fourth *Saeima*, in the midst of the global depression in 1931, exacerbated the situation.

The Farmers' Union and Karlis Ulmanis lost significant electoral support. They, at least through association, were tied to the corruption of cabinet formation and the abuse of power for money that plagued Latvia. The Depression radicalised the electorate which in turn looked for new options. In the countryside, the division between established farmers and the reform-era farmers surfaced in this political conflict. The reform-era farmers, the new farmers (*jaunsaimnieki*), backed their own agrarian party which challenged the Farmers' Union for the right to speak on behalf of Latvian farmers. The Social Democrats, struggling with a communist-backed leftist party, moved sharply to the left to keep their core of worker support. The centre adopted the nationalist rhetoric growing in popularity throughout Europe and further eroded the support of the Farmers' Union. The elections for the fifth *Saeima* set for the autumn of 1934 promised a continued slide for the Farmers' Union with potential growth for Latvia's indigenous fascist party, the *perkonkrusts*.

Faced with a political future of diminishing returns and shrinking status, Karlis Ulmanis began planning a coup by late 1933. He found support in the nationalist ranks of government that shared his concern about the nature of parliamentary democracy in Latvia. The bureaucrats saw themselves in constant struggle with the irrationality of elected politicians. To bureaucrats, elections returned a multitude of plans and agendas, when they believed the state needed united action. The Depression seemed to underline the importance of firm, central authority.[35] The democratic process of building consensus was either too time consuming or unproductive. These officials, who controlled the levers of state, accepted Ulmanis' coup as a vehicle for bringing strong united action to deal with Latvia's problems.

The coup was without significant incident or bloodshed. Close to three hundred politicians were interned in a prison camp for up to two and a half years and many others were under close police surveillance. Overt widespread oppression was lacking. The degree of popular support for the coup is impossible to ascertain, although the lack of popular opposition against the dismissal of parliament suggests apathy. Over the years, several of the progressive supporters of the Ulmanis regime quietly left it and went into passive opposition. The loss of democracy meant the loss of alternative concepts of Latvian society other than those proposed by the state. The nationalists' dream of guaranteeing a Latvian character for the state became government policy. The state gradually assumed more and more control in an attempt to create a central, authoritarian state. After Ulmanis, by his own decree, became State President in 1936, the state pursued the "outward signs of the fascist ... aesthetic, but remained essentially a conservative regime".[36]

The Ulmanis regime spread its control gradually through Latvian society. In a comparative sense, this gradual control can be described as benign, not benevolent. Other authoritarian regimes of Eastern Europe (not to mention the Soviet Union and the Third Reich) crushed opposition and general human rights with violent regularity. Latvia, in contrast, had almost symbolic repression and resistance. Nevertheless, a creeping authoritarianism bent every institution of government and society to the service of the state. The ideals of the democratic republic were replaced by the two slogans of the new regime: the *vadonisprincips* (*fuehrerprinzip*—the infallibility of the supreme leader), and a "Latvian Latvia" (as opposed to the more intolerant Latvia for Latvians). The minority rights protected by minority deputies in parliament were eroded by a series of appointments and decrees. The regime removed minority and leftist politicians, replacing them with sycophants. The school system was similarly cleansed of its minority and leftist teachers. Government censorship of the press, and its selective support of the arts, moulded the cultural content of Latvia. Yet, although the regime often targeted minorities, it was also the only state in the region to ban anti-Semitic writing. As with the political crackdown, the crackdown on minorities was real, but not as hostile as in other states in the region. Finally, the state controlled and orchestrated the economy. The state controlled banks and financial markets, organised labour, set the tone for mercantile operations through the ownership of the largest retail outlets and, through control of agricultural purchases and rural co-operatives, controlled the rural economy as well.[37]

Authoritarianism crippled initial resistance to Soviet and Nazi occupations. The regime established a manner of governance that removed individual input and democratic choice in favour of centralised, authoritarian rule. The regime established state control of the economy and of culture. It dabbled in social engineering that would create obedient Latvians to take the reigns of a Latvian state with a continuously diminishing role for minorities. Nevertheless, the average citizen of Latvia (regardless of nationality) continued to distance himself in terms of prosperity, education, freedoms, and opportunities from those that did not escape from the Soviet Union. By 1940, in almost every measurable category of the quality of life and economic performance, Latvia was leaps ahead of the Soviet state and much of Eastern Europe. The problem, however, was a continued and growing gap between Latvia and the most advanced countries of Western Europe and Scandinavia. Although the regime took pains to show that Latvia was joining Western Europe, sophisticated economists like Arnolds Aizsilnieks and comparative historians like David Kirby refute such claims as exaggerated.[38]

The accomplishments of the state, nevertheless, were very real and palpable to its citizens. In an interview with an American journalist,

Karlis Ulmanis reportedly said that the ultimate defence of the Latvian state in the face of Nazi or Soviet aggression would be the memory of the independence era.[39] Ulmanis predicted that the citizens of Latvia under foreign occupation would remember independence fondly, and fight for its ultimate rebirth. He proved prophetic. The ideal and the accomplishments of independent Latvia survived fifty years of occupation to help rebuild independent Latvia.

The larger geopolitical conditions that favoured the emergence of independent Eastern European states in the aftermath of World War One began to break apart by the late 1930s. An aggressive, irredentist Nazi Germany stated plainly its desire to re-acquire lost territory in Eastern Europe. Particularly damning for the Baltic Republics would be co-operation between erstwhile enemies, Nazi Germany and the Soviet Union. As long as these two states competed with each other for influence in the Baltic region, the Baltic states could manoeuvre between them and preserve their independence. Of particular interest to Latvia was the establishment of closer, and hopefully binding links with Western European countries. In 1939, however, a diplomatic game began to secure the favour of the Soviet Union. The Allies were anxious to recruit the Soviets into an alliance that would stop Hitler from invading Poland, whereas Hitler was determined to guarantee his Eastern flank in case of war over Poland. As Artis Pabriks succinctly wrote:

> Control over the Baltic states was the price Stalin demanded for the support of one side or the other. Nazi Germany appeared to be more flexible than the Western allies when it came to giving up control over countries not in its possession.[40]

The Soviet Union and Nazi Germany carved up Eastern Europe in the now well-known secret protocols of the Molotov—Ribbentrop Pact of August 23, 1939.[41]

On October 5, 1939, the Soviet Union forced the Latvian government to sign a Mutual Assistance Pact that deployed between twenty-five and fifty thousand Soviet soldiers in Latvia. The Latvian state found itself trying to maintain independence within a radically changing international world while simultaneously calming its worried citizens. For example, communist agitation increased significantly after the arrival of Soviet forces. The Latvian political police tried to contain the spread of such agitation without offending Moscow. Similarly, the Finno-Soviet War strained both government and populace. Factions within the army and government agreed with popular sympathies for the Finns in their struggles against the Soviets. The most difficult change to explain in the troubled times following the Molotov—Ribbentrop Pact was the repatriation of the Baltic German population in the winter of 1939–1940.

The Latvian state had an uneasy relationship with its Baltic German minority. On the one hand, the Baltic Germans were the descendants of the traditional enemy of the Latvian peasantry; they were seen as allies of the Kaiser during World War One; they fought Latvia's popular agrarian reform all the way to the League of Nations; and they maintained economic power disproportionate to their numeric strength. On the other hand, the Baltic German presence ensured that Latvia's cultural development followed pan-European trends. Liberal and progressive Germans supported the Latvian state. Baltic German economic experience and skill assisted Latvia's post-war economic reconstruction, and the cultural vitality of the Baltic German community was a vital component of Latvia's society. Following Ulmanis' coup, the progressive wing of the Baltic German community lost influence to a new radical group with close ties to the Nazis. These Baltic Germans welcomed Hitler's call for repatriation to the Reich in the fall of 1939, but many others were not as enthusiastic. For many, Latvia was their home and had been for centuries. The combined pressure of Nazi rhetoric, the threat of Soviet aggression in the Baltic region, and Latvian state encouragement convinced the vast majority of Baltic Germans to leave Latvia. Latvian authorities described the repatriation as a victory in the campaign to create a Latvian Latvia. Repatriation was, however, a national tragedy. Latvia lost a vital component of its society and a link to its past. Most ominously, outsiders decided the fate of 53,000 of Latvia's citizens.[42]

OCCUPATION AND WAR, 1940–1945

In June 1940, the Soviets' interest in the Baltic states returned with plans to incorporate them fully into the USSR. On June 15, Soviet troops attacked a Latvian border post near Maslenki, killing five and kidnapping thirty-seven. The Latvian cabinet ignored the provocation, but on the following day the Soviets delivered an ultimatum that demanded the establishment of a new Latvian government and the free entry of Soviet troops into Latvia. Faced with invasion and an aerial bombardment of its cities, Ulmanis and his Ministers accepted the terms of the ultimatum.[43] Soviet tanks poured into Latvia and the Latvian government helped co-ordinate their movements to avoid incidents with the local population. Political prisoners were released from jails (although Latvia's fascists, the *perkonkrusts*, were not released) and Soviet-orchestrated demonstrations greeted forces in Riga. By June 18, Latvia, like Lithuania and Estonia, was completely occupied by the Soviet Union.

Andrei Vishinsky, the prosecutor of the 1930s Moscow show trials, was sent to Latvia as the Soviet envoy charged with supervising the sovietisation of Latvia. A biologist, Dr Augusts Kirhensteins, was named the new Minister President, and the popular writer (and closet communist) Vilis Lacis was named Minister of the Interior. With the backing of the Soviet army and the supervision of the Soviet embassy, the new government quickly neutralised political opposition and moved Latvia towards full incorporation into the Soviet Union. More than sixty senior civil servants and all newspaper editors were removed from their positions within the first weeks of occupation. From July 14 to 15, mandatory "elections" to a "People's *Saeima*'" were held. The Soviet police arrested all groups and individuals that opposed the Soviet-sponsored "Working People's League" in the elections. On July 17, the authorities announced an almost unanimous victory for the Commu-nist-backed list. Four days later, the false parliament convened and promptly applied for membership in the USSR (in violation of the Constitution of Latvia). On August 5, 1940, the USSR accepted Latvia's application. In a little less than two months, the Soviets used false pretences, a mockery of democratic procedure, political terror, repression, and unbridled propaganda in an attempt to disguise that their occupation of a sovereign state was the free choice of a nation that "voted" to dissolve itself and accept political domination.

The sovietisation of Latvia continued until the Nazi invasion of the Soviet Union in June 1941.[44] Within a month of occupation, Latvia's 800 largest industrial enterprises (including all publishing houses) were nationalised. The banking system merged with the Soviet State Bank. Latvia's currency, the Lats, which had been valued at three per ruble, was exchanged on a one for one basis. Personal savings in excess of 1,000 rubles were confiscated. By the end of the year, all commercial enterprises, as well as houses or apartments exceeding 220 square metres (170 in rural areas), were expropriated. In the countryside, all land exceeding 30 hectares was expropriated and redistributed to the landless in small, economically unviable, ten-hectare plots. Change did not end with political and economic "reform", but included an attack on the social fabric of the Latvian state. The new regime dismissed about 700 teachers and replaced them with poorly educated, but politically reliable, party supporters. Communist agitation, propaganda, "red corners" (shrine-like displays of Soviet heroes and accomplishments), and four hours of Russian language a week were introduced to primary school curricula. University students were expected to learn Russian, the faculty of theology was closed and libraries were purged of the more than 4,000 titles from the official list of banned books and brochures. As the year continued, the authors of banned works were targeted for extermination as well.

The Soviet regime hoped to attract some supporters through their attacks on the Latvian state. Pay increases for workers and land reform, for example, targeted Latvia's poorest workers and peasants. The indigenous Communist Party was minuscule. A few dozen Communist Party agitators released from prison jumped to positions of importance, and thousands of Latvian Soviet citizens (Latvians who had not returned to Latvia following its independence in 1918) came to staff positions within the new administration. The Stalinist purges of the 1930s, however, handicapped Soviet efforts; thousands of high-ranking Latvian Bolshevik loyalists were executed as enemies of the state. The Soviet regime succeeded in co-opting many Latvian Social Democrats into the Soviet administration, but the distrustful Soviet Communist Party did not accept them into the Party. The regime also hoped to find supporters among Latvia's minorities. The minority school system that had come under attack during the Ulmanis era, for example, was expanded.

The radical nature of sovietisation lessened the chance of recruiting supporters for the new system. The co-opted Social Democrats became increasingly uneasy about Soviet intentions and fell under Soviet suspicion as well. Workers and the middle class suffered under Soviet decrees that wiped out their savings and diminished their purchasing power. Stores emptied of all goods as Soviet officers and speculators from Leningrad descended upon goods unavailable in Soviet stores. Minority communities grew disillusioned with Soviet rule as well. Most of Latvia's Russian and Jewish communities were devoutly religious. Official Soviet atheism attacked the conservative religious nature of these communities. With only small pockets of support, the Soviet regime turned increasingly to terror in order to rule the occupied country.

Through 1940 and 1941, terror increased each month as arrests, imprisonment and executions became more and more common (Table 2). Already in July of 1940, the top ministers and civil servants of the Ulmanis regime (including Ulmanis himself on July 22nd) were arrested and tried as enemies of the state.[45] From 1940 to 1941, at least 7,292 people were arrested for "political crimes", about 1,500 of these were executed. On the night of June 13–14, the NKVD arrested thousands in the first mass deportation. More than 14,000 people were deported that evening, either for having been the ruling elite of independent Latvia or for being potential opponents to Soviet rule. About 2,000 of the deported were Jewish, both the devout and the wealthy were enemies of the state to the Soviet regime. The German army invaded on the heels of these mass deportations. Many beleaguered Latvians greeted them as liberators.

Nazi occupation proved little better, and for many much worse. German troops attacked the Soviet Union on June 22, 1941. Blitzkrieg troops reached the Latvian border by the 26th, occupied Riga on the 1st of July, and all of Latvian territory by the 5th. Within six months, the Nazis with the assistance of up to 2,500 Latvians murdered the great majority of Latvia's Jewish population.[46] Afterwards the Riga ghetto and camps across Latvian territory became part of the Nazis' holocaust complex for the extermination of European Jewry, gypsies, homosexuals, Jehovah's Witnesses, socialists, communists, and other "non-desirables". The Nazis were quite able to find hundreds of willing, individual volunteers to guard camps, and even to equip the infamous Arajs Kommandos who participated in the murder of Jews

TABLE 2 POLITICAL ARRESTS DURING THE FIRST YEAR OF SOVIET OCCUPATION

Month and year	Number of arrests
June 1940	20
July 1940	141
August 1940	300
September 1940	291
October 1940	507
November 1940	331
December 1940	236
January 1941	268
February 1941	290
March 1941	281
April 1941	288
May 1941	273
June 1941	3,991
July 1941	75

Source: Viksne, Rudite and Kangeris, Karlis, eds., No NKVD lidz KGB Politiskas Pravas Latvija 1940–1986 (Riga: Latvijas vestures apgads, 1999), p. 972.

across Latvia and perhaps throughout occupied Eastern Europe. Latvia, however, was an occupied country being occupied by another invader. Latvians had no rights, did not participate in the creation of policy and could not lobby for a change of policy. One of the surviving controversies about the Holocaust in Latvia is the question of the interregnum. The idea is that there existed a chaotic time period lasting from days to weeks between Soviet occupation and the establishment of Nazi order. During this interregnum, Latvian forces supposedly organised the murder of Jews independently of Nazi plans. The most thorough study, Andrievs Ezargailis' *The Holocaust in Latvia* however, has not found conclusive evidence to support these claims. Soviet control was complete up to the final hours, and Nazi rule immediately secured all independent armed forces. Nazi troops disarmed and disbanded Latvian partisans with methodical regularity. Nevertheless, the seeming lack of an interregnum does not diminish the role that individual Latvians played in the murder of Latvia's Jews after the establishment of Nazi rule.

The ultimate policy of the Nazis, similar to the Soviets, was the destruction of all traces of an independent Latvian state. Hitler hoped to colonise the Baltic region with Germans, use some Latvians for slave labour and exterminate the rest. However, his long-term plans did not have an opportunity to materialise. Short-term plans for Latvia revolved around two principles: establishing and maintaining Nazi control of the territory, and squeezing every resource, human or material, out of the region for the war effort. Soviet rubles were immediately exchanged at a ten to one rate of exchange with the Reichsmark. Latvians, who lost a significant share of their savings during the Soviet exchange, now lost even more in the German exchange. Food rations were introduced that favoured German administrators over local inhabitants, and all property nationalised by the Soviets became the property of German firms. By September, roughly 25,000 German officials had entered Latvia as the occupation's administrators while at least 35,000 Latvians were deported to Germany as labour.

If the Soviets relied on Latvian Bolsheviks and the disaffected of the Ulmanis era for collaboration, the Nazis looked to co-opt the former elite of the Latvian state. The former bureaucrats, civil servants, business leaders and politicians had watched the Soviet regime dismantle the state that was their lives' work in less than a year. For those who survived that year, particularly after watching many of their colleagues deported in early June 1941, collaboration with the Nazis held the prospect of rebuilding parts of the old Latvian state. The Nazis were keenly aware of the prospect of recruiting Latvian collaborators and of the benefit they would provide in governing the occupied territories and maximising their economic productivity. The Nazis looked for Latvians with enough legitimacy among the population to head a Latvian Self-Administration that would look after daily administrative functions. The Germans needed a devout anti-Bolshevik who would serve Nazi interests before Latvian ones. After failing to persuade some, they settled on a conservative, retired general, Oskars Dankers, who had left Latvia with the repatriating Baltic Germans. Once appointed Director of the Latvian Self-Administration on August 21, 1941, Dankers attempted to convince other Latvian nationalists to work with the German occupation.[47]

By the end of autumn, Dankers convinced many Latvian nationalist leaders to assume posts within the Self-Administration. The Self-Administration was divided into directorates that tried to balance Latvian interests with the demands of Nazi authorities in administrative and judicial arenas. The directorates had limited success in keeping control over cultural and educational policy. They were even less successful in curbing Nazi demands on labour conscription, economic exploitation, military conscription and opposing the Holocaust. As Nazi fortunes waned after the Battle of Stalingrad, the German high command considered mobilisation in occupied territories to improve their situation at the front. The Self-Administration called for Latvian autonomy and equal rights within the formation of Latvian Legions. The German High Command ignored such demands and introduced "voluntary" conscription. Two Latvian officers, General Bangerskis and Colonel Silgailis, called on Latvians to join the Legion on March 22, 1943. The response was disappointing. Although thousands volunteered, thousands more deserted or fled and conscription was needed to draw the 146,000 former Latvian citizens into the Latvian Legions of the German army as the 15th and 19th Waffen-SS Divisions.[48]

Although the Nazis succeeded in co-opting many Latvian nationalists, their rule was not without opposition. Red partisans, a guerrilla wing of the Soviet Army, operated from the woods of eastern Latvia. Similarly, Soviet Latvian officials worked from Moscow to form Latvian divisions within the Red Army. These Soviet divisions consisted of at least 43,000 soldiers who were evacuated former Latvian citizens and Soviet citizens of Latvian origin. They frequently fought against the Latvian Legions of the Nazi army, adding a macabre, forced civil war-like aspect to World War Two for many Latvians.[49]

Some of Latvia's democratically minded intellectuals and politicians foresaw the destruction of the Latvian state under Soviet or Nazi occupation. They established contacts with Sweden and the Western Allies in July of 1943 and on August 13, 1943 established the Latvian Central Council (LCC), headed by Konstantins Cakste, the son of Latvia's first president.[50] The Council had representatives from Latvia's four largest political parties, including the Social Democrats and the Farmers' Union. The Council denounced co-operation with either Soviet or Nazi forces and established contacts with similar organisations in Lithuania and Estonia. In the fall of 1944, many of the Council's leaders were arrested by the Germans and sent to Stutthof concentration camp in Germany (where Cakste died in 1945). The remaining Council leaders attempted to create a military force. German troops disarmed the Latvian renegades, executed eight officers and sent a further 457 soldiers to the Stutthof concentration camp.

Why did Latvians who had worked for independent Latvia work for the Soviet or Nazi occupations? The implicit goals of both occupations went against democratic and nationalist desires for Latvia. The motivation lies within the turbulence of World War Two and the memory of World War One. Those that participated in the Holocaust and Soviet deportations were simply criminals. Those in bureaucratic posts throughout government were a combination of opportunists, survivalists and misguided patriots. They lived through World War One where the most important things for Latvia were the presence of a military force and institutions and personnel that could step forward for Latvian autonomy and independence when geopolitical forces were most opportune. They believed that military service or work in Soviet or Nazi administration could preserve some Latvian territorial integrity or autonomy. Perhaps, if both the Soviets and Nazis lost World War Two, a geopolitical phenomenon similar to the end of World War One could lead to an independent Latvia yet again. These wishes were pure fantasy, but they were the only comforting delusion available to Latvians faced with extermination in the horrors of World War Two.

Soviet troops returned to eastern Latvia in 1944, but did not re-conquer all territory until the armistice. A sliver of German and Latvian forces remained hugging the eastern shore of the Baltic Sea until May 8, 1945. Latvian guerrilla resistance continued until the early 1950s.[51] Tactically, the partisans were doomed. However, they were able to resist for so long because they had the tacit support of the great majority of the population. The early partisans were deserters from the German and Soviet armies as well as soldiers from the Latvian Legion that refused to surrender at the armistice. Over the years of struggle, thousands of Latvians with no previous combat experience joined the partisan movement. After each wave of Soviet repression, a new ripple of survivors or relatives of Soviet victims took up arms in the woods of Latvia. Ultimately, the Soviets succeeded by planting informers within the partisans or through interrogation of captives, discovering the location and nature of the partisan organisation. The Soviet forces broke the Latvian partisans' popular support by breaking the populace at large.

The legacy of occupations, World War Two and the partisan war was a country devastated. The countryside was yet again criss-crossed with barbed wire and bomb craters. Towns, cities and farmsteads were yet again destroyed or damaged. Tens of thousands of people were killed, in exile or crippled. Latvia may have lost 30 per cent of its pre-war population owing to the war, including almost the entire historic populations of the Baltic Germans, Jews and gypsies. Latvia's Jews suffered most. As Modris Eksteins summarised:

> Of the Latvian Jews who went to Russia in 1941, perhaps four thousand—one in three—survived; of the roughly eighty-three thousand who fell into German hands in Latvia, not more than nine hundred survived; and of the more than twenty thousand Western Jews sent to Latvia, only some eight hundred lived through the deportation until liberation. This was the highest percentage of eradication in all of Europe.[52]

David Kirby, on the other hand, remarked on the larger scale: "the sheer volume of numbers of those who perished or suffered frightful deprivations as a result of war and occupation is so overwhelming as to make any attempt at description or enumeration futile."[53] Latvia was a state in ruins, joining Estonia and Lithuania as the only European states not to regain independence after WWII.

SOVIET LATVIA, 1945–1985

Following World War Two, the sovietisation of Latvia entered a new, more aggressive phase. From 1945, the Soviet regime, using the army and the secret police, employed brute force to establish obedience and transform Latvian society. The Soviet regime embarked on a plan of socialist reconstruction in Latvia that called for accelerated industrialisation and collectivisation of agriculture. The first secretary of the Latvian Communist Party, Janis Kalnberzins, hoped that Latvia would quickly be converted into an industrial Soviet republic. By 1947, much of the pre-war industrial infrastructure had been repaired and new industrial enterprises, particularly heavy industry, were planned. Rapid industrialisation just after the depopulation of war meant that industrial manpower had to come from outside Latvia. In 1946, 41,000 people had already arrived in Latvia with over 30,000 settling near the industrial concerns in Riga. Between 1945 and 1955, approximately 535,000 labourers were shipped to Latvia from the Soviet Union. By 1948, almost 45 per cent of all industrial workers were not ethnically Latvian. Rapid industrialisation had altered the ethnic composition of Latvian society.[54]

Post-war sovietisation meant war against the Latvian countryside. After rural support for the partisans became clear, collectivisation became state policy. In 1947, Soviet authorities introduced excessive, progressive taxation on small farms. The new tax rates reached 75 per cent of estimated income, and as peasants were unable to pay, the state seized their livestock, equipment and property. Despite the considerable pressure and the concomitant propaganda in favour of collectivisation, few farmers joined the *kolkhozes*. The Soviet regime claimed that the lack of enthusiasm for collectivisation was due to *kulak* (so-called wealthy peasants) agitation against the *kolkhozes*. From March 25 to 28, 1949, the Soviet regime struck viciously against these *kulaks*, deporting 43,200 of them (including more than 10,000 children) to Siberia.[55] By 1952, 98 per cent of Latvian farmers lived and worked on collective farms.

The sovietisation of Latvian society also included a comprehensive, planned assault on independent Latvian cultural life. The first blow was the loss of the majority of the pre-war intelligentsia through emigration. Fearful of Soviet repression (a justified fear), 120,000 Latvians fled the country in the final stages of World War Two. Within the refugee population were 2,062 teachers, 197 university lecturers, more than half of Latvia's doctors, engineers, architects and Lutheran clergymen.[56] Those that remained were targeted by the Soviet regime for arrest and deportation. All literature, newspapers and journals became closely supervised Communist Party mouthpieces. History was rewritten to stress the Baltic region's "organic" connection to Russia and streets and squares were renamed after Soviet heroes. The Soviet authorities banned traditional Latvian festivals such as Midsummer's Eve. Increasingly, the Russian language was stressed over Latvian.

The Stalinist era from 1945 to 1953 was the height of Soviet terror. Anyone could be arrested, tried and punished for political crimes

TABLE 3 POLITICAL ARRESTS DURING THE STALINIST ERA

Year	Number arrested
1944	3,708
1945	14,702
1946	3,967
1947	2,424
1948	3,131
1949	3,542
1950	2,987
1951	2,427
1952	969
1953	616

Source: Viksne, Rudite and Kangeris, Karlis, eds., No NKVD lidz KGB Politiskas Pravas Latvija 1940–1986 (Riga: Latvijas vestures instituta apgads, 1999), p. 972.

without warning or guilt. The numbers are staggering (Table 3). Not counting the deportation of the *kulaks*, tens of thousands were arrested. Recently, the Historical Institute of Latvia has completed an exhaustive search of NKVD and KGB files to create a "catalogue" of all people tried for political crimes (according to the Soviet Constitution) from 1940 to 1986. The book consists of more than 950 pages of condemned names, their date of birth, place of birth, "crime" and archival location.

The general population of Latvia was suspect in Soviet eyes. There was, however, a tiny minority that supported the sovietisation of Latvia: the surviving Latvian Bolsheviks, cadres from the USSR, and young, true believers in communism. At least 9,000 Communists were transferred to Latvia from 1945 to 1951. The young true believers were Latvian communists baptised during World War Two and indoctrinated with Soviet methods of dealing with dissent and ordering society. They were grateful to the Soviet system for unprecedented upward mobility—the sons of poor peasants and factory workers were placed into universities and technical institutes to raise the percentage of proletarians and lower classes in higher education. In effect, they benefited from a sort of affirmative action for the lowest classes both in educational opportunities and, with loyal service to the Communist Party, in governmental employment. These Latvian Communists, just reaching senior positions within the Party and government, pushed for "national communism" during the era of Khruschev's thaw.

In March 1953, Stalin died and the grip of mass terror and repression on the entire Soviet Union relaxed. The Khruschev thaw took time to take effect, but with the death of Stalin there was a noticeably sharp slowdown in the scale of political terror. In Latvia, political arrests dropped from 616 in the last year of Stalin's life to ninety-two the following year. The Soviet occupation under Khrushchev could not be described as beneficial to Latvians, but terror slowed and space for manoeuvring politically, economically and culturally emerged for a nation that had lived a nightmare for nearly a decade and a half. At first, Khruschev followed a political strategy of finding some rapprochement with the Soviet Union's many nationalities. The change was more than a loosening of political terror and symbolic rapprochement with nationalities. Administrative decentralisation converted many all-Union ministries into Republic ministries and greater economic control was given to regional economic councils. The changes fostered a Latvianisation of administrative, bodies including the Communist Party, and spurred a revival of national, cultural identity.

With the cultural thaw, at least 30,000 Latvian survivors of the Gulag began returning to Latvia in 1954. Among them were a number of Latvian intellectuals who were also permitted to return to creative work. Their lives and the lives of their children, however, were exceedingly difficult—they had a "stain" on their records that would close doors to universities, professions and other opportunities. Nevertheless, the cultural thaw allowed for a cultural identity based on national principles to survive.[57] The thaw brought regime approval to classic Latvian literary figures such as Fricis Barda, Janis Akuraters, Karlis Skalbe and others. Furthermore, a new generation of Soviet Latvian cultural figures emerged as promising national representatives. The ban against Midsummer's Eve celebrations was also lifted. Cultural life, although still severely limited, tolerated a national component.

Soon after the change in nationality policy, a Latvian, Vilis Krumins, was appointed to the post within the Latvian Communist Party responsible for policy on local cadres. Over the next several years, ethnic Latvian content increased considerably in the Communist Party and in local Soviets. On the local level, more than 50 per cent of the regional party leaders, 60 per cent of the *kolkhoz* chairmen, 35 per cent of the directors of enterprises and 30 per cent of engineers were replaced in the mid-1950s. On the republic level, indigenous Latvian representation at the Council of Trade Unions increased significantly, as it did on the Central Committee. The ultimate example of this trend towards "national communism" was the rise of Eduards Berklavs to the post of Vice-Chairman of the Latvian Council of Ministers in 1957.[58]

Berklavs was one of the young, true believers in communism. His beliefs, however, blended the ideology of communism with the interests of the Latvian nation. As Vice-Chairman, he spearheaded a campaign of

Latvianisation that included attempts at slowing labour migration and strengthening the rights of the Latvian language. He argued that Communist agitation in the countryside faltered because Party members could not adequately converse with Latvian peasants. He pushed the non-Latvian Communists to learn Latvian, and increased the amount of Latvian instruction in schools. Likewise, directors of factories were given the right to release employees if they could not demonstrate Latvian proficiency within two years. Berklavs and the "national communists" hoped to use their growing control of the republic's economy to limit the growth of heavy industry. The rapid industrialisation required a tremendous in flow of immigrant labour. The first post-war census conducted in 1959 showed that the number of Russians had increased by 388,000 since 1935. The number of Belorussians had jumped by 35,000 and the Ukrainian population jumped from a nearly non-existent 2,000 to 30,000. The ethnic Latvian population, however, dropped by 170,000. The "national communists" hoped to stop this debilitating in-migration and slow the recruitment of Latvian labour for settlement projects outside Latvia. The "national communists" hoped to channel industrialisation towards industries viable within the manpower resources of the republic. They argued that if the Latvian SSR developed consumer goods and agriculture, Latvia's economic performance would maximise the republic's contribution to the development of the USSR. These rapid changes, however, were unpopular among factions in Moscow, non-Latvian Communists, and the older guard of the Latvian Communist Party.

From 1957, Khruschev's soft nationality policy began to ossify. He had consolidated power and began to view non-Russian Soviet republic elites as a potential threat. In 1958 a new law on education made secondary languages voluntary, effectively neutralising the Russian-speaking population of non-Russian republics' need to learn the local language. Berklavs, and similarly minded leaders from Lithuania, Estonia, Ukraine and the Caucasus, opposed the new reforms. As tensions mounted, a delegation of the Central Committee of the Communist Party of the Soviet Union arrived in Riga in April 1959. Khruschev followed in June and castigated Berklavs for the excesses of "national communists". Supporters of the new mood in Moscow, as well as the old guard from Stalinist days such as Arvids Pelse and Janis Kalnberzins, attacked Berklavs and others. Berklavs was accused of "national isolationism", and he and his supporters were purged. At least 2,000 "national communists" lost their positions, the entire leadership

TABLE 4 POLITICAL ARRESTS DURING THE RISE AND FALL OF THE "NATIONAL COMMUNISTS"

Year	Number of arrests
1954	92
1955	54
1956	43
1957	101
1958	103
1959	44
1960	46
1961	71
1962	73

Source: Viksne, Rudite and Kangeris, Karlis, eds., *No NKVD lidz KGB Politiskas Pravas Latvija 1940–1986* (Riga: Latvijas vestures instituta apgads, 1999), p. 972.

of Latvia's Komsomol was dismissed, and a slight upswing in political arrests stamped out any remaining reforming zeal (Table 4).

In 1963, the new First Secretary of the Latvian Communist Party, Arvids Pelse, demanded the suppression of all expressions of nationalism, narrow localism, national self-separation, idealisation of the past or adherence to reactionary traditions and customs. The Midsummer's Eve celebration was yet again prohibited and many of the classic Latvian literary figures were banned anew. The new generation of literary figures came under increasing attack. A new wave of Russification struck primary schools, and Russian was adopted as the language of communication within the Party, the governmental apparatus and the field of economics. The new leadership returned to the industrialisation policy of the Stalin era. Pelse requested that Moscow transfer new cadres and industrial manpower to Latvia. Ethnic Latvian control of industry dropped quickly and by 1961 less than a quarter of industrial managers were ethnic Latvians. With the purge of national communists and the reversal of many of their programmes, the regime lost touch with the Latvian national community. Serious reform would not recur until the 1980s.

The struggle of the national communists revealed several themes in the development of Soviet Latvia. The supporters of Eduards Berklavs represented a new, Latvian elite that struggled for the survival of the Latvian nation, but were committed to incorporating such an identity into the existing political structure of Khruschev's Soviet Union. After their early reforms, however, the nature of their demands led to conflict between the incompatible Latvian and Soviet interests. The national communists succeeded in venting the agony of the Latvian nation from WWII and Stalinism, and slowed the destruction of the ethnic Latvian nation. Working within the Soviet system, however, the national communists could not re-establish a link with the pre-war nation. After the purge of the national communists, however, the Soviet state could not forge a structural link between local community, organisational ideas and Soviet state institutions.

Augusts Voss became First Secretary in 1966, a post he would hold until 1984. The Voss era mimicked the Brezhnev era on a republic level in many respects. Voss ruled co-terminously with Brezhnev in a long era that has since been characterised with stagnation. The almost twenty years were remarkable for the little amount of political change or initiative, the collapse of the Communist Party in local participation, and the continued modernisation and industrialisation of the state. The Communist Party became an institution foreign to the great majority of Latvians. Roughly one-third of the Latvian Communist Party was ethnically Latvian (the lowest percentage of titular nationality composition in the Soviet Union), and only 5 per cent of Latvians belonged to the Party.[59]

The Voss years, like the Brezhnev years, are enigmatic; the regime stifled reform, but many of Latvia's inhabitants enjoyed a degree of relative prosperity and social peace that had not been seen for nearly twenty-five years. The changing economy and nature of Soviet Latvia spurred the development of two new social forces. On the one hand, the continuing rapid industrialisation and modernisation dissolved traditional values and helped create the cult of the individual in Soviet Latvia. On the other hand, the lack of democratic political rights and a Soviet nationality policy that discriminated against Latvians meant that a widening gap emerged between a Soviet Latvian's expectations and the real opportunities available to him or her. In other words, the average inhabitant may have begun to live better, but he or she was alienated from the regime.[60] Likewise the manner of rapid industrialisation created stagnation in production and efficiency, massive environmental problems and the continuing arrival of Slavic workers.

The continued industrialisation and modernisation from 1960 to 1980 transformed Latvia from a primarily agricultural country to an industrial one. In 1935, only 17 per cent of Latvia's inhabitants were employed in industry, whereas by 1972 that number had jumped to 42 per cent. Simultaneously, those employed in agriculture and forestry dropped from 65 per cent in 1935 to 16 per cent in 1985. These dramatic changes meant significant alterations in the composition of Latvia's cities and countryside. Soviet authorities began consolidating collective farms, hoping to create a more efficient agricultural sector. The number of farms dropped from 1,105 *kolkhozes* in 1960 to 392 by 1975.[61] The number of people employed in agriculture dropped precipitously and the number living in the countryside plummeted. By the early 1980s, only 30 per cent of the inhabitants of Latvia lived in the countryside. Those that remained there lived in bizarre apartment-complex islands within a sea of fields. The logic of the apartment buildings was that the Soviet system brought an equality of amenities to city and country. By living more or less collectively, *kolkhoz* workers could rationalise electrical usage, plumbing and modern luxuries such as cafeterias and laundromats. On the ground, however, the logic failed; services were not on par with the potential comforts of the city. Central planning dictated rural development, and the collective farmers could not initiate or modify the changes taking place in the countryside. These changes targeted the old, single farmstead style of living for destruction, but offered little in its place.

The city pulled people from the collective farms because it offered better standards of living and better opportunities. During the 1960s and 1970s, provincial towns such as Valmiera, Ogre and Olaine grew as large fibreglass, textile and pharmaceutical plants that exploited low wage, unskilled female workers were built. Latvia became one of the most important industrial centres of the Soviet Union.[62] Despite the considerable flow of Latvians from the countryside to the cities, the rapid industrialisation of the Soviet central planners demanded a far larger pool of labourers than Latvia could provide. The in-migration of predominantly Slavic workers continued unabated. From 1961 to 1989, an additional 1,466,700 labourers moved to Latvia. Many stayed for brief periods of time and again left Latvia, but more than 330,000 settled in Latvia permanently, primarily in Riga and the cities of Latgale. The number of ethnic Slavs doubled in most Latvian towns. Ethnic Latvians approached minority status within their own Republic.[63]

The rapid industrialisation and modernisation of Latvia with its concurrent lack of political change led to a rapid increase in social problems. From the 1960s, the crime rate rose steadily, as did suicide, abortion and divorce rates. Birth rates dropped markedly. Often the root problem was alcohol. Alcohol abuse became common on collective farms and on the factory floor. Drinking was involved in three-quarters of all drowning accidents and crimes, in one-half of all automobile accidents and deaths by fire, and was a leading cause of divorce. One-third of male deaths between the ages of twenty and fifty were linked to alcohol.[64] The debilitating alcohol problem is and will continue to be one of the longest-lasting vestiges of Soviet rule in Latvia.

The growing social problems and the general alienation from the regime affected cultural developments in Latvia from 1960 to 1980. Two cultural currents rejected modernisation: the growing interest in folk art, customs and traditions, and the small, underground hippie movement of the early 1970s. These movements showed the growing importance of Latvian cultural contacts with the outside world.[65] Choirs, dance groups, orchestras, artists, writers and scholars travelled abroad much more frequently. Western books and plays were more frequently translated into Latvian. A non-political, cultural revival included record attendance at theatre performances, museum exhibits and film screenings. Despite these impressive accomplishments, Latvian culture was under assault owing to continued Russification. The number of books, journals and newspapers published in Latvian dropped each successive decade until by 1984, fewer than 50 per cent of all books published were in Latvian. Technical books as well as children's books were primarily in Russian, creating a two-pronged Russification attack (from the cradle to the professions).[66] Not surprisingly, by 1979 fewer than 20 per cent of Russians living in Latvia could speak Latvian, whereas close to 60 per cent of Latvians spoke Russian. The pressure of general Russification extended to the mass media where two-thirds of all radio and television broadcasts were in Russian.

The continued Soviet assault on Latvian identity fostered dissent and opposition. Active opposition in the 1970s and early 1980s, however, was confined to individuals or small groups.[67] In the summer of 1971, the "Letter of the Seventeen Communists" was sent to European Communist Parties. Eduards Berklavs was one of the authors of the letter, which protested against the continuing Russification of Soviet Latvia. Most dissent, however, came from small, fringe organisations such as the Latvian Independence Movement, Latvia's Democratic Youth Committee or Latvia's Christian Democratic Organisation. These groups tried to draw attention to the illegality of the Soviet occupation or to the desperate state of the Latvian nation. They were, however, easily marginalised and monitored by the Soviet authorities (Table 5). Few people were aware of their existence, and except for anti-regime graffiti, there was no mass, popular movement against Soviet rule until the mid-1980s.

The legacy of the Brezhnev—Voss era is twofold. On the one hand, ordinary Latvians knew that they were not free to express themselves

TABLE 5 POLITICAL ARRESTS FROM 1963 TO 1986

Year	Number arrested	Year	Number arrested
1963	27	1975	2
1964	21	1976	13
1965	17	1977	16
1966	22	1978	12
1967	24	1979	8
1968	9	1980	22
1969	16	1981	22
1970	17	1982	10
1971	19	1983	24
1972	33	1984	13
1973	18	1985	5
1974	8	1986	6

Source: Viskne, Rudite and Kangeris, Karlis, eds., *No NKVD lidz KGB Politiskas Pravas Latvija 1940–1986* (Riga: Latvijas vestures instituta apgads, 1999), p. 972.

in certain ways, but their daily toil was rewarded with education, health care and entertainment. Latvians still remember with fondness the nearly free access to subsidised sports, entertainment, health care and travel. Granted, ties to the West remained problematic and the divide between Latvian émigrés and Latvians in Latvia continued to be monitored intensely by secret police. Latvians were, however, able to travel extensively throughout the Soviet Union. It may be impossible to visit a Latvian home today without seeing a souvenir of an excursion to the Caucasus or Moscow or to a resort on the Black Sea. This is the nostalgia that so haunts the pensioners of Latvia today. They have the misfortune of timing of birth. A lifetime of work in the Soviet Union is undervalued, if not valueless, in today's Latvia. The other half of the legacy of the Voss—Brezhnev era is that the entire Soviet system was collapsing and contemporary Latvia still struggles with the debris. On a macroeconomic level, the system of perks and benefits coupled with massive military expenditures, the misuse and exhaustion of resources and the decline in productivity was akin to a colony of termites voraciously consuming the foundations of an elaborate wood palace. Unfortunately, today many blame those that addressed and continue to address the rot as the culprits for the fall of the Brezhnev palace. One of Brezhnev's successors, Yuri Andropov, realised the danger of the rot in the foundations. He hoped to shore them up with economic reforms. His plan, cut short by his death, continued with the work of his protégé, Mikhail Gorbachev.

Gorbachev continued prodding the economy with his package of reforms loosely termed *perestroika*. His relaxation of political control meant the return of demands for national principles in the governance of the Latvian Soviet Socialist Republic. The dilemma solved with force in the 1950s returned with a vengeance; reform of empire presented the possibility of the empire's dissolution. Latvians moved from calling for the introduction of popular content in the formation of government policy to rearranging the foundations of the Latvian Soviet Socialist Republic. The first move threw out the Soviet, and the second dismissed Socialist principles, leaving the Republic of Latvia, after a fifty-year hiatus. Rebuilding the Republic, however, would be a daunting task. Reconstruction would demand the correcting of fifty years of Soviet occupation, healing the many tragedies of World War Two and addressing the weaknesses of the independent era. Latvia's troubled twentieth century hangs over its move into a new millennium.

1 "Herder", in *Latvijas Padomju Enciklopedija*, 1984, p. 62.
2 A.J. van Reenan, *Lithuanian Diaspora, Konigsberg to Chicago* (London: University Press of America, 1990), pp. 4–5, as quoted in Artis Pabriks, *From Nationalism to Ethnic Policy: The Latvian Nation in the Present and the Past* (PhD Dissertation for Aarhus University, 1996), p. 42.
3 The most comprehensive nineteenth-century history is Arveds Svabe, *Latvijas vesture, 1800–1914* (Stockholm: Daugava, 1968).
4 See also Andrejs Plakans, *The Latvians; A Short History*, "Chapter 6: A Century of Reforms" (Stanford: Hoover Institute Press, 1995), pp. 80–111.
5 Benedict Anderson, *Imagined Communities; Reflections on the Origin and Spread of Nationalism*, revised and extended version (London: Verso, 1991), p. 116.
6 Anderson, p. 195–196.
7 Margers Skujeneeks, *Latvija: Zeme un eedzivotaji* (Riga: Valsts Statistiskas Parvaldes Izdevums, 1922), p. 273.
8 David Kirby, *The Baltic World 1772–1993: Europe's Northern Periphery in an Age of Change* (London: Longman, 1995).
9 Edward C. Thaden, ed., *Russification in the Baltic Provinces and Finland, 1855–1914* (Princeton: Princeton University Press, 1983).
10 Andrejs Plakans, "The Latvians, The Eighteen Nineties" in Edward C. Thaden, ed. *Russification in the Baltic Provinces and Finland, 1855–1914* (Princeton: Princeton University Press, 1981), pp. 248–267.
11 Andrejs Plakans, *The Latvians; A Short History* (Stanford: Hoover Institute Press, 1995), p. 103.
12 *Ibid.*, p. 107.
13 See Erich E. Haberer, "Economic Modernization and Nationality in the Russian Baltic Provinces 1850–1900", *Canadian Review of Studies in Nationalism*, 12, 1 (1985).
14 Uldis Lasmanis, *Berga bazara un laikmeta labirintos* (Riga: 1997).
15 Anders Henrikkson, *The Tsar's Loyal Germans: The Riga German Community: Social Change and the Nationality Question, 1855–1905* (Boulder, CO: East European Monographs, 1983).
16 Conversation with Herr Donat Frankfurt, August 1995.
17 Margers Skujeneeks, *Latvija: Zeme un eedzivotaji* (Riga: Valsts statistiskas parvaldes izdevums, 1922), p. 452.
18 Skujeneeks, p. 523.
19 For two interpretations of 1905, see Janis Krastins, *1905. gada revolucija Latvija 1905–1907* (Riga: Zinatnu Akademija, 1948); and Alfreds Bilmanis, *A History of Latvia* (Princeton: Princeton University Press, 1951).

20 Ronald Grigor Suny, *The Revenge of the Past: Nationalism, Revolution, and the Collapse of the Soviet Union* (Stanford: Stanford University Press, 1993), p. 64.

21 *The Revolution in the Baltic Provinces of Russia; A Brief Account of the Lettish Social Democratic Workers' Party* (London: Independent Labour Party, 1907) is the best English language primary source.

22 For a general source on Russia: Teodor Shanin, *Russia, 1905–07 Revolution as a Moment of Truth* (London: Macmillan, 1986).

23 Allan K. Wildman, *The End of the Russian Imperial Army* (Princeton: Princeton University Press, 1980).

24 Opinions about the Revolutions of 1917 range wildly. See Richard Pipes, *The Russian Revolution* (New York: Vintage Books, 1990); Sheila Fitzpatrick, *The Russian Revolution, 1917–1932* (New York: Oxford University Press, 1982); N.N. Sukhanov, *The Russian Revolution 1917, A Personal Record* (Princeton: Princeton University Press, 1984).

25 Suny, p. 64.

26 Andrievs Ezergailis, *The 1917 Revolution in Latvia* (Boulder, CO: East European Monographs, 1974).

27 L. Lundin, "The Road from Tsar to Kaiser: Changing Loyalties of the Baltic Germans, 1905–1914", *Journal of Central European Affairs*, X, 1950.

28 Andrievs Niedra, *Tautas nodeveja atminas* [*Memoirs of the Traitor of His People*] (Riga: Zinatne, 1998).

29 Stanley Page, *The Formation of the Baltic States; A Study of the Effects of Great Power Politics upon the Emergence of Lithuania, Latvia, and Estonia* (New York: Howard Fertig, 1970).

30 All WWI damages from Margers Skujeneeks, pp. 207–218, 353–373, 456–466.

31 Hugo Celmins, "Agrara reform", in Alfreds Bilmanis, Julijs Izaks and Lizete Skalbe, eds., *Latvijas Republikas Desmit Pastavesanas Gados* (Riga: Golts un Jurjans, 1928).

32 The Constitution is at the following web-site: http://www.saeima.lv/Lapas/Satuersme_Visa.html or English language version: http://www.saeima.lv/Lapas English/Constitution_Visa.html or see *Constitution of the Republic of Latvia* (Stockholm: Latvian National Foundation, 1984).

33 Aldis Purs, *Creating the State from Above and Below: Local Government in Inter-War Latvia* (Ph.D. Dissertation, University of Toronto, 1998).

34 Alfreds Bilmanis, Julijs Izaks and Lizete Skalbe, eds., *Latvijas Republik Desmit Pastavesanas Gados* (Riga: Golts un Jurjans, 1928).

35 Aldis Purs, *Creating the State from Above and Below.*

36 The sentiment comes from Vejas Liulevicius on Lithuania, but applies equally to Latvia.

37 Arnolds Aizsilnieks, *Latvijas saimniecibas vesture 1914–1945* [*History of the Latvian Economy 1914–1945*] (Stockholm: Daugava, 1968).

38 Arnolds Aizsilnieks; David Kirby, *The Baltic World 1772–1993: Europe's Northern Periphery in an Age of Change* (London: Longman, 1995).

39 Cited in Indulis Ronis and A. Zvinkulis, eds., *Karlis Ulmanis trimda un cietuma: Dokumenti un materiali* [*Karlis Ulmanis in exile and prison*] (Riga: Latvijas vestures instituta apgads, 1994).

40 Pabriks, 106.

41 Lithuania was assigned to the Soviet sphere later, after the partition of Poland between Germany and the Soviet Union.

42 John Hiden, *The Baltic States and Weimar Ostpolitik* (Cambridge: Cambridge University Press, 1987); John Hiden and Tom Lane, eds., *The Baltic and the Outbreak of the Second World War* (Cambridge: Cambridge University Press, 1992).

43 Ilga Gore and Aivars Stranga, *Latvija: neatkaribas mijkreslis* [*Latvia: Twilight of Independence*] (Riga: Izglitiba, 1992); I. Grava-Kreituse, I. Feldmanis, J. Goldmanis, and A. Stranga, eds., *Latvijas okupacija un aneksija 1939–1940, dokumenti un materiali* [*Latvia's occupation and annexation 1939–1940*] (Riga: By the authors, 1995).

44 Alfreds Ceichners, *Latvijas bolsevizacija, 1940–1941* [*The Bolshevization of Latvia, 1940–1941*] (Riga: 1944. Reprint, Gauja, 1986); E. Zagars, *Socialist Transformation in Latvia, 1940–1941* (Riga: Zinatne, 1978).

45 Indulis Ronis, ed., *Karlis Ulmanis Trimda un Cietuma* [*Karlis Ulmanis in exile and prison*] (Riga: Latvijas vestures instituts, 1994).

46 The definitive work on the Holocaust is Andrievs Ezergailis, *The Holocaust in Latvia: The Missing Center* (Riga: Historical Institute of Latvia, 1996).

47 Karlis Kangeris, "The Former Soviet Union, Fascism and the Baltic Question. The Problem of Collaboration and War Criminals in the Baltic Countries", in Stein Ugelvik Larsen, ed., *Modern Europe After Fascism 1943–1980s* (Boulder: Social Science Monographs, 1998); Haralds Biezais, *Latvija Kaskrusta vara. Svesi kungi, pasu laudis [Latvia under the Power of the Swastika. Foreign Lords, our own people]* (Grand Rapids, 1992).

48 An uncritical account of the Legions is A. Silgailis, *Latviesu Legions* (Imanta: 1962).

49 On Soviet Latvian divisions, see A. Drizulis, *Latvijas PSR Vesture* (Riga: Zinatne, 1986). A further 57,000 were mobilised by the Red Army in the last months of 1944. See Kangeris, p. 750.

50 See E. Dunsdorfs, *The Baltic Dilemma* (New York: Robert Speller & Sons, Publishers Inc., 1975).

51 Heinrichs Strods, *Latvijas nacionalo partizanu kars, 1944–1956 [Latvia's National Partisan War]* (Riga: Latvijas vestures instituts, 1996).

52 Modris Eksteins, *Walking Since Daybreak: A Story of Eastern Europe, World War II, and the Heart of Our Century* (Toronto: Key Porter Books, 1999), p. 154.

53 Kirby, p. 366.

54 Juris Dreifelds, "Immigration and Ethnicity in Latvia", *Journal of Soviet Nationalities*, vol. 1, 4, pp. 34–51; A. Drizulis, ed., *Riga Socialisma Laikmeta, 1917–1975 [Riga in the Socialist Era, 1917–1975]* (Riga: Zinatne, 1980).

55 On collectivization, see Kangeris, pp. 752–758.

56 H. Kreicbergs, *Vainigie un Nelaimigie* (Riga: Avots, 1989); J. Krastins, "Komunistiskais Genocids Latvijas Kulturvide", in I. Sneidere, ed., *Komunistiska Totalitarisma un Genocida Prakse Latvija* (Riga: Zinatne, 1992).

57 The Khruschev era in the Baltic states is thoroughly covered in: R. Misiunas and R. Taagepera, *The Baltic States. Years of Dependence 1940–1990* (London: Hurst & Company, 1993).

58 See also: G. Simon, *Nationalism and Policy toward the Nationalities in the Soviet Union. From Totalitarian Dictatorship to Post-Stalinist Society* (Oxford: Westview Press, 1991).

59 See Plakans, *The Latvians; A Short History*; O. Kregere, "Industrializacijas destruktiva politik", in I. Sneidere, ed., *Komunistiska totalitarisma un Genocida Prakse Latvija* (Riga: Zinatne, 1992).

60 Pabriks, "Part 3. The Role of the Latvian Ethnic Group as a Linkage between the Individual and Overall Society", pp. 174–181.

61 P. Zvidrins and I. Vanovska, *Latviesi. Statistiki Demografisks Portretejums* (Riga: Zinatne, 1992).

62 O. Kregere, p. 142; Dreifelds, p. 51.

63 Zvidrins and Vanovska, pp. 56–58; Dreifelds, p. 52.

64 Misiunas and Taagepera, 225; P. Eglite and I. Markausa, "Vairakumtautibu Demografiska Uzvediba Latvijas PSR 70–80. Gadu Mija", in I. Apine, ed., *Socialie Procesi un Nacionalas Attiecibas Padomju Latvija* (Riga: Zinatne, 1987).

65 The hippie community even created Soviet Latvia's only underground film, *Pasportrets*.

66 M. Soikane-Trapane, "Status of the Latvian Language in Present-Day Latvia", in I. Kalnins, ed., *Conference on Security and Cooperation in Europe Follow-up Meeting in Vienna, November, 1986: Soviet Violations in the Implementation of the Final Act in Occupied Latvia* (USA: World Federation of Free Latvians, 1986); J. Dreifelds, "Russification of Press and Publishing in Latvia", in same.

67 See Misiunas and Taagepera; O. Pavlovskis, ed., *Latvian Dissent, Case Histories of the 1983 Soviet Campaign to Silence Political Dissidents in Occupied Latvia* (World Federation of Free Latvians, 1983).

Chapter 2

LATVIA'S POLITICS 1987–1991: THE THORNY ROAD TOWARDS INDEPENDENCE

Latvia, Estonia and Lithuania celebrate the turn of the millennium as democratic independent European nations aspiring to join the European Union and NATO. Just ten years ago, their situation looked quite different. Only very courageous dreamers could have imagined that these countries would ever re-establish their lost statehood. What was wrong with Soviet rule in Latvia? How did Latvians organise their independence movement? Why did they succeed? These are a few of the questions addressed in this chapter.

QUESTIONING DEPENDENCE, 1987–1990

Many observers, from a range of analysts to popular activists themselves, refer to the five-year period from 1987 to 1991 as the *third Awakening*.[1] This period marks the third wave of Latvian popular activity in the past two centuries. If the first two awakenings were the rise of Latvian ethno-cultural awareness and the proclamation of the independent Latvian state in 1918, the third awakening is characterised by the massive popular movement that succeeded in restoring Latvian independence after fifty years of Soviet occupation. In the period from 1987 to 1991, the world witnessed massive upheaval in the Baltic republics of the Soviet Union. On June 14, 1987, a large part of Latvian society awoke like a sleeping beauty to participate in anti-Soviet demonstrations taking place at the Monument of Freedom in the heart of Riga. During the following days, the Soviet media identified "hostile western radio stations" as the "kiss thirsty" prince that caused the national awakening of Latvians. Many Soviet politicians propagandised, as well as truly believed, that there were no reasons for Latvians or other Balts to oppose Soviet authority. Just five years later it was clear that those who believed in the strength and integrity of the Soviet system were wrong. Latvia and the two other Baltic states, Estonia and Lithuania, re-established their independence *de facto* and *de jure* in 1991.

WHAT WAS WRONG?

At the end of the 1980s, civil protests united about 250,000 people in the Latvian Popular Front (LPF) movement. The many demonstrations with thousands of participants gave Latvians a sense of national unity and solidarity that they had not experienced since shortly before the birth of the Latvian nation-state.[2] Two reasons can be given to explain why such a movement originated in the Soviet Baltic Republic and why it succeeded in re-establishing the Latvian state after fifty years of occupation. These two reasons are *opportunity* and *ability*.

The general reforms of the Soviet system provided the opportunity. In 1985, the election of Mikhail Gorbachev as General Secretary of the Communist Party of the Soviet Union (CPSU) marked the launching of the socio-political reforms known as *perestroika*. A combination of economic problems and strains in foreign policy led to these reforms which were meant to safeguard the position of the USSR as a world power. The reforms aimed for a relative democratisation of Soviet society, but emphasis was put on economic advancement.[3] The Soviet leadership succeeded in the first steps towards democracy in Russia, but the country lost its political as well as economic might. Ironically, relative democratisation, which was initially taken only as a means, became the only achieved end in those days. The initial goal, namely economic and political might, was lost, maybe for ever. In the 1980s, however, the changes in the nature of the oppressive regime had one large consequence, an *opportunity* to express opposing views. The Baltic nations, just like many of their Eastern European neighbours, took the chance. The opportunity was there for two reasons.

First, Gorbachev, like most of his predecessors, disregarded, under-estimated and fundamentally misunderstood the existing cultural and national problems of the Soviet system. Gorbachev seems not to have thought that the multinational composition of the Soviet state was an important consideration in his decisions for reform in 1985. From 1985 to 1987, Gorbachev approved of Communist Party declarations that the historical problem of relations among the nations of the USSR had been resolved once and for all.[4] In his book *Perestroika*, Gorbachev argued that the multinational character of the Soviet Union was a sign of strength rather than weakness and that the nationality question was, in principle, solved.[5] He compared the Soviet Union to the United States of America, stating that in the USSR people cannot avoid learning Russian, just as English cannot be avoided in the USA.

Second, Gorbachev's reforms depended upon popular support for the success that would neutralise his opponents from the conservative wing of the Soviet communists. Gorbachev urgently had to deal with the Soviet socio-economic crisis. Growth rates in the economy had declined steadily and dramatically since the 1960s. In the1980s, the USSR experienced an economic depression. Under such conditions the Soviet Union appeared increasingly incapable of maintaining a military establishment and of developing high technology competitive with that of the United States and the West. Gorbachev argued that *perestroika* meant more democracy as well as more patriotism.[6] Apparently Gorbachev understood that without the revival of civil activity and interest in solving the country's problems, his reforms would fail. In the mid-1980s the Soviet leadership correctly appraised the second estimation, but was fundamentally wrong with the first. In other words, the ghost of nationalism was increasingly noticeable in the Soviet Union while the almighty ghost of communism was speedily disappearing. National mobilisation and solidarity was the factor that provided Latvians with the ability to organise themselves around their ethnic values and overthrow the Soviet regime.

At first glance, one might argue that Latvians were well integrated in Soviet society and that their ethnic differences diminished during the last three decades of Soviet rule. Thus, from 1970 to 1979, the percentage of ethnic Latvians speaking Russian as their second language increased from 45.4 to 58.3 per cent. Latvians were the second most bilingual nation in the USSR after the Belorussians (52.3 and 62.9 per cent, respectively).[7] The percentage of Latvians speaking Russian as their native language increased as well, from 19,000 to 28,900 (1.46 to 2.15 per cent).[8] Finally, during the 1970s and 1980s, the number of ethnically mixed marriages increased to approximately 25 per cent of all marriages.[9] This data alone, however, could lead to misleading conclusions. Although the use of Russian as a second language increased enormously owing to Russification policies, the adoption of Russian as a native language increased very little.

If we regard language as the main source of ethnic identification among Latvians during the Soviet period, the 2.15 per cent of Latvians whose native tongue was Russian does not seem a convincing argument for assimilation. The percentage rate of mixed marriages also loses significance if one bears in mind that the majority of children (about 71 per cent) from these mixed marriages declared themselves to be Latvian.[10] Contrary to accepted opinion, during the last twenty years of Soviet rule in Latvia, the Latvian population increasingly consolidated around ethno-cultural values such as language and culture. This happened because Latvians found it increasingly difficult to pursue their individual and collective goals within the overall Soviet society. The attitude of Latvians, and the Balts at large, can be

TABLE 1 DECREASE IN NUMBER OF ETHNIC LATVIANS AMONG ENGINEERS AND POLICEMEN (PER CENT)

Year	Engineers	Policemen
1970	46	39
1979	41	36
1989	39	34

Source: Zvidrins, P. and I. Vanovska, *Latviesi. Statistiski Demografisks Portretejums* (Riga: Zinatne, 1992).

described as a *social* and *psychological* rejection of everything Russian and consequently of everything Soviet. Consider a few examples. Since approximately 60 per cent of ethnic Latvians were able to use Russian relatively freely, it could be argued that language proficiency would not play a decisive role in choosing one's education or profession, even if the language of instruction was Russian. Ethnic Latvians, however, preferred to study subjects taught mainly in Latvian that would give them a future chance of working in Latvian surroundings, frequently without any kind of financial considerations in mind. Thus, until the mid-1980s, students of Latvian origin preferred to study at the Academy of Art, the Academy of Music, the Academy of Agriculture and the University of Latvia. In each of these institutions, ethnic Latvians made up more than 55 per cent of all students. At the same time, in the Institute of Civic Aviation, where the language of instruction was Russian, only 1.4 per cent of the students were of ethnic Latvian origin.[11] It should be added that ethnic Latvians were seemingly subject to an unwritten prohibition against holding leading posts in shipping, aeronautics or railroad services. These professions were considered to be of a half-military nature in the Soviet Union. Neither were ethnic Latvians ever proportionally represented among many other professions, such as engineers or policemen. Moreover, their proportions were decreasing (Table 1).

Similarly, ethnic Latvians were considerably under-represented in administration. In Latvia, only 28 per cent of those in the Soviet administrative network were ethnic Latvians.[12] The situation did not improve when Mikhail Gorbachev became First Secretary of the CPSU. Recruitment to the Soviet elite was a two-step process: first one had to become a member of the Communist Party, and then advance through the nomenclature system.[13] Latvians were under-represented on both levels. From the 1950s, Latvia's Communist Party had the lowest percentage of titular ethnic members in the entire Soviet Union. Even Lithuania and Estonia had greater representation within their Communist Parties (Table 2).

TABLE 2 COMMUNIST PARTY SIZE AND ETHNICITY 1960–1990 (THOUSANDS OF MEMBERS AND CANDIDATES)

	Latvia		Estonia		Lithuania	
	Size	% of Latvians	Size	% of Estonians	Size	% of Lithuanians
1960	66	35	33.4	49	54.3	58
1965	95.7	39	54.8	52	86.4	64
1970	122.4	–	70.2	52	116.6	67
1975	140.0	–	81.5	52	140.2	68
1980	158.0	–	95.3	51	165.8	69
1985	170.8	–	107.6	50	190.6	71
1989	–	–	–	–	209.5	71
1990	177.4	39	110.3	50	–	–

Note: Owing to the low percentage of ethnic Latvians in the Communist Party, data on the ethnicity of the members in Latvia was never published.
Source: S. Høyer, E. Lauk, and P. Vihalemm, *,Towards a Civic Society. The Baltic Media's Long Road to Freedom. Perspectives on History, Ethnicity and Journalism* (Tartu: Nota Baltica Ltd, 1993).

Most of the ethnic Latvian communists were outside the decision-making nomenclature. In 1986, for example, only four out of thirteen Latvian Central Committee members were Latvians.[14]

In the 1980s, Latvian political and professional misrepresentation overlapped with real, as well as imagined, economic and cultural deprivations. In order to legitimate the Soviet occupation of the Baltic Republics, Soviet ideologists frequently argued that Soviet rule brought to Latvia a flourishing economy and technological revolution. According to official Soviet data, production in Latvia increased by 4,600 per cent compared to 1940. Similarly, GNP increased by 1,150 per cent, and social labour productivity increased by 1,009 per cent.[15] Latvia together with Estonia and Lithuania enjoyed the highest living standards within the USSR. In 1985, the per capita consumption in the Baltic states exceeded the all-Union average by 12 to 28 per cent.[16] The "economic successes" of the post-war years were a major element in the Soviet claim to have a legitimate right to rule in Latvia.

Despite this relative success, however, many Latvians did not find Soviet achievements quite so glorious, for several reasons. First, despite the impressive statistics, in the 1980s, average Latvians felt increasingly deprived economically because it seemed to them that they did not receive an equitable share of what they annually delivered to the whole Union market. Soviet Latvia contained only one per cent of all the inhabitants of the Union, but produced 29 per cent of all passenger wagons for rail transportation (including all the coaches for electric trains), almost half of all telephone receivers, every fourth radio receiver, 22 per cent of all trolley-cars, 10 per cent of all synthetic fibre, 12 per cent of all washing machines and many other goods at a percentage rate several times exceeding the ratio of Latvia's inhabitants to those of the Soviet Union.[17]

Knowing their "importance" to the Soviet economy, Latvians frequently complained that if not for the "greedy union" with the centralised, socialist economy, they would be much better off. Starting from the end of the 1960s, this argument became increasingly popular as people experienced shortages of food and clothing despite relatively well advanced production. Economic difference between the Soviet centre and Latvia was one of the rational sources of centrifugal force making many Latvians feel sceptical about the Soviet system and urging them to search for alternative perspectives.

Second, once the Soviet economy started its decline, it became increasingly difficult for the Soviet propaganda apparatus to use economic arguments against anti-Soviet ideas. Opponents of the Soviet system in Latvia argued that when assessing the development of the Soviet Latvian economy during the post-war years, one must take into account what would have happened if the Soviet Union had not incorporated Latvia by force. The Latvian émigré economist Viksnins argued, for example, that if Latvia had not been occupied, and if Marshall aid had been available to Latvians, then quite probably at least Finnish income levels would have been reached.[18] Thus, the ideological argument entered into the discussion about Latvia's future.

Third, the modernisation that required the import of labour threatened Latvians with becoming a minority in their homeland, and their culture and language with becoming quaint, ethnic artefacts.[19] In the period from 1961 to 1989, about 332,000 newcomers from other parts of the Soviet Union settled permanently in Latvia.[20] Added to nearly half a million immigrants that came to Latvia in the first ten post-war years, the immigrant population constituted nearly one-third of Latvia's population at the end of the 1980s. Along with increasing Russification in the workplace and the decrease of the actual usage of the Latvian language in many spheres of social activity, the cultural and linguistic argument gained weight in the vocabulary of the opponents of the Soviet system.

The permanent pressure of linguistic and cultural Russification, immigration of Slavic populations to Latvia, lack of possibilities of equal representation, and cultural misrecognition were the arguments that pushed the average Latvian further and further away from Soviet politics and government. Under these circumstances Latvian politics from 1987 to 1991 was increasingly determined by one major popular goal, independence. Self-determination and the restoration of the independent Latvian republic was seen as a panacea to all the above-mentioned social, cultural, economic and political problems.

MOVEMENT FOR INDEPENDENCE

Although opposition to the Soviet regime in Latvia existed throughout the post-war period, organised mass opposition can be found only in the first ten post-war years and then again starting from the second half of the 1980s. If in the late 1940s and early 1950s opposition centred on the national partisans in the Latvian forests, then in the 1980s opposition was embodied by a mass organisation, the Latvian Popular Front (LPF), adhering to strictly peaceful and political means instead of armed resistance.

Chronologically, as well as substantively, the movement for Latvia's independence can be divided into three phases. The first phase was the birth of environmental and human rights organisations in Latvia. The second phase was distinguished by the creation of mass popular movements with the participation of intellectuals. In the third phase, a number of communist nomenclature members joined the Latvian Popular Front.

During the *first phase*, a small group of artists, craftsmen and intellectuals that had been active in various folklore and cultural groups initiated a grass roots campaign of preservation of historical buildings. Dilapidated, rural churches were the initial targets, and renovations included the replacement of crosses, a historic symbol opposed by the atheist Soviet regime. The campaign grew into the Environment Protection Club (Vides Aizsardzibas Klubs—VAK). Later, the organisation served as a basis for the Latvian Green Party which, unlike most of Europe's greens, was quite nationalist orientated, thus combining the post-modern values of "thinking green" with the nationalist values of "thinking Latvian". Significantly, the banner of VAK resembled the Latvian national three-stripe flag. Instead of red, their banner had green (a black and white photograph of one, such as a newspaper photo, however, was indistinguishable from the other). The odd combination of nature protectionists and nationalists can be explained by the philosophy of the first VAK members, which considered society and people as a part of nature. This approach stressed the mutual interdependence of people and nature. Consequently, oppressed Latvian culture came under the imagined protection of the Latvian Environment Protection Club.

In 1986, when Soviet authorities planned to construct a hydroelectric complex on the Daugava River close to Daugavpils, a sudden wave of fierce opposition led by Latvian intellectuals and greens appeared. In October 1986, the young journalist Dainis Ivans and his computer specialist colleague Arturs Snips published an article in the cultural journal *Literatura un Maksla* (*Literature and Art*). They raised cultural and environmental issues about the dam, as well as questioning the economic logic of the massive, long-planned project.[21] Their point of view was quickly supported by more than 700 letters and 30,000 signatures sent by the population at large to the journal. Facing such unexpected popular resistance, the USSR's Council of Ministers cancelled the project.

The campaign against the dam was not only about ecology and the rich cultural heritage of the Daugava River Valley. It was also about the centralised manner in which decisions were made. The central Moscow government did not even pay lip-service to the local authorities and population. The campaign was also about opposition to continued labour in-migration since such a complex construction site would inevitably lead to the immigration of many Slavic workers to Latvia. The local Latvian population was not willing to continue accepting decisions made without their consent that could possibly threaten their collective identity and environment. At the same time, Gorbachev's leadership provided the people with the opportunity to express themselves with less fear of repression. The movement against the dam can be seen as the first success story of Latvian collective action against Soviet authorities. Popular mobilisation provided a taste of democracy and whetted the appetites of those who craved more. Thus, the first cobblestones were laid in the road to a mass independence movement.

In the same year, three workers from the port of Liepaja in western Latvia founded Helsinki-86, a human rights watch group. The group declared that their objective was to "monitor how the economic, cultural, and individual rights of our people are respected".[22] The organisation consisted of about ten people, but its presence made the KGB panic. The group was placed under permanent surveillance. Repression and threats almost disrupted Helsinki-86's call for a popular demonstration to commemorate the Soviet deportations of Latvian citizens in June 1941. On June 14, 1987, the first of the "calendar demonstrations" with about 5,000 participants took place at the Monument of Freedom in Riga.[23] In the same year, two other calendar demonstrations marked the anniversaries of the Molotov–Ribentrop Pact and Latvian Independence Day on August 23 and November 18 respectively.

The period from October 1986 to March 1988 was the first phase of the popular movement for independence, marked by ecological protests and the calendar demonstrations. During this phase, the mass activities were mainly of a near-spontaneous nature organised by former dissidents, political prisoners and people outside the established elite circles. While being trailblazers in the fight for human rights and national independence, these individuals frequently lacked the skills needed for detailed public work. Therefore, in the following years the leadership of the movement for independence and democracy in Latvia was taken over by a number of establishment intellectuals, and later by members of the communist nomenclature. Therefore, the sad historical generalisation that every revolution abandons its initiators also applies to the Latvian national awakening of the 1980s.

The *second phase* of pro-independence activities began in March 1988 when the creative unions, organisations that united the majority of Latvian intellectuals, called for a discussion of the "tragic consequences" of Stalinism and how intellectuals should deal with them.[24] On June 1 and 2, 1988, Latvia's intellectuals met to discuss contemporary social and economic problems and demanded the public unveiling of the so-called "white spots" of history. In the 1980s, this term referred to everything that official Soviet propaganda avoided or pretended did not exist. Among these issues were the Soviet occupation of the Baltic states, the Soviet–Finnish war, and Soviet repression. During the conference, Mavriks Vulfsons, a journalist, political analyst and an old communist, stated openly that in 1940 Latvia was violently occupied by Soviet military forces. This could be considered the first *official challenge* of the legitimacy of Soviet power in Latvia. With this announcement, Vulfsons opened a Pandora's box, initiating a broad discussion on the legality of Soviet authority in Latvia and the two other Baltic countries. The announcement triggered tremendous political upheaval in society that was soon dubbed the *"Awakening"* by journalists.[25] Suddenly, popular opposition to the Soviet regime became a just cause in the eyes of the Latvian majority. Increasing numbers of individuals took part in the "calendar demonstrations" which became a predictable popular activity throughout the Baltic Republics over the next three years.[26] Calendar demonstrations marked the anniversaries of the darkest moments of Soviet occupation in Latvia and its two neighbouring countries. Simultaneously, this list was enlarged by the anniversaries of the independent pre-war republic. The public, after a nearly fifty-year hiatus, celebrated November 11th (Heroes' Day) and November 18th (Latvia's Independence Day). The participation of the Latvian intellectual elite, starting from 1988, gave these commemorative events additional legitimacy in the eyes of the broad public as well as helping to protect the participants from violent mass repression.

In 1988, with increasing public activity and the lack of serious mass repression by Soviet authorities, two mass organisations were founded. On June 17, 1988, the Latvian National Independence Movement (LNNK) was founded as the *first national mass movement* demanding the restoration of an independent Latvia. Eduards Berklavs, one of the senior purged national-communists of the 1950s, reappeared on the public scene as the head of the LNNK. In the following years this organisation grew into one of Latvia's first and largest political parties. In October 1988, the more moderate Popular Front of Latvia (LTF) was created under the guidance of Latvian intellectuals.[27] Within a year, LTF membership grew to about 250,000, uniting more than 10 per cent of Latvia's population in the country's largest *democratic movement*. The initial goal of LTF was broad political and economic autonomy for Latvia. In 1989, however, the changing political situation, as well as the political pressure of LNNK, pushed the LTF leadership to demand full independence.

With the intellectuals' involvement in the mass movement, the Latvian Communist Party, Latvia's territorial branch of the Communist Party of the Soviet Union (CPSU), faced their own political dilemmas. This period can be defined as the *third phase* of the independence movement. The first dilemma was the question of a willingness to reform the party organisation along democratic principles. The second dilemma concerned the push for Latvia's independence. This included a number of questions related to historical interpretations and Russian–Latvian relations in general. The Latvian Communist Party (LKP) could not avoid these issues because a majority of the intellectuals holding LTF membership were also members of the Communist Party. Thus, at the founding congress of the LTF, approximately one-third of its delegates were Communist Party members.[28] Also, the LKP had to establish a stand on what was happening in the Latvian Soviet Socialist

Republic for their own survival. Using Laitin's terminology, the "old elites", in order to survive, had to choose either to support the newly growing political forces or to continue to collaborate with their former "lords" in the imperial centre.[29] Latvia's communists were divided over this choice. Some opted in favour of the Popular Front while others remained faithful to Moscow. From 1988, the Latvian Communist Party consisted of a reform wing and a conservative wing, yet the structural split of the party came only in 1990. With little chance of compromise between the two wings, the reform communists at a meeting on February 24, 1990, called for a founding congress for an independent Latvian Communist Party. A few days earlier, the reformists, who composed about one-third of the Communist Party, had been expelled from the original LKP by the Moscow loyalists.

It must be noted that the majority of the reform communists opting for co-operation with the popular movements were ethnic Latvians, while the remaining loyalists were mainly of Slavic origin. The active participation of the national communists in the popular movements was a *matter of self-identification* as well as a question of *rational choice*.[30] The option to go "in step with the nation" meant receiving a "pardon" for the "mistakes" these individuals had committed when joining the Communist Party. Formerly, they had been rejected or simply avoided by their compatriots because of their political decision to bolster the Soviet regime in Latvia. The rational part of the choice, on the other hand, was connected with the calculation that with every year of *perestroika* the Soviet system was becoming more democratic. This meant that the popular voice would increasingly get more power, and by opposing the popular movement the ruling position of former elite members would be undermined. The reform communists also appeared useful to the Popular Front. They were personally familiar to Gorbachev and his circle, and they knew the language of power used in Moscow. The reform communists, like Anatolijs Gorbunovs, were the Latvians with whom Moscow could negotiate. Therefore, the marriage between the Popular Front and reform communists appeared mutually acceptable, at least for the first years. The relationship also seemed profitable for the central authorities since, with the help of the reform communists, they had a feeling of control over the mass movements in the Baltic Republics. The triangle was beneficial for all those involved.

While the reform communists made common cause with the Popular Front, the conservative wing of Latvia's Communist Party did not lose complete public support. In response to the increasing popularity of the

Popular Front, an alternative mass movement, Latvia's International Working People's Front (Interfront), was established in 1988. Interfront was strongly supported by the conservative communist leadership in Moscow and Riga as well as by the KGB. Interfront mainly united aged Russian speakers and former Soviet army officers. Interfront's aim was to avoid the erosion of Communist Party rule and oppose any changes that threatened the privileged status of Russian-speaking immigrants. Interfront, however, with its old-style rhetoric, was more an object of ridicule than real political concern for most Latvians.[31]

In general, the non-Latvian population of the country was divided in their attitude towards the political activities taking place in Latvia. Though the programme of the LTF was relatively moderate, among the members of the movement, anti-immigrant and anti-Russian sentiments were quite widespread. Many Russian speakers had strong qualms about the movement, whose programme was frequently presented in nationalist and anti-Soviet vocabulary. The Russian-speaking masses were not willing to yield their privileged status on language issues. As a result, ethnic polarisation obtained a structural frame, with most ethnic Latvians in the Popular Front while Interfront gained support from Russian speakers.

A number of non-Latvians, however, supported the Popular Front, especially ethnic minorities whose families lived in Latvia before the occupation and who understood the growing Latvian aspirations for independence owing to their own pre-war experience in independent Latvia. At the same time, the leadership of the Popular Front moved quickly to support minority cultural autonomy. In 1988, the Popular Front helped to initiate eighteen National Cultural Associations (for Jews, Ukrainians, Belorussians, gypsies and others). Then, in co-operation with the Communist Party leadership, the Popular Front helped organise a Nationalities Forum. The forum provided an opportunity for all groups to air their grievances. The participants passed three resolutions expressing support for Latvia's sovereignty, concern about the ecological situation, and support for the efforts of minority groups to preserve their cultures.[32] Along with these developments, in September 1989 the first Jewish school in the Soviet Union was opened in Latvia, symbolising the tendencies of multicultural politics within the Latvian Popular Front.[33]

Minorities saw the chance to promote their own interests while co-operating with Latvians and the Popular Front. The Russian immigrant population was divided in their attitude towards the suddenly (at least for them) resurgent Latvians. The local conservative communist leadership frequently did not receive direct orders from Moscow on how to deal with mass movements demanding independence. At the same time they lacked the courage and imagination to deal with such large popular masses unified in common political action themselves. Their indecision and inactivity worked in favour of Latvia's Popular Front, which gained increasing public support with each passing day.

FROM OPPOSITION TO POSITION

In October 1989, during the second LTF Congress, 24 per cent of the 1,046 delegates to the congress were Communist Party members. Moreover, 45 per cent of the 100-member LTF Council were party members, which showed the increasing influence of the reform communists on the Latvian popular movement.[34] The communist content definitely provided more legality to the Popular Front, making it more predictable as well as respectable, at least in the eyes of the Moscow central authorities. Despite the fact that leaders of the LTF were generally more moderate than the membership at large, their room for manoeuvring was constrained by the fact that ordinary members of the Front increasingly insisted on national independence, and sometimes even opposed any collaboration with authorities of the Soviet Union. Therefore, at the end of the 1980s, it is possible to distinguish between two simultaneous ways of trying to achieve independence in Latvia. The first was the *parliamentary way* while the second was the *legalistic attempt* to restore Latvian statehood.[35] Curiously, in the Baltic case of the 1980s, the legal and parliamentary methods were not the same.

Parliamentary method

The parliamentary method of regaining independence meant the use of the existing Soviet institutions in order to increase Latvia's political and economic autonomy.[36] Usually this was done by replacing the existing Soviet bureaucracy with more pro-Latvian oriented individuals or by the ideological reorientation of the existing Soviet civil servants. At the end of the 1980s, this method appeared to be quite successful and there was a chance to reach most of the LTF's goals by using the existing structures of the regime. In March 1989, the Latvian Popular Front won the majority of seats allocated for Latvian delegates to the all-Union Congress. In the same year, the Latvian Popular Front was victorious in municipal elections and was thus able to use these structures to further the Front's political agenda. By 1989, the Communist Party's monopoly of power in Latvia had been broken.[37]

The parliamentary method proved peaceful and constitutional according to the Soviet understanding; however, it did not follow a traditional understanding of democratic procedures. The main problem of Soviet elections was that the Soviet authorities regarded the "people" as everybody who was living in occupied Latvia, including the large garrison of Soviet troops. Consequently the electorate consists not only of citizens of the former Republic of Latvia, but also of all the post-war Soviet immigrants, their descendants, and the military posted in Latvia. A large part of the Latvian population feared that the participation of these individuals in the elections would not only undermine the legality of such elections, but also endanger Latvia's prospects for independence since the majority of the above-mentioned social groups opposed any increase in Latvia's autonomy. Moreover, there was a fear that if the participation of the Soviet immigrant population were institutionalised, the legal continuity between the pre-war Latvian Republic and re-established modern Latvia would be broken. The opposition of the majority of Soviet immigrants to a proclamation of independence,[38] the loss of the legal continuity of Latvian statehood and the possibility that Latvia would remain a political satellite of Moscow while being formally independent were the major concerns of the opponents of the parliamentary method of regaining independence.

This question gained even more importance when the elections to the Latvian Supreme Council approached in 1990. If the Popular Front lost these elections it could not proclaim independence and Soviet authority in Latvia would be simultaneously legitimised. Taking into account that Latvia had the largest Russian-speaking minority among the Baltic Republics as well as a sizeable Soviet garrison that would vote against independence, these fears seemed well founded. The consequence of such an outcome would be a split of the united Baltic approach to co-operation for the sake of common independence for all three Baltic nations.

Legal attempt to restore independence
In February 1989, the most radical wing of Latvian political activists began establishing Citizens' Committees to avoid the possible corruption of the principle of independence. The Latvians borrowed the idea from the Estonians who were faced with the same problem while aiming to become an independent nation. The Committees registered all those

who were citizens of Latvia before occupation in June 1940, and their descendants. June was the last month of the independent Latvian state before Soviet occupation. The idea behind the Committees was that those registered would elect delegates to a congress that would decide the future of Latvia.[39] The legitimacy of the congress was based upon the argument that only the citizens of the former, independent, legal republic (and their offspring) have *de jure* rights. Those who entered the Baltic states as officials of the "occupying power" did not.

By the beginning of 1990, this new movement had gathered 900,000 signatures of people who wanted to be citizens of an independent Latvia; in effect, carrying out an unofficial referendum. The "Citizens Movement" pushed increasingly for a rapid break with the Soviet Union and the disfranchisement of Soviet immigrants.[40]

The confrontation between the two movements began at the end of 1989, with the approach of elections to the Supreme Council. The majority of the Popular Front leadership supported the parliamentary approach with the goal of gaining broad autonomy for Latvia, while the Committees argued for independence through a legalistic and immediate solution to Soviet occupation. The Committees threatened to boycott the elections in order to avoid further legitimisation of Soviet authority. At the same time, the Citizens' Committees were unable to provide a reasonable and peaceful alternative on how to re-establish independence because the Soviets did not seem to be yielding to the nationalists' demands. Shortly before the crucial elections, the Popular Front and the Committees came to an agreement that the Committees would not boycott the elections while the Popular Front, in case of victory, would demand complete national independence.

Citizens' Committees existed from 1989 to 1993, until the election of the 5th *Saeima* (Parliament) of Latvia, providing an alternative to the parliamentary method of re-establishing independent Latvian nationhood. The Citizens' Committees together with the Popular Front also played an important role in the development of civil society and democratic understanding within Latvian society. They provided healthy competition between the two independence movements and they served as a guardian of the principle of continuity of Latvian statehood. Without the Committees, the Popular Front leadership would have faced more difficulties in keeping its backbone in negotiations with Moscow. The influence of the reform communists might have also been larger if not for the radical demands of the members of the Committees.

The compromise between the Popular Front and the Citizens' Committees achieved on the eve of the Supreme Council elections of 1990 secured an LTF victory in the Council and placed the independence movements in the parliamentary position. During the elections to the Supreme Council (March to April 1990), the LTF obtained 134 of 200 seats. Anatolijs Gorbunovs, the high-ranking reform communist, was elected Chair of the Supreme Council while Ivars Godmanis, a leader of the Popular Front, became Prime Minister.[41] The Popular Front and the reform communists took institutional power in Latvia in less than a two year-period from the inception of the Popular Front.

The next step of the newly elected Latvian Supreme Council was to decide how to re-establish independence. Before making this decision, the Latvian Popular Front seriously considered the experience of its neighbours, Lithuania and Estonia. The Lithuanian Supreme Council passed its unilateral Declaration of Independence on March 11, 1990. The Soviet Union responded with an economic blockade to "show" Lithuanians what independence meant in reality. The confrontational tactics chosen by the Soviet leadership appeared to be fruitless because the Lithuanian decision was based on moral and psychological grounds while the Soviet response tried to use rational arguments. From today's point of view it is clear that the only outcome of the Soviet blockade was an even stronger Lithuanian demand for independence. Estonian legislators, on the other hand, chose a rather more careful way by declaring independence, while also announcing a transition period until the constitutional institutions of the Estonian state could be created.

The Supreme Council of Latvia decided to follow in the Estonians' footsteps and on May 4, 1990 passed the awkwardly titled "Declaration About the Renewal of the Independence of the Republic of Latvia". Simultaneously, the Council declared that independence would come into force only after the end of a transition period.[42] Hoping to avoid direct confrontation with Moscow, the Latvian declaration, similar to the Estonian declaration, was a conditional one calling for a "transition period" of an indeterminate length leading to eventual, complete independence from the Soviet Union. 138 Council members, 134 of whom were Popular Front activists, voted for the declaration. Fifty-seven conservative communists and Interfront members walked out of the assembly and did not participate in the vote. The electoral success of the pro-independence forces was striking, taking into account that almost half of Latvia's population in 1990 were Russian speakers. Thus, it is possible to argue that an increasing number of non-Latvians opted for independence as well as for the political programme of the Popular Front. The period from 1990 to the end of 1991 can be regarded as the honeymoon in relations between various ethnicities inhabiting Latvia. According to the Latvian sociologist Brigita Zepa, 85 per cent of ethnic Latvians and 26 per cent of non-Latvians opted for an independent Latvia in 1990.[43]

THE UNBEARABLE LIGHTNESS OF INDEPENDENCE

The time from the declaration of independence in 1990 to contemporary politics is distinguished by three periods in Latvian politics. The first period was from May 1990 to August 1991 when theoretically Latvia existed as an independent state, but legitimate power could not be exercised owing to the Soviet Union's intransigence and refusal to negotiate independence. The second period started after the unsuccessful coup in Moscow in August 1991, after which Latvia became independent *de facto* as well as *de jure*. The beginning of the third period can be set with the election of the fifth *Saeima* (Parliament) in 1993, the first truly independent elections after more than fifty years of Soviet occupation. During these three periods, the people of Latvia went through a tremendous experience in democracy, economic hardship and social transition. Very few, if any, of the voters could imagine what was in store for them over the next few years. Nevertheless, the Baltic states were among the most ardent supporters of the socio-political transition among their Eastern and Central European fellows. Continued Latvian support can be attributed to connections with the interwar Republic, anger, pain and resentment from fifty years of Soviet occupation, and a passion for making up for those lost decades. For many Latvians, however, as well as non-Latvians, independence did not fulfil their dreams of a paradise on the shores of the Baltic Sea.

Dual government from 1990 to 1991

In 1990, despite the Soviets' refusal even to discuss Latvia's independence, the newly elected government and its Cabinet of Ministers headed by Ivars Godmanis decided to proceed, more or less, as if Latvia were already independent. The main duties of the government were to introduce new legislation and to create new state institutions that would correspond to the needs of an independent state. In general terms, in the "Declaration of 4 May 1990 concerning the Renewal of Independence of the Republic of Latvia", the Supreme Council provided guarantees of fundamental rights. On the same day the highest Latvian legislative body proclaimed Latvia's accession to fifty-one international human rights instruments. These included forty-eight declarations, conventions and resolutions drawn up in the United Nations or among its specialised agencies.[44] Latvian authorities also declared that in their legislative activity they would be guided by the relevant documents adopted by the Council of Europe and the European Parliament. The primacy of the fundamental principles of international law over national law was also recognised. The legislative process itself, however, was slowed by Latvia's new legislators' lack of democratic experience. Contacts with Western countries were still weak and hindered by Soviet-controlled borders. There was also little co-ordination between various governmental institutions thereby undermining the process of transformation from dependence to democracy.

In the institutional sphere, major changes were also initiated. The government established a Customs Department and created custom points at the Latvian border. A new prosecutor's office was also created because the existing office refused to obey the new government, continuing to follow orders from the central Soviet authorities in Moscow. Thus, in 1990, Latvia had two state Attorneys General. Severe problems existed also in the Ministry of Interior. Its Riga department declared that it did not recognise the Declaration of Independence.[45] Consequently the work of the police was paralysed for at least a year. The government, however, managed to establish the Bank of Latvia, a crucial first step towards economic independence. Although many changes were initiated during the first year of theoretical independence, many institutions continued to function in the old Soviet manner or simply narrowed their activities while waiting for a political solution to the existence of two governments, the Latvian and the Soviet.

Although the transition period was relatively peaceful (compared to the bloodshed of the disintegration of the Soviet Union in other republics), ordinary citizens experienced quite a bit of panic and concern about their security and well-being. By the end of 1990, it was obvious that Latvia, Estonia and Lithuania were going to renew their independence. The radical forces within the CPSU, the Soviet military and the Latvian Communist Party decided to use force to halt what they called the disintegration of the Soviet Union. Soviet paramilitary units (OMON) stationed in Latvia struck with violence at random targets of the new Latvian state and its civilians. The OMON attacked and burned customs posts, kidnapped individuals and generally tried to convey an intimidating presence in Latvia. When the US-led Allied attack on Iraq commenced in January 1991, many expected the OMON and other reactionary forces (KGB, CP) to strike against the Popular Front government and the independence movement. On January 2, 1991, at the request of the Communist Party of Latvia, OMON troops occupied the Press Building, effectively halting the popular media for a short while. On January 20th, the same units attacked Latvia's Ministry of Interior. In the ensuing firefight, five people were killed and twelve others wounded. A week earlier, on January 13th, the Soviet army used tanks to attack a major Lithuanian TV station in Vilnius. Thousands of unarmed Lithuanians surrounded the TV station and tried to defend their freedom with bare hands. The TV station, however, was occupied, leaving fourteen killed and hundreds injured. The Communist Parties of all three Baltic republics organised "National Salvation Committees", and called for the resignation of the elected parliamentary governments.

On January 13th at 4.45 a.m., Dainis Ivans, the Supreme Council member and Popular Front leader, appealed to the nation in Latvian and Russian to defend government buildings against attack.[46] The streets of Riga quickly clogged with tens of thousands of people constructing barricades in anticipation of an all-out assault. People stayed wearily at the barricades for several weeks, but the OMON and the Communist Party did not press the battle, stunned by both international reaction and popular resistance in all three Baltic countries. Significantly, a sizeable minority at the barricades were Russian speakers. Ilga Apine, a Latvian scientist, has rightly called this period the honeymoon of ethnic relations. Indeed, in the newspapers it was possible to find articles with the following headlines: Latvians and Russians—On One Side!; All Together—Against Militarism!; People of Latvia—Against the Attempts of Moscow to Intervene in Latvian Affairs.[47] In the shadow of the January events, many Russians had to make a choice between the communist reaction on one side and Latvian nationalism on the other. Taking into account that Latvian nationalist politics at that time included a portion of democratic and liberal values, Russians chose what seemed to them the lesser evil, namely co-operation with Latvian independence seekers. It seemed that many Russians living in Latvia were repulsed by the Soviet resort to force and decided that they feared reaction generally more than Latvian nationalism particularly. When the Latvian government held a plebiscite on independence, on March 3, 1991, 73.68 per cent of the voters opted

TABLE 3 SUPPORT FOR INDEPENDENCE ACCORDING TO ETHNICITY (%)

	1989	1991
Latvians	55	94
Russians	9	38

Source: Zepa, B., "sabiedriska Doma Parejas Perioda Latvija: Latviesu un Cittautiesu Uzskatu Dinamika (1989–1992)", Latvijas Zinatnu Akademijas Vestis, No.2 (1992).

for independence, a figure that included a large number of Russians (Table 3).[48]

Besides uniting most of the population in favour of Latvia's short-term political objectives, the period of dual government was important in Latvian politics for another reason. According to Levits, during this period the cornerstone of the future national bureaucracy with its heavy dependence on former Soviet technocrats and reform communists was laid.[49] The Communists argued that in order to create an efficient and functioning government, there was a need for "professionals" which they saw as being themselves. At the same time it was not taken into account that the reform communists and the Soviet administration, in general, did not possess the skills needed for democratic government. During the January crisis of 1991, the reform communists were the most scared of popular defensive activities and the building of barricades, frequently arguing the need to find an easier compromise with Soviet authorities.[50] The popular movement with its human sacrifice of 1991, however, was the catalyst that paved the road to Latvian independence. At the beginning of 1991, Latvians mentally stopped living as citizens of the Soviet Union because the regime was killing and humiliating its own people.

The period of dual government ended suddenly in August 1991, when a coup d' état organised by reactionary communists and the conservative forces within the military failed in Moscow. Latvian authorities used this confusion and on August 21, 1991, declared the end of the transition period towards independence. The goal was reached almost four years after the first anti-Soviet demonstration took place in Riga—a speed and accomplishment not imagined in those first fateful hours.

1 J. Stradins, Tresa atmoda (Riga: Zinatne, 1992).

2 Artis Pabriks, From Nationalism to Ethnic Policy: The Latvian Nation in the Present and the Past (Aarhus: PhD diss., 1996), p. 163.

3 B. Nahaylo, "Nationalities", in M. McCauley, ed., The Soviet Union under Gorbachev, (England: Macmillan Press, 1987: 94); O. Norgaard, "Gorbatjovs Reformstrategi", Politica, 20, 1 (1988), p. 14.

4   R. Lukic and A. Lynch, *Europe from the Balkans to the Urals. The Disintegration of Yugoslavia and the Soviet Union* (Sipri: Oxford University Press, 1996), p. 129.

5   M. Gorbacovs, *Parkartosanas un Jauna Domasana. Musu Valstij un Visai Pasaulei* (Riga: Avots, 1987), p. 103.

6   *Ibid.*, p. 29.

7   G. Simon, *Nationalism and Policy toward the Nationalities in the Soviet Union: From Totalitarian Dictatorship to Post-Stalinist Society* (Oxford: Westview Press, 1991), p. 397.

8   F. Rajevska, "Kritika asvestenija burzuaznimi ideologami roli i znatsjenia dvuhjazitisija v SSSR", in F. Rajevska, ed., *Nacionalas Attiecibas un Ideologiska Cina*, (Riga: LVU, 1982), p. 50.

9   P. Zvidrins and I. Vanovska, *Latviesi. Statistiski Demografisks Portretejums* (Riga: Zinatne, 1992), p. 84.

10 J. Dreifelds, "Immigration and Ethnicity in Latvia", *Journal of Soviet Nationalities*, 1, 4 (1990), p. 60.

11 Zvidrins and Vanovska, p. 100.

12 *Ibid.*, p. 103.

13 A. Steen, *Recirculation and Expulsion: The New Elites in the Baltic States* (University of Oslo: Department of Political Science, 1994), p. 4.

14 Misiunas and Taagepera, p. 206.

15 G.J. Viksnins, "The Latvian Economy: Performance and Prospects", in I.Kalnins, ed., *Conference on Security and Cooperation in Europe Follow-up Meeting in Vienna, November 1986: Soviet Violations in the Implementation of the Final Act in Occupied Latvia* (USA: World Federation of Free Latvians, 1986), p. 123.

16 G.E. Schroeder, "Nationalities and the Soviet Economy" in eds., *The Nationalities Factor in Soviet Politics and Society*, L. Hajda and M. Beissinger, (Oxford: Westview Press, 1990), p. 51.

17 Viksnins, p. 122.

18 *Ibid.*, p. 120.

19 A. Veisbergs, "The Latvian Language—Struggle for Survival", *Humanities and Social Sciences*, 1 (1993), p. 32.

20 Dreifelds, p. 52; Pabriks, p. 168.

21 *Literatura un Maksla*, October 14, 1986.

22 O. Eglitis, *Nonviolent Action in the Liberation of Latvia* (Cambridge, MA: The Albert Einstein Institution), p. 9.

23 O. Bruvers, "1987. 14. Junijs Riga", *Latvija Sodien*, 15 (1987), pp. 1–10.

24 J.A. Trapans, "The Sources of Latvia's Popular Movement", in ed., *Towards Independence: The Baltic Popular Movements*, J. Trapans (Oxford: Westview Press, 1991), p. 31.

25 Eglitis, p. 14.

26 A. Plakans, "Democratization and Political Participation in Postcommunist Societies: The Case of Latvia" in K. Dawisha and B. Parrot *The Consolidation of Democracy in East-Central Europe*, eds., (Cambridge: University Press, 1997), p. 170.

27 Trapans, 35.

28 *Latvijas Tautas Fronte. Gads Pirmais* (Riga: 1989), p. 65.

29 D.D. Laitin, "The National Uprisings in the Soviet Union", *World Politics*, 44, 1 (1992), p. 146.

30 Pabriks, p. 194.

31 N. Muiznieks, "Latvia: Origins, Evolution and Triumph" in eds., *Nation and Politics in the Soviet Successor States* J. Bremer and R. Taras, (Cambridge: University Press, 1993), p. 197.

32 *Ibid.*

33 D. Bleiere, *Latvijas notikumu hronika* (Riga: N.I.M.S., 1996), p. 17.

34 W.C. Clemens, *Baltic Independence and Russian Empire* (Basingstoke: Macmillan, 1991), p. 208.

35 E. Levits, "1990. gada 4. maija Deklaracija par Neatkaribas Atjaunosanu", in eds., *Latvijas valsts atjaunosana 1986—1993* U. Bormanis and V. Kanepa, (Riga: 1998), p. 215.

36 Levits, p. 215.

37 *Ibid.*, p. 217.

38 In June 1989, 55 per cent of ethnic Latvians and only 9 per cent of non-Latvians opted for independence (Zepa, 1992, p. 23).

39 Trapans, p. 41.
40 Muiznieks, p. 199.
41 Misiunas and Taagepera, p. 335.
42 Levits, pp. 222–224.
43 B. Zepa, "Sabiedriska Doma Parejas Perioda Latvija: Latviesu un cittautiesu uzskatu dinamika", *Latvijas Zinatnu Akademijas Vestis*, 2 (1992), p. 23.
44 5th Saeima-Republic of Latvia Standing Commission on Human Rights (1993), p. 32.
45 T. Jundzis, "Neatkarigas valsts aparata veidosana un tiesibu reformas", in eds., *Latvijas valsts atjaunosana 1986–93*, U. Bormanis and V. Kanepa, (Riga: 1998), p. 240–241.
46 Ivans, p. 321.
47 I. Apine, "Nomenclature. The Peculiarities of Latvia", in *The Transition towards Democracy: Experience in Latvia and in the World, Nov. 12–14, 1992* (Riga: University of Latvia, 1994); Pabriks, p. 208.
48 Muiznieks, p. 200.
49 Levits, pp. 223–224.
50 Ivans.

Chapter 3

LATVIA'S DEMOCRACY EXAMINED: 1991–1999

What kinds of political dilemmas did Latvians face after re-establishing their independence? What are the major problems in contemporary Latvian politics? What are the perspectives for Latvian democracy? These are a few of the questions that we would like to address in this chapter. First, we will examine the post-awakening period when a new elite formed, and political parties were institutionalised through the first independent parliamentary elections of 1993. Next, contemporary ethnic dilemmas will be discussed. The chapter will end with speculations about the weaknesses and strengths of Latvia's political system.

THE POST-AWAKENING (1991–1993)

The morning after is usually the morning of a hangover. In the two years after re-establishing independence, many Latvians had to cure their post-awakening hangover. These two years of independence can be characterised as the end of the Baltic revolution. The first victims were the leaders of the singing revolution who left of their own accord or were forced to leave Latvian politics. Their time was over, just as the time of the Popular Front and its policies was over as well.

Dainis Ivans, the spiritual and intellectual leader of the Popular Front, was among those who on his own initiative left the Presidium of the Supreme Council as a protest against the corruption of the principles of the popular movement. According to Ivans, in those early days the first post-awakening elite was formed by uniting the former Soviet *nomenklatura* members with young career seekers.[1] Latvian politics of these years seemed dominated by one sole principle—the search for money and power. The formerly united Popular Front split into various sections, preparing the ground for Latvia's emerging political parties and interest groups. Among the main political events of the post-awakening period it is relevant to distinguish:

- party and elite formation;
- fifth *Saeima* elections.

Party and elite formation

In his memoirs, Ivans recalls that the post-awakening political culture changed drastically. Formerly, many decisions had been based on public discussions or on close consultation with the public, thus keeping the distance between the average citizen and the Popular Front leaders measurable in inches, not miles.[2] After the restoration of independence, however, political decision-making frequently moved to "dark rooms" where it has largely remained, thus contributing to the alienation of the masses from the ruling èlites. Among the first of these "power organisations" was *Club 21*, which united many former and future leaders of the Latvian elite. According to Ivans, every member and supporter of the Club was promised an office in the future state or private institutions in unofficial talks.[3]

Ivans' memoirs do not seem exaggerated bearing in mind that *Club 21* served as the basis for *Latvia's Way (Latvijas Cels)*, the party that won the relative majority of seats (36 out of 100) in the elections to the fifth *Saeima*. From 1993 onward, the party has been Latvia's kingmaker and has yet to be in parliamentary opposition. Latvia's Way (LW) was formed in January 1993 by a group of MPs formally belonging to the parliamentary faction of LTF. Among the leading figures was Anatolijs Gorbunovs, Supreme Council Chair and ex-ideological secretary of the Latvian Communist Party.[4] According to Cerps, this party united about 70 per cent of Latvia's most popular politicians, thus being compared to "a team of stars". In its nascent years, there were several well-known Latvian émigrés among the LW.[5] They gave the party the image of an open-minded, liberal political force able and willing to lead Latvia through reform. In the following years, the role of the émigré Latvians drastically decreased among the top political elite, in part because these individuals were not and could not be considered as insiders by the former Soviet nomenclature who were still the unchallenged leaders of Latvian politics. The role of the reform communists did not decrease during the post-awakening period because they proved flexible enough to cope with independence. After the restoration of independence, former CP members constituted a sizeable share of the new Latvian elite. According to Anton Steen, former CP members constituted 63 per cent of the new Latvian elite in 1993. The former CP members' hold on the central administration and the municipalities specifically was even greater; they accounted for 66 and 71 per cent of these institutions respectively.[6]

Latvia's Way combined within itself a large number of former *nomenclatura* representatives, career seekers and some moderate

Popular Front members. Because of or despite this combination, the party is among the longest-lasting and most successful political parties in contemporary Latvia. Only a few other political forces can match Latvia's Way's record of longevity in parliamentary representation. These parties on the right of the political spectrum are the *Latvian National Independence Movement* (LNNK) and *For Fatherland and Freedom* (TB). LNNK was born three months before the Popular Front in July 1988.[7] From its inception, this political movement declared its aim to be the restoration of democracy and Latvia's independence. In the post-awakening period, LNNK represented Latvian nationalist philosophy and interests. In parliamentary elections, their support has come from voters who looked for an alternative that was similar to Latvia's Way, slightly more nationalistic and less associated with the former Soviet *nomenklatura* and reform communists. Further to the right than the LNNK in the Latvian political spectrum is For Fatherland and Freedom. This party emerged from the organisers and supporters of the Citizens' Committees. In contemporary Latvia, this party represents the far right in Parliament. In the elections to the sixth, and particularly the seventh parliaments, they nominated a joint block of candidates with LNNK. In this merger of the nationalist right, the more extreme TB has taken the controlling lead of the far right in Latvia's politics.

Generally speaking, once the Communist Party's monopoly on power collapsed and after the Popular Front splintered with its achievement of independence, political parties sprouted like mushrooms after a warm, autumn rain in post-Soviet Latvia. Many had already emerged before 1991. Thus, in July 1989, Latvia's Social Democratic Workers' Party (LSDSP) was re-established based upon roots in the strong social democratic movement of the pre-war period. Although social democracy had a long and honourable history in Latvia, its contemporary popularity was corrupted by fifty long years of communist rule. The LSDSP tried to establish its independence from the Communists, even becoming a full-fledged member of the Socialist International before Latvia's Declaration of Independence of May 4, 1990.[8] In the spring of 1990, a number of former reform communists founded Latvia's Democratic Labour Party (LDDP). Juris Bojars, a former KGB major and splendid orator, led the party, but a law forbidding KGB personnel to run for office handicapped the party's future parliamentary campaigns. Rounding out the parties left of the political spectrum was *Harmony for Latvia—Revival for the Economy* (SLAT). Janis Jurkans, Latvia's first Minister of Foreign Affairs who had been sacked from his post because of his liberal views on granting citizenship to the former Soviet immigrants, led this party.

In the same year, several members of Latvia's Popular Front and the earlier environmental movement VAK established Latvia's Green Party (LZP). Latvia's Communist Party also re-formed as the *Equal Rights Movement*. Its main political argument throughout the 1990s was the defence of the rights and interests of Russian speakers in Latvia. The Movement's popular image has always stressed its close association to pro-Soviet, and later pro-Russian, conservative communist circles. Few of Latvia's voters trusted this party, seeing it as a possible threat to Latvia's independence. Along with the above-mentioned parties and political movements, the pre-war Latvia's Farmers' Union (LZS) and Democratic Centre Party (DPC) were re-established. A Christian Democratic Union (KDS) was also founded on tenuous links to the pre-war Republic. Despite these parties' claims to be the successors of pre-war Latvian parties, their connections were primarily symbolic. Symbols and programmes alone were insufficient. The organisers and members of Latvia's many parties had little democratic or parliamentary experience; moreover, they had the huge burden of the Soviet past on their shoulders.

Fifth Saeima elections
In the discussions about electoral procedures, the new parliament-to-be was designated as the fifth *Saeima*. The previous democratically elected parliament in Latvia was the fourth *Saeima* of 1931. The Supreme Council also re-established the power of Latvia's Constitution of 1922, thus legitimising the renewed republic with a connection to the interwar parliamentary tradition. In 1992 the Law on the Election of the *Saeima* was introduced. According to this modified version of the electoral law of June 9, 1922, the whole country was divided into five electoral regions (Riga, Vidzeme, Zemgale, Kurzeme and Latgale). The electoral law created a system of proportional representation.[9] A 4 per cent threshold, however, was introduced to avoid the pre-war weakness of the Latvian parliamentary system where too many small parties gained seats and caused eventual political instability.

The array of groupings and parties that eventually put forward candidate lists numbered twenty-three and contained virtually every identifiable viewpoint on the Latvian political spectrum.[10] The elections to the fifth *Saeima* were held on June 5 and 6, 1993. Almost 90 per cent of all eligible voters participated in the elections. A further 18,000 émigré citizens voted in foreign countries from Canada to Australia. Electoral rules stipulated that only pre-war citizens and their descendants could vote; thus Soviet immigrants were excluded from the democratic process, causing widespread domestic and international protest. Many of the twenty-three contesting parties were ideologically similar; politicians' egos and inability to build consensus created artificial divisions and barriers. Other parties appealed to single issues and were little more than special interest lobbies. A connection to the past through adopting the name of pre-war parties or by using ageless former members of these pre-war parties as political tools legitimised many political forces in the first years of contemporary Latvia. In order to gain popularity as well as political legitimacy, relatives of popular pre-war politicians, as well as émigré Latvians, were included at the top of many emerging parties. Guntis Ulmanis, the grandnephew of Karlis Ulmanis, for example, rose from relative obscurity to become one of the newly re-formed Farmers' Union candidates. Latvia's Way highlighted its connection to the interwar Republic with Gunars Meierovics, the son of Latvia's first Minister of Foreign Affairs, Zigfrids Anna Meierovics. The Christian Democrats hoped that the émigré Aivars Jerumanis, as their candidate, would become the first post-occupation president of Latvia.

Of the twenty-three parties, eight overcame the 4 per cent threshold and claimed their seats in the fifth Latvian Parliament.

The newly elected *Saeima* convened on July 6th 1993. The *Saeima* promptly elected a presidium and the state president. Anatolijs Gorbunovs, the former reform communist leader, became the Parliament's Speaker, a position he had also held in the Supreme Council. Guntis Ulmanis became the fourth elected President in the history of the Republic of Latvia (his granduncle, Karlis Ulmanis, assumed the post in 1936, two years after he had released parliament). The issue of

TABLE 1 DISTRIBUTION OF SEATS IN THE 5TH *SAEIMA*, 1993

Party/Political group	Per cent of vote	Seats
Latvia's Way	32.3	36
Latvia's National Independence Movement	13.4	15
Harmony for Latvia-revival for the Economy	12	13
Latvia's Farmers Union	10.6	12
Equal Rights Movement	5.8	7
For Fatherland and Freedom	5.4	6
Christian Democratic Party	5	6
Democratic Centre Party	4.8	5

Source: Diena, June 8, 1993.

citizenship and the future of the Russian speakers in Latvia caused a permanent tension that shaped Latvian politics throughout the next five years.

ETHNIC POLICY AND THE POLITICAL PERSPECTIVES OF LATVIA

From the awakening of the mid-1980s to the late 1990s, Latvia's politics centred on two major political cleavages. The first was the independence cleavage, the second the ethnic cleavage. Both cleavages were closely intertwined. Latvia, of course, also developed other cleavages, such as: left–right cleavages, rural–urban cleavages, elites–people cleavages, and generational cleavages. These cleavages, however, became more important at the end of the 1990s. The first two cleavages, independence and ethnic politics, dominated most of Latvia's political discourse of the decade. Therefore, we shall first discuss the ethnic cleavage before turning to other cleavages.

Citizenship—The core issue of ethnic cleavage

From the party formation period of the early 1990s to the 1998 referendum on the citizenship issue, nationalist rhetoric played a central role in Latvian politics. In order to acquire legitimacy, almost every newly emerging political force felt a need to reassure voters that their party would follow a hard line towards Soviet immigrants, thus securing national independence as well as Latvian collective identity. At first, the "proof" of a "real patriot" was a conservative initiative in the sphere of citizenship, but later, conservative views on language and education policy also became symbolic. In 1991, most of the nationalists as well as the former reform communists opted for the exclusive citizenship policy which restricted political participation to descendants of the pre-war Republic. As a result, the majority of Latvia's ethnic Russians were disenfranchised, and the ethnic honeymoon which began in 1990 ended abruptly.

On October 15, 1991, the Supreme Council passed a resolution "On the Renewal of Republic of Latvia's Citizens' Rights and Fundamental Principles of Naturalisation".[11] The law argued that despite the long-standing internationally illegal annexation of Latvia by the USSR, a body of Latvian citizens had continued to exist. According to the resolution, all the pre-war citizens of Latvia and their descendants automatically became citizens of the restored republic. An absolute majority of ethnic Latvians could claim to have some kind of legal connection to the pre-war republic, but the majority of ethnic Russians were unable to do so. Consequently, in 1991, Latvia's population was split into two groups. One group was made up of citizens, of whom about 78 per cent were ethnic Latvians.[12] The other group consisted of non-citizens, former Soviet-era immigrants and their descendants of whom an absolute majority were ethnic Russians, Belorussians and Ukrainians.

For persons who could not claim to have any legal relation to the previous body of Latvian citizens, the Supreme Council established several fundamental principles of naturalisation and stated that naturalisation would begin no sooner than July 1, 1992—the date when all residents of Latvia were supposed to have registered.[13] In reality, naturalisation began only in 1995 since the citizenship law enabling former Soviet immigrants to naturalise was passed only in July 1994. The law can be regarded as the most conservative among the Baltic states because, not considering the language requirement and a knowledge of the basics of Latvia's history and its Constitution, the potential applicants were divided into groups according to age and status. Thus instead of the principle of individuality, a collective principle was applied to all potential applicants.[14] The law was liberalised only in October 1998 after a national referendum.

In the period from 1991 to 1994, the post-war immigrants found themselves in a legal limbo. They tried to guess which rights and freedoms they still possessed. Fears that they might, for example, be denied re-entry into the country after a sojourn abroad seemed reasonably well founded because for the first years of independence, a number of rights, such as the freedom to return to Latvia, were reserved for citizens.[15] Paul Kolstoe rightly argues that shortly after independence those non-natives who had supported Baltic independence found themselves at a crossroads.[16] Most felt unjustly treated by the Latvian politicians and Popular Front activists who before independence had promised equal treatment and equal rights for all minorities. After independence, it seemed that Latvia's politicians departed from their promises. Restrictions connected with the lack of citizenship suddenly affected more than 600,000 people. After the citizenship law was passed in 1994, ethnic relations deteriorated in the public sphere owing to the restrictive nature of this law. However, in the private sphere ethnic tension was rarely seen.

Language policy
Changes in the Language Law, as well as mandatory language tests, were another reason why inter-ethnic passions heated in the newly independent Latvia. In 1989, under pressure from the popular move-

ments, the Latvian Supreme Council had adopted a Language Law that introduced a practical bilingualism of Latvian and Russian. The law envisaged a three-year transition period during which non-Latvians working in the state sector would have to learn varying degrees of Latvian according to their position. For example, managers and senior officials were supposed to be able to communicate in Latvian, and doctors had to be able to talk with their patients.[17] Along with the regulations on the Latvian language, the law contained provisions encouraging the use of other minority languages.

In 1992, the linguistic transition period stipulated in the law of 1989 ended. At the same time, the political compromise of legal equality between the Latvian and Russian languages lost its importance because Latvia had become an independent state. There was no reason to continue to pay tribute to the Russian language as the *de facto* official language of the Soviet Union because the Soviet Union no longer existed. At the same time, it is possible to argue that the legal equality between the Latvian and Russian languages did not guarantee practical equality. Most ethnic Latvians were able to communicate in Russian while the majority of Russians were unable to understand Latvian. Therefore, official bilingualism only prolonged an actual linguistic inequality in Latvia. In 1992, the Latvian authorities adopted several amendments to the Language Law of 1989 in order to change the hierarchy of languages. Latvian became the principal language in the government and administrative bodies. The law allowed the use of other languages, but in most cases required translations into Latvian if a participant in a public activity requested it. While guaranteeing the right to be educated in Latvian, the law declared that residents of other nationalities living in Latvia had the right to be educated in their native language (Article 10). The Latvian state also assumed the financial responsibility of supporting minority schools in Latvia. According to the law, state-financed universities would teach primarily in Latvian from the second year of studies (Article 11). Thus, students who had not mastered Latvian were granted a period of transition in order to improve their language skills.

Despite the official rhetoric, the amendments to the Language Law of 1989 acted like a completely new language law, altering the hierarchy of languages by proclaiming Latvian the *lingua franca* in the territory of Latvia. Along with the adoption of amendments to the Language Law, a flurry of other laws and regulations concerning the use of languages were adopted throughout 1992. In May 1992, the Council of Ministers passed regulations concerning the "Official State Language Proficiency Certification" establishing three levels of language proficiency—a basic level, an intermediary, and an advanced level. In June 1992, the Place Name Commission was created with the purpose of promoting the preservation, restoration and precise usage of historical place names characteristic of the Latvian cultural environment.[18] In July 1992, a law "On Additions to the Latvian Code on Administrative Violations Concerning the Official State Language Issues" was adopted clarifying the responsibility of persons and organisations in cases of violations of the language law.[20] In July, the Council of Ministers passed a resolution on the "Naming and Renaming of Railway Stations, Ports, Airports and Geographic Objects". Finally, during November 1992, the resolution on the "Official State Language Usage in Titles and Information" was passed.[21]

Both aspects of the ethnic policy of the post-awakening period, namely the citizenship issue and linguistic policies, were based on rational political reasons and justifications. It is much easier, however, to justify the Latvian linguistic strategy of the early nineties than the policy regarding citizenship. As far as linguistic policy is concerned, Latvian authorities introduced a number of legal acts aimed at reviving the use of the Latvian language in all spheres of public life. At the same time, the linguistic policy of the early 1990s encouraged the public use of minority languages in schools as well as in other institutions. One might of course argue that since more than 30 per cent of the population uses Russian as its native tongue, Russian should be introduced as a second official language. In our view, however, this argument does not hold. By granting equal rights for public use to Russian as well as Latvian, the beneficiaries of such a policy would be Russian speakers who would continue to be monolingual. The Latvian language would be used situationally, and would not regain its lost position of *lingua franca* in Latvia. A two official language policy would continue a *de facto* discrimination of the Latvian language.

As far as citizenship policy is concerned, it is difficult to justify Latvia's policy. For legal and symbolic reasons, it was important to stress the continuity of the contemporary Republic of Latvia and the interwar Republic by renewing the same base of citizenry. Granting automatic citizenship to Soviet-era immigrants would have disrupted this legal continuity. But once independence was achieved, the dilemma of non-citizens should have been more actively redressed. The only reason to delay the solution of the legal status of the Soviet immigrants was the hope that many of these people would leave Latvia and settle in other parts of the former Soviet Union. Indeed, some non-citizens emigrated,[22] but most saw Latvia as their home and chose to remain. An acceptable solution for the citizenship issue in late 1991 would have been the quick declaration of reasonable principles of naturalisation based on an applicant's pledge of loyalty to the new state as well as some knowledge of Latvian. Such a solution would have gone hand in hand with the earlier promises of the Popular Front leadership and was the most popular solution to Latvia's residents (see Figure 3.1). The figure represents the three alternatives for citizenship. The first and third alternatives represent the radical demands of Latvian nationalists and Russian conservatives respectively. The central choice corresponds to the previous political rhetoric of the Popular Front and many of the reform communists. More than half of all ethnic Latvians, as well as one-third of the minorities, supported the alternative of the centre. The radical alternatives, on the other hand, found support in one or another ethnic group exclusively. In total numbers, the first alternative was supported by 42 per cent of the total population, the third alternative by 64 per cent of residents, while the middle alternative found support among 84 per cent of Latvia's population.

Figure 1   The three most favoured citizenship models

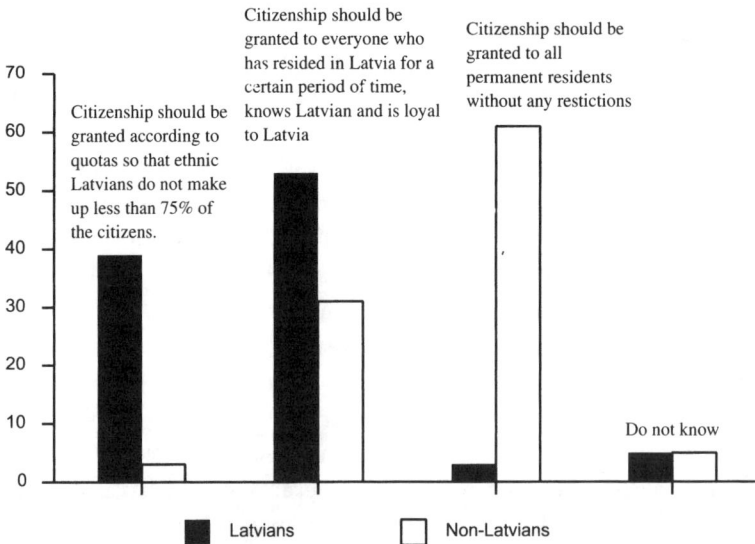

Source: *Diena*, October 7, 1993.

Why did Latvia's new politicians not choose the most popular option? According to Poulsen the reason behind the restrictive legislation lies in the inability of the new political elite to make unpopular decisions.[23] Much of the emerging political elite had a reform communist background and lacked moral authority. They felt a desperate need to prove their loyalty to their doubting voters by choosing a nationalist stand. At the same time, reform communists profited most from this political decision. Many administrative and economic posts that had been occupied by Soviet immigrants were now open only to citizens—Most likely open to ethnic Latvians, and even more likely to the former reform communists.

In the early 1990s, Latvians built state authority around the culture and language of the core nation, thus following the traditional patterns of nation-state building splendidly described in the writings of Anthony Smith and Ernest Gellner. In the Baltic case in general, and in the Latvian case in particular, the dilemma of the nation-state building process was that the expansion of the Latvian state, culture and language overlapped with the population's general demand for more freedoms and democracy. Consequently, the setting of collective borders and state authority had to be adjusted to the increasing demands of individuals and minorities. Therefore, Latvian nationalism, as an indivisible ideological shadow of any nation-state building process, had to be associated and intertwined with the basic principles of human and civil rights and freedoms. The inability to solve the above-mentioned dilemma between nationalism and liberal democratic principles was the foundation of ethnic tensions between the authorities and sizeable Russian-speaking minorities who justly felt that nation-building was carried out at their expense while disregarding the above-mentioned principles. As well as these objective grounds for the end of the ethnic honeymoon, there were also a number of subjective reasons feeding potential ethnic tension. Minority complaints were frequently psychologically grounded. For many Baltic Russians, even at the end of the 1990s, it was difficult to imagine the status of the minority after fifty years enjoying the privileges of the majority. Likewise, the minority syndrome of Latvians hindered their ability to cope with the responsibilities of being the state's majority. Objectively, Latvia's ethnic honeymoon ended because rational as well as psychological arguments were against this "marriage" in the early 1990s. The Latvian majority was afraid that granting the full scope of political and civil rights to Soviet immigrants would threaten Latvian ethnic identity as well as challenge the recently regained political independence. In turn, Latvian elites frequently used this sentiment to gain personal profit and diminish professional competition by successfully hindering non-ethnic Latvian participation in politics and administration.

The Russian claims that in 1991 the Latvian state divided people into first and second class citizens can only be regarded as partially true. The Latvian government indeed institutionalised the line between the former Latvian citizens and Soviet immigrants. The origins of this division go back to the Soviet occupation when Soviet immigrants were made partially responsible for the injustices committee by their state.

CONTEMPORARY ETHNO-POLITICAL TRENDS AND PERSPECTIVES

After the election of the fifth *Saeima*, for some time it was difficult to define a watershed in political events that challenged the process of transition. Within ethnic relations as well, the deadlock did not end with the adoption of the citizenship law of 1994. Owing to the restrictive nature of the new law and a degree of political apathy among many non-citizens, the naturalisation process proceeded very slowly. From 1995 to 1998, about 200,000 Soviet immigrants were eligible to apply for citizenship. Up to the end of 1996, less than 17,000 made this choice.[24] Between 1996 and 1998, several attempts were made to liberalise the citizenship law. All of these attempts, however, met with fierce opposition from the Latvian nationalist wing which threatened the fall of cabinet or other governmental instability if the law was liberalised.

Finally, on June 22, 1998, the Latvian Parliament adopted sixteen amendments to the Law on Citizenship, lifting the so-called window system and permitting every resident the right to apply for citizenship without any restrictions. The amendments were considered in accordance with the recommendations of the OSCE as well as the European Union (EU). Nationalist deputies of Parliament, however, began a constitutional manoeuvre that suspended the law pending a signature campaign in support of a referendum on the matter.[25] On October 3, 1998, the referendum was held, and more than 53 per cent of the voters supported the liberalised citizenship law.[26] Along with the liberalisation of the citizenship law went the preparation of the State Programme of Social Integration in Latvia. The Programme draft was prepared in late 1998, and aired at public hearings in the spring of 1999. The hearings themselves were a new element in Latvia's unfolding democracy. They were organised by the Latvian state and the Soros Foundation in Latvia, while also sponsored by the United Nations Development Programme

(UNDP). After three months of heated public debate, the Programme was reworked and improved, taking into account various criticisms and suggestions. The Programme is expected to be passed in 2000 after fighting its way through fierce nationalist opposition and the indifference of many other politicians.

The liberalisation of the citizenship law, as well as the preparation of a governmental programme for social integration, can be regarded as the first signs of a possible improvement in ethnic relations since the end of 1991. While it is still too early to gauge the full impact of the amendments, they appear to have contributed to a significant increase in interest in naturalisation among non-citizens. In 1998, 4,439 persons were naturalised, a notable increase from the previous year. After the amendments were passed, 2,049 applications for naturalisation were submitted in November and December alone.[27] Despite the liberalisation of citizenship provisions, a large number of non-citizens will continue to reside in Latvia for the next decade and potentially longer. Therefore, more must be done to improve and define the status of non-citizens in Latvia by giving them more rights, freedoms and certainty. At the end of 1998, the Latvian parliament began consideration of a new law, "On the Status of a Stateless Person in the Republic of Latvia".[28]

It is still too early to argue that ethnic relations have improved in Latvia during 1998 and 1999. It is possible, however, to claim that after being the hottest political issue for seven years, since 1998 the citizenship issue has receded in importance in the ethnopolitical discourse of Latvia. Successful solution of the ethnic problems inherited from Soviet occupation now depends on several factors.

First, successful ethnic relations depend on the success of the launch of the social integration programme. The success of the programme is dependent not only on political support, but also on public involvement in its implementation. Both sides, Latvians as well as Russians, are still quite sceptical about social integration. Russians, for example, fear losing their national identity if social integration equals assimilation. Latvians, on the other hand, face a difficult transition from a minority syndrome to accepting difference. Latvian national identity throughout its history has depended upon a zealous protection of ethnic identity against a ruling power (Baltic German, Tsarist, Nazi or Soviet) bent on destroying Latvianness. Now, Latvians must grow to accept the responsibility of being the state and accommodating its minorities. The change in perception is slow to come to many Latvians. Even in 1998 when asked *Would you support your child if they chose to marry*

TABLE 2   THE ATTITUDE TOWARDS LATVIA JOINING RUSSIA

Join Russia	Yes	No	Hard to say
Citizens—Latvians	6	88	6
Citizens—Non-Latvians	26	51	23
Non-citizens	35	44	21

Source: The Programme for Studies and Activities "Towards a Civic Society". Report, The Results of 1st and 2nd Stages, Baltic Data House, Riga, 1998.

someone of a different nationality?, only 56 per cent of citizens answered that they would support their child's choice. Among non-citizens, support would be given by 90 per cent of parents.[29]

Second, different political forces should not misuse ethnic solidarity in order to fight for votes. In the unfolding situation where more and more non-ethnic Latvians will obtain Latvian citizenship such a policy can lead to two consequences. If the ethnic argument is used broadly by political parties, the state will increasingly polarise along ethnic lines. If the existing ethnic Latvian-oriented parties, however, understand that the many new citizens are their potential voters, they might attempt to avoid nationalist rhetoric. Right-wing parties concerned with fiscal conservatism, for example, may find electoral support as a multi-ethnic, pro-business party. As far as ethnic Russian-oriented parties are concerned, their credibility is still undermined in the eyes of many Latvians who frequently see them as a potential fifth column of Russia opposing Latvian independence. Among many ethnic Russians, the national independence of Latvia is still not perceived as a value. According to data from 1998, about 35 per cent of non-citizens would support Latvia joining Russia while only 6 per cent of Latvians would consider such an option.[30] Eight years after regaining independence, the proportions of pro-Latvian, anti-Latvian, and neutral-oriented former Soviet immigrants remain much the same. Forty-four per cent of non-citizens were against Latvia joining Russia, while 21 per cent were ambivalent about their choice (see Table 2).

Third, state authority needs to be rediscovered with popular participation that facilitates the search for successful solutions to possible ethnic tensions. Indeed, how can Latvians claim to have some moral authority and the right to ask former Soviet immigrants for a pledge of allegiance, when they themselves are sceptical about their state, government and the rule of law? In 1998, for example, only about 21 per cent of Latvia's citizens trusted parliament compared to 79 per cent trusting television and 82 per cent trusting schools.[31] Of course, trust and allegiance are reciprocal and parliament and Latvia's politicians have to demonstrate that they serve their country and not narrow interests or their own pocketbooks.

Fourth, the success of social integration is dependent upon the political will of various parties. At this stage, only one party, the nationalist Fatherland and Freedom party, has an established conviction. They are against the integration of Soviet immigrants. The rest of the leading political parties pay lipservice to the integration process, but do little to grab political or financial support for this obvious need in Latvian society. There are several reasons explaining this behaviour: the lack of resources, the fear of losing nationalist voters, a lack of political responsibility, and the low interest of politicians in the integration policy compared to the more financially lucrative privatisation programme.

The successful social integration of various ethnic groups in Latvia is possible only if the democratisation and liberalisation of the country continue at a rapid pace. Only in this case might ethnic cleavages lose their dominating role in Latvian politics. The party system, however, still remains largely undeveloped and personalities rather than political principles frequently dominate politics.

CONTEMPORARY POLITICAL VALUES, PARTIES AND POLITICS

According to Guillermo O'Donnell, reasonably free and fair elections provide a means of vertical accountability or a presence of democracy in many transition countries. At the same time many new polyarchies, including the post-communist Baltic countries, suffer from the weakness of horizontal accountability, which implies a corresponding weakness in the *liberal* and also the *republican* components.[32] In other words, the political transition cannot be fully successful if democracy, liberalism and republican philosophy do not shape the transition process equally. Democracy facilitates equality, liberalism promotes individual rights and a commitment to freedom in society while republicanism supports the constitutional state and the rule of law. Latvia, unfortunately, does not have a surplus of democracy, freedoms or rule of law. In 1998, the World Bank listed Latvia's primary post-independence problems as a relatively low level of civil participation, an absence of interest in politics, a lack of knowledge about basic civil freedoms and rights, and substantial levels of corruption.

Since regaining independence, Latvia has held three parliamentary elections. The first post-Soviet parliament (fifth *Saeima*) was elected in

1993, the sixth *Saeima* was elected in 1995 and the seventh *Saeima* was elected in 1998. The results of the last three parliamentary elections confirmed that the political participation of Latvia's inhabitants is slowly declining; the number of those who take part in elections is decreasing. In 1993, 89 per cent of the electorate voted in the fifth *Saeima*, while only 72 per cent of citizens took part in the election of the sixth *Saeima*.[33] In the election of the seventh *Saeima*, almost 72 per cent of voters took part, but this relatively high number is deceptive. Simultaneous with the parliamentary elections was the referendum on citizenship, which drew considerable attention and participation. Moreover, about 75 per cent of citizens and 90 per cent of non-citizens in Latvia are not members of any political organisation. In 1998, 71 per cent of citizens and 89 per cent of non-citizens had not taken part in any political activity over the past year.[34] Weak civic participation can be explained by the traditions inherited from previous authoritarian regimes when public opinion did not really matter. It also can be explained by the present economic hardships which do not leave much spare time or energy for public activities. Finally, public and political activities are the privileges of free and independent citizens, but in contemporary Latvia many lack a complete understanding of their rights and freedoms. Many Latvian residents still have a very weak grasp of the shape of traditional values of the modern Western world. For instance, classical works such as Locke's *Letter of Toleration* or *The Federalist Papers* are still waiting to be translated into Latvian in order to be available to a broader public.

The general state of society also influences the contemporary Latvian party system, its development as well as the responsibilities of its elites towards their voters and society in general. According to Scott Mainwaring, the conventional criteria by which party systems are usually compared are the number of parties and the degree of ideological polarisation. Mainwaring also speculates about the level of institutionalisation of parties, which is a substantial criterion for countries in transition.[35] Latvia's parties can be easily critiqued according to these criteria.

Table 3 shows that while submitted lists did not decrease radically from one election to another, the number of elected parties did. Partly, this can be explained by the threshold which was increased to 5 per cent before the 6th *Saeima*. The parties offer surprising results when analysed according to the left/right spectrum.

Table 4 clearly shows that the parties of the right have dominated Latvian politics during the last three elections. This dominance,

TABLE 3 NUMBER OF PARTY LISTS AND THEIR SUCCESS IN THE PARLIAMENTARY ELECTIONS (1993, 1995, 1998).

Parliament	Submitted lists	Represented in the parliament
5th	23	8
6th	19	7
7th	21	6

Source: Parliamentary elections.

however, did not make the Latvian political system more stable. From 1993 to 2000, Latvia experienced eight different governments (see Table 5). This characteristic of Latvian politics seriously undermined the electorate's trust in the parliament, parties and democracy *per se.*

Another obvious characteristic of the Latvian party system is the relatively weak representation of left-leaning parties. One of the main reasons for the electoral weakness of the left is the Latvian Soviet past. The average voter tends be sceptical about anything leftist because the policies of the former regime discredited a general leftist orientation. Another reason is practical; to proceed with market economic reforms, voters believe that they should be right orientated in order to bring about continued change. It would be misleading, however, to assume that Latvian voters lack leftist sentiments. Instead, it would be correct to say that many Latvian voters think left, but vote right. Many voters still assume that a leftist orientation automatically means an orientation towards Russia. Finally, it should be noted that the left–right spectrum

TABLE 4 PARTIES ACCORDING TO THEIR ORIENTATION[35]

Saeima	Right	Left	Centre	Difficult to say
5th	5	2	1	
6th	4	1	–	2
7th	3	1	1	1

TABLE 5 LATVIAN PRIME MINISTERS AND THEIR PARTY AFFILIATION

Parliament	Prime Minister	Party affiliation/orientation
5th	Valdis Birkavs	Latvia's Way (centre/right)
5th	Maris Gailis	Latvia's Way (centre/right)
6th	Andris Skele	No affiliation
6th	Andris Skele	No affiliation
6th	Gundars Krasts	For Fatherland and Freedom/LNIM (right/nationalist)
7th	Vilis Kristopans	Latvia's Way (centre/right)
7th	Andris Skele	People's Party (right/conservative)
7th	Andris Berzins	Latvia's Way (centre/right)

in Latvia is distorted not only by economic or moral understandings, but also by attitudes towards the past, national independence and Soviet immigrants. Many left–wing voters are Russian speakers who support closer ties with Russia and equal citizenship rights for former Soviet immigrants. They may, however, be economically quite centre oriented. The surprising performance of the Social Democrats and the heirs of the Equal Rights Movement in the elections to the seventh *Saeima* may point to a gradual re-emergence of the left in Latvia. These politicians, however, continue to deal with the popular perceptions that they are closer to Moscow than the European Union. Because of the ideological confusion of the Latvian political system almost half of Latvia's inhabitants (38 per cent of citizens and 49 per cent of non-citizens) could not assess their own political views according to the traditional left–right scale.[36] Last but not least, the Latvian left's popularity is also influenced by the current problems of many of the West's SD parties.

Another characteristic of the Latvian party system is that only a few parties have survived through all three independence-era elections. Thus, party and political continuity in Latvia are relatively weak. In the seventh *Saeima*, three of the six parties are represented in parliament for the first time, while two parties (People's Party and New Party) were created about half a year before the elections. The People's Party's youth did not stop it from becoming the largest party in parliament. The same happened in the elections to the sixth *Saeima*, when the party *Saimnieks* received the most votes. In the elections to the seventh *Saeima*, however, the same party lost its representation in parliament and perished from the Latvian political scene. A similar fate befell the *National Movement for Latvia* or Siegerist's party which became the third largest party in the sixth *Saeima* (16 seats out of 100) by successfully manipulating marginalised and desperate voters. Just a few years later, in the elections to the seventh *Saeima*, the party failed to overcome the 5 per cent threshold. Barely two years after the elections, the *New Party* (*Jauna Partija*) is already experiencing serious divisions within their parliamentary faction.

The frequent rise and fall of parties is characteristic of an unstable party system. In turn, it contributes to further political instability in general. The political allegiance of voters remains unclear; in 1998, almost half of Latvia's residents could not give a concrete answer about their political choice.[37] Mainwaring argues that in transitional societies, not only are there many parties and ideological polarisation, but the strength of party institutionalisation is important. The unstable politics of Latvia's parliamentary elections demonstrate that party institutiona-lisation in Latvia is still rather weak and developing. Four of Mainwaring's theoretical arguments could be used to support this conclusion:[38]

- Party stability is rare. Only some parties remain on the scene for more than a few years. In Latvia, some parties suffer precipitous declines, while other parties enjoy sudden electoral upsurges.
- Many parties are weakly rooted in society. Only a minority of citizens regularly vote for the same party. Instead, citizens vote according to candidates or, if they vote because of the party label, they easily switch party preferences. In the Latvian case, the People's Party won the most seats in the seventh *Saeima* largely because of its leader, Andris Skele. The Social Democrats collected many votes owing to their populist duo of Juris Bojars and Janis Adamsons, while Latvia's Way continues to win (through three elections) largely because of the appeal of Anatolijs Gorbunovs.
- Mainwaring's third argument about the weak legitimacy of parties in the eyes of individuals and groups can also be found in Latvia. For instance, 8–11 per cent of inhabitants support the view that the best option to make weighted political decisions in elections would be just one political party in Latvia.[39]
- An important characteristic of Latvian political parties is that parties have few resources. Parties are the creation of, and remain at the disposal of, individual political leaders or more recently business interests. Intra-party processes are not well institutionalised. If, for example, a party loses its most popular leader, the party faces immediate decline and a loss of support of a considerable number of voters. With a few exceptions, parties have serious problems finding replacements for their departing leaders because parties are organised around a small group of elites that have only marginal contacts with the average party members. This was clearly the case with Cevers and Siegerists and their respective parties *Saimnieks* and the National Movement for Latvia. The future of the People's Party would also be questionable if the popular Andris Skele suddenly retired from politics.

The present Latvian party system and politics in general are shaped by many of the same problems that other Eastern and Central European states in transition are facing. What separates Latvia and the two other

Baltic states from the rest of the countries in transition are independence and ethnic cleavage. Both cleavages are the result of the fifty-year Soviet occupation and its consequences. In other words, Latvia, Estonia and Lithuania, unlike their Eastern and Central European counterparts, were much more effectively cut off from all foreign cultural, economic and political contacts. Latvia experienced a greater degree of control by the Moscow authorities than any Eastern Bloc country. At the same time, the interwar experience of independent statehood was an argument that set Latvians, Estonians and Lithuanians apart from other Soviet Republics. The memory of independence helped the Baltic peoples maximise their efforts for independence, and also helped them organise themselves once independence was achieved.

In the 1990s, Latvians faced two relatively controversial political goals at once. On the one hand, they had to liberalise their society, implement market reforms and democratise their political system. On the other hand, they had to create an effective administrative and executive state apparatus as well as clarify their political and other collective identities. The first goal was more connected with the degree of individual freedoms and rights, while the second was more about collective values and state coercion. In most other Eastern and Central European countries, state bureaucracies were more or less in place. Thus these societies could devote more attention than the Balts to the questions of democratisation, liberalism and market economy.

The two cleavages mentioned above, independence and ethnic tension, will not leave Latvian politics in the near future. There are, however, several reasons to think that their importance might decrease with time. First, the institutionalisation of the political system continues successfully despite its dark sides. There appears less need for the political employment of ethnic solidarity in present-day Latvia, even if it continues to be the strongest source of collective identity. Second, the ethnic integration programme has finally been accepted as necessary by most of Latvia's political parties. Latvian authorities seem to have understood that they have to strengthen democracy, increase the level of public participation and integrate their diverse populations if they hope to keep Latvia as an independent political and administrative entity well into the twenty-first century. Third, the economic aspects of transition are starting to play an increasing central role in Latvian politics. Fourth, even if the ethnic question continues to be among the top priorities, issues of economic and European integration can overshadow it. As Latvia is integrated into common European and global structures, ethnic and independence cleavages will weaken. Latvia's current international relations, its continuous quest for a guaranteed permanence of Latvian national independence and other related questions of national security are all aspects that contribute considerably to ethnic tension in domestic politics and undermine the balanced development of the country's political system.

1 D. Ivans, *Gadijuma karakalps* (Riga: Vieda, 1995), p. 368.
2 *Ibid.*
3 *Ibid.*
4 U. Cerps, "The Left-Wing Parties in Latvia and their Performance in the 1993 Parliamentary Elections", in Jan Ake Dellebrant and Ole Norgaard, eds., *The Politics of Transition in the Baltic States. Democratization and Economic Reform Policies* (Sweden: Umea University, 1994), pp. 89–90.
5 Among them, the most prominent was Gunars Meierovics, a former leader of the World Federation of Free Latvians and son of the famous Latvian inter-war politician and foreign minister Zigfrids Anna Meireovics. Participation of well-known émigré Latvians strengthened LW's legitimacy in the eyes of many voters as well as among émigré Latvian voters (Authors).
6 A. Steen, *Recirculation and Expulsion: The New Elites in the Baltic States* (Oxford: Westview Press, 1994), p. 6.
7 V. Bluzma, "Sabiedriski Politiskas Domas un Partiju Sistemas Attistibas", in U. Bormanis and V. Kanepa, eds., *Latvijas valsts atjaunosana 1986–1993* (Riga: LU Zurnala Latvijas Vesture Fonds, 1998), p. 261.
8 Cerps, p. 93.
9 Cerps, p. 87.
10 A. Plakans, "Democratization and Political Participation in Postcommunist Societies: The Case of Latvia", in K. Dawisha and B. Parrot, eds., *The Consolidation of Democracy in East-Central Europe*, (Cambridge: University Press, 1997), p. 261.
11 5th Saeima–Republic of Latvia Standing Commission on Human Rights, 1993, pp. 71–73.
12 *Diena*, February 15, 1995.
13 A. Grigorjevs, "The Ghosts", *Freedom Review*, May–June (1993), p. 12.
14 According to the law of 1994, during the year 1995, members of a citizen's family could apply for citizenship. From January 1, 1996 to 2000, people born in Latvia could apply; after 2001 people born outside Latvia could apply for citizenship. Additionally priority was given to young people. According to this law many well-integrated Soviet immigrants had to wait for years in order just to apply for citizenship while others who perhaps were not prepared to pass the citizenship tests were granted such an opportunity. By applying collective rules to the citizenship issue Latvian authorities alienated many potential loyal citizens and discredited the Latvian legal system in the eyes of many residents (Auth.).
15 P. Kolstoe, *Russians in the Former Soviet Republics* (Bloomington: Indiana University Press, 1995), pp. 124–125.
16 *Ibid.*, p. 128.
17 A. Kamenska, *Valsts valoda Latvija* (Riga: Latvijas cilvektiesibu un etnisko studiju centrs, 1994), p. 7; Veisbergs, p. 34.
18 Official State Language in Latvia, 1992, pp. 53–54.
19 *Ibid.*, pp. 39–42.
20 *Ibid.*, pp. 59–62.
21 In 1990, 8,706 persons left Latvia. Most of them were ethnic Russians, Belorussians, and Ukrainians. In 1991 there were 10,796 persons leaving Latvia, while 46,931 persons left the country in 1992, mainly to go to the territory of the former Soviet Union. After 1992, emigration slowly decreased and in the last years of the 1990s emigration did not exceed a few thousand per year, in fact approaching zero balance (*The Ethnic Situation in Latvia. Facts and Commentary*, 1994, p. 9;

Diena, May 10, 1995).

22 J.J. Poulsen, "Nationalism, Democracy and Ethnocracy in the Baltic Countries", in Jan Ake Dellebrant and Ole Norgaard, eds., *The Politics of Transition in the Baltic States. Democratization and Economic Reform Policies* (Sweden: Umea University, 1994), pp. 15–29.

23 UNDP, 1997, pp. 86.

24 OSCE/ODIHR, 1998, pp. 10–11.

25 *Diena*, October 4, 1998.

26 LCHRES, 1998, p. 41.

27 *Ibid.*, p. 42.

28 BDH, 1998, p. 93.

29 *Ibid.*, p. 63.

30 *Ibid.*, p. 83.

31 G. O'Donnell, "Horizontal Accountability in New Democracies", *Journal of Democracy*, 9, 3 (1998), p. 113.

32 BDH, p. 79.

33 *Ibid.*

34 S. Mainwaring, "Party Systems in the Third Wave", *Journal of Democracy*, 9, 3 (1998), p. 66–68.

35 The fifth *Saeima's* right can be described as Latvia's Way, Latvian National Independence Movement, Latvian Farmers' Union, For Fatherland and Freedom, Christian Democratic Party. The Left was Harmony for Latvia–Revival for the Economy, and Equal Rights Movement. The centre was the Democrats Centre Party. In the sixth *Saeima*, the right was Latvia's Way, Latvian National Independence Movement, Latvian Farmers' Union, and For Fatherland and Freedom. The centre was the Latvian Unity Party. The left was the National Harmony Party. Two parties were difficult to identify. They were the Democratic Party "Saimnieks" and People's Movement for Latvia. In the seventh *Saeima*, the right is Latvia's Way, For Fatherland and Freedom, and People's Party. The New Party can be described as the centre, while For Human Rights in United Latvia is the left. The Social Democrats are difficult to identify, yet they lean towards the left.

36 BDH, p. 71.

37 BDH, p. 75.

38 Mainwaring, p. 68.

39 BDH, p. 73.

Chapter 4

LATVIA'S ECONOMY SINCE 1991

A fitting tribute to Latvia's economic transformations is the rapidly changing nature of scholarly analysis on the state of the country. When the Popular Front first pushed for independence from the Soviet Union, many Western experts countenanced economic patience and restraint: independent Latvia should remain within a Russian ruble zone, for example. A few years later, the émigré political scientist Juris Dreifelds wrote of a Latvia "in transition" both politically and economically. At the end of the decade, however, the Finnish scholar Marja Nissinen confidently writes of Latvia's transition to a market economy, as if transition was an accomplished fact. Nissinen does not naively claim that Latvia is now on the same footing as a Western European economy, but instead argues that since the majority of state enterprises have been privatised (with the rest on their way), and since public opinion supports difficult economic reforms, and since almost every major political party supports the move towards a market economy, we can safely assume that Latvia will not return to the command economy of its Soviet past. Latvia is committed to the Western model of a market economy; this transition is complete. Much, however, remains to be done. The IMF approach to macro economic stabilisation must continue, yet the state needs to address a new range of problems that have developed with transition: continued corruption, dangerous inequities within society, and the future of entire sectors of the Latvian economy are among the most desperate problems. Latvia's economic performance through the 1990s has been impressive, but has also not lacked its calamities. Although Latvia has not yet pulled off an economic miracle, the standard of living of the vast majority of its citizens is demonstrably better than most former Soviet citizens can claim.

Before independence, however, the Latvian Soviet Socialist Republic was economically a cog within the larger command economy of the Union of Soviet Socialist Republics. Market forces played no real role in the structuring of the production of goods and services. Instead, the Soviet economy was a maze of economic plans from the all-Union level in Moscow, the Republic level in Riga and more regionalised councils. Moscow's decisions were always paramount. Industrial managers created a whole degree of mechanisms to deal with central planning. Factories, for example, had to meet production quotas. These quotas could be based on quantity of product, weight of product, or some combination of these and other indicators. In response, the factory manager would stockpile parts, falsify records, or use the flourishing grey market (not quite a legal or illegal market) to meet or overachieve the quotas. Poor quality control standards, rampant corruption and a system that bred poor industrial productivity all contributed to the crisis in Soviet economics. Planned products, from apartment buildings to dishes, were largely in short supply and of poor quality. Shortages and queues became endemic. The planned economy did not respond quickly or adequately to consumer demands. Perhaps the food sector illustrates the economy best of all. Collective farms were largely ignored in terms of investment and they wallowed in inefficiency. Collectively, the farms somehow fed the Soviet Union (with the aid of food imports)—few went hungry, but the quality and diversity of food was dismal. Collective farmers sold food grown on small private plots at markets in the city, but at high prices. Some goods earmarked for the *nomenklatura* were largely denied the common people.

By the mid-1980s, the Soviet leadership realised the gravity of their economic hardship. The Soviet Union relied on foreign grain, despite the possibility of bumper crops. Increased Western military spending drove the Soviet armed forces to consume ever-larger portions of the planned economy. Beginning with Andropov, the Soviet regime toyed with reforms of the Soviet economy. Gorbachev's ultimate campaign for *glasnost* was a by-product of economic reforms. Gorbachev designed *perestroika* to allow industrial managers to tinker with market forces to reform their outdated industries. The regime meant *glasnost* as a public spur to recalcitrant industry managers. Journalists publicised inefficiencies and industrial graft and corruption. Political reform, however, careered ahead of economic intentions. The collapse of the Soviet Union and the democratisation of the former Union developed much more rapidly than economic transformation. Most former Republics are still stuck in a quagmire of state control of the economy. Collectively, the market transition has been a slow and difficult one. The Baltic Republics are the exception. Just as politically the Baltic Republics quickly broke binding ties with the remains of the Soviet Union (pointedly remaining outside the Commonwealth of Independent States—CIS), economically Estonia, followed by Latvia and Lithuania, have moved the farthest and fastest in economic reform. Latvia is now primarily a market economy eager for European Union membership. Latvia's greatest economic challenges are continued reform and economic growth and a more equitable distribution of Latvia's growing prosperity throughout society. If a quarter of Latvia's population still lacks citizenship, perhaps three-quarters still live in poverty.

SUMMARY OF THE ECONOMY DURING THE INDEPENDENCE MOVEMENT

The reasons for the rapid advancement of political reform with slower economic change are relatively sensible. As the Popular Front pushed towards independence, factory managers and industrial planners hedged their bets. Unlike the moral claims of the independence movement, the economy existed in a material world entirely dependent upon the larger Soviet planned economy. Latvia was entirely dependent on energy supplies from other parts of the Soviet Union. Likewise, industrial raw materials came from outside Latvia, while the ultimate destination of the finished products was also outside Latvia. As a result, economic reform waited for the results of political liberalisation, with one significant exception. During the years 1987–1991, the previously small role of Republic-level planning grew in importance. Gorbachev's reforms assumed that greater local control would allow for the flexibility needed to create viable industries. Latvian politicians and factory managers took advantage of the opportunity to stake their claim to Republic-level control as part and parcel of the greater movement towards independence. The change in jurisdictional control of industry was quite marked, as shown in Table 1.

The increasing control of industry within Latvia by Republic-level officials produced positive and negative consequences. Greater local control meant a degree of expertise and experience before the wholesale take over of Union-controlled industry with the collapse of the coup of August 1991. Bluntly, this meant that Latvian industry did not start independence at ground zero in an organisational sense: factories, and

TABLE 1 JURISDICTION OF INDUSTRY IN LATVIAN SSR (BY PERCENTAGE)

Type	1980	1987	1990
All union	37	40	37
Joint union/republic	46	50	21
Republic	17	10	42

Source: *Latvia: An Economic Profile for the Foreign Investor* (Riga: Council of Ministers of Latvia, September 1991), p. 8 as quoted in Juris Dreifelds, *Latvia in Transition* (Cambridge: Cambridge University Press, 1996), p. 113.

Republic planning could both pursue production and management according to local concerns rather than Moscow orders. Unfortunately, these few years could not divorce Latvian industry from its complete dependence on the larger Soviet economic unit. Supplies for production and markets for finished goods disappeared almost overnight, throwing Latvian industry into a tailspin at the dawn of Latvian independence. More importantly, Republic or factory control and planning did not prepare Latvia for the leap into the market.

The rise in Republic-level control of industry also led to the appearance (often justified) of the former Communist Party *apparat's* grip on the commanding heights of independent Latvian industry. Politicians from the nationalist right argued that industrial planners and factory managers, all Communist Party members, benefited first from the transition to Republic control and then the transition to independence. These former communists were allegedly able to work their ties with the old system into wealth at the expense of the people of Latvia. Furthermore, an alleged shady coalition of former party members and organised crime figures dismantled Latvian industry for immediate riches. The heyday seemed to be the "coloured metals" craze that saw the stripping of factories, and even railroads, for their metal content which was then sold on the world market. No doubt, many managers and former communists translated their previous position to comfort in the new order, but this is inevitable in any period of radical transition.[1] The more extreme conspiracy theories, however, are the figments of an industrialised nation coming to terms with industrial collapse in a few short years. Still production lines are a symbol of a popular movement stolen, not the result of transformation to a world market economy. The comparative wealth of the few (particularly politicians, former communists, or factory managers) is assumed to be of crooked origins. The presence of highly visible cases of corruption and embezzlement coupled with the ostentatious lifestyle of Latvia's *nouveaux riches* seems to offer more proof of the plundering of Latvia's wealth and industry. Latvia, however, is relatively fortunate. The manner of privatisation in Latvia demanded much more transparency in transactions than was (and continues to be) the case in much of the former Soviet Union.

Although short-sighted governmental policy or corrupt individuals may have harmed the recovery of the Latvian economy by increments, they are simply symptoms of a difficult recovery, and not the reason for Latvia's economic woes. The Soviet legacy is the hackneyed, but incompletely understood, explanation for the poor starting position of independent Latvia's economy and its continued difficulties. What does this mean exactly? As already stated, the Latvian economy was a single cog within a larger Soviet machine—the cog could not work on its own. The whole Soviet machine, however, barely worked. The machine, to continue the analogy, had not been oiled in decades. As a result, independent Latvian factories entered the competitive world market with obsolete machinery, and the raw materials provided through central planning in Soviet times now had to be purchased at world prices. The sudden high cost of petrol and oil crippled industry generally and the high price of cotton specifically hampered Latvian textiles. Furthermore, products needed to find new markets. Their increased cost and politically motivated pseudo-sanctions from Russia handicapped Latvian goods in eastern markets, and their lack of quality standards, inferior manufacture and poor marketing doomed them in the West. The cost of reinvestment seemed extremely daunting, particularly because Latvia had virtually no homegrown capital. Righting the economy would be arduous and seemed to call for exceptionally unpopular measures.

The Soviet legacy in agriculture was similarly profound. Collective farms were mismanaged and underfunded. Overall, the Soviet regime seemed, by and large, to have given up on collective farms. As a result agricultural productivity lagged. With little incentive to pursue bumper crops, fields went fallow. Latvia is one of the few countries in the world that has more forestland now than prior to World War Two. To compensate for the small yields, caused largely by the lack of individual incentives, collective farms used inordinate amounts of chemical fertilisers. The resulting runoff into rivers and streams left Latvia with extensive rural environmental damage (matched by the poor pollution control standards of industry). Livestock production also suffered from a lack of veterinarians and effective vaccines, leaving much of the stock of independent Latvian farms sickly or diseased.

The Soviet legacy in finances is perhaps the least detrimental because there was less of a legacy. Financial institutions were, after all, odd creatures to a centrally controlled economy. The ruble's inconvertibility isolated Soviet monetary policy from world financial markets. Not surprisingly, the strongest success of independent Latvian economic growth is monetary reform and finance. The new state was not building on Soviet foundations, or left with as many detrimental physical remains of the Soviet era. Here, more than elsewhere in the economy,

Latvia started afresh. At first, many Western economists suggested that the Baltic should remain in the ruble zone (countries of the former Soviet Union that decided to continue using the same currency) to preserve their position within former Soviet economic markets. The Latvian government, as well as those of Lithuania and Estonia, however, pushed ahead with plans to return to the old currencies of the interwar Republics. The currency was a symbol of independence restored, but it also allowed the Baltic Republics to escape the extreme turmoil of CIS economics. If Latvia had remained in the ruble zone, the government would have had no control over the amount of money in circulation and would have been tied inseparably to the vagaries of the Russian economy. With a currency of its own, the Lats, Latvia has charted economic policy to match its political independence. Monetary policy has become the linchpin of Latvia's economic structural adjustments.

Latvia's macroeconomic reforms

After the collapse of the Soviet Union, its former Republics and satellites throughout Eastern and Central Europe were faced with difficult decisions regarding economic reform. Few, perhaps only Belarus, have largely disavowed intentions to pursue a market economy. The debate instead centred on the proper approach towards transition, gradual or "shock therapy". The gradual approach was favoured by more cautious states wary of the market and its effect on their populations. Gradualists argued that sudden, drastic economic reform would overwhelm democratic reforms (people might vote populist) and push industries too far too quickly. Shock therapists, on the other hand, argued for the most rapid pace of reform possible. In many respects, they also distrusted the general population's ability to appreciate the need for painful economic reform. They argued that by "going slow" the pain of economic readjustment was prolonged, and feared that an impatient electorate would change course in the proverbial midstream. Nearly a decade of experience throughout Eastern and Central Europe suggests that as important as the pure economic reasoning to the manner of adjustment was the underlying political economy of the new states. Economic decision-making cannot be divorced from political influences. Latvia has followed the "shock therapy" programme, and together with its other proponents has experienced the most promising economic transformations. The concerns about the electorate, however, seem unfounded. Despite tremendous economic privations, Latvia's citizens have returned pro-rapid transformation parliaments three successive times. At the end of the decade, these reforms seem to be bearing fruit, but "rotten apples" remain in the economic orchard.

Latvia, eager to join the European Union and NATO, has religiously followed the advice of Western economic organisations, particularly the International Monetary Fund (IMF). The IMF's plan for Eastern European states has followed a "liberal-monetarist philosophy" which is based on sound currency, market forces and private property. As Marja Nissinen succinctly summarises:

> The standard approach to economic adjustment is to limit drastically government involvement in the economy through cutting subsidies, liberalising prices, deregulating economic activity and then letting the market do the job. The programmes typically include strong deflationary measures in the form of monetary and fiscal restrictions, aimed at reducing domestic demand. Institutional and structural changes focus on foreign trade liberalisation, the introduction of currency convertibility, privatisation of state-owned enterprises and reform of the financial sector.[2]

Latvia, through numerous governments, has not strayed from this liberal economic path. Its firm acceptance of the market and concomitant rejection of state interventionism has placed Latvia among the top tier of reforming countries in the eyes of international financial organisations. Latvia, for example, enjoys standby credit from the IMF and has already been accepted into the ranks of the World Trade Organisation (WTO). Despite the international praise, however, Latvia's economy with its strengths has also developed chronic weaknesses.

Einars Repse, the Governor of the Bank of Latvia, and his team deserve much of the credit for Latvia's greatest success, the introduction of the Lats.[3] Repse, and the Bank of Latvia (which is largely independent of parliament), followed IMF advice closely throughout 1991–1993, which included an economic crash caused in part by the tight control of the monetary supply (by not granting government credits and ending subsidies). A transitional currency, the Latvian ruble, was released into circulation on May 7, 1992, followed by the Lats on March 5, 1993. The success and seamlessness of the new currencies was due to the overlap between the introduction of one and the removal of the other. For example, both the Soviet and Latvian ruble were equal, legal tender for several months, before the Soviet rubles were gradually removed. A similar process followed with the Latvian ruble and the Lats. As a result, both the Latvian ruble and the Lats enjoyed credibility almost immediately and no panic occurred with the change of money. Once introduced, Repse kept the Lats firmly pegged to an IMF-defined basket of currencies to shield it from the fortunes of a single currency. The Lats is guaranteed 100 per cent by gold and foreign currency reserves. This has led to a stable exchange rate for the Lats (one of Eastern Europe's most stable currencies); a continuing testament to the tight monetary control practised by Repse and the Bank of Latvia.

The strong Lats has benefited imported goods at the expense of domestic products, but has remained relatively popular owing to monetary chaos in Russia. Latvians may grumble that the exchange rate should move a little, but they accept Repse's currency guidance when they reflect that with relatively low inflation and a strong Lats prices remain similar, month after month.[4] News from Russia, on the other hand, mentions hyperinflation and a collapsing ruble. There exists a certain "comparative misery and hardship" factor that cannot be easily calculated but has been instrumental in the acceptance of economic reform in Latvia. As long as the Russian example is unbelievably worse than the Latvian experience, the comparative advantage acts as a coating for the bitterness of economic transformation.

As the years go by, however, this factor of comparative misery and hardship will diminish in importance. As political independence succeeds and sinks deeper roots, people will see Latvia as a separate entity from the former Soviet Union. In time, the Latvian economy will be judged by its own yardsticks, and not by comparison to chaos in Russia. Latvia's continuing economic challenge is to use the transition period to establish a strong economic foundation so that when people reflect on the popular benefits of independence they will not be disappointed.

As already noted, the reintroduction of the Lats has been an unparalleled success in Latvia's recent economic reforms. The Bank of Latvia's tight control of monetary supply has forced relatively balanced governmental budgets. Latvian state budgets have little or no deficit. A sense of financial propriety rules over the discussion of state income and expenditures, which with the exceptions of unforeseen circumstances (the collapse of Latvia's largest private bank in 1995) keeps government spending in line with incoming revenue. Of course, the great challenge has been to increase revenue and pare "superfluous" spending. One bonus has been Latvia's lack of foreign debt to finance. In 1991, Latvia emerged from the Soviet Union without a share of the Soviet debt and thus had greater budgetary leeway than most Western countries (or other

Eastern European countries). Over the years, Latvia has gradually incurred foreign debt, but still not to the levels of high-spending Western countries. The amount saved on interest and payments cushioned the collapse of most other governmental sources of revenue from 1991–1995.

Latvian state revenue comes primarily from taxes and tariffs. Throughout the early 1990s, however, the collection of tariffs collapsed, and the introduction of personal and business taxes was largely ignored. Latvia's porous borders encouraged smuggling and the low salaries of border guards and customs agents fostered bribery and endemic corruption. Contraband alcohol and cigarettes were probably more widely consumed than legitimate ones. Smuggled consumer products such as sugar damaged the state's early economic perform-ance two fold. The unpaid duty robbed the state treasury of desperately needed funds and the low prices of attractive, contraband Western goods overwhelmed Latvia's indigenous production. Likewise, Latvian industry and agriculture overstated the degree of their financial collapse to minimise their tax dues. Official tax returns for the early 1990s must read as if there were hardly a single profitable company in all of Latvia. Businesses were particularly slow to pay accurate payroll taxes. Individuals only began paying near-accurate income tax as late as 1995. Governmental attention and international aid, however, have focused on addressing the basic shortcomings of Latvian government revenue. Tighter border control and customs points have returned tariff revenue to government coffers. Similarly, markets and businesses have switched, by government decree, to computerised cash registers which keep more accurate tallies for proper sales tax collection (Latvia, as a result, has steadily collected more sales and excise taxes). Finally, the Ministry of Finance hired more tax collectors to review more comprehensively the claimed taxes of both businesses and individuals. Incentives for catching tax frauds include departmental rights to a percentage of the taxes from tax fraud cases (there are similar provisions for border and customs officers).[5] The government has also instituted tighter controls of alcohol sales to ensure revenue from the high consumption levels of Latvians. Although tax and tariff policy are still fraught with loopholes and non-compliance, the foundations for sound collection of governmental revenue have been set. All businesses pay a flat 25 per cent tax, and goods and service are taxed at 18 per cent. A social tax targets salaries, fees, royalties and other remunera-tion at 37 per cent (partially paid by the employer, partially by the employee). Other revenue comes from excise tax, personal income tax, property tax, and taxes on natural resources.[6] Simultaneous with the improvement of the collection of governmental revenue, the Latvian government has cut state expenditures.

The budgetary axe has fallen on the bureaucracy and on "expendable" programmes. Latvia has traditionally struggled with a top-heavy government—there were too many government officials and ensuing bureaucratic red tape. A British traveller in the 1920s, went so far as to muse that Latvians struggled for independence in order to move into the positions of petty, despotic governmental officials.[7] Disgruntled entrepreneurs and foreign investors might assume the same of the national awakening of the 1980s. Corruption continues to be a serious problem at the end of the century, from traffic police through all halls of government and business. From the era of the Popular Front government, however, Latvia's new rulers have struggled to cut government personnel—first targeting obsolete or superfluous ministries. Bureaucratic resistance has eroded some governmental reform efforts, but the low salaries of government bureaucrats in the early 1990s succeeded in driving many out of government service. The types of cuts often assumed political significance, as with the Ministry of Culture.

The first Latvia's Way government put the Ministry of Culture on the chopping block, relegating the Minister of Culture to a state minister within the new Ministry of Education, Science, Research, and Culture (all were previously Ministries of their own). The popular composer Raimonds Pauls, who was the Minister, resigned in protest, but governmental funding of culture has largely subsided in the tough economic transition times. The various governments have proceeded with renovations of the Latvian Opera House, and returned to the interwar tradition of awarding governmental medals of honour with stipends to cultural figures, but the generous grants to the arts that the Soviets dispensed have disappeared. An occasional headline about the Academy of Art's inability to pay electrical bills or the financial hard times of previous darlings of Latvian culture shock the nation with the degree of the withdrawal of state support of the arts. A foreign observer, unfamiliar with Latvia, may even conclude that *Soros Funds* is Latvian for the Ministry of Culture, reflecting on the central role George Soros' humanitarian organisation plays in maintaining a vibrant cultural scene in Riga. The government, however, would reply that at present the state is too poor for governmental largesse for culture. Pensions tax the new government, infrastructure improvements are desperately needed, and state-supported culture brings memories of a Soviet art scene not always responsive to popular taste. Latvia's ruling elites believe in the current Western conviction that the forces of the market must dictate art and culture (with support for specific, symbolic nation-building cultural institutions).

Prudence over the state's own financial matters has merged with general price liberalisation as the state has left much of the economy to the forces of the market. Although the market sets most prices, there are administratively determined prices on state housing, energy, transportation, telecommunications, public utilities and medical services.[8] The state is moving towards liberalising these prices, but often faces stiff public reaction. The recent increase in tariffs for Lattelekom (Latvia's telecommunications) met with demonstrations, and many disapprove of government plans to bring petrol prices completely in line with world market prices. Although, on occasion, the government has slowed its remaining liberalisation of prices owing to public pressure, it has not reversed its course. The Lattelekom reaction is indicative. After angry demonstrations by people on fixed incomes who feared losing telephone services, the tariff hike was stopped pending further investigation. After tempers cooled, the original tariff was generally reintroduced.

All of these macroeconomic steps followed by successive Latvian governments led to a strong economic forecast by the end of the 1990s. The tight money supply and government action controlled inflation, foreign debt and budget deficits. Latvia has the lowest inflation rates of the three Baltic states. After the triple-digit inflation rates of the early 1990s, Latvia's economic policy bore fruit in the second half of the decade. By 1996, inflation was a respectable 13.1 per cent, by 1997, 7 per cent, by 1998, 4.7 per cent, and 1999 is forecast to be 3.5 per cent. According to the *Central European Economic Review*, Latvia will have the lowest inflation rate of all of Central and Eastern Europe for 1999.[9] Latvia is not only committed to, but already meets the European Union's Maastricht criteria in regard to foreign debt, budget deficits and overall debt. Similarly, with an eye to the EU, Latvian legislators and officials are busily creating the legal framework not only for a market economy, but also for EU membership. Banking reform, although apparently unsuccessful owing to the many bank failures, is also approaching European standards. Generally, macroeconomic reform has been successful, but problem spots remain: privatisation, foreign direct investment, and sector restructuring and reorientation.

LATVIA'S CONTINUING ECONOMIC CHALLENGES

The generally successful governmental efforts at laying the economic foundations of the new state fall disappointingly short in privatisation. The Soviet regime, upon incorporating Latvia into the Soviet Union, wasted little time in nationalising businesses and property in 1940–1941. The drive to collectivise agriculture was all but complete by the end of 1949. The Supreme Council controlled by the Latvian Popular Front declared its intent to return confiscated property to the rightful owners or their descendants. Restitution became a central pillar and, although politically important, has handicapped the economic effectiveness of privatisation to date. Returning property to individuals slowed owing to multiple applicants for the same properties, the ensuing court cases and the need for archival authentication. Arcane property laws further muddled the question of ownership with distinctions between separate ownership of land and the buildings upon the land. This privatisation mess confused the legal status of businesses on lands in the midst of litigation, and stalled the privatisation of these businesses. Similarly, the government at times hesitated in privatising large, potentially profitable businesses. Further contributing to the chaos of privatisation is Latvia's tortured experience of privatisation certificates (or vouchers), which are supposed to involve the people in privatisation, and thus capitalism and the new state.

The great problem in privatisation has been the frequent changes in direction. Marja Nissinen devotes a brilliant thirty pages to almost every aspect of privatisation in the authoritative *Latvia's Transition to a Market Economy*. Nissinen describes how the political and social need for restitution and vouchers lacked economic sense and ultimately failed to provide "justice" to the former owners at the same time. Furthermore, the voucher system required almost three years of legislative construction, further hampering privatisation. Likewise, the government vacillated about the method of divestiture, who would execute privatisation, who would be targeted for acquisition, and what would be needed for acquisition. By 1995, the result was a slow, tortured process that seemed to support popular notions of an independence movement robbed by shady underworld figures and former Communist *apparats*. The uncertain real estate market continued to dampen investment and reconstruction. The third turnaround in privatisation approach began to move more steadily, but the arrival of the Skele cabinet in 1996 sped the pace of privatisation vigorously. Nissinen enthusiastically argued that: "The year 1996 became a turning point in mass privatisation. The main bulk of enterprises were privatised in 1996–97, while the target was set to complete the privatisation by mid-1998."[10] Soon after, however, the Skele cabinet fell and privatisation slowed as Latvia entered a parliamentary electoral season. The Kristopans cabinet further slowed the pace of privatisation (amounting to the last remaining large, but potentially very lucrative, enterprises) in the first half of 1999. A new Skele cabinet from the summer of 1999 promises to finish privatisation, but it is now clear that this is a process that will continue into the twenty-first century. More so than in any other economic reform, political interests (specific regional and business interests) have impinged upon privatisation in Latvia. The debate has often been heated and acrimonious. Insinuations of corruption and megalomania have lowered Latvia's politics to a turf war between economic interests and frustrated the completion of Latvia's economic transition.

The lack of clarity about privatisation may be one cause of Latvia's continuing difficulty in attracting foreign direct investment (FDI). Government policy has attempted to create a climate for investment. Latvia's currency convertibility laws should encourage such investment. The Lats is freely convertible, individuals (resident and non-resident) can open bank accounts in either Lats or in foreign currencies, and they can enter or leave the country with any type of currency. Non-residents can repatriate all of their capital or dividends and there are considerable tax incentives (such as free economic zones) to encourage FDI. An independent, governmental body, the Latvian Development Agency, has been specifically created to aid FDI. The real estate market, however, remains a concern. Latvia's legislators have been slow to allow foreign ownership of land (several political parties are against it, and others create different types of property—urban and rural). Furthermore, although tax rates have been simplified, in practice the bureaucratic red tape involved in opening and operating a business can be so excessive that only the most devoted investor will commit or stay. Generally investment concentrates in transport, communications and finances; this includes the lucrative modernisation of Latvia's telecommunications by a British–Finnish consortium. A few industries have also drawn investment, but almost solely in Riga. FDI has done little to solve (perhaps has exacerbated) Latvia's growing regional disparities. Although many of Latvia's economic indicators are as strong as other Eastern and Central European countries, Latvia fails to attract similar rates of FDI. Per capita foreign direct investment in

Latvia is only 20 per cent of investment in Hungary, and 30 per cent of that in Slovenia. Even Estonia, which has been more impressive than Latvia in economic terms, still attracts more than twice as much FDI per capita as Latvia.[11] The lack of foreign investment is all the more troubling because one of the foundations of the IMF-backed economic policy that Latvia has followed is the expectation that such a macroeconomic climate will draw foreign investment that will then help the state grow and prosper. Latvia itself does not have the capital needed to create a state and economy on par with its European expectations. If foreign investment does not increase, Latvia's economic aspirations will be partly unrealised.

The trickle of foreign direct investment and the continuing problems with privatisation were reflected in the performance of the Riga Stock Exchange. If, in a general sense, an elected parliament is a mirror to a country's political life, in a market economy, the stock exchange is a similar mirror to the economic world. The Riga Stock Exchange followed the rapid privatisation of 1996 and 1997 with quick growth, high turnovers per session and growing index values (including a Baltic first, a Dow Jones Riga Stock Exchange index value). The Riga Stock Exchange Price Index (RICI) indicated the rise and crash of the exchange. The index base value (100 points) was set on April 2, 1996 and reflects the price changes of the shares within the index. By the end of 1996, that index had quadrupled, while by the end of 1997, the index showed seven fold growth. The collateral damage of the Russian crisis dropped the index near the original mark at 185.7 by the end of 1998.[12] Of all Central and Eastern European states, only the Russian Stock Exchange had a worse 1998 in terms of performance. The Riga Stock Exchange continued to be quiet through 1999 with minute daily trading volumes, adopting a wait-and-see approach to recovery from the Russian crisis, the fall out from Latvian banking problems, and the on going privatisation drama. To further underline the exchange's weakness, of the sixty-plus companies listed on the exchange, the great majority of that small daily trading is of nine companies' stocks. Plans for merging the Vilnius, Riga and Tallinn Stock Exchanges to create a larger body that would benefit each component part with the larger volume have been as inconclusive as plans for greater Baltic unity in the political sphere.

With the stock exchange generating little capital, Latvia's native financial institutions, the commercial banks, are an equally weak sector of the national economy. The first years of independence were rocky ones for the private financial sector of the Latvian economy, which started from scratch. Savings banking began on a particularly unsure footing. The Soviet savings banks transferred all of their funds to Moscow and following Latvia's independence these funds were never returned, effectively robbing Latvia's citizens of all of their accumulated capital. The Latvian Bank and government stepped in and guaranteed the shortfall. Fortunately for Latvian banking, people did not rush for their deposits. Rapid inflation in the early years of the Republic eased the load of carrying the missing Soviet savings money. The near crisis, however, defined the patterns of banking in Latvia through the 1990s: a cycle of near crises, crises and recovery.

Most dangerous to the nascent banking industry was 1995, the year of insolvent banks. The roots of the banking crisis lay, in part, in the lax rules and regulations for private banks. Originally, new banks needed little money to begin operations and were not required by law to conduct independent audits and financial reviews. Some small banks simply opened their doors, collected people's and businesses money and disappeared. More often, unstable banks played a dangerous gamble that began with high interest rates to attract depositors. Increasing numbers of depositors worked, in part, as a pyramid scheme providing the money for the original depositors. These banks gambled that continuing expansion would last until the Lat exchange rate crashed, effectively covering the high interest rates for later depositors. The Bank of Latvia, however, did not waver on keeping a strong Lats, and prodded by international advice clamped down on banking rules. The arrival of Coopers & Lybrand and other Western auditing firms brought the crisis to a head and the Bank of Latvia declared dozens of banks insolvent. The crisis quickly moved from a few smaller banks throughout the sector owing to a complex network of interbank credits. The banks were dependent on oil capital from Russia and other former Soviet states, and when much of this capital was withdrawn owing to the growing crisis, most banks "faced a liquidity crisis".[13] The largest private bank in Latvia, with one-fifth of all deposits, *Banka Baltija*, closed its doors in the spring of 1995.[14] The crisis caused a 10 per cent drop in Latvia's GDP for 1995, mushroomed the government's budget deficit for the year, and sidetracked Latvia's economic growth.

Banks, however, rebounded from the crash of 1995. Bank consolidation brought the number of banks from sixty down to thirty-one, with five major banks (Unibanka, Parex Banka, Rigas

Komercbanka, Rietumu Banka and Hansabank-Latvia). A wary public was still uncertain about the security of their deposits, but there seems to be a general acceptance of banking as a necessity of modern times. The introduction of debit cards, credit cards, ATM machines and other modern banking conveniences also wooed depositors into the private banking industry. Ironically, the strength of the Lats, which doomed the gambling private banks before 1995, brought international capital back to Latvian banks. The capital, by and large, comes from the former Soviet Union and includes a great deal of ill-gotten gains and unreported income. Latvian banks are relatively more secure than the banks of the former Soviet Union, the strength of the Lats guarantees deposits do not devalue and Latvian banks are far from the prying eyes of tax collectors of the former Soviet Union. Some Latvian banks have not been bashful in promoting this offshore banking advantage – one bank even uses the motto "We are closer than Switzerland" before listing its services, which include numbered accounts.

The 1995 bank crisis seemed to repeat itself in 1998–1999, although the causes were different and the short-term effects were much more contained. In the autumn of 1998, when Russia devalued the ruble, defaulted on its domestic debt and suspended trading on its Stock Exchange, Latvia's banks were caught "heavily exposed". Ten of Latvia's thirty-one banks had more than 10 per cent of their assets in either the Russian Stock Exchange or in other Commonwealth of Independent State (CIS) securities. Latvian banks tried to hold out for a Western bailout of Russia, but when none was forthcoming, the weakest banks fell. Rigas Komercbanka, Latvia's fifth largest, and considered one of the safer banks, closed its doors. The long-term effects of the Russian crisis continue to strain Latvia's banks. As late as the summer of 1999, Standard and Poor's Corp. concluded, "Latvian banks' direct Russian exposure was 'the highest of any banking system in the world".[15] Unlike the bank crisis of 1995, however, most banks avoided the domino effect. The crisis also spurred a new round of bank consolidation and an incursion of Western bank interventions (primarily Swedish and Norwegian). Even the closed Rigas Komerc-banka entered difficult and protracted restructuring negotiations and reopened in late 1999 as Pirmabanka. After the Russian crisis, bank analysts routinely talk of diversified investments in safer climates and the eventual consolidation of Latvian banking into three or four banks.

The Russian crisis reverberated beyond just banking and through Latvia's industrial and agricultural sectors of the economy. After the devaluation of the Russian ruble, Latvian agriculture and industry were cut off from their main market. Latvia's gross domestic product fell by 1.9 per cent in the last quarter of 1998, and a further 2.3 per cent in the first quarter of 1999. Industrial output dropped by 14.5 per cent. Companies were generally caught between Russian and CIS clients that could no longer afford Latvian products (and were behind in payments) and the almost prohibitive cost of retooling for Western markets.[16] The effects of the crisis are like the proverbial glass of water that optimists see as half full, but pessimists see as half empty. Either the crisis shows the shortcomings of the past decade's restructuring and reorientation and the economy's continued dangerous reliance on Russia or it shows how well Latvia has divorced itself from its tumultuous neighbour in a relatively short period of time.

Industry in the 1990s, is a good example of half-completed reconstruction and reorientation with some sectors devoted almost solely to the West and others still reliant on the countries of the former Soviet Union. Prior to independence, Soviet-era industrialisation needed cheap, Soviet (but not Latvian) raw materials and Soviet markets—both of which disappeared in 1991. Independent Latvia was left with an industrial infrastructure that the state could not support. Machine building, metallurgy, electronics, electrical power and chemicals dominated the industrial sector, as did military production.[17] Consumer goods and services were particularly disregarded in the command Soviet economy. After independence, only consumer goods and services grew as the remainder of industry in Latvia went into a steady freefall from 1991 through to 1993. By 1994, some growth in production figures re-emerged, but industry showed an increase in manufacturing only in 1996. Production is not expected to be at 1989 levels until well into the twenty-first century.

Industry has retooled towards pursuits that do not require the extensive raw materials that Latvia does not possess. Electronics, telecommunications and automobiles are likely to disappear from Latvian industrial production. The Latvija minibus production lines quickly stopped, as did other automotive industrial pursuits. The once-massive State Electronics Factory (*Valsts Elektroniska Fabrika*, VEF) which produced a lion's share of Soviet telephones and other electronic goods is now defunct. During the state's negotiations with British and Finnish telecommunications providers for the overhaul of Latvia's telephone services a place for VEF was sacrificed. The introduction of modern, digital services continues (as proof, Latvia is awash in web-sites, and Riga has high-tech telephone booths), but the renovations are produced almost entirely abroad. Logic would have dictated that government negotiations for telecommunications overhauls in the 1990s would have included a role for (and modernisation of) Latvia's telephone industry. The result would have been industrial production for the immediate domestic need, and perhaps the seeds of an industry that could participate in the restructuring of the former Soviet Union's telecommunications system.[18] The government contract, however, failed to include a significant role for VEF, and admitted the monopoly of Western parts and technology. The site of the former VEF factory, which in Latvia's previous independence built the world's smallest cameras, is scheduled to become a shopping arcade.

By the late 1990s, industrial production focused on export goods based on Latvia's resources or on "skills-based" production. Juris Dreifelds, for example, highlights pharmaceuticals, biotechnology and furniture as somewhat successful (and potentially vibrant) sectors of the economy.[19] Furniture and the timber industry in general is an ideal success story for Latvian industry. Initial wood exports were of raw timber (Latvia is one of the few countries to have more forests in 1999 than in 1939), but as skills, capital and the search for new markets developed, more and more exported wood was finished and treated lumber. The logical extension would be the growth of a skilled furniture industry. Information technology is another Latvian industrial success story. Soviet education's strength in mathematics and the remnants of the Soviet military industrial complex in Latvia have left Latvia with a well-educated and highly skilled information technology labour force. Several Latvian software companies have grown from this base to be competitive in the Western market. Pharmaceuticals and biotechnology are well poised in industrial restructuring, for similar reasons. Some factories, such as the fibreglass factory in Valmiera, successfully attracted Western investment and produce for Western markets. These factories weathered the Russian crisis easily.

Most factories, however, had failed to reorient by 1998. The great obstacle, however, has been that while successful sectors of industry restructured, the temptation towards the old export orientation remained. The networks of trading and contacts were more familiar in the former Soviet Union, and the markets were less demanding. Latvian goods were much more competitive in Russian and CIS markets than in Western Europe. The Russian crisis in 1998 and the subsequent freeze in orders affected Latvia's industry regardless of its restructuring.

The crisis was a new catalyst in reorienting towards Western markets, but this change is exceedingly difficult. The example of pharmaceuticals is telling. The drug manufacturer, Olainfarm, restructured its Soviet plant but continued selling primarily to Russian markets. With the Russian crisis, they do not have the needed capital to reorient their goods to new markets. Through 1999, the company fell back on the short-term solution of bartering with Russian buyers to survive the crisis. The hope in pharmaceuticals and throughout industry is for an infusion of Western capital. Dreifelds concluded insightfully that "industry requires capital for modern equipment, and intensive retraining of its workers and managers for the development of various skills required in a competitive market-based capitalist economy".[20] The slow rate of foreign direct investment is all the more serious considering the precariousness of the industrial sectors of Latvia's economy.

Agriculture is in a grimmer predicament than industry. Despite the dismantling of *kolkhozes* and *sovkhozes*, and the privatisation of agricultural land, agriculture remains in a state of crisis in contemporary Latvia. In the early 1990s, many Latvians who received previous property as restitution returned to the land. They were motivated in part by fears of coping with the rapid structural changes in the cities. On the farm, at least they could feed their families. These returnees to agriculture quickly became designated as the new *Jaunsaimnieki*, alluding to the 1920s agricultural reform. There are similarities in the conditions of the new farmers of the 1920s and of the 1990s.[21] Agrarian reform in the 1920s distributed plots of land too small for wheat production, and so agriculture focused on pork and dairy products. Nearly a third of new farmers in the 1920s began their farming lives without a home; between a quarter and a half had no barns, inventory or livestock. Most of these farmers fell quickly into debt. The hardships of making the farms work pulled new farmers out of social life, limiting their reading of newspapers and participation in social organisations and agricultural aid courses. Most shockingly, the need for the labour of children kept school attendance down during agricultural seasons. In 1993, the new generation of new farmers echoed the conditions of the 1920s. One quarter did not have homes, 42 per cent had no electricity, 52 per cent no tractors, and 85 per cent no telephones. Farm size was yet again too small for affordable grain production, and private farmers turned to the production of pork, dairy and poultry. Some government assistance came in the form of artificially high grain prices and compensation for the rise in diesel fuel prices. Nevertheless, private farmers have faced difficulties in getting credits from private banks owing to their lack of collateral, and many have fallen into debt. Lack of transport and telephones isolated the new farmers of 1993, and yet again the need for the labour of children drove down school attendance. New farmers in the 1920s, and new farmers in the 1990s, translated their numeric strength into political clout by backing political parties devoted to their concerns (Farmers' Union and New Farmers' Party in the 1920s, the Farmers' Union and others in the 1990s).

The similarities, however, end with the difficult lives of new farmers in the 1920s and 1990s. The agricultural growth of the 1920s is unlikely to repeat, for numerous reasons. The world market of the 1920s and 1930s favoured Latvia's pig farmers and dairy producers. Following the massive destruction of World War One, the demand for pork and butter was nearly unquenchable. Latvia patterned itself on the successful Danish dairy model and although never reaching such levels of efficiency, sold products easily and steadily on the world market. The Depression slowed foreign orders for a few years, but by the second half of the 1930s, Western Europe was purchasing heavily in order to build up reserves in case of war. The 1990s, and the foreseeable future, have not been as kind to Latvian agriculture. Latvian farm products are not entering an agricultural market vacuum, rather a sophisticated, competitive market of global agribusiness. Despite the subsidies that Western governments give their own agricultural sectors, Western-run international financial institutions such as the IMF pressure developing countries, like Latvia, to cut government support to farmers in reverence to the power of the free market. As a result, Latvian agricultural goods cannot compete with the mechanised production, high quality and superbly marketed agricultural goods of the West, even in Latvia itself. The only salvation for Latvian agriculture is either a share of production for the home market, or speciality products on the world market (entrepreneurs' schemes range from organic foods for Scandinavia to Christmas trees for Germany). Regardless, agriculture will be a junior player in Latvia's economy, as statistics already warrant. Nearly 20 per cent of Latvia's inhabitants make their living in agriculture (although without cutting lumber on the side, many could not survive), but agriculture produces less than 10 per cent of Latvia's GDP. The harshest critics of agriculture would go so far as to say that the most productive role the new private farms play is the maintenance of 20 per cent of the labour force in the countryside, and therefore out of the pools of unemployed in the cities.

Agriculture, fishermen and food processing felt the effects of the Russian crisis of 1998. Much of Eastern Latvia's agricultural production and food processing had been sold to Moscow and St Petersburg, but with the crisis, orders ended (and outstanding debts were not paid). Unemployment grew markedly throughout eastern Latvia as food processing (and industry generally) laid off workers. The region around the town of Rezekne recorded Latvia's worst unemployment, nearing the 30 per cent mark. Fishing communities suffered similarly. Fishermen themselves were out of work throughout Latvia by September 1998. By November, fish processing plants and other associated industries were silent. In the region around the town of Talsi (which has fishing communities) unemployment jumped month by month through the autumn of 1998. At 4.3 per cent in August, unemployment rose to 7.6 per cent by November—the rise in this unemployment was focused in a few fishing villages with nearly 30 per cent out of work.[22]

As with industry, however, sectors or individuals within agriculture that have modernised their farms and reoriented themselves away from complete dependence upon Russian and CIS markets are prospering. Throughout Latvia, industrious, modernising farmers have successfully wooed foreign investment and created model farms. Janis Varpa, a farmer in the agricultural community of Ledurga, is a good example of the future of Latvian agriculture. Through a combination of aid from international veterinary companies (or rather their representatives in Latvia) and others, Varpa introduced 250 Holstein Friesian dairy cattle to his farm at an expense of nearly 150,000 Lats. Improvements did not end with the introduction of high-quality cows; appropriate dairy technologies from barn and stall design to milking machines to proper veterinary medical care were also introduced. The dairy cattle produce much more milk than Latvia's standard cattle, and are better suited to Latvia's climate. Farmers from across Latvia have followed the progress of Varpa's model farm with great interest.[23] Similar model farms have introduced high-quality beef cattle (destined for Riga's finest restaurants), and quality sheep stock. Some agricultural communities have benefited from direct investment from Western agricultural communities, bypassing the bureaucratic intricacies of the national scale. Model farms point to the ultimate future of Latvian agriculture—more productive and modernised agriculture that will require far fewer employed hands. Agriculture will continue to develop towards the example of most industrialised states where roughly 5 per cent of the population works in agriculture. In other words, the trend towards modernising and rationalising agriculture will continue until farming is an established and productive part of the economy (although never the driving force behind the economy). Latvia's challenge will be to accommodate the many people now working in agriculture who will become "superfluous" with the development of the sector.

Trade is often trumpeted as the potential saviour of the Latvian economy. Historically, Latvia has three ports that are ice-free all year round, Riga, Ventspils and Liepaja. The recent return of Liepaja to the fold, after decades of use by the Soviets as a submarine base, should expand trade through Latvia's ports. The slow, but steady, construction of the Via Baltica, a highway linking Warsaw, Vilnius, Riga, Tallinn and Helsinki, should expand overland trade as well. Bustling trade between East and West, with Latvia as the prism through which all trade flows, is a dream of future Latvian prosperity. The nature of trade, however, entails potentially explosive questions about the orientation of Latvia's economy and has been the root of political discord between economic interests favouring industrial reorientation and restructuring and interests favouring Latvia as a transit trade state.

Contemporary Latvia is geographically situated to pursue trade with the former Soviet Union. The Western powers and Russia are eager to expand trade. Russia is dependent on Western aid and the continued development of trade on the world market. Before the crisis of 1998, Russia threatened Latvia with cutting trade, but this is no longer an option for financially desperate Russia. Latvia can be assured of continued transit of Russian oil through Ventspils. Ventspils, a small port before WWII, was massively developed during the Soviet era, and became the Soviet Union's primary European port for the export of oil and oil products. It is the world's twelfth largest port and the busiest on the Baltic Sea.[24] The city of Ventspils has parlayed this transit trade into the wealthiest region of Latvia. Ventspils's mayor, Aivars Lembergs, is one of Latvia's most important political and economic figures. As a result, the "Ventspils interests" believe that the restructuring of Latvia's economy should not include a reorientation away from the oil fields of Russia. The "Ventspils interests" believe that guaranteeing the continued transit of oil should be Latvia's economic and political priority.

Latvia's policy makers, however, need to balance their approach to international trade. They should diversify trade relations as much as possible (a policy in construction, as shown by the bevy of trade agreements that the Latvian government has signed with nations all over the world) in order to minimise the weight of any single trading partner. The Russian crisis showed the changing nature of Latvia's foreign trade. Germany surpassed Russia as Latvia's leading export and import partner. Great Britain rose to a surprising second place in exports. To underline Latvia's slow but steady progress towards realignment, in 1998 almost 56 per cent of Latvia's imports were from EU states, only 15.6 per cent from CIS countries. Exports were similar, with 56 per cent headed for EU states and less than 20 per cent to the CIS.[25] Ideally, Latvia should try to develop a trade policy that even takes advantage of the opening of competing ports (Tallinn, and a proposed port near St Petersburg) and caters to growing economic centres around the Baltic Sea.[26] A diversified trade balanced between port and highway traffic dealing with many types of goods from many nations would also buffer the Latvian economy against the sudden drop of a single product, or exporter/importer. Likewise, an extensive trade system would reinvigorate more of Latvia's economy and move away from the current trend of economic growth in Riga (and a select other few), and stagnation everywhere else.

Tourism also highlights the trend of economic concentration in Riga, with little elsewhere. Nevertheless, the expansion of tourist destinations suggests the possibility of extending economic growth throughout the rest of Latvia. Throughout the Soviet period, tourism blossomed on the shores of the Gulf of Riga. The resort-city of Jurmala, long known for its exclusive spas and resorts, became the playground of the Soviet *nomenklatura*. Many Soviet army officers and governmental leaders retired to Jurmala and continue to reside in its environs. After the collapse of the Soviet Union, the tourist waves from the East ebbed. New waves of tourists, however, are slowly gaining in strength. Scandinavian tourists escaping the high prices of alcohol in their homelands have made the Baltic Republics a popular weekend getaway. Latvian émigrés continue to return to Latvia to visit relatives, reclaim property, attend cultural festivals and even occasionally to open businesses. German tourists, particularly the descendants of Latvia's Baltic Germans, have also visited Latvia en masse to find their lost roots. Other tourists, including the ever-present backpackers scouring Eastern Europe for bargains and "off-the-beaten path adventures", have made Riga the less known Prague. Western hotels, from Best Western to the luxury SAS-Radisson, have opened to meet the shortage of high-quality Western hotel rooms. The growth of services that accommodate Western tourists has definitely aided Latvia through its economic hardships.

The extent of growth for the tourist industry, however, is still unclear. Latvia has, without a doubt, a multitude of treasures to offer the Western tourist. Riga is a beautiful, turn of the century architectural gem with the best collection of *Jugendstil* buildings in the world. The bluntness of Soviet Intourist hospitality has disappeared, replaced with a range of friendly, private restaurants, clubs and pubs. Outdoor patios and beer gardens spill throughout Old Riga in the summer and entrepreneurial Latvians have picked up English and German surprisingly quickly. Outside Riga, the beaches of the Gulf of Riga and Baltic Sea are, by European standards, often deserted and beautiful. There is a healthy smattering of medieval castles, renaissance palaces and quaint towns throughout Latvia that offer the tourist more to see than just Riga. One recurring vision is Latvia as a sportsman's paradise with fresh water fishing in the Gauja National Park and game hunting in Latvia's forests. Western tourists and Latvian entrepreneurs alike are already exploiting all of these options, but capital is again a problem. For Latvian tourism to develop, the country as a whole, and points of interest specifically, have to be marketed relentlessly in a crowded market of tourist destinations. A co-ordinated plan with enticing airfares, packaged vacation plans and aggressive ad campaigns needs to be centrally co-ordinated and continued year after year to ensure a healthy crop of yearly tourists. The government would be ideally situated to encourage and lead such a campaign, but that seems unlikely given its financial precariousness and its attraction to free market principles.[27] The danger is that Latvia as a Western tourist destination will become faddish or a hidden gem. Both options are fine for the Western tourist, but economic death sentences for an industry hoping to revitalise Latvia's economy. Furthermore, the early, naive hope that tourism could be the engine behind Latvian economic recovery is simply extinguished by Latvia's northern climate which restricts waves of tourists to a few summer months.

The fads of Europe and the West will dangle over the fate of Latvia's economy in more than just the tourist industry. Latvia's economic development must begin at home, but can always be undermined by "external shocks". Historically this was the plight of Latvia in the interwar years; the greater European crash initiated Latvia's depression from 1929–1933, and the partition of Europe between Nazi Germany and Soviet Russia determined Latvia's fate. In the new millennium, the fate of Latvia hangs by the threads of the world market. The nations of Africa, for example, can describe to Latvia in detail how quickly modest reforms and accomplishments can be erased by an oil crisis.[28] The fate of Latvia, and many others, can be little more than collateral damage in manipulations of the world market by financial speculators, or regional trading blocs. Moreover, there is a tendency of Western business to think in cyclical terms and therefore create cyclical conditions, often in countries rushing to join the club of the most prosperous nations. All too often it seems that the darlings of Western economists just as quickly plummet out of favour (see the Czech Republic or Thailand). In part, nations' economies stumble of their own accord, but Western investment seeks out potential windfalls and just as quickly abandons them for other, new potential windfalls. This is the trap of the West for states like Latvia, particularly now that the free market ideology is touted as "the most promising institutional arrangement for world-wide prosperity that history has ever seen".[29] The danger becomes all the more real scanning the non-Western nations of the world. The IMF and the World Bank have been intricately involved in the financial planning of governments around the world for over twenty years, but few states seem likely to catch the West. Of course, every country has its own reasons for success or failure, but the economic dilemma for Latvia is the pursuit of the free market at what price. The government's pursuit of macroeconomic reform puts the state's role in solely business and financial terms. Latvia's elite could follow every *laissez-faire* option and by the whim of an "external shock" gain nothing. The consequences would be all the more serious when one stops to consider the non-financial issues that beg for government interference.

HUMAN SIDE OF ECONOMIC REFORM

Latvia faces a multi-faceted demographic crisis, from cradle to grave. A recent study conducted by Healthcare Europe concluded that only Russia had a worse health record than Latvia in all of Europe.[30] Table 2 shows that the huge gap between Latvia and Sweden is not just in economic terms.

Latvia has long had negative population growth, and perhaps more significantly more abortions per year than live births. The degree of negative growth is troublesome for several reasons. First, it is an indicator of the despair people feel at bringing new lives into the present situation. Likewise, a country's inability to reproduce its own

TABLE 2   A COMPARISON OF LATVIAN AND SWEDISH HEALTH STATISTICS

	Latvia	Sweden
Infant deaths per 1,000 births	15.2	4.0
Life expectancy, men	64.0	76.5
Life expectancy, women	76	81.5
Deaths per 1,000 inhabitants	13.6	10.5
Births per 1,000 inhabitants	7.6	10.1
Natural increase	–6	–0.4
Tuberculosis per 10,000 inhabitants	68.0	5.6
Abortions per 1,000 births	1141	337
Accidental deaths per 100,000 men	260.8	57.2
Accidental deaths per 100,000 women	64.2	23.6

Source: Dace Plato, "Latvijas iedzivotaju veseliba – viena no sliktakajam Eiropa", *Diena*, Dec. 5, 1998.

inhabitants will mean hosts of transitional problems ranging from jobs to health care when the youngest generation matures. The smaller number of people translates to a smaller number of taxpayers per pensioner, already a crisis for Latvia's government. Latvia's pensioners are a vocal, voting minority that pressures the government incessantly to raise their meagre pensions. To be sure, their current pensions do not provide a living wage, but they nevertheless consume 30 per cent of the state's budget. Pensioners' calls for doubled benefits would amount to the ransom of future generations. What makes matters worse is that until recently Latvia kept Soviet laws for retirement that saw women retire at fifty-five and men at sixty. Attempts to raise the age of retirement were watered down by the pensioners' lobby, so that the age of retirement will be slowly increased by half a year per year until levelling at a universal sixty-five. A future with even fewer taxpayers and more pensioners would be insurmountable under present economic conditions.

Contemporary Latvia needs to address the crisis in the health of its population. Life expectancy is already shockingly low for an industrialised country. The Latvian government needs to raise the quality of health care throughout life, but does not have the needed resources. Latvia must seek creative alternatives, perhaps more vigorously embracing preventive health care. An ounce of prevention is better and cheaper than a pound of cure, particularly when one considers that so many of Latvia's health concerns are easily addressed through prevention. Alcoholism, for example, plagues Latvia and is responsible, in part, for the particularly low life expectancy for men. Aggressive anti-smoking and anti-alcohol campaigns can only bring future savings in financial and societal returns. Reproductive health care and pre-natal care could drive down the reliance on abortion as birth control, and contribute to healthy children and mothers.

Finally, the state needs to remember that projects and services that lose money still benefit society. The arts, sports and education all return dividends in intangible ways. Arts and sports not only enrich people's lives, but also comment on the human condition, in this case the condition specific to Latvia. Identifying the Latvian condition is a necessity in forging national identity and attachment across multi-ethnic Latvia. Finally, in this era of nearly universal acclaim for capitalism and a global free market, Latvia should remember that economic advice comes and goes. Deficit spending was once popular, now a sin. State ownership and centralisation, once heralded, now fall before the craze for privatised decentralisation. Above all of these trends in theories, one economic truth seems to stand out. Vigorous and constant investments in education and health care are acknowledged universally as the only guarantors (if all else goes well) of a strong economy and a healthy society. As Latvia's political economy now stands, these are hardly prioritised. Latvia has suffered a serious brain drain from 1990 to 1998 as between four and five hundred academics and scientists have left the country (primarily owing to terrible wages in Latvia).[31] Of all of the Eastern and Central European states noted as potential frontrunners for EU membership, Latvia invests the least in education. Less than 0.25 per cent of GDP went to education in the 1999 budget. Lithuania, Estonia, the Czech Republic and Poland roughly doubled Latvia's investment, and Slovenia invested three times more than Latvia per capita. In the entire region, only Albania, Macedonia and Bulgaria invested less in education.[32] The OECD has further underlined that there is a growing gap between the quality of schools in Riga and elsewhere in the country.[33] Government action turned to education in 1999, budgeting more money for school repairs and highlighting the sector for increased state investment,[34] but it may be too little too late to slow the crisis in education. The Skele government's decision to convert university education to tuition for all may be financially prudent, but it will add to the growing gap between the "haves and have-nots". Although a degree of collapse in healthcare and education was inevitable given the huge economic changes of the past decade, macroeconomic reform is undermining one of the few resources that makes Latvia competitive, a well-educated and healthy workforce.

Latvia's economic transformation has been generally successful. Macroeconomic reform particularly has been impressive. As Marja

Nissinen commented, however: "Until the present day, it has been easy to legitimise giving priority to the rebuilding of the economy over and above the expansion of the social sector, in spite of widespread poverty."[35] Latvia is, at times, in danger of forgetting that its successful macroeconomic reform is a means to an end, and not an end in itself. The reforms are meant to create the conditions for prosperity and growth. The wealth, comfort and materialism of a new middle class is not enough. Latvian policy makers need to remember that not only must they create the preconditions for prosperity, but also they should use prosperity for continuing growth and the greater good. Education and health care should become top priorities. At present, foreign aid and humanitarian organisations (from PHARE to the Peace Corps) pick up the slack for certain projects, and the government responds to others. Many foreign programmes, however, will soon leave the Baltic States. Latvia's new challenge will be to tackle the human side of economic reform as aggressively as it did the transition to the market. If Latvia tries this approach, its inhabitants will continue to have faith in the new system. If it does not, however, the growing disparities between regions and classes could become explosive.

1   For example, after Ivars Godmanis was soundly defeated in the elections to the *Saeima* in 1993, he left public life, and joined the board of the lucrative SWH Corporation in the summer of 1993. In 1998, he returned to parliament and became the Minister of Finance. The crossover between politicians and big business which is commonplace in the West was a new innovation at the time and viewed very cynically by a disgruntled public. For a discussion of ties between organised crime and new elites see Stephen Handelman, *Comrade Criminal; Russia's New Mafiya* (New Haven: Yale University Press, 1995).

2   Marja Nissinen, *Latvia's Transition to a Market Economy: Political Determinants of Economic Policy* (London: Macmillan Press,1999), p. 62. Chapter 10, "The Nationality Question" is an authoritative analysis of the ethnic dimension of Latvia's economic reform.

3   "Latvia's World Class Banker" in *The Baltic Times*, August 28–September 3, 1997, p. 14. The article emphasises that the December 1996 issue of *Euromoney* listed Repse as the only East European in the top ten of the fifty financial leaders of the twenty-first century.

4   *Diena*, July 22, 1997. A public opinion poll showed Repse as the most respected public figure among citizens with 40 per cent of respondents choosing him. PM Andris Skele was second with 21.4 per cent.

5. Conversation with customs officer, Spring 1995, and Juris Dreifelds, *Latvia in Transition* (Cambridge: Cambridge University Press, 1996), p. 119.

6   Nissinen, p. 71.

7   Arthur Ruhl, *New Masters of the Baltic* (New York: E.P. Dutton & Company, 1921).

8   Nissinen, p. 64.

9   "Economic Barometer", *Central European Economic Review*, March 1999, p. 26–27.

10 Nissinen, p. 81.

11 *Central European Economic Review*, March 1999, pp. 26–27.

12 See the official web-page of the Riga Stock Exchange: http://www.rfb.lv

13 Nissinen, p. 73.

14 See William Hallagan, "The Evolution of the Latvian Banking Market", *Journal of Baltic Studies*, 28, 1 (Spring 1997), pp. 65–76.

15 Burton Frierson, "Baltic States struggle to replace Russia", *Globe & Mail*, September 2, 1999.

16 *Ibid.*

17 Nissinen, p. 74. Nissinen estimates that in the 1980s, 15 per cent of Latvia's labour force was employed in military production.

18 *Dienas Bizness*, April 25, 1997; April 28, 1997 reported meetings between PM Andris Skele and the British telecommunications company Cable & Wireless about decreasing the Lattelekom monopoly of services (Lattelekom). Two months later *Diena*, June 12, 1997, reported the Privatisation Agency's plans to privatise the state's shares of Lattelekom, which will complete the full circle of the state's abandonment of involvement in the telecommunications industry. See Juris Dreifelds, *Latvia in Transition* (Cambridge: Cambridge University Press, 1996), pp. 129–130 for more details about the VEF telephone connection.

19 Juris Dreifelds, *Latvia in Transition*, p. 130.

20 Juris Dreifelds, *Latvia in Transition*, p. 130.

21 The data on the *jaunsaimnieki* comes from Hugo Celmins, "Agrara reforma", in *X. gada svetki gramata* (Riga: 1928); data on current farmers from Juris Dreifelds, *Latvia in Transition* (Cambridge: Cambridge University Press, 1996): pp. 131–133.

22 Sarma Kocane, Janis Trops and Dace Plato, "Bezdarba skartajas pasvaldibas situacija pasliktinas", *Diena*, November 27, 1998; "Bezdarbnieku skaits jau ir parsniedzis 100 tukstosu robezu," *Diena*.

23 Gunita Ozolina, "Zemnieks grib razot konkuretspejigu pienu", *Diena*, May 4, 1999.

24 Nissinen, p. 76.

25 "Nozimigaka partnere areja tirdznieciba bijusi Vacija", *Diena*, January 19, 1999.

26 See the discussion on the "Amber Gateway" in Chapter 5.

27 *Diena*, June 3, 1997 and *Business & Baltics*, June 10, 1997 both reported that over 1,700,000 tourists visited Latvia in 1996, spending 118.5 million Lats, 38 million of which ended up in the state's budget The EU, however, highlighted that Latvia needed to spend much more to promote Latvian tourism abroad.

28 Combinations of a drop in export prices and the energy crises of the 1970s, for example, crippled many African economies. Some are only now recovering from these "external shocks". See "Emerging Africa", *The Economist*, June 14, 1997.

29 Jeffrey Sachs, "The Limits of Convergence; Nature, nurture and growth", *The Economist*, June 14, 1997, p. 22.

30 "Latvijas iedzivotaju veseliba—viena no sliktajam Eiropa", *Diena*, December 5, 1999.

31 Gunita Nagle, "Zinatnieki emigre uz neatgriesanos", *Diena*, Nov. 27, 1998.

32 Gunita Nagle, "Austrumeiropa Latvijas zinatnei viszemakais finansejums", *Diena*, Nov. 17, 1998.

33 Gunita Nagle, "Izglitibas problemas nerisina, atsaucoties uz finansu trukumu", *Diena*, April 21, 1999.

34 Baiba Melnace, "Valsts investicijas—infrastrukturai", *Diena*, Feb. 15, 1999.

35 Nissinen, p. 271.

Chapter 5

THE FOREIGN POLICY OF LATVIA

There are few nations whose international affairs contain as many broken threads as Latvia's. This has inevitably left obvious traces on Latvian foreign policy goals and tactics. Latvia was born as a nation-state at the beginning of the twentieth century, yet will still enter the next millennium as a political youngster that has to find and define its role in the international community. There are three main reasons for this relative youth. The first is the relatively late construction of the nation-state (1918). The second is the more than 50 years of Soviet occupation. Finally, the third reason is Latvia's challenging geopolitical situation, which has yet to offer a stable and long-lasting solution to national security. Therefore, in Latvia, the elements of foreign policy are constituted by historical background and accumulated experience.

This chapter will address several questions about Latvia in international politics. What kinds of realities determine contemporary Latvian foreign policy? What are the major goals of this policy? Are these aims achievable? Can Latvian foreign policy be called successful? How does the international society accept Latvia? First, the origins of Latvian foreign policy will be outlined. Next, its main determinants will be analysed and foreign policy orientation will be revealed. Finally, the successes and failures of Latvian foreign policy will be summarised.

ORIGINS AND BASIC GOALS OF LATVIAN FOREIGN POLICY BEFORE 1991

Latvia has been a frequent battleground for various European powers trying to dominate the Baltic Sea and its shores in the distant and more recent past. The Baltic geographic situation has caused the area to be a zone of contact and conflict between the West and Russia for several centuries. In previous centuries, Germans, Swedes and Poles who subdued the local population and struggled with a rising Russian Empire to control the Baltic region represented the West. This history, however, matters most in how it strengthens recent interpretations by contemporary policy makers. What matters most for contemporary Latvian politicians and society is the experience of the twentieth century.

At the beginning of the century, when Latvians started to search for political participation and autonomy they were denied self-rule and liberal reforms by the Russian Tsar. In 1917, Latvian representatives failed to negotiate autonomy with the short-lived Russian provisional government as well. The Russian policy was summarised by the representative of the Russian middle class Miliukov who said: "If autonomy is granted to the Latvian people, it must also be granted to the Lapps and Samoeds."[1] The situation did not improve after the October Revolution. Lenin paid lip-service to Latvian political independence, but tried to establish Soviet authority over the Baltic lands. Independence came to Latvia only after the Western Allies (Britain, France, USA) decided to support independent Latvian statehood politically and militarily. In 1919, British warships helped Latvian forces defend Riga from a renegade German and White Russian siege. Allied political and military support was also crucial for Latvian efforts in the war with Soviet Russia.

In 1920, Latvia concluded a Peace Treaty with Soviet Russia. According to the treaty, Russia forfeited all of its historical claims over Latvian territory for ever.[2] "For ever" lasted only twenty years, before Latvia was occupied and incorporated into the Soviet Union. The contemporary Russian analyst Pikayev argues convincingly that the Baltic States and Russia possess a long history of mutual mistrust[3] that heavily influences Latvian foreign affairs at the end of the twentieth century. Moreover, the Russian factor is the main determinant of contemporary Latvian foreign policy as far as national security is concerned.

The origins of contemporary Latvian foreign policy can be traced before the re-establishment of independence. At the end of the 1980s, the emerging Popular Front presented their understanding of the future of Latvia and its foreign policy directions.[4] The Popular Front leadership assumed that its political success depended not only on domestic politics, but also on the international environment. Foreign policy was largely determined by the political peculiarities of the struggle for democratisation and self-determination. Foreign affairs and domestic politics were intertwined. Latvia's Popular Front attempted to reach three main goals in foreign relations. First, to draw international attention to the forgotten Baltic States and their movement for independence. Second, to gain international support and assistance in this fight for independence. Third, to prepare the ground for Latvia to join different international organisations that could facilitate the achievement of independence.[5] The programme of the Latvian Popular Front (LPF) of 1988 declared (Article 3) that the Front supports an increase in political, economic and cultural contacts between the countries of the Baltic Sea.[6] The programme also supported Latvia's membership in such organisations as the UN and UNESCO.

Pre-independence success in foreign policy can be traced to popular movement activities like the "Baltic Chain" whose ten-year anniversary was celebrated throughout the Baltic States in 1999. On August 23, 1989, the Popular Fronts of the three Baltic countries organised a chain of people holding hands. The chain meandered roughly 700 kilometres from Tallinn through Riga to Vilnius (the three Baltic capitals), marking the fiftieth anniversary of the Molotov–Ribbentrop Pack of 1939 that led to Soviet occupation of the Baltic states. About 1.5 million people from all three nations participated in this peaceful resistance, which drew international media attention to the Baltic states and their quest for independence.

Until 1990, however, Latvian foreign policy can be described more as the ideas and informal activities of LPF members than as a consistent foreign policy. Basic state factors like national government, independent economy and national territory were still absent. The situation started to change after the Popular Front received a majority of seats in the Supreme Council and declared a transition period towards independence on May 4, 1990. During the period from the declaration of independence on May 4, 1990, to August 1991 (the day of actual independence), Latvian politics had two goals within the frame of international relations. First, the newly elected authorities attempted to negotiate national independence with Soviet authorities. Second, they initiated the process that would eventually achieve international recognition. Latvians soon realised that they were in a Catch-22 situation. Western countries refused to recognise Latvia before Moscow did. Moscow, however, refused to discuss independence with Latvia *per se*. Additionally, until mid-1991, Latvia did not have a functioning Ministry of Foreign Affairs.

In July 1990, the Latvian Council of Ministers established a Ministry of Foreign Affairs with a staff of thirty-two people. The facilitation of *de facto* independence was declared as the Ministry's main goal.[7] In the transition period towards independence, as far as foreign relations are concerned, the best co-operation existed between Latvia and the Nordic countries, as well as with the other two Baltic States. A common political achievement inspired by the Nordic countries was the opening of Baltic information bureaus in several Nordic capitals at the end of 1990.[8] After the re-establishment of independence, these bureaus served as Latvian, Estonian and Lithuanian embassies. Also, the Nordic Foreign Ministers met with their Baltic counterparts at an official level. Likewise, Baltic representatives were frequent guests at the meetings of the Nordic Council of Ministers. No other Soviet republic enjoyed similar support.

The first foreign policy agreements of Soviet-ruled Latvia were made with the two other Baltic countries experiencing the same political developments as Latvia. On April 12, 1990, the Agreement of Economic Co-operation between Latvia, Estonia and Lithuania was signed.[9] The Baltic States also adopted the Declaration on Unity and Co-operation which renewed the Treaty on Unity and Co-operation among the three Soviet-occupied nations. The Treaty had legally been in force since 1934. The Council of the Baltic States was also created. Before 1991, however, there was no official opinion concerning Latvia's international orientation after regaining independence. The Latvian authorities were not willing to challenge Moscow before real independence was attained, and thus all of their efforts were aimed at providing a peaceful transition to independent statehood.[10]

DEVELOPMENT OF FOREIGN POLICY AFTER 1991

After the abortive Moscow coup in August 1991, the Latvian government was quite unprepared to deal with foreign policy strategy and duties. After regaining independence, Latvia, Estonia and Lithuania were poorly prepared for their new roles as independent international actors. Unlike Eastern and Central European countries exiting the Soviet bloc, Latvians had limited diplomatic experience as well as very limited personnel and material resources. In foreign and security policy, Latvia started from scratch.[11] Additionally, compared to other former Soviet satellites, Latvia had weaker domestic state institutions, less international recognisability, absence of a national economy and the presence of a large minority of Russians strongly oriented towards Russia. Latvia's immediate problems were, therefore, international recognition, the establishment of diplomatic contacts and representation in the appropriate international forums—the UN, the OSCE, Council of Europe, the EU and NATO. Other Soviet bloc states had, at least, UN and Council of Europe representation. Latvia, on the other hand, had too many un-adopted acts, non-created institutions and systems; all of these were needed immediately. The Baltic politicians were also unprepared for independent foreign policy making. Despite the term "re-emergence" in international politics, Latvia faced problems similar to those countries that are newcomers in international politics (like the other Soviet republics). From the very beginning, however, Latvian diplomacy refused to compare itself with the rest of the former Soviet Union, for several reasons. First, Latvian officials did not want to undermine the already weak continuity between the first republic and the renewed Latvia. Second, Latvia wanted to gain international recognition of the occupation in 1940. Third, Latvia's ambitions were much larger than those of other former Soviet republics because of its past independence. Fourth, politicians and people wanted to detach themselves from the dishonourable and violently imposed Soviet past and identity.[12]

In the international sphere, Latvians saw themselves as closer to their Eastern and Central European nations recently freed from Soviet dominance rather than as one of the Soviet republics. In certain respects the parallel with the post-communist states in Eastern and Central Europe is valid. Just after the collapse of the Soviet Union, all of these countries shared a common foreign policy, to become a part of the rest of Europe by establishing strong political, economic, social and cultural links with Western Europe.

The withdrawal of Russian troops had the highest priority in Latvian foreign policy during the first years of independence. In 1991, there were between 500 and 600 military installations of varying sizes and importance in Latvia. More than 51,000 Soviet troops were stationed in Latvia. Riga was the headquarters of the Baltic Military District.[13] By the spring of 1992, it was clear that getting the Russian troops to leave would prove more difficult than originally hoped. The head of the Russian delegation categorically denied admitting that Latvia was occupied in 1940, and tied the withdrawal of troops to other requirements such as the granting of citizenship to all Soviet immigrants living in Latvia.[14] Russia hoped to keep the intelligence-gathering installation near Ventspils, the phased array at Skrunda and the naval harbour in Liepaja for several years. The Latvian authorities refused most of these demands. The transition period for the Skrunda radar station, however, was permitted. The final agreement allowed some base functioning until August 1998. The last person serving at Skrunda left at the end of 1999.

Success with Russian troop withdrawal was only reached after Latvians succeeded in drawing international attention to the issue and thereby gaining considerable support from Western sources. In

November 1992, the UN General Assembly passed a resolution requiring the withdrawal of foreign troops from the Baltic states.[15] The troop withdrawal was completed in 1994, three years after Latvia was declared to be an independent country. Foreign, especially American, assistance played an enormous role in the negotiations between Latvia and Russia. Latvian politicians were not experienced in foreign politics, and they lacked a source of bargaining, namely power. "Empty-handed" diplomacy continues to be the most common characteristic of Latvian diplomacy and it frequently determines diplomatic failures or successes. If an issue becomes important for some of the world powers, Latvian goals might succeed. Otherwise, Latvia has only a slight chance for reaching its foreign policy goal. In the case of Russian troop withdrawal, the issue was of particular interest to its Nordic neighbours and the European Union as well as to the United States who intensively assisted Latvian aspirations.

MAJOR TRENDS IN LATVIAN FOREIGN POLICY STRATEGY

Even though Latvia, like Estonia and Lithuania, is continuously eager to emphasise its European rather than Soviet heritage, its foreign policy strategy included this approach only several years after regaining independence. Only in 1995 did Latvian authorities adopt a statement clarifying the main foreign policy goals fot the period to 2005.[16] The main goal is strengthening the country's independence so that it is never lost again. The statement argues that this goal can only be reached through integration into European security, political and economic structures as well as through active participation in global political processes. The statement formalised Latvia's long-standing aspirations for possible future membership in the European Union and NATO. A domestic basis for this strategy was distinguished. The development of parliamentarian democracy, domestic stability and sustained market economic reforms were defined as the domestic foundations of this strategy.

The foreign policy document argued that membership in the EU would realistically secure the survival of the Latvian nation and state. Therefore, relations with the European Union member states were stressed as a priority of foreign policy. At the same time political and economic co-operation among the Baltic countries, as well as with countries such as the Nordic states, Poland and Germany, were stressed. Relations with the United States of America were also declared to be a priority of foreign policy. As far as relations with Russia were concerned, the policy statement declared that Latvian interests lay in the development of a normal relationship based on internationally accepted legal norms and obligations. This foreign policy was based on the assumption that regained Latvian independence was not irreversible. Despite the end of the Cold War and the collapse of the Soviet Union, Baltic region security still had to be defined. Understanding Latvia's security concerns requires some insight into Latvia's relations with its neighbours. This explains the possible reasons for Latvian nervousness about their national security. Latvia's relations with various countries and international organisations need to be analysed according to their declared and actual degree of importance to Latvia.

Following the classification of national interests elaborated by RAND researcher James Thomson,[17] Latvia's interests can be divided into:

- vital interests;
- essential interests;
- general interests.

Vital interests are defined as the international events that have an immediate and radical influence on a country's security perspectives and political, economic or social well-being. In the Latvian case it would be equal to the threat of losing national independence by leaving the country in a security vacuum. Essential interests are determined by international developments that might radically influence the political environment around the particular country, undermine its international position and have reasonable influence on domestic politics. In turn, general interests are determined by activities that influence a country as a regional or global actor. Offences against these interests have either long-term or minor effects on the country.

For Latvia, vital interests can be defined as interests that guarantee the independence and security of the country. Latvia's foreign policy statement declares that its vital interests can be reached through integration into the European Union and NATO. Consequently, Latvia's vital interests are connected with a strong US presence in Europe, and a stable and democratic Russia without geopolitical and territorial claims.

Latvia's essential interests are the development of good bilateral relationships with the countries of the Baltic Sea region, active participation in the Baltic and North European region, and co-operation with Germany and Russia.

Latvia's general interests are the stability, prosperity and peacefulness of Europe (including the territory of the former Soviet Union), the increase of the role of small states in the international environment, and collaboration with the Latvian Diaspora as an important lobby for the Baltic region. After determining this initial framework, let us examine how Latvia's particular relations with neighbouring countries promotes or hinders the implementation of these interests.

RELATIONS WITH BALTIC NEIGHBOURS

Latvia's relations and co-operation with its closest neighbours, Estonia and Lithuania, can be described as good but still underdeveloped. Baltic co-operation has a historically rooted sentiment and is based on many common trends in their socio-political experiences. Ethnically, Latvians share the same origin as Lithuanians, being the only two nations of the Baltic linguistic group within the Indo-European group. Historically, however, Lithuanians are closer to Poland, while Estonians and Latvians share Germanic and Scandinavian influences. Apart from these long-term historical, cultural or linguistic ties, Baltic co-operation has developed most strongly when all three nations faced a common threat to their national identity or security.

The first attempts to deepen political and military co-operation among the Baltic states goes back to the first years of the interwar period when all three Baltic nations were establishing their states and fighting common enemies. The search for Baltic unity started as early as 1918 and 1920, when Baltic military and political leaders met to discuss mutual concerns aimed at winning independence from the Soviet Union.[18] Despite several attempts, there was no real success in creating Baltic unity until the mid-thirties when in 1934, the Baltic Entente was established. The Baltic Entente agreement provided for periodic conferences and set aside special problems.[19] Despite the good ideas behind it, the Baltic Entente never really became an institution of deep political, military and/or economic co-operation among the Baltic states similar to those of the Nordic states and Benelux co-operation. All three countries at times lacked the will to overcome their national ambitions and problems for the sake of broader Baltic co-operation. Without a doubt, Latvia can be described as the most ardent promoter of Baltic unity in the interwar period.[20] The attempts, however, came to an end with the Soviet occupation of the Baltic Republics in 1940.

Baltic co-operation saw its revival once more during the Baltic struggle for independence in the late 1980s, reaching its zenith during the famous (and above-mentioned) Baltic Chain. In 1991, following the interwar experience, the leadership of all three countries established the Baltic Assembly, composed of twenty deputies from each of the three parliaments. The first Assembly session took place in Riga in 1992. On October 31, 1993, the Baltic Council and the Council of Ministers of the Baltic states were created.[21] The purpose of the Baltic Council is to facilitate trilateral co-operation between the legislative and executive branches of the three states. The Council of Ministers, which started to function in June 1994, is to oversee and implement inter-governmental and regional co-operation.[22] On August 13, 1999, the Secretariat of the Baltic Assembly was established in Riga.

In the first years of its existence, the Baltic Assembly directed its activities towards troop withdrawal from all three countries. After the Russian troop withdrawal from Latvia, Estonia and Lithuania, one of the major Baltic co-operative efforts became the attempt to join NATO and improve Baltic security. In August 1993, all three Baltic presidents met in Latvia and agreed to work on the establishment of common peace-keeping units.[23] In 1994, the BALTBAT (Baltic Battalion) was established with strong support from the Nordic countries, Great Britain and the United States. In the second half of the nineties, Baltic peace-keeping troops served in missions in different parts of the former Yugoslavia. Baltic military co-operation has also developed in the spheres of naval and common air defence. Thus, Baltron, the joint naval squadron, was created as well as Baltnet, a regional air surveillance network centred in Kaunas, Lithuania, with direct assistance from the US. In 1999, the Baltic Military Academy was established in Tartu, Estonia to educate young officers from all three Baltic countries. Symbolically, the academy flag consists of the three colours of the Baltic nations as well as the blue colour of NATO.[24] Collaborating in matters of national defence, the Baltic countries are preparing themselves for admission into NATO.

Western support for Baltic military and defence co-operation should be understood in the context of support for Baltic reintegration in European and transatlantic organisations by making Baltic troops compatible with Western standards. Therefore military and security co-operation among the Baltic countries is difficult to assess without considering the Latvian–Nordic and Latvian–US components of such co-operation.

Contemporary Baltic co-operation also has its economic aspects. In November 1997, the Baltic countries signed an agreement on the abolition of non-tariff barriers in mutual trade.[25] Many unresolved economic problems, however, continue to exist between Latvia and the two other Baltic countries. In the relatively short, post-Soviet period, Latvia has already experienced a "herring war" and "pork war" with Estonia, as well as an "oil war" and "egg war" with Lithuania. The cross-border movement of labour is still unresolved, and the common Baltic market has not been established in spite of ardent proposals from the Latvian side particularly. In July 1999, Andris Skele, the Latvian Prime Minister, offered to improve economic and political co-operation between the states by developing a common Baltic border while diminishing the relevance of inter-Baltic borders. Such suggestions, however, do not always find support among Estonian and Lithuanian authorities. Undoubtedly, Latvia as the geographical centre could benefit the most from inter-Baltic co-operation. Thus, Latvians are most interested in Baltic co-operation.

The Russian crisis of 1998 pushed the Baltic states closer because of the lost eastern markets. In 1999, for the first time in post-Soviet history, Latvian trade with Estonia and Lithuania exceeded its trade with Russia and CIS countries.[26] Despite these impressive statistics, the Baltic nations frequently underestimate the need for Baltic economic and social co-operation in their eagerness to knock at the door of the European Union. Benefits gained from regional co-operation would translate in international circles as the Baltic nations' ability to make rational decisions and create successful foreign policy. Finally, Baltic integration would prepare the Baltic states for broader integration in European structures. There is, as of yet, no clear indication that the Baltic states can truly co-operate without reservation. Latvia and Estonia, for example, continue to have difficulty concluding acceptable border treaties with Russia. Unlike the matter of troop withdrawal, Western supporters cannot contribute much to Baltic attempts to solve this issue. In fact, why should they if Latvia and Estonia themselves cannot work in concert?

The similar nature of all three Baltic countries in the external political environment accounts for the solidarity in national security at times of political tension. Similar economies, however, frequently force the states into unneeded competition, instead of co-operation for the sake of the common good. Baltic co-operation is most fruitful in the sphere of military and security co-operation. Security aspects are the only spheres in which it is difficult to imagine contradictions or counter-productive competition among the Baltic countries. In the military sphere, Balts understood what they were unable to comprehend in the social and economic spheres: mutual co-operation can bring them closer to the European Union, just as military co-operation brings them closer to NATO.

Latvian–Estonian and Latvian–Lithuanian relations are frequently on the level of general interests. Baltic co-operation alone cannot solve the vital problems of any of the countries, but raising Baltic co-operation to the essential level of national interest would increase the Baltic states regional leverage. Thus, Baltic co-operation would serve Baltic essential interests. Presently, Baltic co-operation is implemented only on the general level, whereas common sense would suggest that Baltic co-operation is of essential interest to each of the Baltic states. Baltic hesitancy is probably due to policy makers' short-sighted estimation that the Baltic states are too uninfluential internationally, and that they are somehow competing with each other for similar positions in Europe and the world. Competition and short-sightedness should be replaced by co-operation and long-term political planning within Baltic relations.

NORDIC COUNTRIES AND BALTIC SEA REGION CO-OPERATION

According to the national foreign policy statement, strengthening Latvian–Nordic co-operation is one of the main directives of Latvian foreign policy. Within the Baltic Sea Region (BSR), increasing integration is taking place. The Scandinavian countries were the first nations to recognise Baltic independence in 1991. Before actual independence, Latvia, as well as Estonia and Lithuania, signed a co-operation agreement with Denmark.[27] The Baltic nations gained friends and ardent supporters among such influential Scandinavian politicians as Carl Bildt, Uffe Ellemann-Jensen and many others. From 1988, Denmark and Iceland were the most passionate advocates and supporters of Baltic independence.[28] This support was manifested on two levels. First, Denmark and Iceland supported the Baltic states in various international contexts. Second, these countries promoted Baltic access to such international organisations as CSCE (OSCE). In fact, the original Baltic foreign policy strategies went hand in hand with the interests of their Scandinavian neighbours. Finland, Sweden, and to a lesser degree Norway, combined pro-Baltic sentiments with an interest in not undermining the economic or security issues of the Nordic–Soviet relationship. If, in the last pre-independence years, Nordic unity towards the Baltic states showed differences of opinion,[29] the situation changed after the Baltic countries regained their independence. The question was no longer how to support the Baltic states without infringing on the fragile balance in the political game in Moscow.[30] The new situation allowed Sweden, and especially Finland, to pursue a much more open policy towards their Baltic neighbours.

Nordic countries, after Great Britain, were the first countries to allow visa-free travel to Baltic citizens. Scandinavian countries were also the first countries to offer various educational exchange programmes for Baltic scholars, students and politicians in order to integrate them as soon as possible into European society. Economic support and other initiatives of the Nordic countries are increasingly channelled to Latvia and the other two Baltic countries through bilateral or multilateral channels, particularly through Nordic and EU programmes. Western military support to the Baltic countries also has a significant Nordic component. Latvia, along with the other two Baltic states, is declared to be in the sphere of *essential interests* to the Nordic countries.[31] Nordic interest in the political developments of the Baltic countries can be understood in conjunction with the domestic policies of the Nordic countries themselves. The Danish, Swedish, Finnish and Norwegian political actors base their attitudes towards Latvia on the assumption that the success or failure of the Baltic states to improve their national security, strengthen market economies and develop democracy will have direct implications on Nordic policies. Therefore, relations with the Baltic states are on the priority list of Nordic foreign policy agendas. In March 1992, Denmark with the support of Germany evolved an alternative strategy where instead of Nordic–Baltic co-operation the scheme of the Council of the Baltic Sea States (CBSS) was offered in Copenhagen.[32]

Today, the Council of the Baltic Sea States consists of Denmark, Sweden, Iceland, Norway, Finland, Latvia, Lithuania, Estonia, Poland, Germany and Russia. Initially, the CBSS tried to focus on non-controversial issues such as environmental protection, the sharing of information and knowledge, regional co-operation and the development of civil society.[33] Issues such as regional security, troop withdrawal from the Baltic, and the rights of Russian speakers were occasionally addressed. The CBSS was one of the first international organisations in which Latvia was an equal, international member. In 1997, Latvia acted as president of the CBSS, and in January of 1998, Riga hosted the CBSS ministerial meeting that brought Germany's Chancellor Helmut Kohl and Russia's Prime Minister Victor Chernomyrdin to Latvia. Neither of these politicians had previously visited Latvia. The CBSS appeared to be a moderating body able to influence relationships between countries of the Baltic Sea region. The ministerial meeting, for example, dealt with energy, crime and regional co-operation—issues important for the Baltic countries and Latvia in particular.

Baltic Sea regional co-operation is important for Latvia because of its political and economic perspectives. First, the regional population is roughly 80 million people and it is one of the most vital economic regions of Europe. Second, regional co-operation brings together nations from previously different political systems that after the end of the Cold War are pursuing similar goals. The Baltic nations are geographically at the heart of this region, giving them a chance to prove their international value. Ojars Kalnins, the former Latvian ambassador to the United States, proposed the idea of an *Amber Gateway*, thus attempting to define Latvia's role in the new international and regional context. As Dzintra Bungs argues,[34] Latvia can be defined as the only Baltic country because Estonia is tightening its relationship with Finland, and Lithuania is approaching Poland and Central Europe. Therefore, Latvia's orientation and identity are frequently questioned. Latvians feel close to both Baltic countries, but they can identify neither with Central Europe nor with Finland. Latvian identity lies within the Baltic Sea region and this aspect is the core idea of the *Amber Gateway*. The project proposes a Latvian initiative to increase Baltic–Nordic, Baltic–Baltic region and EU–Russian co-operation in which Latvia and the other Baltic states would play an important role. The *Amber Gateway* is an ambitious plan to turn the negative aspects of the strategic importance of the Baltic states into a positive advantage. The *Amber Gateway* project, if implemented, would also be advantageous from the perspective of "soft security". Regional interdependence between the Baltic and Nordic countries, and potentially Germany as well, would increase. It would also provide Russia with a better chance to co-operate with the expanding European Union while co-operating with the Baltic States on an equal basis. Similarly, US–Nordic co-operation in security issues would give economic and political support for the whole North European security architecture. Finally, regional co-operation would benefit the European Union itself, and its regional development projects. At present, however, the *Amber Gateway* is more an idea than a reality owing to the inactivity of the Latvian bureaucracy, and a lack of support for the Latvian initiative from Estonia and Lithuania.

Baltic–Nordic and Baltic Sea regional co-operation is an important aspect of Latvian foreign policy that is still relatively undeveloped. The economic, cultural, social and security advantages of regional co-operation could be multiplied if a more active policy from the member countries were to be followed. Increasing regional activity would be one of the few arenas in which longer-term Latvian policy could avoid "empty-handed" diplomacy and offer its partners something tangible. For example, by increasing regional interdependence the Baltic countries, or at least Latvia and Estonia, could become equal partners with other Nordic countries. This would place the Baltic states as vital Nordic interests, instead of simply essential ones. Greater regional co-operation might also improve Latvian–Russian relations. Such co-operation could also increase Latvian (Baltic) value in the eyes of the Americans who already possess geopolitical interests in the North European region. Finally, successful regional co-operation could facilitate German interest in the Baltic region.

GERMANY, PRO OR VERSUS LATVIA

According to the Latvian foreign policy statement, intensive political, security and economic co-operation with Germany is among Latvia's political priorities.[35] From the Latvian side, priority to Latvian–German relations is due to Germany's leading role in the expansion of the EU, and because of its leading economic and political role in Europe generally. Baltic–German and Latvian–German relations experienced a renaissance after the Baltic states regained independence in 1991. In September 1991, the German Foreign Minister, Hans-Dietrich Gen-scher, visited the Baltic countries and expressed Germany's support for the once again independent nations.[36] In 1992, Genscher with Uffe Ellemann-Jensen proposed the creation of the CBSS, which granted the Baltic states new international opportunities. In 1994, the German Foreign Minister Klaus Kinkel declared that Germany had assumed the role of Baltic advocate in the EU.[37] Despite the promises of high-ranking German officials, the reality of Latvian–German relations seems somewhat greyer.

Germany does actively support Latvia, and the two other Baltic states, as far as market reforms and monetary assistance and expertise are concerned, which no doubt helps Latvian integration into European structures. Thus, from 1992 to 1998, about 245 million German marks had been allocated to PHARE and other EU aid programmes within the Baltic states.[38] Germany is also one of Latvia's major trade partners, particularly after the Russian economic crisis of 1998 forced many Latvian enterprises to reorient to markets other than Russia. In 1999, for example, Laima, the largest Latvian candy producer, exported more than 22 per cent of its production to Germany, compared to 5 per cent in 1988.[39] Compared to Nordic support to the Baltic, however, Germany looks quite different. Germany's assistance is primarily cultural and economic, frequently avoiding political support. Second, German assistance is channelled through regional co-operation programmes. Latvia receives most of its German assistance from the *Laender*. On the national level, Germany has been hesitant to grant support to the Baltic States or Latvia. Germany was one of the last European countries granting a visa-free regime for Baltic citizens visiting Germany. Compared to Baltic–Nordic relations, where prime ministers, presidents and members of royal families visit the Baltic states more than once, high-ranking German officials seldom are guests in the Baltic states (except in the aforementioned framework of the CBSS when Kohl visited Latvia). The German President, Roman Hercog, visited Latvia only in May 1999, the last month of his office.

Germany's priority, apparently, is relations with Russia, as well as EU and NATO expansion to its nearest neighbours: Poland, Czech Republic and Hungary. Germany's well-being and security are directly connected with its Central European neighbours; the Baltic states do not play a crucial role. Central European affairs are vital interests to Germany, while relations with Russia are of essential importance. The Baltic states are generally important for Germany, but as Lothar Ruhl, the former state secretary of the German Defence Ministry, claimed, Germany is following the formula "neither isolate nor provoke Russia".[40] Taking into account the present Baltic–Russian relationship, it is difficult for Germany to claim to be a Baltic advocate and maintain the above-mentioned formula.

LATVIAN–RUSSIAN RELATIONS: WHAT ABOUT THE FUTURE?

Latvian–Russian relations inherited the enormous burden of political mistrust owing to the events of the past fifty years, which saw the Latvian state effectively deconstructed, and Latvian society decapitated through deportations and numerous mass repressions organised by the

Soviet regime. Since the restoration of independence, relations between Latvia and Russia can be described as cool and sometimes even unfriendly. The withdrawal of Russian troops was Latvia's main foreign policy goal in its relations with Russia during the first years of independence. Russia attempted to tie troop withdrawal to citizenship rights for former Soviet immigrants in Latvia.[41] Bilateral negotiations between Latvia and Russia started in 1992. Under the Helsinki agreement of 1992, Russia agreed that the two questions of Latvia's "discriminatory" policy against Russians, and troop withdrawal should not be linked. Russia was obliged to withdraw troops unconditionally.[42] The mainstream Western position was to demand quick Russian troop withdrawal from the Baltic, and strongly oppose any linkage between troop withdrawal and Latvian or Estonian citizenship policies. Moreover, the US Senate voted on July 1, 1992, to tie American aid to Russian troop withdrawal from the Baltic. The bill stipulated that US aid would be restricted to humanitarian aid if within a year Russian troop withdrawal had not made significant progress.[43]

Latvia was also bound by international obligations to guarantee the human rights of the former Soviet citizens. This, however, did not include the automatic provision of Latvian citizenship to Soviet immigrants. David Atkinson, the Chairman of a Committee of the Council of Europe pointed out, while recommending that the Baltic countries become members of the European Council, that it was impossible to grant citizenship automatically to all those who came to a country in the period of occupation. Such blanket citizenship would be "unparalleled in history".[44] In order to monitor the political developments concerning Soviet immigrants in Latvia, however, CSCE (OSCE) and the Latvian government agreed to host an OSCE Mission, which opened in November 1993. The OSCE presence has multiple implications on Latvian politics. First, it can counter unreasonable Russian claims about human rights abuses by providing impartial expertise. Second, it assists the Latvian legislature to draft various minority/human rights related laws. Third, it serves as a mediating institution between the ethnic communities of Latvia. Fourth, the OSCE itself is an international organisation that fosters Latvia's international participation. Through the structures of the OSCE, Latvian observers themselves participated in missions in the former Yugoslavia.

Another important foreign policy issue for Latvia was the border treaty with Russia. Unlike foreign troop withdrawal, at the end of 1999

Latvia (like Estonia) had still not resolved the issue of its Russian border. According to Aivars Stranga, the border treaties are not yet signed because of at least two miscalculations on the Latvian side.[45] First, Latvia (like Estonia) initially insisted on references to the Peace Treaty of 1920 between Latvia and Russia. The reference would give Latvia a claim for territorial reunification with the region of Abrene, which was stripped from Latvian territory without Latvian consent in 1944. Similar territorial problems affect Estonian–Russian relations. Second, Latvia and Estonia did not co-ordinate their efforts to find a common compromise with Russia in their negotiations. Thus, Russia's policy towards the Baltic states could follow the formula "divide and rule". Third, the domestic constellation of power in Latvia frequently favoured nationalist parties whose foreign policy concepts were frequently based on unrealistic assumptions. As long as the borders between Russia, Latvia and Estonia remain undefined, Russia can fulfil two goals: obstruct Latvia's and Estonia's progress towards the EU, and facilitate contradictions between the two countries.[46]

The contemporary Latvian foreign policy statement suggests a normalisation of relations with Russia that would create a mutually acceptable and beneficial international environment based on international rights and observance of international obligations.[47] The practical arrangements of these goals are still far from coming true, due to several reasons on both sides. As far as the Latvian side is concerned, as Stranga observed, Baltic politicians need to learn to work as a team when common interests are at stake. Latvia also suffers from an inability to overcome its bitter historic experiences and formulate its policy to an agenda that subdues all other interests in favour of the main goal—membership in the EU and NATO. Latvian policy makers must understand that the border treaty depends on their negotiation skills with Russians, not the use of the West as mediating forces. Finally, the improvement of Latvian–Russian relations is very dependent on changes in the Russian attitude towards the Baltic states in general.

Since the beginning of Latvian independence in 1991, Russian diplomacy has stressed that the Baltic region is of vital interest to Russia and has opposed Latvia's integration into Western security structures. Both of these decisions are dictated by the Russian wish to avoid the formation of a uni-polar world with the United States as the major political actor. Russia also hoped to maintain Russian influence in the former territory of the Soviet Union. According to Russian foreign policy guidelines published in February 1997, Russia's strategic objectives were considered threatened if the Baltic states joined NATO.[48] Russian opposition to Baltic membership in NATO was repeated in the foreign policy analysis in October 1997 prepared by a group of senior Russian politicians, academics and businessmen.[49]

Besides Russia's geopolitical objections to NATO expansion in the Baltic states, Russian diplomats argue that minority issues and ethnic tension in the Baltic states is a reason why the West should stay out of the Baltic. In order to add weight to this reasoning, Russian authorities always stress their particular interest in the fate of "Russian speakers" in the Baltic. This is meant as a warning to Western governments about including the Baltic in NATO; Russia suggests that NATO membership would be a package of problems. First, Russia would consider it an offence to its own national interests. Second, Russia would be able to destabilise Baltic domestic politics through the sizeable Russian minority.

Russian policy concerning possible NATO expansion in the Baltic states is based on a combination of arguments of fear and reason. Are these arguments well grounded and convincing? As far as ethnic tension is concerned, Russia will always be able to organise a demonstration of a few hundred or thousand of Russia's supporters in Riga or Tallinn. It is doubtful, however, that it would cause more than a serious threat to regional stability. All the while, continuing Baltic integration into the European Union, improving economies compared with the economic disasters faced by Russia, and the ongoing building of civil society will with each passing year lessen the chance that Latvia's non-Latvian population would actively support Russia's strategic interests in the Baltic. The Baltic Russians will have nothing to win by supporting Russia, while they will gain increasing personal advantages in remaining loyal to their countries of residence.

Russia's second argument about a possible deterioration of Western–Russian relations in the case of NATO expansion in the Baltic is based upon the Russian assumption that the Baltic is a vital sphere of its interests. Russia might favour Baltic membership in the EU, while opposing membership in NATO.[50] Russia understands that it would benefit from a direct border with the EU. The Baltic states could promote co-operation between the European Union and Russia. In other words, Russia would like to have a rich neighbour, but one without a lock on its door.

Russia's definition of the Baltic as a region of its vital interests is ideological rather than strategic. Russian treatment of the Baltic as the "near abroad", official non-recognition of the occupation of the Baltic States in 1940, and difficulties in building relations based on equality and mutual respect suggest the psychological problems Russia has in accepting the independence of its former satellites of Latvia, Estonia and Lithuania. Carl Bildt observed that by using the term "near abroad" for the Baltic states, Russia is claiming rights beyond those in conformity with international law.[51] The Russian approach can also be illustrated by the phrase of the spokesman of the Russian president who, speaking about NATO enlargement in 1998, demanded that "Latvia ... at least should be loyal to Russia's interests."[52] This is but one example showing that as late as 1999, Russia was unable to break with its imperial past. Carl Bildt, the former Swedish Prime Minister, correctly observed that Russian policy towards the Baltic states is a litmus test for the country's wish to reform itself into a modern democracy.[53] A democratic Russia most probably would not claim the Baltic as within its vital sphere of interests. The situation did not improve after Putin's election to the presidency of Russia. Moreover, precisely after Putin's election there was an increase in the aggressive tone of Russia's diplomatic language towards Latvia, and the Baltic States in general.

In the optimal case, having good relations with Baltic countries will become Russia's vital interest. The best source of such good relations is mutual trust and common interest. While the old understanding of Russia as a dominating power in the Baltic region continues to rule the minds of the majority of Russian politicians, mutual trust cannot develop. As the Latvian analyst Aivars Stranga commented, Russian policy is based on its old-fashioned "zero-sum" game application in Baltic politics.[54] In other words, at the end of the millennium Russia had not exchanged the idea of geopolitical competition for a co-operative peace strategy.

Russian rhetoric concerning vital interests, therefore, is based on a desperate attempt to oppose increasing US influence in Europe in general, and in Northern Europe and the Baltic in particular. Nevertheless, the economic benefits of Western aid needed to be guaranteed. In the words of Dmitry Trenin, especially after the Kosovo crisis, Russia distinguishes between the "good West" (European Union) and the "bad West" (United States and NATO).[55] Therefore, Baltic–Russian relations are dependent on two mutually intertwined factors: the Baltic states' ability to develop a dialogue with Russia, and the growing encroachment of the EU and NATO in the Baltic.

LATVIA, THE UNITED STATES AND NATO EXPANSION IN THE BALTIC

Latvians, like other Balts, have justifiably looked to the United States as an active advocate of their integration into the international sphere, as well as a serious security guarantee for Latvia. This evaluation of the US–Latvian partnership has historical and contemporary reasons. After the Soviet occupation of the Baltic states in 1940, the US Administration, resorting to the Stimson Doctrine, promised never to acknowledge the annexation. Post-war administrations stuck to this decision and reiterated Latvia's right to independence.[56]

After independence was regained, President Bill Clinton visited Latvia in 1994 and gave a public speech at the Monument of Freedom in the heart of Riga. This occasion could be compared to President Kennedy's visit to Berlin during the Cold War era, and it confirmed to the Balts that Americans have a strategic interest in Baltic development. As a result, European integration and the presence of the United States in Europe are two major factors that provide security and stability on the continent.[57] After the collapse of the Soviet Union, the United States of America retained its major role in Europe, and became the only world power. Baltic attempts to gain US support and to establish a permanent American presence in the Baltic are an understandable component of Latvia's foreign policy. The Latvian, as well as Baltic, leadership understands that the Americans may have the decisive voice that opens or closes the NATO door to the Balts. Latvians, Estonians and Lithuanians consider membership in NATO as the only guarantee of their political independence and security.

American policy has contributed to creating such an image among the Balts. Indeed, since 1991, the United States has played a critical role in helping the Baltic states implement democratic and free market reforms, and strengthen their security and sovereignty. The United States was crucial during Russian troop withdrawal from Latvia, Estonia and Lithuania from 1991 to 1994. Over the first six years of independence, the United States provided the Baltic states with over 136 million dollars through the Support for East European Democracy Program (SEED). In 1994, the United States established the Baltic–American Enterprise Fund, capitalised at 50 million dollars, to promote the growth of small and medium-sized businesses in these countries. More than 500 Americans have served in the Peace Corps in the Baltic states since 1992. In 1998, the Baltic–American Partnership Fund was created with equal contributions from the US government and the Soros Open Society Institute. The Fund will have 15 million dollars for the promotion of civil society and non-governmental organisations in Latvia, Estonia and Lithuania. This litany of aid and assistance visibly demonstrates an American presence in the Baltic states.

The United States has also helped Latvia, Estonia and Lithuania to meet NATO membership requirements. According to Kaufman, the US has tried to apply its general foreign policy principles to its relations with the Baltic countries. The Clinton Administration's basic strategy document, *A National Security Strategy of Engagement and Enlargement*, consists of three main components. The first plank is a combination of strong defensive capabilities for the promotion of co-operative security measures. The second plank calls for open foreign trade markets to spur global economic growth. The third is the promotion of democracy. The American understanding of the Baltic region is similar to the Scandinavian view expressed by Carl Bilt, Niels Helveg Petersen and Uffe Ellemann Jensen, that "stable, democratic, prosperous and secure Baltic nations are key to regional peace".[58]

Despite the United States' general support for the Baltic states concerning NATO, this support is not guaranteed. Within American politics there are those for, and those against, Baltic membership in NATO. The "pro-NATO camp" is associated with Madeleine Albright, Ronald Asmus and Zbigniew Brezinski. Among the more "careful" Baltic supporters is William Perry who in September 1996 announced that the Baltic states are "not yet" ready for NATO membership. Although Perry carefully emphasised that "not yet" did not mean "never", the United States and NATO officials have repeatedly turned down Baltic requests for an announcement on which countries would eventually receive NATO invitations. This hesitancy has aggravated Baltic fears of being left out.[59] Latvian and Baltic diplomats reacted understandably after the Czech Republic, Poland and Hungary were announced as the first possible members of NATO. Povilas Gylys, the Lithuanian Foreign Minister, summarised Baltic fears that political instability is imaginable in the region owing to the possible emergence of a security vacuum or so-called grey zones. This geopolitical situation reminds Balts uneasily of pre-WWII Europe. The Baltic states perceive such a possibility as their main security risk.[60]

To diminish such Baltic fears, the US began the creation of a Baltic–American action plan in 1996. In mid-1997, the US began work on *A Charter of Partnership among the United States of America and the Republic of Estonia, Republic of Latvia, and Republic of Lithuania*.[61] In October 1997, while commenting on NATO enlargement, Deputy

Assistant Secretary of State Ronald Asmus said that Europe would not be secure until this part of the world was secure.[62] The Presidents of the United States and all three Baltic states signed the Charter of Partnership on January 16, 1998. The Charter made clear that the Baltic states are a part of the US vision for a new Europe, and that the Baltic states would not be left outside NATO or discriminated against owing to their history or geography. The Charter also noted that the United States has a "real, profound, and enduring" interest in the security and independence of the Baltic states.[63] The Charter, however, does not pre-commit the United States to Baltic membership in NATO, nor does it offer back-door security guarantees. Nevertheless, the Charter can be considered a considerable success in cementing the Baltic–US relationship. The Charter successfully made the Baltic states feel closer to the North Atlantic Treaty Organisation, and diminished their fears of being left in a grey zone of insecurity.

In April 1999, at a NATO Summit in Washington DC, the NATO leaders reaffirmed their enduring commitment to the Open Door policy. They also outlined, in the Membership Action Plan, what aspiring countries have to do to strengthen their candidacies. In the summit communiqué, the prospective new members were named in groups. The first group consisted of Romania and Slovenia, both countries almost accepted at the last NATO summit in Madrid in 1997. These countries have relatively important strategic positions, and draw strong support from France and Germany. The Baltic states were mentioned in the second group, which can be considered a Baltic success bearing in mind the fierce Russian opposition as well as the Baltic states' relatively small lobby group in international politics. From the NATO member countries, only Denmark and Iceland favoured Baltic membership. Danish international weight, however, is not equal to that of Germany or France.

In the Baltic countries, many hopes are placed on the NATO summit of 2002. Slovenia and Lithuania seem the most plausible future participants of NATO enlargement. By admitting them, NATO would expand in the southern as well as northern regions. Additionally, Lithuania is the most advanced among the Baltic states concerning military reforms and appropriate budgets. Lithuania is also the only Baltic state that has a border with a NATO state, Poland.

Western and Baltic politicians, however, have to bear in mind that NATO expansion into the Baltic would face initial, intense Russian opposition. On April 27, 1999, for example, Reuters reported that the

Russian Defence Minister Sergeyev said that Russia would never agree to the Baltic states joining NATO. He argued that such a move would be a great threat to Russia. Sergeyev also said that Russia would have to reshape its military doctrine accordingly by 1999. Bearing this in mind, many Western politicians and academics are guarded in their support of NATO enlargement in the Baltic. Thus, the analyst Kaufman argues that NATO's possible expansion into the Baltic is based on flawed US foreign policy that prefers a realist approach to co-operative peace measures in the region.[64] This approach argues that Europe and the world cannot create a lasting security system without Russia. This is common thinking among NATO member countries as well.[65] Western–Russian co-operation, however, cannot be based on the assumption that the Baltic states are the ultimate and vital sphere of Russian interest or are within Russia's sphere of control. This approach contradicts the modern values of international relations. Likewise, Baltic democracy and security are as vital an interest for Europe as are Russia's geopolitical claims over the region. The possible NATO expansion into Northeast Europe should not be seen as a policy that excludes Russia. EU and NATO enlargement would create several complex consequences for Russia. Kaufman's fundamental mistake, like other Western opponents of NATO enlargement, is his inability to see the potentially positive impact of NATO enlargement.

Stuart J. Kaufman argues that while US policy contains rhetoric about building co-operative peace on the European continent, the US follows the realist logic of geopolitical competition with Russia while keeping co-operative peace ideology in reserve for better times.[66] This is the only logical policy towards an unpredictable Russia, facing democratic, economic and identity crises. The Russian identity problem and its weak association with democratic values is one of the main reasons for Russian–Baltic tensions and Russia's opposition to NATO enlargement. Russia has to learn to see the Baltic states as a region of co-operation and integration instead of a source of insecurity.[67] The security of Russia cannot be built on the permanent insecurity of the Northeast European region. Such a high cost for security can be neither endurable, long lasting nor associated with a co-operative peace strategy. Baltic security, as well as North European security, cannot wait until democracy and stability become part of the Russian political tradition.

For political co-operation there must be at least two parties. Until Russia is able to change its attitude from the "zero-sum" game to co-operative strategy, the realist approach in the US foreign policy is the only feasible solution to Baltic as well as North European security dilemmas. While pursuing such a strategy, however, the West should not forget about the need to keep doors open for mutual co-operation with Russia. The West, as well as the Baltic states, should increasingly contribute to democratic reforms in Russia. Increasing US financial assistance to joint Baltic–Russian projects would be one means to this end. The United States should also promote the border agreement between Russia and Estonia and Latvia, as well as pressure Russia to denounce and apologise to the Baltic states for the occupation of 1940. In other words, the realist or stick policy should be complemented with co-operative or carrot policies towards Russia. Echoing the words of Danish Foreign Minister Niels Helveg Petersen, the EU and NATO are the two defining organisations of European security architecture. Only together can they provide long-lasting stability and peace.[68] NATO provides the realist portion of the co-operative strategy in step with the eastward expansion of the European Union.

EU ENLARGEMENT AND LATVIA

Since regaining independence in August 1991, Latvia has striven assiduously to become a full-fledged member of the European Union.[69] Latvia is motivated by two major factors to join the EU. First, EU membership is believed to provide Latvia with "soft" security guarantees. Second, Latvia would benefit from the economic expertise, trade and aid that the EU offers. Latvia's estimation of the benefits of membership in the EU goes hand in hand with the official philosophy of the European Union itself. According to Niels Helveg Petersen, the Danish foreign minister, the European Union was originally created to provide peace and security to Europe through integration, and political and economic co-operation. After the fall of the iron curtain, enlargement represented a historic opportunity to unite most of the European countries in one organisation. In June 1993, the European Council laid down the political and economic criteria to be met by applicant countries.[70] In the political sphere, the applicant countries must observe fundamental democratic rights. They should not have internal political instability or conflicts with their neighbours, and they must have a functioning market economy. Added to the Copenhagen criteria, the applicants must also fulfil the Madrid criteria, namely the establishment of a functioning state administrative system.

In Strasbourg in the summer of 1997, the European Commission recommended that the EU open accession negotiations with five former Eastern bloc countries: the Czech Republic, Estonia, Hungary, Poland and Slovenia. The announcement was part of *Agenda 2000: For a Stronger and Wider Union*, a comprehensive treatise on EU development into the 21st century.[71] Estonia was the first to be admitted to start negotiations because it was considered the most advanced of the Baltic countries. Estonia's success had a twofold influence on the other two Baltic countries. First, by including Estonia in pre-accession talks, EU authorities admitted politically that Estonia's past inclusion in the Soviet Union did not matter. This created an impression in the Baltic states that all three countries were finally viewed as equals with other Eastern and Central European post-communist states. As noted by Danish political scientist Nikolaj Petersen, this was not always the case.[72] Second, Estonian success underlined Latvian and Lithuanian domestic problems and facilitated regional competition in a positive sense. The inclusion of Estonia in the first round of talks, however, also had negative consequences for Latvia and Lithuania. Many Latvians felt that they had been rejected and left at the gates of the EU without any particular reason.[73]

In November 1998, Valdis Birkavs, Latvia's Foreign Minister, stressed that among the second group candidate countries, Latvia was the only country whose European Commission progress report was strong enough to warrant 1999 as a possible date for the beginning of negotiations about accession.[74] Latvian politicians widely expect that in December 1999, at the end of the Finnish EU presidency, an invitation will be forwarded to Latvia. Despite this optimism, Latvia must still proceed with reform in order to meet the Copenhagen and Madrid criteria. There are many unsolved social, political and economic issues that Latvia must address. Latvia's economy is still relatively small and weak; there are high levels of institutionalised corruption, weak border control, and human rights problems in prisons. Not meeting the Copenhagen and Madrid criteria, however, is not the sole determinant for inclusion in the EU. Those states that joined the EU in the 1980s, for example, did not fulfil all of these conditions. Geo-strategic aspects dominated earlier decisions. The Iberian countries, for example, have taken a long time to adjust to life in the EU, and Greece has arguably still not done so. In this sense the EU has already given itself one "weak underbelly". Some might say that Central and Eastern Europe would constitute another.[75]

In the existing European situation, Germany's will for enlargement appears to be the key factor.[76] Germany used its authority to promote the accession of Poland, Hungary and the Czech Republic to the European Union because these countries were considered vital to German interests. Additionally, the Czech Republic and Hungary are the most advanced former Eastern bloc candidates. In turn, Poland has sizeable markets, impressive recent economic performance and historical weight in securing Germany's support for EU membership. Germany is, however, sensitive about Russia. While Russia has not opposed Baltic membership in the EU, Germany may be hesitant to antagonise Russia. German support for Baltic membership is also dictated by economic factors within the EU, one of which is their considerable contribution to the EU budget in times of economic uncertainty. In EU enlargement in the Baltic, Baltic–Nordic co-operation plays a crucial role as well. Three of the five Nordic countries are members of the European Union, and thus the Baltic EU lobby has considerable strength. EU enlargement in the Baltic Sea region can be seen as a complex issue that ultimately could facilitate regional security, economic prosperity and increased co-operation with Russia.

CONCLUDING REMARKS ON LATVIA'S INTERNATIONAL POSITION

After 1991, the international system transformed from a two-superpower system into a one-superpower system. This implied a new security environment and political strategy for the Baltic states.[77] Under these circumstances, Latvia has been largely preoccupied with two tasks: first, to re-enter and define its role in the world of independent nations; second, to develop a mutually acceptable relationship with a neighbouring Russia. The main goal of contemporary Latvian foreign policy remains much the same as at the beginning of the twentieth century. It is to secure the irreversibility of independence.[78] Latvia's foreign policy analysts believe that this goal can be achieved through Latvia's integration into the EU and NATO. These two organisations would guarantee economic integration in a common Europe, as well as provide political security. Latvia's main goals of sovereignty and security can only be achieved through integration in European and global structures. Indeed, the paradox emerges that if Latvia and the two other Baltic countries are determined to secure their national independence, the best way is through regional and global integration. Only integration and interdependence can provide the "empty-handed" Latvian diplomats with bargaining power. The same holds for Russia. If Russia wishes to regain power and glory, it should orientate itself towards co-operation instead of dreaming about returns to imperial domination.

When NATO was established, the first Secretary General of NATO coined the simple shorthand for NATO's mission: "to keep the Russians out, the Americans in and the Germans down". International affairs specialist Heurlin now offers today's shorthand: "to keep Americans and Russians in, and Germans and the new members down".[79] Latvia's regional and strategic interests, however, would read: "to keep Americans and Nordic countries up and in, while keeping Russians and Germans down and in". Baltic security is largely dependent on regional security, and so the above-mentioned formula should apply to the rest of Europe.

The latest developments in the Baltic region have shown that the most considerable support for Latvia has come from their Nordic neighbours and the United States of America. North Europe is the region critical to the new US–Russian relationship. Simultaneously, Baltic security and prosperity are critical to the Nordic countries. American and Nordic countries are among the most ardent supporters of Baltic integration partly for ideological reasons, partly for their own national interests. Along with Nordic–American co-operation, Baltic co-operation is crucial for Latvia as well. Without deeper Baltic co-operation it is difficult to imagine broader regional integration. Latvian foreign policy should devote more attention to aspects of Baltic–Nordic and inter-Baltic co-operation. Without a doubt, integration into NATO and the EU is of utmost importance to Latvian prosperity and security. The president of Iceland, while visiting Latvia, however, outlined that Nordic–Baltic co-operation can be an influential tool in the future of European politics.[80] Latvian attempts to integrate into the European Union and NATO should be complemented with a conceptual strategy and political activities that also aim at a deeper regional integration.

Foreign policy will never be fully successful if it does not rest on sufficient and successful domestic government. To benefit from and to be successful with international co-operation and economic, political and military assistance, Latvia has to work on several domestic issues as well. First, through tireless work, Latvia must improve its relations with Russia on a regional and national level. Second, Latvia must accelerate the integration of its Soviet-era immigrants. Third, Latvia must make its defence capabilities compatible with NATO countries. Finally, Latvia must proceed with market reforms and the growth of democracy, including the strengthening of human rights and freedoms. These last aspects are enormously important because small countries contribute to the global society from an authority built on comparative wealth and individual freedom, not military power. Thus domestic policy is indivisibly intertwined with foreign policy aspirations. At the end of the twentieth century, Latvian integration into the EU and NATO is the first, and perhaps only, chance the country has had to make itself and the whole Baltic Sea region externally secure, internally free and prosperous. One need not be a prophet to claim that without Baltic security, European security is without guarantees as well.

1   A. Spekke, *History of Latvia* (Stockholm: Zelta Abele, 1951), p. 340.
2   The second paragraph of the 11 August 1920 peace treaty stated that "Russia without exception recognises the independence, freedom and sovereignty of the Latvian state and willingly and for all time disclaims any sovereign rights which may have belonged to Russia with respect to the Latvian people and land whether they be based on the judicial system of the former state, or on international agreements which for the purposes of this agreement shall be deemed to be null and void for all future time" (Duhanovs, Feldmanis, Stranga, 1994, pp. 8–9).
3   A.A. Pikayev, "Russia and the Baltic States: Challenges and Opportunities", in Birthe Hansen and Bertel Heurlin, eds, *The Baltic States in the World Politics* (Great Britain: Curzon Press, 1998), p. 133.
4   Vaares, p. 21.
5   *Latvijas Tautas Fronte Pirmais Gads*, pp. 3–23; Vaares, p. 21.
6   *Latvijas Tautas Fronte*, p. 4.
7   T. Jundzis, "Atgriesanas starptautiskaja aprite—atjaunotas Latvijas arpolitika", in U. Bormanis and V. Kanepa, eds., *Latvijas valsts atjaunosana 1986–1993* (Riga: LU zurnala Latvijas vesture fonds, 1998), p. 296.
8   Vaares, p. 28.
9   *Ibid.*, p. 22.
10 Jundzis, p. 297.
11 I. Faurby, "Baltic Security: Problems and Policies Since Independence", in N. Petersen, ed., *The Baltic States in International Politics* (Copenhagen: The Danish Institute for International Studies, 1993), p. 51.
12 N. Petersen, "Introduction", in N. Petersen, ed., *The Baltic States in International Politics* (Copenhagen: The Danish Institute for International Studies, 1993), p. 3.
13 Jundzis, p. 349; Faurby, p. 71.
14 Jundzis, p. 351.
15 *Ibid.*, pp. 353–354.
16 *Latvijas Vestnesis*, February 10, 1995.
17 Z. Ozolina, "Baltic-Nordic Interaction, Cooperation and Integration", in Atis Lejins and Z. Ozolina, eds., *Small States in a Turbulent Environment: The Baltic Perspective* (Riga: Latvijas arpolitikas instituta, 1997), p. 63.
18 Crowe, p. 7.
19 Crowe, p. 24; von Rauch, pp. 180–184.
20 Stranga and Feldmanis, pp. 85–86.
21 Dz. Bungs, *The Baltic States: Problems and Prospects of Membership in the European Union* (Germany: Stiftung Wissenschaft und Politik. Forschungsinstitut fur Internationale Politik und Sicherheit, 1998), pp. 93–95.
22 *Ibid.*
23 Sics, pp. 54–55.
24 *Diena*, August 24, 1999.

25 Bungs, p. 98.

26 *Diena*, August 24, 1999.

27 Bungs, p. 101.

28 O. Norgaard, "Danish Policies Towards the Baltic States", in Nikolaj Petersen, ed., *The Baltic States in International Politics* (Copenhagen: The Danish Institute of International Studies, 1993), pp. 156–157.

29 *Ibid.*

30 *Ibid.*, p. 159.

31 Ozolina, p. 63.

32 Norgaard, p. 160.

33 Bungs, p. 99.

34 *Ibid.*, pp. 92–93.

35 LAI, p. 41.

36 Samorodni, p. 136.

37 Bungs, p. 102.

38 *Ibid.*, p. 104.

39 *Diena*, August 17, 1999.

40 Bungs, p. 103.

41 Faurby, pp. 52–53.

42 *The Economist*, February 5, 1994; Zhuryari, p. 79.

43 A. Park, "Ethnicity and Independence: The Case of Estonia in Comparative Perspective", *Proceedings of the Estonian Academy of Sciences, Humanities and Social Sciences*, 44, 2 (1995), p. 318.

44 *Ibid.*

45 Stranga, pp. 199–200.

46 *Ibid.*, p. 201.

47 LAI, p. 38.

48 Bungs, p. 110.

49 *Nezavisimaja Gaseta, October 28, 1997.*

50 *Diena, August 18, 1999.*

51 K. Bildt, "Baltic Security and European Stability", in Atis Lejins and Pauls Apinis, eds., *The Baltic States on Their Way to the European Union, Security Aspects* (Riga: Latvijas arpolitikas instituta, 1995), pp. 44–45.

52 *Diena*, March 9, 1998.

53 Bildt, p. 51.

54 Stranga, p. 188.

55 *Diena*, August 18, 1999.

56 Vaares, p. 30.

57 Bungs, 1998, p. 121.

58 S.J. Kaufman, "The Baltic States in Post-Cold War U.S. Strategy", in Birthe Hansen and Bertel Heurlin, eds., *The Baltic States in the World Politics* (Great Britain: Curzon Press, 1998), p. 49.

59 S. Girnius, "Back in Europe, to Stay", *Transition*, 3, 6 (1997), p. 7.

60 P. Gylys, "Integration Policy of the Three Baltic States", in *The Baltic Dimension of European Integration. A Conference at Riga 24–25 August 1996* (Riga: Latvijas arpolitikas instituta and Royal Danish Embassy, 1996), p. 16.

61 Bungs, 1998, p. 122.

62 Bungs, 1998, p. 121.

63 The White House, October 19, 20, 21, 22, 1998.

64 Kaufman, pp. 46–65.

65 B. Heurlin, "NATO, Security, and the Baltic States: A New World, A New Security, A New NATO", in Birthe Hansen and Bertel Heurlin, eds., *The Baltic States in the World Politics* (Great Britain: Curzon Press, 1998), p. 76.

66 Kaufmann, p. 46.

67 LETA, April 20, 1998.

68 N.H. Petersen, "The Role of the European Union: 'Soft' Security?", in *The Baltic Dimension of*

European Integration. A Conference at Riga 24–25 August 1996 (Riga: LAI, Royal Danish Embassy, 1996), pp. 92–93.

69 Bungs, 1998, p. 7.
70 Petersen, p. 93.
71 Bungs, p. 9.
72 Petersen, p. 10.
73 Bungs, p. 9.
74 *Latvija un Eiropas Savieniba*, November, 1998.
75 H. Grabbe, "The EU's Enlargement Strategy" in *The Baltic Dimension of European Integration. A Conference at Riga 24–25 August 1996* (Riga: LAI and Royal Danish Embassy, 1996), p. 57.
76 *Ibid.*
77 Heurlin, p. 65.
78 *Latvijas Vestnesis*, February 10, 1995.
79 Heurlin, p. 72.
80 *Diena*, June 11, 1998.

Chapter 6

CONCLUSIONS

The post-independence hangover is often more dangerous to new states than the movement to achieve independence. Latvia is a particularly good case in point of the disappointments and disillusionment of a mass movement. At the apogee of the Latvian Popular Front's campaign for the reestablishment of independence, 94 per cent of the ethnic Latvian community supported them.[1] Many of those 94 per cent did not realise what they were creating, or what to expect from the wheels set in motion by the collapse of the Soviet Union. Nationalist aspirations and dreams of greater freedoms merged painlessly with the opulent prosperity that the West seemed to offer. The hangover came from the disparity between the hopes for the future in 1991, and the everyday difficulties of 1999. The performance of Latvia in a few short years, however, has been quite impressive. The ongoing challenge for Latvia is to involve its citizenry (and non-citizenry) in continued progress and tend to state responsibilities that do not return profits.

The Latvian economy nearly stumbled to a stop from 1991 to 1993, but has since rebounded, quite enthusiastically in some sectors. The economic indicators that Latvia has posted from 1996 would make much of the former Soviet Union jealous. Even the comparative ease with which Latvia struggled through the fallout of the Russian crisis of 1998 points to how far the country has distanced itself from the CIS. The transition from a command to market economy is firmly rooted in post-Soviet Latvian soil. The economy, however, continues to have weak points, many of which are structural difficulties that will be difficult to overcome. The agricultural sector lags behind the rest of the economy. Unemployment figures are creeping upwards and its geographical distribution shows economic depression in Latvia's east. Underemployment is also troubling. Who is unemployed, underemployed or poorly paid is even more troublesome. The poor economy of Latvia's east also translates as unemployed non-Latvians that weakens non-Latvian allegiance and identification with the Latvian state. The weaknesses of the economy can reopen the sores of ethnic politics. Latvian farmers disgruntled with agricultural collapse support Latvian national extremists, while unemployed Russians in the East support political forces for a return to communism and the imagined glory days of the Soviet Union.

Latvia's recent changes in citizenship laws have calmed tensions with the non-Latvian community. Latvia, however, must not assume that the issue is solved. The state should continue aggressive integration programmes to bring its nearly 700,000 non-citizens into Latvian citizenship. The danger now is that the Russian-speaking community is indifferent, not antagonistic, to the new state. Latvia needs to involve all of its residents in state–building and the future of the country. Involving citizens and residents does not mean compromising ethnic survival or assimilating minorities, but means an acceptance and trust in the people to create their own state.

The fundamental idea that the state should reflect society and not mould it is largely foreign to Latvia's historical experience. The "tyranny of antecedent circumstances" haunts contemporary Latvia with a lack of civil society and frequent reliance on the notion of *prikaz*, not procedure. *Prikazi* were simple executive orders that held little recourse to appeal in Tsarist Russia. Tsarist officials administered and ruled by what Lenin called approvingly the transmission belt system of government. The transmission belt carried a central decision to all other levers of government, which then implemented it without input or criticism. This approach to governance united the Ulmanis era, and Soviet and Nazi occupations. Government was one-way administration, not a nexus between popular local input and national goals. The Republic of Latvia at the turn of a new century is faced with the old challenge—to overcome the state's historic distrust of society and to create responsive local and national government, and not central administration in a new guise.

Bureaucrats and governments are not solely responsible for Latvia's penchant to slip into state control of society. Everyday people on the one hand believe in a larger power's control over every negative event, and on the other hand look for the salvation of a strong man. Society effectively tries to remove its own blame and/or responsibility. Conspiracy theories abound in contemporary Latvia even in explaining popular achievements. The growth of civil society and resistance to encroaching central control of society needs a critical public that accepts failure, and its part within failures. Without this element, the public blames one ambiguous entity, and looks to another to set thing straight.

Minorities, and not just ethnic ones, are easy first targets of a state turning on its citizens. In the interwar years, for example, doubts lingered about the loyalties of Baltic Germans, Russians, Poles and Jews. The apparent ability of ethnic parties to make and break coalition governments gave rise to the idea that the Latvian state was held hostage to minority demands, when minority rights gradually, yet continuously, eroded. Ethnic communities were not the only targets. The Social Democrats, for example, were seen almost as traitors because of their refusal to sing the national anthem on particular occasions and their stated long-term goal of socialism. Viewing the problems of the state through lenses coloured with ethnic minorities particularly, and minorities generally, is a mirage. The citizens of the state do not fall into such neat compartments, nor will they ever. The Latvian nationalist parties in the 1920s and in the 1990s did not and do not even receive the majority of ethnic Latvian votes. Similarly, minority communities do not vote as a unified bloc. Economic interests, moral issues, geography, sex, age and a host of other variables influence all citizens of the state and blur the political distinctions between citizens. A forced, bi-polar view of society, the us vs. them paradigm, seems to be a natural human response, but it is wrong. The undermining of minority rights is simply the undermining of all citizens' rights, and a step towards the return of authoritarianism and central control.

The ethnic collage of Latvia is almost a mosaic. The strong ethnic Latvian majority of the interwar years was as artificial as the strong Russian minority is now. At the turn of the twentieth century, the countryside was overwhelmingly Latvian, and the cities were not. World War One brought death, waves of refugees leaving Latvia and the evacuation of Tsarist industry and administration. The eve of World War Two and the war itself raised the Latvian percentage even higher. The Baltic Germans "repatriated" to the Reich in the fall of 1939 and Latvia's sizeable Jewish population were murdered in the Holocaust. Artificial changes in demographic patterns over the first fifty years of the twentieth century raised the Latvian percentage of the country. Equally artificial demographic changes in the second half of the twentieth century whittled away the ethnic Latvian share of Latvia almost to a minority. Deportations and the in-migration of hundreds of thousands of Slavic workers brought the ethnic Latvian share of the population to its all-time nadir. Ethnic percentages are again shifting. Similar to World War One's evacuation of industry, the removal of Soviet troops has led to the simultaneous exodus of attached Slavic families. Higher mortality rates and lower birth rates in the Slavic communities of Latvia continue to accelerate the growth of the ethnic Latvian percentage of the population.[2] Latvia may well settle into a Latvian percentage similar to just before World War One. Of course, most other people are now Russian, instead of a wide variety of equally weighted minorities.

A firm commitment to minority rights in the face of the state is a firm commitment to citizens' rights in the face of the state. There is no reason to believe that Latvia cannot promote Latvian ethnic identity, while still acknowledging the multi-ethnic composition of the Republic of Latvia. The strongest Latvia is a united Latvia, but not a false unity based on the suppression of differences of opinion. The state needs unity based on the acceptance of difference, and ethnic, religious, ideological and/or sexual preference.

The importance of a citizen's rights in relation to the state is not an endorsement for the diminished role of government and the state. Nor is it an unqualified endorsement of the rights of the individual at the expense of all else. Despite the West's current mania about the all-powerful forces of the free market and the running of government as business, the role of the state must be significant in Latvia for the foreseeable future. Reform towards a market economy is essential, but the state needs to work for growth with equity. Latvia's poorer cities and regions or its disadvantaged inhabitants cannot be left alone to the vagaries of the market. The state must step forward to help ease the burden of economically depressed areas of the country.

Latvia must be governed as a state of citizens and not as an abstract variable in economists' blueprints. Economic policy, the state and its citizens need above all else to conduct themselves with common sense and a sense of the public weal. The common sense needed by government and citizens is the ability to look critically at any programme, party, economic plan or military alliance. The first ingredient of this common sense is a lively and informed public debate about the nature of the state. No economist, politician, adviser or expert can know what Latvia needs; that knowledge can come only from an airing of concerns and public reflection. A democratic state with guarantees of civil liberties and mutual involvement and participation between citizens and government will help ensure that such debate will lead to public consensus. Common sense, however, also dictates that the state and its citizens realise that there is no easy path to salvation. Bank crises, for example, are not reflections of conspiracies, but poor financial planning and lax state supervision. With common sense, the state corrects its banking regulations and citizens do not store their savings in their mattresses. Instead they scrutinise the surviving financial institutions and invest more carefully. The state punishes negligent bankers to set an example to future financial speculators and to reinforce the rule of law. The banking crisis of 1995 showed that the foundations of common sense exist in Latvia. The outlined steps were followed almost precisely. Such common sense and public debate need to be applied to all matters, be they admission to the European Union, to NATO, the implementation of IMF guidelines, or the state's role in the commonweal.

Commonweal is an almost obscure term in contemporary political and economic debate dominated by terms such as deficit and debt cutting, free trade and privatisation. Common sense allows for a place for fiscal conservatism, but mediates away from turning it into the sole purpose of state and government. Latvia cannot afford to create a welfare state like those of Western Europe, but that does not mean that government should abandon social welfare. Public debate can build consensus to prioritise the state's involvement in investments for the future. Education and health care seem likely choices for the most intensive investment. Common sense should temper popular aspirations about the limits of the commonweal.

All states have an inherent tension or stress unique to that country. This fundamental dilemma is the source of creative and destructive energies. The crisis of the state, for example, fuels the artistic manifestations of the nation, as well as the political and economic remedies designed to smooth differences. These are the best of times. The same tension, however, produces the dark side of a state, when the divisions are not addressed, politics turn hostile, economics discriminatory, and art propaganda-like. These are the worst times. The United States, for example, is defined by race, Canada by the anglophone/francophone divide, the United Kingdom by class, Germany the limits of order and Russia its relationship with the West. Latvia's inherent tension may appear to be the question of minorities, but that is just a symptom of the real divide. Latvia is defined by the tension between state and citizen. At its best moments, the creative energies and impulses of a nation are pooled to bridge the gap between state and citizenry to create a working whole—the agrarian reform of the 1920s, or the movement for independence in the late 1980s. The dark side, however, is when the balance is lost and the state rides roughshod over the citizen as in the case of the Ulmanis regime, collaboration with foreign occupation and the continuing desire for simple solutions and strong leaders. Latvia and her citizens must use public debate to build a consensus around common sense that uses the state to benefit the commonweal, but does not unleash the state over its citizens.

1 Juris Dreifelds, *Latvia in Transition* (Cambridge: Cambridge University Press, 1996), p. 71.
2 All of Latvia, however, is demographically collapsing. There are 188,000 fewer people in Latvia in 1997 than in 1991. Total population of Latvia has fallen by 7 per cent in eight years. The birth rate is far below the death rate, and 1,813 fewer babies were born in 1996 than in 1995. *Diena*, June 9, 1997.

BIBLIOGRAPHY

Aizsilnieks, Arnolds. *Latvijas saimniecibas vesture 1914—1945* (History of the Latvian Economy 1914–1945). Stockholm: Daugava, 1968.

Anderson, Benedict. *Imagined Communities; Reflections on the Origins and Spread of Nationalism.* New York: Verso, 1991.

Apine, I. "Nomenclature. The Peculiarities of Latvia". *The Transition towards Democracy: Experience in Latvia and in the World, November 12–14, 1992,* Riga: University of Latvia, 1994.

Balabkins, Nicholas and Aizsilnieks, Arnolds. *Entrepreneur in a Small Country: A Case Study Against the Background of the Latvian Economy, 1919–1940.* Hicksvile, NY: Exposition Press, 1975.

Biezais, Haralds. *Latvija: Kaskrusta vara. Svesi kungi, pasu laudis* [Latvia under the Power of the Swastika. Foreign Lords, Our own People]. Grand Rapids, 1992.

Bildt, K. "Baltic Security and European Stability". In *The Baltic states on Their Way to the European Union, Security Aspects,* edited by Atis Lejins and Pauls Apinis. Riga: Latvijas Arpolitikas Instituta, 1995.

Bilmanis, Alfreds, Izaks, Julijs, and Skalbe, Lizete, eds. *Latvijas Republikas Desmit Pastavesanas Gados.* Riga: Golts un Jurjans, 1928.

Bilmanis, Alfred. *A History of Latvia.* Princeton: Princeton University Press, 1951.

Birons, A., ed. *Latvijas stradnieki un zemnieki 1905.–1907. g. revolucija* (The Latvian peasants and workers in the revolution of 1905–1907). Riga: Zinatne, 1986.

Bleiere, D. *Latvija. Notikumu Hronika.* Riga: N.I.M.S, 1996.

Bluzma, V. "Sabiedriski Politiskas Domas un Partiju Sistemas Attistiba". In *Latvijas Valsts Atjaunosana 1986–1993,* edited by Uldis Bormanis and Vija Kanepa. Riga: LU Zurnala "Latvijas Vesture" Fonds, 1998: 361–390.

Boikova, Tatjana. "Competitiveness of Latvia's Economy: A Pre-requisite for long-term growth". *Humanities and Social Sciences Latvia,* 4, 21: 93–103.

Bruvers, O. "1987. Gada 14. Junijs Riga". *Latvija Sodien,* No. 15. USA: The World Federation of Free Latvians, 1987: 1–10.

Bungs, Dz. *The Baltic States: Problems and Prospects of Membership in the European Union.* Germany: Stiftung Wissenschaft und Politik. Forschungsinstitut fur Internationale Politik und Sicherheit, 1998.

Ceichners, Alfreds. *Latvijas bolsevizacija, 1940–1941* (The Bolshevization of Latvia, 1940–1941). Riga: 1944. Reprint, Gauja, 1986.

Cerps, U. "The Left-Wing Parties in Latvia and Their Performance in the 1993 Parliamentary Elections". In *The Politics of Transition in the Baltic States. Democratization and Economic Reform Policies,* edited by Jan Ake Dellebrant and Ole Norgaard. Sweden: Umea University, Department of Political Science, 1994: 85–105.

Cice, Ausma. "The Employment of Latvia's Economy: A Pre-Requisite for Long-Term Growth". *Humanities and Social Sciences Latvia,* 4, 21: 104–124.

Clemens, W.C. *Baltic Independence and Russian Empire.* Basingstoke: Macmillan, 1991.

Constitution of the Republic of Latvia. Stockholm: Latvian National Foundation, 1984.

Corrsin, Stephen B. "The Changing Composition of the City of Riga, 1867–1913". *Journal of Baltic Studies* 13 (1982): 19–39.

Dreifelds, Juris. "Russification of Press and Publishing in Latvia". In *Conference on Security and Cooperation in Europe Follow-up Meeting in Vienna, November, 1986: Soviet Violations in the Implementation of the Final Act in Occupied Latvia,* edited by I. Kalnins. USA: World Federation of Free Latvians, 1986: 68–76.

—— "Immigration and Ethnicity in Latvia". *Journal of Soviet Nationalities,* 1, 4 (1990): 34–51.

—— *Latvia in Transition.* Cambridge: Cambridge University Press, 1996.

Drizulis, A., ed. *Riga Socialisma Laikmeta, 1917–1975* (Riga in the Socialist Era, 1917–1975). Riga: Zinatne, 1980.

Drizulis, A., ed. *Latvijas PSR Vesture.* Riga: Zinatne, 1986.

Duhanovs, M., Feldmanis, I., Stranga, A. *1939. Latvia and the Year of Fateful Decisions,* Riga: University of Latvia, 1994.

Dunsdorfs, E. *The Baltic Dilemma.* New York: Robert Speller & Sons, Publishers Inc., 1975.

Eglite, P. and Markausa, I. "Vairakumtautibu Demografiska Uzvediba Latvijas PSR 70–80. Gadu Mija". In *Socialie Procesi un Nacionalas Attiecibas Padomju Latvija,* edited by I. Apine. Riga: Zinatne, 1987.

Eglitis, O. *Nonviolent Action in the Liberation of Latvia.* Cambridge, Massachusetts: The Albert Einstein Institution, 1993.

Eksteins, Modris. *Walking Since Daybreak: A Story of Eastern Europe, World War II, and the Heart of Our Century.* Toronto: Key Porter Books, 1999.

Ezergailis, Andrievs. *The Holocaust in Latvia: The Missing Center.* Riga: Historical Institute of Latvia, 1996.

—— *The Latvian Impact on the Bolshevik Revolution.* Boulder, CO.: East European Monographs, 1983.

—— *The 1917 Revolution in Latvia.* Boulder, CO.: East European Monographs, 1974.

Faurby, I. "Baltic Security: Problems and Policies Since Independence". In *The Baltic States in International Politics,* edited by Nikolaj Petersen. Copenhagen: The Danish Institute of International Studies, Jurist-og Oekonomforbundets Forlag DJOF Publishing, 1993.

Fitzpatrick, Sheila. *The Russian Revolution 1917–1932.* New York: Oxford University Press, 1982.

Germanis, Uldis. "The Rise and Fall of the Latvian Bolsheviks". *Baltic Forum* 3: 1 (1988): 21–5.

Girnius, S. "Back in Europe, to Stay". *Transition,* Vol. 3, No. 6 (1997): 7–10.

Goba, Alfreds. *Dienas gramata* (Diary). Unpublished, 1918.

Gorbacovs, M. *Parkartosanas un Jauna Domasana. Musu Valstij un Visai Pasaulei,* Riga: Avots, 1987.

Gordon, Frank. *Latvians and Jews between Germany and Russia.* Stockholm: Memento, 1990.

Gore, Ilga and Aivars Stranga. *Latvija: neatkaribas mijkreslis. Okupacija* (Latvia: Twilight of Independence). Riga: Izglitiba, 1992.

Grabbe, H. "The EU's Enlargement Strategy". In *The Baltic Dimension of European Integration. A Conference at Riga 24–25 August 1996.* Riga: Latvijas arpolitikas instituta and Royal Danish Embassy, 1996: 41–48.

Grava-Kreituse, I., Feldmanis, I., Goldmanis, J., Stranga, A., eds. *Latvijas okupacija un aneksija 1939–1940, dokumenti un materiali* (Latvia's occupation and anexation 1939–1940). Riga: by the authors, 1995.

Grigorjevs, A. "The Ghosts". *Freedom Review,* May–June, 1993: 12–14.

Gulans, Peteris. "Transformation and the Development of an Economic Model". *Humanities and Social Sciences Latvia,* 4, 21: 56–86.

Gylys, P. "Integration Policy of the Three Baltic States". In *The Baltic Dimension of European Integration. A Conference at Riga 24–25 August 1996.* Riga: Latvijas arpolitikas instituta and Royal Danish Embassy, 1996: 14–40.

Haberer, Erich E. "Economic Modernization and Nationality in the Russian Baltic Provinces 1850–1900". *Canadian Review of Studies in Nationalism,* 12, 1, 1985.

Hamm, Michael F. "Riga's 1913 City Election: A Study in the Baltic Urban Politics". *Russian Review,* XXXIX, 4.

Handelman, Stephen. *Comrade Criminal; Russia's New Mafiya.* New Haven: Yale University Press, 1995.

Henrikkson, Anders. *The Tsar's Loyal Germans: The Riga German Community: Social Change and the Nationality Question, 1855–1905.* Boulder, CO.: East European Monographs, 1983.

Heurlin, B. "NATO, Security, and the Baltic States: A New World, A New Security, A New NATO." In *The Baltic States in the World Politics,* edited by Birthe Hansen and Bertel Heurlin. Great Britain: Curzon Press, 1998: 65–86.

Hiden, J.W. *The Baltic States and Weimar Ostpolitik.* Cambridge: Cambridge University Press, 1987.

Hiden, John, ed. *The Baltic and the Outbreak of the Second World War.* Cambridge: Cambridge University Press, 1992.

Hiden, John, and Thomas Lane. *The Baltic Nations and Europe: Estonia, Latvia, and Lithuania in the Twentieth Century.* New York: Longman, 1991.

Høyer, S., E. Lauk and P. Vihalemm, eds. *Towards a Civic Society. The Baltic Media's Long Road to Freedom. Perspectives on History, Ethnicity and Journalism.* Tartu: Baltic Association for Media Research/Nota Baltica Ltd, 1993.

Human Rights in Latvia in 1998, Latvian Centre for Human Rights and Ethnic Studies, Riga: 1999.

Ivans, D. *Gadijuma Karakalps.* Riga: Vieda, 1995.

Jerans, P., ed. *Latvijas Padomju Enciklopedija* (Latvia's Soviet encyclopedia). 10 vols. Riga: Galvena enciklopediju redakcija, 1981–1987.

Jundzis, T. "Tiesibu Reformu Loma Neatkaribas Atjaunosana". In *Latvijas Valsts Atjaunosana 1986–1993,* edited by U. Bormanis and V. Kanepa. Riga, 1998: 151–165.

Jundzis, T. "Neatkarigas valsts Aparata Veidosana un Tiesibu Reformas". In *Latvijas Valsts Atjaunosana 1986–1993,* edited by U. Bormanis and V. Kanepa. Riga, 1998: 239–258.

Jundzis, T. "Konfrontacija ar Latvijas Neatkaribas Pretiniekiem". In *Latvijas Valsts Atjaunosana 1986–1993,* edited by U. Bormanis and V. Kanepa. Riga, 1998: 278–295.

Jundzis, T. "Tiesibu Sistemas Reforma". In *Latvijas Valsts Atjaunosana 1986–1993,* edited by U. Bormanis and V. Kanepa. Riga, 1998: 315–333.

Jundzis, T. "Atgriesanas Starptautiskaja Aprite-Atjaunotas Latvijas Arpolitika". In *Latvijas Valsts Atjaunosana 1986–1993,* edited by Uldis Bormanis and Vija Kanepa. Riga: LU Zurnala Latvijas Vesture Fonds, 1998: 347–361.

Kamenska, A. *Valsts Valoda Latvija.* Riga: Latvijas Cilvektiesibu un Etnisko Studiju Centrs, 1994.

Kangeris, Karlis. "The Former Soviet Union, Fascism and the Baltic Question. The Problem of Collaboration and War Criminals in the Baltic Countries". In *Modern Europe After Fascism 1943–1980s,* edited by Stein Ugelvik Larsen. Boulder: Social Science Monographs, 1998.

Karnite, Raita. "Globalization and the Latvian Economy: Observations and Conclusions". *Humanities and Social Sciences Latvia,* 4, 21: 4–33.

Kaufman, S.J. "The Baltic States in Post-Cold War U.S. Strategy". In *The Baltic States in the World Politics,* edited by Birthe Hansen and Bertel Heurlin. Great Britain: Curzon Press, 1998: 46–65.

King, Gundar. *Economic Policies in Occupied Latvia.* Tacoma, WA.: Pacific Lutheran University Press, 1965.

Kirby, David. *The Baltic World 1772–1993: Europe's Northern Periphery in an Age of Change.* London: Longman, 1995.

Kolstoe, P. *Russians in the Former Soviet Republics.* Bloomington: Indiana University Press, 1995.

Krastins, Janis. *1905. gada revolucija Latvija: 1905–1907.* Riga: Zinatnu Akademija, 1948.

Krastins, J., ed. *Riga 1860–1917.* Riga: Zinatne, 1978.

Krastins, J. "Komunistiskais Genocids Latvijas Kulturvide". In *Komunistiska Totalitarisma un Genocida Prakse Latvija,* edited by I. Sneidere. Riga: Zinatne, 1992.

Kreicbergs, H. *Vainigie un Nelaimigie.* Riga: Avots, 1989.

Kregere, O. "Industrializacijas destruktiva politik Prakse Latvija". In *Komunistiska totalitarisma un Genocida Prakse Latvija,* edited by I. Sneidere. Riga: Zinatne, 1992.

Laitin, D.D. "The National Uprisings in the Soviet Union". *World Politics,* 44, 1 (1992): 139–177.

Lasmanis, Uldis. *Berga Bazara un laikmeta labirintos.* Riga: 1997.

Latvia Human Development Report, Riga: UNDP, 1997.

Latvia Human Development Report. Riga: UNDP, 1998.

Latvija un Eiropas Savieniba, November, 1998

Latvijas Tautas Fronte. Gads Pirmais. Riga, 1989.

Latvijas Tautas Frontes Programma. Latvijas Tautas Frontes Statuti, Riga: Avots, 1988.

Levin, Dov. "Arrests and Deportations of Latvian Jews by the USSR during the Second World War". *Nationalities Papers* 16 (1988): 50–70.

—— "The Jews and the Sovietization of Latvia, 1940–1941". *Soviet Jewish Affairs* 5 (1975).

Levits, E. "1990.gada 4. maija Deklaracija par Neatkaribas Atjaunosanu". In *Latvijas Valsts Atjaunosana 1986–1993,* edited by U. Bormanis and V. Kanepa. Riga, 1998: 207–237.

Lievan, Anatol. *The Baltic Revolution: Latvia, Lithuania, Estonia, and the Path to Independence.* New Haven, CT.: Yale University Press, 1993.

Loit, Aleksander, and John Hiden, eds. *The Baltic in International Relations between the Two World Wars.* Stockholm: University of Stockholm, 1988.

Lundin, L. "The Road from Tsar to Kaiser: Changing Loyalties of the Baltic Germans, 1905–1914". *Journal of Central European Affairs*, X, 1950.

Lukic, R. and Lynch, A. *Europe From the Balkans to the Urals. The Disintegration of Yugoslavia and the Soviet Union.* Sipri, Oxford University Press, 1996.

Mainwaring, S. "Party Systems in the Third Wave". *Journal of Democracy*, 9, 3 (1998): 51–67.

Misiunas, R. and R. Taagepera. *The Baltic States. Years of Dependence 1940–1990.* Expanded and Updated edition. London: Hurst & Company, 1993.

Muiznieks, N. "Latvia: Origins, Evolution and Triumph". In *Nation and Politics in the Soviet Successor States*, edited by J. Bremer and R. Taras. Cambridge: Cambridge University Press, 1993: 3–29.

Nahaylo, B. "Nationalities". In *The Soviet Union under Gorbachev*, edited by M. McCauley. England: Macmillan Press, 1987: 73–97.

Niedra, Andrievs. *Tautas nodeveja atminas* [Memoirs of the Traitor of His People]. Riga: Zinatne, 1998.

Nissinen, Marja. *Latvia's Transition to a Market Economy: Political Determinants of Economic Reform Policy.* London: Macmillan Press, 1999.

Nørgaard, O. "Gorbatjovs Reformstrategi". *Politica*, 20, 1 (1988): 10–26.

Norgaard, O. "Danish Policies Towards the Baltic States". In *The Baltic States in International Politics*, edited by Nikolaj Petersen. Copenhagen: The Danish Institute of International Studies, Jurist-og Oekonomforbundets Forlag DJOF Publishing, 1993: 155–175.

O'Donnell, G. "Horizontal Accountability in New Democracies". *Journal of Democracy*, 9, 3 (1998): 112–127.

Ozolina, Z. "Baltic-Nordic Interaction, Cooperation and Integration". In *Small States in a Turbulent Environment: The Baltic Perspective*, edited by Atis Lejins and Z. Ozolina. Riga: Latvijas Arpolitikas Instituta, 1997: 113–147.

Pabriks, A. *From Nationalism to Ethnic Policy: The Latvian Nation in the Present and the Past.* University of Aarhus: Department of Political Science, 1996.

Page, Stanley W. *The Formation of the Baltic States: a Study of the Effects of Great Power Policies on the Emergence of Lithuania, Latvia, Estonia.* Cambridge: Harvard University Press, 1959; New York: Howard Fertig, 1970.

Park, A. "Ethnicity and Independence: The Case of Estonia in Comparative Perspective". In *Proceedings of the Estonian Academy of Sciences. Humanities and Social Sciences*, No. 44, 2 (1995): 302–323.

Pavlovskis, O., ed. *Latvian Dissent, Case Histories of the 1983 Soviet Campaign to Silence Political Dissidents in Occupied Latvia.* World Federation of Free Latvians, 1983.

Petersen, N. "Introduction". In *The Baltic States in International Politics*, edited by Nikolaj Petersen. Copenhagen: The Danish Institute of International Studies, Jurist-og Oekonomforbundets Forlag DJOF Publishing, 1993: 9–21.

Petersen, N.H. "The Role of the European Union: 'Soft' Security?" In *The Baltic Dimension of European Integration. A Conference at Riga 24–25 August 1996*, Riga: LAI, Royal Danish Embassy, 1996: 90–100.

Pikayev, A.A. "Russia and the Baltic States: Challenges and Opportunities". In *The Baltic States in the World Politics*, edited by Birthe Hansen and Bertel Heurlin. Great Britain: Curzon Press, 1998: 133–161.

Pipes, Richard. *The Russian Revolution.* New York: Vintage Books, 1990.

Plakans, Andrejs. *The Latvians: A Short History.* Stanford: Hoover Institution Press, 1996.

—— "The Latvians". In *Russification in the Baltic Provinces and Finland 1855–1914*, edited by Edward C. Thaden, 207–84. Princeton, NJ: Princeton University Press, 1981.

—— "Democratization and Political Participation in Postcommunist Societies: the Case of Latvia". In *The Consolidation of Democracy in East-Central Europe*, edited by K. Dawisha and B. Parrott. Cambridge: University Press, 1997: 245–289.

Poulsen, J.J. "Nationalism, Democracy and Ethnocracy in the Baltic Countries". *The Politics of Transition in the Baltic States. Democratization and Economic Reform Policies*, edited by Jan Ake Dellebrant and Ole Norgaard. Sweden: Umea University, Department of Political Science, 1994: 15–31.

Purs, Aldis. *Creating the State from Above and Below: Local Government in Inter-War Latvia.* Ph.D. diss., University of Toronto, 1998.

Rajevska, F. "Kritika asvestenija burzuaznimi ideologami roli i znatsjenia dvuhjazitsija v SSSR". In *Nacionalas Attiecibas un ideologiska cina,* edited by F. Rajevska. Riga: Latvijas valsts universitate, 1982: 41–60.

Rauch, Georg von. *The Baltic States: The Years of Independence, 1917–1940.* Berkeley: University of California Press, 1974.

Reenan, A.J. van. *Lithuanian Diaspora, Konigsberg to Chicago.* London: University Press of America, 1990.

Remnick, David. *Lenin's Tomb: The Last Day of the Soviet Empire.* New York: Vintage Books, 1994.

—— *Resurrection: The Struggle for a New Russia.* New York: Random House, 1997.

Republic of Latvia Parliamentary Election and National Referendum 3 October 1998, Assessment Mission, OSCE/ODIHR Election Observation, October 3, 1998.

Rimsevics, Ilmars. "How to Survive Economic Globalization? Latvia's Case". Humanities and Social Sciences Latvia, 4, 21: 87–92.

Ronis, Indulis, ed. *Karlis Ulmanis Trimda un Cietuma* (Karlis Ulmanis in exile and prison). Riga: Latvijas vestures Instituts, 1994.

Sabiedribas Integracija Latvija (projekts). Valsts programmas Koncepcija, Riga, 1999.

Sachs, Jeffrey. "The Limits of Convergence; Nature, nurture and growth". *The Economist,* June 14, 1997, 22.

Saul, John Ralston. *Voltaire's Bastards: The Dictatorship of Reason in the West.* Toronto: Penguin Books, 1993.

Schroeder, G.E. "Nationalities and the Soviet Economy". In *The Nationalities Factor in Soviet Politics and Society,* edited by L. Hajda and M. Beissinger. Oxford: Westview Press, 1990.

Scruton, R. *A Dictonary of Political Thought.* United Kingdom: Pan Books in association with The Macmillan Press, 1983.

Shanin, Teodor. *Russia, 1905–1907; Revolution as a Moment of Truth.* London: MacMillan, 1986.

Sics, U. *Latvija Cela uz NATO.* Riga: Latvijas Arpolitikas Instituts, 1996.

Silgailis, A. *Latviesu legions.* Imanta: 1962.

Simon, G. *Nationalism and Policy toward the Nationalities in the Soviet Union. From Totalitarian Dictatorship to Post-Stalinist Society.* Oxford: Westview Press, 1991.

Singleton, Fred. *The Economy of Finland in the Twentieth Century.* University of Bradford: 1986.

Skujeneeks, Margers. *Latvija: Zeme un Eedzivotaji.* Riga: Valsts Statistiskas Parvaldes Izdevums, 1922.

Soikane-Trapane, M. "Status of the Latvian Language in Present-Day Latvia". In *Conference on Security and Cooperation in Europe Follow-up Meeting in Vienna, November, 1986: Soviet Violations in the Implementation of the Final Act in Occupied Latvia,* edited by I. Kalnins. USA: World Federation of Free Latvians, 1986.

Spekke, A. *History of Latvia,* Stocholm: Zelta Abele, 1951.

Steen, A. *Recirculation and Expulsion: The New Elites in the Baltic States.* University of Oslo: Department of Political Science, 1994.

Steinbuka, Inna. "Competitiveness of the Latvian Economy and Economic Convergence with the EU". *Humanities and Social Sciences Latvia,* 4, 21: 33–55.

Stradins, J. *Tresa Atmoda.* Riga: Zinatne, 1992.

Stranga, A. "Baltic–Russian Relations: 1995–Beginning of 1997". In *Small States in a Turbulent Environment: The Baltic Perspective,* edited by Atis Lejins and Z. Ozolina. Riga: Latvijas Arpolitikas Instituts, 1997: 184–238.

Strods, Heinrichs. *Latvijas nacionalo partizanu kars, 1944—1956* [Latvia's national partisan war]. Riga: Preses nams, 1996.

Sukhanov, N.N. *The Russian Revolution 1917, A Personal Record.* Princeton: Princeton University Press, 1984.

Suny, Ronald Grigor. *The Revenge of the Past: Nationalism, Revolution and the Collapse of the Soviet Union.* Stanford: Stanford University Press, 1993.

Svabe, Arveds. *Latvijas vesture 1800–1914* (History of Latvia, 1800–1914). Stockholm: Daugava, 1958.

Svabe, Arveds, ed. *Latvju enciklopedija* (Latvian encyclopedia). 3 vols. Stockholm: Tris Zvaigznes, 1950–1951.

Taylor, C. "The Politics of Recognition". In A. Gutmann (ed.), *Multiculturalism and "The Politics of Recognition*, edited by A. Gutmann. Princeton, New Jersey: Princeton University Press, 1992.

Thaden, Edward C., ed. *Russification in the Baltic Provinces and Finland, 1855–1914*. Princeton: Princeton University Press, 1981.

The Revolution in the Baltic Provinces of Russia; A Brief Account of the Lettish Social Democratic Worker's Party. London: Independent Labour Party, 1907.

The Programme for Studies and Activities Towards a Civic Society. Report. The Results of 1st and 2nd Stages, Baltic Data House (BDH), 1998.

Trapans, Andris. "The Latvian Communist Party and the Purge of 1937". *Journal of Baltic Studies*, XI, 1: 25.

Trapans, J.A. "The Sources of Latvia's Popular Movement". In *Toward Independence: The Baltic Popular Movements*, edited by J.A. Trapans. Oxford: Westview Press, 1991: 25–43.

Vaares, P. "Dimensions and Orientations in the Foreign Policies of the Baltic States". In *The Baltic States in International Politics*, edited by Nikolaj Petersen. Copenhagen: The Danish Institute of International Studies, Jurist-og Oekonomforbundets Forlag DJOF Publishing, 1993: 21–41.

Veisbergs, A. "The Latvian Language-Struggle for Survival". *Humanities and Social Sciences*, 1 (1993): 27–40.

Viksne, Rudite and Kangeris, Karlis, eds. *No NKVD lidz KGB Politiskas Pravas Latvija 1940–1986: Noziegumos pret padomju valsti apsudzeto Latvijas iedzivotaju raditajs* [From NKVD to KGB Political Trials in Latvia 1940–1986: Index of Citizens of Latvia Accused of Criminal Activites Against Soviet Power]. Riga: Latvijas vestures instituta apgads, 1999.

Viksnins, G.J. "The Latvian Economy: Performance and Prospects". In *Conference on Security and Cooperation in Europe Follow-up Meeting in Vienna, November, 1986: Soviet Violations in the Implementation of the Final Act in Occupied Latvia*, edited by I. Kalnins. USA: World Federation of Free Latvians, 1986:, 120–134.

Wildman, Allan K. *The End of the Russian Imperial Army*. 2 volumes. Princeton: Princeton University Press, 1987.

Zagars, E. *Socialist Transformation in Latvia, 1940–1941*. Riga: Zinatne, 1978.

Zalite, E. "Rupniecibas uznemumu reevakuacijas gaita un problemas 20.–30. Gadi". *Latvijas Vestures Instituta Zurnals*, 2, 7 (1993): 69.

Zepa, B. "Sabiedriska Doma Parejas Perioda Latvija: Latviesu un Cittautiesu Uzskatu Dinamika (1989–1992)". *Latvijas Zinatnu Akademijas Vestis*, 2 (1992).

Zhuryari, O. "The Baltic Countries and Russia (1990–1993): Doomed to Good-Neighbourliness?" In *The Foreign Policies of the Baltic Countries: Basic Issues*, edited by Pertti Joenniemi and Juris Prikulis. Riga: Centre of Baltic–Nordic History and Political Studies, 1994: 75–87.

Zvidrins, P. and I. Vanovska. *Latviesi. Statistiski Demografisks Portretejums*. Riga: Zinatne, 1992.

5th Saeima – Republic of Latvia Standing Commission on Human Rights. *The Republic of Latvia, Human Rights Issues*. Riga: 1993.

PERIODICALS

Diena [Day], daily, Riga.

Dienas Bizness [Day's Business], daily, Riga.

Economist, weekly, London.

Fokuss [Focus], weekly, Riga.

Humanities and Social Sciences Latvia, quarterly, Riga, 1993-.

Jaunakas zinas [The Latest News], daily, Riga, 1918–1934.

Journal of Baltic Studies, quarterly, USA., 1971–.

Labrit [Good Morning], daily, Riga, ?–1995.

Latvijas Vestnesis [Latvian Herald], daily, Riga, Official Newspaper of the Latvian Government.

Latvijas vesture [Latvia's History], quarterly, Riga.

Lauku Avize [Rural Newspaper], weekly, Riga.

LETA, Latvia News Service.

Likumu un valdibas rikojumu krajums, later Likumu un ministru kabineta noteikumu krajums [Collection of Laws and Decrees], Riga, 1919–1940.

Literatura un Maksla [Literature and Art], weekly, Riga.

Neatkariga Cina [Independent Struggle], daily, Riga.

Nezavisimaja Gazeta, daily.

The White House press releases.

Vakaras Zinas [Evening News], daily, Riga.

Valdibas Vestnesis [The Government Herald], daily, Riga, 1918–1940.

WEB-SITES

If the Internet is the information superhighway, it is a highway under construction. The great difficulty in measuring the worth of web-sites is the inability to guarantee how often they are updated, and if they even continue to exist at the given addresses. Nevertheless, here is a short list of web-sites specializing in Latvia:

http://www.alausa.org
    American Latvian Association
http://www.bns.lv
    Baltic News Service
http://www.grida.no
    Maps
www.mfa.gov.lv
    Ministry of Foreign Affairs
http://www.bank.lv
    The Bank of Latvia
www.iem.gov.lv
    Ministry of the Interior
htttp://www.latnet.lv
    Education and Science Network
http://www.riga.post.lv
    Latvia's Post
http://www.saeima.lv
    Latvia's Saeima
www.inyourpocket.com
    Riga in your pocket
http://www.radioswh.lv
    Radio Station SWH
http://www.usis.bkc.lv
    US Mission to Latvia
www.latnet.lv/wwwsites/government/governmental.html
    Guide to government web-sites.
www.delfi.lv
    Partal to Latvian news and events
www.diena.lv
    Latvia's largest newspaper

Other international organization web-sites also have information on Latvia, or on topics that affect Latvia significantly, such as:

http://www.un.org/
    The United Nations Official Home Page

http://www.nato.int/
   NATO's home page
http://www.oecd.org/
   The Organization for Co-operation and Development Home Page
http://www.sipri.se/
   The Stockholm International Peace Research Institute Home Page
http://www.embassy.org/
   The Electronic Embassy Home Page
   and
www.balticstudies-aabs.lanet.lv
   The Association for the Advancement of Baltic Studies.
www.riga.lv/minelres
   Minority Electronic Resources

Index

1920 Peace Treaty, 120, 135
Abortions, 38, 114
Abrene, 135
Academy of Agriculture, 48
Academy of Art, 48, 98
Academy of Music, 48
'A Chartership of Partnership among
    the U.S.A. and the Republic of
    Estonia, Republic of Latvia and the
    Republic of Lithuania', 139–40
Adamsons, Janis, 85
'Agenda 2000: For a Stronger and
    Wider Union', 143
Agrarian reform, 17, 107, 153
Agreement of Economic
    Co-operation, 122
Aizsilnieks, Arnolds, 22
Akuraters, Janis, 34
Albright Madeleine, 139
Albania, 115
Alcoholism, 38, 114
Amber Gateway, 131
Anderson, Benedict, 3
Andropov, Yuri, 40, 90
Apine, Ilga, 63
Apparats, 92, 100
Arajskommandos, 27
Asmus, Ronald, 139–40
Atkinson, David, 134
Attorneys General, 62
August 21, 1991, 64

Baltic-American Action Plan, 139
Baltic-American Enterprise Fund, 138
Baltic-American Partnership Fund,
    138
Baltic Assembly, 127
BALTBAT (Baltic Battalion), 127

Baltic Chain, 121, 127
Baltic Council, 127
Baltic Entente, 126
Baltic Germans, 1–11, 14, 17, 24, 31,
    111, 151
Baltic Military Academy, 127
Baltic Military District, 123
Baltic Sea Region (BSR), 129
Baltron, 127
Bangerskis, General, 29
Banka Baltija, 103
Bank of Latvia, 62, 95–6, 103
Barda, Fricis, 34
Belorussians, 35, 47, 56, 73
Berklavs, Eduards, 34–6, 39, 54
Berzins, Andris, 83
Best Western Mara, 111
Bildt, Carl, 129, 137, 139
Birkavs, Valdis, 83, 143
Bolsheviks, 12–15, 26, 28, 33
Bolshevism, 9, 11
Brest-Litovsk Treaty, 13–14
Brezhnev, Leonid, 37, 39–40
Bribery, 19, 97
Brzezinski, Zbigniew, 139
Bojars, Juris, 69, 85
Bulgaria, 115
Bungs, Dzintra, 131

Cabinet of Ministers, 61
Cakste, Janis, 10
Cakste, Konstantins, 30
Calendar demonstrations, 53–4
Central European Economic Review,
    98
Cevers, Ziedonis, 85
Chernomyrdin, Victor, 131
Christian Democratic Union, 70, 71

Citizen's Committees, 58–60, 69
Citizens' Movement, 59
Citizenship, 72–3, 75–6, 78, 80, 123, 134, 150
Citizenship Law of 1994, 78–9
Civil parishes (*pagasti*), 16
Clinton, Bill, 138
Club-21, 68
Collective Farms (*Kolkhozes*), 32, 34, 38, 90, 93, 107
Collectivization, 100
'Coloured metals', 92
Commonwealth of Independent States (CIS), 90, 94, 104–6, 111, 128
Communist Party of the Soviet Union (CPSU), 46, 48, 54, 62
Constituent Assembly, 13, 15–20
Coopers & Lybrand, 103
Copenhagen criteria, 142–3
Corruption, 19, 90, 98, 101, 143
Council of Europe, 62, 122
Council of Ministers, 74, 121
Council of Ministers of the Baltic States, 127
Council of the Baltic Sea States (CBSS), 122, 130–2
Coup of May 15, 1934, 19–21
Customs department, 62
Czech Republic, 113, 115, 133, 139, 143–4

Daily Page (Dienas lapa), 6
Dankers, Oskars, 29
Daugava River, 9, 12, 16, 52
Daugavpils, 8, 52
'Declaration about the Renewal of the Independence of the Republic of Latvia' (May 4, 1990), 60–1, 69, 121
Democratic Centre Party, 70–1
Denmark, 129–30, 140
Depression, 21, 47, 108, 112
Divorce, 38
Dreifelds, Juris, 89, 106–7
Dual power, 13
Duma, 10, 12

Egg war, 128
Eksteins, Modris, 31
Ellemann-Jensen, Uffe, 129, 132, 139
Émigrés, 32, 40, 68, 71, 111
Equal Rights Movement, 70–1, 84
Estonia, 13, 30–1, 35, 45, 48–9, 58, 60, 62, 86, 90, 94, 102, 115, 122, 124, 128–32, 134–5, 138, 143
European Commission, 143
European Council, 134, 142
European Parliament, 62
European Union, 45, 78, 84, 91, 95, 99, 111, 115, 122, 124–5, 128–30, 132–3, 135–7, 142–6, 153

Farmers' Union, 14, 18, 20–1, 30, 70–1, 108
February Revolution, 13
Federalist Papers, 82
Finland, 5, 10, 13, 129–31
Foreign Direct Investment (FDI), 99, 101–2
For Fatherland and Freedom (TB), 69, 71, 81, 83
France, 2, 120, 140
Free Economic Zones, 101

Gailis, Maris, 83
Gaujas National Park, 112
Gellner, Ernest, 77
Genscher, Hans-Dietrich, 132
Germans, 1, 15, 119
Germany, 15, 16, 23, 108, 111, 124–5, 130–3, 140, 144, 153
Glasnost, 90
Godmanis, Ivars, 60–1
Gorbachev, Mikhail, 40, 46–8, 52, 55, 90–1
Gorbunovs, Anatolijs, 55, 60, 68, 85
Great Britain, 111, 120, 127, 130
Greece, 143
Gulag, 34
Gylys, Povilas, 139
Gypsies, 27, 31, 56

Hansabanka-Latvija, 103
Harmony for Latvia-Revival of the
    Economy (SLAT), 69, 71
Healthcare Europe, 113
Helsinki, 110
Helsinki-86, 52
Helsinki Agreement of 1992, 134
Hercog, Roman, 133
Herring war, 128
Heurlin, B., 145
Hippie movement, 39
Historical Institute of Latvia, 33
Hitler, Adolph, 23, 28
Holocaust, 27–30, 151
Hungary, 102, 133, 139, 143–4

Iceland, 129–30, 140, 145
Imperial Germany, 14
Industrialisation, 31–2, 35–8, 105
Institute of Civic Aviation, 48
International Monetary Fund (IMF),
    89, 95–6, 102, 108, 113, 153
Interregnum, 28
Ivans, Dainis, 52, 63, 67–8

January 2, 1991, 63
Jaunsaimnieki, 20, 107
Jerumanis, Aivars, 71
Jews, 7, 26–7, 31, 56, 151
Jugendstil, 7, 112
Jurkans, Janis, 69
Jurmala, 111

Kalnberzins, Janis, 31, 35
Kalnins, Ojars, 131
Kaufman, S.J., 139, 141
Kaunas, 127
KGB, 33, 52, 56, 63, 69
Kinkel, Klaus, 132
Kirby, David, 22, 31
Kirhensteins, Augusts Dr., 25
Khruschev, N., 33–6
Kohl, Helmut, 131
Kolstoe, Paul, 73
Krasts, Gundars, 83

Kristopans, Vilis, 83, 101
Krumins, Vilis, 34
Kulaks, 32–3
Kurzeme, 9, 12, 17, 70

Lacis, Vilis, 25
Laima Chocolate, 133
Laitin, D.D., 55
Language law, 73–4
Latgale, 2, 16, 38, 70
Latgalian, 11
Lattelekom, 99
Lats, 25, 94–6, 101, 104
Latvia's Central Committee, 49
Latvia's Christian Democratic
    Organisation, 36
Latvia's Communist Party (LKP), 23,
    25–6, 31–7, 48, 53–5, 57–8, 60,
    62–3, 69–70, 92
Latvia's Constitution of 1922, 17–18,
    25, 70
Latvia's Democratic Labour Party
    (LDDP), 69
Latvia's Democratic youth
    Committee, 30
Latvia's Development Agency, 101
Latvia's Green Party, 51, 70
Latvia's International Working
    People's Front (Interfront), 56, 60
Latvia's Popular Front (LPF), 45, 51,
    54–61, 63, 67, 70, 76, 84, 91, 98,
    100, 120–1, 149
Latvia's Way, 68, 71, 83, 85, 98
Latvian Association of Riga, 4
Latvian Central Council, 30
Latvian Independence Day
    (November 18, 1918), 14, 53–4
Latvian Independence Movement, 39
'Latvian Latvia', 29, 31
Latvian National Army, 14–15
Latvian National Council, 14
Latvian National Independence
    Movement (LNNK), 54, 69, 71
Latvian Opera House, 98
Latvian Refugee Association, 12

Latvian Rifles, 12–13
Latvian ruble, 95
Latvian Self-Administration, 29
Latvian Song Festival, 4
League of Nations, 15, 17
Ledurga, 109
Legalistic Attempt, 58-61
Lembergs, Aivars, 110
Lenin, V.I., 12, 120
'Letter of Seventeen Communists', 39
Letter of Toleration, 82
Levits, Egils, 64
Liepaja (Libau), 5, 8, 15, 52, 110, 123
Limbazi, 8
Literatura un Maksla, 52
Lithuania, 7, 13, 30, 31, 35, 45, 48–9,
    60, 62–3, 86, 90, 94, 115, 122,
    124, 126–31, 138, 140, 143
Lithuania's Declaration of
    Independence, March 11, 1990,
    60
Locke, 82

Maastricht criteria, 99
Macedonia, 115
Madrid criteria, 142–3
Mainwaring, Scott, 82, 84–5
Maslenki, 24
Meierovics, Gunars, 71
Meierovics, Zigfrids Anna, 14, 71
Merkel, Garlieb, 2
Membership Action Plan, 140
Midsummer's Eve, 2, 32, 34, 36
Miliukov, Pavel, 120
Minister President, 14, 18, 25, 83
Ministry of Culture, 98
Ministry of Finance, 97
Ministry of Foreign Affairs, 14, 121
Ministry of the Interior, 62, 63
Mitau (Jelgava), 5, 8
Molotov-Ribbentrop Pact, 24, 53, 121
Monument of Freedom, 45, 53, 138
Moscow, 12, 29, 35–6, 40, 55–62,
    84, 86, 89, 92, 109, 121–2, 130
Mutual Assistance Pact, 23

'National Communists', 34–7
National Cultural Associations, 56
National Movement for Latvia, 84–5
National Salvation Committees, 63
Nationalities Forum, 56
Nazi occupation, 27, 30, 150
'Near abroad', 137
New Current (*Jaunstravnieki*), 5-6
New Farmers' Party, 108
New Party, 84
Nissinen, Marja, 89, 95, 100, 115-16
NKVD, 7
Nomenklatura, 49, 53, 67, 69, 90, 111
North Atlantic Treaty Organisation
    (NATO), 45, 95, 122, 124–5, 127,
    129, 135–42, 144–6, 153
Niedra, Andrievs, 15
Nordic Council of Ministers, 122
Nordic countries, 121, 127, 130, 132,
    145
Norway, 129–30
November 11th (Heroes' Day), 54

October Manifesto, 9-10
October Revolution, 13–14
O'Donnell Guillermo, 81
Official State Language Proficiency
    Certificate, 75
Ogre, 38
Oil war, 128
Olaine, 38
Olainfarm, 107
OMON, 62–3
'On the Renewal of Republic of
    Latvia's Citizens' Rights and
    Fundamental Principles of
    Naturalisation', 72
'On the Status of a Stateless Person in
    the Republic of Latvia', 79
OSCE (CSCE), 78, 122, 129, 134

Pabriks, Artis, 23
Parex Banka, 103
Parliamentary method, 57–8
Partisans, 29–31, 51

Pauls, Raimonds, 98
Peace Corps, 116, 138
Pelse, Arvids, 35–6
Pensions, 114
People's Party, 83–5
People's *Saeima*, 25
Perestroika, 40, 46–7, 55, 90
Perkonkrusts, 21, 25
Perry, William, 139
Petersburg Newspaper (Peterburgas avize), 5
Petersen, Niels Helveg, 139, 142
PHARE, 116, 133
Pikayev, A.A., 120
Pipes, Richard, 13
Place Name Commission, 75
Plakans, Andrejs, 6
Plebiscite on Independence (March 3, 1991), 63–4
Privatisation, 99–101
Prvatisation certificates (vouchers), 100
Poland, 13, 23, 115, 124, 130, 133, 139, 140, 143, 144
Poles, 7, 119, 151
Police, 9, 48, 62, 98
Pork war, 128
Poulsen, J.J., 77
Putin, V., 137
Prikaz, 150

RAND, 125
Red Army, 15
Referendum of October 3, 1998, 72, 78
Reform Communists, 55, 57, 59–60, 64, 68–9
Repatriation, 24
Repse, Einars, 95–6
Restitution, 100, 107
Revolution of 1905, 9, 11
Rezekne, 109
Rietumu Banka, 103
Riga, 1, 4–8, 12–13, 15–16, 25, 27, 38, 56, 62, 64, 70, 89, 98, 109–10, 111–12, 115, 120–1, 127, 130, 136

Rigas Komercbanka (Pirma banka), 103–4
Riga Stock Exchange, 102, 104
Riga Stock Exchange Price Index (RICI), 102
Romania, 140
Ruble zone, 89, 94
Ruhl, Lothar, 133
Russia, 13, 16, 46, 83, 104–5, 110–11, 119–20, 122, 124–5, 128, 130, 133–8, 141, 144–5, 153
Russian Empire, 1, 4, 6, 8
Russian Provisional Government, 13, 120
Russians, 7, 26, 35, 63, 73, 79–80, 151
Russian Stock Exchange, 104
Russification, 5–6, 8, 11, 36, 9, 47, 50

Saeima, 18–20, 59, 61, 67–8, 70–1, 78, 81, 82, 84–5
Saimnieks, 84–5
SAS-RadissonDaugava, 111
Secretariat of the Baltic Assembly, 127
Serf emancipation, 2
Sergeyev, 141
'Shock therapy', 94
Siberia, 32
Siegerist, J., 84
Silgailis, Colonel, 29
Skalbe, Karlis, 34
Skele, Andris, 83, 85, 100–1, 115, 128
Skrunda, 123
Sloka, 8
Slovenia, 102, 115, 140, 143
Smith, Anthony, 77
Snips, Arturs, 52
Social Democrats, 6, 10, 12, 14, 18, 20, 26, 30, 69, 84–5, 151
Socialist International, 69
Socialist reconstruction, 31
Soros, George, 98
Soros Foundation, 78
Soros Open Society Institute, 138
Soviet deportations, 33, 52
Soviet-Finnish War, 23, 53

Soviet immigrants, 32, 35, 38, 50, 52, 58, 70–3, 75, 77–8, 80–1, 84, 123
Sovietisation, 25, 31, 32
Soviet occupation, 24, 30, 34, 45, 49, 53–4, 59, 61, 86, 119, 121, 123, 150
Soviet State Bank, 25
Soviet Union, 16, 21–3, 25, 27, 37, 38, 40, 47–8, 50, 76
Stalin, Josef, 33
Stalinism, 32, 36, 53
Standard and Poor's Corp., 104
State Electronics Factory (*Valsts elektroniska fabrika*, VEF), 105–6
State President, 18, 21
'State Programme of Social Integration in Latvia', 78
Steen, Anton, 68
Stimson Doctrine, 138
Stranga, Aivars, 135, 137
Strasbourg, 143
Suicide, 38
Suny, Ronald, 9, 13
Support for East European Democracy Program (SEED), 138
Supreme Council, 74, 100, 121
Sweden, 30, 113, 129–30

Tallinn, 102, 110–11, 121, 136
Talsi, 109
Tariffs, 97
Tartu, 3, 5, 127
Taxes, 97–8
The First Ten Years of the Republic of Latvia, 19
'Third Awakening', 45
Third Reich, 21
Thomson, James, 125
Treaty of Unity and Co-operation, 122
Trenin, Dmitry, 137
Tsar Alexander I, 2
Tsar Nicholas II, 9

Ukrainians, 35, 56, 73
Ulmanis, Guntis, 71

Ulmanis, Karlis, 14, 18, 20–1, 23–4, 27, 71
Ulmanis regime, 153
UNESCO, 121
Unibanka, 103
United Nations (UN), 62, 121–22
United Nations Development Programme (UNDP), 78
United Nations General Assembly, 124
United States of America (USA), 46–7, 120, 124–5, 127, 135, 138-142, 145, 153
University of Dorpat (Tartu), 3, 5
University of Latvia, 48

VAK Environment Protection Club, 51–2, 70
'*Vadonisprincips*', 22
Valdemars, Krisjanis, 4
Valmiera, 38, 106
Varpa, Janis, 109
Ventspils (Vendau), 5, 8, 110, 123
Via Baltica, 110
Vidzeme, 9, 17, 70
Viksnins, George, 50
Vilnius, 63, 102, 110, 121
Vishinsky, Andrei, 25
Von Herder, Johann Gottfried, 1
Voss, Augusts, 37, 39–40
Vulfsons, Mavriks, 53
Vyborg Declaration, 10

Warsaw, 110
Working People's League, 25
World Bank, 81, 113
World Trade Organisation (WTO), 95
World War One (WWI), 11–12, 16, 23–4, 30, 108, 151–2
World War Two (WWII), 24–33, 41

Young Latvians (*Jaunlatviesi*), 3–5, 9–10

Yugoslavia, 127, 134

Zemgale, 12, 16–17, 70

Zepa, Brigita, 63
Zelbstschutz, 10

Lithuania
stepping westward

Thomas Lane

Lithuania

Postcommunist States and Nations

Books in the series

Belarus: A denationalized nation
David R. Marples

Armenia: At the crossroads
Joseph R. Masih and Robert O. Krikorian

Poland: The conquest of history
George Sanford

Kyrgyzstan: Central Asia's island of democracy?
John Anderson

Ukraine: Movement without change, change without movement
Marta Dyczok

The Czech Republic: A nation of velvet
Rick Fawn

Uzbekistan: Transition to authoritarianism on the silk road
Neil J. Melvin

Romania: The unfinished revolution
Steven D. Roper

Lithuania: Stepping westward
Thomas Lane

Latvia: The challenges of change
Artis Pabriks and Aldis Purs

Estonia: Independence and European integration
David J. Smith

Bulgaria: The uneven transition
Vesselin Dimitrov

FOR NICK, MAX, ANA AND BARBARA

TABLE OF CONTENTS

Chronology   ix

Preface   xiii

Introduction   xvii

Map of Lithuania   xxxix

Section I: Revival and Repression 1914–1985   1

1   Independent Lithuania between the Wars   1

2   Sovietization 1940–1985   49

Section II: Independence and the Politics of Transition
1985–1999   87

3   The Achievement of Independence 1985–1991   87

4   Government and Politics in Independent Lithuania   131

5   The Lithuanian Economy after Independence   163

6   Lithuania's Foreign and National Security Policy   199

Bibliography   225

Index   235

CHRONOLOGY

11th Century AD	Lithuania first mentioned in chronicles.
1200	The Pope summoned a crusade to convert the Baltic pagans to Christianity. The Teutonic Order of Knights was the Church's main instrument.
1236–63	Reign of Mindaugas, who unified the Lithuanian tribes in defence against the Teutonic Order.
1316–41	Reign of Grand Duke Gediminas; vigorous defence against the Teutonic Order combined with territorial expansion into Russian lands.
1323	Beginnings of the construction of Vilnius by Gediminas.
1385	Marriage of Grand Duke Jogaila to Queen Jadwiga of Poland: beginnings of Lithuania's conversion to Christianity as a result.
1392–1430	Reign of Grand Duke Vytautas: under him Lithuania achieved its greatest power and widest territorial extent, from the Baltic to the Black Sea.
1410	The Lithuanians and Poles inflict crushing defeat on the Teutonic Order at Zalgiris (Grunwald, Tannenberg).
1569	Treaty of Lublin between Poland and Lithuania establishes the Polish-Lithuanian Commonwealth in the face of the growing challenge of Moscow.
1697	Polish made the official language of the Commonwealth.
1714–80	Life of Kristijonas Donelaitis, one of the most celebrated Lithuanian poets.
1795	Third Partition of Poland, leading to the incorporation of Lithuania in the Russian Empire.
1830	Revolt against Russian rule begins the process of intense Russification.
1832	University of Vilna (Vilnius) closed down.
1861	Emancipation of Lithuanian peasantry.
1863	Second major revolt against Russia.
1863–1904	Intense opposition by Lithuanians to the policy of Russification: book smuggling from Lithuania Minor (in East Prussia) into Lithuania to circumvent the ban on books in the Latin script.
December 1905	Meeting of Lithuanian Assembly in Vilnius accompanying the revolution in Russia.
1911	Death of M.K. Ciurlionis, Lithuania's greatest painter and a distinguished composer.
1914	First World War begins.
1915	Occupation of Lithuania by the German Army.
1918	On 16 February Lithuania declared its independence, despite German obstruction, after Russia withdrew from the war.
1920	The Soviet-Lithuanian Peace Treaty signed.
October 1920	The Vilnius region seized by Polish General Zeligowski.
1920–25	Major land reforms initiated.
1922	Foundation of a university in Kaunas, which became the Vytautas Magnus University in 1930.

1922	Adoption of the Lithuanian constitution.
January 1923	Klaipeda (Memel) territory incorporated by force in Lithuania.
1926	Signing of the Soviet-Lithuanian Non-Aggression Treaty.
1926	Political coup; Antanas Smetona became President and remained so until 1940.
1934	*Entente* between the three Baltic states.
1938	A new constitution introduced, with greater powers for the presidency.
1938	Diplomatic relations established with Poland under threat of force.
March 1939	Germany seizes Klaipeda (Memel) from Lithuania and incorporates it into the Reich.
23 August 1939	Signature of the Molotov-Ribbentrop Pact.
28 August 1939	First Secret Protocol of the Molotov-Ribbentrop Pact allocating Lithuania to the German sphere of influence.
28 September 1939	Second Secret Protocol of the Molotov-Ribbentrop Pact transferring Lithuania from the German to the Soviet sphere.
10 October 1939	Pact of Defence and Mutual Assistance between Lithuania and the Soviet Union.
14 June 1940	Ultimatum from the Soviet Union to the Lithuanian government.
15 June 1940	Soviet occupation of Lithuania began.
21 July 1940	People's Assembly applied for membership of the USSR after fraud and intimidation in the election.
July 1940	Beginnings of Sovietization of Lithuania.
3 August 1940	Supreme Soviet of the USSR granted Lithuania's request for incorporation in the Soviet Union.
13/14 June 1941	Mass deportation of Lithuanians to the Soviet Union.
22 June 1941	Hitler attacked the Soviet Union and a Lithuanian uprising against the Soviet occupation forces took place.
June/July 1941	Lithuania occupied by German troops.
1941–43	The Holocaust: killing of virtually the entire population of Lithuanian Jews by the Nazis.
1944	Red Army reconquered Lithuania.
1944	Imprisonments, executions and mass deportations of Lithuanians to Siberia and the Arctic north resumed.
1944	Anti-Church campaign began, and continued through the following decades.
1944–52	The rise and fall of the Resistance movement.
1946–51	Collectivization campaign succeeded.
1953	Death of Stalin.
1956–57	Beginning of Khrushchev's 'thaw'.
1965	Brezhnev ends limited autonomy of the republics.
1972	The *Chronicle of the Catholic Church in Lithuania* began to circulate clandestinely.
1976	Formation of the Lithuania Helsinki Group.
1978	Formation of the Catholic Committee for the Defence of Believers' Rights.
1970s and 1980s	Slowing down of Soviet economic growth.
1985	Mikhail Gorbachev became General Secretary of the Communist Party of the Soviet Union; launched *perestroika* and *glasnost*.

1987	The 'Four Man Proposal' in Estonia proposing economic self-management for the Soviet republics became a theme of Lithuanian reformers.
1987–88	'Calendar' demonstrations took place.
3 June 1988	Foundation of *Sajudis*, the Lithuanian movement for reconstruction, in Lithuania.
Summer 1988	A series of mass rallies in support of *Sajudis* occurred.
October 1988	*Sajudis*'s founding congress.
March 1989	Elections to the Congress of People's Deputies: *Sajudis*-backed candidates swept the board.
May 1989	Communist Party of Lithuania endorsed the principle of Lithuanian sovereignty and condemned the Molotov-Ribbentrop Pact.
23 August 1989	Massive human chain linked up the three Baltic capitals in a demonstration against the Molotov-Ribbentrop Pact's secret clauses.
3 November 1989	Liberal citizenship law passed.
December 1989	Congress of People's Deputies eventually accepted the existence of the secret clauses of the Molotov-Ribbentrop Pact and declared them illegal under international law.
December 1989	The Communist Party of Lithuania separated itself from the CPSU and declared Lithuania a multi-party state.
January 1990	Gorbachev's visit to Lithuania a failure.
24 February 1990	Multi-party elections to Supreme Soviet of Lithuania: overwhelming victory for *Sajudis*-backed candidates.
11 March 1990	Supreme Soviet declared the sovereignty of the Lithuanian state.
March 1990	Formation of the first *Sajudis* government.
March 1990	Soviet sanctions against Lithuania began.
June 1990	Agreement to begin talks: end of sanctions.
December 1990	Gorbachev proposed strengthening central power at the expense of the republics: conservatives called for crackdown on opposition to the Soviet state.
13 January 1991	Soviet troops seized the TV tower and centre in Vilnius: 15 people shot or crushed by tanks.
9 February 1991	Referendum in Lithuania produced overwhelming support for independence.
29 July 1991	Russia recognized the independence of Lithuania.
19 August 1991	Attempted coup by conservatives against Gorbachev: failure owing mainly to resistance of Boris Yeltsin, president of the Russian Federation.
August/ September 1991	Lithuania received international diplomatic recognition and was admitted to the UN.
August 1991	Lithuania takes full responsibility for own affairs as an independent state; Landsbergis remained Chairman of the Lithuanian Supreme Council (Parliament) and therefore head of state; Vagnorius was Prime Minister.
November 1991	Prices of foodstuffs and manufactured products freed.
1 May 1992	The *talonas* introduced to supplement the Russian rouble as the medium of circulation.
May 1992	Landsbergis defeated in his attempt to create an executive presidency.
July 1992	Prime Minister Vagnorius is forced to resign by the Lithuanian Supreme Council.

October 1992	New Constitution approved.
25 October 1992	The Lithuanian Democratic Labour Party (LDDP—the former Communists) won a major victory in the parliamentary elections.
February 1993	Victory in the Presidential election to Algirdas Brazauskas of the LDDP.
1993–96	Government's economic reform programme implemented.
25 June 1993	Lithuanian currency, the *litas*, introduced.
31 August 1993	Russia completed the withdrawal of its troops from Lithuania.
4 January 1994	Lithuania applied for membership of NATO and joined the Partnership for Peace.
1994	Baltic Council of Ministers formed.
1 April 1994	Currency Board system introduced linking the *litas* to the US dollar and controlling inflation by strict limits on the money supply.
April 1994	Visit of President Walesa of Poland to Vilnius to sign the Lithuanian-Polish State Treaty.
1 January 1995	Free trade agreement between the EU and Lithuania.
January 1995	Most Favoured Nation treaty with Moscow implemented.
June 1995	Association Agreement with the EU signed.
December 1995	Lithuanian Bank Crisis.
October 1996	LDDP overwhelmed in parliamentary elections: victory for the right wing and centre parties, the Conservatives, the Christian Democrats and the Centre Union.
September 1997	Lithuania not included in the list of six states to begin negotiations for EU entry.
October 1997	Treaty with Russia on borders.
Autumn 1997	Presidential election campaign culminating in the election of Valdas Adamkus, a Lithuanian-American, defeating Arturas Paulauskas.
1998	Europe Agreement with the EU came into effect.

PREFACE

Lithuania is a small country with a long history. She restored her independence in August 1991 after a half century of submission to Soviet and Nazi totalitarianism. Some 73 years earlier, in February 1918, she had reclaimed her independence for the first time after more than a century of Russian rule. An independent mediaeval state, Lithuania became a partner with Poland in the Polish-Lithuanian Commonwealth in the late fourteenth century, a relationship which endured until the third partition of Poland-Lithuania by the surrounding great powers in 1795.

The generation of Lithuanians who regained independence for the second time in 1991 were deeply influenced by the history of their people. Soviet rule had severed their connections with western and central Europe, and their dearest wish, and the anchor of their foreign policy, was to 'return to Europe', whose history they had shared for centuries. Returning to their historical and cultural roots, however, did not mean an uncritical restoration of what had existed before. Just as the first re-establishment of independence did not restore the old Commonwealth, so the second revival did not recreate the republic of the inter-war years, despite strong pressure from some Lithuanians to do so. Shaped by their history in so many ways, Lithuanians have not been determined by it. They have adapted, and will continue to adapt, to the rapidly changing external security and political environment. What is unchanging, however, is a determination never to submit to tyranny, nor to be drawn back into a Russian sphere of influence. Alongside that, Lithuanians are resolved, even in a period of unprecedented change, to maintain their cultural identity.

This work may be described as a contemporary history of Lithuania; its major focus is on the last fifteen years of the twentieth century. However, the leading participants in the dramatic events which occurred in this period, such as Vytautas Landsbergis, the unlikely hero of the independence movement, were acutely aware of the history and culture of their country, which they had struggled assiduously to keep alive during the Soviet occupation. To understand the actions, policies, and values of the Lithuanian leadership and the people they represented during the campaign for independence requires some knowledge of the forces which shaped them. Accordingly this book deals quite extensively with the earlier history of Lithuania in the twentieth century, and throws a brief backward glance at the main themes of Lithuanian history from the mediaeval period to the First World War.

Nonetheless, it is the independence movement, the subsequent transition to democracy and the market, and the re-integration of Lithuania into western political, economic and security organizations which is the core of this book. There have been a number of scholarly histories of Lithuania in the twentieth century in English. Those which cover the contemporary period almost all conclude with the achievement of independence. The aim of this book is to take the analysis as far as possible into the processes of economic modernization and democratic maturation after independence. This longer perspective provides the opportunity to assess the direction and success of the domestic and foreign policies of successive post-independence Lithuanian governments. An intriguing complication of this period is the growing recognition that membership of the European Union, to which Lithuania aspires, will inevitably diminish her recently-recovered Lithuanian sovereignty. How will such a contradiction be received by the Lithuanian people?

Most histories of Lithuania in English have been written by members of the Lithuanian-American community. No one writing about Lithuania in this century can fail to be indebted to the distinguished work of Alfred Erich Senn and the late V. Stanley Vardys. I have also been strongly influenced in my understanding of recent Lithuanian history by the writings of Tomas Venclova and Aleksander Shtromas, both dissidents during the Soviet period, whose perceptions and humanity are a source of admiration. Many other historians, economists and political scientists have helped to clarify my views on the course of Lithuanian history, and their works are recorded in the bibliography. A number of these are from the Nordic countries which have been deeply interested in, and committed to, the process of transition in the Baltic states for the last fifteen years, and have strong historical links with the eastern Baltic. Yet there is a case for a history of Lithuania to be written from the vantage point of Great Britain which, over a long period, sustained economic and political links with Lithuania, and never recognized the occupation of Lithuania by the Soviet Union in international law. Moreover, the Lithuanian community in Britain, with its strong representation in the City of Bradford, helped to keep alive the memory of Lithuanian history and culture during the darkest years of the Cold War. A British historian may hope to bring to the writing of Lithuanian history both admiration and detachment, a recognition of the heroism and endurance of the Lithuanian people combined with reservations about aspects of contemporary Lithuanian nationalism, respect for the idea of the ethnic Lithuanian nation combined with a recognition of the importance in the Lithuanian past and present of ethnic pluralism. And no one writing about Lithuania can fail to reflect on the centuries-old relationship with neighbouring Poland. My particular debt to the work of the doyen of historians of Poland in English, Norman Davies, is gratefully acknowledged here.

My curiosity about the Baltic states has been nurtured over the years by my colleagues in the Baltic Research Unit at Bradford, and by a succession of visiting fellows and researchers from the region. My greatest intellectual debt is to my friend and colleague John Hiden who has encouraged my interest by innumerable conversations over the past fifteen years, and has demonstrated by his own energetic example how to combine the writing of perceptive history with the study of contemporary developments. I have also learned a great deal from discussions with other colleagues in the Baltic Research Unit: David Smith, Leonidas Donskis and Darius Furmonavicius. I am especially grateful to David Smith for his careful reading of the manuscript. As well as pointing out errors and obscurities, he made some welcome suggestions about the structure of the book. Others who have influenced my understanding of Lithuanian history and foreign relations, particularly relations with Poland, either by their writings or their conversations, include Alfonsas Eidintas, Stasys Vaitekunas, Elzbieta Stadtmüller, Yves Plasseraud, R.G. Hurner, and David Kirby. It goes without saying that I take full responsibility for the final version of this book.

I am grateful to the staff at Harwood Academic Publishers for commissioning this book, and for their sympathetic, helpful and tolerant approach, especially when the book began to slip behind schedule.

Finally, I am delighted to record my gratitude to my family, from whom I have, as usual, received enormous support. Jean, as always, has been a model of understanding and encouragement and Nick, Max, Ana and Barbara, to whom this book is dedicated, have been interested and stimulating from first to last.

INTRODUCTION

In the nineteenth century it was commonplace to divide nationality groups into 'historic' and 'non-historic' categories. What distinguished one from the other was the past existence of statehood. In central and eastern Europe the Poles, Czechs and Hungarians were prominent examples of historic nations whose states had been destroyed and territories absorbed by neighbouring powers. Following the nineteenth-century growth of national self-consciousness, with its associated demands for national self-determination, the historic nations became the prime candidates for the restoration of statehood. It was generally assumed that the remaining 'non-historic' nations would achieve some form of autonomy within the structure of existing or soon-to-be-created states, and gain protection for their rights under minority legislation. The Lithuanians were believed to fall into this category.

The victorious Entente Powers who created the map of post-war Europe during the peace negotiations at Versailles in 1919, expected that Lithuania would again become part of Russia, as it had been since the Third Partition of Poland-Lithuania in 1795, whatever the outcome of the Russian civil war. Despite Lenin's rhetoric about self-determination, the Bolsheviks would have incorporated Lithuania and the other Baltic nations into Soviet Russia had they not been repulsed in the various wars on the western borders of Russia after the Armistice of 1918. Poland, its statehood recreated in the peace settlement and aspiring to great-power status in central Europe, assumed that Lithuania would be re-incorporated in it, either as part of a reconstituted and federally-organised Polish-Lithuanian Commonwealth or as an integral part of a unitary Polish state.

For the most part Poles did not take seriously Lithuanian claims to self-determination, believing that the Lithuanians were being deceived or manipulated by Germany, which preferred an independent Lithuania to one incorporated in Poland. After all, Lithuania was small and economically backward, its large estate owners were Polish or Polonized Lithuanians who identified with Poland, its peasantry were ethnically divided between Lithuanians, White Russians and Poles, and its commercial activities were carried on by Jews. It was conceded that there was a small and growing professional middle class and intelligentsia, but this was not taken seriously as the embodiment of Lithuanian national aspirations. For these reasons it was widely held that Lithuania could not sustain independence and would need the 'protection' of Poland. Polish attitudes towards their Lithuanian cousins were, on the one hand, patronising and condescending, and on the other, protective and affectionate. In some cases the relationship was even closer, as blood brothers aligned with one side or the other in the 'family' dispute.

The nineteenth-century Lithuanian national renaissance had created very different perceptions and aspirations among the Lithuanians. Starting with the objective of preserving and developing Lithuanian language and culture, Lithuanian nationalist leaders, by the time of the Russian Revolution of 1905, were seeking to achieve autonomy within a constitutional and democratic Russia. Towards the end of the First World War, however, the ambition for autonomy was replaced by a drive for independence. Lying behind this change were a number of factors: betrayal by Russia of promises made during the revolutionary turmoil of 1905; the defeat of both Russia and Germany in the First World War and their subsequent marginalisation in European politics for a crucial period; the skilful opportunism of the embryonic Lithuanian leadership combined with the publicity and financial assistance given to the cause by the émigré Lithuanian communities; and the rapid and effective mobilisation of military forces by the Lithuanian government to defend the infant state. Added to this was the reluctance of the Polish leadership to annex Lithuania by force for fear of League of Nations' sanctions, and, in Piłsudksi's case, out of a genuine, if exasperated, affection for his native Lithuania.

Underlying Lithuanian success in achieving independence was a deeply-held conviction among the nationalist middle class leadership, resting on wide support among the peasantry, that the Lithuanian case for independence was no weaker than that of the so-called historic nations. Indeed, in the face of western indifference and Polish antagonism, they vigorously asserted their claim that Lithuania *was* an historic nation. It followed that Lithuania should be seen as *restoring* its independence, just as Poland was doing. This interpretation may not have made many converts outside Lithuania, but it steeled the resolve of the Lithuanians in the face of almost insuperable difficulties after the First World War.

One of the keys to Lithuanian history is, in fact, the determination of most Lithuanians to survive in the face of extreme oppression by foreign rulers. Indeed, the harsher the oppression, notably after the great anti-Tsarist revolt of 1863 or again in the post-1945 period, the more resolute the Lithuanians seemed to become. By contrast, the Union of Lublin of 1569 which united Poland and Lithuania in a Common-wealth, achieved a marked Polonization of the Lithuanian upper classes and some Polonization of the peasantry by a combination of the seductiveness of Polish culture and the practical benefits of adopting the Polish language. But even then the Lithuanian language survived among some of the peasantry and lesser nobility, Lithuanian folk culture resisted assimilation, and Lithuanian-language books were published. The collective memory, usually passed down in oral form from generation to generation, sustained a sense of Lithuania's unique achievements in medieval Europe. Systematic study of Lithuanian culture and history in the eighteenth century by linguists and folklorists ensured that its distinctiveness and value became widely known among scholars. Attempts to extinguish this culture by the Tsarist governments in the nineteenth century only stimulated the resolve to defend it on the part of scholars, publicists, clergy, poets and historians. One may not agree with Senn's claim that at the beginning of the twentieth century the word 'Lithuanian' only 'referred to a language spoken by a people living on the South-East shore of the Baltic where there had once been a medieval state', yet the continued existence of that language was itself a tribute to the resistance of the Lithuanian people in the face of generations-long attempts to destroy their identity.[1]

The subterranean persistence of Lithuanian culture over centuries places the assimilationist attempts of the Soviet era in perspective. The period of Lithuanian independence between the wars reinforced the foundations of Lithuanian culture through the growth of education, the encouragement of the arts, and the dissemination of scholarly research into the Lithuanian past. By contrast the Soviet occupation of Lithuania in 1940-41 and again after 1944 was a deliberate and brutal attempt to destroy all that was meant by the word 'Lithuania', at first by arrests, imprisonment and executions of the leadership groups, and then by mass deportations of hundreds of thousands of ordinary Lithua-nians. Despite pessimistic assessments that Lithuanian culture was doomed to disappear along with the Lithuanian state, resistance in various forms was maintained. As in the nineteenth century, the saving grace for Lithuanians was the tenacity of the collective memory in the face of a repressive official ideology. For most of their existence since the medieval period, therefore, Lithuanians have retained the integrity of their mental world. This has provided them with both the tools for resistance and an ambition to shape their own future without external direction. An understanding of post-Communist Lithuania requires, therefore, some insight into the Lithuanians' mental world, their historical perceptions, and their collective memory.

THE MEDIAEVAL EXPERIENCE

By the 11th century AD the various Baltic peoples who had settled along the south-eastern shores of the Baltic Sea between the Vistula and Daugava rivers had become distinguishable from each other, and the name Lithuania appears for the first time in the chronicles of the period.[2] At this time the Lithuanians were largely concentrated in the Nemunas (Niemen) river basin. To the West and North respectively were their fellow Balts, the Prussians and the Letts (later Latvians). All were pagan tribes sharing a common geographical origin, supposedly in the region of the Caspian Sea. The Lithuanians engaged in some trade but remained largely isolated in their remote forests and swamps. This isolation came to an end following efforts to convert them to Christianity. In 1200 the Pope summoned a crusade for this purpose and two orders of knights were formed to execute the task: the Knights of the Cross in present-day Poland, and the Knights of the Sword in Riga, among the Letts. The Teutonic Order, to adopt the name given to the two organizations after their merger, was unscrupulous in its use of force and terror as a means of conversion. Indeed, extermination was often a substitute for conversion, notably in the crusade against the Prussians after 1230. A less bloodthirsty route was taken among the Letts to the north. When the two orders of Knights tried to link up along the shoreline between Prussia and the centres of the Lettish tribes to the north the Lithuanians themselves were directly threatened. This danger provided the stimulus to unite the hitherto separate Lithuanian tribes.[3]

Although the threat from the Teutonic Knights was the major stimulus to unity, another contributory factor was the desirability of joint military expeditions against various Russian princes, and the requirements of defence against the advancing Tatar forces in the East.[4] During the next two centuries, under the vigorous leadership of a succession of princes such as Gediminas, Algirdas, Kestutis and Vytautas, Lithuania maintained the dual policy of vigorous defence against the Teutonic Order in the West and territorial expansion into Russian lands in the East, the latter being effected by a series of victories against the Tatars and a policy of judicious intermarriages with Russian princely families.[5] Among the Lithuanian leaders of this period

Gediminas made serious attempts to consolidate Lithuania's position in the West and to end her isolation by establishing relations with various European countries. As part of this process he extended a welcome to foreign artisans, craftsmen, agriculturalists and traders, and established trade links with the Hanseatic League. Not the least of his achievements was to found the town of Vilnius around his great castle, construction of which began in 1323. This became and has remained for Lithuanians their capital city and most important cultural centre.[6]

Following the joint reigns of Gediminas's two sons, Algirdas and Kestutis, who ruled for over three decades after 1345, there occurred one of those fundamental decisions which shape a country's history and which affected the course of European development. Jogaila, the son of Algirdas and a Russian mother, considered contracting a dynastic marriage with a Russian princess and converting to Orthodoxy. But the relative weakening of Lithuania's position *vis-à-vis* the Teutonic Order persuaded him instead to ally with Poland, which offered him marriage with Jadwiga, their newly-crowned Queen. The famous agreement of 1385 had a price. Jogaila had to consent to become a Christian, to convert his country to Christianity, and to join Lithuania to Poland in perpetuity. This settlement had two major political consequences: it aligned Lithuania with the Catholic West rather than the Orthodox East; and it marked the beginning of the Polonization of Lithuania and its conversion to Christianity.[7]

The significance of this decision was well understood in Lithuania. Vytautas opposed his cousin Jogaila's marriage and resisted attempts by the Polish leadership to annex Lithuania. Under a treaty of 1392 with Jogaila, Vytautas became the grand duke of Lithuania but he was unable to sever ties with Poland owing to Polish resistance.[8] Under his leadership, however, Lithuania achieved its greatest power and widest territorial extent, from the Baltic to the Black Sea and from Moscow to Poland. In the process it engaged in decisive battles with the Tatars and the Teutonic Knights, the historic Lithuanian-Polish victory over the latter at Zalgiris (Grunwald, Tannenberg) in 1410 marking the end of their persistent challenge to the Lithuanian state.[9]

Vytautas was only the first of many Lithuanians who wanted to reverse the dynastic union with Poland. This course was not dictated by hostility to, or ignorance of, the Latin West; indeed Vytautas spoke both German and Latin and had travelled in Western and Southern Europe. He was concerned above all for Lithuanian independence which he believed could only be achieved if the Grand Duchy were elevated into a kingdom, with himself as crowned monarch. Independence would grant Lithuania equal status with Poland and free it from Polish intervention. Vytautas's ambition was to be denied both by Polish obstruction and his own death.[10]

The reign of Vytautas the Great was the high-water mark of the mediaeval Lithuanian state. Nineteenth-century Lithuanian nationalists invoked the names of Vytautas and his predecessors in claiming historic status for the Lithuanian nation. Moreover, the Lithuanian state was not only historic but had an imperial character; its territory was extremely extensive, and it ruled over Russians, Belarussians, Jews and Tatars, who together outnumbered the ethnic Lithuanians by nine to one. The growth of trade and city populations attracted merchants, bankers and artisans of many different nationalities. Lithuania was not, however, a well-integrated state, lacking a centralised administrative system to compensate for its national and religious diversity. It adopted a tolerant attitude to the religion, language and customs of its territorial acquisitions. Moreover, it utilised the old Slavonic form of Russian as the language of its diplomacy. All the signs are that Russians would increasingly have assumed positions of leadership in the Lithuanian state, which 'would probably soon have become Russified and reverted to the Byzantine civilization of the East', had it not been for dynastic union with Poland.[11]

The leaders of the nineteenth-century Lithuanian cultural renaissance invoked both the traditions of the mediaeval Lithuanian state and Lithuania's linguistic and cultural inheritance. It was probably the long period of independence and pagan religion which enabled Lithuanian peasant culture to put down such deep roots which not even the forced conversion to western Christianity was able to eradicate. The dynastic union of 1385, furthermore, left Lithuania with its own administrative system, code of laws, army and treasury, thus ensuring that the independence of the state was not immediately undermined.[12]

The persistence of ethnic Lithuanian culture was attributable to a combination of language and religion. It is widely appreciated that Lithuanian is one of the the oldest Indo-European tongues, studied by philologists for the insights it offers into the evolution of language. 'If you wish' said Goethe 'to hear how our ancestors spoke, listen to the Lithuanians'. It has close ties to Sanskrit and some affinity with Greek and Latin. Although it was not written until after the Reformation in the sixteenth century, and the oldest literary document in Lithuania is reputed to be a translation of a hymn which appeared in 1545, it remained the language of most Lithuanian peasants through the centuries, despite the adoption of Polish by the gentry and nobility. It was also the language of the *dainos*, folk songs connected with ancient Lithuanian rites, some of which were incorporated into Christian church services, and were 'discovered' by folklorists in the eighteenth and nineteenth centuries when they were preserved in written form.[13]

Though the conversion to Christianity sometimes encouraged attempts to destroy pagan survivals, generally there was a degree of tolerance towards the old religion. Lithuanian paganism's central characteristic was a veneration for nature which it held to be sacred. Various elements in nature were personified in the form of deities and were propitiated by the rituals of worship.[14] Christianity seems to have been careful to incorporate aspects of the pagan religion into its own rites and symbols. Pagan altars were moved into Christian churches, the pagan shrine at the bottom of Gediminas hill in Vilnius was converted into a cathedral, the Christian cross incorporated the signs of the sun, moon and stars, and some of the dramatic pagan rituals and ceremonies were introduced into Christian religious services. Experts on ancient Lithuanian mythology suggest that the Christian All Souls' Day was welded on to the existing pagan ritual of celebrating the holiday of dead souls.[15]

POLAND'S JUNIOR PARTNER

In the sixteenth century the rise of Muscovy shifted the balance of power in Eastern Europe. Seeking to bring all Russian lands within its ambit Moscow represented a potent threat to Poland-Lithuania, which had absorbed many Russian principalities in the course of its expansion. The Muscovite challenge provided the main inducement for the Poles and Lithuanians to strengthen their dual state by signing the famous Treaty of Lublin in 1569.[16]

The Lublin settlement provided for a joint elective monarch as head of state and joint meetings of the parliaments of the two countries. Lithuania retained its own laws, law courts, currency, army and treasury, the title of Grand Duke for the monarch, and the great seal of state. The significance of Lublin has been disputed by historians. On the one hand the continuation of separate institutions meant that the two states could have almost independent existences. The title of the state was Commonwealth of the Two Nations (*Rzeczpospolita Oboiga Narodov*) in which the two partners enjoyed equality, even though the *de facto* state language was Polish. But though there was formal constitutional equality, in practice Lithuania suffered a loss of individuality becoming, in Norem's words, the 'minor partner' in the relationship. In fact Lublin, so Vardys believed, ultimately reduced Lithuania to a dependent 'provincial status'.[17]

Arguably the acceleration of certain long-term social trends after Lublin was ultimately more responsible for the supremacy of Poland in the Commonwealth than the treaty itself. The generations-long process of cultural 'Polonization', reinforced by inter-marriage, led to the Lithuanian nobility and gentry voluntarily accepting Polish culture and language. This process more or less obliterated the cultural differences between the Polish and Lithuanian upper classes. The fact that the Polish language was not made the official language of the Common-wealth until 1697, more than three centuries since the dynastic union between the states, testifies to the voluntary nature of the cultural assimilation. From then on, if not before, it was assumed in Poland that Vilnius (or Wilno in Polish) was as Polish as Warsaw. With some exceptions, the Lithuanian upper classes now saw themselves primarily as Poles, albeit of a Lithuanian ethnic background. Polonization had introduced them to a more developed and sophisticated culture, greater wealth, superior manners, the attractions of a higher civilization, and an entrée into the western world of the Renaissance.[18]

Despite these advantages for the Lithuanian upper classes, the Commonwealth's survival as an independent state was questionable in the face of grave internal political weaknesses. Muscovy's remorseless determination to acquire Slavic lands on its western and southern frontiers, and attacks from the North by Protestant Sweden, exploited these weaknesses to the full. Then, in the last third of the eighteenth century, Russian, Austrian and Prussian territorial ambitions were fulfilled with the conquest and division of Poland-Lithuania as a result of the three Partitions. Although Lithuania retained its name for a while, after the revolt against Russian rule in 1830 it became part of the Northwest Territory of the Russian Empire and lost its separate political identity.[19]

THE UNCOVERING OF LITHUANIAN CULTURE

In the late eighteenth and early nineteenth centuries knowledge and appreciation of Lithuanian language and culture became better known among small groups of scholars, philologists and folklorists. The original stimulus to this investigation of Lithuanian peasant culture was practical in nature; it was linked to the Protestant Reformation and a movement to improve the material conditions of the peasantry who were still enchained by serfdom. It first took root in Protestant Lithuania Minor, not in the Catholic Lithuanian heartland, and had a most important role to play in the Lithuanian cultural renaissance of the late nineteenth century.[20]

The emphasis in Protestantism on direct access to the Word of God through the Bible encouraged the Lutheran church to translate religious works into Lithuanian. In 1547 Luther's Catechism was translated, to be followed at intervals by translations of the Gospels and, in 1701, the entire New Testament. The University of Koenigsberg, founded in 1544, trained ministers of Lithuanian background for pastoral work.[21] The Reformation stimulated a new interest in Lithuanian language and culture which laid the foundations for the Pietist and Romantic movements in the late eighteenth and early nineteenth centuries. One of the most celebrated poets in the Lithuanian language, Kristjonas Donelaitis (1714–1780), who was born in Lithuania Minor and spent his life there as a Protestant minister, took advantage of the opportunity offered to theological students by the University of Koenigsberg to enrich his knowledge of Lithuanian. In his narrative poem, 'The Seasons', which, to quote Harrison, 'abounds in local and ethnographic colour', he creates a vivid picture of nature and country living and depicts the everyday life of the Lithuanian serfs in Lithuania Minor. He speaks up against the increasing threat of Germanization, praises Lithuanian folk culture, and urges the peasants to hold on to what they have inherited from their parents. It is often said that Donelaitis was the first writer in Lithuanian to reveal the richness and power of the language. His work, which was known about before it was published several decades later in 1818, quickened the development of a Lithuanian national consciousness.[22]

Donelaitis's writings in Lithuanian did not stand alone in the eighteenth century; there were a number of books printed in Prussian Lithuania, notably grammars, a dictionary, and collections of folksongs. And in the early nineteenth century scholars at the University of Vilnius undertook research into Lithuanian folklore, customs and history. The most notable of these figures was Simonas Daukantas (1793–1864), who published an anthology of Lithuanian folklore, studied the culture of ancient Lithuania (*The Character of the Ancient Lithuanians*), and wrote three historical works on Lithuania up to the Lublin Union.

Vytautas Vanagas believes that for Daukantas the past 'was a source of moral power from which he derived the strength to raise national consciousness'. It was axiomatic that the ultimate objective was to achieve political independence which alone could guarantee protection of the national identity.[23]

The uncovering of Lithuanian peasant culture was accelerated by the German Romantics, Lessing, Herder and Goethe, who became interested in Lithuanian folk songs, poetry, customs and history which were being recorded, translated and published by scholars at the University of Koenigsberg. The Romantics were struck by the beauty and richness of Lithuanian folk rites and the songs which accompanied them. Moreover, the Lithuanian language, which had been widely considered as merely a peasant dialect, now gained a new standing as an ancient Indo-European language.

This rediscovery of Lithuanian folk culture was significant in interpreting Lithuanian history and creating a vision of the future. The romantics held that folk culture was the original expression of a people's artistic activity and the 'repository' of its history and myths. It offered the most reliable evidence of national character and identity. It was a measure of a people's uniqueness. Folk culture could therefore be used by political activists to subvert the corrupt contemporary order and to build the foundations for its replacement. Folk culture was, as it were, a prelapsarian phenomenon offering inspiration to rebels and dissidents in the existing alien and oppressive regimes. The 'rescue' of this culture and its substantial reinforcement over almost two centuries produced in Lithuania an intellectual and cultural steadfastness strong enough to withstand the severest repressions of the Tsarist and Stalinist periods.[24]

RUSSIFICATION—AND LITHUANIAN RESISTANCE

The great Polish-Lithuanian revolts against Tsarist rule in 1830 and 1863 marked the inception of an intense policy of Russification in the territory of the former Commonwealth. The general thrust of the repressive measures was anti-Polish; indeed some attempt was made in the 1860s to conciliate Lithuanian sentiment and to split the Lithuanian peasants from their Polish landlords. But any positive effects of this policy were completely nullified by other heavy-handed Russification measures. Russia's repressive measures after the uprisings were characteristic: hangings, sentencing to penal servitude, and internal exile for the leaders. 45,000 Polish families were transported to Russia and Russian colonists were established in their place. The Polish-language University in Vilnius (Vilna in Russian) was closed down in 1832, many of its academic staff were imprisoned or deported, and the holdings of its library were dispersed in Russian collections. Polish schools were shut and all education had to be conducted in the Russian language. Later, Poles were taxed more heavily than Russians, and the use of the Polish language in papers, plays and on shop signs, for example, was prohibited. Church property was confiscated and churches and religious houses were closed. The Orthodox Church was authorized to implement a Russification policy.[25]

The rationale behind the policy of extreme repression was to extirpate Polish culture and to 'liquidate the Polish question' from the Lithuanian territory. It was hoped that these measures would guarantee political stability and, by separating Lithuanians from Polish influence, make them more responsive to assimilationist policies, or at least strengthen the links between Lithuanians and Russians.[26]

The policy of divide and rule was, however, totally undermined by an egregious error on the part of the Tsarist government, namely the decision to prohibit the publication of Lithuanian books and journals in Roman type and to insist that Cyrillic print be used instead. This law aroused implacable opposition among the new and growing Lithuanian intelligentsia, a number of whom were bishops and priests of the Roman Catholic Church, since it seemed to be an attempt by the Russians to convert the Lithuanians to Orthodoxy. For believers the Roman alphabet had a profound religious significance since it helped to distinguish the Latin church from its Orthodox rival. This measure fortified the Lithuanian national renaissance of the later nineteenth century.[27] Renaissance, however, implies education and literacy. The Lithuanian lands, defined for our present purposes as Lithuanian majority areas, namely the province of Kovno (Kaunas) and five districts of the Vilna (Vilnius) and Suwalki provinces, were populated almost exclusively by peasants. The vast majority of the population of the larger towns were ethnic Poles, Jews and Russians, meaning that there was virtually no ethnic Lithuanian middle class or proletariat. In the Russian Census of 1897, for example, only 14 Lithuanians were identified as merchants. The leaders and supporters of the renaissance, therefore, did not emerge from a well-established urban middle class but from the first generation of the peasantry to acquire a formal education. The appearance of this group, many of whom received their higher education in Russian universities, was dependent on the abolition of serfdom in 1861. It is worth remarking that many of the early members of the nationalist movement in Lithuania, as von Rauch noticed, came from the province of Suwalki, which had been allocated to Prussia during the Partitions, and where the peasants received their freedom under Napoleonic rule as early as 1807.[28]

The emancipation in Lithuania was achieved on more favourable terms for the peasantry than in the rest of the Russian empire. This was part of the Russian strategy of divide and rule by splitting off the Polish landlords from the Lithuanian peasantry during the uprising of 1863.[29] The emancipation act helped some of the Lithuanian peasantry to become more prosperous. They offset their redemption debts by improved agricultural methods, the draining of swampland, the establishment of cooperatives, and the purchase of landlords' land, often using remittances from relatives in the large Lithuanian-American community in the United States for this purpose (a community which had grown significantly as a result of substantial flows of immigration from Lithuania). It is likely that most peasant holdings were large enough to support the owners and their families. It was from this relatively prosperous section of peasant society that a group of men emerged, educated either in the seminary or in Russian universities, acquainted with their country's history and language, and trained in medicine, law, engineering and theology. It was they who became the leaders of the Lithuanian cultural renaissance and who formed the political, administrative, judicial and cultural elites in the independent Lithuanian state after 1918. Coming from a peasant background the new leaders were egalitarian in sympathies, reformist, democratic and, above all, uninterested in the Polish landlords' objective of restoring the old Commonwealth. Instead, they wanted self-determination for Lithuania within its ethnographic boundaries, free from both Russian and Polish domination. But their first task was to raise Lithuanian national consciousness.[30]

There were already substantial foundations on which to build. The University of Vilna (Vilnius), suppressed after the 1832 revolt, had been a centre of the study of Lithuanian history, culture, language and law under leading scholars such as Lelewel, Onacewicz and Danilowicz. These studies continued between the two uprisings, exemplified by the work of the Romantic writer and linguist Stanevicius, the historian and lexicographer Poska and the historian and littérateur Daukantas, who tried to recreate the life of the ancient Lithuanians in *The Character of*

Ancient Lithuanians and Samogitians. These writers' purpose was to deepen understanding and love for Lithuania and to increase Lithuanian national awareness. One of the most significant figures in the Lithuanian renaissance after 1863 was Bishop Motiejus Valancius. Described as the 'greatest Lithuanian personality in the nineteenth century' he bridged the divide created by the 1863 revolt and established the strategy of Lithuanian resistance to intensified Russification. An historian and a writer of spiritual and secular works aimed at the Lithuanian peasants, he was most influential during the period of intense Russification after 1863. Unyielding in his opposition to the attempted imposition of the Cyrillic alphabet, Valancius placed the Catholic Church at the head of the opposition to Russification and hastened the identification of Lithuanian Catholicism with national resistance. This is not to say that all nationalists were religious believers, but the Church, in opposing the denationalization of Lithuania, created the tradition of Catholic resistance to Russification, whether in its Orthodox or atheistic guise.[31]

Valancius' aim was not only to stimulate resistance but also to meet the religious, practical and moral needs of the Lithuanian peasants by increasing the level of literacy and by disseminating religious and moral texts in the Lithuanian language. He encouraged his clergy to become active scholars and writers, and converted some religious seminaries into centres for the study of the Lithuanian language and culture. The Church's leading role in opposing Russification enhanced its reputation in the eyes of the Lithuanian peasantry. This encouraged able sons of Lithuanian peasants to become priests, weakening the long-standing connections between the Church and Polonism.[32]

Since it was impossible to publish works in the Latin script in Lithuania after the Tsarist ban, Valancius made himself responsible for planning and organizing the smuggling into Lithuania of books, journals, prayer books and almanacs in the Latin print from the Lithuanian community in East Prussia. The Roman Catholic clergy played a significant role in this clandestine operation which lasted for four decades until the ban was withdrawn in 1904. The purpose of these works was to increase religious devotion, to teach literacy and disseminate information about Lithuania's history and culture, as well as offering a critique of Russian rule.[33]

With the passage of time the role of the Church in book smuggling became relatively less important, as secular works became more numerous. East Prussia offered a kind of safe haven for Lithuanian book publishing, a home for the Society of the Friends of Lithuania, which supported publication and provided the resources to enable Lithuania to survive the intense Russification campaign. Book smugglers have rightly entered the ranks of Lithuanian national heroes. Punishments for those caught were severe but there were always other volunteers to take the place of those arrested. Hundreds of thousands, not to say millions of copies of publications found their way across the border from the East Prussian town of Tilsit, which was the centre of Lithuanian publishing.[34] Between 1891 and 1902 alone the Russian customs seized 172,000 items, but their strike rate was apparently not high. The majority of publications ended up in secret hiding places all over Lithuania, used by peasant families and in clandestine Lithuanian language schools. By focusing on language, history and culture, myth and folklore, these works inculcated a knowledge and appreciation of Lithuanian identity, inspired the growth of national consciousness, and developed embryonic feelings of political self-direction among the Lithuanian people. The political and cultural perceptions of at least two generations of Lithuanians were sharpened by this imported literature.[35]

The most famous of these periodicals was *Ausra* (*Dawn*) which appeared between 1883 and 1886. Founded (and edited until 1885), by Juonas Basanavicius, one of the great figures of the Lithuanian renaissance, it nourished among its readers a love for the culture, language and history of Lithuania. It pressed for the establishment of Lithuanian language schools and use of Lithuanian in local administration. In attempting to recapture the distinctiveness of Lithuania's past it celebrated the paganism of the period before Christian conversion. Its advocacy of economic development and egalitarian principles appealed to the peasantry but dismayed the Polish landlord class. In the atheism or agnosticism of its editorial board, its idealisation of paganism, and its perception that the Roman Catholic Church was the agent of Polonisation in Lithuania the journal alienated most of the clergy who, until then, had been the main defenders of Lithuanian culture against Russian persecution. On the other hand, a small but growing number of clerics followed the logic of nationalism by trying to remove the Church from Polish control.[36]

Another journal, *Sviesa*, also published in Tilsit between 1887 and 1890, was more sympathetic to the existing social order though, like *Ausra*, it advocated improvements in the agricultural economy and promoted economic development. Unlike *Ausra* it identified much more with the Catholic Church and the Polish-speaking landowning class who it hoped would adopt the Lithuanian language once more. Far more radical was the journal *Varpas* (*The Bell*) published in Tilsit between 1889 and 1905. Edited by Vincas Kudirka it expressed the opinions and values of radical and secular nationalists. Like the other journals it sought to raise national consciousness but made the clear distinction between nationality and religion. Unlike *Ausra* it was clearly political with an increasingly socialist orientation. It adopted a very critical stance towards the Tsarist authorities and ridiculed the Russian bureaucracy. At the centre of its demands was the restoration of the Latin text for Lithuanian publications. Other journals published outside Lithuania included *Apzvalga* (*Review*), *Tevynes Sargas* (*Guardian of the Fatherland*), *Naujienos* (*News*), and *Ukininkas* (*Peasant*).[37]

By the 1890s it was clear that there was a split in the ranks of the renaissance between secular nationalists and the priesthood, the latter on the whole being opposed to a Lithuanian national state. The secular forces drew their strength from the new Lithuanian intelligentsia who had been educated in Russian universities and had grown away from the Catholicism of the Lithuanian peasantry from whom they had mainly sprung. The Tsarist regime, in offering help and encouragement to young Lithuanians to enter Russian universities, hoped to Russify them, weakening their attachment to both Lithuania and Poland. Instead, these young people joined Lithuanian student societies, learned about their own language, history and culture, and graduated as lawyers, doctors, and other professionals. This was the first generation of politically active nationalists to repudiate Russification and Polonism. Also a number of them became socialists or revolutionary radicals and were among the first to join the Social Democratic Party of Lithuania in 1896. In their opposition to privilege and inequality they represented the inclinations of their peasant forebears.[38]

It was not easy for these educated Lithuanians to find employment in the Lithuanian provinces since ethnic Lithuanians were banned from the civil service, and opportunities in law and medicine were hard to obtain. At the end of the nineteenth century perhaps 1,500 educated Lithuanians remained in other provinces of Russia. They became a very valuable source of educated personnel for Lithuania after independence. Nevertheless, an increasing proportion of these graduates returned to work in their own country and supported the nationalist movement. In 1901 the Lithuanians received St. Michael's Church in Vilnius for services in the Lithuanian language and it became a centre for Lithuanian activities. At this point, as Eidintas has shown, the movement for the restoration of the Lithuanian state as the successor of the medieval Grand Duchy began to take hold. However, the first demands of the Lithuanian movement were for autonomy within the Russian empire, involving Lithuanian language education, an end to the ban on the Roman alphabet, freedom for the Roman Catholic religion, and civil and political rights.[39]

POLITICAL RESURGENCE

The opportunity for the Lithuanian national movement to express itself politically and to achieve some of its most pressing demands arose as a result of the Russo–Japanese War of 1904. In the face of external conflict and internal economic difficulties the Tsarist government needed to conciliate some of its domestic opponents. Moreover, it finally recognised that the ban on the Roman alphabet for Lithuanian books had failed in its main purpose of Russifying the Lithuanian population. Accordingly the ban was removed and the use of the Lithuanian language in primary schools was permitted. This gave an immediate stimulus to the national renaissance. A number of newspapers began to be published in the Lithuanian provinces and one of them, *Vilniaus Zinios* (*Vilna News*), issued a call for a meeting of a Lithuanian Assembly in Vilnius in December 1905. Almost 2,000 delegates were chosen from all parts of Lithuania, from each of the embryonic political parties, and from Lithuanians living in other parts of Russia. This impressive gathering represented a landmark in Lithuanian nationalism, marking the transition from vague national aspirations to a concrete set of political demands.[40]

The chief demand of the manifesto issued by this assembly was a call for political autonomy within a Russian federation. The main resolution read as follows:

'Only self government will satisfy the aspirations of the Lithuanian people. Lithuania must therefore be resuscitated within her ethnographic boundaries as an autonomous State in the Russian Empire. Her relations with other Russian States must be established upon a federative basis. Vilnius will be the capital of the country and the seat of parliament. The latter will be elected by general, secret and direct ballot, in which women will also participate ...'

'The Lithuanian language is the official language. The schools must be the nurseries of the Lithuanian spirit and must be directed by teachers freely chosen. Good wishes for further success shall be expressed to all Lithuanians who, in the Vilnius government, are fighting against Polonization.'

This resolution shows that Lithuanian self-assertion, which had developed in the Russification period, was directed against Polish as well as Russian attempts to assimilate Lithuanian culture. After the restoration of independence in 1918 Polonization succeeded Russification as the greatest threat to Lithuanian cultural identity, as it had been before the Partitions.[41]

The Assembly decided to pursue the objective of autonomy by non-violent means, namely a refusal to pay taxes and to undertake military service, the withdrawal of children from Russian schools, and participation in strikes. Demands were also made for Lithuanian to be the language of local government and of instruction in all, not just primary, schools. However, when attempts were made to pursue these tactics the Russian government cracked down and many activists were imprisoned or exiled while others escaped abroad. Nonetheless, until the outbreak of war in 1914, the Lithuanians raised national self-awareness through various publications and patriotic organizations, promoted economic development, and increased living standards. The Stolypin reforms, the introduction of the Lithuanian language in primary schools, a greater freedom of association and expression, and a license to establish cultural and scientific societies such as the Lithuanian Scientific Society (organized by Basanavicius in 1907) and the Society of Fine Arts, contributed to strengthening the national movement.[42] The ending of most press restrictions led to a proliferation of new journals—there were 22 in 1914, most of them weeklies—which tended to heighten national feeling. This movement was not now discouraged by the Tsarist government; belatedly St. Petersburg began to recognise that Lithuanian national sentiment was a counterbalance to Polish influence.[43]

Evidence of Lithuanian aspirations for autonomy took a very visible form in the Russian Duma to which Lithuanian deputies were elected. A Lithuanian–Jewish voting bloc ensured that seven Lithuanian deputies were elected in the first two Dumas and four in the two subsequent ones. All these deputies came from the Kaunas and Suwalki provinces since the population of Vilna province was mainly Polish, Belarussian and Jewish. The Lithuanian deputies sat in a separate group in the Duma in order to preserve their independence from the Polish deputies in the nationalities bloc. Their work for Lithuanian autonomy bore no immediate fruit but parliamentary experience was useful preparation for independence after 1918.[44]

The easing of Russian repression once again brought to the forefront Lithuanian–Polish relations. The new Lithuanian intelligentsia and professional middle class rejected the Polish vision of restoring the old Commonwealth. For the Poles, who outnumbered Lithuanians in Vilna province, Lithuanian nationalism was an artificial construct, inspired and fomented by Russians (later by Germans), with no deep roots in Lithuanian society. Separation of Lithuania from Poland would, they believed, weaken Catholicism in Eastern Europe in its contest with Russian Orthodoxy. They claimed that the Lithuanian peasantry was not interested in the national movement. This dubious assertion ignored the evidence of peasant hostility to the Polish landowning class. It underestimated, too, the obvious appeal of land distribution in favour of the peasantry which the Lithuanian nationalists were advocating.[45]

The Poles clearly misjudged the depth and weight of Lithuanian national feeling. The renaissance had done its work well. The Lithuanians could see little virtue in the former Commonwealth and had no wish to revive it. In Anne Applebaum's words, the Lithuanians accused the Poles of 'cheating' them out of their medieval kingdom, and Polish aristocrats of cheating them out of their language. Lithuanian culture was authentic, Polish culture an alien imposition. Lithuanian nationalists idealized opponents of the Commonwealth, men such as the aristocratic Radvilas brothers and those Lithuanian nobles who had gathered at Valkininkai in 1700 to oppose Polish forms of administration. But of more immediate importance were the Lithuanian peasantry. They had largely escaped Polonization, they still spoke Lithuanian, and they retained their own customs and traditions. They had succeeded in resisting the centuries-long attempt by Catholic priests to Polonize their flocks, notably by prohibiting the use of Lithuanian in the vernacular parts of church services. Many of the clergy were ethnically Polish and ethnic Lithuanians who aspired to the priesthood complained of discrimination in entry to the seminary at Vilna. In 1906 the Lithuanian clergy petitioned the Pope for the separation of the Polish and Lithuanian churches and the creation of a Lithuanian archbishopric. There was no response. Polish dominance opened the clergy to attacks by secular groups, such as free thinkers and anti-clericals, for betraying Lithuanian culture. It was a short step to blame the eclipse of the Lithuanian nation on the Polish-led Christian conversion of the country in the fourteenth century.[46]

By 1914 the Lithuanian national movement was maturing rapidly. The decision of the Russian government to engage in war with Germany in 1914 ultimately brought down the Romanov regime and opened up the possibility of the subject nations achieving the autonomy or independence they had long craved. Although a condition of independence was the collapse of the imperial structures in Central Europe, there were still many obstacles in the path. Lithuanians had to fight hard to win, and then to consolidate, their independence, as the next chapter will show.

1. Alfonsas Eidintas, Vytautas Zalys, Alfred Erich Senn, ed., Edvardas Tuskenis, *Lithuania in European Politics: The Years of the First Republic 1918–1940* (Basingstoke and London: Macmillan, 1997), p.1.

2. Between the 7th and 5th centuries BC various related tribes had settled in the south-east Baltic region. They were referred to by Tacitus as Aestii, or Aestians, and were later denominated as Balts. They were praised for their farming skills; Adam of Bremen lauded their customs and criticised them only for not being Christian: E.W.Polson Newman, *Britain and the Baltic* (London: Methuen, 1929), p.102; E.J. Harrison, ed. and compil., *Lithuania 1928* (London: Hazell, Watson and Viney Ltd., 1928), pp.23–30; Third Interim Report of the Select Committee on Communist Aggression, House of Representatives, Eighty-Third Congress, Second Session 1954, *Baltic States: A Study of their Origin and National Development, their Seizure and Incorporation into the U.S.S.R.*, third Reprint Edition, William S. Hein and Co. Inc., Buffalo, New York, 1972, pp.64–66.

3. The Grand Duke Rimgaudas was the first to achieve some measure of unity but the real unifier of the Lithuanians and the founder of their state was Mindaugas (1236–63). He inflicted a notable defeat on the Knights of the Sword in 1237 and later converted to Christianity, though after his assassination in 1263 his successors reverted to paganism: Owen J.C. Norem, *Timeless Lithuania* (Chicago: Amerlith Press, 1943), pp.30–36; Romuald Misiunas and Rein Taagepera, *The Baltic States: Years of Dependence 1940–1990* expanded and updated edition (London: Hurst & Co., 1993), pp.2–3; Select Committee on Communist Aggression, pp.64–66.

4. Bronis J. Kaslas, *The Baltic Nations—the Quest for Regional Integration and Political Liberty: Estonia, Latvia, Lithuania, Finland, Poland* (Pittston, Pa.: Euramerica Press, 1976), p.56; Newman, p.103; E.J. Harrison, *Lithuania Past and Present* (London: T. Fisher Unwin Ltd, 1922), p.41.

5. Norem, p.30–37; Harrison, *Lithuania Past and Present*, p.41.

6. Kaslas, pp.55–56; Misiunas and Taagepera, p.3.

7. Misiunas and Taagepera, p.3; Simas Suziedelis, 'Lithuania from Medieval to Modern Times: A Historical Outline' in V. Stanley Vardys ed., *Lithuania under the Soviets: Portrait of a Nation, 1940–1965* (New York, Washington, London. Frederick A. Praeger, 1965), pp.4–6; V. Stanley Vardys and Judith B. Sedaitis, *Lithuania, the Rebel Nation* (Boulder, Colorado; Westview Press, 1997), p.12.

8. In 1401 Jogaila officially confirmed Vytautas' authority and title of Grand Duke for life, retaining for himself a nominal title of supreme duke and the right to inherit the Grand Duchy of Lithuanian: *Lithuania: an Encyclopedic Survey* (Vilnius: Encyclopedia Publishers, 1986), p.94.

9. Newman, p.103; Misiunas and Taagepera, p.3; Suziedelis, pp.4–5; Norem, pp.42–7.

10. The Poles feared the probable separation of the two states following Vytautas' coronation and tried to persuade the Pope that this division would weaken Latin Christianity in its struggle with Orthodoxy: Harrison, *Lithuania Past and Present*, pp.43–6; Kaslas, pp.58–9; Vardys and Sedaitis, pp.12–14.

11. E.J. Harrison, *Lithuania's Fight for Freedom* (New York: Lithuanian American Information Center, 1952), pp.8–9; Kaslas, pp.58–9; Suziedelis, pp.5–6; Norem, pp.42–7; Select Committee on Communist Aggression, p.66; Robert H. Lord, 'Lithuania and Poland', *Foreign Affairs*, vol.1, 1923, pp.39–40.

12. Harrison, *Lithuania's Fight for Freedom*, p.5.
13. Kaslas, p.12; Harrison, *Lithuania 1928*, pp.128–31.
14. Examples of these deities were Perkunas, the God of Thunder and Lightning, Zemininkas, the God of the Earth, and Garija, the God of Fire: Norem, p.9; Harrison, Lithuania's Fight for Freedom, pp.5–7.
15. *The Baltic Independent,* 4–10 November, 1994; Norem, p.44; Harrison, *Lithuania 1928*, pp.114 and 131.
16. Another anxiety was the anticipated extinction of the current Jagellonian line which would have eliminated the only political link between the two countries: Kaslas, p.59; Suziedelis, p.5.
17. Suziedelis, p.5; Kaslas, p.59; V. Stanley Vardys, 'Lithuanians', in Graham Smith ed., *The Nationalities Question in the Soviet Union* (London: Longman, 1990, rept. 1992), p.73; Misiunas and Taagepera, p.3; Norem, p.65.
18. The division between a Polish-speaking landlord class and a Lithuanian-speaking peasantry was widened by the imposition of serfdon in the sixteenth century. In the nineteenth century this introduced a class element into the Lithuanian nationalist movement: Misiunas and Taagepera, p.4; Anne Applebaum, *Between East and West: Across the Borderlands of Europe* (London and Basingstoke: Papermac, 1995, first published Pantheon Books, 1994), pp. 45–8; Newman, p.106; Alfred Erich Senn, *The Emergence of Modern Lithuania* (New York: Columbia University Press, 1959), pp.4 & 17; Kaslas, p.62; Vardys and Sedaitis , p.13; Georg von Rauch, *The Baltic states: the Years of Independence. Estonia, Latvia, Lithuania 1917–1940* (London: Hurst & Co. 1974), p.3.
19. Suziedelis, pp. 5–6; Newman, p.107; Select Committee on Communist Aggression, p.68.
20. For centuries Lithuania Minor had been part of East Prussia, with its capital city at Koenigsberg, called Karaliaucius in Lithuanian. Ethnic Lithuanians were concentrated in the areas round Tilsit, in the north-east part of the country. The landowners were German, the peasants Lithuanian. During the Reformation the population was converted to the Lutheran form of Protestantism. After the Partitions the Russian Empire annexed all of Lithuania except Lithuania Minor which remained part of East Prussia: James D. White, 'Nationalism and Socialism in Historical Perspective', in Graham Smith ed., *The Baltic States: the National Self-Determination of Estonia, Latvia and Lithuania* (London: Macmillan, 1994); Senn, p.2; Norem, p.72.
21. In the Commonwealth, too, there was a brief interest in Protestantism when Reformers were invited to visit Lithuania by Prince Radvilas among others, leading to the opening of some schools and printing presses. But this development was cut short by sectarian conflict and the Counter-Reformation which regained all the ground lost for the Catholic Church.
22. In his criticism of the landowners, his moral condemnation of serfdom, and his concern to improve the spiritual and material condition of the peasants, Donelaitis was close to the Pietist movement which favourably contrasted the virtuous life of the ordinary people with the frivolity and uselessness of the aristocracy: White, p.20; see also James D. White 'National Movements in the Baltic Provinces', *International Politics*, vol.33, no.1, March 1996, p.68; Vardys and Sedaitis, p.16; Senn, p.14; Norem, p.60; August Rei, *The Drama of the Baltic Peoples* (Stockholm: Kirjastus Vaba Eesti, 1970), p.31; Harrison, *Lithuania 1928*, p.104; Vytautas Vanagas, 'Ancient Written Literature and the Beginning of Belles Lettres' in Vytautas Kubilius ed., *Lithuanian Literature* (Vilnius: Vaga Publishers and the Institute of Lithuanian Literature and Folklore, 1997), pp.40–46.
23. Vytautas Vanagas, 'Literature at the Crossroads of Enlightenment and Romanticism', in Kubilius ed., pp.61–2; Norem, p.79.
24. Harrison ed., *Lithuania 1928*, p.131; Rei, pp.31–2; White 'National Movements', p.68.
25. One example of this was the compulsory baptism in the Orthodox Church of the children of Catholic-Orthodox Church of the children of Catholic-Orthodox marriages: Vardys 'Lithuanians', p.73; Misiunas and Taagepera, p.6; Foreign Office Peace Handbooks, vol.VIII, *Poland and Finland* (London: HMSO, 1918–19), pp.43–4; Kaslas, pp.61–2; von Rauch, p.18; Harrison, *Lithuania Past and Present*, pp.56–9; Theodore Weeks, 'Lithuanians, Poles and the Russian Imperial Government at the Turn of the Century', *Journal of Baltic Studies*, vol.XXV, no.4, Winter 1994, pp.291–2.

26. The system of repression was in fact applied more thoroughly in the Lithuanian part of the former Commonwealth since de-Polonisation was deemed to be more feasible there than in Poland itself: Weeks , p.294; White, 'Nationalism and Socialism', pp.18–19; Vardys and Sedaitis, p.16

27. Weeks, p.292; White, 'National Movement', p.75; Harrison, *Lithuania Past and Present*, p.58.

28. Von Rauch, p.17; Eidintas, Zalys, Senn, p.16; Rei, p.28.

29. Notably redemption payments were reduced by 20 per cent from the levels originally set, peasants were given an unconditional right to purchase land, and land allocated to the peasants under the Emancipation Act (endowment land) became individual not communal property as in the rest of Russia: Peace Handbooks, pp.126–7.

30. Peace Handbooks, p.127; Kaslas, p.63; Rei, p.29; Select Committee on Communist Aggression, p.70; Eidintas, Zalys, Senn, pp.12–13.

31. Juozas Jakstas, 'Lithuanian to World War I', in Albertas Gerutis ed., *Lithuania 700 Years* (New York: Manyland Books, 1969 rept 1984), pp.112–17, 124–5; White, 'National Movements', p.76; Thomas Remeikis, *Opposition to Soviet Rule in Lithuania 1945–80*, (Chicago: Institute of Lithuanian Studies Press, 1980) (reprinted extract from *Ausra* No.1 on Valancius), p.374.

32. Remeikis, p.373; White 'National Movements', p.75; Vardys and Sedaitis, p.17; von Rauch, p.20.

33. Vardys, 'Lithuanians', p.73; White, 'National Movements', p.77; Kaslas, pp.63–4.

34. This unofficial literature has been likened to *samizdat* writing which flourished in Soviet Lithuania in the 1970s. However, the number, variety and volume of clandestine publications was infinitely greater than in the later period.

35. Kaslas, pp.63–4; Stanley W. Page, *The Formation of the Baltic States: A Study of the Effects of Great Power Politics upon the Emergence of Lithuania, Latvia and Estonia* (Cambridge, Mass: Harvard University Press, 1959), p.2; Senn, p.9; Marite Sapiets, 'Religion and Nationalism in Lithuania', *Religion in Communist Lands*, vol.7, no.2, 1979, p.77; Harrison, ed., *Lithuania 1928*, p.110.

36. Page, p.2; von Rauch, p.21; White, 'National Movements', pp.78–80.

37. White, 'National Movements', pp.78–80; Page, p.2; Harrison, ed., *Lithuania 1928*, p.112; Harrison, *Lithuania Past and Present*, p.62.

38. White, 'National Movements', pp.75ff.

39. Eidintas, Zalys, Senn, p.14; Harrison, *Lithuania Past and Present*, pp.65–7.

40. Kaslas, pp.63–5; Page, p.8; von Rauch, pp.22–3; Alfred Erich Senn, *The Emergence of Modern Lithuania*, (New York: Columbia University Press, 1959) pp.9 and 11–13; White, 'Nationalism and Socialism', pp.29–31.

41 White, 'Nationalism and Socialism', pp.29–31; Kaslas, pp.63–5; Senn, *The Emergence*, pp.9–13; Peace Handbooks, pp.47–8; Page, pp.8–9; Select Committee on Communist Aggression, p.69; Eidintas, Zalys, Senn, pp.18–19.

42. The Scientific Society, for example, carried out systematic studies of Lithuanian folklore, and published annually collections of *Dainos*, popular traditions and customs, and ancient myths and legends.

43. Vardys and Sedaitis, pp.19–20; Kaslas, p.65; Harrison, *Lithuania Past and Present* pp.71–4; Eidintas, Zalys, Senn, p.19–20.

44. Eidintas, Zalys, Senn, p.19; Senn, *The Great Powers*, p.6.

45. Peace Handbooks, pp.60–62;

46. idem, pp.60–62; Weeks, pp.296–8; Applebaum, pp. 51–2; Norem, p.68; Alfred Erich Senn, *The Great Powers, Lithuania and the Vilna Question 1920–28* (Leiden: E.J. Brill, 1966), pp.3–4; Harrison, *Lithuania Past and Present*, pp.51–2.

Map of Lithuania

Section I

REVIVAL AND REPRESSION, 1914–1985

Chapter 1

INDEPENDENT LITHUANIA BETWEEN THE WARS

Mikhail Gorbachev's attempt to modernise and revivify the Soviet economic system through the processes of *glasnost* and *perestroika* loosened the controls imposed on Lithuania since the Soviet occupation of the state in 1940. Western commentators were surprised by the speed, energy and ingenuity of the Lithuanians who, like 'greyhounds in the slips', were impatient to seize their opportunity for self-determination. Their response to Gorbachev's challenge revealed to a somewhat bewildered world the abject failure of the Soviet policy of assimilation. At the same time, it showed that the memory of independence in the inter-war period remained very much alive in popular consciousness. The sense of a separate and distinctive Lithuanian identity survived both Stalin's savage attempts to destroy it and the persistent efforts of his successors to replace it by a new Soviet allegiance. Just as the Lithuanian national renaissance prepared the ground for the restoration of independence in 1918, so the experience and achievements of the inter-war years strengthened the foundations of Lithuanian identity. Stalinist repression served to reinforce rather than to undermine those foundations.

At the time of Gorbachev's accession to power the Lithuanian people remained keenly aware of their history, culture, and practical achievements, despite the passage of several generations since the independence period. The confidence stemming from this awareness, combined with their revulsion against the Soviet system, enabled them to strike out in the direction of self-determination. Despite all attempts by the Soviets to eradicate or re-interpret the Lithuanian collective memory, the population held fast to its history, using it like a compass to plot its course to independence. Some familiarity with the history of Lithuania since 1914 is therefore crucial to an understanding of the second restoration of independence in 1990–91.

THE FIRST WORLD WAR AND LITHUANIAN INDEPENDENCE

The Lithuanians made great strides towards independence during the First World War despite the destruction and upheavals of the period. The eclipse of their powerful neighbours and putative overlords, Germany and Russia, created the opportunity for them to move beyond their pre-war ambition for autonomy towards the restoration of independence. But they faced formidable obstacles, not least the opposition of their powerful neighbours Germany, Russia and Poland. Moreover, support for their cause from Great Britain and France was slow to emerge, and it was not until December 1922 that Lithuanian independence was recognized *de jure*.

After the occupation of the Lithuanian provinces by the German army in 1915, German policy, as described by Eidintas, was one of annexation, colonization and Germanization. Germany's intention, in replacing Russia as the dominant power, was to impose the German language and German culture on its new province, thus avoiding a revival of Polish cultural influence there. In the cause of Germanization, Lithuanian-language newspapers and journals were suppressed, use of the Lithuanian language in official business or in schools was prohibited or restricted, public meetings were banned, and Lithuanian place names were changed or modified. Later in the war, when it became clear that there would not be a German victory, German ambitions for Lithuania inevitably became more modest, involving the creation of a formally independent state but one which was still dependent on Germany.[1]

Lithuania could expect no greater sympathy for its aspirations from Russians of whatever political complexion. The Whites, who opposed the Bolsheviks in the Russian civil war of 1919–20, refused to accept the independence of the Baltic states. At most, they might have been prepared to concede a form of self-government, the precise nature of which would have been determined by a Russian constituent assembly. The Bolsheviks' attitude was ultimately more relevant to the future status of Lithuania. Their declaration of 15 November, 1917 assured all the nations of Russia of their right to self-determination. It soon became clear, however, that they would not deny themselves the right to influence these nations either by agitation or political action. Lenin and Stalin wanted to keep Russian territories united, and expected the various nationalities to re-integrate voluntarily in the Soviet Russian state. This is what the principle of self-determination meant in practice.[2]

The Lithuanians knew that they would have to struggle for their independence against both the Germans and the Russians. The same was true of their relations with the Poles. During and immediately after the war Polish official representatives did not anticipate an independent Lithuania. On the contrary, when Germany and Austria-Hungary announced in November 1916 that a Polish state would be created at the end of the war, the Polish Temporary Council assumed that this would be a Polish-Lithuanian state, a reconstruction of the former Polish-Lithuanian Commonwealth which, as we have seen, expired with the Third Partition in 1795. The Poles promoted the idea that Lithuania was simply a backward Polish province which should logically be united with Poland since it was unfit to achieve independence. They conceded that, as part of the greater state of Poland, Lithuania would be granted 'a generous measure of self-government'. At the back of this political manoeuvring was the fear that Lithuania would become a client state of either Germany or Russia without Polish overlordship. Indeed the Poles were convinced that Lithuania's desire for independence was simply the result of German pressure. The desirability of a secure Polish access to the sea through the Lithuanian port of Klaipėda (formerly Memel) also influenced Polish attitudes. Solving the intractable problem of frontier delimitation between Poland and Lithuania in the ethnically complex region of eastern and south-eastern Lithuania was another potential gain from the incorporation of Lithuania in Poland.[3]

Faced with the territorial ambitions of three large neighbours the first Lithuanian governments could have had few illusions about the difficulty of achieving independence. Acute economic dislocation and lukewarm attitudes from the western powers presented further obstacles. It is difficult to believe that the success of the post-independence governments in the face of considerable odds was not an inspiration and an example to later generations of Lithuanians, who had to endure Soviet repression and then the risks and dangers of seizing independence in the teeth of Gorbachev's opposition. The war and occupation produced an embryonic representative body in Lithuania, the Executive Council, made up of five people (including Antanas Smetona, later President of Lithuania) from various political groupings. This Council not only defended the interests of Lithuanians, but also began to develop the political agenda for the Lithuanian people. The German and Austro-Hungarian governments' pledge in 1916 to restore an independent Polish state, followed by President Wilson's speech in January 1917 in favour of self-determination, and then the United States' entry into the war in April 1917, encouraged Lithuanian nationalists to pursue the goal of independence very energetically.[4] They received indispensable support in this quest from the large Lithuanian communities in the United States and Russia, which not only provided relief for Lithuanian war victims but advocated independence for Lithuania.[5]

External support, however sympathetic, was insufficient to achieve independence for Lithuania in the face of opposition from powerful neighbours. Ultimately Lithuanians had to rely on themselves and to wait for an improvement in the international climate. The latter came about, Page argues, as a result of the more difficult military situation in the winter and spring of 1916–17, German perceptions that concessions to Lithuanian nationalists would weaken Polish influence in the region, and the possibility that the February Revolution in Russia would encourage independence demands among the subject nationalities. These considerations disposed the German government to make concessions to Lithuanian demands for self-determination. While the Lithuanians pressed for a fully representative elected assembly, German opposition resulted instead in the selection of an assembly from the local leaders in each district and from all the political parties. The so-called Vilnius Assembly which met from 18–22 September, 1917 was composed of 220 delegates, many of them prominent and well-known professional men.[6]

The Assembly's resolutions began the process which led to Lithuanian independence. The delegates called for an independent state within ethnographic boundaries, a guarantee of the cultural rights of the national minorities, and the election of a constituent assembly to lay the political foundations of the independent state. A National Council or *Taryba* of 20 persons was chosen as an executive body. For its part the Assembly promised to enter into special relations with Germany at some future date in return for support for Lithuanian independence at the peace conference.[7]

In the face of indifference from Russia and the Entente Powers, Lithuania had no option but to negotiate with Germany. At the same time Germany hoped to use the Assembly and the *Taryba* to draw Lithuania into a state of dependence, believing that an independent Lithuania would be helpless in the face of Russian and Polish machinations. Negotiations with the German Foreign Ministry eventually led to the recognition of Lithuanian independence provided that close military, economic, trading, and monetary ties to Germany were maintained. On 11 December, 1917 the 'first Lithuanian Declaration of Independence' was agreed though it was never made public. The

Lithuanians expected that Germany would recognize their independence and permit them to take over the administration of the state.[8]

This expectation was not fulfilled. Frustrated by German obstructionism the Lithuanian *Taryba* decided to publish its Declaration of Independence on 16 February, 1918. This stated:

> 'The Lithuanian Council, as the only representative of the Lithuanian nation, basing itself on the right of national self-determination and the decision of the Vilnius conference of September 18–23, 1917, proclaims that it is re-establishing an independent, democratically ordered Lithuanian state with Vilnius as its capital and that it is separating that state from any state ties that have existed with other nations.'[9]

What differentiated this from the December declaration was the omission of the provision calling for close links with Germany, and the inclusion of a reference to a Constituent Assembly to establish the legal basis of the Lithuanian state. Both declarations, however, referred to the 'reconstitution' of the Lithuanian state, clearly demonstrating the influence of the late-nineteenth-century national renaissance.[10] The immediate task of the Lithuanian leadership was to establish an independent and responsible government for Lithuania. This was not easy since the country was still under German occupation, and the German military leadership still aspired to unite Lithuania with Germany, either by annexation or by linking it to the Prussian crown, with the object of obtaining strategic and economic advantages as well as living space for German settlers. The treaty of Brest-Litovsk between Germany and the Bolsheviks in March 1917 strengthened the German grip on Lithuania since the Russian side surrendered its territorial claims. German recognition of Lithuanian independence, therefore, had very little practical significance. However, German defeat on the western front put paid to its hopes of taking over Lithuania. The appointment of Prince Max von Baden as German Chancellor in October 1918 marked the end of German intransigence on the Lithuanian question.[11] On 15 October, 1918 Prince Max announced that countries occupied by Germany had the right to self-government. Five days later Lithuanian representatives received permission to take over the administration of the country and by 2 November a provisional constitution had been adopted. Under it the *Taryba* became the State Council with legislative powers, and a three-man Presidency led by Antanas Smetona was established. The Presidency nominated Augustinas

Voldemaras, who had returned from Russia along with Martynas Yčas, to be Prime Minister on 11 November, 1918. It was decided that a Constituent Assembly should be convened as soon as possible to agree on a permanent constitution. Six Belarussians and three Jews were invited to join the Council of State and the two communities were represented by one minister each in the cabinet. The Polish community refused to participate.[12]

DEFENCE OF LITHUANIA AGAINST EXTERNAL ATTACK

Governments under the Provisional Constitution had to carry the heavy responsibility of organizing the defence of Lithuanian territory against Bolshevik, White Russian and Polish aggressors. The energy released by the achievement of independence, combined with a robust sense of national pride, generated strong powers of resistance to external attack. Only in the Vilnius region were Lithuanian forces ultimately unsuccessful. Effective defence depended in part on a realistic assessment of the boundaries of the new state. Lithuanians sensibly rejected the reconstitution of the historic state of Lithuania with its extensive territories in White Russia and Ukraine. They preferred an ethnographic Lithuania, with an outlet to the sea, and its capital at Vilnius. In broad terms this meant a state embracing the former Russian provinces of Vilna, Kaunas and Suwałki together with the city of Grodno (Gardinas in Lithuanian), and the area of Klaipėda or Lithuania Minor, which had been part of East Prussia for many centuries. These territorial claims were widely known in 1919 except for the claim to Klaipėda (Memel), which was not publicized while Lithuania depended on German assistance against the Bolshevik invasion in early 1919. Although the contest for Klaipėda was delayed until 1923 the remainder of ethnographic Lithuania was subject to invasion and attempted occupation, first by the Soviet Russian army, then by the Poles, and finally by the adventurist forces of Bermondt-Avalov. Although Lithuania's survival as an independent state was remarkable, she nevertheless failed to establish her preferred frontiers as a result of Polish opposition to the idea of ethnographic Lithuania.[13]

In fact, it is unlikely that Lithuania could have survived her wars for independence without German help. Whether it liked it or not the Lithuanian government was forced to rely on Germany during 1919. Germany was the only one of the major powers to recognise Lithuanian independence. Under the terms of the Armistice, it was agreed that the Germans would hold the eastern front until the victorious powers requested withdrawal. The Entente's aim was to ensure that the Bolshevik forces did not occupy territory evacuated by Germany. The presence of German troops was therefore Lithuania's best defence against Soviet invasion. However, when German forces began to withdraw prematurely, the Bolshevik army took control of the abandoned territory. In Vilnius an embryonic Communist Party announced the formation of a provisional revolutionary government ahead of the arrival of the Red Army. The Bolshevik forces arrived on 5 January, 1919, when a Soviet Socialist Republic was established in Lithuania.[14]

When Piłsudski, the Polish leader, who came from a Lithuanian Polish family, offered the Lithuanians help, they would accept it only on condition that Poland recognised an independent Lithuanian state with its capital at Vilnius. This the Poles consistently refused to do. Piłsudski advanced his favoured solution to relations between the two countries— a reconstituted Polish-Lithuanian federation, in effect the restoration of the Treaty of Lublin of 1569. Lithuanian nationalists of different parties, who were largely the products of the Lithuanian national renaissance, could see in the federation idea only the recreation of Polish cultural and political dominance in Lithuania, and rejected it. Nor were they induced to take a more positive line towards Polish claims by the constant incursions into Lithuanian territory by Polish forces in breach of several armistice lines negotiated with the help of the Entente powers in the summer and early autumn of 1919. An attempted Polish coup in Kaunas in August 1919 further soured relations.[15]

Fortunately for the Lithuanians the German army halted its retreat in the vicinity of Kaunas. This gave the Lithuanians the necessary breathing space to establish their own volunteer military units and by February 1919 they were starting to push the Soviets back in the direction of Vilnius. By late Spring this small but effective Lithuanian army was poised to capture Vilnius, only to be pipped at the post by a Polish army under Haller which took control of the whole of the region around Vilnius and placed it under Polish administration.

Happily for the Lithuanians, the Bolshevik forces were driven out by the late summer of 1919. The Bolsheviks' failure was not simply the result of military defeat or partisan warfare. Their basic weakness was a complete inability to convert the Lithuanian population to communism. Their advocacy of farm collectivization was anathema to the Lithuanian peasantry who wanted land redistribution and private ownership. A possible source of support, the industrial working class, was too small to have significant political influence even if it had not been alienated by Soviet abolition of existing civic institutions and their replacement by soviets. The atheistic doctrines of the communist government did not enamour them to the staunchly Roman Catholic population. Requisitioning of foodstuffs by the Soviet army, the compulsory use of Russian in government offices and official documents, and the nationalisation of the most important manufacturing concerns aroused serious opposition among the Lithuanian population. If there was a dominant ideology in Lithuania after the war it was nationalism, not communism.[16]

The final enemy of Lithuanian independence in 1919 was the notorious force of Bermondt-Avalov, ostensibly supporting the Russian White forces under General Yudenich, but in reality seeking to undermine the independence of the Baltic states and to restore the rule of the Baltic Germans in Estonia and Latvia. These so-called Russian forces were largely composed of Germans equipped and paid from German sources but 'dressed as Russians'. Towards the end of November 1919 these mercenaries, now 50,000 strong, occupied large parts of northern and western Lithuania and threatened the independence of the Lithuanian state. But on 21 and 22 November they were defeated at Radviliskis and Šiauliai and were ordered to leave Lithuania by an Inter Allied mission under General Niessel.[17] This marked the end of the first phase of Lithuanian independence. Boundaries had been at least partially established, invading armies, with the exception of the Poles in the south east, had been driven out, a small but effective Lithuanian army had been established and had proved itself in a series of battles against numerically stronger forces, a civilian administration had been created and a constituent assembly was due to meet in 1920 to establish a permanent constitution for the country. Although the Lithuanians had long-term anxieties about their security and their relations with their powerful neighbours, they now had the opportunity to construct their own state, free from the immediate fear of invasion. They took this opportunity with considerable energy and inventiveness in the face of formidable obstacles.

THE ECONOMIC INHERITANCE OF INDEPENDENT LITHUANIA

The Lithuanian economy had considerable potential though at the time of the first census in 1923 Lithuania was a small and impoverished state. This potential was substantially realised in the inter-war period as Lithuania's record of economic achievement shows. Including Klaipėda but excluding the Vilnius region, Lithuania's population was 2.17 million. Soviet sources suggest that it had increased to almost 2.9 million in 1939 as a result of natural increase and the incorporation of part of the Vilnius region in October 1939.[18] Approximately three quarters of the population worked in agriculture in 1923, and only about 10 per cent in industry and handicrafts, transport and communications, and commerce and credit. These proportions had not changed very much by 1939 when industrial workers were still only 8 per cent of the work force. Throughout the interwar period Lithuania remained an essentially agrarian economy though urban dwellers increased their share of the population from 13 per cent to 27 per cent between 1913 and 1940. In its ethnic composition the population of the towns was mainly Jewish, Polish and Russian. In the former Kovno province, for example, only about 9 per cent of the urban population consisted of ethnic Lithuanians. Overall, ethnic Lithuanians constituted about 80 per cent of the population in March 1939, the Jews 7 per cent, the Germans 4 per cent, Poles 3 per cent and Russians just over 2 per cent. After the loss of Klaipėda and the return of the Vilnius region in 1939 the proportion of Jews and Poles increased and that of the Germans fell. Moreover, if the Lithuanians had achieved their ambition of establishing an 'ethnographic' Lithuanian state, the proportion of ethnic Lithuanians would have been significantly smaller. Allowing for inaccuracies in the Russian census of 1897, the figures suggest that, in the Kovno (Kaunas) province and in the five districts of Vilna and Suwałki provinces with Lithuanian majorities, only about three-fifths of the population was ethnically Lithuanian.[19]

A contemporary observer of Lithuania, E.W. Polson Newman, noted that it was impossible to visit Kaunas without realising that Lithuania was a very poor country. Similarly, the British Consul in Danzig, who visited Memel (Klaipėda) in 1922, wrote to the British Foreign Secretary, Lord Curzon, that on crossing between Memel and Lithuania 'one finds oneself in Russia'. The fields were poorly and wastefully cultivated, the roads unrepaired, the houses, cottages and hovels dilapidated. The people were 'not a bad looking lot of primitive peasantry ... but the scene left a sense of pathos in the mind'. It was a pity, he concluded, that they had been left so long under Slav domination.[20]

Indeed, the Russian legacy combined with the effects of the war had bequeathed to the new state a backward and devastated economy. The

Tsarist regime had impeded economic development though there were a few signs of progress in the decade before 1914. A class of large landowners of Polish, Polonized Lithuanian, or Russian origin dominated the agrarian economy. Some 3,000 owners held 26 per cent of the land; in 1912 there were some 700 estates of over 2,700 acres. In all about 48 per cent of the land was held in private estates in 1905, compared with 45 per cent held by peasants under the terms of the emancipation act of 1861 (so-called *nadyel* land).[21]

After the Stolypin reforms peasant plots in the open fields began to be consolidated into individual farms, opening up the possibility of greater enterprise and innovation in agricultural methods. Peasants held the greatest share of livestock, and new foreign breeds were introduced to improve stock and encourage dairy farming. Manufacturing activity was minimal; most enterprises engaged in food and raw material processing such as distilling, brewing, flour milling, saw milling and woollen textiles. Major exports were foodstuffs and raw materials which were exchanged for manufactured goods, machinery, coal and cotton textiles.[22]

Economic development and good government were both handi-capped by high levels of illiteracy and a shortage of educated and experienced personnel. In the early period of the republic there was a paucity of qualified people capable of running government departments and conducting state business. Some of the best were sent abroad as diplomats. Moreover, the determination to eliminate Polish influence from the country combined with the redistribution of land 'skimmed the country of its aristocracy' and reduced the number of those qualified for public service.[23]

During the First World War economic backwardness was exacer-bated by the devastation inflicted by the rival armies of Russia and Germany. When the Russian army withdrew in 1915 it destroyed all buildings of strategic value. Large factory units which could be dismantled and most movable machinery were transported to Central Russia. During the fighting townships, villages, and farms were destroyed. Tens of thousands of wooden houses and other buildings were burned down, leaving their inhabitants without homes and food. The Russians deported large numbers of people living near the front line, compelling them to burn their farms and abandon their crops. When the army retreated, it was accompanied by large numbers of civilians who could not face the prospect of a German occupation. In their turn the Germans requisitioned supplies, mobilized the work force, and Germanized the population. German exploitation of land in the war was 'imperishable in the memory of the people'. Huge quantities of foodstuffs were sent to Germany or consumed by the German army, and large areas of forests were damaged or felled indiscriminately. It has been calculated that out of 214,000 farms, 92,000 were ruined; some 57,000 frame buildings were burned; during the entire war some 90,000 horses, 140,000 head of cattle, 767,000 sheep and pigs were requisitioned. By the end of the war the number of farm animals had fallen by not far short of a half.[24] Young Lithuanian males, who were forced to labour in the forests and on the land to supply the German army with the necessary food and timber, were harshly treated and underfed. At the same time the shortage of labour as a result of conscription into the Russian army, and the disruption of the supply of artificial fertiliser and manure during the war, adversely affected agricultural productivity. Those who aroused the displeasure of the *Ober Ost* were arrested and often deported to Germany. Press censorship, the compulsory teaching of German in schools, and restrictions on free movement were all put at the service of Germanizing the population.[25]

LITHUANIAN ECONOMIC DEVELOPMENT

Lithuanian economic development in the inter-war period took place within the structure of a largely agrarian economy. Although industrial output grew by some 350 per cent between 1928 and 1939, and the number of workers in Lithuanian industries rose by about 600 per cent in the period 1913–39, there were still only about 40,000 industrial workers in 1939, and only 34,000 if those working in Klaipėda are deducted from the total. Lithuania's property ownership was of a mixed character, embracing a very substantial and successful cooperative sector, a private sector which received considerable government stimulus and investment, and an increasing number of state-owned enterprises. Successive governments had little interest in a planned economy until the Finance Ministry was given strong interventionist powers in May 1940 as a result of the war-time economic emergency.[26]

Macro-economic policy during the inter-war period was remarkably consistent. Fiscal, budgetary and monetary policies were, and remained, extremely cautious and prudent. The new currency, the *litas*, was fixed to gold and was never devalued in the period despite the pressures of the world economic depression in the early 1930s. Successive governments ensured that the budget was balanced in virtually every year. The foreign debt remained small, and overseas trade was in surplus for almost the whole period. The balance of payments also benefited considerably from remittances from Lithuanian communities abroad, mainly in the United States. This economic strategy kept inflation down, though the policy of tariff protection, which was designed to support infant industries, raised domestic commodity prices. The consistency of economic policy was mostly attributable to Prime Minister Tūbelis, brother-in-law of President Smetona, who 'preferred autarchy to foreign debt'. The government compensated for the relative shortage of foreign investment by itself investing heavily in domestic enterprise, both co-operative and private. Under Smetona the state invested up to around 60 per cent of the capital in joint stock companies, mainly in agricultural industries. Particularly important was the government's active role in modernizing the country's communications infrastructure—railways, roads, telecommunications and harbour facilities—after years of Russian indifference. Equally important were land reclamation projects in the countryside, such as the draining of swamp lands.[27]

Within this broadly stable economic strategy the nationalist government of President Smetona, which took power in 1926, decided to promote an export-based agricultural economy with a major emphasis on meat, dairy products and poultry. Lithuania had too few basic raw materials and skilled workers to develop an efficient heavy industrial sector. Moreover, the world economic depression after 1929 offered few market opportunities for new industrial producers. The concentration on foodstuffs, by contrast, yielded excellent returns. Butter output increased 12 times between 1928 and 1939, and 90 per cent of it was exported. The number of dairy cattle doubled as did the milk yields per cow. Similarly there was a considerable increase in the number of pigs and a notable improvement in the quality of Lithuanian bacon and eggs, much of which found its way to the markets of western Europe, and in particular to the breakfast tables of Great Britain. It was precisely this strategy of encouraging foodstuffs production which offered a major stimulus to the growth of Lithuanian industry. The policy received a significant boost from 1935 when the agricultural terms of trade improved and farm purchasing power increased. The success of this policy was quite remarkable. Whereas world industrial production in 1939 had fallen to 74.8 from a 1929 figure of 100, the comparable figure for Lithuania was 140.7. Lithuania benefited from its specific tariff system as compared to the more commonly used *ad valorem* alternative, since this increased the level of protection during the period of falling world commodity prices. And its exports of processed foodstuffs remained relatively buoyant, reflecting improvements in quality combined with relatively low prices.[28]

Food and raw material processing was at the centre of the Lithuanian industrial economy during this period. For example, the meat, dairying and sugar industries constituted 41 per cent of Lithuania's industrial output by value in 1939, compared to the next largest industry of textiles with around 15 per cent of the total. Other major industries were clothing, metallurgy, woodworking, and paper. Lithuanian industries produced basic consumer goods for the domestic market, and relied on domestic sources for two thirds of the raw materials consumed. The generation of energy depended mainly on domestic sources of supply, notably wood and peat, only 40 per cent of raw materials for the energy sector coming from abroad.[29]

In international trade, grain exports, which had been the most important element just before the first world war, were displaced by dairy products. Other significant exports were timber, bacon, eggs and flax. Overall, in 1939, food products accounted for around 53 per cent of Lithuanian exports, compared with 17 per cent in 1925, and the next largest contribution of around 30 per cent came from raw materials and semifinished goods. Lithuania's major imports included machinery, motors, coal, cement, cotton fabrics and scientific instruments. Lithuania experienced a severe decline in foreign trade as a result of the Great Depression of the early 1930s from which she only began to recover after 1935.[30]

Her main trading partners in the inter-war period were Germany and the United Kingdom. Over 66 per cent of Lithuanian exports went to Germany and the UK, while 53 per cent of imports came from there. After Hitler began to impose sanctions on Lithuanian trade in the 1930s in order to exert pressure over Klaipėda, the UK succeeded Germany as the major market for Lithuanian goods and the biggest exporter to the Lithuanian market. Independence cut Lithuania off from its former Russian markets and, largely as a result of Soviet policy, they did not recover, Lithuanian exports to the Soviet Union never reaching more than about 10 per cent of total exports before 1939. A major feature of the economy of independent Lithuania, therefore, was the reorientation of its international trade towards the markets of the West, which in turn imposed demands for value added and high quality products. The parallel with the post-Soviet period in Lithuania is striking.

Lithuanian economic achievements, though creditable, should be put in perspective. On the one hand Lithuanians exported as much butter per head as the Irish and the Dutch, though not as much as the Danes, the Estonians and the Latvians.[31] Equally the Lithuanian standard of living was higher than Poland's and substantially greater than the Soviet Union's. Yet Lithuania's exports and imports by value per capita were relatively undeveloped and her economy was heavily dependent on the export of processed farm products to a relatively limited number of west European countries. Productivity in agriculture, though improving, remained lower than in the other Baltic states and was only two thirds to three quarters of the average European figure in 1931–35. This suggests that there was relative over-population in the countryside. The reality, however, was that there was a shortage of farm labour combined with unemployment among former farm labourers who had moved to the towns in search of other work. Only continued improvements in agricultural output per worker and per hectare could bring Lithuania up to European levels.[32]

The weaknesses in Lithuanian economic performance between the wars should not be allowed to obscure one of the great successes in this period, namely the co-operative movement's 'decisive' contribution to economic development. Co-operatives existed in the Lithuanian provinces of Russia before the First World War, though their development was restricted by the Russian government. After independence a major stimulus to the development of co-operatives came from the Union Law of the Co-operative Societies in 1919, which was supported by the Ministry of Commerce and Industry. Most co-operatives benefited from tax exemption, and the producer co-operatives were assisted by government investment, though this was not an unmixed blessing since it was accompanied by government interference. Credit and consumer co-operatives grew rapidly and fulfilled important functions, particularly in the peasant economy after the extremely important land reforms, which are discussed in the following section. The redistributive measures strengthened the middle and small peasant farms, and co-operative principles encouraged mutuality and self-government. Consumer, credit and insurance co-operatives along with producer co-operative organizations, numbered approximately 1,300 in 1939 with around 200,000 members. It was, however, the producer co-operatives which enjoyed the most stiking success and made the most significant contribution both to economic development and the efficient distribution of goods in both domestic and international trade.[33]

The earliest and most important of these was *Lietūkis*, the Central Union of Agricultural Co-operative Societies, established in 1923 to help peasant farmers with the marketing of their grain, flax and wood products, and to facilitate their purchases of agricultural machinery and supplies, such as fertilizers, salt, cement etc. *Lietūkis* dominated Lithuania's grain export business. It was also responsible for the import of 80% of agricultural machinery. *Lietūkis* ran experimental farms where improvements in animal breeding, the treatment of grains, and harvesting techniques were pioneered. Until 1926 *Lietūkis* was also responsible for co-operative dairies. It then helped to form an organization called *Pienocentras*, the Central Union of Dairy Co-operatives, which set up dairies and creameries throughout Lithuania to purchase milk and eggs from farms. It established a virtual monopoly in the export of Lithuanian dairy products and eggs. Like *Lietūkis* it actively promoted improved production methods so that Lithuanian dairy products could compete in quality and price on world markets. Last in the trio of major producer co-operatives was *Maistas*, though this was a mixed enterprise in which almost half the shares were owned by *Lietūkis* and the rest by the government. *Maistas*'s purpose was to purchase stock from farmers and to process meat for domestic and export markets to the highest standards. *Lietūkis* and *Pienocentras*, and to a lesser extent *Maistas*, embodied the traditional co-operative principles of democratic control and self-help. However, they combined these with a strategy of investing in the best production and marketing methods to improve and promote Lithuanian products in export markets.[34]

SOCIAL REFORMS

Accompanying these economic changes were far-reaching social reforms. The driving force behind Lithuanian independence, as we have seen, were the sons and grandsons of peasants who were determined to establish a state where democratic principles were embodied in the constitution and laws. The co-operative movement was just one example of democracy in practice. Another was the far-reaching reform of landholding, embodied in three major laws between 1920 and 1925, which had economic, social and political objectives. The first aim of the laws was to achieve greater equality in land distribution and to assuage the deep hunger for land on the part of the peasantry. Forty per cent of land was in the hands of a relatively small group of landlords, 20 per cent being held by only 450 gentry families. This was widely regarded as an 'intolerable inequity'. Secondly, it was apparent that a large proportion of this land employed backward methods of cultivation; many of the landlords were absentees with little or no interest in improved methods, and one third of this land lay fallow. The Russian, Polish or Polonized Lithuanian landlord class had generally opposed independence and could therefore expect no favours from the new ethnically Lithuanian 'bourgeois' government. Finally, redistributive reform was essential to weaken the attraction of communist propaganda among the rural proletariat.

Estates granted by the Tsarist government were confiscated in their entirety and other large landholders were restricted to 200 acres, later raised to 325 acres. The land pool thus created, some 1.8 million acres, was allocated in part to poorer and landless peasants. The object was to settle some 35,000 to 40,000 families on small plots and to increase the landholdings of some 26,000 existing smallholders. On average, new farmers received about 23 acres and existing small landowners an additional 8 acres. The number of landowners grew by 18 per cent while 13 per cent of peasants added to their holdings. The new owners, except for demobilised soldiers, were expected to pay for this land in interest free instalments over a period of 36 years.[35]

Did the laws achieve their objectives? Simply as a measure of redistribution the legislation created, in Hope's words, 'a social revolution' in Lithuania. The old aristocratic structures of wealth and power based on land were seriously weakened and largely replaced by 'a classless nation of propertied small farmers'. This band of middle-sized family farms lay at the heart of improving and increasingly prosperous Lithuanian agriculture. About 50 per cent of land was in holdings of from 37 to 125 acres and only 0.6 per cent of farms were more than 250 acres in extent. At the same time this redistribution strengthened the loyalty of the peasants to the Lithuanian republic and deflated the appeal of radical communistic proposals in the rural areas. The reforms also confirmed the existing trends towards the establishment of individual farmsteads away from the collectivism of the village, which gave a stimulus to individual enterprise. The result was shown in greatly improved agricultural methods and much increased yields. There was generally higher productivity on individual farmsteads than on the large estates, particularly in livestock raising and poultry. Hence the redistribution in itself, coupled with the break up of the villages in favour of individual farmsteads, effected an increase in output. But, in addition, there was an across-the-board improvement in agricultural methods used in grain production, animal breeding and dairy output. This resulted from a combination of farmers' entrepreneurial talents, government incentives and protection, and the technical expertise of the co-operatives.[36]

It would be a mistake, however, to present the Lithuanian agrarian economy as a tale of unqualified success. The agricultural census of 1930 showed that there were some 53,000 farms of between 2.5 and 12.5 acres. Simutis calculated that farms up to about 30 acres in size were generally too small to produce a surplus for the market, and these represented 56 per cent of all farms in Lithuania. By contrast there were some 127,000 farms above 30 acres which produced a surplus and supplied the bulk of the agricultural exports. As population increased the size of farms tended to decrease, placing a larger proportion in the less prosperous and vulnerable category.[37] Small farmers also suffered from the loss of customary rights after the ending of communal agriculture, such as the right to graze cattle on the common pasture. By 1930 there were some 150,000 agricultural labourers, many of whom moved to the towns in search of more congenial and better paid work. Alongside the relatively prosperous and successful band of medium sized farms there was a growth of dwarf holdings and of landless labourers; it was among these classes that there arose economic and political discontents in the 1930s. Critics of the governments have suggested that too little was done to cushion these poorer sections of the rural community.[38]

There was little criticism, however, of the government's educational, cultural and social reforms. Some simple statistics indicate the scale of the educational achievements of independent Lithuania. In 1913 only 15 per thousand of the population attended school compared with a figure of 116 per thousand in 1931/32. The budget of the Ministry of Education almost tripled between 1923 and 1937. The number of students attending grammar schools rose four fold between 1920 and 1939 and the number of primary schools tripled. The ethnic minorities established their own schools with state support. For example, 93 per cent of Lithuanian Jewish children were taught in Yiddish or Hebrew language schools in 1925. The illiteracy rate had fallen to 12 per cent on the eve of the Second World War, having been between one third and one half at the beginning of the 1920s. Adult education was expanded through the co-operatives and by the attachment of special classes to elementary schools, resulting in 497 such courses attended by 30,000 students in 1927. Technical and vocational education was enhanced by the creation of special agricultural schools, technical and trade schools, and schools of commerce. The loss of the Vilnius region to Poland meant that the ancient University of Vilnius could not become a Lithuanian institution. Consequently the Lithuanian government founded a university in Kaunas in 1922 which was given the title of Vytautas Magnus University in 1930. The government created other higher academic institutions such as the Agricultural Academy, the School of Veterinary Medicine, the Art School and the Conservatoire. Taken together these institutions created a literate, educated population with appropriately qualified professional, scientific, and technical personnel.[39]

Complementing the educational reforms was a generous programme of state investment in the arts. National pride demanded that the new Lithuanian state should have a flourishing cultural life in which Lithuanian composers, artists and performers would have the opportunity to exercise their talents and to present to the Lithuanian public through the theatre, on the opera and ballet stage, in the art galleries, and through the public buildings themselves the best examples of world and Lithuanian art and architecture. This policy resulted in the state financing the construction of two theatres, an opera and ballet company, two symphony orchestras and museums and galleries, notably a gallery to display the works of Lithuania's greatest painter (and composer) M.K. Čiurlionis who died tragically young in 1911.[40] Kaunas itself was transformed from a small and undistinguished Russian garrison town into a worthy capital city with impressive public buildings. During the interwar period Lithuania enjoyed a major cultural revival which immeasurably strengthened its national identity and inspired confidence in the literary, artistic and organizational talents of the Lithuanian people. It enabled them to show to themselves and the world that they did not deserve the derisive Polish epithet of 'peasant'. This thriving cultural life was matched by a proliferation of newspapers, weekly and monthly magazines, and scientific, professional and scholarly journals. Some 157 periodicals were published in Lithuania in 1937, up from 112 in 1928. In 1914, by contrast, only 22 periodicals had been published. Each of the ethnic minorities had its own publications. Evidently the thirst for information, comment and debate had been deepened by the growth in literacy and general education since independence. Similar progress was made in the provision for social welfare and health care, and this was maintained during Smetona's regime after 1926. State schemes for sickness, accident and unemployment insurance were initiated, and conditions at work were established under International Labor Organization norms. An eight-hour day was introduced in manufacturing industry, and pension laws were enacted.[41]

GOVERNMENT AND POLITICS BEFORE THE COUP OF 1926

The constitution of 1922 provided the framework for the conduct of government and politics in the early inter-war period. The Constituent Assembly, elected by universal suffrage in April 1920, aimed to create a constitution which would give the Lithuanian people, in Rei's terms, 'the opportunity to express their sovereign will in a most direct and unhampered way'. The democratic and egalitarian convictions of the 'founding fathers' were the result of a number of shaping influences: the peasant origins of most of the members of the Assembly and their consequential desire to reconstruct Lithuanian society in the interests of their constituents; the impact of Russian radicalism on many of the Russian-educated Lithuanian intelligentsia; the effect of the February Revolution in Russia on Lithuanian refugees who had fled before the German advance in 1915; the fear of dictatorship; and the example of West European constitutions such as the Weimar Republic's. The result was a highly democratic form of government in which the legislature was dominant, the executive was weak, and the President largely a figurehead. The legislature was a single chamber elected by universal suffrage under a system of proportional representation. MPs were elected to the legislative assembly, or *Seimas*, for three years and they in turn elected the President of the Republic. The *Seimas* had the power to impeach the president and to dismiss him from office by a two-thirds majority. On the other hand, the President could dissolve parliament and call an election, which moderated parliamentary powers but put his own position at risk. Although the President appointed the Prime Minister, the Cabinet was responsible to the *Seimas* and in fact completely subordinate to it. Moreover, the President could take no action without cabinet approval.[42]

As could be expected, this highly democratic constitution guaranteed civil rights and individual freedoms, though there remained in place some legal restraints on the freedom of assembly and the press. These were directed mainly against communists. It also offered guarantees of the rights of minorities in the state. Challenged by Sapieha, the Polish

Foreign Minister, to base Lithuania's policy towards its national minorities on 'the principles of equity and justice', the Lithuanian Foreign Ministry retorted that Lithuania was 'a profoundly democratic state' which had 'never wronged or will wrong the rights of national minorities'. Each was given full autonomy in the administration of its religious, educational and cultural institutions and could draw upon its share of the national educational budget for this purpose. Initially two ministries, for Jewish and Belarussian affairs respectively, were established but these were discontinued in 1923. The Lithuanians, however, went beyond the minimum guarantees of minority rights laid down by the League of Nations, even though they were not prepared to continue the offer of separate legislative jurisdiction to the Jewish community after 1924. Nonetheless, the Jews in common with the other minorities of Germans, Poles, Belarussians etc. maintained their own schools, publications, cultural activities and linguistic rights, and continued to elect their representatives to the *Seimas* until the coup of 1926. In the parliamentary period the Jewish community elected between three and six deputies, the Polish between two and four and the Germans one or two until the 1926 election when an extra five representatives were elected. It was not unreasonable for the Lithuanian state to insist that the Lithuanian language be taught in minority schools, that a good knowledge of Lithuanian be required for entry into Lithuanian higher education, and that street names and public notices should be in Lithuanian (though in the case of notices more than one language could be used). As E.H. Carr reported to the British Foreign Office, parliamentary speeches could be made in other languages than Lithuanian but they could not be entered in the records of debates unless a translation was provided.[43]

The democratic politics resulting from the Lithuanian constitutional structure was, as Hiden and Salmon have noted, 'all too easy to caricature'. The drafters had very little practical experience of politics, except of the oppositional or revolutionary variety, and adopted a scheme based on western models which was not necessarily appropriate for Lithuania. The census of 1923 noted an illiteracy rate of 44%, which posed problems for the effective conduct of democratic politics. The constitutional model adopted provided for a multi-party democracy, but the resulting proliferation of parties made it difficult to establish workable coalitions. In the elections of 1923 12 parties put up candidates and 10 of them were represented in the Seimas. In the first seven years of independence Lithuania had 11 cabinets. These frequent cabinet crises, it has been argued, created a constant climate of political uncertainty and tended to discredit democracy. This was compounded by a constitutional structure which failed to provide checks and balances between the three arms of government.[44]

But despite frequent changes of government there was some continuity; for most of the period up to 1926 governments were dominated by a coalition of Christian Democrats and Populists and there was a degree of continuity among ministers and civil servants. Moreover, this 'happy anarchy', as it has been called, produced fundamental and beneficial economic and social reforms, for example in landholding and education. The weakness of the 'civic culture', to use Taagepera's term, was an unavoidable handicap, but its major impact lay not so much in the proliferation of parties as in the failure of the right wing parties and the military to accept the judgment of the electors in 1926. Evidently ethnic nationalism combined with rightist fears of the political left was stronger in some influential circles such as the army and to some extent the Church than the commitment to democratic processes.[45]

There were four parliamentary elections from 1920 to 1926 before the period of authoritarian rule began. Three parties or party blocs dominated the elections, and between them formed the majorities in the *Seimas*, sometimes with the help of the representatives of the minorities. The Christian Democratic bloc composed of the Christian Democrats (CDs), the Farmers' Union and the Labour Federation, constituted the largest grouping. It was a reformist party, appealing both to rural and urban constituencies including professional people and rural labourers and smallholders. But the cement which held it together was a deep loyalty to the Roman Catholic Church. Priests were very active in its leadership and among the rank and file; as a British observer noted, the second vice chairman of the Seimas was a clergyman, and two of the leaders of the CDs were priests, one of them at the same time being a leading banker. Priests and the lay leadership of the CDs tended to come from the same socio-economic background of relatively prosperous peasant families. The party's support held up well until 1926. It won 59 out of 112 seats in the election of 1920, 38 out of 78 in 1922, 40 out of 78 in 1923 but only 30 out of 85 in 1926.

The second major party was the Party of Rural People, more commonly called the Populists, which was liberal in orientation and on the Centre Left. Its economic and social reform programme, though somewhat more radical, was similar to that of the CDs, and it drew its support mainly from small landholders. Its major conflict with the CDs was over the relations between church and state. The CDs had fought successfully for compulsory religious education in state schools, for financial support for the Church from the state, and for church control over the registration of births, marriages and deaths. No civil marriage or divorce was permitted in Lithuania during the parliamentary period. The Populists, being anti-clerical, opposed the CDs on these issues, and this was the major source of conflict between them. It proved impossible for them to form a coalition after 1924.

The Social Democratic Party (SDP), a Marxist party committed to parliamentarism and reform, drew its support mainly from the urban workers. It had a more radical economic and social agenda than the Populists though, like them, it rejected the power of the Church in politics. It remained a minority party, at the peak of its success in 1926 winning only 17% of the vote. To the left of the SDP were the Communists who were proscribed in 1919 but reappeared in the 1920s behind a front organization. On the far right was a party which had very little electoral success but enjoyed high visibility. This was the Nationalist Party which emerged in 1924 after the merger of two existing groups. Some of the leading figures of the Lithuanian independence movement were members of this party but had been rejected by the electorate subsequently. They included Smetona, Voldemaras, Yčas and Basanavičius. They fell outside the partisan consensus on land and other reforms and were very critical of the parliamentary system. They made no secret of their admiration for strong government and a powerful military. Their opportunity came in 1926.[46]

The CDs experienced a loss of support as a result of the economic recession and the publicising of corruption in the government and administration. They also suffered from the Concordat between Poland and the Vatican which in effect recognised Poland's claims to the Vilnius region. Since this was the most salient issue in Lithuanian politics, the CDs had to take the blame for this failure. The new government was a coalition of Populists, the SDP and the Minorities parties. It proceeded to alienate the church, the army and the mass of devout Lithuanians.

The coalition government of 1926 formulated a programme which reflected the priorities of its different elements. The Social Democrats, for example, wanted the removal of some legal restraints on the freedom of assembly and the press dating from 1920, which had been aimed mainly at the communists. They also demanded an amnesty for those imprisoned under these laws. The Populists' priorities were addressed by proposals for cutting the budgets for religious teaching in schools, introducing civil registration of births, and reducing the size of ecclesiastical landholdings. The proposal to increase the number of Polish language schools pleased the Polish minority. But the decision to cut defence expenditure and to retire some senior army officers, combined with the signing of the Soviet-Lithuanian Non-Aggression Treaty in 1926 (despite the fact that this had been initially negotiated by the preceding Christian Democrat government) aroused alarm in military circles. Political opponents accused the new government of moving leftwards in both domestic and foreign policy, and betraying the new nation by its concessions to minorities and political extremists. The continued failure to realise Lithuanian ambitions over Vilnius remained a festering sore.[47]

The Christian Democrats saw the opportunity to remove their opponents from power, the Nationalists to establish an authoritarian form of government. Their mood was illustrated by the apocalyptic words of Antanas Smetona who described the 'swarms' of unemployed as 'dark mobs, led by Russianized chieftains ... insisting upon a Bolshevik government' and threatening 'to deprive Lithuania's peaceful and industrious inhabitants of their property'. Rightist university students demonstrated against the 'Polonization and Bolshevization' of the country and the 'cosmopolitanism and internationalism' of the coalition government (referring to its leftist and minorities' sympathies). In the circumstances, Piłsudski's coup in Poland in 1926 offered the army an example to follow. The Lithuanian coup ushered in an era of authoritarian government which lasted, with some modifications, up to the outbreak of the Second World War.[48]

INDEPENDENT LITHUANIA UNDER SMETONA

The coup of 1926 was engineered by young army officers in Kaunas on the pretext of saving the republic from a communist plot. The army leadership, which was close to Smetona, took care to maintain a pretence of legitimacy and continuity. President Grinius was persuaded to appoint Augustinas Voldemaras Prime Minister and then to resign. Smetona was then elected President by the *Seimas* in the absence of the Populists and Social Democrats who boycotted the session. The *Seimas* was itself dissolved in the Spring of 1927. The Christian Democrats played a passive but supportive role during the coup and then voted for

Smetona as head of state. He reciprocated by appointing them to the cabinet. At this point, however, the Christian Democrats and the Nationalists began to part company. The former supported the coup to remove the leftist government but not to destroy the 1922 Constitution. They expected a speedy return to parliamentary government. The Nationalists had no intention of obliging them. Having dissolved parliament they refused to call new elections and in 1928 they decreed a new constitution which was to be ratified within the next ten years. The general effect of this document was to weaken the legislature and to strengthen the executive. To preserve the illusion of continuity the constitution proposed in 1928 and the adopted constitution of 1938 were regarded as amendments to the 1922 constitution.[49]

Defending the transfer of powers to the head of state, critics claimed that the 1922 constitution was inappropriate for Lithuania at that stage in its socio-economic and political development, and that the dominance of parliament over the executive failed to restrain excessive partisanship and to encourage compromise and toleration. It was rational to propose constitutional changes which would alleviate some of the popular impatience with the parliamentary system. The leaders of the coup and the Christian Democratic leadership both wanted a safe right wing nationalist government, but they differed over the means to achieve it. Smetona went down the authoritarian route, while the Christian Democrats wanted to retain a parliamentary system with stronger safeguards against anarchy and 'partocracy'.[50]

One of the commonest accusations levelled against the Smetona regime by the Soviet government after the Soviet annexation of Lithuania in 1940 was its fascist character. It followed that the incorporation of Lithuania into the Soviet Union in 1940 and again in 1944 represented a liberation from fascism and its replacement by 'progressive' forces. However, this characterisation of Smetona's government as fascist is highly questionable. Determining the true character of Smetona's regime is important for an assessment of Soviet actions in 1939–40. It also has profound implications for present-day Lithuanians who have to ask themselves whether their form of government after 1926 was fascist or authoritarian. If the latter, could it be justified as a corrective for an extreme form of parliamentarism which Lithuania was too politically immature to operate effectively?

Historians of Lithuania from the Baltic expatriate communities in the West have been very sensitive to the charge of fascism which, on the whole, they have rejected. Taagepera, for example, noted the 'mild authoritarianism' of the Smetona government; Eidintas asserted that 'none of the basic elements of totalitarianism was present in Lithuania'; others have referred to the benignity rather than malignancy of the regime, and to its semi-democratic, semi-dictatorial and temperate character. Smetona, they argue, distanced himself from the fascist model and condemned national socialism publicly. Others claim that Smetona saved Lithuania from fascism in 1929 by dismissing, and later imprisoning, his Prime Minister Voldemaras who had developed pro-Nazi tendencies and allegedly received financial subsidies from Nazi Germany.[51] Smetona is also defended from charges of fascism because his personality and temperament made him a most unlikely candidate for totalitarian leadership. He was mild-mannered, a 'conservative humanist', who read Plato in the original each day, a man who was reluctant to use excessive force, and who scorned racism and anti-Semitism. His political success and survival depended as much on his manipulative and management skills as on the use of violence, repression, demagoguery, or bewitching oratory.

His apologists do not, however, attempt to absolve him from the charge of authoritarianism and political dictatorship, nor do they reject the argument that his ideology, rather than conduct, was substantially influenced by fascism. He admitted admiration for Italian fascist principles, and adopted the cult of the leader, taking the title Leader of the People (*Tautos Vadas*), and refusing to tolerate opposition. But some of the essential qualities of a Mussolini or a Hitler, such as megalomania, oratorical power, demagoguery, effortless histrionic performances, crude and indiscriminate violence, and 'gangsterism' were absent in Smetona. Even his title seemed modest and low key, compared with the challenging and contemptuous *Duce* or *Fuehrer*.[52]

If the 'leadership principle' was not implemented in a rigorous way, and if Smetona was, as right-wing critics like the *Vairininkai* alleged, too fastidious about the use of force, too reluctant to realise in practice the fundamental principles which he espoused, nevertheless the political class whose activities he had suppressed could present a major indictment of his regime, could indeed accurately describe it as an authoritarian and arbitrary form of government underpinned by a fascistic ideology. For example, no political parties apart from Smetona's own Nationalist Union were allowed to function openly. Initially many opposition leaders were arrested, though arrests had virtually ceased by 1938. At the same time, however, many opposition politicians continued to act 'informally' and unofficially, retaining prominent jobs in the co-operatives, in education and in other occupations. Moreover, there was no attempt to eliminate physically the opposition, apart from the Communists who suffered from wholesale arrests and repressions. This did not please the more militant members of the nationalist movement who wanted to see more recourse to police measures and forced labour camps.[53]

Press censorship and limitations on the right of assembly were imposed by the regime. Although a modicum of press freedom remained, new restrictions were imposed in 1935 and 1936, but it cannot be claimed that the regime operated a 'watertight' censorship controlling all sources of information and 'prefabricating' data.[54] Trade unions lost their independence and were placed under state control. The regime received strong support from the army, the *quid pro quo* being substantial allocations to defence in the national budget, about a quarter of the total expenditure. Smetona could also call on the support of an armed militia, the *Sauliu Sajunga*, composed of Lithuanian veterans and members of youth organizations, whose virulent nationalism was sometimes frustrated by Smetona's reluctance to give them a free rein.

In practice, a basic pluralism of social and civic organizations was permitted. The court system, for example, remained independent, cities and towns retained self-government, and a market economy dominated by co-operatives and private enterprises continued to exist, supported by government subsidies and protection. Literally thousands of private economic, professional, social and cultural organizations functioned independently of, and without interference from, the Nationalist government. Religious tolerance remained a prominent feature of national life and generous state subsidies to religious denominations and minority language schools were continued. The administration actively supported the Jewish community which enjoyed a thriving cultural and religious life.

Even though the educational and cultural activities of the minority communities were largely unaffected by the Smetona regime, economic and social developments had an adverse impact on the Jewish community in particular. For example, the increasing role of the co-operatives in Lithuania's foreign and domestic trade reduced the share of Jewish merchants in these sectors. In 1923 Jews controlled about three-quarters fo Lithuania's trade but by 1936 the number of Jewish-owned commercial enterprises had declined by one-tenth compared with a roughly 300 per cent growth in ethnic Lithuanian firms.

Similarly, the traditionally heavy presence of Jews in higher education was diminishing as increasing numbers of ethnic Lithuanians qualified for entry into the university and academies as a result of better educational provision in schools. The proportion of Jews in higher education had fallen to 15 per cent in 1935, though this was still double the proportion of Jews in the population as a whole. During the 1930s there was increasing criticism from the ethnic Lithuanian business sector of the still powerful role of Jews in trade and industry, and the relatively limited opportunities for ethnic Lithuanians in these sectors. Anti-semitic sentiments were more frequently expressed, even though Smetona himself emphasised that ethnic minorities were fellow citizens, not foreigners, and that their cultures must be respected.[55]

Ethnic Poles were the second largest minority community. Relations between them and ethnic Lithuanians were affected by the hostility over the Vilnius region which existed between Poland and Lithuania during the inter-war period. Lithuanians were sensitive to the treatment of their fellow ethnics by the Polish authorities in Vilnius, particularly in the field of education, and were tempted to respond in kind. For example, a law was passed in 1936 forbidding children to attend Polish language schools in Lithuania unless both parents were ethnic Poles. This was part of a general movement to discourage enrollment in these schools. But Polish minority rights under the constitution and the laws were not circumscribed during the period.[56]

Smetona's regime embodied an extreme authoritarian nationalism. In Smetona's ideology, individualism was subordinate to the demands of the nation; discipline, and conformity to the national will, superseded personal freedom; everything alien to the Lithuanian 'soul', such as parliamentary democracy, was rejected; national consciousness and love of the fatherland were paramount; the leader's will was supreme; the key to national survival was to mobilize and drill the nation like an army. In short, the nation was likened to an organism, a 'monolithic body', in which human needs and personal responsibility, not human rights, took precedence. In his last major speech Smetona painted an idealised picture of the Lithuanian people working together in solidarity towards one end, to ensure the vigour and health of the nation.[57]

But, it may be concluded, neither Smetona's ideology nor the political realities of his regime could be summed up as fascist. Employing a set of criteria drawn up by Norman Davies to distinguish fascist regimes, we find that Smetona's government failed most of the questions on the totalitarian, fascist-communist test. Even where it conformed in general to fascist norms it lacked the rigorous and ruthless applications of its ideas characteristic of fascist states. August Rei's description of authoritarian developments in the Baltic states deserves to be recalled in this context. 'Even though these deviations from the democratic form of rule ought in justice to be regretted, it is easy to show that they give no occasion for upbraiding the Baltic peoples, let alone vilifying their young states as lairs of the blackest reaction.' In sum, Lithuania's experience of genuine fascism had to await the Nazi and Soviet occupations.[58]

This conclusion is reinforced by political developments during the last three years of independence. A new constitution was introduced in 1938, prompted in part by economic and political unrest in South-West Lithuania and the loss of support for the regime, evidenced by the faster growth of Christian Democratic organizations than Nationalist ones. The Polish ultimatum to Lithuania to establish diplomatic relations in 1938 and the loss of Klaipėda to Germany in March 1939 reduced the prestige of the government and brought the political opposition back to life. While some Christian Democrats supported authoritarianism provided it was given direction by the Roman Catholic Church, others, along with Populists and Social Democrats, demanded more representative government and a restoration of civil rights. Taking advantage of these more favourable circumstances the Christian Democrats and the Populists joined together in an anti-nationalist campaign. The new constitution of 1938, though providing for a stronger presidency and a weaker legislature than in the 1922 version, at least provided the opportunity for a new legislature to be elected in the same year. The new constitution also reaffirmed the commitment to equality before the law, irrespective of race and religion.[59] In order to retain power Smetona had to widen the basis of support for his regime. A new cabinet under General Černius was appointed in March 1939 with four members from the Christian Democratic and Populist parties, although the parties were still formally outlawed. Press restrictions were eased and acts of clemency led to the freeing of some political prisoners. Public meetings were held which offered serious criticism of the Nationalist Party. There was a sharp rise in political tension in the immediate pre-war years so much so that some historians believe that the democratic movement to remove authoritarianism would have been unstoppable if the war had not intervened. On the other hand, there was a growth of right wing extremism and a determination both to hang on to power and to become more resolute in its exercise. Predictably there was a sharp growth in anti-Semitic activity at the end of the decade. It is uncertain how the struggle between democratic and authoritarian forces would have been resolved.[60]

In reviewing the economic, social and cultural progress of Lithuania in the interwar period, Vardys summarised her achievements as follows; she restructured her economy, implemented a radical land reform, created an effective educational system from elementary schools to higher education, initiated a system of social security, and sponsored literature, music and the arts. There was a major infrastructural programme, involving new roads and railways, the draining of swamps, the construction of impressive public buildings including schools and hospitals. The economic growth rate between 1925 and 1939 of almost 5 per cent per year was surpassed by the growth of industrial production of 7.5 per cent per year. Agricultural output grew at a more modest but quite impressive 2 per cent per annum. This financed a rise in overall prosperity though there were sections of the population such as unemployed urban workers and agricultural small holders who were sunk in poverty and were tempted by communist ideas. Finally, the foundations of Lithuanian national culture became more solidly based, and pride in being Lithuanian was enhanced in the light of national achievements. Lithuania's existence in a threatening international environment for the whole of the inter-war years further heightened national awareness and strengthened determination to protect the country's independence.

LITHUANIA IN INTERNATIONAL RELATIONS BETWEEN THE WARS

The borders of the independent Lithuanian state were determined, not by the League of Nations, but by the use of force, either on the part of Lithuania itself or by one of its more powerful neighbours.[61] This forceful resolution of border problems increased tension and instability, damaged good neighbourly relations, and undermined attempts to establish a regional security system. The question of 'where was Lithuania?' was not resolved in the inter-war period, certainly not to the satisfaction of the Lithuanians. For example, the Vilnius region was lost to Lithuania following the notorious attack by the Polish General Zeligowski in October 1920. Part of it was regained in October 1939 as a result of the Soviet invasion of Poland the previous month. Similarly, in January 1923, an attack on Memel (Klaipėda) sponsored by the Lithuanian government effected the transfer of the territory to

Lithuania, only for it to be returned to Germany in March 1939 as a result of Nazi threats to use force. For almost the whole of the inter-war period, therefore, Lithuania was at odds with its two powerful neighbours, Poland and Germany. Although, formally, its relations with the Soviet Union were friendly, Lithuania recognised that its freedom of action was circumscribed by Soviet suspicions and apprehensions.

Attempts to define the territorial extent of the Lithuanian state were complicated by the several meanings to be attached to the term Lithuania. Was it historic Lithuania, the territory of the old Grand Duchy; was it ethnographic Lithuania which, the Lithuanians claimed, was partly inhabited by people of Lithuanian stock who had become denationalized and 'weaned away' from the old language; or was it linguistic Lithuania, the area where Lithuanian language was over-whelmingly spoken? The official Lithuanian claim, adumbrated in Paris and then maintained throughout the inter-war period, was for an ethnographic Lithuanian state composed of several former Tsarist provinces—Kovno, most of Vilna, and parts of Suwałki and Grodno, along with so-called Lithuania Minor, that is the Memel district of East Prussia.[62]

The newly independent Polish state could not accept the Lithuanian claims. After Polish troops had captured Vilna from the Bolsheviks in the Spring of 1919, Piłsudksi assumed that Lithuania, within its historic boundaries, would again form part of the Polish-Lithuanian Common-wealth on the basis of the Treaty of Lublin. Dmowski and Paderewski at Versailles adopted an annexationist position under which an ethno-graphic Lithuania would be united politically with Poland but given a reasonable amount of autonomy. Each of these standpoints was diametrically opposed to the Lithuanian view, expressed in the Lithuanian Constitution, that the old political ties (that is, with Poland), were not to be restored.[63]

The dispute was partially resolved by the Polish decision to recognise Lithuanian self-determination. However, the precise boundaries of an independent Lithuania were a major and unresolvable bone of contention. The main problem was the disposition of the former Russian province of Vilna. The Lithuanians claimed Vilna (Vilnius) as their historic capital and used Russian census data to show that ethnically the Poles had no claim to the region; quite the contrary, the White Russians (Belarussians) were slightly more than 60 per cent of the population compared with the Poles's share of about 8 per cent. German and Polish census material, however, suggested that more than half the population claimed to be of Polish nationality. Moreover, Lord claimed that less than half of the population of ethnographic Lithuania actually spoke Lithuanian. Conflicting data and unyielding nationalist claims made it impossible to reach a solution satisfactory to both sides, despite the later efforts of the tireless League of Nations intermediary Paul Hymans.[64]

Earlier, the League had been drawn into the dispute on the ground. In 1919 it drew three lines of demarcation between Polish and Lithuanian troops, the last of which, in December, separated 'indubitably Polish areas from those with a predominantly non-Polish population'. Vilnius lay on the non-Polish side of the line. Again, the Treaty of Moscow, the Soviet-Lithuanian peace treaty of 1920, drew a frontier which placed the Vilnius district on the Lithuanian side.[65] Finally, the League Council suggested a provisional demarcation line in September 1920 which gave Vilnius to Lithuania. Fighting continued until the agreement at Suwałki between Lithuania and Poland on 7 October, 1920 which drew a line of demarcation which was incomplete but indicated that the Vilnius area would be part of Lithuania. Three days later Zeligowski seized and occupied the Vilnius region and drove out the Lithuanian forces. The League made strenuous efforts to encourage the two sides to reach agreement; negotiations, chaired by Hymans, considered compromise proposals including the cession of Vilnius to Lithuania in exchange for a permanent political, economic and military association between the two states. When no agreement was forthcoming, the Polish authorities in Vilnius organized a plebiscite which was boycotted by the non-Polish elements in the population, though the result was taken by the Polish side to justify incorporation of Vilnius in Poland. The Allies seized this opportunity to allocate Vilnius to Poland on 15 March, 1923.[66]

A propos of this decision the historian, E.H. Carr, then an official of the British Foreign Office, witheringly remarked that the League 'threw up the sponge and the Allied Powers decided to compound Zeligowski's felony'.[67] That is perhaps too dismissive; the repeated refusals of both sides to heed the League's calls for 'a supreme gesture toward peace, consent and conciliation' tested the League's patience up to and beyond its limits. Fierce national pride and an unyielding determination to achieve national objectives on both sides made it impossible to reach agreement, and the League had to recognize that fact. Balfour, the British delegate, observed that no one listening to the two sides could believe that their main objective was to reach agreement. When politicians were suspected of being too conciliatory, popular opinion refused to allow them to make concessions, and backed up their opposition by threats of violence. Consequently, Lithuania refused to open diplomatic relations with Poland and the border remained closed until 1938, when Poland took advantage of the international situation to demand that diplomatic relations be restored.[68]

Just as the Vilnius question poisoned relations between Lithuania and Poland between the wars, so the Memel (Klaipėda) problem kept relations between Lithuania and Germany in a state of tension which became increasingly severe at the beginning of the 1930s. The Memel area had been part of Lithuania until the eighteenth-century partitions of Poland-Lithuania, which allocated the territory to Prussia. At the end of the First World War the Allied Powers separated the area from Germany, the assumption being that it would ultimately be joined to Lithuania. There were two reasons for this. The first was that a majority of the population spoke Lithuanian or a form of it, even though rather more of the population in the town of Memel itself spoke German.[69] The second reason for attaching the Memel to Lithuania was economic. The port of Memel was the only outlet to the sea of an independent Lithuania and this access would be crucial for Lithuanian economic development after the war. Germany's possession of Koenigsberg made Memel of far less economic significance to her than to Lithuania.[70]

However, Memel could not be transferred to Lithuania until her independence had been recognized. As a temporary measure, therefore, Memel was turned over to French administration. Partly as a result of Franco-Polish pressure, international opinion began to lean in favour of granting Memel the status of free city, with open access for Lithuanian trade. Fearing that Memel was slipping away from them, a Lithuanian force of irregulars seized the district in January 1923, and in February the Conference of Ambassadors agreed to accept the transfer, the conditions for which were formalised in the Memel Convention of 8 May, 1924.[71] After Hitler came to power Nazi support in Klaipėda grew, the German elements became increasingly irredentist, and Germany imposed economic sanctions against Lithuanian trade. Tension became acute; disturbances were frequent; and in March 1939 Hitler demanded the transfer of Klaipėda to Germany. The Lithuanians had no option but to accede.[72]

The broader significance of the Vilnius and Klaipėda issues lay in their impact on Lithuania's security between the wars. Although there were difficulties in the German-Lithuanian relationship over Klaipėda, Weimar Germany preferred to see Klaipėda in Lithuanian hands to being under the influence or control of Poland. An anti-Polish Lithuania offered protection for East Prussia, and Polish-Lithuanian antagonism prevented the formation of a Baltic bloc, which neither Germany nor the Soviet Union wanted. Despite the tensions and difficulties over Klaipėda, the Lithuanian and German governments were able to reach agreements in a number of areas in the late 1920s, particularly on trade. The very high proportion of Lithuanian exports which went to Germany strengthened Germany's influence on Lithuania.[73]

So far as relations with the Soviet Union were concerned, the Soviets were the only power to support Lithuania's claim to Vilnius. Moreover, Latvian officials concluded from their contacts with leading Lithuanians in 1930 that the Lithuanian government had no fear of being absorbed or conquered by the Soviet Union owing to the absence of a common border. They assured themselves also that Germany would not permit a Soviet encroachment on Lithuania. In contrast, Estonia and Latvia regarded Germany and the Soviet Union as major threats to their independence, and looked to Warsaw as a balance against Moscow and Berlin. This widened the gap between them and Lithuania. In any case, there was little they could do to help the Lithuanians attain their objectives in Vilnius and Klaipėda. In short, the Vilnius issue prevented the creation of a comprehensive pan-Baltic league composed of the three Baltic states, Finland and Poland during the most favourable circumstances at the end of the First World War.[74]

However, it would be wrong to place the entire blame for this failure on the shoulders of the Lithuanian-Polish conflict. First advanced by the heads of the Estonian delegations in London and Paris and presented to the British Foreign Office in November 1918, the League idea was an ambitious attempt to establish wide-ranging co-operation among Baltic Sea states, the Scandinavian, the Eastern Baltic (Finns, Estonians and Latvians) and the Southern Baltic (Lithuanians and Poles). It envisaged co-ordination of foreign and defence policies, agreements on communications and commerce, the establishment of a Baltic Economic Council and an arbitration covention. The overall objective was to ensure the security of the member states and to guarantee freedom of trade and communciations in the Baltic Sea. This potentially powerful grouping would have stood in the way of any German or Russian attempt to gain control of the Baltic region. It would have meant that 'Imperialist Russia and Imperialist Germany [would] be separated from each other by a dam of living peoples ... If Russian and German imperialism reach power and form an alliance it will threaten with destruction the independence of Finland as well as of the border states' [of Estonia, Latvia, Lithuania and Poland]. This was the conclusion of an article inspired by Foreign Minister Holsti of Finland. A series of conferences between 1920 and 1922 at Helsinki, Bulduri and Warsaw, however, failed to bring the putative League into effect.[75]

Obviously the Lithuanian-Polish conflict bitterly divided two of the potential members and led to Lithuanian withdrawal from the conferences. But the Scandinavians never showed any desire to participate owing to their unwillingness to shoulder military obligations in the Eastern and Southern Baltic area. The Finns, though participating in all the conferences, ultimately rejected an agreement which might have entangled them in Poland's conflicts with her neighbours. Finland's repudiation of Holsti's policy marked her realignment with the Scandinavian states, particularly Sweden. Other factors working against the realisation of the Baltic League were the Soviet Union's bitter hostility to it and the antipathy towards regional security organizations at a time, 1922, when the prestige of the League of Nations still stood high.[76]

The anchor of Lithuania's foreign policy particularly under Voldemaras was that 'the key to Vilnius lay in Moscow and Berlin'. By 1934 this was no longer the case. Lithuania was 'altogether isolated' and had no alternative but to re-orient its foreign policy. The advent of Hitler and Germany's consequential withdrawal from the League, followed by the trial of Nazi leaders in Klaipėda in 1934, irreparably damaged Lithuanian-German relations. Moreover, the German-Polish Non-Aggression Pact of January 1934 removed the possibility of Germany's supporting Lithuania against Poland. Two years earlier, a similar non-aggression agreement between Poland and the Soviet Union meant that the key to Vilnius could no longer be found in Moscow either.[77]

This loss of support against Poland turned Lithuania's thoughts in unfamiliar directions. Under the direction of Stasys Lozoraitis, the Foreign Minister, Lithuania explored ways in which the relationship with Poland could be improved while agreeing to disagree on Vilna. These attempts came to nothing in 1935 owing to mutual suspicions and mistrust. Similarly the failure of the Eastern Locarno proposal removed another foreign policy option for Lithuania. But a parallel attempt to establish an agreement between the three Baltic states resulted in the Treaty of Entente and Collaboration signed in Geneva on 12 September, 1934. This Baltic entente, though fragile and imperfectly developed, formed the basis of Lithuania's foreign policy until the Second World War.[78]

An Estonian-Latvian defence alliance had existed since 1923. In 1934 both states felt threatened by Germany, and German minorities there were somewhat receptive to Nazi propaganda. In the circumstances the governments in Tallinn and Riga accepted Lithuania's proposal to widen the existing alliance into a tripartite entente. The three states were to confer periodically with a view to co-ordinating political, diplomatic and economic policies, but it was agreed that the Vilnius and Klaipėda questions would be excluded from the agreement. No plans were made for a military alliance, however, since Estonia refused to participate. The very close similarity in the three states' economic structures made economic co-operation difficult, and cultural collaboration, also provided for in the agreement, foundered on the absence of a common language.[79]

The entente states felt that their salvation lay in neutrality. Alignment with one of the great powers in the region would ensure hostility from the other. The best way of evading involvement in great power conflicts was to stay uncommitted and aloof. Accordingly all three passed neutrality laws on the Swedish model in the belief that neutrality would help to guarantee their independence. The Baltic states' failure to strengthen their neutral status by forming an effective military alliance convinced Moscow that the policy imperilled Soviet security by tempting fascist aggression in the region. Molotov thought that neutrality was 'too insecure, too unreliable' and Zhdanov gave a characterically brutal warning of the likely consequences: '...if these tiny peoples allow big adventurers to use their territories for big adventures, we shall widen our little window onto Europe with the help of the Red Army'. The great fear was that Germany might turn the Balts into 'involuntary accomplices in an anti-Soviet crusade'.[80]

The dramatic changes taking place in Nazi Germany created a deep sense of insecurity in the Soviet Union. Moscow pursued a variety of policies designed to strengthen security on her western borders, notably the signing of non-aggression pacts with Finland, Poland, Estonia, and Latvia in the early 1930s and the renewal of the 1926 pact with Lithuania in 1934. These were accompanied by attempts to impose frontier guarantees on the Baltic states and Finland, and to establish an eastern Locarno, the latter being sabotaged by Germany and Poland. In the last five years of peace, the Soviet Union relied on the conclusion of mutual assistance pacts with its Baltic neighbours. These were intended to prevent the Soviet nightmare—a third power taking advantage of Baltic weakness to impair or infringe Baltic independence.[81]

WAR AND ANNEXATIONS

In the months before the Nazi attack on Poland on 1 September, 1939, the Great Powers and the Baltic states engaged in complex diplomatic manoeuvres to defend their individual interests. Lithuania strenuously pursued the policy of non-alignment, formulated in her neutrality declaration of January 1939 and in her non-aggression treaties with the Soviet Union and Germany. Neutrality opened Lithuania to the accusation of being either pro-German or pro-Soviet. Her resolve was pressed to the limit by Soviet and German hints that if she came off the fence she might be rewarded with Vilnius. The Soviet Union was, not unreasonably, nervous about the security of its north-west borders. Moscow believed that the policy of neutrality was inadequate defence against possible aggression and that the independence of the small Baltic states could easily be compromised. Were they not too weak to prevent other great powers using them for their own advantage? Soviet nervousness was increased by the Nazi occupation of the Klaipėda region. Shortly afterwards Litvinov warned that any agreement between the Baltic states and third parties which infringed Baltic independence would be intolerable to the Soviet Union, not least because it would violate the terms of the Baltic-Soviet non-aggression pacts. It was becoming clear that the Soviet Union would be satisfied with nothing less than control of Baltic foreign policy, thus rendering the Baltic states' neutrality null and void.[82]

Soviet determination to assert control of the eastern Baltic was at the centre of its negotiations with France, Great Britain and Germany in the summer of 1939. London and Paris were keen to involve the Soviet Union in extending joint guarantees to several Central and East European states including, on Soviet insistence, the Baltic states. But the British were resistant to pressing Soviet guarantees, and Soviet military bases, on the unwilling Baltic states, and were disinclined to accept Soviet definitions of direct and indirect aggression. As R.A. Butler reported to the House of Commons, the British were being asked to agree to limit the independence of the Baltic states, which they were not prepared to do. Despite this assurance an Anglo-French military delegation was still engaged in negotiations in Moscow when the Molotov-Ribbentrop Pact was signed on 23 August, 1939.[83]

Moscow's conclusion of a pact with Germany seemed like a *volte-face* but in fact it represented a tactical rather than a strategic change. The Soviets saw an opportunity to re-establish control over the Baltic states through agreement with Germany once it became clear that their objectives might not be achieved in negotiations with the western powers. Nor was this tactical possibility adopted at the eleventh hour; rather it had been adumbated as early as March 1939 when Stalin said there were no reasons for conflict between Germany and the USSR and, moreover, that he was prepared to maintain peaceful relations with 'aggressive' states so long as they did not interfere with the Soviet Union. Litvinov was then replaced by the less anti-German Molotov. This signal was observed by Berlin and encouraged Hitler to resume the interrupted negotiations later in the summer. Moscow, in short, wanted a deal which would guarantee her control over the Baltic states, and it did not matter very much whether the deal was with Germany or the West.[84]

The first secret protocol of the Molotov-Ribbentrop pact of 28 August, 1939 excluded Lithuania from the Soviet sphere of influence, but this omission was rectified in the second secret protocol of 28 September, which transferred Lithuania from the German sphere in return for some Soviet-controlled Polish territory. This agreement gave Moscow the chance to establish military defences on the territory of the Baltic states and to place Soviet military, air force, and naval units in forward positions.[85]

By a combination of threats and blandishments the Soviet Union was able to conclude Pacts of Defence and Mutual Assistance with each of the Baltic states in turn, concluding with Lithuania on 10 October, 1939. This agreement provided for the establishment of Soviet military and naval bases and the stationing of some 20,000 troops on Lithuanian territory. Stalin and Molotov stressed the danger posed by Lithuania's long land frontier with Germany and the increased security for Lithuania arising from the deployment of Soviet troops there. To sweeten the pill Stalin guaranteed Lithuanian independence, promised not to interfere in the internal affairs of the country, and finally offered Lithuania her heart's desire of Vilnius, which the Soviets had recently occupied during the defeat of Poland.[86]

David Kirby has suggested that the Lithuanian government need not have collaborated with the Soviets to such an extent, and that by agreeing to the Mutual Assistance Pact it compromised the country's future existence. The latter part of the statement is undoubtedly true; any subsequent military threat from Moscow would have been impossible to resist given the large number of Soviet troops already in the country. Moreover, despite Soviet assurances that their bases would not affect the sovereignty of Lithuania, Urbšys and other ministers recognised the falsity of this claim. The Pact quite clearly precluded any Lithuanian attempt to effect an alliance with another state. It also destroyed Lithuania's policy of neutrality.[87]

Kirby's criticism of Lithuanian compliance with Moscow's demands is more questionable. Urbšys offered a number of objections in his discussions in Moscow, not least that the proposed pact would impair the friendly relations between Moscow and Kaunas which had existed since the Moscow treaty of 1920, and would create mistrust between the two states. But he and the entire government were in an impossible position. Moscow had taken care to obtain agreements with Estonia and Latvia first, and were able to use these in negotiations with Lithuania. There was little doubt that Stalin would use force if necessary; this had already been threatened by Voroshilov, the Soviet Defence Commissar, during the negotiations with Estonia. Moreover, before the pact was agreed, Urbšys approached Berlin to verify if German support would be forthcoming in the event of Lithuanian resistance. The Molotov-Ribbentrop Pact ensured that it would not.[88] Smetona was also aware that further resistance to Soviet demands would create the unwelcome impression that Lithuania was acting as Germany's ally. It is therefore difficult to see how any action on the part of Lithuania after 23 August, 1939 would have affected the outcome. Lithuania was in an impossible position and emerged from it with a modicum of dignity.

In retrospect it is clear that Moscow was looking for the opportunity to establish complete control of the Baltic states. It chose the occasion of Hitler's military successes on the western front, when the world's attention was focused on the breaching of the Maginot line and the entry of German forces into Paris.[89] With an eye to world opinion, however, Stalin played out an elaborate and chilling charade; the bear could not be seen simply to swallow the mouse, the mouse had to be shown to be threatening the bear, and then persuaded to ask to be swallowed for its own good. Russia still maintains the fiction that Lithuania and the other Baltic states voluntarily voted for entry into the Soviet Union. Even Stalin could see the glaring contradiction between blatant use of force to absorb the independent states and repeated Soviet promises to respect the independence and sovereignty of these states, going back in Lithuania's case to the 1920 Treaty of Moscow, the non-aggression pact of 1926, and the mutual assistance pact which promised 'never to infringe upon the sovereign rights [of Lithuania], with particular pertinence to their political structure and social and economic organization'.[90]

The charade began in May 1940 when the Soviets charged the Lithuanian authorities with abducting two missing Soviet soldiers stationed in Lithuania and drawing them into espionage activities.[91] A judicial investigation led by a commission of senior jurists under the minister of Justice, Antanas Tamošaitis, concluded that there was no evidence to justify Soviet charges and that the Soviet soldiers were simply deserters.[92] The second stage of the charade began when Lithuanian Prime Minister Merkys was summoned to Moscow for an interview with Molotov. There he faced the grotesque accusation of creating an anti-Soviet military alliance out of the Baltic Entente, and therefore of violating the Mutual Assistance Pact. His Minister of the Interior and Director of State Security were also accused of provocative anti-Soviet actions. Despite a placatory assurance from Smetona that Lithuania had been entirely honorable in its adherence to the Pact, the Lithuanian government received an ultimatum on 14 June, 1940 that it must agree by 10 a.m. on the following day to form a government capable of ensuring the proper fulfilment of the Pact. It must also arrest and try the two named officials, and accept additional units of the Red Army on Lithuanian soil to enforce the Pact. Similar demands were made on Latvia and Estonia shortly afterwards. 'Whatever your answer' said Molotov 'our army will march into Lithuania tomorrow'.[93]

The Lithuanian government accepted the Soviet ultimatum and Soviet troops rolled in at 10 a.m. on 15 June, 1940. There was no resistance. The intense cabinet debate concluded that the shedding of blood would have no useful purpose, given overwhelming Soviet force both on the borders and inside the country. Smetona argued for a short, symbolic resistance after which the whole government should go into exile. He was overruled, and left the country. Some believed that Smetona's record of dictatorship made it impossible for his government to mobilize a vigorous opposition to Soviet aggression. Others asked about the purpose of so much military expenditure in the period of independence if an armed defence could not be mounted. The absence of overt resistance gave the Soviets the opportunity to claim that Soviet rule was acceptable to the population. Although many Lithuanians were no doubt glad to see the end of Smetona, they were quite unprepared for the severe brutality of Soviet rule; if they had been, there might have been more resistance to the occupation.[94]

Indeed, the population and the leadership were probably misled by the soothing noises emanating from Moscow's emissaries in Lithuania, hinting at the maintenance of some form of independence or autonomy.

The new Communist Interior Minister, Mečys Gedvilas, gave an assurance that the 'essential fundamentals of our country have not been changed. No one threatens rightful property or wealth'. The Soviets desperately wanted to avoid a bloodbath, and to convince the world that Lithuania and the other two Baltic states were changing their governments constitutionally and voluntarily. Vladimir Dekanozov, Deputy Commissar for Foreign Affairs, and the Kremlin's newly-appointed special envoy to Lithuania, set about re-organizing the government. Asserting that Smetona's flight was in effect a resignation, which it was not, Dekanozov was able to compel Merkys, as Smetona's deputy, to nominate Justas Paleckis, a left wing journalist, as acting president and Prime Minister. His deputy and Foreign Minister was a popular writer, literary scholar and opponent of Smetona, Vincas Krėvė-Mickevičius. The initial composition of the government was a mixture of liberals and left-wingers, including four communists, one of whom, Mečys Gedvilas, was in a crucial position. His presence along with that of Krėvė-Mickevičius reassured the populace that the new government did not intend to introduce a Soviet regime, still less to incorporate Lithuania in the Soviet Union. It was put about that the Red Army's presence was simply to defend Lithuania against external attack.[95]

Krėvė-Mickevičius was soon disillusioned. Dekanozov was the real ruler of Lithuania and government decrees soon bore a characteristic Soviet stamp. The constitutionally-elected *Seimas* was dissolved, political parties suppressed except for the Communist Party, elected local government officials, police commanders, senior army officers and civil servants dismissed, and the police force replaced by a specially recruited militia. Army rank and file were made to participate in 'spontaneous' demonstrations and undergo political education. Believing that there must be some mistake, and that the Kremlin did not know what its subordinates in Lithuania were up to, Krėvė-Mickevičius demanded an interview with Molotov. At the end of it Molotov's brutal frankness had shattered Krėvė's illusions. The occupation of the Baltic region was, he said, an historical necessity for the Russian state. Small nations were destined to disappear, and Lithuania would 'have to join the glorious family of the Soviet Union'. Moscow's next task was to demonstrate to the world Lithuania's voluntary acceptance of this historical necessity.[96]

It chose to do so by means of a characteristically Soviet electoral process and a cowed 'representative' assembly. The existing evidence points to the fraud, intimidation and unconstitutionality at the heart both of the election and the subsequent decision of the Assembly to seek Lithuanian membership in the USSR. Energetic Soviet efforts to convince the world that the process was both democratic and legitimate had little success at the time, still less subsequently. The reasons are obvious. There can be no doubt that changing the electoral laws by decree was unconstitutional; that the rejection of more than one candidate for any parliamentary seat was undemocratic; that the nomination of candidates exclusively from the Communist-dominated Working People's Leagues was authoritarian; that the nomination of some candidates without their knowledge was effrontery. Threats and violence were used against any one who attempted to resist these breaches of democratic election rules, and the ubiquity of the Red Army was a constant warning against resistance. Just before the election some 2,000 prominent opposition figures were arrested. The corruption of the process was further demonstrated by the failure of the government to compile lists of eligible electors; everyone was encouraged to vote including, one report suggested, ineligible teenagers. Anyone not turning out on election day was damned as an enemy of the people. Since the passports of voters were stamped, these enemies could be identified. The authorities combined an evident craving for the appearance of legitimacy with a cynical contempt for western democratic processes. Playing the game in this way suggests that more open steps to annex Lithuania might have met with widespread popular resistance.[97]

Undoubtedly there were some in the population who welcomed the actions of the new government, for example those who were attracted by the proposal in the electoral platforms to confiscate so-called large landholdings and redistribute them among landless peasants. But no one voted for the incorporation of Lithuania in the Soviet Union since this was not part of the official platform. In fact, very few voted at all, casting doubt on the legitimacy claimed by the authorities for the elections. In time-honoured Soviet fashion, the turnout was claimed to be 95.5 per cent with 99.2 per cent voting for the official candidates. So low was the poll on the first day that the polling stations were opened again on 15 July, but to judge from contemporary sources perhaps no more than 30 per cent turned out over the two days. In any case it was alleged that no effort was made to count the votes and the results were announced in the London press 24 hours before the polls closed. The essential point is that the composition of the People's Assembly, as it came to be called, was determined by Dekanozov, not the people. More important still is that even if the electoral process had been beyond reproach, the Assembly would still not have had the constitutional power to change the form of government, nor to vote for Lithuania's incorporation in the Soviet Union.[98]

When this pliable Assembly met it quickly carried out its master's demand to apply for admission to the Soviet Union and to establish a Soviet Socialist form of government. It was made clear that no opposition vote would be tolerated. On 3 August, 1940 Lithuanian independence was formally extinguished by a vote of the USSR Supreme Soviet to grant Lithuania's request for incorporation. After emphasising the United States' long-term interest in the independence of the Baltic states, Sumner Welles, the U.S. Under-Secretary of State, condemned the 'predatory activities' carried out by force or the threat of force, or any form of intervention on the part of one state ... in the domestic concerns of any other sovereign state ...' Two decades later the Council of Europe's report on the Baltic States commented that Moscow was 'trying to persuade the free world that the Lithuanian nation ... joined the group of Soviet-enslaved peoples by its free volition ... In fact, however, Soviet propaganda has failed to produce international legal act or contractual provision in support of its contention.'[99] The words 'Soviet-enslaved' used in an official report by a body devoted to the protection of human rights provide the theme of the next chapter.

1 Simas Suziedelis, 'Lithuania from Medieval to Modern Times: An Historical Outline', in V. Stanley Vardys ed., *Lithuania under the Soviets: Portrait of a Nation, 1940–1965* (New York, Washington, London: Frederick A. Praeger, 1965), p.7; E.W. Polson Newman, *Britain and the Baltic* (London: Methuen, 1929), p.109; Stanley W. Page, *The Formation of the Baltic States: A Study of the Effects of Great Power Politics upon the Emergence of Lithuania, Latvia and Estonia* (Cambridge, Mass.: Harvard University Press, 1959), pp.27, 32; Georg von Rauch, *The Baltic States: the Years of Independence. Estonia, Latvia, Lithuania 1917–1940* (London: C. Hurst & Co., 1974), pp.39–40; Third Interim Report of the Select Committee on Communist Aggression, House of Representatives Eighty-Third Congress, Second Session 1954, *Baltic States: A Study of their Origin and National Development; their Seizure and Incorporation into the USSR*, third reprint edition (Buffalo, NY: William S. Hein and Co. Inc, 1972), p.74.

2 Suziedelis, pp.8–9; Alfonsas Eidintas, Vytautas Žalys, Alfred Erich Senn, ed., Edvardas Tuskenis, *Lithuania in European Politics: The Years of the First Republic 1918–1940* (Basingstoke and London: Macmillan, 1997), pp.34–5; Alfred Erich Senn, *The Emergence of Modern Lithuania* (New York: Columbia University Press, 1959), pp.28, 95; Page, pp.58–60.

3 Both Roman Dmowski and Ignace Paderewski, the official Polish delegates to the Peace Conference, advocated the renewal of the 'Ancient Union existing between the two nations': Suziedelis, p.8; Page, p.170; Senn, pp.42, 94; E.J. Harrison, *Lithuania Past and Present* (London: T. Fisher Unwin, 1922), pp.87–8; F.W. Pick, *The Baltic Nations: Estonia, Latvia and Lithuania* (London: Boreas Publishing Co. Ltd., 1945), pp.75, 77.

4 Suziedelis, p.8; Select Committee on Communist Aggression, p.75; E.J. Harrison ed. and compil., *Lithuania 1928* (London: Hazel, Watson & Viney Ltd., 1928), p.34; Eidintas et al., p.26; Bronis J. Kaslas, *The Baltic Nations—The Quest for Regional Integration and Political Liberty: Estonia, Latvia, Lithuania, Finland, Poland* (Pittston, Pa.: Euramerica Press, 1976), p.66.

5 Eidintas et al., p.26; Page, p.41; von Rauch, p.41; Harrison, *Lithuania Past and Present*, pp.85–6; Pick, p.59; Select Committee on Communist Aggression, pp.72–3.

6 Page, p.30 Harrison, *Lithuania Past and Present*, p.89.

7 Suziedelis, pp.10–11.

8 op cit., pp.11–12; Eidintas et al., pp.28–9; Select Committee on Communist Aggression, p.79.

9 Select Committee on Communist Aggression, p.80; Suziedelis, p.12.

10 Suziedelis, p.13; Page, pp.52–4 von Rauch, p.42; Eidintas, pp.29–30.

11 Select Committee on Communist Aggression, p.80; Harrison, *Lithuania Past and Present*, p.92; Owen J.C. Norem, *Timeless Lithuania* (Chicago, Ill: Amerlith Press, 1943) pp.93, 95; Eidintas et al., p.33; Senn, pp.35 ff.; Page, p.90.

12 Voldemaras was educated in Russia and became a university teacher there. He was a brilliant linguist and a very able man, but he proved to be a volatile and unpredictable colleague with strong authoritarian tendencies: Suziedelis, p.14; Senn, *The Emergence...*, pp.47–49, 100; Kaslas, p.68; Newman, p.110.

13 Suziedelis, p.15; Newman, pp.110–11; Kaslas, p.65; Senn, *The Emergence...*, pp.128–30; Page, pp.145, 165.

14 Senn, *The Emergence...*, pp.54, 59, 62–70; Alfred Erich Senn, *The Great Powers, Lithuania and the Vilna Question 1920–28* (Leiden: E.J. Brill, 1966), pp.15–16.

15 Senn, *The Great Powers*, pp.18–20; Senn, *The Emergence...*, pp.74, 77; von Rauch, pp.51–4, 100–01; Newman, pp.110–11; Kaslas, p.68; Page, p.96; Suziedelis, p.17.

16 Eidintas et al., p.36; Senn, *The Emergence...*, pp.79–80; Page, pp.130–32.

17 Newman, p.111; Kenneth Bourne, D. Cameron Watt and Michael Partridge, General Editors, *British Documents on Foreign Affairs: Reports and Papers from the Foreign Office Confidential Print*, Part II, From the First to the Second World War. Series F. *Europe, 1919–1939*, vol.59, John Hiden and Patrick Salmon eds., *Scandinavia and Baltic States, January 1919–December 1922*, (University Publications of America, 1996), Memorandum, Political Intelligence Department, Foreign Office, 1 October, 1919, pp.8–9; Senn, *The Emergence...*, p.144–45; Robert Ziugzda, 'Lithuanian in International Relations in the 1920s', in John Hiden and Aleksander Loit eds., *The Baltic in International Relations between the Two World Wars*, (Stockholm: Centre for Baltic Studies, University of Stockholm, 1988), pp.59–60; Page, pp.145 & 165; von Rauch, p.69.

18 The loss of Klaipėda to Germany in March 1939 meant a population reduction of 153,793, but this was more than compensated by the increase of 457,500 following the incorporation of part of the Vilnius region in October 1939: Leonas Sabaliunas, *Lithuania in Crisis: Nationalism to Communism, 1939–1940*, (Bloomington and London: Indiana University Press, 1972), p.3; V. Stanley Vardys and Judith B. Sedaitis, *Lithuania, the Rebel Nation* (Boulder, Col.: Westview Press, 1997), p.41; Select Committee on Communist Aggression, p.160; Anicetas Simutis, *The Economic Reconstruction of Lithuania after 1918*, (New York: Columbia University Press, 1942), p.12.

19 Simutis, pp.12–14; Eidintas et al., p.16; Vardys and Sedaitis, p.41; von Rauch, pp.84–5.

20 Newman, p.114; Consul Fry to Lord Curzon, 3 August, 1922, Report on the Situation in Memel, Doc. 161, in Hiden and Salmon eds., p.316.

21 Many of the large landowners were absentees, with no interest in improvements. Agricultural methods were generally antiquated, land was inadequately drained and there was little use of organic or chemical fertilisers. The principal crop in 1912, taking up more than 70 per cent of the cultivated land, was cereals; other important crops were potatoes, pulses, flax and hemp, and sugar beet. Communications were primitive; the best roads, and the railways, were constructed mainly to serve military and strategic purposes, and were inadequate for the economic needs of the country: Peace Handbooks, issued by the Historical Section of the Foreign Office, vol.VIII, *Poland and Finland* (London: HMSO, 1918–19), pp.108–12, 124, 126–28; Kaslas, p.117; V. Stanley Vardys, 'Independent Lithuania: A Profile', in V. Stanley Vardys ed., *Lithuania under the Soviets: Portrait of a Nation, 1940–65* (New York, Washington, London: Frederick A. Praeger, 1965), pp.22–23.

22 Peace Handbooks, p.124.

23 Sir Tudor Vaughan to Lord Curzon, 30 January, 1923, in Hiden and Salmon eds., vol.60, p.8.

24 48 per cent of the cattle, 44 per cent of the pigs, 30 per cent of the sheep, and 38 per cent of the horses had been commandeered for the armies or for the German civilian population: von Rauch, pp.25 and 39–40; Norem, pp.92–3; August Rei, *The Drama of the Baltic Peoples*, (Stockholm: Kirjastus Vaba Eesti, 1970), pp.125–6; Harrison, *Lithuania Past and Present*, p.132.

25 Harrison, *Lithuania 1928*, pp.33–4; Eidintas et al., p.22; Max Muller to Lord Curzon, 23 March, 1921, in Hiden and Salmon eds., vol.59.

26 Vardys and Sedaitis, pp.40–41.

27 op. cit, pp.41–2; Sabaliunas, pp.99–101; Sir Tudor Vaughan to Lord Curzon, 30 January, 1923 in Hiden and Salmon eds., vol 60, p.8; E.J. Harrison, *Lithuania's Fight for Freedom* (New York: Lithuanian American Information Center, 1952), pp.12–13; Vardys, 'Independent Lithuania', pp.26–7; Norem, pp.130, 240; Harrison, *Lithuania 1928*, pp.7–8.

28 Norem, pp.241–3; Simutis, pp.70–2; Vardys, 'Independent Lithuania', pp.25–6.

29 Simutis, pp.72–80; Eidintas et al., pp.117–20.

30 Simutis, pp.83–91; Select Committee on Communist Aggression, pp.158–9.

31 However, the rate of growth of the Lithuanian dairy industry was much greater than that of Estonia and Latvia since Lithuania had started from a much lower base of production in 1918.

32 Vardys and Sedaitis, pp.40–2; von Rauch, p.127; Pick, p.96.

33 Sabaliunas, pp.92–7; Simutis, pp.32–6.

34 Simutis, pp.36–42; Norem, p.140; Vardys, 'Independent Lithuania', p.26.

35 Both the Social Democratic and Christian Democratic Parties disapproved of compensation for landlords, Vardys, 'Independent Lithuania', pp.23–4; Newman, pp.115–16; Sabaliunas, pp.4, 66-74; Norem, 110–12; Rei, pp.137–8; Eidintas et al., pp.45–7; Simutis, pp.25–8; V. Stanley Vardys and Romuald Misiunas, *The Baltic States in Peace and War 1917–1945* (University Park and London: Pennsylvania State University Press, 1978), p.70.

36 Eidintas et al., pp.47–8; Simutis, p.29; Nicholas Hope, 'Interwar Statehood: Symbol and Reality' in G. Smith ed., *The Baltic States: The National Self-Determination of Estonia, Latvia and Lithuania* (London and Basingstoke: Macmillan, 1994), pp.44, 47–8.

37 In the 1930s this rural subsistence population increased, property sales rose, and around 20 per cent of those who had received land as a result of the reforms had to give it up.

38 Select Committee on Communist Aggression, p.157; Eidintas et al., pp.47–8; Sabaliunas, pp.72–4; von Rauch, pp.90–1.

39 Harrison, *Lithuania's Fight for Freedom*, pp.13–14; Vardys and Sedaitis, pp.42–3; Harrison, ed., *Lithuania 1928*, pp.95–100; Eidintas et al., pp.129–31; Hope, pp.53–5.

40 An authoritative biography of Ciurlionis was written by Vytautas Landsbergis, the leader of the independence movement in Lithuania in 1990–91 and a distinguished musicologist.

41 Harrison ed., *Lithuania 1928*, pp.113, 131–33; Pick, pp.99–100; Harrison, *Lithuania's Fight for Freedom*, p.13; Vardys, 'Independent Lithuania', p.24.

42 It is worth noting that the Social Democrats and the Populists abstained on the vote on the Constitution and it was approved only by the vote of the Christian Democrat and Jewish deputies: Vardys, 'Independent Lithuania', p.31; Sabaliunas, pp.3–4; Vardys and Sedaitis, p.31; von Rauch, pp.76–80; Select Committee on Communist Aggression, pp.120–4; Eidintas et al., pp.43–45.

43 M. Purickis, Foreign Minister of Lithuania to Prince E. Sapieha, Foreign Minister of Poland, 24 July, 1920, in République Polonaise, Ministère des Affaires Etrangères, *Documents Diplomatiques concernant les Relations Polono-Lithuaniennes* (December 1918–September 1920), Warsaw, 1920, p.41; Eidintas, pp.133–7; Vardys and Sedaitis, pp.38–9; von Rauch, pp.136–8.

44 Vardys and Sedaitis, p.31; Kaslas, pp.101–2; Hope, pp.48–9; John Hiden and Patrick Salmon, *The Baltic Nations and Europe: Estonia, Latvia and Lithuania in the Twentieth Century* (London and New York: Longman, 1991), p.5.

45 Vardys and Misiunas, pp.66–7; Rein Taagepera, 'Civic Culture and Authoritarianism in the Baltic States 1930–1940', *East European Quarterly*, vol.VII, No. 4, 1973, pp.407–8.

46 Eidintas et al., pp.51-2; Sabaliunas, pp.5–8; Vardys and Sedaitis, pp.31–5; von Rauch, pp.97–9; Vardys, 'Independent Lithuania', pp.31–4.

47 Vardys, 'Independent Lithuania', p.34; Hope, p.62; Sabaliunas, pp.6–7; Eidintas et al., pp.53–5; Sir Tudor Vaughan to Austen Chamberlain, in Hiden and Salmon eds., vol.60, pp.365 & 386; Vardys and Misiunas, pp.70–71.

48 No attempt was made to ratify the revised constitution; Smetona was re-elected President under a special law in 1931 but a new parliament was not elected until 1936. This drew up a new constitution which was adopted in 1938: Norem, p.133; Eidintas et al., pp.57–8; Select Committee on Communist Aggression, pp.128–30; Hiden and Salmon eds., vol.60, p.386; von Rauch, pp.120–1; V. Stanley Vardys, 'The Rise of Authoritarian Rule', p.68.

49 In 1926 he had founded and developed a paramilitary organization called Iron Wolf which Smetona abolished after Voldemaras's dismissal. More popular among younger army officers than Smetona, Voldemaras offered a real challenge to Smetona's government: Vardys and Misiunas, pp.76–77; Pick, p.91.

50 Taagepera, p.408; Eidintas et al., p.58, note 36; Sabaliunas, xix; Saulius Suziedelis, 'Alfonsas Eidintas, Vytautas Žalys, *Lithuania in European Politics: The Years of the First Republic 1918–1940*, *Lithuanian Foreign Policy Review* 98/1, p.115; Vardys and Misiunas, p.77; Misiunas and Taagepera, pp.12–13.

51 Sabaliunas, p.140; Vardys, 'Independent Lithuania', pp.36–7.

52 Norman Davies, *Europe, a History* (Oxford, New York: Oxford University Press, 1996), p.947.

53 Eridintas et al., pp.133–6.

54 op.cit., pp.136–7.

55 Sabaliunas, pp.32–5, 140–2.

56 Davies, pp.945–48.

57 Vardys, 'Independent Lithuania', pp.37–8; Sabaliunas, pp.41–60.

58 Vardys, 'Independent Lithuania', p.38; Sabaliunas, pp.114–28; Vardys and Sedaitis, p.37; Suziedelis, p.115; von Rauch, p.199.

59 This begs the question of the status of the national minorities who Smetona himself declared were 'not foreigners', but fellow Lithuanian citizens.

60 V. Stanley Vardys 'Lithuanians' in Graham Smith ed., *The Nationalities Question in the Soviet Union* (London: Longman, 1990, rpt.1992), p.74; Norem, p.153.

61 The border with Latvia should be exempted from this comment.

62 Ministry of Foreign Affairs of Lithuania to Wasilewski, Polish Minister Plenopotentiary to Kovno, 6 August, 1919, République Polonaise, Ministère des Affaires Etrangères, p.31.

63 Rei, p.66; Senn, *The Great Powers*, p.12; Senn, *The Emergence...*, pp.91–4.

64 Ziugzda, p.20; Senn, *The Great Powers*, pp.72–6; Vardys, 'Independent Lithuania', p.40.

65 It was agreed, however, that the delimitation of this territory between Lithuania and Poland would be decided by the two states themselves, an arrangement confirmed in the 1921 Treaty of Riga between Poland and the Soviet Union.

66 Baltic States Annual Report, 1923, in Hiden and Salmon eds, vol.60, p.171; E.H. Carr to Austen Chamberlain, 12 October, 1926, Hiden and Salmon, p.349–50.

67 Carr, loc.cit., p.350.

68 Vardys, 'Independent Lithuania', pp.40–1; Senn, *The Great Powers*, pp.76–7.

69 The language of administration and education was German, however, and there was a great incentive to learn German for business reasons. The German census of 1910 suggested that there was a fairly even division between German and Lithuanian speakers and this was more or less confirmed by the Lithuanian census of 1925 which counted 59,337 Germans, 37,625 Lithuanians and 38,404 Memellanders (described as ethnically Lithuanian, but identifying themselves as 'local' and with a loyalty to the German political parties in the district).

70 Kaslas, pp.79–81; Vardys, 'Independent Lithuania', p.42; David M. Crowe, 'Great Britain and the Baltic States, 1938–39' in Vardys and Misiunas eds., pp.111–12.

71 This provided for a degree of autonomy for Klaipėda under Lithuanian sovereignty. It was conflict between the German-dominated legislature and the Lithuanian government that caused so much strain in relations between Germany and Lithuania subsequently. The Klaipėda Germans never accepted becoming Lithuanian citizens and took every opportunity to detach the district from Lithuania. The German government, through its consulate, frequently appealed to the League

about the violation of the Memel Convention and the 'oppression' of the Klaipėda Germans by the Lithuanian government.

72 Kaslas, pp.82–4; Consul Fry to Curzon, 3 August, 1922, in Hiden and Salmon eds., vol.59, p.309; J.W. Headlam-Morley, in Hiden and Salmon eds., vol.60, pp.86–110; von Rauch, pp.197–9; Norem, pp.206–10; Eidintas et al., pp.160–65.

73 Eidintas et al., pp.62–3.

74 Sabaliunas, pp.13–15; Hugh I. Rodgers, *Search for Security: A Study in Baltic Diplomacy, 1920–1934* (Hamden, CT.: Archon Books, 1975), pp.75–6; Vardys, 'Independent Lithuania', p.39; Eidintas et al., pp.101–2; Česlovas Laurinavičius, 'The Baltic States between East and West: 1918–1940', *Lithuanian Foreign Policy Review*, 98/1, pp.88–90.

75 Rei, pp.170–1; Acton to Curzon, 9 December, 1919, in Hiden and Salmon eds., vol.59, pp.17–18; von Rauch, pp.107–9; Sabaliunas, p.18; Walter C. Clemens, Jr., *Baltic Independence and Russian Empire*, (Basingstoke: Macmillan, 1991), pp.46–7.

76 von Rauch, pp.108–10; Rodgers, pp.20–8, 32–8.

77 The key was there but the cost was prohibitive, as was shown in October 1939 when the Baltic states were compelled to sign mutual security pacts with the Soviet Union: von Rauch, p.176; Clemens, pp.47–50; Vardys, 'Independent Lithuania', pp.44–5; Eidintas et al., pp.139–41.

78 Eidintas et al., pp.61, 150–54; Pick, pp.116–18; Sabaliunas, pp.19-20; Kaslas, pp.172–8, 202–4, 219–20.

79 Edgar Anderson, 'The Baltic Entente: Phantom or Reality?', in Vardys and Misiunas eds., pp.129–31; Rodgers, pp.96–101.

80 von Rauch, pp.189–92; Kaslas, p.220; Rei, p.234.

81 Vardys and Misiunas eds., pp.99–101; von Rauch, pp.174–78; Rei, pp.250–1.

82 Eidintas et al., pp.167–8, 171; David Kirby, 'Incorporation: The Molotov-Ribbentrop Pact', in G. Smith ed., *The Baltic States: the National Self-Determination of Estonia, Latvia and Lithuania* (Basingstoke: Macmillan, 1994), p.70; Rei, pp.250–1; Select Committee on Communist Aggression p.213; the problem is clearly stated in David Vital, *The Survival of Small States: Studies in Small Power/ Great Power Conflict* (London: Oxford University Press, 1971), p.112.

83 Bronis J. Kaslas ed., *The USSR-German Aggression Against Lithuania* (New York: Robert J. Speller & Sons, Publishers, Inc., 1973), p.110; Bohdan Nahaylo and Victor Swoboda, *Soviet Disunion: A History of the Nationalities Problem in the USSR* (London: Hamish Hamilton, 1990), p.81.

84 Nahaylo and Swoboda, p.82; Rei, p.249, note, 251.

85 The establishment of Soviet control might have been avoided if Kaunas had accepted the German invitation to attack Poland at the onset of the war. Although Vilnius would have been gained, the price would have been a German protectorate and the loss of any western sympathy.

86 Misiunas and Taagepera, pp.15–16; Heino Arumae, 'Moscow's Point of View of the 1939–40 Events in the Baltic States', *Rahva Haal* (The People's Voice), 14 February, 1991; Nahaylo and Swoboda, p.84; Kirby, pp.73–4; von Rauch, p.212. In view of the long-standing foreign policy priority of restoring Vilnius to Lithuania, it might be supposed that the Soviet offer of Vilna Vilnius (albeit a relatively small part of the total Vilna territory) would have been readily accepted at whatever cost to Lithuanian sovereignty. This was not the case. Smetona himself wondered whether Vilnius was worth having if the price was so high, but recognized that refusing would not prevent Soviet troops entering Lithuania: Eidintas et al., p.171.

87 Kaslas ed., *The USSR-German Aggression against Lithuania*, p.147; Robert von Dassonowsky-Harris, 'The Philosophy and Fate of Baltic Self-Determination', *East European Quarterly*, xx, no.4, January 1987, p.498–9; Eidintas et al., pp.171–3.

88 The Lithuanian government's unavoidable submission to Soviet demands was paralled by a recognition that its freedom of action in foreign policy had been severely constrained by the Mutual Assistance Pact. Nonetheless, negotiations with Germany eventually resulted in a trade treaty in April 1940 providing for German purchase of between 60 per cent and 70 per cent of Lithuanian exports. Discussions were held with the Estonian and Latvian governments on closer co-operation in a number of fields. These were later used by Moscow as the pretext for one of its more absurd allegations about sinister Baltic collaboration against Soviet interests: Misiunas and Taagepera, p.17; von Rauch, pp.218–22; Kaslas, *The USSR-German Aggression...*, pp.186–8.

89 The collapse in the West also brought home to Stalin that Hitler would be ready for eastern aggression much earlier than he had anticipated. Hitler's invasion of Belgium, the Netherlands and Luxemburg also provided Moscow with the opportunity to reiterate its theme that 'the neutrality of small states which do not have the power to support it is a mere fantasy. Therefore, there are very few chances for small countries to survive and to maintain their independence': Misiunas and Taagepera, p.18; Nahaylo and Swoboda, p.84; Kirby, p.75.

90 But Stalin's imperative was to end the status quo: 'What was decided in 1920 cannot last for ever' were his words to the Latvian Foreign Minister Munters: Arumae, in *Rahva Haal*, 13 February 1991; Clemens, p.54.

91 The Lithuanian government went to extreme lengths to satisfy the Soviets that no abduction had taken place. It set up a high-powered investigation, sought full co-operation with the Soviet military, and initiated a vigorous search for the missing men. It received no collaboration from the Soviet authorities.

92 Tamosaitis was later tortured to death in Kaunas prison after the Soviet annexation, a grim reminder of the price to be paid for judicial integrity in a soviet state: Misiunas and Taagepera, p.18; Arumae in *Rahva Haal*, 14 February, 1991; Kaslas, *The USSR-German Aggression*, pp.180–5.

93 The charge of creating an anti-Soviet military alliance rested on a bizarre interpretation of articles in a new journal, the *Revue Baltique*, published in Tallinn in 1940, designed to strengthen economic, social and cultural co-operation between the Baltic states. An article by Merkys on the Baltic entente was seized on by Molotov to accuse him of forming a secret military alliance between the three states, cemented by frequent contacts between military chiefs. This Soviet interpretation was evidence of either acute paranoia or complete cynicism in Moscow. The German Foreign Office believed there were no grounds for Soviet suspicions: Kaslas, *The USSR-German Aggression*, pp.21–2, 186–8, 209, 220–1; Kirby, p.75.

94 Certainly government and people in January 1991 were conscious of the submission of 1940 and were determined not to repeat it. As a leading *Sajudis* figure, Romualdas Ozolas, stated at the time, 'I will be a person whose children will not be able to accuse me that again, as in the 1940s, not a single shot was fired [in the defence of independence]' Vardys and Sedaitis, pp.49–50; Alfonsas Eidintas, 'The Meeting of the Lithuanian Cabinet 15 June 1940' in John Hiden and Thomas Lane eds., *The Baltic and the Outbreak of the Second World War* (Cambridge: Cambridge University Press, 1992), pp.165–73.

95 Kirby, p.78; Misiunas and Taagepera, p.21.

96 Gorbachev was still claiming, some time after conceding the illegality of the Molotov-Ribbentrop Pact in 1989, that Lithuania and the other Baltic states had supported their incorporation in the Soviet Union in 1940. Considering the amount of evidence to the contrary this was extraordinary: Misiunas and Taagepera, pp.25–6.

97 op.cit, pp.26–8; Nahaylo and Swoboda, p.85; British Foreign and Commonwealth Office, Background Brief, 'The Baltic States: 40 Years under Soviet Rule', FO 973/111, September 1980; Harrison, *Lithuania's Fight for Freedom*, pp.26–8; Kaslas, *The USSR-German Aggression*, p.229; von Rauch, p.225; V. Stanley Vardys, 'Aggression, Soviet Style, 1939–40' in Vardys ed., *Lithuania under the Soviets: Portrait of a Nation 1940–1965* (New York and London: Praeger, 1965), pp.54–7.

98 Jurgis Glusauskas, the police chief of Kaunas, who observed the poll in Mariampole, reported from exile in 1953 that several hours before the polls closed the turn out was only 5–10 per cent. The NKVD official in charge then ordered his subordinates to record higher percentages than were actually cast. Within an hour the proportion rose to between 40 per cent and 60 per cent and just before the close of voting the figures reached 99.2 per cent. He believed that in reality no more than 15 per cent voted, which is in line with other estimates: Kaslas, ed., *The USSR-German Aggression*, pp.22, 230–1; Harrison, *Lithuania's Fight for Freedom*, pp.27–9; Kirby, pp.78–9.

99 Misiunas and Taagepera, p.30; Harrison, *Lithuania's Fight for Freedom*, p.28; Kaslas, *The USSR-German Aggression*, pp.23, 414–5; Vardys, 'Aggression, Soviet Style', pp.57–8; John Alexander Swettenham, *The Tragedy of the Baltic States: A Report Compiled from Official Documents and Eyewitnesses' Stories* (London: Hollis and Carter, 1952), pp.50–1.

Chapter 2

SOVIETIZATION 1940–85

The annexation of Lithuania by the Soviet Union was followed by the implacable and remorseless demolition of the political, economic, social and cultural institutions established by the independent Lithuanian state in the inter-war decades. In their place was introduced all the apparatus of the Soviet system. Simultaneously, the élite of the Lithuanian government, political parties and civil service, the senior military and police officers, and the leaders of business and cultural life were arrested and sentenced to deportation or execution. Those who could fled the country. The loss of so much experience and talent was incalculable. The accompanying political repression and economic devastation, coupled with the onslaught on Lithuanian education and culture, was traumatic for the Lithuanian population. The rather passive acceptance of the annexation which had prevailed among the Lithuanian population in the summer of 1940 was now regretted. Perhaps part of this acquiescence arose from the goodwill which Moscow had earned in Lithuania as a result of its support over Vilnius in the inter-war decades. Goodwill was eroded by the Mutual Assistance Pact and fatally undermined for most people by the experience of occupation, which shattered the illusion that Sovietization might be tolerable or short-lived.

Although the communist government was cautious in implementing its policy of collectivization in agriculture, in almost all other respects it pressed ahead unhesitatingly with its programme. As a result the Lithuanian population was left in absolutely no doubt about the meaning of Sovietization. The result was dangerously counter-productive. For the next half century Lithuania resisted Moscow's rule. The resistance took many different forms, from sustained guerrilla warfare to the waving in public of Lithuanian flags, from samizdat publications to demonstrations at sports arenas. But at no time could the Soviets feel that Lithuanians had been won over either to Marxism-Leninism or to Soviet institutions and values.

Historians have tended to apply the term resistance mainly to the Resistance, the great guerrilla war which was at its height in Lithuania from 1944 to around 1950. But resistance both preceded and followed this dramatic episode. It existed in the first year of Soviet occupation,

1940–41, during the subsequent Nazi rule, and again after the end of guerrilla warfare. This persistent rejection of the Soviet system was attributable to the strong sense of identity of the Lithuanian people which was created out of many different mutually reinforcing elements. Among these was the pride in medieval statehood, the memory of resistance to an earlier Russianization, and the cultural renaissance of the nineteenth century. This identity drew strength from the power and tenacity of religious beliefs and the sense of achievement arising from the creation of an independent state. It was further nourished by the struggle of the infant Lithuanian state to survive in a hostile environment, in particular by the constant challenge of Poland's physical and cultural presence. The long historical perspectives of Lithuanians equipped them with the patience to endure and the hope that they would ultimately achieve their objectives.

This strong commitment to the national identity was strengthened, not weakened, by Soviet repression. Ivinskis has noted the evidence of popular resistance even in the first year of annexation. Anti-Soviet leaflets were circulated and underground publications continued to appear. Secret opposition organizations were formed, notably among students and junior academic staff. There was a good deal of passive resistance such as the boycotting of many political activities sponsored by the regime. 'Primitive' Soviet customs and behaviour were ridiculed. In October the Lithuanian Activist Front (LAF) was formed to organize resistance against the Soviet occupation and to campaign for the restoration of Lithuanian independence. But time and opportunity were lacking for developing full-scale opposition; it was the Nazi occupation which provided the breathing space and the resources for this. On the re-entry of the Red Army into Lithuania in 1944 the Lithuanian population was steeled to resist. Their earlier experience of Sovietization left them in no doubt as to what was in store.

The Sovietization process has been described and analysed many times. It is not necessary here to enter into great detail about the Lithuanian version. But to pass over it too abruptly would be to lessen its significance in Lithuanian history and obscure its shaping power over the Lithuanian mentality. Although space precludes an exhaustive discussion, the more significant elements in the process should be identified.

The essence of the Sovietization process was to destroy and to construct. Institutions and people were destroyed in order to break the power of the preceding 'bourgeois' society. In their place came Soviet

Russian systems of government, economy and society, and loyal Soviet citizens to run them. This process is exemplified by the new institutions of government approved by the People's Assembly on 21 July, 1940. Formally the Council of Ministers and the Supreme Soviet ruled the country. The Chair of the Presidium of the Supreme Soviet, Justas Paleckis, was in effect head of state. But the essential purpose of these bodies was to ratify and implement decisions taken elsewhere, either in Moscow or by the Lithuanian Communist Party, whose General Secretary was Antanas Sniečkus. These government and party bodies replaced the institutions of government of independent Lithuania. Similarly, Soviet laws superseded Lithuanian, Russian civil and criminal legislation being adopted by December 1940. A large influx of Communist officials from other parts of the Soviet Union replaced the government personnel of the old regime. Other replacements came from members of the native communist party, blue collar workers, both Gentile and Jewish, who believed that job opportunities and access to higher education would increase under communist rule. Obviously Jews saw the Soviet regime as preferable to a German occupation. But the Soviet government was hostile both to Zionists and to Jewish capitalists, and Jews suffered heavily from arrests and deportations. Unfortunately, the association was established between Jews and communism among many ethnic Lithuanians, and this led to Lithuanian participation in pogroms after the Soviets were expelled and to involvement, the full extent remaining unclear, in the destruction of the Jewish community in Lithuania by the Nazis.[1] In stating this there is no intention of shifting responsibility for the Holocaust from where it truly belongs, nor to condemn the whole of Lithuanian society for what was possibly the work of a relatively small number of individuals. This aspect of life under the occupations will be discussed more fully later.

The term 'replacement' in connection with personnel is, of course, a misnomer. Most of these Lithuanian officials and prominent persons were, in Sovietspeak, eliminated. Around 2,000 people in these categories were arrested just before the July elections and given 8-year terms in labour camps in the Soviet Arctic or Siberia. Anyone considered to be an obstacle in the path of Sovietization or who showed signs of independence was dubbed an 'enemy of the people' and arrested. Perhaps several hundred persons a month were seized in this way. Those who had opposed or disapproved of the new regime were executed. Others were given long sentences at hard labour. One calculation puts the number imprisoned in the year after the annexation at 12,000. Some of those in prison at the time of the Nazi attack on the Soviet Union on 22 June, 1941 were executed before the Soviet withdrawal; many of them showed marks of brutal torture. About 1,400 bodies were subsequently recovered.[2]

The second supplementary stage in the destruction of people involved the planned mass deportations. The objective was to remove from Lithuania what were described as counter-revolutionary elements. The planned numbers were so large, however, (reported to be 700,000) that it must be assumed that an additional objective was to generate such fear and insecurity among the population that social control and the weakening of Lithuanian identity would be facilitated. In fact the execution of this project was only just begun before the German invasion in July 1941. Some 34,000 were deported by freight wagons on the night of 13/14 June; husbands and wives were separated, the men going to forced labour camps, the women and children to collective farms and economic enterprises in remote parts of the Soviet Union.[3]

Compared with the pain and suffering inflicted by these brutal measures, other aspects of Sovietization seem to pale into insignificance. The establishment of collective ownership of industrial enterprises, small workshops, trading companies, banks, shops, restaurants, hotels and housing, though predictable, was executed in a crude and damaging manner. Leaving aside the absence of compensation for the former owners, the neglect of existing management and technical expertise combined with the reckless speed of the nationalization process both damaged the economy and lowered the standard of living of employees.[4] Replacement of former owners by inexperienced and inexpert communist trusties led to the introduction of the usual forms of communist personnel management combining terror, high piecework quotas, the use of informers, and political indoctrination. Trade unions were abolished and workers lost many of their former rights and benefits.

The Lithuanian economy took on Soviet characteristics with startling speed. The exchange rate between the *litas* and the Russian ruble led to the depreciation of the *litas* by between 300 per cent and 500 per cent. Armed with appreciated rubles the Russians descended on the shops and stores like clouds of locusts, emptying them within two months. Lithuanians joined in the rush to buy, some to hoard, others to speculate. Confiscations of goods and equipment for other parts of the Soviet Union accentuated shortages, which were further exacerbated by the fall in production and distribution following nationalization. After a panic run on the banks, a moratorium was placed on withdrawals.

Deposits virtually ceased, and in January 1941 all savings in excess of 1,000 rubles were nationalized. Housing above a certain area of floor space was appropriated and owners of better apartments in city centres were forced to evacuate them in favour of Soviet officials. A housing shortage was worsened by the slowdown in construction. All in all this was plunder on the grand scale, leading, as was intended, to the reduction of the Lithuanian standard of living to Soviet levels, and to the destruction of the property-owning classes.[5]

Since Lithuania was predominantly an agricultural economy, the relatively small industrial and commercial sector could not be expected to offer significant resistance to Soviet measures. The mass of small- and medium-sized farmers, however, had to be treated with circumspection. Precipitate collectivization would have triggered a mass revolt at a time when the Soviet Union was attempting to strengthen security on its western frontiers. Consequently agricultural policy, though severe, was tempered by pragmatism.[6] Redistribution and requisitioning were the most salient features of Soviet farm policy in this period. Farms were restricted to 75 acres; land in excess of that was placed in a state land bank for redistribution. Some 28,000 owners were affected, and around 75,000 smallholders or landless agricultural workers received small allocations from the land pool, too small in fact to support a family, as was admitted. Moreover the short-term loans made available to the new owners were too small to provide adequate finance for the new farms. The loss of land by the larger farms reduced the supply of fodder for existing herds with the result that two to three times the usual number of cattle were slaughtered in the winter of 1940–41. Arguably the intention of this measure was to create a large number of impoverished peasantry for whom collectivization might ultimately seem a desirable option. At the same time the medium and larger farms were subjected to a requisitioning regime under which 30–50 per cent of produce was delivered to the state at about one-sixth of the market price. Simultaneously the terms of trade between agriculture and industry worsened dramatically, so that agricultural purchasing power fell by 45 per cent in 1940. The existing private landholding structure was being prepared for its eventual demolition and alignment with Soviet norms.[7]

Sovietization brought fundamental changes to education and culture. The new regime changed the curricula and textbooks in schools and higher education to ensure conformity with Soviet ideology. The aim of educational and cultural institutions was to substitute for the norms of independent intellectual inquiry a total conformity with Soviet values and ideology. The annexation initiated a fierce ideological struggle against individualism, free thought, independent research and the ideas and values of the bourgeoisie. Indoctrination by means of compulsory classes in Marxism-Leninism and dialectical materialism was a priority in all educational institutions and in workplaces. Many members of the intelligentsia lost their posts and, in some cases, their lives and liberties. Press control became severe. Books were proscribed and many writers became silent, writing only 'for the drawer'. Most cultural bodies were disbanded and replaced by new Soviet institutions. The largest and most powerful independent institution in Lithuania, the Roman Catholic Church, was subjected to a series of measures designed to break its power and to remove its influence in society. All government financial support to the church ceased, many seminaries were closed, and the clergy lost their pensions. The clergy and their congregations were taxed more highly and were subjected to higher rents and charges. The responsibility for the registration of births, marriages and deaths was transferred to the state, and church holidays were abolished. Although measures to separate church and state were justifiable, the campaign against the Church was only in its first stage and the indications of a relentless 'war against God' were already apparent, and were to take full effect during the second Soviet occupation after 1944.[8]

GERMAN OCCUPATION 1941–1944

It will be recalled that a resistance organization, the LAF, was founded in the autumn of 1940, with cells established in all parts of Lithuania. By the Spring of 1941 estimates suggest its membership was about 36,000. Its leadership had made contact with an important unit of the Front in Berlin, founded by a group of Lithuanian exiles who had escaped to Germany at the beginning of the Soviet annexation. The group's leader was Colonel Kazys Škirpa, the former Lithuanian Ambassador in Berlin. Elaborate directives were sent to Lithuania on how to prepare a revolt against the occupier. The German government tolerated the Front's activities, including its liaison with cells in Lithuania. It hoped to establish a fifth column there when it launched its invasion of the Soviet Union under the codename Operation Barbarossa.

The LAF and the Germans, it turned out, agreed on only one thing, the speediest possible defeat of the Soviet forces in Lithuania. But whereas the LAF's aim was to establish an independent Lithuania, Hitler intended the country to be a mere colony of the German Reich.

There is no doubt that the Lithuanian uprising, which coincided with the launch of Barbarossa on 22 June, 1941, considerably eased the German military task—the LAF rapidly took over Kaunas, Vilnius and many other towns, seized the radio station, harassed the withdrawing Soviet forces, and mobilized some 100,000 resistance fighters, three times its own membership—and enabled the LAF to establish a Provisional Lithuanian Government before German military units arrived in the capital. The Provisional Government tried all it knew to establish itself but the Germans had other plans and slowly squeezed the life out of it. It ceased to exist in August. From then on Lithuania became part of the German province of Ostland, destined to serve the material and manpower needs of the Reich and to become a location for German agricultural colonists.

Although this was a cruel disappointment to Lithuanians, the uprising had shown the falsity of Soviet claims that Lithuania's becoming part of the Soviet Union was a voluntary act. Moreover, the mass participation in the revolt which cost around 10,000 Lithuanian casualties, with an estimated 4,000 killed, compensated for the failure to offer any resistance to the Soviets in June 1940, and sustained the self-respect of the Lithuanian people. Heroic resistance to the Nazi and the succeeding Soviet occupations has justifiably been a great source of national pride. There is no intention here to demean in any way the bravery and self-sacrifice of the Lithuanian people in the harshest of times, but it would be a serious omission not to refer to the allegations of Lithuanian participation in the destruction of the Lithuanian Jews by the Nazis. Indeed, failure to give proper consideration to the allegations has left a cloud of suspicion hanging over Lithuania. The precise role of Lithuanians in the Holocaust needs to be determined, not only as a matter of historical record, but as a contribution to the contemporary debate. What many Lithuanians regard as wild and exaggerated charges can only be rebutted, if rebutted at all, by a careful examination of the facts.

At the beginning of the Nazi occupation there were around 200,000 Jews in Lithuania. The Jewish community had existed in Lithuania for centuries and relations between ethnic Lithuanians and Jews had been peaceable. Vilnius was one of the great centres of Jewish culture, scholarship and spirituality. Occupationally the Jews were heavily concentrated in commerce and petty manufacturing; they were also heavily urban, and generally more highly educated than the ethnic Lithuanian population. There were, however, many Jews who remained poor, some of whom believed in radical political solutions to economic and social problems. The Soviet occupation offered the opportunity for some Jews with communist sympathies to join the new administrative structures and, with Jewish members of the NKVD, to participate in the Sovietization of Lithuania, including the purges of the Lithuanian élites. The Lithuanian Communist Party was disproportionately composed of Jews. In the eyes of some Lithuanians an equivalence was established between Jews and communist persecutors. At the same time there was probably a disproportionate number of Jews in the Soviet-organized deportation of 1941.[9]

The Nazi occupation offered the opportunity for revenge. In some areas Lithuanian mobs committed atrocities against the Jews. Although the Lithuanian Provisional Government tried to restrain Gestapo activities and protested against mass executions of Jews, it also issued a number of pro-Nazi and anti-Jewish statements. Evidence of 'whole-hearted concurrence' by local authorities serving under the Germans in crimes against the Jews, write two historians, 'is difficult to establish'. They concede, however, that the earlier Soviet executions and deportations, by removing Lithuanian community leaders and many members of the intelligentsia, weakened those elements capable of arguing for restraint in anti-Semitic activity.[10]

Tomas Venclova has challenged contemporary Lithuanians to face up to the fact that the manifestoes of the LAF set out a programme of ethnic and racial purification, and that the programme was published in the press after the June uprising. Objections that purification meant expulsion not extermination are implausible, he believes, given some of the very explicit published statements.[11] Venclova's point is that accusations of Lithuanian collaboration in the Holocaust will not stop, nor will the extent of such collaboration be known, until full investigations of every alleged incident have been carried out. Extenuation or excuses or the rewriting of history will simply fuel suspicion and prolong the controversy.[12] In the interests of fairness it must be remembered that not insignificant numbers of Lithuanians risked their lives to save Jews, and that the Roman Catholic hierarchy and many individual priests consistently opposed German policies and tried to save Jewish lives.[13]

The non-Jewish population greeted the German invasion with relief and were initially optimistic that the clock could be turned back to the economic and political conditions which had obtained in the pre-Soviet period. Such hopes were illusory. It soon became clear that the Nazi authorities were determined to exploit Lithuania's physical and human resources to the limit for the exclusive benefit of the Reich. Long-term plans as expressed in *Generalplanost* in 1942 envisaged the deportation east of up to 85 per cent of Lithuanians deemed incapable of being Germanized. It was intended that those remaining would assimilate with the German, Dutch, and Danish agricultural colonists who would be settled in Lithuania during and after the war.[14] In the short term the population of Lithuania was to provide for the manpower needs of wartime Germany either as industrial workers in the Reich or as support troops in the German military and security services. The German authorities met with very little success in their attempts to implement this mobilization policy. With the help of the clandestine press and radio the Lithuanian underground persuaded the population not to volunteer for service in German Construction Battalions or in German war industries. By means of passive resistance, bribery, or going into hiding, Lithuanians undermined the German recruiting drives. Their objective was to preserve the largest possible population in Lithuania for resistance to a renewed Soviet occupation. There were inevitably some volunteers, notably from former Lithuanian army personnel who joined Nazi Defence Battalions, and quite large numbers were seized in indiscriminate round-ups.[15]

The Nazis were singularly unsuccessful in their attempt to form a Waffen-SS legion, only 286 people volunteering to serve. The Germans blamed the dissuasion carried out by the underground media, and carried out reprisals against the intelligentsia whom they held responsible.[16] The one German success was to organize an independent Territorial Defence Force under General Povilas Plechavičius, a popular Lithuanian commander. Around 30,000 Lithuanians volunteered for this force. When the German authorities tried to incorporate the force in the SS, it disbanded on Plechavicius's instructions and disappeared with its weapons and uniforms into the forests where it formed the nucleus of the anti-Soviet freedom fighters, who waged an eight-year-long guerrilla war against restored Soviet rule.[17]

The policy of mobilization was accompanied by an intense exploitation of physical resources and the suppression of Lithuanian culture. Lithuania was seen as an important provider of agricultural products. Her hopes of a return to private enterprise and a relaxation of draconian Soviet requisitioning were dashed. Special German companies assumed direction of nationalized Soviet firms, land appropriated from larger farmers by the Soviets was not restored, and deliveries of agricultural commodities were stipulated. Worse still, the policy of colonization of Lithuanian farms by Germans was actively encouraged by the Central Colonization Office, and by July 1943 some 4,700 German families, around 20,000 people in all, had been settled on Lithuanian farms, whose owners were expropriated.[18]

German policy, particularly the savage policy of reprisals against non-co-operation or resistance, created a significant underground opposition to Nazi rule. The underground successfully communicated with the Lithuanian population by means of a very active press. The principles of the Atlantic Charter were widely quoted and admired. In 1943 the various strands of opposition to the Nazi regime, with the exception of the Communists, united in a Supreme Committee for the Liberation of Lithuania, which became the *de facto* provisional government of Lithuania. By the time the Red Army reconquered Lithuania in 1944 most of the elements of the resistance movement were in place: armed soldiers from the Territorial Defence Force now acting as guerrilla fighters, a press and underground organization, a clandestine radio, and an expectation that armed resistance would be supported by the western allies against the Soviet occupation forces.[19]

This anti-Soviet resistance was far more active and forceful than the opposition to the Germans. This reflects two considerations. Firstly, most Lithuanians were more hostile to the Soviets than to the Nazis, perhaps because Moscow was more destructive of the Lithuanian identity and culture, and of its intelligentsia. Secondly, the Lithuanians did not wish to lose too many men in active opposition to the Germans lest their capacity to resist the Soviets be undermined. They saw 'no advantage in weakening Hitler against Stalin'.[20] However, many Lithuanians lacked confidence that the Soviet regime could be resisted or overthrown and left for the West in the face of the Soviet advance. Some 70,000 Lithuanians fled, a high proportion of them professional people and skilled workers. In addition to the approximately 175,000 Jews who were killed by the Nazis, calculations of other victims of the German occupation through reprisals, executions, and deaths in auxiliary military services suggest a total of 16,000, along with a further 36,000 deported to slave labour or concentration camps in Germany.

THE RESUMPTION OF SOVIET RULE

A little under a year before the Soviet army returned to Lithuania the political parties issued a joint declaration predicting a still more terrible wave of extermination should the Soviet Union resume its occupation of the country. The first victims would be the most active and vital elements of the population. The Soviet objective would be, as it had been during 1940–41, the destruction of the Lithuanian nation. Accompanying this chillingly accurate prediction was a defiant declaration from the Supreme Committee of Liberation, the essence of which was taken up by *Sajūdis*, the Popular Front movement in the Gorbachev period. This affirmed that the sovereign state of Lithuania would not disappear as a result of occupation; rather, its constitution would remain in force though the functioning of its government institutions would be temporarily suspended. A statement from the Committee in September 1944, when the Soviet forces had taken over most of Lithuania, predicted that the Lithuanian people would 'resist all attempts to re-introduce the undemocratic Soviet regime in Lithuanian and [would] defend themselves against all attempts to deport the Lithuanian masses to the remotest regions of the Soviet Union. This struggle of the Lithuanian people is a fight for its liberty, for its right to an independent life, for its very survival'.[21]

Unlike the states of western Europe which were subject to Nazi occupation during the war, Lithuania could not now expect to return to an independent existence. The peculiarly terrible fate of the Baltic states was to experience not one but three successive occupations, the object of which was, as the Committee of Liberation predicted, the destruction of the Lithuanian identity and the complete integration of the Lithuanian republic into the Soviet Union. For 70 years of the twentieth century Lithuania has been occupied by a succession of repressive regimes. Through its extraordinary powers of resistance it has preserved its identity and recaptured its independence, but only after the most painful experience of terror, persecution, tyranny and personal tragedies.

Space does not permit a detailed analysis of the second period of Soviet rule in Lithuania; in any case, the story has been told many times before. But it is important to discuss the broad contours of the subject so that the movement for independence at the end of the 1980s can be placed in context. The major aim of Moscow's policy was Sovietization, by which is meant the most complete integration into the Soviet Union of the Lithuanian soviet republic, the denationalizing of the Lithuanian identity, the introduction of the command economy directed from Moscow, the total collectivization of the means of production, distribution and exchange, the imposition of a totalitarian form of government, and the elimination of civil society. These objectives were to be achieved by any appropriate methods, including the most stringent and brutal. The methods changed over time, particularly after the death of Stalin in 1953. After that the methodology was composed of variations on the theme first advanced by Stanley Elkins in connection with slavery in the United States, 'accommodation and resistance'. Use of the term accommodation suggests that the methodology provided some elbow room for the preservation of Lithuanian national consciousness. The second major theme of the discussion is how the all-pervading sovietization failed to destroy the Lithuanian identity, and allowed it to flourish as soon as the conformist pressure was reduced under Gorbachev's policy of *glasnost* and *perestroika*. The experience of Lithuania illustrates in a remarkable way the power of national identity to survive the most rigorous attempts to destroy it over many decades.

The period until Stalin's death witnessed the most severe and violent efforts to impose the Soviet system. The first aim of the incoming Soviet troops and security forces was to establish control among a population whose absence of enthusiasm was tangible. The NKVD was quickly reintroduced, headed by Russian personnel. An office was established in almost every town and the local population was carefully screened to identify 'enemies of the people'. There followed a major purge and many arrests, notably of every former village elder, followed by very long sentences in prisons and labour camps or deportation to Siberia or Northern Russia. Many who believed they would be victims of this process went into hiding or joined the freedom fighters in the forests. Deportations had the paradoxical effect of removing many actual or potential opponents of the regime while at the same time generating more resistance, which was followed in turn by further deportations.[22]

The process of imposing Soviet authority was made difficult by a shortage of trustworthy Lithuanian personnel. Moreover, the professional class, many skilled workers and the intelligentsia had either been killed, deported, or evacuated when the German army withdrew. The security organs, the Communist Party, the bureaucracy, the judiciary and economic enterprises had to be staffed by personnel brought in from Russia, many of whom had no knowledge of Lithuanian. As a result the language of government and management had to be Russian which made for difficulties.[23] By the autumn of 1944 some of the leaders of the former Communist government had returned to Lithuania, including Sniečkus, Gedvilas and Paleckis, who resumed their former positions. But Moscow retained a very tight grip for some time through the Lithuanian Bureau of the Central Committee of the

Soviet Communist Party. It also appointed a powerful envoy in Mikhail Suslov, fresh from a successful pacification campaign in the North Caucasus, to take charge of an Organization Bureau of the Lithuanian Central Committee charged with eliminating resistance in Lithuania.[24]

The policy of the government was to prepare ethnic Lithuanian personnel for responsible positions in the Party, bureaucracy, judiciary, security services, and the economy. These were composed of idealistic communists, who believed in the radical reconstruction of society on the basis of Marxism-Leninism, careerists, and people who conformed either by temperament or for safety. Many of these recruits were relatively uneducated and were fast-tracked through full time secondary school courses or evening classes. Others received a higher education or training at party schools for positions in government and Party. However, the small numbers of ethnic Lithuanians joining the Party shows that careerists represented a tiny fraction of the total population. Most Lithuanians either joined, or sympathised with, the resistance movement. They had already seen that the Communists had not changed their spots; all political parties except one had been banned, there was no freedom of speech, or association or habitation. Civil society had been suppressed. For most there was no alternative to active or passive resistance.[25]

DEPORTATIONS

The terrible experience of the deportations confirmed the worst apprehensions of the population, forcing the more active opponents into guerrilla warfare and hundreds of thousands, particularly in the countryside, into aiding the 'forest brothers' in every possible way. There is a contrast between the mass deportation which occurred in June 1941 and those which took place between late 1944 and 1949. In the first instance lists of individuals to be deported were grouped into 14 categories, ranging from members of political parties and officials, to persons who had travelled abroad or who were active in Lithuanian parishes. In the later waves of deportees, the criteria were far more vague and could be interpreted very loosely to embrace anyone who might be regarded as a 'counter-revolutionary' or 'enemy of the people', in a word anyone suspected of lacking enthusiasm for the regime, or offering it overt or covert opposition. Consequently anyone and everyone was at risk, particularly anyone who fitted the very flexible definition of kulak, in essence a landholder, however small, who opposed collectivization or who objected to any aspect of farm policy.

The experience of deportation had a powerful impact on Lithuanian consciousness, not only on the deportees themselves who survived the most severe physical conditions in Siberia, Kazakhstan or the Arctic North and ultimately returned to Lithuania after the death of Stalin, but also on their relatives, friends and neighbours who witnessed their arrests and departures and later learned about the nature of their appalling experiences. Estimates suggest that the mass deportations alone between 1944 and 1949 totalled 350,000 people. To that figure must be added tens of thousands who were deported from Lithuanian prisons after investigation and summary trial in secret.[26] This means that at least ten per cent of the Lithuanian population was forcibly transferred to other parts of the Soviet Union where their life expectancy was very low and where only the youngest, fittest and luckiest survived the cold, hunger and disease for more than two years. Whole families were deported, grandparents, the sick and dying, pregnant women and those nursing infants. Husbands were separated in the trains from their wives and children, and were sent to different destinations. Hardly anyone in Lithuania could have been ignorant of what was occurring, and all must have appreciated that the Soviet authorities were waging a war of extermination, of ethnic cleansing against Lithuania. As one Soviet apparatchik put it, there would be a Lithuania without Lithuanians.[27]

Just to give the figures shows the enormity of the operation. In late 1944 there were around 30,000 deportees; in August/September 1945, an estimated 60,000; February 1946, perhaps 40,000; late 1947, 70,000; May 1948, 70,000; March 1949, 40,000 and summer 1949, another 40,000.[28] But eloquent as these figures are they cannot express the full horror of the experience which was described so graphically in Zoë Zajdlerowa's *The Dark Side of the Moon*. The deportations were not preceded by arrests and interrogations. Instead, people were awoken in the middle of the night, given a short time to pack food and clothing and then transported to the nearest trains where they were loaded into freight trucks for the journey east. There were no specific charges and no indication given of the length of the deportation. But when one considers the conditions under which the journey was made and the extreme harshness of the camps no one could reasonably doubt that 'the intention of the Soviet government was that these people should not survive'.

'After midnight, whole convoys of sledges loaded with families who had been arrested went past the door all night. There were thirty-three degrees of frost ... Columns of similar sledges passed throughout the whole of the following day. The neighbours stood outside their houses and prayed. The parish priest stood on a knoll by the church, holding out a great Cross towards those who were taken past.

As the trains drew out those left behind tried desperately to cling to them somehow, even to hold them back with their hands. There was, however, nothing to hold on to. The doors in the centre of each car closed tightly like a vice, and were secured by an immense iron bar which lay all the way across. The clang of these iron doors coming together was a sound impossible to forget. From the gratings fluttered down showers of white scraps, atoms of paper on which were written names and addresses, last messages begging not to be forgotten, broken sentences and prayers.'[29]

Deportations and the collectivization of agriculture were closely linked processes. The later deportations were intended to break the resistance among farmers to collectivization, and in this they succeeded. The cost in life, output and agricultural efficiency was enormous.[30]

COLLECTIVIZATION

From 1946 on a vigorous campaign in favour of collectivization began. It was hoped that the measures so far adopted to squeeze the private farms would persuade the landholders to collectivize voluntarily. However, despite much cajoling and many party directives, only about 4 per cent of Lithuanian farms were part of collectives at the beginning of 1949. The deportations, increasingly centred on recalcitrant farmers, combined with a new land reform law in early 1949, accelerated collectivization so that by the end of 1949 62 per cent of farms had been collectivized and by January 1951 the figure had reached 89 per cent. The pressure for collectivization seemed to be most effective in the areas where guerrilla operations were infrequent. Collectivization, deportations and the resistance were closely connected. The fear of deportation was a stimulus to farmers to enter collectives, and collectivization, by killing off the private farms, denied the guerrillas sources of food, shelter and other assistance which they had customarily enjoyed. It was a decisive factor in ending the guerrilla war.[31]

The collectives lived down to their reputation. They were poorly organized and badly managed. Decrees to greatly increase their size compounded the problems of management. Many collective farmers, indignant at no or derisory payments for their work on the collective, concentrated their main energies on their small private plots or took up work outside the collective. Others simply migrated to the towns in search of better pay and conditions. So great had been the dislocation and disorganization that overall output and productivity per worker suffered a dramatic decline, particularly in cereal production in the period 1950–55. Output in 1950 was lower than a decade earlier. The sown area continued to decline until 1958. The numbers of livestock suffered catastrophic falls. Output from private plots was a disproportionately high percentage of total agricultural output, reflecting the difference in energy and commitment of farmers when tending to their 'own' land. In 1989 21 per cent of meat and 30 per cent of eggs were produced on private plots. However, it took many years for Lithuanian agriculture to recover. Eventually it reverted to the agricultural patterns of the inter-war period, with a strong concentration and efficiency in meat and dairy produce, 40 per cent of which it exported to the rest of the Soviet Union where agricultural productivity was considerably lower than in Lithuania. But Lithuanian agricultural productivity was dismayingly below that of western Europe and the United States (something like 16 per cent of the American figure). If the progress of Lithuanian agriculture in the inter-war period had been maintained post war it is doubtful whether, by 1989, there would have been significant differences in productivity between a non-Soviet Lithuania and its Scandinavian and west European neighbours.[32]

RESISTANCE

The ultimate collapse of the guerrilla resistance cannot be described as a failure. The memory of it restored the sense of national self-respect which had nose-dived after the annexation of 1940. Irrespective of disputes about whether the war was unduly prolonged or essentially futile, the achievements of the guerrillas in resisting the considerable force of Soviet security troops thrown against them over an eight year period, with insignificant outside help, strengthened patriotic sentiments and national pride.

The membership of the guerrilla forces provided convincing evidence that this was not a class war but a nationalist struggle. People from all types of occupations and all social classes were involved. There were, of course, deserters, opportunists and criminals in their ranks, but the evidence suggests that the vast majority were people from small and medium-sized farms, blue collar workers, students, former soldiers in the Lithuanian army, some priests, and members of the intelligentsia. Life as a freedom fighter was nasty, brutal and short with very few pickings for those who might be attracted by the thought of plunder. It attracted the young, the idealists and those who were escaping arrest and deportation.[33]

The nucleus of the resistance was composed of the members of the Lithuanian Territorial Defence Force set up by the Nazis in 1944 which melted into the forests with its arms when the Nazis broke their agreement. But these forces were quickly supplemented, after the second Soviet occupation, by Lithuanian deserters from the Red Army, by people who feared arrest or deportation, and by those who wished to atone for acquiescing in the first Soviet occupation and had been encouraged by the successful uprising against Soviet troops and installations during the early days of Operation Barbarossa. Members of farm families threatened by collectivization and natural rebels who found life under the Soviets intolerable also joined. Many were influenced by calculations, false as it turned out, that they would be supported by the western powers when the war-time alliance with the Soviet Union ended, as they predicted it would. These optimists placed excessive faith in the Atlantic Charter; their hopes were encouraged by Churchill's iron curtain speech at Fulton and by the Truman Doctrine. In short, they did not think that they would have to fight for long.[34]

The estimated number of guerrillas at any one time between 1945 and 1949 was around 30,000. There was a high rate of attrition, the average life of a guerrilla being around two years. Casualties, arrests, and returns to civilian life reduced numbers which were constantly augmented by new recruits. It has been calculated that over the eight years from 1944 perhaps 100,000 people were involved in guerrilla activity. Add to these the tens of thousands of civilian supporters who offered shelter and supplies to the guerrillas, who liaised with them, provided information from government offices and distributed under-ground papers, and it becomes evident that a very significant share of the Lithuanian population was engaged in one way or another in the freedom movement. Soviet figures suggest that some 20,000 guerrillas were killed in anti-guerrilla operations and as many members of the security forces. Soviet operations were directed by top people such as Beria's deputy and successor General Kruglov. Some 70,000 Interior Ministry (MVD) and Ministry of State Security (MGB) troops were thrown into anti-guerrilla operations, and there was a back-up of eight

Red Army divisions, local militia and 'special extermination squads' when supplementary forces were required. The fact that the guerrillas held out for so long is testimony to their enormous courage, resourcefulness and endurance, the support of the local population, particularly the independent farmers, and their ability to supply information (some acquired through radio monitoring and through agents in government offices) to the people through the underground press. However, the war of attrition took its toll; by 1951 the number of guerrillas had fallen to an estimated 5,000 and by the end of 1952 to around 700.[35]

As popular support declined, and in the continuing absence of foreign supplies and bases, it was inevitable that the guerrilla war would be lost. Arms and ammunition were in increasingly short supply and the mass deportations accompanied by collectivization deprived the underground of essential civilian co-operation. Guerrilla seizure of food supplies from collective farms began to alienate their erstwhile supporters since compulsory deliveries still had to be made to the state from a diminished total. The persistence of the Soviet regime and a more accurate perception of the policy of the West convinced many that the time had come to end open resistance. In 1952 an order was given to this effect, but opposition continued on a diminished scale until 1956 when Snieckus announced the end of the 'nationalist bandit underground movement'.[36]

The question remains whether the guerrillas were, as the Soviets described them, extreme nationalists, fascists, counter-revolutionaries and agents of western intelligence. The evidence suggests that they were not, although their ultimate objectives were not entirely clear. Obviously they were guilty of one of the Soviet charges, that of being counter-revolutionary. They wanted to restore an independent demo-cratic republic in the tradition of western liberal democracy, but they were understandably vague on detail. It would seem that they did not support the revival of the pre-war authoritarian system. That they were nationalists is obvious. Indeed, the Resistance had a profound impact on Lithuanian national consciousness in a number of ways.[37]

For example, the sheer savagery and brutality of Soviet repressive methods against the guerrillas, notably the use of torture, banishment to the gulag, the destruction of the homes of suspects, and the callous and inhuman treatment of the corpses of the partisans, left a lasting mark on the memory of the survivors and was passed down in families, among friends and in communities through later generations. As Sniečkus later remarked, these 'violations of Soviet legality' alienated the population, 'deepened national solidarity' and harmed the Sovietization process. Moreover, even the Lithuanian Party recognized that the population held an 'affectionately patriotic image' of the resistance movement, which would have to be overcome if the people were to be converted to the Soviet system. Another significant consequence of the prolonged period of guerrilla warfare was the retardation of economic development directed from Moscow. Investment was slowed, reaching in per capita terms only one quarter of the level in Estonia and one half that of Latvia. Consequently Lithuanian industrial growth was delayed compared with that of its neighbours. When it began to catch up in the late 1950s the institutional context was different, providing more republican control over planning and immigration. This was a key difference between the post-war experience of Lithuania and that of Estonia and Latvia. The preservation of the ethnic Lithuanian character of the population in a proportion similar to that of the pre-war period strengthened the sense of national identity in the country.[38]

SOVIETIZATION AFTER STALIN

While it may be argued that the Resistance reinforced nationalist sentiments, after its suppression Sovietization was pursued by other, more subtle and more persistent means. Many Soviet experts in the West expressed surprise when nationalist activities re-emerged in the Soviet republics during the Gorbachev period. The surprise was the consequence of a misunderstanding of the situation in the republics partly brought about by too great a readiness to believe Soviet pronouncements on the subject, and partly through a shortage of information. In retrospect, however, there were enough indications of national resistance to Sovietization for different conclusions to have been drawn. In that case, the nationalist renaissance under Gorbachev would have been seen as a continuation and intensification of existing trends rather than a novelty.

And yet it is hardly surprising if the experts were misled. The capacity of the Baltic and other peoples in the Soviet Union to resist the 'uncompromising Bolshevik implacability', to use Khrushchev's words, of the Sovietization process was as unexpected as it was remarkable.[39] Unexpected, that is, if one's historical perspective was foreshortened, but still remarkable even with a broad understanding of the capacity of the Lithuanians to survive denationalizing processes. The attitudes of the Lithuanian people in the 1980s and 1990s, their approach to politics, their negotiations with the Russians over troop withdrawal, their affirmation of their European values and so on, are a product of the various strands of their history. Not the least significant of these strands was the prolonged process of Sovietization which did not crush the Lithuanians but shaped their consciousness in significant ways.

What did Sovietization mean in practice for the Lithuanians? After the end of the Resistance and the death of Stalin in 1953, the resulting 'thaw' seemed to offer the opportunity for the development of the cultural life of the constituent republics of the Soviet Union. But towards the end of the decade Khrushchev began to reassert Russian control, reimposing centralization in economic, cultural and social life. The 1961 Soviet Party programme refers to the achievement of the complete unity of the Soviet nations and the elimination of national cultural and linguistic differences. What is quite curious in the policy statements of the Soviet Communist Party under Khrushchev and Brezhnev is the dualism which they express. On the one hand they affirm and reaffirm the policy of Sovietization, on the other they pronounce with satisfaction that the national question has been solved. When Andropov admitted that the road to 'fusion' or assimilation was still a long one and that nations would survive for a long time, it was a recognition that earlier claims of success in Sovietization were hollow. In other words, the policy of Sovietization needed constant re-affirmation because it was not working, despite the most intense and prolonged efforts.[40]

The official Soviet ideology on nationality relations was that, under the influence of Lenin's enlightened policies, national cultures would flourish while at the same time experiencing integration or rapprochement (*sblizhenie*). The dialectical relationship between these two trends would eventuate in fusion, or assimilation (*sliianie*), of all the national cultures into a new Soviet whole. A Soviet people would be created, united by a common ideology ('the socialist content of the cultures of the USSR') and a common language.[41] The Party's repeated exhortations to the different nationalities to integrate more closely show that its faith in 'fusion' through the dialectical process was not complete. In practical, as opposed to ideological, terms the new Soviet constitution of 1977 attempted to strengthen national union. While maintaining the theoretical right of secession of the individual republics the Constitution 'neutralized' that right by defining the Soviet Union as a 'unitary state', the 'state of the entire people'. Nonetheless the federal form of the constitution was maintained, providing a legal framework for the expression of republican interests. The interpretation of this constitution, and its proposed amendments by the Soviet leadership, became a question of critical political importance during Gorbachev's contest with the Baltic republics in the early 1990s.[42]

Moving away from these ideological and 'high policy' considerations, we should consider what Sovietization actually meant for Lithuania in the three decades after Stalin's death. And was Sovietization another form of Russification? In broad terms Sovietization involved the imposition of a totalitarian system in politics, economics and social life, the extensive compulsory use of the Russian language at the expense of Lithuanian, the undermining of Lithuanian culture and national identity and the substitution for it of a superior 'international' culture, Russian, and finally the introduction of Russian and other non-Lithuanian personnel into Party, government and economic posts. Each of these aspects is worth comment since in totality they represent a system which was quite alien to the Lithuanians and against which they struggled. These struggles cost them dearly in the form of imprisonment, consignment to the Gulag or psychiatric hospitals, and isolation from the wider world of European and world civilization of which they felt part.

Sovietization subordinated all social organizations to Party and State. What we have come to call 'civil society' ceased to exist. Free speech and freedom of assembly were of course eliminated, and artistic expression controlled. Private and voluntary organizations had no role and were banned. Individualism and personal responsibility had no place in a system which encouraged the population to look to the state for provision of all its needs. When the state failed, as it frequently did, to provide these essentials the population had to waste time and energy and become extraordinarily resourceful, or corrupt, to provide for itself. But generally the scope for initiative was very restricted, leading to the claim that people became 'infantilized', irresponsible and dependent. At the same time everyone was encouraged under pain of sanctions to join state- and Party-sponsored organizations. Expressions of disillusionment or opposition were reported by informers to the security organizations. The eyes and ears of the state were everywhere. People withdrew into their families and cultivated, so far as they could, their private lives.[43]

If *sblizhenie*, still more *sliianie*, were to be achieved, there would have to be a unifying language through which all Soviet people could communicate and gain access to world culture. Since Moscow was the centre of empire, Russian was inevitably the language of unification. Language policy was absolutely central to the Sovietization process. Until the death of Stalin Lithuanian had the primary role as the language of education, literature and even the press. But Russian language teaching became compulsory in school at an early age and was as important as instruction in the Lithuanian language. Nonetheless the language of instruction remained Lithuanian throughout the educational system. Under Khrushchev, however, Russian became the 'primary language of all Soviet peoples' and legislation in 1958–59 offered parents the right to choose the school language, which in effect made Lithuanian optional. This marked the beginning of a process of shifting the emphasis in language teaching in favour of Russian. In 1961 the programme of the CPSU referred to the 'wiping out [of] national differences, chiefly the linguistic ones' with the aim of making Russian the 'common language of co-operation' of all peoples of the Soviet Union and 'the bridge to world culture'. Under Brezhnev Moscow strengthened its control of education in the republics. In 1978–79 legislation was introduced to increase the amount of Russian language teaching in Lithuania and the other republics still further. Russian was now to be introduced into nurseries and kindergarten; and in the schools more resources were to be used to increase the effectiveness of the teaching. Another school reform law in 1984 again emphasised the need to improve teaching of Russian. But no reciprocal obligations were placed on Russians in Lithuania or the other republics to learn the indigenous language. In effect there were two types of schools: those where the language of instruction was Russian and Lithuanian was optional; and other 'national' schools where instruction was carried out in Lithuanian but where considerable time was devoted to compulsory Russian language teaching.[44] Accompanying this legislation was a campaign to strengthen the Russian language press and concomitantly to sustain the decline in the proportions of books, journals and periodicals in Lithuanian. Russian became the language of government and administration, business enterprise, and culture.

The downgrading of the Lithuanian language was intended to strike a major blow at Lithuanian culture and identity. Other attacks took the form of the destruction of Lithuanian monuments, the erosion of traditions and the weakening of Lithuanian culture by the rewriting of history, the revision of textbooks and limitations on Lithuanian language publishing and radio and TV broadcasts. Leading editors and cultural figures were dismissed from their posts. The major denationalization campaign was conducted against the Roman Catholic Church, the most powerful upholder of Lithuanian identity and the only institution to retain its ideological and organizational independence from the Soviet regime.[45]

Since the first Soviet occupation the Church had come to be seen as the only national body capable of fighting to defend Lithuanian culture, values and traditions. Accordingly Lithuanians remained extraordinarily loyal to it despite persecution and discrimination.[46] This loyalty was reinforced by the manner in which priests and bishops shared in the sufferings of their flocks; executions, deportations and imprisonment of priests were commonplace, particularly but not exclusively in the post-war decade. Accordingly, the Kremlin correctly identified the Church as a major obstacle to Russification and the arch opponent of Soviet ideology. It therefore launched a menacing campaign against the Church from the time of the occupations of 1940 and 1944. 'The war against God' was conducted by different methods during the next four decades, but the fundamental hostility to the Church was unremitting throughout the period.[47]

Under Stalin the Lithuanian church was subjected to state-sponsored terrorism. Bishop Borisevičius was executed in 1947, Archbishop Reinys of Vilnius along with two other bishops was deported. He died in Vladimir prison, the other two returned to Lithuania in 1956 after the amnesty but were broken in health and never resumed their duties, dying shortly afterwards. Estimates suggest that 78 priests were executed after 1944 and 180 deported. It is probable that most of these punishments were inflicted owing to priestly refusal to condemn the Resistance and to support the collectivization campaign. A further loss of priests and bishops had taken place as a result of the second Soviet occupation in 1944 when some 250 clergy emigrated to the West and another 150 to Poland. In 1939 the number of priests was put at 1439 but after the deportations, executions, and emigration the numbers fell significantly but recovered to some 900 in 1960 after more priests entered their vocation and others returned from the labour camps. By 1975 the number had fallen further to 765 and by 1985 it was down to 677. The government's attack on the Church was not confined to personnel. All church property, including the Churches themselves, was nationalized, all seminaries but one were closed, as were all monasteries and convents. Vicarages and attached land were confiscated and clergy had to find alternative accommodation. Taxation on Churches was punitive and the clergy sank into extreme poverty, subsisting on the charity of their parishioners. A government appointee, the so-called Commissioner for Religious Affairs, supervised Church administration and demanded that the bishops seek his approval for decisions connected with ordinations, appointments and transfers of clergy. At the parish level the regime insisted that control should pass from the priest to elected committees subordinated to local Soviets. The authorities tried to cement this system of control by breaking the links between the Lithuanian church and the Vatican.[48]

After the death of Stalin the authorities concluded that the continued use of terror was unproductive and would not break the spirit of the older generation of believers. A more tolerant attitude was adopted and coercion was downgraded. This allowed the Church to regain some strength and Khrushchev decided on a new and more militant campaign based on moral pressure, intensified atheistic education and propaganda, and a sustained attempt to wean young people away from religious influences. Local governments were exhorted to clamp down on the illegal teaching of religion to children. The right to teach religion was not permitted under the Soviet constitution, only the performance of religious rituals, whereas the right to atheistic education and propaganda was recognized. Priests were not permitted to prepare children for the first communion, only to examine them individually, not in groups. Those wishing to be confirmed had to travel long distances. Other obstacles were erected in the path of believers, such as compulsory work on Sundays. Children of believers were picked on at school and subjected to ridicule. The number of priests in training was restricted to 24 in 1965–66, rising to 30 subsequently (down from a total of 80 in 1958), but not enough new priests were being ordained annually to maintain existing numbers. The publishing of prayer books and catechisms was either forbidden or permitted in totally inadequate numbers. Religious papers or radio and tv broadcasts were banned although the media were filled with atheistic articles and programmes asserting the incompatibility of religion and science. When Archbishop Steponavičius of Vilnius refused to ensure that his priests conformed to the prohibition on religious teaching and resisted state interference in matters of internal church discipline, he was exiled from his diocese in 1961 and was not allowed to return for decades. The authorities believed that under the impact of these repressive measures the Church in Lithuania would be dead in two decades.[49]

POPULATION MOVEMENTS

Moscow no doubt calculated that Sovietization would be facilitated by the movement of Russians and other nationalities, such as Ukrainians and Belarussians, into the Baltic states. Yet there were other explanatory factors behind the immigration, otherwise the volume of immigrants would have been similar in each of the states. However, in proportion to total population immigration was much heavier in Estonia and Latvia than in Lithuania.[50] This was accounted for by a number of factors, namely the relatively low level of investment in Lithuania after the war, the lack of a more developed industrial base, and a significantly higher birth rate. The relative decline in the Lithuanian birth rate in the 1960s accounts for the increase in immigration dating from that period. But though the absolute numbers of immigrants rose substantially, the share of ethnic Lithuanians in the population remained constant at around 80 per cent during the next two decades.[51] This was a remarkable achievement given the enormous population losses in Lithuania in the decade of the 1940s. Tarulis calculated that there was a loss of population of around one million people between 1939 and 1959, resulting from deportations, executions, the Holocaust, voluntary emigration, and the resulting reduction in the numbers of births.[52] The resistance war and the consequential slow rate of economic growth ensured that immigration would not fill the gap left by population losses. The relatively high rate of natural increase fulfilled this function.

It has often been suggested that high levels of immigration into Estonia and Latvia were part of a policy of 'purposeful imperialist Russianization'. This would be consistent with the high profile Sovietization policy of successive governments in Moscow. On the other hand, relatively low levels of immigration in Lithuania suggests that at least the policy of Sovietization was flexible and could be pursued by a range of measures of which immigration was a desirable but not an indispensable part. It may be that immigration was a 'side effect' or 'an incidental rather than a deliberate consequence' of Soviet economic policies. Arguments supporting this view suggest that a number of factors were at work in increasing immigration in Estonia and Latvia. Among them were the higher standard of living there compared with the rest of the Soviet Union, the heavy Soviet investment in already existing industries such as shale oil and metal manufacture which could make a significant contribution to Soviet reconstruction after the Second World War, the possibility of jumping the queue for housing, and the opportunity for retired Soviet military officers to settle in the area. Above all, the low birth rate in Latvia and Estonia created a demand for industrial labour which could only be filled by immigrants from elsewhere in the Soviet Union, particularly at a time (the two post war decades) when there was a high rate of natural increase in the Soviet Union. This is not to say that Sovietization was not the ultimate goal, merely to emphasise that there were other plausible reasons for immigration supplementing the Sovietization idea.[53]

Recruitment of Russians and other nationalities in Lithuania was particularly heavy in government and among economic managers, especially in the post war decade. A shortage of experienced and loyal Lithuanian Party members meant that Russians and Russified Lithuanians from other parts of the Soviet Union played a central role in policy-making. Lithuanians were usually given the top positions in administration but there were numerous Russians in high Party and government positions who were there to provide expertise and supervision, and to ensure that Moscow directives were carried out. The KGB was controlled directly from Moscow through senior operatives in Lithuania who were almost always Russians. In local government too Russians had a major voice; in 1945–46 three-fifths of the chairmen and vice-chairmen of city and district governments were Russian. Obviously greater numbers of Russians in Lithuania would strengthen the use of the Russian language. No doubt it was also expected that intermarriage would become more common and that the children of such marriages would adopt Russian culture and identity. In practice mixed marriages were rare and where they occurred the majority of children retained the Lithuanian identity and language and went to national schools.[54]

THE FAILURE OF SOVIETIZATION

When Crèvecoeur asked in the 1770s 'Who then is the American, this new man?' he believed he had identified a people with a unique identity. By contrast, while the Soviet leadership claimed they had created a new Soviet person from the raw material of the Russian empire, their aspirations had exceeded their achievements. Whereas the Americans had voluntarily adopted a new identity and developed it under the influence of the North American physical environment, the 'Soviet people' was a conception imposed on the population of the Soviet empire by force and intimidation, and was in reality only skin deep. Stalin admitted as much in the Second World War when he reverted to the symbols of Russian patriotism to inspire opposition to the Nazis. For Lithuanians to become Soviet persons they had to become Russophiles, atheists and materialists. They had to be fluent in the Russian language. They had to reject their 'bourgeois national' identity, and accept the Communist system of production and exchange, Communist criteria in artistic and cultural endeavours, and Communist morality in which the ends justified the means. In addition they were required to accept the suppression of free speech and individualism in deference to the supreme wisdom of the Party and the collective good. Finally, they had to surrender independence of judgement in the pursuit of knowledge and absorb in a mechanical fashion the dogmas of the Party. In short, they had to become alienated from their real selves, their backgrounds and their convictions. In 1976 a dissident journal *Aušra,* the re-incarnation of the nineteenth century journal of the same name, described what Sovietization meant, namely the age-old process of Russification but with a new ideological content, Marxist-Leninism. This reality was concealed under the terms 'internationalism', 'Soviet patriotism', 'the Soviet people'. What this meant for the Lithuanians was the suppression of their national culture, and a brutally sustained attempt to 'obscure, de-emphasise and slander' their nation's past. That it did not succeed was testimony to the deep roots of the Lithuanian identity and to the people's determination, under extreme pressure, to hold on to their culture and language.[55]

THE LITHUANIAN ECONOMY UNDER THE SOVIETS

Sovietization Stalin-style proved to be incompatible with economic growth. The guerrilla war, the massive deportations and collectization did not provide a stable context for investment. After 1945, the Soviet Fourth Five Year Plan provided the next lowest level of investment per capita in Lithuania of all the Soviet republics.[56] The rate of industrial growth in Lithuania in the 1940s and early 1950s was much slower than in Estonia and Latvia. By the late 1950s, however, growth was accelerating and industrial output rose quite steeply in the next decade. The rate of output increase began to decline in the following two decades; nonetheless, in the 1980s the industrial sector was responsible for about 60 per cent of GNP. The surpluses of rural labour coupled with a higher rate of natural population increase than in Estonia and Latvia reduced the demand for labour from other parts of the Soviet Union. In terms of industrial output figures the Soviet achievement was notable. By 1973 industrial production was 39 times higher than in 1940, although allowances have to be made for the dislocation in economic life in that year caused by occupation and war. As we have seen Lithuania was primarily an agrarian economy in the inter-war period. By 1990, however, it was industrialized, with industry and construction accounting for some 64 per cent of gross national product, agriculture some 25 per cent, transport almost 4 per cent and commerce 6.7 per cent.[57]

The composition of industrial output reflected the development of already existing industries and the introduction of new ones. A proportion of new investment was determined by locational theory but some could only be explained by Moscow's desire to integrate Lithuania ever more tightly into the Soviet economy, whatever economic absurdities might result.[58] Industries such as woollen textiles, food processing, fishing and fish processing, furniture, clothing, and construction materials were developed from an existing base. However, investments in new industries became increasingly important, notably in machine construction, precision tools, chemicals, energy and wood products. In general the emphasis was placed on labour intensive rather than resource-based industries and on consumer rather than producer goods. But in the 1960s the chemical industry experienced rapid development and Lithuania became an important producer of plastics, artificial fibres, synthetic polymers and chemical fertilisers. Plastic products supplied a substantial local market in the Lithuanian machinery and precision instruments industries.

These new industries, particularly chemicals, were heavy consumers of energy, most of which in the form of natural gas and oil had to be imported from elsewhere in the Soviet Union. To meet this shortfall there was considerable investment in the energy industry in Lithuania both in conventional electric power stations and in the massive nuclear power plant at Ignalina in north–east Lithuania. The Lithuanian Party and government was able to influence the location of new industry, drawing on the finding that industrial plants established in small towns or rural areas could on average depend on the locality for almost 90 per cent of their labour needs; only the remaining 10 per cent had to be recruited in other republics. Accordingly as many new enterprises as possible were established in or near small towns or provincial cities.[59]

The existence of a relatively skilled and well-trained workforce was an important factor in decisions about industrial location. The generally high levels of education in independent Lithuania provided the basis for a substantial investment in education and training in the Soviet period to provide the skilled workers, technicians and professionals required by the developing economy. The composition of the work force changed significantly under the impact of economic development. For example, the proportion of agricultural workers fell by more than half (to around 30 per cent of the total work force) between 1939 and 1970. During the same period the proportion of urban inhabitants more than doubled to 45 per cent and by 1989 had further increased to 68 per cent. The urban population had a lower proportion of ethnic Lithuanians (69 per cent) than the population as a whole (approximately 80 per cent); by contrast ethnic Russians were better represented in the urban population than the rural. This led to fears that urbanization might weaken the Lithuanians' sense of identity. However, the situation did not become acute. Although the number of Russian speakers in Lithuania quadrupled by the 1980s, their overall proportion of the population did not exceed around 20 per cent, substantially below the figures for Estonia and Latvia (38 per cent and 48 per cent respectively) which aroused widespread apprehension in those countries.[60]

The dominant sector in the pre-war economy, agriculture, made a much smaller contribution to total output of goods and services in the Soviet period. Agriculture sustained a massive setback as a result of the collectivization campaign, which led to the deportation of huge numbers of skilled and experienced farmers. The remaining farmers were organized and re-organized into the new collectives, so that by the beginning of 1951 almost 90 per cent of private farms had been absorbed. Bad management, low morale and passive resistance on the part of the workforce, and poor working conditions resulted in a 60 per cent decline in agricultural output in 1951 compared with pre-war. The number of dairy cattle fell from 848,000 to 504,000 in 1951. The area under cultivation continued to decrease until the end of the 1950s. Crop yields diminished. Compulsory state deliveries at much below market prices and ludicrously low rates of pay, or no pay at all, for collective farm workers induced them to concentrate their best efforts on their private plots at the collectives' expense. Output did not recover for more than a decade. Recovery was based on a reversion to the specialization of production which had characterized Lithuanian agriculture before the war, namely a concentration on dairy, meat and eggs production. Eventually Lithuania became the most efficient producer of meat and dairy products in the Soviet Union. But a very significant contribution to total production of meat (21 per cent) and eggs (30 per cent) and other crops still came from the farmers' private plots. In 1975 the private plots produced 39 per cent of total agricultural output, a far higher proportion than their share of arable land. By Soviet standards Baltic agriculture was a great success story, and the productivity of farm workers in Lithuania was some 47 per cent higher than in the rest of the Soviet Union.

Another major contribution to the success of the collectives derived from food processing enterprises established on the farms such as mills, canneries and mineral water bottling. Farms also diversified into the manufacture of products such as furniture, metalwork and barbed wire. But despite the fact that rural incomes were often higher than urban the quality of life in the country was not as appealing owing to the shortage of modern housing, a lack of running water and a proper sewage system, and the absence of paved streets. Even in the relatively successful area of labour productivity the Lithuanian achievement was quite modest by international standards. In sum, Soviet collectivization was a major obstacle to Lithuania's keeping up with agricultural development in western Europe and Scandinavia.[61]

Although, as we have seen, integration (*sblizhenie*) of the Soviet republics into the Union was very imperfect, it was most complete in the field of economic development. Moscow's policy was to create intense economic inter-dependence between the republics, for markets, raw materials, finished goods and energy.[62] In 1982 Lithuania exported 80 per cent of its industrial production to other Soviet republics and imported almost 90 per cent of goods and supplies from there. There was a high degree of centralization in economic decision-making, apart from a brief interlude under Khrushchev when some devolution in the management of industry was permitted. By the 1980s five-sixths of all industrial enterprises in Lithuania were controlled either by all Union ministries in Moscow or by Union and republican ministries combined. The Lithuanian government itself directed only about 15 per cent of industry. Certainly all the key industries and the majority of large factories were controlled from Moscow, which meant that all the major decisions were taken there. The development of inter-republican enterprises in the 1970s reinforced the process of integration, not least by confirming Russian as the operational language, and by encouraging labour mobility between republics on the basis of professional qualifications.

BENEFITS AND COSTS

It is generally accepted that owing to, or despite, Soviet economic policy Lithuania and the other Baltic republics achieved a standard of living that was the highest in the Soviet Union. Salaries were above the national average and manufacturing and farming enterprises were held up as models for emulation elsewhere in the Union. Per capita state expenditure on education, health and other social services was higher in the Baltics resulting in, to take just one example, infant mortality rates at less than half the Soviet average. Rapid economic growth in the 1950s and 1960s was the basis for this relative prosperity, but the pace of growth slowed in the 1970s, and with it increases in living standards. Per capita incomes were higher in Estonia and Latvia than Lithuania owing to Lithuania's less developed industrial base pre-war, the delayed investment post-war, and the higher proportion of the work force in agriculture. Still, Lithuanian incomes were above the Soviet average by about 9 per cent in 1989, the incomes of collective farmers were substantially higher than the national average, and there was generally a higher quality of life. Moscow's claims that the difference was accounted for by subsidies from the centre were disputed by Lithuanians.[63]

Improvements in standards of living inevitably produced calls for an improvement in standards across the board, for example in the supply and quality of consumer goods, in better and more available housing and in a whole range of services. The points of comparison for Lithuanians were not the rest of the Soviet Union but Scandinavia and the pre-war republic. The latter was somewhat idealised but there was truth in the contention that the gap between the Lithuanian and Scandinavian economies had widened considerably since the war and that there was little prospect of relative improvements under Soviet communism. This consciousness of relative deprivation partly accounts for the avidity with which the Lithuanians and their Baltic neighbours seized on the opportunties presented by Gorbachev's *perestroika* to reshape their economies.[64]

Frustration with the constraints on economic development was less effective in mobilizing public opinion in the pre-Gorbachev period than a growing awareness of the damage inflicted on the natural environment by developments in industry and agriculture. The fact that most of the damage was done by industrial plants created and controlled by Moscow focused the indignation not on local agencies but on the all-Union authorities. This provided the opportunity for an alliance between the government in Vilnius and environmental protection groups, composed very heavily of members of the scientific and cultural intelligentsia. The importance of nature in Lithuanian religious and cultural traditions, and the deep affection of Lithuanians for the natural world, made it inevitable that a strong wave of protest would emerge against the violation of Lithuania's heritage.[65] Some environmental campaigners were punished by the security forces for their involvement but generally speaking protesters attracted support, albeit indirectly, from the government in Vilnius.

The roots of the problem lay in the decision by Moscow to abandon *sovnarkhoz* in 1965 and to return to the branch system of managing industry, under which decisions on planning, location and investment were taken by central ministries in Moscow and not by the territorial authorities i.e. the republican governments. In practice this meant neglect of republican views on the location of plants and the best use of local resources. Moreover, there was little inducement for central ministries to allocate investment funds for environmental protection since this would reduce the rate of return on capital[66] Environmental damage had become so serious by the late 1970s that the Vilnius Party Committee urged Moscow to transfer funds for environmental protection from all-Union ministries to territorial governments. What was so alarming was not only the pollution itself but the apparent total indifference of Moscow to Lithuanian anxieties, and the determination to pursue existing policies without regard to local public opinion. This was widely seen as an attack on the Lithuanian heritage and a threat in the long run to Lithuanian identity.[67]

What then was the nature of the environmental problem which caused so much concern in Lithuania and mobilized so many to protest? Water and air pollution were the major worries. Most rivers were badly polluted by untreated effluent. Of particular concern was the major Lithuanian river, the Nemunas, which was heavily polluted and flowed into the ecologically sensitive delta area and then into the shallow waters of the Courish lagoon. Untreated industrial waste flowed directly into rivers or into municipal sewers, and accounted for about 70 per cent of all raw sewage discharged by those sewers. In addition, only one of the five major cities in Lithuania possessed a complete sewage treatment plant. Sewage treatment in rural areas was rudimentary. Agriculture was another major source of pollution: chemical fertilisers, manure, insecticides and herbicides were major pollutants of surface and ground water, with run-offs into streams and rivers. All streams in the coastal zone were heavily polluted and this in turn contributed to the pollution of the lagoon and the Baltic Sea itself. The heavy presence of chemical fertilisers promoted the growth of algae and the depletion of oxygen in the water.[68]

Much of this pollution was cumulative, from hundreds or thousands of small outlets. There were, however, some major enterprises which were so damaging, either potentially or actually, that they attracted considerable opposition and marked the beginning of successful protest. One of the most notable was the proposal to build an oil refinery at Jurbarkas on the Nemunas River. This threatened major air and water pollution and dire consequences for the Courish lagoon and the fishing industry of the area. Twenty one leading scientists, economists and writers successfully protested against the proposal and a new site at Mazeikiai was chosen.[69]

Although this threat to the lagoon was removed there were other major polluters in the area, notably the Klaipeda Cellulose and Paper plant which discharged massive amounts of waste into the harbour, enterprises in the towns of Sovetsk and Neman in the Kaliningrad region which discharged effluent into the Nemunas, illegal oil discharges from ships in Klaipeda harbour, and pollutants from the fish processing plant. The substantial expansion of the chemical industry in Lithuania contributed substantially to water and air pollution. Notable examples of polluters in this industry were the chemical plant at Panevezys and the fertiliser works at Jonava.[70]

Air pollution was a serious threat to public health. As late as 1980 two thirds of all industrial establishments in the five largest Lithuanian cities had installed no pollution control equipment, despite promises to do so. A good example of this negligence was the Mazeikiai plant. Where such equipment existed it often did not work effectively. Vilnius, for example, suffered intensely from air pollution; the concentration of nitrogen dioxide, carbon monoxide, acetone, hydrochloric acid and ammonia, sulphuric acid, formaldehyde and phenol was some 24 per cent above the maximum permissible level.[71]

The growth of the economy through the 'sixties and 'seventies led to a doubling of energy requirements every four years. Growing environmental problems with conventional fossil fuel or hydro electric power generation led to the decision to construct a giant nuclear power station at Ignalina on Lake Druksiai in North East Lithuania. This was projected to be the largest nuclear installation in the world. Construction began in the late 1970s. It was anticipated that the installation would have some adverse effects on the local environment. However, these paled into insignificance after the disaster at Chernobyl since Ignalina was of the same type of nuclear installation and was presumed to be equally dangerous. As a consequence the Lithuanian government blocked the use of funds for a third reactor and there was considerable popular hostility to any further development. Public perceptions about the environmental dangers caused by indifferent and unaccountable Moscow ministries were among the first and most important generators of constructive action, as opposed to protest, among the Lithuanian population. For example, a Nature Protection Association was formed and between 1982 and 1984 the Lithuanian Academy of Sciences, in collaboration with other organizations, prepared a scheme to protect the natural environment and to preserve 'the priceless resources of our country' in the period up to the year 2,000.[72]

Evidently the balance between accommodation and resistance was shifting in the 1980s. The calculation of costs and benefits, even on an economic level, was showing an increasing deficit as the pace of economic growth diminished. Comparisons between the West and the Baltic states became more and more unfavourable to the Soviet system. The pollution of the environment by careless and indifferent Soviet economic planners deeply offended influential sections of the Lithuanian population, including senior figures in the Lithuanian Communist Party itself, who were sensitive to the need to protect the Lithuanian countryside and its architectural heritage, so important to Lithuanian identity. Hence, when Mikhail Gorbachev came to power in 1985 with a reforming programme, there was enormous pent-up demand for change waiting to be released in Lithuania. The big question was, would Gorbachev's reforms help the Lithuanians to take control over their own lives, their economy, and their environment by breaking the power of the Soviet system.

1 David Kirby, 'Incorporation: the Molotov-Ribbentrop Pact', in G. Smith ed., *The Baltic States: the National Self-Determination of Estonia, Latvia and Lithuania* (Basingstoke and London: Macmillan, 1994), pp.79–80, Alfonsas Eidintas, Vytautas Zalys, Alfred Erich Senn, ed., Edvardas Tuskenis, *Lithuania in European Politics: The Years of the First Republic 1918–1940* (Basingstoke and London: Macmillan, 1997), pp.188–9; Evald Uustalu, Appendix: Events after 1940, in August Rei, *The Drama of the Baltic Peoples*, (Stockholm: Kirjastus Vaba Eesti, 1970), p.318; V. Stanley Vardys and Judith B. Sedaitis, *Lithuania, the Rebel Nation* (Boulder, Col: Westview Press, 1997), p.53.
2 Among the most prominent political figures deported were Voldemaras, Merksys, Urbsys and Stulginskis. The future Prime Minister of Israel, Menachem Begin, was deported shortly after the annexation. Of the ten surviving Lithuanian prime ministers, six were deported, and four escaped to the West: Vardys and Sedaitis, p.54; Georg von Rauch, *The Baltic States: the Years of Independence, Estonia, Latvia, Lithuania 1917–1940* (London: C. Hurst & Co., 1974), pp.226–8; Bronis J. Kaslas ed., *The USSR-German Aggression against Lithuania* (New York: Robert Speller & Sons, Publishers, Inc., 1973), p.23; Aleksandras Shtromas, 'The Baltic States as Soviet Republics: Tensions and Contradictions', in Smith ed., p.87; Romuald Misiunas and Rein Taagepera, *The Baltic States: Years of Dependence 1940–1990* (London: Hurst & Co, 1993),

pp.39–40; Uustalu, p.318; Eidintas et al., Footnote 18 to Chapter 6; John Alexander Swettenham, *The Tragedy of the Baltic States: A Report Compiled from Official Documents and Eyewitnesses' Stories* (London: Hollis & Carter, 1952), pp.134–7.

3   Eidintas et al., Footnote 19 to Chapter 6; Uustalu, pp.320–3; Kaslas, p.23; Shtromas, p.86; Misiunas and Taagepera, pp.40–41.

4   By early 1941 nationalization was all but complete; it involved not only privately-owned firms but the co-operatives as well, which had absorbed individual craftsmen. Where the expertise of the owner was absolutely indispensable for the continued operation of the enterprise, he was retained until an adequate substitute could be found: Misiunas and Taagepera, p.34; Kaslas, p.247; Uustalu, pp.314–5; Swettenham, pp.32–8.

5   Swettenham, pp.96–101.

6   Uustalu, pp.313–4.

7   This policy did most damage to the largest and most productive farms and the better land on these farms was targeted for confiscation: Swettenham, pp.79–83; Misiunas and Taagepera, pp.35–6; Bohdan Nahaylo and Victor Swoboda, *Soviet Disunion: A History of the Nationalities Problem in the USSR* (London: Hamish Hamilton, 1990), p.87.

8   Misiunas and Taagepera, pp.36–8; Swettenham, pp.103–9.

9   Misiunas and Taagepera, pp.61–64.

10 op. cit., p.61; Eidintas et al., pp.189–91.

11 Venclova quotes from an editorial in *Naujoji Lietuva*, 4 July, 1941: 'The greatest enemy of Lithuania and other nations was and in some places remains a Jew ... Today, as a result of the genius of Adolf Hitler ... we are free from the Jewish yoke ... A New Lithuania, after joining a New Europe of Adolf Hitler must be clean from Jews ... To exterminate the Jewry and Communism along with it is a primary task of the New Lithuania' Tomas Venclova, 'A Fifth Year of Independence: Lithuania, 1922 and 1994', *East European Politics and Societies*, vol.9,. no.2, Spring 1995, p.365.

12 For example, Skirpa's claim that a 1940 appeal, which was couched in Nazi vocabulary and included some anti-semitic sentiments, was designed to win over the German government to support LAF-sponsored resistance in Lithuania, or another suggestion that Lithuanians were powerless to resist under the German occupation or to raise their voices against the Holocaust. See Venclova, p.366; Kaslas, pp.312–3.

13 Among these were the parents of Vytautas Landsbergis, and Bishop Vincantas Borisevičius whose protection of Jews did not save him from execution by the Soviets 1947, despite Jewish pleas on his behalf: E.J. Harrison, *Lithuania's Fight for Freedom* (New York: Lithuanian American Information Center, 1952), p.71; Zvi Segal, 'Jewish Minorities in the Baltic Republics in the Postwar Years' in Dietrich Andre Loeber, V. Stanley Vardys and Laurence P.A. Kitching, eds., *Regional Identity under Soviet Rule: the Case of the Baltic States* (Hackettstown, NJ: Publications of the Association for the Advancement of Baltic Studies, 1990), p.229

14 Alfred Rosenberg, the Minister for Eastern Occupied Territories, was instructed to 'transform the [Baltic] region into part of the Greater German Reich by germanizing racially possible elements, colonizing Germanic races, and banishing undesirable elements'. Kaslas, p.321.

15 Perhaps 20,000 persons joined these battalions in the expectation that they would serve only in Lithuania. However, they were sent East to provide support behind the front line, which often involved them in anti-guerrilla operations and civilian control tasks. This activity has sometimes involved them in accusation of committing atrocities against the local populations: Misiunas and Taagepera, p.57; Shtromas, p.91; K.V. Tauras, *Guerrilla Warfare on the Amber Coast* (New York: Voyages Press, 1962), pp.27–8.

16 Leading intellectuals were sent to concentration camps, and the universities of Vilnius and Kaunas were closed in 1943: Misiunas and Taagepera, pp.54, 58; Shtromas, p.91; Thomas Remeikis, *Opposition to Soviet Rule in Lithuania 1945–80* (Chicago, Ill: Institute of Lithuanian Studies Press, 1980), p.268

17 Harrison, pp.44–6; Zenonas Ivinskis, 'Lithuania during the War: Resistance against the Soviet and the Nazi Occupants', in V. Stanley Vardys ed., *Lithuania under the Soviets: Portrait of a Nation 1940–1965* (New York: Praeger, 1965), pp.83–4; Misiunas and Taagepera, p.59.

18 It was planned to bring in another 20,000 settlers in 1944: Harrison, p.34

19 Harrison, pp.49–51; Vardys and Sedaitis, p.57; Ivinskis, pp.77–9.

20 An opposite view is expressed in the following quotation which makes no distinctions between Germans and Russians who are 'as inseparable as the Siamese twins ... The Russians might be using more unpolished policy, which is more characteristic of the East, while the Germans might be executing their dirty work in a more subtle, more intelligent way ... But even this difference is fading out ... All means are equally good to both the Germans and the Russians as long as they lead to the achievement of their purpose which is to incorporate Lithuania ... into Russia or Germany, as the case may be. The Russians as well as the Germans are of the opinion that this can be accomplished only when Lithuania is inhabited either by Russians or Germans, but not by Lithuanians', quoted from *Nepriklausoma Lietuva* (Independent Lithuania) of 15 June, 1943, in Harrison, p.48

21 Harrison, p.76.

22 Harrison, pp.52–3; Vardys and Sedaitis, pp.60–1; Uustalu, pp.349–51.

23 In 1947 around 30 per cent of ministers were Russians and another 13 per cent were Russian Lithuanians. Most deputy ministers were Russians. Lithuanian officials were usually supervised by non-Lithuanian assistants. In 1945 the Lithuanian Communist Party was only about 18 per cent ethnic Lithuanian, a percentage which had grown to 38 per cent by 1957.

24 Misiunas and Taagepera, pp.77–83; Vardys and Sedaitis, pp.59–62.

25 Harrison, pp.52–3; Shtromas, pp.94–6.

26 Harrison, pp.62–7.

27 It has been calculated that 80 per cent of the deportees never returned to Lithuania after the amnesty of 1955. Most of them had perished in the gulag: Uustalu, pp.353–5.

28 Vardys 'Lithuanians', p.75; Walter C. Clemens, Jr., *Baltic Independence and Russian Empire* (Basingstoke: Macmillan, 1991), pp.56–7; Shtromas, pp.93–4; Misiunas and Taagepera, pp.73–4.

29 Zoe Zaidlerowa, ed. John Coutouvidis and Thomas Lane, *The Dark Side of the Moon* (Hemel Hempstead: Harvester Wheatsheaf, 1989), pp.63–5.

30 Tauras, pp.66–8; Harrison, pp.74–6; Remeikis, p.227; Misiunas and Taagepera, pp.94–5.

31 Misiunas and Taagepera, pp.97–107; Tauras, pp.67–9.

32 Vardys and Sedaitis, pp.65–6.

33 Harrison, pp.54–8; V. Stanley Vardys, 'The Partisan Movement in Postwar Lithuania', in Vardys ed., pp.94–5.

34 Vardys, 'The Partisan Movement ...', pp.94–5; Tauras, pp.91–5; Misiunas and Taagepera, pp.84–7; Shtromas, pp.92–3; Vardys and Sedaitis, p.82; Remeikis, pp.43–7.

35 Shtromas, p.93; Misiunas and Taagepera, pp.90–4.

36 Misiunas and Taagepera, pp.93–4; Shtromas, p.93.

37 Vardys and Sedaitis, p.84; Remeikis, pp.58–63.

38 Remeikis, pp.63–4; Vardys and Sedaitis, p.84; Vardys, 'The Partisan Movement', p.108.

39 Nahaylo and Swoboda, p.141

40 op.cit., pp.141, 153–4, 221.

41 At the 24th Congress of the CPSU in 1971 Brezhnev announced that 'a new historical community of people—the Soviet people—had emerged' which was united in 'monolithic solidarity' by ideology and shared experience. Nonetheless the Party would do all it could to promote the further drawing together of nations, suggesting that the old slogan 'nationalist in form, socialist in content' was still applicable: op.cit. pp.153–4; Clemens, pp.62–4; Jan Ake Dellenbrant, 'The Integration of the Baltic Republics into the Soviet Union', in Loeber et al. pp.101–4; Andrejs Urdze, 'Nationalism and Internationalism: Ideological Background and Concrete Forms of Expression in the Latvian SSR', in Loeber et al. pp.355–8

42 Clemens, pp.62–4; Nahaylo and Swoboda, p.210.

43 Vardys and Sedaitis, pp.71–4.

44 It is unclear how this separation of schools was supposed to foster integration : Sergei Zamascikov, 'Soviet Methods and Instrumentalities of Maintaining Control over the Balts', in Loeber et al., pp.94–5; Zvi Segal, 'Jewish Minorities in the Baltic Republics in the Postwar Years', in Loeber et al., pp.236–8; Nahaylo and Swoboda, pp.140–2; Uustalu, pp.382

45 Misiunas and Taagepera, pp.211–13; V. Stanley Vardys, 'Modernization and Baltic Nationalism', *Problems of Communism*, XXIV, 5, 1975; Vardys, 'Lithuanians', pp.77–8; British Foreign and Commonwealth Office Background Brief, 'The Baltic States: 40 Years under Soviet Rule', FO 973/111, Sept. 1980; Vardys and Sedaitis, pp.72–4; Silvia P. Forgus, 'Manifestations of Nationalism in the Baltic Republics', *Nationalities Papers*, vol.VII, No.2, Fall 1979, p.205.

46 Dissident sources suggested that there were one million believers (about 47 per cent of the population) in Lithuania in 1977; others claimed that the proportion of believers was as high as 75 per cent of the population in 1969. Vardys thought that probably one and a half million people participated in religious life in the 1980s: V. Stanley Vardys, 'The Role of the Churches in the Maintenance of Regional and National Identity in the Baltic Republics' in Loeber et al., p.157

47 Dennis J. Dunn, 'The Catholic Church and the Soviet Government in the Baltic States 1940–1941', in V. Stanley Vardys and Romuald J. Misiunas eds., *The Baltic States in Peace and War 1917–1945* (University Park and London: The Pennsylvania State University Press, 1978), p.155; J. Savasis, *The War against God in Lithuania* (New York: Manyland Books, 1966), pp.32–42; Marite Sapiets, 'Religion and Nationalism in Lithuania', *Religion in Communist Lands*, vol.7, no.2, 1979, pp.78–81; V. Stanley Vardys, 'Recent Soviet Policy toward Lithuanian Nationalism', *Journal of Central European Affairs*, vol.XXII, 1963, pp.325–6.

48 Uustalu, p.364; Vardys 'Lithuanians', pp.75–6; Harrison, p.72, Remeikis, pp.107–8; Letter to Pope Pius XII from the Catholics of the Lithuanian Republic quoted in Remeikis, pp.485–91; Savasis, pp.26–7; Vardys in Loeber et al. eds., pp.157–8

49 Remeikis, pp.108–11, 448, 527–8; Vardys and Sedaitis, pp.85–6; Savasis, pp.52–58.

50 In relation to their pre-war populations Estonia received three times as many immigrants as Lithuania between 1950 and 1989, Latvia more than twice as many: Ole Norgaard et al., *The Baltic States after Independence* (Cheltenham: Edward Elgar, 1996), pp.170–1.

51 The Lithuanian rate of natural increase was 10 per thousand in 1967 and 8.7 in 1970 compared with an Estonian rate of 6.7 in 1960 and 3.2 in 1972. This was accounted for by a combination of higher birth rates and a smaller number of abortions. The rate of growth of natural increase in Lithuania fell from 1.5 per cent per annum to 0.6 per cent per annum between 1960 and 1980. This meant a weakened local labour supply in the 1980s reflected in an upward movement in immigration between 1980–84. Overall immigration accounted for only 12 per cent of Lithuanian population growth in the earlier period of 1959–70: Vardys, 'Modernization and Baltic Nationalism', pp.39–40; Misiunas and Taagepera, pp.281–2; Norgaard, pp.171–3; Rein Taagepera, 'Baltic Population Changes, 1950–1980', *Journal of Baltic Studies*, vol.XII, no.1, 1981, pp.37–44; Rein Taagepera, 'The Population Crisis and the Baltics', *Journal of Baltic Studies*, vol.XII, no.3, 1981, pp.239–41; Thomas Remeikis, 'The Impact of Industrialization on the Ethnic Demography of the Baltic Countries', *Lituanus*, vol.XII, Part 1, 1967, pp.33–41.

52 Tarulis calculated that if the pre-war birth rate had been maintained through the period 1939–59 the population would have been 3.7 million instead of the actual 2.7 million in the Soviet census of 1959: Albert N. Tarulis, 'A Heavy Population Loss in Lithuania', *Journal of Central European Affairs*, no.4, January 1962, pp.452–64.

53 Taagepera, 'Baltic Population Changes', pp.41–4; Taagepera, 'The Population Crisis', pp.241–2; Tonu Parming, 'Population Processes and the Nationality Issue in the Soviet Baltic', *Soviet Studies*, vol.XXXII, no.3, July 1980, pp.401–2; Misiunas and Taagepera, p.281.

54 Both at this period and later, when economic immigration began to increase, immigrants had higher average levels of formal education than ethnic Lithuanians. This is not inconsistent with Shtromas's contention that Russian immigrants were mainly industrial workers in the lowest socio-economic positions: Vardys, 'Modernization and Baltic Nationalism', pp.40–1; Zamascikov, p.95; Dellenbrant, pp.110–11; Vardys and Sedaitis, p.69; Aleksandras Shtromas, 'Prospects for Restoring the Baltic States' Independence: A View on the Prerequisites and Possibilities of their Realization', *Journal of Baltic Studies*, vol.XVII, no.3, Fall 1986, p.262; Remeikis, 'The Impact of Industrialization', p.35

55 Swettenham, pp.103–14; Tauras, pp.100–02; Vardys, 'Recent Soviet Policy', pp.328–30; *Ausra*, No.2, February 1976, quoted in Remeikis *Opposition to Soviet Rule*, pp.405–9.

56 The amount per capita was four times as high in Estonia and twice as high in Latvia: Remeikis, *Opposition to Soviet Rule*, p.63

57 Zamascikov, pp.91–3; Vardys and Sedaitis, p.66; Vardys 'Lithuanians', pp.76–7; Vardys, 'Modernization and Baltic Nationalism', pp.37–8, 40; Misiunas and Taagepera, pp.234–5.

58 For example, the largest cast iron foundry in the Baltic region was required to export 55 per cent of its output to Russia in 1980 while at the same time metal industries in Lithuania had to import cast iron from producers a thousand miles away in other parts of the Soviet Union. Numerous examples of such practices can be found: Vardys and Sedaitis, p.67

59 The adoption of this policy had been facilitated by the *sovnarkhoz* strategy under Khrushchev which gave individual republics greater power to determine economic development. For industrial development see Vardys and Sedaitis, pp.6–8; Remeikis, *Opposition to Soviet Rule*, pp.81–4; Remeikis, 'The Impact of Industrialization', pp.39–41; Henry Ratnieks, 'The Energy Crisis and the Baltics', *Journal of Baltic Studies*, vol.XII, No.3, Fall 1981, pp.246–52; Augustine Idzelis, 'Locational Aspects of the Chemical Industry in Lithuania: 1960–1970', *Lithuanus*, vol.XIX, no.4, 1973, pp.51–2, 59–60; Benedict V. Maciuika, 'The Role of the Baltic Republics in the Economy of the USSR', *Journal of Baltic Studies*, vol.III, no.1, 1972, pp.19–20.

60 Vardys, 'Modernization and Baltic Nationalism', pp.37–8; Uustalu, p.360.

61 Misiunas and Taagepera, pp.106–7, 230–33; Vardys and Sedaitis, pp.64–5; Remeikis, *Opposition to Soviet Rule*, p.64

62 Lithuania, for example, had a virtual monopoly of the production of household electric metres and TV channel changers in the Soviet Union, produced about half the baths and fittings for Soviet housing, and a disproportionate share of other products. In turn it was dependent on other parts of the Soviet Union for its oil, natural gas, coal, cotton and wood for manufacturing. Vardys and Sedaitis, pp.67–8.

63 Moscow argued that higher standards of living were the result of Lithuania's annual trade deficit which was supported from the centre. Vilnius countered this argument by reference to Lithuanian transfer payments to Moscow, to Lithuanian expenditure on housing, health and pensions for Russian immigrants, and to the severe costs of deportations and forced labour in other parts of the Soviet Union. Vardys and Sedaitis, p.68; Zamascikov, pp.92–3; Itzchok Adirim, 'Realities of Economic Growth and Distribution in the Baltic States', and Armin Bohnet and Norbert Penkaitis, 'A Comparison of Living Standards and Consumption Patterns between the RSFSR and the Baltic Republics', both in Loeber et al., eds., pp.289–90 and 301–4; Michael Binyon, *Life in Russia*, (London: Panther, 1983), p.300; Clemens, p.66

64 Parming, p.408

65 In 1959 the Lithuanians were the first group in the Soviet Union to press for an environmental protection law and created a Lithuanian Nature Protection Association: Augustine Idzelis, 'Responses of Soviet Lithuania to Environmental Problems in the Coastal Zone', *Journal of Baltic Studies*, vol.X, no.4, 1979, pp.304–5; Vardys and Sedaitis, pp.69–71.

66 It might produce a high social rate of return but this would not be recognized in the Ministries' accounts

67 Remeikis, *Opposition to Soviet Rule*, pp.83–4; Misiunas and Taagepera, pp.240–1; Augustine Idzelis, 'Institutional Response to Environmental Problems in Lithuania', *Journal of Baltic Studies*, vol.XIV, no.4, 1983, pp.300–01; Idzelis, 'Response of Soviet Lithuania', pp.303–4

68 Idzelis, 'Response of Soviet Lithuania', pp.299–302; Idzelis, 'Institutional Response to Environmental Problems', pp.296–8

69 It was believed that the environmental damage on this site would be less; to help meet the concerns of environmentalists the most up-to-date methods of pollution control were promised. The refinery was completed in the early 1980s: Idzelis, 'Locational Aspects of the Chemical Industry', pp.53–5; Misiunas and Taagepera, pp.238–9; Remeikis, *Opposition to Soviet Rule*, p.83.

70 Idzelis, 'Response of Soviet Lithuania', p.301–2.

71 Idzelis, 'Institutional Response', pp.298–9.

72 Augustine Izdelis, 'The Socioeconomic and Environmental Impact of the Ignalina Nuclear Power Station', *Journal of Baltic Studies*, vol.XIV, Fall, 1983, pp.250–3; Vardys and Sedaitis, p.71

Section II

INDEPENDENCE AND THE POLITICS OF TRANSITION,
1985–1999

Chapter 3

THE ACHIEVEMENT OF INDEPENDENCE, 1985–1991

Few people in the 1960s and '70s were bold enough to forecast the demise of the Soviet Union. Moscow, it was generally believed, still posed a formidable security threat and a major ideological challenge to the West. There were no indications that the Kremlin was about to loosen its hold on power, either in the Soviet Union or in East Central Europe, even though there was some evidence of systemic weakness and long-term relative decline. The emergence of Mikhail Gorbachev in 1985 as General Secretary of the Communist Party of the Soviet Union (CPSU) raised hopes in some quarters that the decline could be reversed. Indeed Gorbachev showed that a new leader could make a significant difference, but not in ways that could have been predicted. In fact, the reform process initiated by Gorbachev developed its own anti-system dynamic and defied the best attempts of the Kremlin to re-assert its control.

The Soviet Union had not one but several Achilles' heels, the economy, the environment, and nationalities policy being among the most conspicuous. Arguably, the decades-long policy of Sovietization was the Soviet Union's greatest failure. As soon as Gorbachev loosened the controls the various Soviet nationalities began to reassert themselves amid calls for self-determination and the restoration of sovereignty. Although there was undoubtedly a connection between Gorbachev's assumption of power and the increased visibility of national movements, it would, nevertheless, be a mistake to over-emphasise the discontinuity of the mid-1980s. After all, a formidable dissenting movement had existed in the Lithuanian republic for two decades. What the world saw was a small group of active dissenters who were prepared to sacrifice themselves for their varied causes. For them imprisonment, consignment to psychiatric hospitals and loss of career were to be expected. What it did not generally see was that the active dissenters were the tip of the iceberg, enjoying the covert, and occasionally the open, support of large sections of the population. Gorbachev did not bring the dissenters and the reformers into life. But unintentionally he gave them the opportunity through *glasnost*, to form mass movements and to express their views openly before the world's media. 1985 did not, therefore, represent a clean break with the past. Soviet rule had never been regarded as legitimate; aspects of it such as the denial of religious rights, the restriction of human freedoms, and the rejection of national self-determination had been increasingly challenged by the dissident movement. However, the population had been divided over tactics between the very small minority who wished to challenge the regime overtly, and the pragmatic majority who wanted to work within the permissible limits to achieve change, or to resist unwelcome Soviet-inspired policies, or to conserve what they could of the Lithuanian landscape, natural resources, and architectural heritage. Even the Lithuanian Communist Party leadership used what influence or obstructive capacity it had to oppose some of the most damaging Muscovite policies. After Gorbachev, this division between the dissenters and the pragmatists became increasingly artificial as the limits on the permissible widened significantly.

SPONTANEOUS PROTESTS AND ORGANIZED DISSENT

The defeat of the resistance movement in the early 1950s enforced a degree of quiescence, if not acceptance, on the Lithuanian population. We cannot agree, however, with one historian's claim that there was 'total outward compliance' since for three decades after 1955 there were spontaneous mass protests in Lithuania against Soviet rule.[1] Sometimes these took the form of demonstrations of solidarity against Soviet oppression elsewhere, notably in Hungary and Czechoslovakia in 1956 and 1968 respectively. Occasionally sporting contests between Lithuanian teams and teams from other parts of the Soviet Union produced anti-Soviet demonstrations. Soviet attempts to celebrate anniversaries, notably the twentieth anniversary of Soviet rule in Lithuania in 1960, met with hostile demonstrations. When Romas Kalanta, a 19-year-old student, set fire to himself in 1972 as a protest against Soviet rule, there were violent and prolonged protests by university students and high school pupils. Sometimes demonstrations were accompanied by the singing of independent Lithuania's national anthem, the waving of the national flag, or the placing of flowers on national memorials. These episodes, to be sure, were relatively infrequent, and could be interpreted by the outside world as isolated events of no long-term significance.[2]

These outbreaks of open dissent were paralleled in the 1960s by an underground dissenting movement. This expressed itself in civil disobedience, the production of *samizdat*, and the use of mass petitioning. The movement's most important objectives were to protect the rights of religious believers, to advocate the principle of national self-determination, and to advance the cause of human rights. Protection of Lithuanian culture and the physical environment were also important aspects of the movement.[3]

A major impetus to the dissenting movement was the government's attack on the activities and self-government of the Roman Catholic Church in Lithuania, and the threat posed to its long-term survival by government restrictions. Archbishop Steponavičius had been exiled in 1961 for his refusal to implement government demands that priests be prevented from teaching religion to children in small groups, and that the state take over the appointment of priests. The government also proposed to make a drastic reduction in the number of priests in training so that there would be insufficient newcomers to replace those dying or retiring. These restrictions became law in 1966. Resistance initially took the form of petitions, the most notable being one to Brezhnev, General Secretary of the Communist Party of the Soviet Union (CPSU) in 1971, which was signed by 17,000 people against the denial of religious freedom and freedom of conscience. The signatories claimed that this denial violated both the Soviet constitution and international conventions to which the Soviet government was a party.[4]

Although these petitions had a useful publicity and morale-building effect, they were unsuccessful in changing Soviet policy. Dissidents began to look for more effective means of advancing their interests. They chose the model of *samizdat* borrowed from the Russian human rights movement. The *Chronicle of Current Events*, which started circulating in Moscow in 1968, was the obvious model for the *Chronicle of the Catholic Church in Lithuania* which began in 1972. The association between Lithuanian and Russian dissidents dated from the 1960s when Sakharov, Ginzburg, Kovalev and other Russian dissidents campaigned for the release of Baltic political prisoners. The *Chronicle of Current Events* first published news of dissent in Lithuania in 1970. Later, Russian dissidents in Moscow were able to make available to western correspondents copies of Lithuanian *samizdat*.[5] The objective of the *Chronicle of the Catholic Church in Lithuania* was at first simply to record abuses and violations of religious rights but it soon widened its scope to report on human rights abuses in general. It restricted this role after 1976 when the Lithuanian Helsinki Committee was formed and several other *samizdat* were in circulation.[6]

The growth of the religious rights movement was enhanced by the election of Cardinal Karol Wojtyla of Poland as Pope John Paul II in 1978. This inspired Lithuanian as well as Polish Catholics. The new Pope explicitly supported Roman Catholics in the Soviet Union, convincing them that their interests would not be subsumed in a Vatican policy of *ostpolitik*. Perhaps inspired by this an organization called the Catholic Committee for the Defence of Believers' Rights was formed in Lithuania in 1978 to document breaches of religious rights, such as dismissal from employment for attending religious services or prohibition on entry to university for religious students.[7]

Although dissent first took organized form in the religious rights movement it quickly became more diversified when a separate movement for the defence of national and human rights, composed of secular and religious activists, emerged in the 1970s. It is not always easy to distinguish between religious and national rights. The Roman Catholic Church in the nineteenth century was an important defender of national rights against Russianization. It was axiomatic for most Lithuanians that destruction of the Church's influence was a prerequisite for denationalizing the state. The Catholic religion was widely believed to be an indispensable element of the Lithuanian national identity and the Catholic Church the only remaining institution capable of defending Lithuanian culture and traditions. Hence the unremitting government campaign against the Church was a vital element in Sovietization. Nonetheless, a separate movement for the defence of human and national rights emerged in the mid-1970s, composed of secular as well as religious activists. After its formation the *Chronicle* began to focus exclusively on the defence of religion. The *Chronicle* was joined by around a dozen other *samizdat* publications having both secular and religious orientations. At the heart of the new movement was the Lithuanian Helsinki Group formed in 1976, one of a number established in the Soviet Union. The Final Act of the Conference on Security and Co-operation in Europe (CSCE) held in Helsinki in 1975, in effect offered the Soviet Union international *de facto* recognition of its territorial gains in the Second World War in return for the observation of human rights and the rights of peoples to self-determination.[8] The Lithuanian Helsinki Committee was established by a group of individuals to monitor the Soviet government's adherence to its undertakings. It maintained links with similar committees in other parts of the Soviet Union and issued public statements on alleged human rights violations. It linked the struggle for national rights to the Helsinki process, calling the Soviet Union to account for failing to observe the internationally accepted standards to which it had solemnly subscribed.[9] The ecumenical character of the Lithuanian group is shown by its founder members which included two in the Catholic tradition, the Reverend Karolis Geruckas and Viktoras Petkus, two secular figures, Tomas Venclova and Ona Lukauskaitė-Poškienė, both poets and leftist intellectuals, and Eitan Finkelstein, a Jewish scientist and a leader of the Jewish emigration movement.[10]

The formation of the Helsinki Committee was paralleled by the flowering of the *samizdat* movement. Two took the titles of dissident journals published during the Lithuania renaissance, *Aušra* (1975) and *Varpas* (1977). Other notable publications were *Perspektyvos*, *Dievas ir Tėvynė* (*God and Fatherland*), and *Laisvės Šauklys* (*The Upholder of Freedom*). Unlike the *Chronicle*, a number of them were forced to stop publication and their editors arrested and sentenced. Yet there were more dissident publications and demonstrations in Lithuania in relation to population than in any other Soviet republic. There was also much unofficial publishing of books and periodicals for religious needs, including thousands of copies of prayer books.[11]

By the late 1970s the dissident movement in Lithuania was growing significantly and the Kremlin was beginning to see it as a danger which had to be snuffed out. Protests against the Afghan war, messages of solidarity to Lech Wałęsa, whose activities in the Polish opposition were closely followed in Lithuania, many petitions and protests seeking religious rights, culminating in a well-supported petition in 1979 demanding the observance of rights to education in the Lithuanian language, all represented a challenge to the Kremlin which could not be ignored. The last straw was a statement published in 1979 signed by representatives of all three Baltic states and supported by Russian dissidents including Sakharov, demanding that the Soviet and both German governments publish the full text of the Molotov-Ribbentrop Pact on its 40th anniversary, and requesting that the signatories to the Atlantic Charter condemn the Pact and its outcome. What infuriated the Kremlin was the accompanying demand that the Pact be nullified and Soviet troops withdrawn from the Baltic republics. As Shtromas has convincingly argued, this marked a 'new departure' in dissident politics since it went beyond calls for human and religious rights to demands for the restoration of Baltic independence. The independence movements from 1988 to 1991, Shtromas suggests, sprang from this significant 1979 petition. The petition was reinforced by a call in *samizdat* on Christmas Day 1983 that a European Parliament resolution demanding decolonization in the Baltic states be placed on the agenda of the CSCE. In fact, the challenge of dissent over Afghanistan and human and national rights was now so formidable that a full-scale crackdown was imposed by the Kremlin, even though this meant a damaging blow to the Helsinki process.[12]

The number of demonstrations and *samizdat* publications, coupled with the severity of the Soviet repression, suggests that the dissident movement in Lithuania was not confined to a small minority of intellectuals but embraced blue collar and farm workers as well. The key to achieving working-class participation was religion, and it was the Catholic rights movement which above all distinguished Lithuanian dissent from other dissident groups in the Soviet Union.

The environmental movement also mobilized support for dissent. Its campaign against water and atmospheric pollution broadened to include defence of the architecture, historical monuments and landscape of Lithuania. A decision in 1967 to replace thousands of farmsteads with new rural settlements generated intense discussion about the impact of such radical change on the identity and traditions of the Lithuanian people, particularly since the purest forms of Lithuanian architecture were believed to be found in rural areas. Since environmentalism could often be pursued without an overt challenge to the regime, it encouraged participation in the reform movement of those who were primarily conservators, and who pursued their objectives through existing institutions, including even the Communist Party itself.[13] It has been pointed out that the reform movement *Sajūdis*, founded in 1988, sprang from these within-system reformers rather than from the dissidents. Ultimately, however, even this kind of in-system reform involved the wider issue of Lithuanian self-determination and Lithuanian relations with Moscow.[14]

The Lithuanian mentality on the eve of Gorbachev's reforms reflected, on the one hand, a robust sense of identification with western culture and Lithuanian traditions and, on the other, frustration at the Kremlin's continuing power to strangle Lithuanian aspirations. The extent of involvement in reform on the part of individuals depended on a number of factors such as personality or career calculations. Environmental or ethnographic groups might be perceived as offering less of a political challenge to the regime than human and religious rights activists or advocates of curriculum reform in schools and universities, particularly in the disciplines of history or philosophy. Wherever reform or dissent involved obstruction of the process of Sovietization it was regarded as too risky by conservative reformers and was left to the system rejectionists. However, the latter were not totally isolated from the rest of society since they attracted broad sympathy for their general objectives.[15]

LITHUANIA ON THE EVE OF *PERESTROIKA*

There was considerable potential for change in Lithuania in the immediate pre-Gorbachev period. Few of the younger generation of Party members were either militant ideologues or total conformists. Some were criticised for allowing nationalists into positions of influence in education and publishing. Their adaptation of Kremlin commands to Lithuanian realities were 'acts of considerable sophistication and finesse'. Between the dissidents and the Party was the majority of Lithuanians who did not challenge the regime openly, or did so on rare occasions, or on particular issues such as the environment. Though they disagreed on tactics and objectives, what the vast majority of Lithuanians shared was a determination to defend and preserve Lithuanian national culture, identity and traditions. They were, in Shtromas's term, conservationists. They resisted where they could, they accommodated when they had to. They could probe the limits of Moscow's tolerance. But they could not continue to oppose if Moscow was determined to force an issue. Only the dissidents were prepared to face the ultimate consequences of the pursuit of principle, but many more privately agreed with their stand.[16]

Compared with the other two Baltic states Lithuania was in a rather stronger position to resist Kremlin pressures. Immigration of Russians was relatively low and ethnic Lithuanians predominated in the population, even in the cities. The birth rate among Lithuanians was higher, and in the 1970s and 1980s there was still a rural labour surplus to draw on to meet the needs of the industrial workforce. The Lithuanians could also gain confidence from their long history as a separate state.[17] Like the Estonians and Latvians, the Lithuanians were beginning to enjoy increased western contacts, particularly with their expatriate communities in Western Europe and North America through family visits in both directions. Lithuanians also benefited from broadcasts from the Vatican and the greater sense of participation in the universal Church after the election of John Paul II. There were also scholarly and cultural exchanges, and greater availability of western books, films, theatrical productions and popular music. This was reassuring confirmation that they were part of the Western tradition, and at the same time reinforced their alienation from Soviet Russian culture and ideology. Lithuanian organizations in the United States were able to influence American foreign policy, helping to persuade the United States government to finance broadcasting to the Baltic states and to make occasional supportive statements for the Baltic cause. This kept alive the Lithuanian dream of independence. Further encouragement was derived from the Solidarity movement in Poland and from evidence of declining Soviet confidence in dealing with opposition in its Eastern European satellites. Moreover, Soviet aggression in Afghanistan highlighted the nature of the Moscow regime and strengthened Baltic interpretations of their status as occupied and colonized countries.[18]

Two factors in particular impinged on Lithuanian minds in the first decade of the 1980s, illustrating the Soviet ability to restrict Lithuanian freedoms and ambitions. The first was a threat to the Lithuanian language and, by implication, to Lithuanian identity. The defence of religious and human rights in the 1970s had led to counter-measures from the Kremlin, including the imposition of Russian language teaching in kindergarten as well as the first year of primary school. There was great pressure, too, for the increased use of Russian in the media, public administration, scholarship, the schools, and business enterprises. Lithuanians fought back by preserving an extra year of school education compared with the Soviet norm, thus restoring the time available for Lithuanian language and cultural teaching. The population enthusiastically supported the use of the native tongue in schools, broadcasting, and academic life. The intimate connection between language and identity was never forgotten.[19]

Secondly, Lithuanians had been able to enjoy the fruits of rapid economic growth in the 1960s. This put them and the other Baltic states at the top of the economic ladder among the Soviet republics. But, beginning in the 1970s and continuing into the 1980s, the rate of growth began to slow and shortages of goods in the shops became a regular feature of life. Food shortages were blamed on exports of food to other parts of the Soviet Union, raising charges of colonial exploitation by the centre. Housing was also in short supply. The relative lack of consumer goods meant that there was too little to buy and no alternative but to save. Consequently Lithuanian savings accounts were bulging. This situation caused frustration rather than happiness, showing that the Soviet economy was incapable of achieving a steadily improving standard of living for all its inhabitants. At the same time the gap between Lithuanian and Scandinavian standards of living continued to widen. For many, Soviet control was just tolerable if the economy prospered. If not, the advantages of economic autonomy seemed increasingly appealing.[20] To sum up, a number of factors combined to form a potent mixture in the Lithuanian psyche— economic frustration, cultural apprehensions, widespread popular disenchantment with communism, hostility to Sovietization, increased contacts with the outside world, a new post-Stalin generation, activist dissent, environmental anxiety, the memory of independence and the strength acquired from surviving Stalinist repression. The introduction of *perestroika* and *glasnost* into this unstable compound presented Lithuanians with an unexpected opportunity which they seized with energy, inventiveness and courage.

GORBACHEV, *PERESTROIKA* AND *GLASNOST*

The demand for independence was the outcome of a short but very intense process, the culmination of a movement of national re-awakening which emerged very cautiously but quickly flowered under the exhilarating conditions of the new-found freedom of expression. For most Lithuanians the movement began in 1988 as an attempt to find ways of making more effective Gorbachev's concept of *perestroika*. In less than two years, however, the movement was responsible for the declaration of independence issued by the Supreme Soviet (or parliament) of the Lithuanian republic. While there were differences about tactics among the supporters of the declaration, all agreed on the principle that Lithuania should reclaim the sovereignty removed from it illegitimately in the Molotov—Ribbentrop pact and by the events of 1940. Our central task is to account for the radicalization of the reform movement in Lithuania in those two years. This can be explained by reference to the interaction between a number of political forces which, in varying combinations and under the impact of events, headed inexorably in a radical direction. Among these forces were the moderate reformers who founded *Sajūdis*, the more radical Kaunas faction in *Sajūdis*, the intransigent radicals of the Lithuanian Freedom League who sought independence from the outset, the Lithuanian Communist Party which split under the pressure of events, the pro-Soviet group

Yedinstvo (unity), composed mainly of Russian and Polish-speaking inhabitants of Lithuania and, finally, the hitherto all powerful CPSU led by Gorbachev which was ditched by its leader in an increasingly desperate struggle for survival. A second linked problem is to explain the success of the Lithuanian independence movement. How, in other words, the claim for independence expressed in March 1990 was translated into reality within the short period of 18 months.

In attempting to reform the Soviet economy Gorbachev was building on the rather shallow foundations laid by his mentor Andropov. Recognizing the depth of Soviet problems and acknowledging that his early attempts at reform were yielding few rewards, Gorbachev embarked on a new course whose destination proved to be both unexpected and unwanted. As an idealistic communist Gorbachev had little tolerance or understanding of nationalism which he saw as one of the products of a bourgeois mentality and therefore destined for extinction. Yet it was his policies that stirred a repressed nationalism into life in almost all parts of the Soviet Union, and particularly in the republics of the peripheral Baltic and trans-Caucasus regions. Why did policies intended for the purpose of economic revival produce these vigorous movements for national self-determination? And how was it possible for them to succeed in a state which had exercised the most iron grip on nationalist manifestations over the previous seven decades?

In embarking on the path of reform Gorbachev had a clear objective in view, which was to revive and modernize the ailing Soviet economy. It was clear to the leading figures in Party and government that the economy was stagnating, that incomes per capita had stopped rising, and that the burden of the military industrial complex was becoming increasingly insupportable. The question was how to achieve a programme of socialist renewal, because at that stage there was no intention of introducing a free market economy. Social problems were also evident, such as widespread corruption, malingering in the work-place, alcoholism, and demoralisation of the labour force. Shortages, coupled with very high levels of savings, produced increasing discontent among consumers. Gorbachev's first attempts at economic reconstruction (*perestroika*) were not dissimilar from those of Andropov, namely, efforts to increase labour discipline, to dismiss dead wood in the government and party, and to energize industrial managers and central planners. When these yielded limited results, Gorbachev flirted with the idea of the self-financing of enterprises, removing from them some of the planning targets in the hope that decentralisation and the introduction of very limited market mechanisms would stimulate growth. But these attempts to galvanise the economy not only achieved little in the first two years of Gorbachev's rule but foundered on the rock of the 'conservatism, venality and sloth' of the Communist Party itself and on the half-baked nature of the reforms.[21]

Glasnost or openness was introduced as a means of energising the reform process. It was hoped that freedom of expression would stimulate ideas and create a culture of complaint against lethargic and unresponsive bureaucrats who were obstructing desirable change. *Glasnost* would help the Party leadership mobilize popular sentiments to sweep away resistance to *perestroika*. But, of course, freedom of expression is indivisible. Criticism in the newly enfranchised press and in public gatherings was not confined to reactionary party bureaucrats but focused on a multitude of issues on which public discussion had been suppressed. The water which had been lapping around the top of the dam now poured over in a massive flood sweeping Gorbachev into frantic improvisations to save what he could of the Soviet Union. It led him, for example, to seek constitutional changes which would ultimately sideline the Communist Party itself, and to initiate discussions on replacing the command economy by a free market system.[22]

In Lithuania, as in the other Baltic republics, open discussion focused on many important issues, not least the topics highlighted for years by the dissidents, such as the illegitimacy of Soviet rule and the persecution of the Roman Catholic Church. Indeed, it was this very strong tradition of dissent which ensured that *glasnost* was more widely exploited in the Baltic republics than in most other parts of the Soviet Union. In Lithuania *glasnost* was utilised not mainly to support *perestroika* (though the possibilities of economic acceleration were seized on) but to question the very basis of Soviet rule and to seek political and cultural freedom. So far from being strengthened by *perestroika*, the Soviet Union was shaken by the myriad pressures for change. Natan Sharansky declared that Gorbachev had got himself into a terrible dilemma—he could save the empire or the economy, but not both. By his chosen methods for invigorating the economy he actually unleashed the passions for self-determination among the subject republics. If he wished to save the Union he would have to turn the screw of repression which he had only just loosened. This, in turn, would have been the kiss of death for *perestroika*. Almost to the end of his tenure of power Gorbachev insisted that the Union was necessary for the success of *perestroika* and that republican independence or even wide-ranging autonomy would kill it. He could never admit that democratisation meant the end of empire.[23]

Reform in the Baltic states was also energised by the priority given to the region in the Kremlin's post-*perestroika* economic planning. The government believed, correctly, that the Baltic region was particularly receptive to change and that it could show the rest of the Soviet Union how to apply reform principles. Lithuania's experience of a market economy and her trading contacts with the West in the interwar period, the higher levels of productivity and skills than in most parts of the Soviet Union, and the fact that the Baltic republics had been 'laboratories for [economic] experimentation' in the recent past prepared them for leading roles in Soviet economic reform. By encouraging the Baltic peoples to seize newly available opportunities, Moscow hoped that the Baltic states would provide a model for the rest of the Soviet Union. But the states themselves were determined to use these opportunities to pursue their own interests. They would test Soviet tolerance to the limit by claiming the maximum possible independence in economic management.

It was four Estonians who in 1987 first extended the idea of enterprise cost accounting to the republic economies themselves (the so-called 'four man proposal), giving the republics self-management at the expense of the central Moscow ministries. Arguably, local control of economic policy, including fiscal, monetary, trade and investment policy, would improve economic efficiency. Implementation of policy would also require republican ownership of land, natural resources, and all state property. The ideas behind this movement spread from Estonia to Lithuania and were adopted by *Sajūdis* in the summer of 1988 as part of its broad reform programme. For Lithuanian reformers, as opposed to *perestroika* supporters in Moscow, economic self-management was not only a means of improving economic output but a decisive step towards local autonomy. Reformers quoted Article 76 of the Soviet constitution in their support: that each Union republic had the obligation 'to ensure comprehensive economic and social development' on its territory. The principle of economic self-management was conceded by the central authorities in the summer of 1989 and enacted by the Supreme Soviet in November of that year, for introduction in January 1990. It seems probable that this concession was designed to avert the possibility of the Lithuanian Communist Party's breaking its connections with the CPSU.[24]

THE RADICALIZATION OF REFORM

The early, essentially mild, expressions of reform were the starting point of a process which generated a profound challenge to the integrity of the Soviet Union, at first by means of claims for republican sovereignty and then by demands for the restoration of independence. Why this occurred can be explained by reference to the interaction between significant participants in the political process.

A fundamental condition of these developments was the growing freedom of expression both in the press and in public meetings which was a consequence of *glasnost*. In Lithuania the media reported and commented on a number of events which aimed to commemorate important historical episodes and to bear witness to Lithuanian national identity. In 1987 and 1988 a number of so-called 'calendar' demonstrations were organized to commemorate significant dates in Lithuanian history, notably 23 August, the date of the Molotov-Ribbentrop pact, and Independence Day on 16 February. Given its stern resistance to Sovietization the Lithuanian Freedom League (LFL) not surprisingly took advantage of *glasnost* to organize these early demonstrations calling for the annulment of the Molotov-Ribbentrop Pact and the restoration of Lithuanian independence.[25] For the more 'moderate' reformers, Senn has argued, the LFL was too militant, and for the LFL, the moderates were too cautious. The LFL was against participating in elections to 'illegitimate' Soviet institutions since participation meant collaboration with the occupier. *Sajūdis*, on the other hand, believed in using existing institutions to further the cause of reform. Most Lithuanians agreed with them. However, in the course of the next two years the moderate reformers and even the great majority of the Communist Party of Lithuania (CPL) were pushed to the right by events, and ultimately accepted the LFL's independence programme though not its uncompromising refusal to cooperate with communists or ex-communists.[26]

The growing freedom of expression and the radicalism of the LFL were not the only contributions to the swelling tide of reform. There were three other notable examples, each of which was rooted in, and heavily influenced by, the Lithuanian cultural and scientific intelligentsia. The first was the environmental movement which, as we have seen, had been actively fighting damage to Lithuania's physical environment for some years. Since most of the environmental degradation resulted from Soviet central planning and was mainly associated with Moscow-controlled industries it was inevitable that this movement would have decentralizing overtones. Of particular concern was the future of the Ignalina nuclear power station and whether a third reactor should be built in the aftermath of Chernobyl in 1986. The possible expansion of an already heavily polluting chemical industry and plans for oil drilling off the coast were other critical issues for environmentalists. A second important contribution to the process of reform was the attempt by writers, artists and journalists to throw off government controls on free expression in the creative arts.[27] Finally, the intelligentsia began to call for a new history, free from orthodox communist interpretations and exploring anew all those 'black holes' in the Lithuanian past which had been the subject of Communist lies and distortions. Linked with this attempt to recapture Lithuanian history for free research and open debate was the movement to preserve and restore historic monuments, which bore witness to Lithuania's pre-communist past, and which had often been the target of communist destructive impulses or malign neglect. These interlocking movements aimed to preserve Lithuania's physical environment, her historical identity and her national culture. Clearly *glasnost* was producing results which Gorbachev could not have expected or wanted. However, had they simply provided the opportunity for the population to let off steam while at the same time stirring the economy into more vigorous life, a reformist Kremlin might have tolerated them.[28]

The Lithuanian Communist Party (CPL) under two conservative general secretaries, Griškevičius until December 1987 and Songaila until autumn 1988, was not so sophisticated, attempting to contain *glasnost* and using the army and police to sweep the streets on the occasion of the calendar demonstration on 16 February, 1988. The CPL was determined to resist what it called the 'nationalist and clerical extremists'. It is probable that the embryonic reform movement could have been crushed at that early stage had it not been for the protection it received from Gorbachev and his Muscovite reformers. As late as May 1988 a demonstration to commemorate the May 1948 deportations was dispersed by the police. But the longer the movement survived without serious adverse consequences, the more confident it became, and its activities became more daring and experimental.[29]

An influential but hitherto relatively silent minority began to participate openly in reform in the summer of 1988. These were the closet reformers in the CPL, who now came out and added their voices to the calls for reform. Together with non-communist representatives of the artistic and scientific intelligentsia they believed that the pace of reform was too slow and a new ginger group was required to speed things up—all in the name of *perestroika*. Accordingly they formed an Initiative Group to develop the Lithuanian Movement for Reconstruction (*Lietuvos Persitvarkymo Sajūdis*), commonly called *Sajūdis*. Among the leading lights of *Sajūdis* were the musicologist Vytautas Landsbergis, the philosopher Arvydas Juozaitis, the physicist Zigmas Vaišvila and the economist Eduardas Vilkas. From the the second rank of the Communist Party came Kazimiera Prunskienė and Bronius Kuzmickas, both academics, and Romualdas Ozalas, a philosopher and essayist. The purpose of the organization was to support *perestroika*. It declared its moderation by refusing to admit dissidents and by welcoming members from the CPL. As a general reform movement it offered a home for the members of various single issue reform strands such as environmentalism, heritage, religious freedom and human rights. The participation of CPL members aroused suspicions in the Lithuanian Freedom League that the movement was a tame Party front designed to direct reform into innocuous channels. The increasing radicalism of *Sajūdis* showed this suspicion to be misplaced. Instead of being a 'harmless outlet for nationalist feelings' the movement became their focus.[30]

There is clear evidence of radicalization in the summer and autumn of 1988. We know that national identity had remained strong through the years of sovietization. It took a little time for the mass of people to realise that *glasnost* and *perestroika* offered them an unexpected opportunity to express their deep sense of national identity. With each newspaper article and public demonstration the sense of solidarity and shared ideas became more intense. And the absence of repression by the regime encouraged more daring expression of long-held convictions. Gorbachev used the movement against the conservatives in the CPL but it soon became clear that *Sajūdis* would not stop at the boundaries set by him, and that his objectives and theirs were radically different. But by this time his benign tolerance of the movement had allowed it to grow to such proportions that it could not be easily steered. The putative puppet had achieved a life of its own. Various commissions were set up by *Sajūdis* to formulate policies for approval by its founding convention. They expressed the ideas of moderate reformers: to make the bureaucracy accountable and to end its privileges; to save the cultural heritage; to open up historical research and discussion; to institute the rule of law; to introduce economic self-management; to preserve the Lithuanian environment, and to recognise Lithuanian as the official language of the state. These ideas were discussed and elaborated at a series of giant demonstrations during the summer of 1988 on the occasions of the Lithuanian delegation's departure for, and return from, the 19th CPSU Congress in Moscow, and of the commemoration of the Molotov-Ribbentrop pact, which attracted a crowd of over 200,000.[31]

However, new demands were raised at the later rallies, taking the reform movement into unfamiliar territory. Linked to a request to publish the text of the secret clauses of the Molotov-Ribbentrop pact was a demand for the declaration of republican sovereignty. This demand was accompanied by requests for the rehabilitation of deportees and the release of all political prisoners. All three partly reflected the influence of the LFL. But of more immediate significance was the dawning realisation that Gorbachev's notion of *perestroika* was different from that of the Lithuanian reformers. At the CPSU congress he had demurred from offering the degree of economic autonomy being demanded by the Balts. Moreover, he insisted that reform would have to be implemented within the existing structure of the Soviet state. This was a theme taken up by Gorbachev's close colleague and emissary to Lithuania, Alexander Yakovlev, who visited the republic in August 1988 to assess the situation. While giving clear signals to the CPL to made concesssions to national consciousness, and stating that the Kremlin was on the side of reform, he cautioned against allowing nationalism to imperil *perestroika*. He shared with Gorbachev the belief that reform should not be permitted to break up the Union. His visit gave enormous confidence to *Sajūdis* in its dealings with the CPL, convincing it that the tide was flowing strongly in its direction, but also making it aware that there were clear limits to the Kremlin's tolerance.[32]

The radicalization of reform picked up pace in the autumn of 1988 under the impact of a series of events. The first of these, a violent break up by the police of a demonstration led by the LFL in Gediminas Square in Vilnius on 28 September, had profound consequences. It was interpreted as an attempt by the conservatives in the CPL to reassert control. But instead of achieving this, it threw together the radical LFL and *Sajūdis*, permitting their leaders to share a platform for the first time. The radicals' analysis of the situation and their political prescriptions began to sound much more convincing. Secondly, as a result of impassioned popular opposition to the actions of the authorities, the leadership of the CPL, seemingly encouraged by Moscow, replaced Songaila as General Secretary by Algirdas Brazauskas, hitherto secretary for industrial affairs. Brazauskas was experienced, flexible, courageous and popular, in fact the most popular politician in Lithuania for many years. His actions suggested that he intended to commit the CPL to a reform programme. He seemed to share the objective of *Sajūdis*, namely to achieve sovereignty for Lithuania, but disagreed over timing and tactics. In the short term, however, his actions were constrained by a conservative central committee. His elevation was welcome to *Sajūdis*, since it meant a more sympathetic ruling party and an indication that the obstacles to reform were weakening.[33]

A third factor in the radicalization of reform was the strategy of Gorbachev himself. On the one hand he claimed *Sajūdis* as a positive force in support of *perestroika*; on the other he believed that the strengthening of reform would ultimately make Lithuanian self-determination irrelevant. But then, on the very weekend of *Sajūdis'* founding congress in October 1988, he introduced his proposals for constitutional reform which shook the Lithuanian reform movement and committed it explicitly to the restoration of Lithuanian sovereignty. Gorbachev's proposals involved some remarkably progressive features by Soviet standards, but also clarified his thinking about the possibilities of republican autonomy or self-determination. In brief Gorbachev proposed a new central parliament composed of a Congress of People's Deputies which would meet periodically, and a new Supreme Soviet elected by the Congress which would meet continuously as an active legislative body. There would also be a new executive presidency, which was destined to be filled by Gorbachev himself. *Sajūdis* believed that Gorbachev had reneged on his promise to the 19th Congress of the CPSU in June to propose greater autonomy to the republics, transferring to them some of the powers of the Central Committee of the CPSU. The Lithuanian reformers also condemned the absence of equal representation for the republics in the central legislature. More serious still was the fear that the new constitution removed the right to secede under Article 72 of the existing Soviet constitution. Even when they were reassured on this point the Lithuanians were told that the central government would have the power to rescind any constitutional changes proposed by the republics and to impose what were called 'special powers of administration' when it saw fit. These proposals outraged the Lithuanian population and provoked a petition signed by 1.8 million people calling for the proposed amendments to be postponed for further consideration. There was now no doubt at all that Lithuanian expectations of sovereignty and self-determination did not form part of Gorbachev's *perestroika* programme. He had made it abundantly clear that his purpose was to strengthen the central administration's authority to carry out the desired reforms. Decentralization in his view would weaken that capacity.[34]

Sajūdis and the Kremlin were now on a collision course and *Sajūdis* could not back down, nor did the centre have the power to compel it to, short of resorting to a Stalinist level of terror, which was entirely contrary to the objectives of reform and totally destructive of rapprochement with the West. The opposing positions had been clarified by the *Sajūdis* founding congress in Vilnius on 22–23 October, 1988. The congress was composed of over one thousand delegates, the overwhelming majority of whom were university graduates, mainly artists, writers, professional people, and scientists. It also included members of the LFL who pressed for radical change, and a new grouping within *Sajūdis* known as the Kaunas faction, from the fact that many of its members were from Kaunas or the provinces. Reports suggested that it was strongly anti-Soviet and in the van of the independence movement. It was undoubtedly another radicalising influence on reform in Lithuania.[35]

At the founding congress, Landsbergis emphasised that *Sajūdis* was a movement, not a party, though it was composed of various groups which could evolve into parties. It was committed to peaceful change and to the parliamentary process. Its platform was very broadly-based, including many of the proposals being discussed in the summer concerning the protection of Lithuanian culture, language and identity and the establishment of self-management. It added a number of other planks such as an end to the current organization of military service in the Soviet army and a guarantee of the rights of minorities. The really significant new development, however, was the proposal which made very explicit the differences between *Sajūdis* and Moscow, whatever the professions of support for Gorbachev's *perestroika*. This was the claim for sovereignty, meaning in practice the supremacy of Lithuania's laws over Soviet laws. A sovereign Lithuania could permit Moscow to have responsibility for functions such as foreign affairs, defence and all-Union budgets, but this arrangement would be dependent on agreement between the Kremlin and Lithuania. Lithuania could delegate these powers to the centre or withdraw them if they were exercised unsatisfactorily. The republics could also choose secession from this **voluntary** federation. For *Sajūdis*, reform could best be achieved in the context of republican sovereignty; for Gorbachev, a strengthened centre was essential to achieve his reform objectives.[36]

The two sides now contested for popular support. *Sajūdis* hoped that it could carry the CPL with it in favour of republican sovereignty and present a united front to Gorbachev. In this it was disappointed. In November the Lithuanian Supreme Soviet, which was dominated by the Communists, failed to follow the Estonian example and declare Lithuanian law sovereign.[37] The indignant *Sajūdis* response was that Lithuania's will was the highest law, that Soviet law could be vetoed by the Lithuanian parliament, and that Lithuania wanted the right to legislate for itself within the Soviet Union. Failure to achieve its objectives at the Supreme Soviet was not soothed by significant government concessions to nationalists, concessions which would have been hailed as major achievements only a few months before.[38]

The situation in late November 1988 showed that by exercising influence on the CPL Gorbachev had put a brake on the Lithuanian march to sovereignty. The CPL, which had been drawing closer to *Sajūdis*, had been compelled to distance itself from the movement. In the event radicalization did not come to a halt owing to the very process which Gorbachev had himself devised a month earlier, namely the initiation of a new electoral process to create reformed Soviet institutions. Elections to the Congress of People's Deputies were scheduled for March 1989 and *Sajūdis* intended to endorse candidates standing for Lithuanian sovereignty. This was an excellent opportunity to test the support of public opinion for the *Sajūdis* position and at the same time show the meagre popular endorsement for the Communist candidates. The election campaign widened the debate to the entire electorate, taking it out of the hands of the small circle of intellectual and cultural leaders who had hitherto shaped and dominated the discussions. The results showed beyond any doubt that the electorate shared the position of *Sajūdis*. Candidates endorsed by the movement, communist and non-communist alike, won 36 out of the 42 seats.[39] This result gave enormous impetus and encouragement to the movement. No one could now convincingly assert that reform in Lithuania was the preserve of a few extremist intellectuals and dissidents. The whole population had demonstrated their engagement, which went beyond the rallies of the previous summer, impressive though they were.[40]

The defeat had a salutary effect on the CPL. On the one hand, there were old-fashioned demands to take control of the situation and to restore order. On the other, there was a growing recognition that the survival of the Party depended on adjusting its position to the popular mood and drawing closer to the reform movement. The stark choice was between communism or Lithuanian nationalism. An increasing number of members began to choose the Brazauskas line of endorsing the principle of Lithuanian sovereignty and condemning the Molotov-Ribbentrop pact. In May 1989 the Lithuanian Supreme Soviet followed this lead by condemning the pact and asserting the Lithuanian right to self-determination.[41] It also adopted the law on economic self-management, giving Lithuania control of its own economy subject to negotiations with the centre. By this action the CPL effected a virtual consensus in Lithuania on the question of Lithuanian self-determination and self-dependence. However, it should be emphasised that there were relatively few voices at that stage calling for Lithuania to exercise its sovereignty by claiming complete independence from the Soviet Union. For example, the Baltic Assembly of the Popular Fronts meeting in May refrained from calling for secession since 'complete self-dependence' could be secured within the Soviet Union.[42]

THE RESTORATION OF INDEPENDENCE

The decision to restore independence arose out of events at the Congress of People's Deputies in Moscow. At least four major factors contributed to this decision: the acceptance by the Congress of the illegality of the secret clauses of the Molotov-Ribbentrop pact; the decision of the CPL to secede from the CPSU and to take its place in a multi-party system in Lithuania as the only way to restore its political position; the visit of Gorbachev to Lithuania in January 1990, which exposed the profound differences between his thinking and that of the Lithuanians; and, finally, the overwhelming support of the Lithuanian electorate in the elections to the Lithuanian Supreme Soviet in February/March 1990 for candidates committed to seeking independence. Each of these factors requires a few words of explanation.

The Congress gave Gorbachev the chance to accept a new union treaty which would turn the Soviet Union into a voluntary federation.[43] But Gorbachev turned down the opportunity, refusing to shift from his objective of strengthening the union as a means of entrenching *perestroika*, and implicitly rejecting the demand for self-dependence except in the field of economic autonomy. For the Balts a key objective was to persuade the Congress to investigate the official documents on the Molotov-Ribbentrop pact and to endorse their own conclusions as to its illegality. This was not a straightforward matter; the Kremlin was initially reluctant to admit even the existence of the secret protocols since the issue was potentially explosive.[44] After persuasion from supporters of the Baltic case such as Roy Medvedev and Yuri Afanasyev, and a vigorous campaign by the Baltic delegations, the Congress eventually agreed to establish a commission under the chair of Alexander Yakovlev. On 20 July, 1989 the Commission concluded that the pact did indeed contain secret clauses to allocate the Baltic states to the Soviet sphere of influence and called on the Congress to invalidate them as illegal under international law. In December 1989 it finally did so. Yet Gorbachev still refused to accept that the Baltic republics had not joined the Soviet Union voluntarily. In support of this view *Pravda* published the original texts of the appeals by the assemblies of the Baltic states to join the Soviet Union in 1940. Yakovlev, who finally came round to accepting the illegality of the secret protocols, defiantly asserted that the protocols were not the basis for the 'legal and political status of the Baltics'. Both Yakovlev and Gorbachev implicitly based their case on the elections of July 1940 which, as we saw earlier, were sham elections based on force.[45]

The shabbiness of the case presented by the Kremlin and the exposure of the secret clauses by a commission of the Soviet parliament itself gave enormous moral authority to the popular fronts in the Baltic states. It propelled them into the next stage of their quest for self-determination. On 22 August a commission of the Lithuanian Supreme Soviet declared the 1940 incorporation of Lithuania into the Soviet Union to be invalid and on the following day, 23 August, two great events occurred which shifted the Baltic reform movement on to a new plane. Firstly, a massive and spectacular 'human chain' linked up the three Baltic capitals in an impressive show of protest against the incorporation of 1940. Secondly, *Sajūdis* courageously drew the logical conclusion from the invalidation of the secret clauses and called on Lithuania to become independent again. Although it desisted from claiming immediate independence it made the incontestable point that relations between the Soviet Union and Lithuania should be based in future on the treaty of 12 July, 1920, which was a treaty between two **independent** states. The response of Kremlin hawks, an implicit threat of retaliatory violence, strengthened the determination of *Sajūdis* to press ahead in its search for independence. Even *The Guardian*, a strongly pro-Gorbachev British newspaper, shared the logic of the *Sajūdis* position. 'The Soviet Constitution' it declared 'says that the Soviet state is a unit of free and equal peoples which have united of their own free will.' If they had not, the union was not voluntary, and if nations were kept under constraint 'then that is no longer a union'; it was a superpower, an empire or a prison of nations.[46]

The second major reinforcement for the independence drive came from the decisions of the CPL in December 1989. Having belatedly supported the declaration of Lithuanian sovereignty, the Party again found itself adrift of public opinion after the events of August. It therefore became imperative to adopt policies which met the public mood and to cut itself off from its greatest source of unpopularity, namely its connection with the CPSU. The first stage of its re-orientation was to convert Lithuania into a multi-party state and to drop the hitherto sacrosanct leading role of the Communist Party. Deprived of its Leninist rationale as the advance guard of the proletariat, the Party could only gain legitimacy through a victory in properly constituted elections. It therefore proposed a democratically elected Supreme Soviet, for which elections would take place in February 1990. In this way the Party lost its constitutional monopoly of power and a multi-party system was legalized.[47]

The Party leaders, especially Brazauskas, recognised that it would be the kiss of death for the Party to fight an election while still part of the CPSU. There had been a great exodus of members from the Party. In local elections in Latvia and Estonia in December 1989 there had been massive defeats for the Communist candidates. Accordingly, after much soul-searching and opposition from the Kremlin, the CPL voted overwhelmingly to separate itself from the CPSU and to declare its independence. The significance of this decision was not lost on observers, either in Lithuania or Moscow. The declaration of the newly independent Party that its most important objective was the achieve-ment of an 'independent democratic state' suggested that this was the first practical step taken by Lithuania on the way to independence. The Party had, by this action, re-attached itself to the mainstream of opinion and narrowed the distance between itself and the *Sajudis*-backed parties. Most significantly, it cast off the cloak of lies which had surrounded it since the loss of independence and threw itself on the mercy of the electorate. Although it had no hope of winning, its decisions in December constituted a necessary first step in the process of rehabilitation.[48]

Gorbachev's visit to Lithuania in January 1990 was another important element in shifting public opinion in favour of independence. His intention in going to Lithuania was to convince the Lithuanian public that he was in favour of self-determination up to the point of secession. He reiterated that independence would pose a serious threat to *perestroika* and that the best solution was to create a real federation. Preparations were being made in Moscow, he announced, to draft a new constitution for the Soviet Union which would give the 15 republics wide-ranging political and economic independence and even the means to secede from the federation. It was evident to many of his hearers in factories, on the streets, among the political class and on TV that Gorbachev was playing for time. There were doubts about his sincerity and his timing. His proposals were vague and belated. Could the Lithuanians delay their own preparations for independence on the strength of such vague generalizations from a man whose attitudes and achievements on the nationalities question were the weakest part of his otherwise remarkable record?

There were also informed rumours circulating about the shape of the new Soviet constitution. The proposed executive presidency, which was seen as a vehicle for Gorbachev to maintain real power as his former power base, the CPSU, crumbled, was believed to have been granted emergency powers which could be used against the self-determination of the republics. And since the president was granted power to define a state of emergency, to interpret and suspend the constitution, and to rule by decree, the new constitution could hardly be considered a bulwark of republican rights. As for the possibility of secession, Landsbergis was right to be sceptical that the proposed new law would offer a genuine and speedy opportunity to quit the Union. Leaving aside the obvious fact that a secession law was irrelevant to a republic which had been occupied by force, critics also feared that the law would prevent rather than facilitate secession.[49] They believed that Gorbachev was following a strategy of agreeing in principle to the possibility of independence while making time for the full implications of the step, both economic and otherwise, to sink in. Presumably he hoped that by placing so many obstacles in the path of divorce the partners would decide to continue living together. He once again misjudged the public mood. He was not trusted, his record did not command confidence, and he was seen as an opportunist whose major concern was to maintain his hold on power and keep the Union intact. If the Soviet Union could admit aggression in Afghanistan and Czechoslovakia, asked Landsbergis, why could it not do the same regarding Lithuania? What hope was there of a meeting of minds when Gorbachev could pronounce in all sincerity, in face of the findings on the Molotov-Ribbentrop pact and the evidently fraudulent elections of July 1940 that 'Some people have doubts about whether Lithuania is part of the Soviet Union or not. This simply is not serious'. It was therefore not surprising that he made few converts, and that slogans among the Vilnius crowds during his visit called for independence and the withdrawal of the Soviet army. In sum, Gorbachev's visit was a failure for him, and confirmed for any waverers that independence was the only option for Lithuania.[50]

Finally, the elections to the Lithuanian Supreme Soviet which took place in three rounds starting on 24 February, were an unequivocal endorsement of independence. All parties supported it except the rump of the Lithuanian Communist Party which had remained loyal to Moscow and was mainly backed by ethnic Russian and Polish electors. sajūdis-backed candidates picked up 72 out of the first 90 seats to be decided in the first round and made further gains in the second round. The final result showed just under 100 of the 141 seats were held by candidates backed by *Sajūdis* on a turnout of 75 per cent. Since *Sajūdis* candidates now held a minimum of two-thirds of the seats in the Supreme Soviet, the latter had the power to approve the necessary constitutional changes needed for independence. Whether it would do so immediately, or in what form, was open to question.[51]

There was an intense debate among the factions as to the correct course of action. The radical nationalists in *Sajūdis* were all for an immediate declaration, the more moderate majority were more cautious. Landsbergis, who was elected chair of the Parliament in preference to Brazauskas, initially reserved his judgement though, in Lieven's words, 'he denounced the spirit of subservience'. There were a number of factors which swung the debate in favour of an immediate declaration. Some argued that voting against immediacy might be interpreted as a vote against the declaration itself, and nobody wished to be caught in that position. There was acute apprehension that action needed to be taken before the new Congress convened in Moscow on 12 March and conferred on Gorbachev the new presidential powers which would have given him the legal right to block independence. A new law on secession would have a similar legal effect, whereas an immediate declaration would fall under the right to secede guaranteed in the existing Soviet constitution.[52] An additional consideration was the realisation that, after a prolonged debate in which all the possible avenues were explored, a refusal to declare independence would be equivalent to complicity in the existing system, especially in the light of the now accepted illegality of the Molotov-Ribbentrop pact. Another persuasive argument was that independence would enable Lithuania to negotiate with Moscow on equal terms regarding such matters as troop withdrawals and other aspects of Soviet disengagement. It was quite widely believed that western states would recognise and support Lithuanian independence. Although reservations were expressed by Brazauskas, among others, concerning the need to establish economic independence before declaring political independence, and the import- ance of keeping in step with Estonia and Latvia on this crucial issue, they were overborne by the general mood of sober determination among the legislators. The suggestion that there was an agreement among the deputies to accept Kazimiera Prunskienė as Prime Minister, Brazauskas as one of the deputy prime ministers, and a number of former communist ministers as members of the new government in return for support in the independence vote among the waverers does not lack plausibility. In what were expected to be difficult negotiations with Moscow and with the possibility of economic sanctions on the horizon, the participation in the government of the experienced Brazauskas and other Communist officials was seen as advantageous to maintaining consensus in difficult times.

So, on 11 March, 1990, the fateful decision to 'restore the exercise of the sovereign powers of the Lithuanian state' was taken in the Lithuanian Supreme Soviet. The republic changed its name from the Lithuanian SSR to the Republic of Lithuania and suspended the Soviet constitution. Although *de facto* independence was declared, observers considered that the declaration could best be understood as an attempt to establish a legal basis on which to negotiate the practical transfer of powers rather than the prelude to the unconditional seizure of these powers. Some members of *Sajūdis* apparently believed that the declaration would persuade the western powers to accord *de facto* as well as *de jure* recognition of Lithuania's independence. In this they were to be rapidly disillusioned.[53]

THE REALIZATION OF INDEPENDENCE

It is not to deny or diminish the importance of Lithuania's own contribution to the achievement of independence to say that the key to success lay in Moscow where a change in the balance of political forces ensured that new political structures would emerge on Soviet territory. The new disposition of power was not caused by the failed coup of August 1991, led by the nexus of army, KGB and military industrial complex. Rather, the coup's outcome symbolized the new balance and confirmed that power had unequivocally passed from the centre to the republics, which in turn would choose whether to associate together in a new association of states.

This discussion will therefore emphasise the struggle for power in Moscow between the conservatives, who wished to maintain a strong Union, and the radicals who believed that only the republics had the will and capacity to implement the reforms. The contest between the two sides fluctuated, with first one and then the other gaining the advantage and with Gorbachev trying to find a firm footing somewhere between. His manoeuvring in these eighteen months suggests that his primary aim was to preserve the Union and to maintain himself in power. But by August 1991 it was apparent that the most he could achieve was a severely weakened Union and a much reduced role for himself. His forced alliance with the radicals clarified how far the balance of power had changed. This situation was, in turn, unacceptable to the conservatives who believed that a forceful attempt to restore order and to halt the disintegration of the Union would meet with popular acquiescence. Their failed coup in August 1991 torpedoed Moscow's opposition to the decentralisation of power and to the restoration of independence for Lithuania and the other Baltic republics.

Shifts in the balance of power in Moscow were critically important, but equally indispensable to the achievement of independence was the Lithuanians' own unyielding determination and their adoption of a strategy of 'active non-resistance'. In addition, decisions in Moscow were influenced by the demands of great power relationships and by international opinion; the responses of Washington, London, Paris and Bonn to events in the Soviet Union could not be ignored by the Kremlin so long as it relied on Western assistance to achieve its economic and political reforms. It is true that Western governments wanted to avoid destabilizing the Soviet Union or undermining Gorbachev's position. Nevertheless, it was made quite clear to him that certain policies, such as the use of force in the Baltic states, would have an adverse effect on Soviet-Western relations and on the West's economic support for restructuring.

One of the most important factors tilting the balance of forces in Moscow towards the decentralizers was the state of the economy. Economic regeneration was one of the major objectives of *perestroika*, and one of its biggest failures. The depth of the Soviet economic crisis showed that Gorbachev's attempts at reform had failed to reverse, indeed had worsened, the long term decline in economic performance.[54] Falling production, rising foreign debt, an increasing trade deficit with the West, and runaway inflation were accompanied by shortages of food and consumer goods. Attempts at stabilisation such as the decision to double or treble the price of basic foods, including bread and flour, with effect from 1 July, 1990 and to remove state control from many areas of the economy, alarmed rather than reassured the population and led to panic buying and further shortages of goods in the shops. The future brought no better prospects. There was sharp disagreement between the reformers, who believed that reform had not gone nearly far enough, and the conservatives who thought it had gone too far and was mainly responsible for the current problems. While the political debates went on, the budget deficit was expected to rise four fold between 1990 and 1991 and national income to fall by at least 11 per cent in 1991. To place these figures in broader perspective, the standard of living of Soviet citizens was only about one fifth of the level in the United States and was expected to fall further in comparison. The record and prospects for the Soviet economy discredited Gorbachev's view that economic reform could only be successful if led from the centre. Indeed, they illustrated the bankruptcy of economic reform under Gorbachev's stewardship and inevitably led to renewed pressure for devolution of economic decision-making to the republics.[55]

An even more potent influence in shaping the balance of political forces was the emergence of Boris Yeltsin, a reformer, as the powerful leader of the Russian Federation. Yeltsin's impact in offering a powerful challenge to the centre, in supporting the other republics in their desire for independence or for a looser Union, and in bringing to bear in Soviet politics the full economic, military and political weight of Russia, was profound. Yeltsin's influence can be clearly identified in several important developments in 1990 and 1991.

The first of these was Moscow's new-found willingness to accept a moratorium on Lithuania's declaration of independence on unexpectedly generous terms. Admittedly, the adverse impact of economic sanctions on the Soviet economy itself influenced the decision. Moreover, pressure from the West for a settlement may have had some effect. But the major factor was Yeltsin's election as President by the Russian parliament. Until then Lithuania's plight was unenviable. After her declaration of independence on 11 March, she had been subject to a range of Soviet sanctions designed to bring her to heel.[56] Accompanying these actions was a non-stop barrage of threats and promises from Gorbachev, ranging from flat denunciations of the illegality of Lithuania's action and refusals to engage in negotiations on secession, to promises of a dialogue on a whole range of problems but within the framework of the Soviet constitution. However, the promised dialogue was conditional on the immediate annulment of 'illegal acts' and a return to the situation on 10 March, 1990. Underlying Gorbachev's approach was his view that 'problems should be resolved by reforming the Federation, not dividing it'.[57]

On 17 April, 1990 the heralded economic sanctions on Lithuania were imposed by the Kremlin. Since Lithuania was part of an integrated economic system centred in Moscow it was calculated that sanctions would quickly bring it to its knees by depriving it of essential supplies and markets. For example, only 4.5 per cent of all Lithuanian production was exported outside the Soviet Union and Lithuania supplied only 5 per cent of its energy consumption, the rest being imported from other parts of the Union. Accordingly, the cutting off of oil deliveries and the reduction of gas supplies by 80 per cent was designed to have a crippling effect on the Lithuanian economy. Almost equally harmful was the reduction in supply of key raw materials.[58] The effects of economic sanctions were severe. Fuel stocks for transport and domestic and industrial heating were rationed to prevent their rapid exhaustion, and industries dependent on the import of raw materials were hard hit. The Lithuanian oil refinery at Mazeikiai and a major fertiliser factory at Jonava were closed shortly after sanctions were imposed. Other enterprises soon joined them. Collective farmers stopped or limited the use of farm machinery owing to fuel shortages. On 21 April Brazauskas said that there were only 12 days supply of fuel left and it would be impossible for the Lithuanian economy to continue for an extended period under the blockade. If sanctions had continued into the summer there would have been a severe threat to the harvest and to Lithuania's food supplies.[59]

Yet, owing to the integrated nature of the Soviet economy, sanctions on Lithuania served to deprive other parts of the Union of products made exclusively or mainly in Lithuania, such as TV tubes, vacuum cleaner parts, electrical measuring devices etc. The Soviet economy as a whole certainly suffered from the sanctions policy. It is also true that western pressure on both Lithuania and the Kremlin had an effect in persuading the two sides to agree to a moratorium. Nevertheless, it is probable that Yeltsin's election as chair of the Russian Supreme Soviet in May 1990 and Russia's declaration of sovereignty in June were decisive in compelling Gorbachev to modify his policy of economic sanctions on Lithuania.[60] Yeltsin had been persuaded by reformers in the Russian Federation to blame continued Soviet economic decline on the 'imperialist policies of the centre' and to demand economic and political sovereignty for Russia, with the same rights to self-government and economic autonomy as the other republics were demanding. His election, albeit by a small majority, gave added weight to his demands for radical economic reform and devolution of power. The massive importance of the Russian Federation in the Soviet economy had a decisive effect on the balance of power between the centre and the republics. Russia produced 91 per cent of total Soviet oil output and accounted for half of its population. In the light of Russia's declaration of sovereignty the Soviet Commission on Compliance with the Constitution announced it would no longer bring cases against any of the republics for non-compliance. In these changed circumstances Gorbachev's demand that Lithuania return unconditionally to the Soviet constitution made little sense.[61]

Moreover, it was doubtful whether the economic sanctions could continue to be effective in view of Yeltsin's meeting with Landsbergis on 2 June to discuss direct cooperation between Russia and Lithuania involving trading links and energy supplies. Yeltsin confirmed that if the Baltic states were to opt for independence Russia would be the first to sign treaties with them.[62] The upshot of this promise was a weakening of Gorbachev's position, a willingness on his part to bring the policy of sanctions to an end, and a commitment to begin negotiations (a term Gorbachev had hitherto refused to employ) with Lithuania on the restoration of sovereignty. All this was a far cry from Moscow's original demand that Lithuania fully revoke her declaration of independence and return to the Soviet constitution.[63]

Yeltsin's second critical intervention on behalf of Lithuania occurred in January, 1991 after the notorious killings and woundings by Soviet paratroops and militia forces of unarmed civilians at the Lithuanian TV tower. The significance of this event and the importance of Yeltsin's actions should be considered in the context of an attempted counter revolution by the Kremlin conservatives in the autumn of 1990. Yeltsin's success in June 1990 in focusing discussions with Gorbachev on the formulation of a new Treaty of Union leading to the creation of a so-called Union of Sovereign States had alarmed the old guard. Gorbachev was displaying his characteristic agility in espousing devolution of power as a means of harnessing the support which

Yeltsin had captured. His apparent support for the Shatalin 500 day plan for economic transformation gained the approval of the radicals, but also mobilized conservative opposition. In the view of these critics, the plan gave too much power to the republics in implementing economic reforms. And the introduction of a Presidential Council in March 1990 had demonstrably failed to strengthen presidential power. Hence the conservative counter-attack in the autumn of 1990 aimed to destroy devolution, create a reinvigorated centre and ensure that economic reform was effected by Union agencies and not by the republics.

When it became clear to Gorbachev that negotiations on a new union treaty were unlikely to produce agreement between the centre and the republics he once again felt impelled to assert Soviet authority and conciliate conservative critics. In late October, therefore, he proposed drastic curbs on the powers of the republics and asserted the precedence of Soviet legislation over republican laws until the new Union Treaty had been concluded. When this treaty was finally published at the end of November it was obvious that it would not win support from the republics. In proposing that Soviet laws should be paramount and the bulk of economic policy remain under central control, Gorbachev showed his preference for a unified state with a single market, single currency, and central control of foreign, defence, fiscal and monetary policies. Still worse from the republics' point of view, secession from the Union would be governed by the law passed in the Spring of 1990, which placed severe obstacles in the way. Compared with this centralizing proposal Yeltsin wanted a looser union built on bilateral agreements between the individual republics and between the republics and the centre, with much greater decentralization of powers.

Gorbachev's ambition to further consolidate presidential power produced a series of proposals to the Congress of People's Deputies meeting in December 1990 which included a strengthened Federation Council, a new security council composed of leading ministers, the new post of Vice-President, a cabinet and a prime minister. The security council was composed of conservative adherents of the Union such as Yanaev, the Vice-President, Pugo the new Minister of the Interior in place of the more liberal Bakatin, Yazov the Defence Minister and Pavlov, who shortly became Prime Minister. In a stunning development Foreign Minister Eduard Shevardnadze, one of the keenest proponents of *perestroika* and an old ally of Gorbachev, resigned along with Yakovlev, once Gorbachev's closest colleague, alleging that the 'vengeful and merciless' conservative forces had made a prisoner of Gorbachev.[64]

There was some evidence to support this view. In addition to the draft Union Treaty and the strengthening of the central government Gorbachev was granted emergency powers to restore order and bring the dissident republics into line. Hawkish communists of the younger generation such as Blokhin and Alksnis, the founders of militant right-wing faction, *Soyuz*, proposed the imposition of a state of emergency, the dissolution of all parliaments and the establishment of a committee of national salvation with wide powers. In December a letter from 53 conservatives called for a crackdown on opposition to the Soviet state and for urgent measures to be taken against separatism and inter-ethnic strife, a reference to alleged violation of the human rights of the Russian-speaking minorities in the Baltic states.[65]

The conservative reaction came to a head in January 1991 and placed the Lithuanian reformers in great peril. All the signs pointed to the use of emergency powers by the centre and there was every chance they would succeed in destroying Lithuanian aspirations for independence. In fact, as the world saw on its TV screens, the events of 11–14 January, 1991 in Vilnius, showed the dramatic failure of Moscow's attempts to seize power in the Baltic states. There were a number of reasons for this, including the fearless non-violent resistance of the Lithuanian people and the condemnation of the West which, contrary to Soviet predictions, was not so distracted by the Gulf War that it ceased to pay attention to events in Lithuania. Moreover, the coup lacked ruthlessness; no one was prepared to take responsibility, least of all Gorbachev who subsequently showed himself to be either ignorant or guilty but incompetent. Nevertheless, in the atmosphere of uncertainty about whether to risk massive casualties by an attack on an unarmed crowd in front of the Lithuanian parliament and about the reliability of Soviet troops in carrying out such actions Yeltsin played a very significant role. Not for the last time he was able to demonstrate courage and an ability to influence the course of events by decisive action. From his position as President of the Russian Federation he criticised Gorbachev for the use of force and called on Russian soldiers not to allow themselves to be used against the rule of law in the Baltic states. His support for Lithuania and the other Baltic states, as shown both by his words and by his courageous flying visit to Estonia to show solidarity with the independence movements, discredited the argument that Baltic independence would be harmful to the interests of Russia. His condemnation of the attack on the TV tower helped to swing Russian democrats behind the Balts. The failure of the coup in Vilnius once again changed the balance of forces in the Soviet Union and dealt a body blow to the conservatives. In the eyes of the West Yeltsin's actions established him as a major political figure with whom they might have to deal. There was now a greater readiness in western thinking to envisage new political arrangements on the territory of the Soviet Union, such as a loose, and voluntary, confederation of member states with independence for those republics which did not wish to join.[66]

There were two other occasions when Yeltsin's interventions decisively weakened conservative forces and prepared the ground for the disintegration of the Soviet Union. The first was his decision to gain popular approval for direct elections for the post of Russian president. His victory in the subsequent election in the Spring of 1991 enormously strengthened his prestige and authority, and gave him a clear moral advantage over Gorbachev who had never faced the voters. This advantage was particularly evident in negotiations on the draft union treaty. Yeltsin and his fellow republican leaders brought their influence to bear in radically amending the original proposals of December 1990. The continued decline in the economy and the strengthening of democratic forces after the failed repression in the Baltic states alerted Gorbachev to the new political balance. For him, even a weakened Union was better than no Union at all and perhaps he saw that no meaningful economic reform could emerge from the centre, given the dominance of conservatives in the Security Council.[67]

Fortified by his election victory Yeltsin was now a towering political figure with whom Gorbachev had to negotiate as an equal. His major demand was for a voluntary confederation, in which the sovereign republics would delegate to the centre only those powers they chose not to keep for themselves. In practice most powers hitherto held at the centre would be vested in the republics. He believed that the centre was incapable of reform because it was in the grip of three powerful conservative forces, the Army, the KGB and the military-industrial complex, which opposed change because it would damage their interests. The only way to implement reform was for the republics to escape the embrace of the centre. The negotiations during the late Spring and Summer of 1991 reflected these aspirations.[68] The final draft of 14 August divided powers between the centre and the republics, conceding to the centre responsibility for defence and foreign policy but giving the republics control of their economies and resources. The centre would retain a supervisory role over military industrial plants but the republics obtained the sole right of taxation and gained a share of the Union's gold, foreign currency, and diamond reserves.[69]

But what about the ambitions of republics such as Lithuania which demanded independence? In July Gorbachev said that republics which did not join the confederation would be offered some alternative form of cooperation such as a common economic space. Yeltsin's role in achieving this outcome was profoundly important. At the same time the conservatives were seeking Gorbachev's scalp and Pavlov warned that the country faced complete chaos after 20 August when the Union Treaty was due to be signed.[70]

The attempted coup of 19 August, led by the country's Vice-President, the heads of the armed forces and the KGB, and the minister of the interior, was a desperate attempt by the conservative military and police nexus to destroy the new Union and to shift the balance of power away from Russia and the other republics back towards the centre. It is probable that the plotters hoped or expected that their actions would be greeted by relief, or at least acquiescence, on the part of the Soviet people. Indeed, although the reformers had powerful support, there was still very strong backing for the conservatives and the outcome of the struggle was uncertain. It was at this point that Yeltsin made his fourth very decisive contribution to the independence of Lithuania, publicly and courageously resisting the illegal action of the conspirators and urging the army not to support the coup. The army, unsure of which authority, Soviet or Russian, to obey, and tired of taking responsibility and unpopularity for the actions of incompetent politicians, refused to support the conspirators. When the coup failed the victory was for the principle for which Yeltsin had been fighting, republican sovereignty and autonomy. When Gorbachev returned from detention he had to recognize that the shift in power was now massively towards Yeltsin and decentralization. For Lithuania and the other Baltic republics the failure of the coup enabled them to seize the chance of independence and to gain diplomatic recognition from the international community.

Indeed, in the eighteen months since their declaration of independence, the Lithuanians had shown that they were eminently capable of the prudent exercise of power. There were a number of occasions during that period when decisions in Vilnius could have resulted in a loss of international confidence, or have provided Moscow with a plausible case to intervene without damage to Gorbachev's international reputation. Yeltsin's defence of the republics was made easier by the measured and reflective policies adopted in Vilnius and by the calm determination of the people to defend their republic by non-violent and non-provocative measures. Intelligent self-restraint was the defining quality of the Lithuanians' response to Moscow's hostility, and it won them international sympathy.

This is not to say that the government in Vilnius was without divisions or conflict. The first administration formed after the declaration of independence was composed of many ministers who had been members of Soviet governments or Communist Party apparatchiks. Their continued presence was testimony to the very small pool of experienced professional politicians outside the former Communist Party. The Prime Minister, Kazimiera Prunskienė, was an exception in that, although she had been a member of the Party, she was an academic and a founding member of *Sajūdis*. Algirdas Brazauskas, the former General Secretary of the Party and now leader of the post-Communist Democratic Labour Party (LDDP) became a deputy Prime Minister along with Romualdas Ozolas, a member of *Sajūdis*.[71] The government was one source of authority, but it was overshadowed by the recently-elected Parliament (still called Supreme Soviet) with its large *Sajūdis* majority under the chairmanship of Vytautas Landsbergis. Never a Party member, Landsbergis was a founding member of *Sajūdis* and a leading figure in the Vilnius intelligentsia. As the presiding figure in the parliament Landsbergis had to steer a course between the radical Kaunas faction, one of whose most prominent figures was Česlovas Stankevičius, and the generally more moderate Vilnius circle. He could not forget that in the country Brazauskas was a more popular candidate for Chair of the Parliament than he himself. The Parliament saw itself as the legitimate leadership of the country. It suspected the government of being too eager to come to an agreement with Moscow. The Parliament and the government therefore eyed each other warily, and tensions strongly characterized their relationship until the coup attempt in Vilnius on 13 January, 1991.[72]

The key figure was Landsbergis. Standing on the authority of Parliament of which he had been elected leader, he was resolute in his resistance to the Soviet economic sanctions. Memories of 1939 were present in the minds of many Lithuanians, and Landsbergis echoed these in his declaration that 'Moscow can annihilate us, they can set up another puppet regime here but they cannot kill our wish for freedom. There can be no turning back on sovereignty'. At the same time he believed that the type of resistance could be crucial in maintaining public support and in winning over international opinion. Consequently he repudiated talk of armed resistance emanating from some of the more radical quarters.[73]

While Landsbergis's position was firm on the question of resistance, it was not clear what his attitude would be to the proposed moratorium on Lithuania's declaration of independence which was actively discussed during May and June 1990. Prunskienė had been under pressure from the United States and European governments to try to reach a settlement with Moscow. Eventually she became convinced that a discussion on a moratorium with Gorbachev within tightly agreed conditions was imperative. The softening of Gorbachev's own position on negotiations with Lithuania persuaded her that an acceptable agreement could be achieved. Her proposal confirmed the suspicions of the radical *Sajūdis* deputies that the government wanted to keep Lithuania in the Soviet Union and to participate in discussions on the structure of a new federation. Landsbergis, who was slow to make up his mind, ultimately made a decisive contribution in swinging parliamentary opinion behind the moratorium. He re-affirmed Lithuania's right to independence, but was prepared to recommend negotiations with Moscow as a means of consolidating that independence. The Lithuanian parliament's declaration summarised his position: '... the Lithuanian parliament continues to express the sovereignty of the Lithuanian nation. Its goal is to rebuild Lithuanian independence. For this the Parliament desires negotiations with the Soviet Union'. The Lithuanian radicals recognized their defeat; Landsbergis had no doubt noted that public opinion favoured the government's position on the moratorium and that international opinion actively supported it.[74]

It was agreed that the moratorium would not come into effect until negotiations with Moscow began and would extend for only 100 days. In the event only a few preliminary discussions took place between the two sides. As we have seen, in the autumn of 1990 Gorbachev was busy preparing the new Union treaty and was falling under the influence of the Soviet conservatives. Prunskienė confirmed radical suspicions by favouring Baltic participation in drafting the new Soviet constitution. She reasoned that by retaining some sort of constitutional link with the Soviet Union the republics would avoid undue economic disruption and the possibility of ethnic conflict. Her popularity with the Parliament declined further as a result. She was also criticised for the slow pace of economic reform and privatisation in Lithuania, and for her government's proposal in December 1990 to increase the price of food, which achieved the remarkable feat of uniting in opposition the leftist ethnic Russian and Polish workers and the *Sajūdis* radicals. When the price increases were revoked by Parliament Prunskienė resigned.[75]

It was this political conflict in Lithuania that persuaded the conservative forces in Moscow to target Lithuania as the first republic to experience the restoration of Soviet rule under the shadowy leadership of a 'committee of national salvation'. Military intervention was timed to coincide with the expiration of the ultimatum to Saddam Hussein, when it was assumed that the West's attention would be absorbed by the Gulf War. Paratroops and special forces were sent in to Vilnius, crowds of Russian speakers poured on to the streets shouting pro-Soviet slogans, there were strikes in the factories and small bomb explosions in the city, and deserters from the Soviet army were actively pursued. The familiar Soviet trap was about to be sprung; a breakdown of law and order would be declared and a self-appointed pro-Moscow committee would invite in Soviet forces to establish a loyal government and restore the Soviet constitution. The TV tower was targeted first of all for occupation by troops owing to the alleged anti-Soviet broadcasts of Lithuanian TV. But an unarmed crowd was rapidly mobilized to surround the building and, in trying to enter, the troops killed thirteen people and injured very many more. At this point there was a general awareness that the next target of attack would be the centre of Lithuanian resistance, the Parliament building, where Landsbergis had his office. This was to be Landsbergis's finest hour. He became the embodiment of Lithuanian resistance: determined, cool under pressure, courageous and patriotic, he summoned the Lithuanian people to act as a human shield round the parliament building. Their response was as impressive as his own conduct. His insistence on passive resistance was followed to the letter, giving Soviet troops no excuse to storm the building. In demonstrating their obvious commitment to independence and their willingness to sacrifice themselves for the cause they showed supreme political maturity and unflinching determination. It was these events that identified Landsbergis as a great patriot and national leader and the Lithuanian people as worthy of the independence which they claimed. On the other hand, undisciplined or violent behaviour by the Lithuanians would have forfeited the admiration and respect they justly acquired.[76]

Similarly, at the time of the August coup, when once more Interior Ministry troops carried out aggressive actions in Vilnius and other cities, thousands gathered in front of the parliament building, as determined as in January to defend it. Once again Landsbergis was resolute, refusing to recognise or collaborate with any illegal or puppet government and recommending Lithuanians to pursue civil disobedience. When asked why Soviet forces did not continue with their military action in Vilnius in January, Landsbergis identified the resistance of the people, the defence of key buildings, the unity in the government, and strong international opposition as being the key factors. He could have added that the January events unified Lithuania and strengthened relations between parliament and government. Popular opinion rallied round the leadership of Landsbergis and increased western contacts with the Baltic republics.[77]

Two other developments strengthened Lithuania in its quest for independence: the referendum on independence, and the decline in Gorbachev's international reputation. The decision to hold a referendum had been resisted by Lithuania but the change of mind was precipitated by Gorbachev's decision, as part of his campaign for a new Union treaty, to seek popular endorsement on 17 March, 1991, for preserving the Soviet Union as a 'renewed federation of equal Soviet republics'. The Lithuanians had major reservations about this proposed referendum; how high would the turn out and the majority need to be to give Gorbachev the mandate he was seeking; would Soviet troops have the right to vote in any republic in which they were stationed and could they be moved around to strengthen the pro-Soviet vote; would an overall Soviet majority in favour overrule an overwhelmingly negative vote in a small republic; would results be falsified? Lithuania, along with 5 other republics, decided not to participate. Instead, it resolved to hold its own referendum on 8 February, 1991, which asked the simple question: 'Do you want Lithuania to be an independent democratic republic?'. There was an overwhelming yes vote. On a turnout of 85 per cent, 90 per cent voted in support of independence. This made it impossible for the Soviet Union to claim a shred of legitimacy for rule in the region and removed all doubts about the representativeness of Parliament's decision to declare independence on 11 March, 1990. Gorbachev's own poll, by contrast, provided him with quite modest support; only a bare majority in the Union as a whole, 54 per cent, on a 68 per cent turn out, was in favour of preserving the Union. Both the turn out and vote in support were significantly lower in Lithuania.

International opinion, which had requested a referendum in Lithuania, was left in no doubt about the virtual unanimity of

Lithuanians in support of independence. At the same time, however, their attachment to Gorbachev as the moderate Soviet leader with whom they could do business, was significantly weakened by the January events. Landsbergis said that the order for the attacks came from 'someone in Moscow'. Until Gorbachev provided an answer as to who that person was, the responsibility rested with him. Gorbachev was in an impossible position; if he accepted responsibility his reputation in the West, which was his main asset in internal Soviet politics, would be in tatters; if he did not, this would show that he was not in control of his own government. Either way his reputation as a supporter of the constitution and the rule of law, and as the effective and responsible leader of his country, was irreparably damaged. Pictures of Gorbachev receiving the Nobel Peace Prize contrasted vividly with photographs of Soviet tanks advancing into crowds of unarmed civilians. Both the referendum and Gorbachev's fall from grace had important implications for the balance of forces in the Soviet Union which, as we have seen, resulted in a much looser confederation and the possibility of republics choosing not to join.

Domestic politics in the Soviet Union and the evolution of relations between the centre and the republics were not conducted in a vacuum. The West had a major interest in the course of events. On the one hand most western countries had refused to extend *de jure* recognition of the Soviet occupation of the Baltic states, and continued to do so. On the other hand, they would not accord diplomatic recognition to Lithuania after its declaration of independence because recognition depended on Lithuania's achieving 'real independence', which meant a viable state with control over its external policy and its territory. Western interests, furthermore, were tied up with Gorbachev. Whatever the reservations about his domestic policies, the West could not bring itself to place Lithuanian interests above the necessity of maintaining the momentum of East-West relations, which depended on Gorbachev's continued tenure of power. It was axiomatic that continued reform in the Soviet Union, progress towards arms control, nuclear disarmament, consolidation of reform in East Central Europe, and German reunification had a far higher priority than the independence of Lithuania.[78] This did not mean that the West would be completely indifferent to Soviet policy towards the republics. Discreet pressure was put on Gorbachev to moderate Soviet actions in Lithuania after the latter's declaration of independence. It was made clear that the imposition of an economic blockade, still worse the use of force, would damage US-Soviet relations, particularly in the area of arms control treaties and trade agreements. Gorbachev was receptive to these warnings since he had an interest in maintaining Western good will and cooperation, not only because this strengthened his political position but also because the West offered indispensable assistance to his economic modernization programme.

As time passed, however, the respect and gratitude which the West had extended to Gorbachev for his achievements and for his cooperation were increasingly offset by his swing towards the conservatives in the autumn of 1990, and by the continued decline of the Soviet economy. The turning point in western relations with Gorbachev came with the use of force in Lithuania, which contrasted with the dignified and restrained behaviour of the Lithuanian people and leadership. A frost descended on US-Soviet relations, the summit scheduled for February was placed in doubt, the attempted coup was condemned as a retrograde step, and continued western aid was placed under review. The British Foreign Secretary, Douglas Hurd, commented acidly that 'simply because the world has its eyes on the crisis in Kuwait, the Soviet Union should be left in no doubt that it cannot commit deeds in a corner without the world noticing'. The January events not only unified Lithuania, but shifted the balance of western opinion in the direction of new structures in the Soviet Union. Support for Gorbachev now had to be weighed against the emergence of Yeltsin and the possibility that western interests might now best be served by decentralization of power, including the independence of those republics who wished it.

The shift in the balance of western opinion was consolidated by the attempted coup in Moscow on 19 August, 1991. While Gorbachev was kept a prisoner in his holiday villa in the Crimea, the conservatives in his government attempted to defeat the movements for the independence of the Soviet republics, and to reassert central control. This proved to be the last throw of the dice by the unreconstructed conservative forces. Their failure in the face of fierce opposition both at home and abroad opened the door for the Baltic republics to claim their independence, and to seek recognition of this independent statehood from the rest of the world. Within a short time Lithuania's *de facto* as well as *de jure* independence had been recognized by many states, including the newly-independent Russian Federal Republic, and by the United Nations. Jubilation in Lithuania at this achievement was matched by a sober recognition of the economic, political and strategic challenges facing the newly-restored republic. Yet, while not even the most far-seeing of politicians could have been totally prepared for the rigours of independence, the tenacity which Lithuanians had shown in the twentieth century in the face of totalitarian repression would be likely to serve them equally well in overcoming the political and economic challenges of the post-Soviet period.

1 Aleksandras Shtromas, 'The Baltic States as Soviet Republics: Tensions and Contradictions', in G. Smith ed., *The Baltic States: The National Self Determination of Estonia, Latvia and Lithuania* (Basingstoke: Macmillan, 1994), p.101

2 Romuald Misiunas and Rein Taagepera, *The Baltic States Years of Dependence 1940–1990* (London: Hurst and Co., 1993), pp.251–3; Bohdan Nahaylo and Victor Swoboda, *Soviet Disunion: A History of the Nationalities Problem in the USSR* (London: Hamish Hamilton, 1990), p.126; V. Stanley Vardys and Judith B. Sedaitis, *Lithuania: the Rebel Nation* (Boulder, Col: Westview Press, 1997), p.88; Aleksandras Shtromas, 'Prospects for Restoring the Baltic States' Independence: A View on the Prerequisites and Possibilities of their Realization', *Journal of Baltic Studies*, vol.XVII, no.3, Fall 1986, p.265

3 A Soviet law of 1968 allowed petitioning to the government without fear of arrest. This encouraged thousands to participate and mass petitions became an important aspect of dissent: Vardys and Sedaitis, pp.85–6

4 This petition was smuggled out to United Nations' Secretary-General Kurt Waldheim for onward transmission to Brezhnev. Two further large petitions were published in 1973 and a massive 150,000 people petitioned in 1979 against the authorities' take-over for non-religious purposes of a newly-completed church in Klaipeda in 1961: Thomas Remeikis, *Opposition to Soviet Rule in Lithuania 1945–1980* (Chicago, I11: Institute of Lithuanian Studies Press, 1980), p.114

5 This information was published in the West and re-transmitted to Lithuania by means of western broadcasts, thus ensuring much wider publicity and reducing the sense of isolation in the country: Misiunas and Taagepera, p.255; British Foreign and Commonwealth Office, Background Brief, 'The Baltic States: 40 Years under Soviet Rule', September 1986; Vardys and Sedaitis, p.86; Remeikis, pp.114, 130

6 The *Chronicle* was produced clandestinely for 17 years from 1972 to 1989. All told there were 81 issues. Each issue had some 40–70 single spaced typed sheets. The KGB put enormous effort into tracing those responsible for its preparation and circulation, and several prominent activists were arrested, tried and imprisoned in 1974. However, they never identified the editor, the Rev. Sigitas Tamkevičius, nor did they interrupt publication. When Tamkevičius was arrested on unrelated charges in 1983 his place was taken by the Rev. Jonas Boruta. Other objectives of the *Chronicle* were to counter the misinformed views of the Vatican about the condition of the Lithuanian Catholic Church and by means of steadfast opposition to the state's violations of religious rights, to counter the impression that Church leaders complied with state policy: Misiunas and Taagepera, pp.255–6; Nahaylo and Swoboda, p.229; Vardys and Sedaitis, pp.86–7; V. Stanley Vardys, *The Catholic Church, Dissent, and Nationality in Lithuania* (Boulder, CO: Westview Press, 1978), p.222

7 An attempt to break the power of the Committee was made in 1983 when two activist priests were sentenced to ten years imprisonment and internal exile. A subsequent protest letter signed by nearly 47,000 people from 71 parishes showed the depth of opposition and the strength of the movement's grassroots

8 Parts VII and VIII of the Act stated: 'The participating States will respect human rights and fundamental freedoms, including the freedom of thought, conscience, religion or belief, for all without distinction as to race, sex, language or religion ... The participating States will respect the equal rights of peoples and their right to self determination ... all peoples always have the right, in

full freedom, to determine, when and as they wish, their internal and external political status ...'. Such an unequivocal statement of principles offered Lithuanian dissidents a 'moral standard of condemnation' of the Soviet system: Bronis J. Kaslas, *The Baltic Nations—the Quest for Regional Integration and Political Liberty: Estonia, Latvia, Lithuania, Finland, Poland* (Pittston, PA: Euramerica Press, 1976), p.2; Remeikis, p.100; Yaroslav Bilinsky and Tonu Parming, 'Helsinki Watch Committees in the Soviet Republics: Implications for Soviet Nationality Policy', *Nationalities Papers*, vol.IX, Spring, Part 1, 1981, p.2;

9 Among Lithuanian dissidents national rights were at least as important as civil rights; they took the view that human rights included the rights of collectivities, particularly the right of nations, to self determination. Operating underground, though involved in *samizdat* publication, was the Lithuanian Freedom League, which assumed the leading role in organizing mass protests in Lithuania in 1987–88: Nahaylo and Swoboda, pp.196–7; Foreign and Commonwealth Office, Background Brief; Shtromas, 'The Baltic States as Soviet Republics', pp.103–4; Remeikis, p.431

10 They suffered the displeasure of the Soviet authorities: Petkus was sentenced to 10 years in prison and five years internal exile in 1978; Venclova was allowed to emigrate but was then stripped of his Soviet passport; Lukauskaite-Poskiene and Finkelstein, who had hosted Sakharov's visit to Vilnius to observe the trial of his fellow dissident, Kovalev, were interrogated by the KGB. Despite this the Helsinki Committee managed to survive through the 1980s: Remeikis, p.145; Vardys and Sedaitis, p.90

11 Between 1965 and 1978 10.3 per cent of all protests and demonstrations in the Soviet Union took place in Lithuania which had 1.3 per cent of the total Soviet population: Shtromas, 'The Baltic States as Soviet Republics', pp.103–6; V. Stanley Vardys, 'Lithuanian National Politics', *Problems of Communism*, July/August vol.XXXVIII, 1989, p.54; Remeikis, pp.130, 153–4

12 During his trial in 1971 in 1971 Simas Kudirka condemned the Soviet government's policies in Lithuania and demanded self-determination for his country: Jan Ake Dellenbrant, 'The Integration of the Baltic Republics into the Soviet Union', in Dietrich Andre Loeber, V. Stanley Vardys and Laurence P.A. Kitching eds., *Regional Identity under Soviet Rule: the Case of the Baltic States* (Hackettstown, NJ: Publications of the Association for the Advancement of Baltic Studies, 1990), pp.112–13; Aleksandras Shtromas, 'The Baltic States as Soviet Republics', pp.105–6; Walter C. Clemens, Jr., *Baltic Independence and Russian Empire*, (Basingstoke: Macmillan, 1991), p.68; Misiunas and Taagepera, pp.270–1; Nahaylo and Swoboda, p.209–13

13 A favourite tactic was to designate churches as architectural monuments to save them from destruction or desecration. Medieval castles such as Trakai or great houses and palaces were also protected and restored as national treasures. Even Snieckus, the long-serving General Secretary of the Communist Party, was believed to have opposed Khrushchev over the restoration of Trakai

14 Augustine Idzelis, 'Institutional Response to Environmental Problems in Lithuania', *Journal of Baltic Studies*, vol.XIV, no.4, Winter 1983, pp.300, 302–4

15 Vardys and Sedaitis, pp.93–4; Shtromas, 'Prospects for Restoring the Baltic States' Independence', p.264;

16 Shtromas, 'Prospects ... ', p.264; Endel-Jakob Kolde, 'Structural Integration of Baltic Economies into the Soviet System', *Journal of Baltic Studies*, vol.IX, no.2, Summer 1978, p.167; V. Stanley Vardys, 'How the Baltic Republics Fare in the Soviet Union', *Foreign Affairs*, vol.44, no.3, April 1966, p.514; V. Stanley Vardys, 'Recent Soviet Policy toward Lithuanian Nationalism', *Journal of Central European Affairs*, vol.XXII, 1963, p.316; Alfonsas Eidintas, Vytautas Žalys, Alfred Erich Senn, ed., Edvardas Tuskenis, *Lithuania in European Politics: The Years of the First Republic 1918–1940* (Basingstoke: Macmillan, 1997), p.194; Ole Norgaard et al., *The Baltic States after Independence* (Cheltenham: Edward Elgar, 1996), pp.37–8

17 It was reported that Lithuanian youth in the early 1960s were proud of Lithuania's independent past and the heroic aspects of her history: Vardys, 'Recent Soviet Policy', p.332; Remeikis, p.87

18 V. Stanley Vardys, 'The Role of the Churches in the Maintenance of Regional and National Identity in the Baltic Republics', in Loeber et al eds., p.158; Norgaard, p.19; Eidintas et al. p.194; Clemens, p.70; Misiunas and Taagepera, pp.242–244

19 Isabelle T. Kreindler, 'Baltic Area Languages in the Soviet Union: a Sociolinguistic Perspective', in Loeber et al., pp.238–42.

20 Norgaard et al., p.19; Misiunas and Taagepera, pp.209–11; Michael Binyon, *Life in Russia* (London: Panther, 1983), pp.45 & 227–8

21 Binyon, pp.393–6; Jan Winiedki, 'Are Soviet-Type Economies entering an Era of Long-term Decline?', *Soviet Studies*, vol.XXXVIII, no.3, July 1986, pp.326–31; *Financial Times*, 12 March, 1990; *The Economist*, 19 January, 1991; Yegor Ligachev, pp.43–51

22 *The Guardian*, 9 January, 1990; *Financial Times*, 12 March, 1990; Clemens, pp.74–81

23 Paul Goble, 'Moscow's Nationality Problems in 1989', in Radio Liberty, *Report on the USSR*, vol.2, no.2, 12 January, 1990; *The Times*, 13 January, 1990, 21 December, 1990, 14 January, 1991; Article in *Daedalus*, reprinted in *The Times*, 11 January, 1990

24 That Moscow's agreement was mainly opportunistic is supported by the flawed implementation of the policy in 1990. In any case, Moscow's introduction of economic sanctions after Lithuania's declaration of independence in March 1990 effectively terminated the policy: Clemens, pp.138, 248–51; Nahaylo and Swoboda, pp.278–9; Misiunas and Taagepera, pp.325–6; Kristian Gerner and Stefan Hedlund, *The Baltic States and the End of the Soviet Empire* (London & New York: Routledge, 1993), p.98; *The Times*, 28 November, 1989

25 Existing in secret since 1978 the LFL took advantage of the new openness to go public in 1987. It was led by some of the leading dissidents such as Antanas Terleckas, Petras Cidzikas and Nijole Sadunaite who placed a marker on the right of the political spectrum: Alfred Erich Senn, *Lithuania Awakening* (Berkeley & Los Angeles: University of California Press, 1990), pp.19–20, 67; Graham Smith, 'The Resurgence of Nationalism' in Smith ed., *The Baltic States: the National Self-Determination of Estonia, Latvia and Lithuania* (Basingstoke: Macmillan, 1994), pp.124–8

26 Senn, pp.67–9, 221–236, 243; Vardys, 'Lithuanians', p.81

27 In December 1987 there were complaints of censorship against the editor of the weekly journal *Literatura ir Menas* (Literature and the Arts) by the Lithuanian Writers' Union. In the previous month the Writers' Union had overthrown its existing conformist leadership

28 Vardys, 'Lithuanian National Politics', p.55; Senn, pp.47–54

29 Vardys, 'Lithuanians', p.79; Vardys, 'Lithuanian National Politics', p.55

30 Vardys, 'Lithuanian National Politics', pp.55–6; Anatol Lieven, *The Baltic Revolution: Estonia, Latvia, Lithuania and the Path to Independence* (New Haven and London: Yale University Press, 1993), pp.224–7; *The Times*, 1 March, 1989

31 Senn, pp.61–5, 124–36; Vardys, 'Lithuanian National Politics', pp.61–2

32 Vardys, 'Lithuanian National Politics', pp.58; Gerner and Hedlund, p.97; Senn, pp.105–12

33 Senn, pp.178–209, 221–2; Gerner and Hedlund, pp.92–3; Vardys, 'Lithuanian National Politics', pp.64–5; *The Times*, 12 January, 1990

34 *The Times*, 16 and 19 November 1988; Vardys, 'Lithuanian National Politics', p.66; Senn, pp.243–5; Norgaard et al., p.27; *The Economist*, 29 October, 1988

35 *The Times*, 3 May, 1990; Senn, p.223–36; Lieven, pp.225–7

36 Vardys, 'Lithuanian National Politics', pp.57–8; *The Times*, 27 February, 1989

37 Brazauskas argued that the proposal would have 'legalized secession' and Moscow would have been forced to rule directly in the republic

38 *The Economist*, 19 November and 26 November, 1988; *The Times*, 6 February, 1989; Vardys, 'Lithuanian National Politics', pp.67–8

39 Two senior figures of the CPL were endorsed owing to their popularity and record, Brazauskas and Beriozov, but many senior figures were defeated, including the Chair of the Supreme Soviet, the Prime Minister and several ministers and CP secretaries

40 Senn, p.251; Vardys, 'Lithuanians', p.84; Vardys, 'Lithuanian National Politics', pp.70ff

41 In declaring Lithuanian sovereignty and the supremacy of Lithuanian laws it reversed the decision of the previous November

42 *The Times*, 19 May, 1989; Vardys, 'Lithuanians', pp.84–5

43 Sakharov made this point, arguing that the republics should enjoy self-determination except in the spheres of foreign affairs, defence and communications

44 Recognition of their existence and their illegality would weaken its claim that the Baltic states had joined the Union voluntarily and confirm Baltic assertions that they did not need to secede formally from the Soviet Union since they had never been part of it

45 Yet Gorbachev still refused to accept that the Baltic republics had not joined the Soviet Union voluntarily. In support of this view *Pravda* published the original texts of the appeals by the assemblies of the Baltic states to join the Soviet Union in 1940 and Yakovlev, who finally came around to accepting the illegality of the secret protocols, defiantly asserted that they were not the basis for the 'legal and political status of Baltics'. Both Yakovlev and Gorbachev implicitly based their case on the elections of July 1940 which, as we saw earlier, were sham elections based on force: Clemens, pp.124–5; Smith, 'The Resurgence of Nationalism', pp.131–3; *The Times*, 2 June, 23 August, 1989; Gerner and Hedlund, pp.63–6; Misiunas and Taagepera, pp.328–9

46 *The Guardian*, 5 January, 1990; Clemens, pp.126

47 This gave the opportunity for various factions existing under the umbrella of *sajūdis* to transform themselves into embryonic political parties some of which had existed in the independence period, such as the Christian Democrats and the Social Democrats, along with more recent creations such as the Greens: Misiunas and Taagepera, pp.324–5; *The Times*, 11 December, 1989; article by Ceslovas Jursens in *The Lithuanian Review*, vol.1, no.1, 15 February 1990

48 *The Lithuanian Review*, vol.1, no.1, 15 February, 1990; *The Times*, 22 and 29 December, 1989, 16 January, 1990; Saulius Girnius, 'The Lithuanian Communist Party versus Moscow' in Radio Liberty, *Report on the USSR*, vol.2, no.1, 5 January, 1990

49 The critics were proved right: the law on secession which was approved by the Supreme Soviet in Moscow in April 1990 raised almost impossible hurdles to republics wishing to leave the Union. There had to be a two-thirds majority in a referendum with a minimum 75 per cent turn out, a wait of five years while the conditions for seceding were negotiated with the central government, and approval by the central legislature and each of the other constituent republics

50 *The Daily Telegraph* 12 January, 1990; *The Times*, 13, 15, 16 January, 1990

51 *The Times*, 5, 17, 26 February, 1990; Gerner and Hedlund, pp.121–4; Misiunas and Taagepera, p.330

52 Although the Lithuanians did not intend to act on the basis of the existing constitution lawyers could claim that an independence declaration fell under existing Soviet law, whereas it would not under a new law of secession

53 Vardys, 'Lithuanians', p.88; Lieven, pp.233–39; Misiunas and Taagepera, p.33; *The Guardian*, 13 March, 1990; *The Times*, 3 April, 1990; *The Times*, 27 February, 9 March, 1990; *The Financial Times*, 12 March, 1990

54 This was confirmed by Abalkin and a team of economic specialists in March 1990 who concluded that economic measures so far implemented were leading Russia into crisis: *The Guardian*, 20 March, 1990

55 *The Times*, 25 and 30 April, 23 and 26 May, 1990; *The Economist*, 19 January, 1991; Gerner and Hedlund, pp.114–17, 157

56 For example, the newspaper printing centre was taken over, Lithuanian deserters from the Soviet army and refuseniks, young men who refused to undertake Soviet military service, were flushed out from hiding by the military, and Moscow began to talk about re-allocating part of Lithuania to neighbouring republics. Lithuanian militias were ordered to hand over all their firearms and Soviet garrisons were heavily reinforced

57 Senn, p.258; Clemens, pp.101–2; *The Times*, 14 and 23 March, 2 and 11 April, 1990; *The Guardian*, 23 March and 6 April, 1990

58 Hanson calculated that purchasing supplies on the world market, even if that were possible, would cost the Baltic states 10 per cent of their Gross National Product since world prices for oil, copper and steel ranged from twice to seven times Soviet prices: *The Times,* 10 and 13 March, 17 April, 1990; Clemens, pp.253–6

59 Brazauskas expected that in May 35,000 people would have been made unemployed by sanctions; in fact the true figures appears to have been 100,000: *The Times*, 21 and 26 April, 26 May, 1990; *The Guardian*, 1 July, 1990

60 *The Times*, 17 and 23 April, 1990

61 *The Times*, 16, 24, 30 and 31 May, 30 June, 1990; *The Daily Telegraph*, 31 May, 1990; *The Economist*, 2 June, 1990

62 He embodied this promise in an agreement of 27 July between the Baltic states and Russia to begin

work on the preparation of treaties without any preconditions: *The Times*, 2 June, 1990; *Estonian Independent*, 6 June, 1990; ELTA *Information Bulletin*, August 1990

63 *The Times*, 30 June, 1990; *The European*, 22–24 June, 1990

64 *The Guardian*, 5 November, 1990; *Estonian Independent*, 29 November, 1990; *The Times*, 11,15,19,20 and 26 November, 3 December, 1990

65 *The Times*, 21 and 27 December, 1990; Gerner and Hedlund, pp.125–7

66 *The Times*, 15 November, 4 and 20 December, 1990; *The Economist*, 5 January, 1991

67 *The Daily Telegraph*, 14 January, 1991; *The Independent*, 16 January, 1991; *Estonian Independent*, 17 January, 1991; *The Economist*, 26 January, 1991; *Atmoda*, 28 February, 1991

68 In a speech in New York Yeltsin singled out Prime Minister Pavlov, KGB chief, Kryuchkov and Defence Minister Yazov as the key antagonists of reform. At about the same time Gorbachev chided Soyuz for being totally divorced from reality'as though under a bell jar': *The Times*, 22 June, 1991

69 *The Economist*, 6 April, 1991

70 Still to be resolved was responsibility for the money supply and macroeconomic policy: *The Times*, 23 August, 1991

71 *The Times*, 25, 27 July, 14 August, 1991; *The Economist*, 3 August, 1991

72 One marked exception to the pattern was the appointment of Algirdas Saudargas, a radical, as foreign minister

73 Lieven, pp.236–9; *The Times*, 3 May, 20 June, 1990

74 *The Times*, 3 May, 1990; *The Independent*, 26 March, 1990

75 *The Times*, 28 April, 18 and 30 June, 1990; Clemens, p.203

76 Lieven, pp.240–1; *The Times*, 2 July, 1990

77 *The Independent*, 9, 10 and 12 January, 1991; *The Times* 14 January, 1991; *Estonian Independent*, 21 February, 1991; Gerner and Hedlund, pp.150–1; Norgaard et al, pp.29–30; Lieven, pp.250–4

78 *The Times*, 11 March, 20 and 21 August, 1991. Lord Home, a former British Foreign Secretary, expressed this sentiment rather brutally: 'a weakened or toppled Gorbachev would be a greater threat to the United Kingdom in the longer term than a crushed Lithuania'.

Chapter 4

GOVERNMENT AND POLITICS IN INDEPENDENT LITHUANIA

The newly independent Lithuania had severed her enforced connection with the Soviet Union and now fulfilled her ambition to rejoin the community of European states. However, she could 'return to Europe' in the fullest sense only if she met the conditions of membership of the various European and Euro-Atlantic organizations, such as the European Union. Acceptance by the community of states meant adopting democratic forms of government and protecting human and minority rights. Accordingly Lithuania approved a democratic Constitution in 1992 and established very liberal citizenship and minorities legislation. External bodies concerned with the protection of human rights such as the Council of Europe and the Organization for Security and Co-operation in Europe (OSCE) gave their seal of approval. Nevertheless, the politics of the post-independence period often reveal a divergence between law and practice. The political class mouthed the vocabulary of democracy but sometimes their actions showed either that they did not understand the meaning of the term, or cynically ignored it. The election of President Adamkus in 1997, in a sense an outsider who had spent almost all his adult life in the United States, epitomized the contrast between two political cultures, each using the same vocabulary but differing radically in their understanding of the words. Adamkus's call for a kind of revolution in the political culture may represent a significant turning point in the politics of post-independence Lithuania. At the least it may help to restore confidence between the people and their political representatives who, at the time of his election, were generally mistrusted and unpopular.[1]

Critics of the practice of Lithuanian politics since independence are open to the charge of failing to recognize the enormity of the challenges faced by the new state. They were too impatient with the stumbling steps of the tyro democrats, and ignored the influence of history on the Lithuanian mentality since 1918, particularly the impact of Soviet rule on attitudes and values and the absence of lengthy periods of democracy to act as a model for contemporary politicians.[2] In the absence of democratic practice for all but four or five years since the first restoration of independence in 1918, it is not surprising that operating a democratic Constitution has proved to be problematic. Unlike the inter-war period, however, there are now external sanctions on behaviour which infringes international standards in the areas of human rights and democratic practice. In the long run, as memories of the Soviet period fade and democratic norms become entrenched, we should expect abuses and violations of western standards to become increasingly infrequent and the balance between rights and responsibilities more exactly struck.

The Constitution offers a good starting point for a discussion of government and politics in post-independence Lithuania. When the Lithuanian Supreme Soviet, on 11 March, 1990, declared the restoration of independence, it also annulled the validity of the Constitutions of the Lithuanian Soviet Socialist Republic and of the Soviet Union. A question of considerable practical and symbolic importance was whether to restore the 1938 Constitution. It was quickly agreed by *Sajūdis* and the Lithuanian Democratic Labour Party (LDDP—the former communists under the leadership of Algirdas Brazauskas) that a new Constitution should be drawn up, despite the symbolic importance of the old Constitution to the notion of *restoring* independence and maintaining legal continuity. This was partly because the 1938 Constitution was authoritarian and unsuited to the times, and partly because its restoration would have enabled the government, had it wished to do so, to exclude Russian immigrants from the franchise. In Lithuania, as we shall see, this was not a controversial issue owing to the relatively small proportion of Russian-speakers in the population. Meanwhile, the 1938 Constitution came into effect as a temporary measure and a Temporary Basic Law was introduced to grant the government the appropriate legal powers to act. Agreement about the need for a new Constitution did not, of course, preclude disagreements about many of its elements during the course of the Constitutional discussions. These disagreements arose out of the mutual suspicions between *Sajūdis* and the LDDP, and focused in part on the powers of the President. *Sajūdis* was apprehensive about the possibility of an LDDP majority in the next parliamentary elections and wanted a strong president of their own movement to push on with reform in face of a government and legislature already dominated by former communists. The LDDP, by contrast, feared that a strong president from the Right would limit the power of an LDDP government to implement its programme. The draft Constitution, which was approved in a referendum at the time of the parliamentary election in October 1992, proposed that the powers of the President should be limited, though they proved to be more extensive than in the neighbouring states of Estonia and Latvia. It also provided, again in contrast to the other Baltic states, for the direct election of the President by popular vote as opposed to a vote of MPs. An overwhelming popular vote for a presidential candidate would increase his authority and influence vis-a-vis the government and parliament, and informally expand his *de facto* powers despite the formal limits on those powers in the Constitution.[3]

In terms of formal powers the President was permitted to have an important role only in shaping and implementing foreign policy, and an essentially consultative and ceremonial role in domestic affairs. He could nominate the Prime Minister who would then formally submit his government for presidential approval. But in the formation of a government it is the support of a parliamentary majority which is critical. The President also has the power to veto legislation which can be overturned by a vote of the deputies. In the case of a parliamentary deadlock the President can call for new elections but only under rather stringent conditions. The Constitution was ratified by a majority of eligible voters; after October 1993 constitutional amendments required a two-thirds majority.[4] It was predictable that after four years of government from 1992–96 the LDDP proposed an extension of the President's powers to call early elections and to appoint the foreign, interior and defence ministers. This change of approach was not unrelated to the likelihood that the LDDP would be defeated in the parliamentary elections of 1996, but might hold on to the presidency in the election of early 1997. It also reflected the experience of Algirdas Brazauskas, president from 1993 to 1997, who believed that election by popular vote and the limited constitutional powers of the president were incompatible.[5]

Lithuania has a unicameral legislative branch. Elections to the *Seimas*, the parliament, take place under a mixed system of proportional representation and direct constituency elections. 70 of the 141 seats are allocated from party lists on a proportional vote. In the election of 1992 a minimum vote of 4 per cent on the national list (2 per cent for parties representing ethnic minorities) was required to qualify for the allocation of seats. The remaining deputies were elected in single member constituencies where the winner was required to gain a minimum of 50 per cent of the vote.[6] The minimum vote of 4 per cent produced a plethora of parties. Consequently the threshold was raised for the 1996 election to 5 per cent for all parties, including those of ethnic minorities; as a result, coalitions of parties were formed in order to break the 5 per cent barrier. Even so, some minor parties were excluded from representation by the small size of their vote.[7]

Until the 1996 parliamentary election the party political spectrum was dominated by two major forces. On the Right was *Sajūdis* which, after the loss of the 1992 election, converted itself from a movement into a right-of-centre party called Homeland Union, or Lithuanian Conservatives. This was a recognition of the fragmentation of *Sajūdis* and the defection of some of its members to other parties. On the Left was the LDDP which held power between the elections of 1992 and 1996. Two parties which had existed in the inter-war period, the Social Democrats and the Christian Democrats, were re-founded after the Lithuanian Supreme Soviet agreed to a multi-party system in December 1989. The legislative stalemate in the summer of 1992 resulting from the splits in *Sajūdis* led to the emergence of a Centre Union to offer an alternative to the two major parties. Other minor parties offering candidates in the 1996 parliamentary election included the Lithuanian Peasants' Party, the Lithuanian Union of Political Prisoners and Deportees, the Lithuanian National Party and the ethnically based Polish Electoral Action. In an attempt to overcome the 5 per cent threshold rule parties looked to form coalitions on the Right and Left; the Homeland Union was successful in forming an alliance with the Christian Democrats, and an association with the Centrists. However, the LDDP's overtures to the SDP were not productive owing to the latter's refusal to associate with what they called a party of capitalists.

Although the LDDP had offered a broadly social democratic platform in 1992, its failure in office to honour many of its electoral promises undermined its socialist credentials with the electorate in 1996. Worse than that was the widespread belief in the corruption of its leading politicians, and the cynical way in which it feathered the nests of the old communist *nomenklatura* at a time when standards of living for ordinary people were in rapid decline. Its period in office confirmed to many electors that the former communists who made up the LDDP were, generally speaking, political opportunists marked by 'cynicism and amoral pragmatism'.[8] It cannot be denied, of course, that the LDDP government had some significant achievements to its credit, particularly in economic and foreign policy, and that its room for manoeuvre was very limited given the external and internal pressures under which it operated. As the LDDP moved towards the centre in its conduct of government, *Sajūdis*, which was in power from the election of 1990 until late autumn 1992, turned sharply right after its party conference in December 1991. For a time its political model seemed to be the inter-war republic. Its attempt to create a strong presidency as a 'controlled dictatorship' with the power to destroy the remnants of communism and to establish rule by decree had uncomfortable echoes of the rule of Smetona. The instincts of many of its members were to support a policy of ethnic nationalism, and this was reflected in the refusal to restore local self-government in the ethnically Polish region of southeast Lithuania until 'the Lithuanian spirit should come back to East Lithuania'. On the other hand, a very liberal citizenship law emerged during the *Sajūdis* stewardship, along with legislation protecting human and minority rights. In the case of both the LDDP and *Sajūdis*, political instincts were moderated both by pragmatism and by the intervention of international agencies in support of liberal democracy and economic reform. *Sajūdis*'s election defeat in 1992 and its failure to win the presidency in February 1993 had a salutary effect. Transforming itself into the Homeland Union with Landsbergis and Vagnorius as its leaders, it dropped its radical nationalist rhetoric and adopted a more moderate and conciliatory approach, and a more liberal economic policy. The radicals in its ranks became a minority with little power. Its determination that Lithuania should join European and transatlantic economic and security organizations imposed on it a clear obligation to modify its nationalist rhetoric in the interests of inter-ethnic harmony.[9]

The party system and politics in general did not enjoy widespread popular confidence. One study showed that more than 60 per cent of Lithuanians supported the creation of a commission of experts to manage the economy instead of elected politicians. There was also a decline in support for democracy as a result of the experience of economic liberalization. Lacking deep roots anyway, democracy was chosen, the argument goes, as the only alternative to the Soviet system and as a way of ingratiating Lithuania with Western governments and international agencies from which it needed help. Democracy was also synonymous with western affluence. When Lithuanians experienced poverty during the economic transition, democracy was tainted by association. On the other hand, the two parliamentary elections of 1992 and 1996 and the presidential contests of 1993 and 1997 effected peaceful and orderly transfers of power from the government to opposition parties. This suggests popular acceptance of democratic norms among both the political élites and the population at large, and a strengthening of the party system. Whether this was the result of a consolidation of democratic practices or an acceptance that democracy was the price to be paid for continued international support is not yet entirely clear. However, a factor working for stability in the political system was the fact that ethnically-based parties have been relatively weak and minorities have voted quite heavily for the LDDP. As Clark has argued, if ethnicity is important in determining party identification, there is often little room for the compromise necessary in a democracy.[10]

In regard to the electoral system, the principle of universal suffrage was accepted and, despite strong nationalist sentiments in certain quarters, Lithuania included ethnic minorities in the political process. Unlike Estonia and Latvia, which produced rather tough citizenship laws for post-war immigrants as a result of demographic imbalance, Lithuania adopted an inclusive policy of offering citizenship to all residents of the state in 1989, the so-called zero option, rejecting the possibility of excluding immigrants from the political community until they had in some way proved their loyalty. Like its northern neighbours, Lithuania awarded automatic citizenship to those who had been citizens in 1940 and their descendants. Finally, as a consequence of the Lithuania-Russia state treaty in summer 1991, citizens of Russia who arrived in Lithuania between the passage of the citizenship law of 3 November, 1989 and the signing of the treaty could also obtain citizenship on the same basis as the pre-1989 residents. Immigrants arriving after December 1991 would not qualify for citizenship until they had passed a language test, demonstrated an understanding of the Lithuanian Constitution, and lived in the country for 10 years. The result of this liberal legislation was that in 1996 over 90 per cent of the inhabitants of Lithuania had become citizens. Since ethnic Lithuanians represented around 80 per cent of the total population, probably more than half of the members of ethnic minority communities became citizens. Membership of the political community was therefore territorially, not ethnically, based. Lithunia's record on citizenship has not attracted criticism from its internal minorities or from international agencies. Since the largest minority in Lithuania is made up of ethnic Russians, the citizenship law was a major factor in establishing better relations between Russia and Lithuania than existed between Russia and the other two Baltic states.[11]

Lithuania's record in protecting human and minority rights, however, has not been free from criticism. Although these rights are guaranteed both in the Constitution and in legislation, their actual enforcement has sometimes fallen short of international standards. The Council of Europe has made a number of criticisms about breaches of good human rights practice. In 1995 a Council representative singled out the 'Soviet-style' operations of the Prosecutor-General's office, which refused to allow arrested persons to consult lawyers before court hearings or to receive visits from relatives. There was also criticism of conditions in Lithuanian prisons, and the high average length of jail sentences compared with normal European practice. In 1998, after a long campaign by the Council, Lithuania declared a moratorium on capital punishment even though public opinion strongly favoured a tough penal regime.[12]

Human rights were also at issue when parliament approved a bill sponsored by Vytautas Landsbergis in 1998 restricting the rights of former KGB employees. The law defined the KGB as a criminal organization and banned all former employees from working in the civil service, in defence, in law and order institutions and in so-called strategic economic enterprises such as banks and private security services, for a period of ten years.[13] President Adamkus vetoed the law on several grounds. Firstly, he believed that decisions about the future of individual KGB officers should be made by the courts, not by a three-person body of politicians. He was dubious about both the legality of restricting employment rights in the private sector and the process of condemning people, not for their individual actions, but on the basis of guilt by association. Moreover, condemning KGB officers without also imposing sanctions on their Communist Party masters was clearly anomalous. Although the Constitutional Court supported the President's view that it was the courts which should decide on the continued employment of KGB officers, it affirmed the constitutionality of the law which emphasised the protection of state security rather than the punishment of offenders. Nevertheless, a scrupulous observance of human rights should ensure that even ex-KGB officers are given full legal protection and have their employment rights safeguarded until found guilty under indictment.[14]

The crime and corruption waves which struck Lithuania in the early years of the transition also had an impact on human rights. A prominent investigative journalist, Vitas Lingys, was murdered for his attempts to expose mafia gangs in Vilnius. The office building of a leading newspaper noted for its investigations of the links between criminals and corrupt politicians was bombed in October 1994. As an emergency measure against violent crime and extortion, parliament approved a law on preventive detention in July 1993, providing for the imprisonment of suspects without charge for a period of two months. It is impossible to say whether or not the law deterred crime but it proved ineffective in achieving convictions since only 43 out of 403 detainees between July and December 1993 were found guilty and only 29 were imprisoned. Finance organizations, whose corrupt activities faced exposure by investigative journalists, made threats against the press, which was also accused by government officials of smearing the reputation of the Lithuanian state. The press criticised the LDDP government for placing itself above the law, and for actively considering the imposition of censorship. Was it simply a coincidence, they asked, that when a journalist revealed a list of alleged LDDP crimes the government placed a ban on wine and beer advertising in the newspapers and proposed to fine any paper responsible for advertisements which were not proved to be completely accurate? These conflicts between politicians and the press seemed to diminish after the change of government in 1996. By 1998 the press and mass media had resisted the challenge to its freedom and was declared by Freedom House, an international human rights group, to be the freest in East Central Europe.[15]

The record of successive governments in protecting the rights of minorities has also been the subject of criticism from minority groups. These will be dealt with at greater length in a later discussion on the Russian, Polish and Jewish minorities. Nevertheless, it is worth noting at this point the positive aspects of minorities legislation. The Constitution and various legal acts offer protection for minority rights and there are many examples of attempts to improve majority/minority relations. Minorities comprise some 20 per cent of the total population of Lithuania. Ethnic Russians constitute approximately 8.2 per cent, Poles 7 per cent and Jews 1.2 per cent. Belarussians and Ukrainians make up between 1 per cent and 2 per cent each. In all there are some 109 nationalities living in Lithuania, among the oldest being the Tatars and the Karaites who were brought to medieval Lithuania and have remained ever since. The Lithuanian Constitution established the right of minorities to protect and support their languages, cultures and traditions. A law on ethnic minorities in 1989 gave members of national minorities the right to learn and be educated in their native language, to practise freely their religion, and to belong to ethnic organizations. They can have their own newspapers, schools and access to TV programmes. The state is required to help minorities develop their cultures and to promote their linguistic and religious identities without discrimination.

A further law in January 1991 provided for the use of the native language alongside the official state language in public offices and on information signs in areas with substantial minority groups of non-Lithuanian speakers. However, in the absence of clear criteria for the application of this law there was ample opportunity for bureaucratic discretion. The language law which made Lithuanian the official state language was, however, quite liberal. Although it enforced the use of Lithuanian by public employees, it did not extend the requirement to the private sector. The deadline for compliance by public servants was generously extended, offering officials adequate time to reach the required standard. We shall see later that a tougher state language law in 1995 aroused the strong opposition of the Polish-Lithuanian community.[16]

Evidently minority rights are firmly rooted in the Constitution and in national legislation. Moreover, there is a long historical tradition in Lithuania of ethnic and religious toleration, which is still alive though damaged by the experience of the Nazi and Soviet periods. However, there are allegations that the administration of these rights is sometimes characterised by bureaucratic indifference or subject to arbitrary interpretations by officials. Public comments can also be insensitive. Former President Brazauskas, whose commitment to the protection of minority rights was admirably firm, admitted that the inflexibility of the state bureaucracy often led to minorities taking offence. The very same terminology was used by Tomas Venclova when he wrote that Poles and Jews in Lithuania, though not mistreated, were often offended. He singled out the tendency of Lithuanian intellectuals to remain silent when there was intemperate public criticism of Polish- and Jewish-Lithuanians.[17] There are, nonetheless, various initiatives designed to strengthen understanding and tolerance between ethnic communities and to undermine negative stereotypes. In 1995 the European Union partially funded the Lithuanian Human Rights Centre to implement an education programme called 'The Integration of Lithuanian Minorities into Lithuanian Society'. One approach was to integrate minority schools into the Lithuanian general education system by concentrating on the shared experiences of pupils and teachers in both majority and minority schools, and on specific problems of minority education. Dispelling the myths surrounding other cultures was a major objective. The introduction of Holocaust studies into the school curriculum was an essential step in helping the Lithuanian communities, particularly the majority community of ethnic Lithuanians, come to terms with their past. In 1997 the Council of Europe sponsored a minorities conference in Vilnius for representatives of 17 East Central European states entitled 'Ethnic Minorities and Education'. Vilnius was chosen owing to Lithuania's 'well-founded' reputation in the area of minorities education. Finally, the Council of Europe approved a plan in 1998 to establish a European Institute for Dispersed Ethnic Minorities in Vilnius, the main task of which was to preserve and popularize the culture of thousands of small ethnic groups, beginning with those of northeast Europe. A leading spirit in these initiatives was Emanuelis Zingeris, a Jewish-Lithuanian deputy and human rights activist, whose work for human and minority rights in the Council of Europe and the *Seimas* has been exemplary. This enumeration suggests that though there are continuing concerns about minority rights in Lithuania, as we shall see when we look at minority groups in more detail, there are interesting attempts to build on Lithuania's centuries-old traditions of tolerance and good relations between ethnic groups.[18]

The politics of the 1990s have been characterized by confrontations between left and right on domestic issues, but on foreign and security policies there has been broad consensus. Although there were differences in emphasis and in detail between the major political groupings on economic policy, particularly on privatization policy, the pressures of the external agencies ensured that governments of either complexion had little room for manoeuvre. Economic policy remained a topic for conflict, but mostly on the level of rhetoric and stylised confrontation. In the main, governments pursued the only policy which would gain the support of international agencies and banks. In foreign policy, similarly, there was broad consensus that Lithuanian security depended on alignment with the West. Russia, it was agreed, would remain an important trading partner and neighbour. However, the parties differed on detailed policies towards Moscow, though there was unanimity on the necessity for the withdrawal of Russian troops, as will be seen in the final chapter.

Despite broad areas of agreement the parties divided sharply on a number of issues. For example, in the first post independence administration of *Sajūdis* led by Gediminas Vagnorius, with Vytautas Landsbergis as Chair of the Supreme Soviet (not to become the *Seimas* until the new Constitution was approved in the autumn of 1992), the government took a strong line on the restitution of private property, a general programme of rapid privatisation, and the de-Sovietization of politics. The latter was intended to 'purge' from public life and positions of responsibility all former political workers in the military, police and civil organizations, along with former Communists. As with the anti-KGB programme of 1998, opponents in the LDDP who were, after all, former Communists and had much to lose from 'de-Sovietization', condemned the guilt by association implied in this programme and criticised the policy for being divisive and inimical to reconciliation after the traumas of the Soviet period. The opposition was so substantial that the policy could not be sustained after the spring of 1992, particularly when investigations revealed that some members of *Sajūdis* had also been KGB informants in Soviet times.[19]

The LDDP also opposed the government's rapid privatisation programme, particularly the intense pace of property restitution and privatisation in the agricultural sector. While *Sajūdis* was intent on destroying every vestige of collectivization and restoring farms to their former owners, the LDDP wanted a slower pace and a more judicious re-organization, which would leave room for larger and more efficient farms and the retention of some collectives. They were afraid that the government's policy would perpetuate inefficient practices and limit the amount of investment available for improvements, a not ill-founded fear. The growing polarization of opinion was reinforced by the results of economic liberalization, which produced hyper inflation and excessive hardship. The opposition seized on *Sajūdis*'s economic record to promote their own policy for protecting living standards and social benefits, and maintaining employment.[20]

As Vagnorius's majority began to crumble in 1992 and an embarrassing stalemate took shape in parliament, the parties agreed to an election in October of that year. The defeat of *Sajūdis* was due as much to energy shortages and loss of traditional export markets as to its economic policy and its perceived incompetence. *Sajūdis*'s harping on the theme of anti-communism and the necessity for de-Sovietization was designed to win votes for a strong new government, with an executive presidency to cut through opposition. These themes did not appeal to the mass of rural voters or to the urban working class, who were suffering from inflation, unemployment and severe declines in living standards.[21] The result was the return to power of the former Communists, who obtained 42.6 per cent of the vote (and an overall majority in the *Seimas*) in comparison with 20.5 per cent for *Sajūdis*, 12.2 per cent for the Christian Democrats, 5.8 per cent for the Social Democrats and 2.4 per cent for the Centre Union. The great voting strength of the LDDP was in the rural areas where they won 31 of the

44 seats as against eight for the parties of the Right. By contrast the Right won in three of the five largest cities, but did less well in Vilnius and Šiauliai. The failure of the various centrist groups to unite in a coalition, and hence to ensure that their votes reached the minimum threshold for representation, also opened the way to the LDDP's victory. The LDDP benefited from the votes of almost all the Lithuanian Russians and about half the Polish electorate. It also projected itself not simply as the defender of living standards but also as a reformist party which would pursue Lithuania's national interests, if necessary in opposition to Russia. Its party organization was superior to that of the other parties; in contrast *Sajūdis* had not yet converted itself into a party and in any case was split between left and right.[22]

The victory of the LDDP in the *Seimas* elections was complemented by the success of Algirdas Brazauskas, the leader of the party, in the presidential election in early 1993, in which he defeated Stasys Lozoraitis, the Lithuanian Ambassador to the United States, by a large majority. Brazauskas had been a popular politican ever since he had shown courage and flexibility as the last leader of the Communist Party of Lithuania before it severed its connection with the CPSU. His position as a deputy premier in the *Sajūdis* administration added a necessary element of administrative competence. A man of stature in a political as well as a physical sense, the voters appeared to trust his experience in government and his more conciliatory approach to Moscow over that of any rival. As to whether he and the LDDP government would turn back the clock by adopting communist policies, or by compromising Lithuanian independence through a conciliatory stance towards Moscow, Brazauskas felt it necessary to assure the electorate that neither he nor his colleagues in government had been communists by conviction. On the contrary, they had recognised that the only way to get ahead was to join the party. The real ideologues, he concluded, were no more than three per cent of the total Party membership.[23]

Voters were evidently prepared to take a chance on the former communists and to believe their electoral promises. What the electorate seemed to want was a modification of the policies pursued by *Sajūdis*, which were devastating the country's economy. They demanded a greater emphasis on social welfare, employment, minimum wages, support for depressed industries, and subsidies for agriculture. The government would no doubt have liked to provide some or all of these things but it inherited an economic programme supported by international agencies, and it found it impossible to do more than tinker with it at the margins. Consequently, its record alienated what might be described as its natural supporters, the former collective farm workers and the factory labour force. By the end of its term of office in 1996 the SDP launched a bitter attack on the LDDP as the betrayer of working class interests and the supporter of capitalism and capitalists. The government's failure to alleviate poverty and solve social problems could not be excused by pleas that external agencies denied it any flexibility in setting economic policy. One of the leading figures in the LDDP, Česlovas Juršėnas, admitted it was paradoxical for a socialist party to pursue market reforms; others on the left would have put it more strongly. The facts were that the gulf between rich and poor had widened and that, shamefully, the LDDP had often identified with the rich.[24]

While critics on the Left denounced the LDDP's failure to deliver its promises, those on the Right criticised its economic record. The government had been too tentative, too conservative, and kept looking over its shoulder at its supporters to try to cushion the impact of its policy on living standards. It denounced its predecessor for being too harsh in implementing reforms. As we shall see in the next chapter, a more radical policy, as in Estonia, actually produced a quicker economic recovery and an earlier improvement in wages, employment and working conditions. For most of the LDDP's period in office economic indices were all negative. It was not until 1995 that the corner was turned. During the election, however, the party boasted of its economic successes: inflation down, the economy growing, privatization successful, exports up, improved quality of production, a stable currency. But the gains were at that stage relatively modest and could not offset the serious decline in living standards of large sections of the population.[25]

Even more damaging to its re-election hopes than its economic record was the widespread perception that the government was corrupt, too close to business, and favoured the former *nomenklatura* in its privatization policy, with the result that many ex-communist managers were able to take control of their newly privatized firms. By November 1995 the Lithuanian press had reported 43 corruption scandals in the government. Three months earlier the leader of the Centre Union, Romualdas Ozolas, began an aggressive anti-corruption campaign in which he made a series of allegations against members of the government, including the Prime Minister. He quoted from a World Bank report that 56 per cent of businessmen admitted paying bribes to government officials. Corruption was made easier, he said, by badly drafted and confusing laws, and ambiguous regulations. There were numerous articles in the press exposing officials' involvement in financial scandals and 'shadowy deals'. However, the President himself seemed to be above suspicion. He defended the government against allegations of corruption, but had to admit that a large number of people had acquired great wealth rather quickly, some had avoided paying taxes, and some had been assisted in their wealth-making activities by state officials. The most notorious example of a member of the government showing insensitivity to the obligations of office was Prime Minister Šleževičius's withdrawal of $30,000 from his bank account the day before the bank crashed, in December 1995. His touching story that he wanted to buy his wife a car for Christmas did not endear him to the mass of small depositors who received no advance warning and whose accounts were frozen after the crash. Although Šleževičius was ultimately forced to resign, he was very reluctant to do so and had to be forced out by the opposition of Parliament, President, and general public. This case along with others suggested that corruption reached right to the top and that the government had lost all credibility.

In the final chapter there will be a discussion of the government's foreign and security policy. Suffice it to say now that its achievements in foreign policy were quite impressive: Russian troops were withdrawn after long negotiations, a deal was agreed with Moscow over military transit through Lithuania, improved relations were negotiated with Poland, and associate membership of the EU was achieved. However, this generally sound progress in implementing a largely bipartisan foreign policy could not offset the scandals and hardship resulting from domestic policy. The result was a general election drubbing at the hands of the right and centre parties in October 1996, and a victory in the presidential election in early 1997 for Valdas Adamkus over the LDDP-backed Artūras Paulauskas.[26]

The unpopularity of the government was reflected in the local election results in 1995 when the LDDP was pushed into second place behind the conservatives of the Homeland Union and only just ahead of the Christian Democrats.[27] Disenchanted by the agricultural policy of the government, some of its rural supporters re-organized the Lithuanian Union of Peasants into an opposition party. Others distanced themselves from the government and either voted against or abstained. In preparation for the general election the following year the political parties attempted to form electoral pacts or coalitions. Parties of the Right were more successful in this than those on the Left. The LDDP vainly appealed to the SDP and the Peasants' Party for an election bloc to maintain the influence of social democratic ideas, should the opposition win. The Peasants' Party declared that rural dwellers wanted a party they could trust. Neither they nor the SDP believed that the LDDP was capable of overcoming the increasing divisions between rich and poor, or of accepting the demands of the Left for more state intervention in the economy. They were unimpressed by the LDDP's claim that it had been forced to compromise its principles by the conditions of international financial aid.[28]

The election was a crushing defeat for the government. The right wing and centre parties, Homeland, Christian Democrats (CDs) and Centre Union won 100 of the 141 seats in the *Seimas*, leaving the LDDP neck and neck with the SDP with a mere 12. Homeland with 71 seats formed a coalition government with the CDs with 16; the Centre Union, though refusing to join the coalition, was prepared to be in partnership with it and was given two ministerial posts. In terms of seats won, Homeland made substantial gains compared with 1992 but the CDs remained static. The largest winner in terms of percentage increase in share of the poll was the Centre Union. By the same criterion, Homeland's share increased by only 10 per cent while that of the CDs fell by some 40 per cent. These figures suggest that there was no great tide of enthusiasm for the parties of the right. Two factors help to explain the election result. The first was the low turnout, only 54 per cent in the first round of the elections. Most of the abstainers were erstwhile supporters of the LDDP, which received 650,000 fewer votes than in the previous election. This clearly reflected the disillusionment of large numbers of electors who had supported the LDDP programme in 1992 as an alternative to the policies of the previous *Sajūdis* government. The LDDP could still plausibly claim that its former supporters abstained rather than voted for the parties of the right. Secondly, a higher percentage of votes went to parties which failed to reach the threshold on the party list vote, thus strengthening those parties that did qualify, mainly the parties of the right.[29]

If the election result is analysed in terms of social class and geographical location, Clark has concluded that the CDs appealed to ardent Catholics on the Right, while Homeland attracted people with strong nationalist sentiments who identified with traditional Lithuanian values. Neither party attracted secular members of the emergent urban middle classes who increasingly were tempted by the Centre Union. Right wing parties fared better in the cities of Kaunas (a centre of nationalism), Klaipėda and Panévėžys while the Left had the advantage in Vilnius, where there was strong minority influence, and Šiauliai, traditionally left leaning. Similarly, the Left polled well in the rural areas of the East and North, the former being heavily populated by minority communities, especially Poles, and the latter formerly a stronghold of the state sector in agriculture, which was still suffering from the economic effects of privatisation. In the southern and western rural areas where the economy was stronger the conservatives attracted better support.[30]

Whilst the election victory was due mainly to disillusionment with the government rather than strong enthusiasm for the opposition parties, the fact remains that the electorate voted more heavily for the right and centre than in the previous election. It is worth examining the reasons for this. At the heart of the Homeland's programme was the dual promise to increase government spending, particularly on social programmes to help the less well off, and to cut taxes. Critics accused Homeland of offering contradictory proposals which simply could not be reconciled. Its response was to argue that faster economic growth would increase government tax revenue. The tax yield would also grow as a result of more efficient tax collection methods. Even with increased spending the budget deficit as a percentage of GDP was also projected to fall from nearly 4 per cent to less than 2 per cent by 1998–2000. The party's aim was to reduce income and corporate taxes as a way of stimulating enterprise and inducing foreign investment. Tax exemptions would also be offered for reinvested profits. Although many conservatives opposed the Currency Board system (which will be fully discussed in the next chapter) and wished to transfer its functions to the Central Bank, they were totally committed to maintaining the stability of the *litas*. There was some understandable popular scepticism on this score in the light of the earlier mixed results of the Central Bank in controlling inflation and maintaining currency stability. In view of the strong support for the Conservatives from the Confederation of Industrialists, scepticism about the new government's monetary policy was probably justified. Many businessmen wanted a devaluation of the currency to help exports. The conservatives were equally determined to ensure that commercial banks were strictly regulated to prevent bank crashes such as the one of December 1995. Although there were question marks over the new government's policies, a central plank in its economic programme, the emphasis on increased foreign invest-ment, was well-judged. A revised and speeded up privatization programme, and the continuation of stabilization policies, were central to this objective. On the political side, the conservatives' promise to introduce an ethics code for public officials as a means of controlling corruption was indispensable if public confidence in government was to be restored.[31]

At the time of writing, when the conservative coalition is only half way through its term of office, it is too early to offer an assessment of its performance. Its fiscal, monetary and privatization policies will be discussed in the next chapter. The Russian crisis had a significant economic impact in Lithuania and the effects of this are still being felt. Yet available data indicated a rate of economic growth of between 5 per cent and 6 per cent per annum by the end of 1998 in comparison with a year earlier, inflation in single figures in 1997 and projected to fall further, a stable banking system, the maintenance of the value of the *litas*, and foreign investment at record heights. On the other hand corruption allegations against both ministers and public officials were once again in the air, despite energetic efforts to purge the customs service and tax inspectorate of dishonest employees.[32]

Honesty and conscientiousness in government, high standards in public life, and the restoration of trust between the political leadership and the people were at the heart of the presidential election campaign in 1997. Many people threw their hats in the ring but the two leading candidates became Valdas Adamkus, who ran with the full support of the Right and Centre parties when Landsbergis was eliminated after the first round, and Artūras Paulauskas, who received the support of the Left. Brazauskas, though still popular, chose not to run, giving his record as a former Communist, his age and the lack of effective power of the Presidency as reasons for his decision. Both candidates were nominally independents, and there was little to choose between their programmes. Paulauskas, a former deputy Prosecutor-General, was supported by the LDDP, which did not think it could find a winning candidate after the withdrawal of Brazauskas. Though young, Paulauskas was associated in the public mind with the former *nomenklatura*. In this respect he was not dissimilar from Alexander Kwasniewski, the President of Poland. Unlike Kwasniewski, his political personality was rather grey and, unlike Brazauskas, he was unable to extend his support sufficiently beyond the boundaries of the Left when it came to the second round of voting. Adamkus, who was part of the

Lithuanian-American *émigré* community, had left Lithuania in 1944 at the age of 17, having been a member of the anti-Nazi underground and also part of a military detachment armed and equipped by the Germans, but not incorporated into the German army. Qualified as an engineer, he rose to become regional director of the US Environmental Protection Agency in the mid-west and, as part of his job, visited the Soviet Union regularly in the 1970s and '80s when he usually made time to return to Lithuania. He was active in *émigré* politics, lobbying for the Lithuanian cause with American politicians. After independence he acquired an apartment in Lithuania and obtained citizenship in 1992, spending some time each year in Lithuania. He helped Lozoraitis in his presidential election campaign in 1992–93. It took him some time to convince the electoral authorities and the courts that he was qualified to run; the polls, however, showed him coming a strong second to Paulauskas and therefore the most likely of the non-left candidates to defeat him. He won the election by the narrow margin of less than 1 per cent on a 74 per cent second round turnout, having increased his share of the poll between the first and second rounds by some 22 per cent compared with Paulauskas's increase of around 4 per cent.[33]

Since their programmes were not dissimilar and since, in any case, presidents have little power in domestic politics, the election turned on personality and morality. Adamkus was widely perceived to be energetic and dynamic despite his advancing years. He had the great qualities for a president of listening to conflicting arguments, acting as a moderator, and negotiating compromises. He was open, tolerant and democratic. He brought with him to a society which was still trying to cast off the remnants of sovietism and authoritarianism a set of liberal-democratic values which were deeply rooted and understood, rather than recently assumed. For many of his supporters the President 'should be a guarantor of justice and democracy, a force who consolidates society and unifies the nation'. Indeed, Adamkus's campaign slogan was 'Accord of the nation', which meant raising the interest of the nation above all party concerns and ambitions, strengthening trust and confidence between the people and their political leaders, and restoring the moral authority of the state. The election, according to Paul Goble, was part of the process of determining what kind of country Lithuania should become. Victory for Adamkus would give Lithuania five more years to escape from its communist past and to develop under the leadership of someone totally committed to democracy, free markets, and integration with the West.[34]

It is far too early to attempt an assessment of Adamkus's presidency. His actions in the first part of his tenure suggested that he was fulfilling the expectations raised during his election campaign. Opinion polls recorded unprecedented levels of popularity and his vetoes of legislation were in almost every case accepted by the *Seimas*. His conflict with Prime Minister Gediminas Vagnorius was an example of a President using the moral authority awarded him by the electorate to fulfil his campaign promises. Vagnorius chose to resign in April 1999 despite commanding a majority in parliament. Adamkus believed that the Vagnorius government was not living up to the standards he, Adamkus, had demanded in the election campaign. His critics accused him of exceeding his powers and acting like an American president. Formally he did not, but he brought to bear his moral authority and his enormous popularity to exert maximum pressure on his unpopular Prime Minister. The extremely popular, effective and 'new look' Mayor of Vilnius, Rolandas Paksas, was persuaded by Adamkus to accept nomination as Prime Minister, and the conservatives, of whom he was a member, agreed to support him despite initial reluctance to do so from Vagnorius and his followers.[35]

Whatever the government in power there were persistent long-term challenges which had to be confronted. We shall see the degree to which successive governments' attention was absorbed by the economics of transition. Of major importance to the identity and reputation of the state was its policy towards minorities, the most visible of which were the Poles, Russians and Jews. Although Lithuania's minorities policies were generally commended by international bodies, and were not so controversial as those of Estonia and Latvia, there has been a history of internal disagreements and resentments, some of which have had international repercussions.

Unlike its two northern neighbours Lithuania has not been involved in conflict with Russia over the treatment of its Russian minority; its major difficulties have arisen over Polish and Jewish issues. Nonetheless, controversy and disagreement have not been absent from Russian minority issues either. It is important to recall that the rights of minorities are firmly established in the Constitution and national legislation, and that the zero option gave citizenship, and therefore voting rights, to all residents in Lithuania in 1989. Controversy arose, however, over the participation of Russophones in supporting Moscow during the campaign for independence. The organization *Yedinstvo* was formed to counter Lithuanian nationalism, and many Russians opposed the Lithuanian declaration of independence in March 1990, supported the attempted coup in Vilnius in January 1991, and were on the side of the Moscow conspirators in August 1991. As a result a county heavily populated by ethnic Russians was deprived of self government for a time as punishment for its disloyalty. Yet the Russian community in Lithuania was not homogeneous. It was composed of Russian families who had lived in Lithuania for centuries, others who had migrated to Lithuania between the wars, and blue-collar workers, technical specialists and security personnel who arrived in the Soviet period. The latter have been described as Russians in Lithuania compared with the earlier settlers, many of whom were well integrated, spoke Lithuanian, looked to Vilnius rather than Moscow, and called themselves Lithuanian Russians. It is difficult to quantify, but it seems that the mass of Russians did not support *Yedinstvo* in its anti-independence campaign. Indeed, more than half the Russian inhabitants are thought to have voted for independence in the referendum of February 1991. It is probable that Russians differed widely in their attitudes to political and economic transition, but citizenship enabled them to pursue their interests through the democratic process. Most of them supported the LDDP; not a single deputy was elected from a Russian ethnic party.[36]

The Russian minority has been particularly exercised by the questions of education and language. At the beginning of the 1990s, however, not a single Russian school had been closed despite the reduction in the Russian-speaking population and the decisions of many ethnic Russian families to send their children to Lithuanian-language schools to ensure they were fluent in Lithuanian. In 1989 about 13 per cent of ethnic Russian children attended schools with Lithuanian as the language of instruction; by 1993 31 per cent did so. At the same time the withdrawal of Russian troops and their families reduced further the number of native Russian speakers in schools. Hence many Russian language schools in Vilnius and elsewhere became undersubscribed and Lithuanian schools became overcrowded and forced to work in shifts. Controversy arose over proposals to rationalise provision by converting Russian language into Lithuanian language schools. It is probably only a matter of time before this rationalisation has to go ahead since the number of ethnic Russian children attending Lithanian language schools is expected to go on rising. At the beginning of the 1990s there remained 42 courses at the University of Vilnius which could be studied in Russian. There was a state drama theatre in Russian, many Russian language papers and magazines, a literary magazine, and Russian language radio and TV programmes. However, Russians in Lithuania also felt threatened during the transition period by the state language law. They demanded the right to use the Russian language in all public offices, and for students in special schools and in higher education to be taught in Russian. Although they failed in this, in practice there was flexibility in using Russian and other minority languages in areas where minorities constituted 'a significant share of the population'. However, the revised language law of January 1995 established Lithuanian as the only state language even in areas where a national minority was in the majority. This was in contradiction to the law on ethnic minorities but addressed a widespread worry that without such a law there would be no inducement for Russian-speakers ever to learn the Lithuanian language, which after all was the language of the state. At the beginning of 1998 the *Seimas* was considering amendments to remove these inconsistencies. Evidently the language law remained a sensitive issue for the minorities, though it seemed to be more of a grievance for the Polish Lithuanians than for the Russian minority, perhaps because the Russian population was more dispersed.[37]

The political leadership of the Polish minority organized its own political party, Polish Electoral Action. As an ethnically-based party, it tended to be inflexible in its relations with the majority parties. It is questionable whether the nationalism of the Polish leadership was fully representative of the rank-and-file since there is some evidence that the majority of Lithuanian Poles were indifferent to the preoccupations of their leaders. At the same time, Lithuanian nationalists were hostile to the Poles on historical and more recent grounds. They did not consider the 'Poles' to be more than polonized Lithuanians who spoke a version of Belarussian with infusions of Polish and Lithuanian words. The Lithuanian Poles, however, declared that their language was a dialect of Polish. The Polish community was concentrated in South-East Lithuania near to the Polish and Belarus borders, mainly in the regions of Šalčininkai (around 82 per cent of the population), Vilnius (67 per cent), Trakai and Švenčionys. Most of the intelligentsia migrated to Poland in 1944 when Soviet occupation ensured that the region would become part of Lithuania. During the Soviet period Polish families found it advantageous in career terms for their children to learn Russian. Owing to their dislike of Lithuanian nationalists and what they imagined would be the cultural consequences for them of Lithuanian independence, the Polish community opposed the aims of *Sajūdis*, taking their opposition as far as declaring autonomy on a number of occasions between 1990 and the August coup in 1991. The Lithuanian Parliament cancelled these measures, but on the last occasion *Sajūdis* temporarily dissolved the local councils in retaliation for their support of the attempted coup. The main representative body of the Polish Community, the Union of Poles, supported a new political party, Polish Electoral Action, which was created in 1994 in response to the recent electoral law stipulating that election candidates could only be nominated by political parties. Although the leaders of the Polish community were in constant conflict with the government, much constructive and 'organic' work was undertaken by members of the Polish community, both to raise living and educational standards and to co-operate effectively with the majority community.[38]

Four issues in particular aroused the active opposition of the Polish leadership to the national government: these were the language law, the changing of electoral boundaries, the enlargement of the city of Vilnius into the surrounding district, and the proposal to raise electoral thresholds in general elections. Polish Electoral Action opposed the 1995 language law, demanding instead that Russian and Polish become official languages in areas of dense Russian and Polish settlement. They linked this with demand for Polish place names to be used alongside Lithuanian ones and for Polish surnames and first names be written in Polish orthography, for example in passports. When it became clear that there was going to be no immediate amendment or supplementation to the language law, the Vilnius district council, in December 1997, decided to permit the use of the Polish language alongside Lithuanian in local government in the area, on the grounds that such a law conformed with the law on ethnic minorities. When the regional government invalidated the decision, the district decided to appeal to the courts to determine the constitutionality of the law. President Adamkus expressed a common fear among the majority of Lithuanian citizens that if minority languages achieved absolute equality of status with the majority there would be no incentive for minorities to learn Lithuanian, which Adamkus believed every citizen should do.[39]

A second controversy arose over the proposal to extend the boundaries of the city of Vilnius, which the Polish community opposed because it was felt that this would interfere with, or even prevent, the restitution of Polish property in the area. Residents were not opposed to expansion *per se* but wanted restitution first.[40] Moreover, Poles had been antagonized in 1992 by the government's decision to redraw electoral boundaries in order to create a number of smaller districts incorporating substantial ethnic Lithuanian populations. This gerrymandering was resented by the Poles, but was an understandable reaction by the majority to what they regarded as the treachery of the Poles during the transition crisis in 1990–91.[41] Finally, the decision before the 1996 general election, to raise electoral thresholds to 5 per cent without any special dispensation for minority parties, was viewed as discriminatory against the Polish community because its effect would be to diminish significantly Polish minority representation in the *Seimas*.[42]

Two factors prevented these disputes escalating into major political crises. The first was the fruitful and friendly co-operation between the Lithuanian and Polish governments, particularly after the signing of the Lithuanian-Polish state treaty in 1994, which inhibited the activities of the Lithuanian Poles.[43] Secondly, ethnic Polish opinion in Lithuania seemed to be both more moderate and more indifferent in its approach to the various issues which exercised the leadership. Although the Lithuanian-Polish community supported Gorbachev's referendum in March 1991 in favour of keeping the Soviet Union together, the turnout was low and unrepresentative. A month earlier more than half the Poles, it has been calculated, voted for independence in the referendum organized by the Lithuanian government. During the first congress of Polish Electoral Action the organization split, the defectors leaving owing to the alleged extremism of the leadership. This suggested that the party could not claim to represent the Polish community, particularly when large numbers of Poles voted for the LDDP, mainly on economic grounds. Since the LDDP was not notable for supporting the Poles in their language and other demands, voting for them implies that economic issues had priority over ethnic ones for significant numbers of Poles. Brazauskas denied that the Lithuanian authorities were curtailing the rights of the Polish minority and claimed that the leadership specialized in making allegations which were not supported by the majority. In 1998 a survey showed that ethnic problems were more interesting to politicians than to the rank-and-file. Only two per cent of Poles believed that intolerance was a problem. Moreover, turnout in elections in Polish constituencies was generally some way below the national average, suggesting no great wellspring of feeling in support of ethnic issues. The establishment of the Polish Congress was a step away from confrontational politics too. Czesław Okinczyc, a parliamentary deputy and president of the Centre for Mutual Understanding, as well as a member of the Polish Congress, believed that stereotypes had to be confronted because there was a damaging inconsistency between the legal rights of minorities and deep-seated attitudes which reflected prejudice and hostility. The aims of the Congress were to improve the economic and social conditions for the Polish community and to encourage closer relations between the majority and minority communities, and between the Polish community and the government, with the object of obtaining practical benefits, especially in the field of education. At the same time the Lithuanian Union of Poles has become more constructive in its approach, with an emphasis on economic improvements, investments in Polish areas, and the fostering of Polish culture and identity. At the end of the decade the signs are good for a period of more harmonious and constructive relations between the Polish minority and the majority communities.[44]

The relations between the ethnic Lithuanian majority and the Polish minority have had, and may continue to have, an impact on the international position of Lithuania, particularly in respect of Poland and the European Union, since poor neighbourly relations could adversely affect Lithuania's progress towards membership of the EU. What might be called the Jewish question in Lithuania, in its impact on the world Jewish community, the Israeli government, and the United States, has the capacity to inflict considerable damage on Lithuania's international reputation. Since the Jewish minority benefits from the minority rights legislation and the protection of the Lithuanian Constitution along with all other minorities, why are Lithuanian-Jewish relations so controversial and emotive?

The heart of the matter relates to history, and the alleged failure of Lithuanians to come to terms with a particularly painful part of it. Jews had lived in the Lithuanian lands for centuries and Vilna was considered to be the northern Jerusalem, not only because of the high proportion of Jews in the population, but because it was an internationally renowned centre of Jewish culture and scholarship. In 1939 there were some 220,000 Jews in Lithuania. Only around five percent of that number survived the war. This was the highest proportion to be killed in any Nazi occupied country. The Nazis were overwhelmingly responsible.[45]

The questions which surfaced after independence were generally unwelcome and potentially embarrassing, and their impact was greater since there had been little attempt to come to terms with this aspect of the war experience in the intervening years. To what extent did Lithuanians themselves participate in the Holocaust? How many of them took part in pogroms even before the arrival of the Nazis in June 1941?

Had war criminals been pardoned by Lithuanian courts in the rehabilitation of Lithuanian political prisoners? Were alleged Lithuanian war criminals not being brought to trial in Lithuanian courts? Was the Jewish emigration of the 1970s, which increased again after *perestroika*, evidence of the inability of the ethnic Lithuanian population to discuss openly, and to try to come to terms with, the role of Lithuanians in the war-time genocide?

Considering the Holocaust first, it is believed that some 40 pogroms took place across Lithuania before the Nazis arrived. Later the Lithuanian security police were charged with having helped the Nazis to round up Jews, and the 12th Police Battalion was allegedly involved in shooting Jews both in Lithuania and Belarus. The question is, how many Lithuanians were actually involved in the genocide and what was the extent of their participation? Jewish critics have alleged that Lithuanians have been reluctant to investigate this question. Until they come to terms with their history, it is argued, Lithuanians will continue to be subject to accusations, some of them wild and exaggerated, and the burden of suspicion will remain. On the other hand, there is a suspicion among Lithuanians that, however thorough the historical research and however conclusive the findings, unless they are unequivocally condemned for their war time record and abjectly apologise for their involvement, their accusers will remain dissatisfied and the issue will rumble on. The conclusions of one piece of research by a Lithuanian historian is unlikely to appeal to the critics, namely that the KGB and its successors sent faked evidence about Lithuanian-American war criminals to the American authorities to incriminate and discredit the Lithuanian *émigré* community. It is this evidence on which allegations of war criminality are at least partially based.[46]

One major problem for Lithuania was that, owing to its isolation behind the iron curtain, it had not appreciated the acute sensitivity in the West to the Holocaust, and the generally accepted opinion that this event was unique in the annals of human atrocities. Statements from official defenders of Lithuania in the early part of the 1990s struck the wrong note, however understandable such sentiments were. It was not enough simply to state in exculpation that Lithuania was a helpless and occupied country when the Holocaust had occurred there. Nor was it acceptable to refer to the number of Jews in the Lithuanian Communist Party and security services which carried out genocide against Lithuanians, as if that acquitted Lithuanians of blame for the Holocaust. As is now known, many Jews were the victims of the Soviets as well as the Nazis, and more Jews were deported than Lithuanians, measured as a proportion of their respective populations. Again, reference by the head of the war crimes prosecution in Lithuania to Jewish interest in the Holocaust in Lithuania as a 'soap bubble' and the Jews' real interest being in the restitution of their property was insensitive, to say the least. Counter charges, however justified in themselves, such as the allegation made by former Lithuanian political prisoners and deportees that no one had been prosecuted for genocide against the Lithuanians and that action should be taken against Jewish participants in Soviet-organized genocide, many of whom lived in Israel, did not remove the Lithuanians' responsibility to investigate thoroughly allegations of Lithuanian participation in the Holocaust. Although Bishop Boruta claimed that the Nazis created the myth of the Lithuanians as Jew killers, he was at least prepared for an open examination of the evidence to establish a complete list of participants in the Nazi and Soviet genocides in Lithuania, perhaps recognizing that simple defensiveness in the face of allegations was inadequate.[47]

Strong moral leadership was given by President Brazauskas, whose speech to the Council of Europe in April 1994 was an eloquent and dignified response to allegations about the Holocaust and the treatment of war criminals. He deplored the extermination of the Jews in Lithuania and regretted that some Lithuanians participated in it. He promised that all war criminals would be exposed and punished. But he also rightly pointed out that many Lithuanians had sheltered and protected Jews from the Nazis, and the Israeli government had recognized and honoured these self-sacrificing people. On a visit to Israel in March 1995 Brazauskas followed the example of Prime Minister Šleževičius the previous year in publicly apologising for the crimes committed by individual Lithuanians against the Jewish people and promising to bring war criminals to justice. But this moral lead was rejected by some Lithuanians, who believed that Brazauskas's apology was a humiliation for Lithuania; who, they asked, would apologise for the crimes against Lithuania? Moreover, they argued, a public apology reinforced the international image of Lithuanians as war criminals. In 1994–95, therefore, defensiveness was still the prevailing mode of response to the allegations of war crimes.[48]

Closely linked to this issue was the Lithuanian government's response to allegations about the rehabilitation of war criminals and the failure to bring charges against them. Tens of thousands of Lithuanians were convicted in Soviet courts after the war and deported. After independence many of these were officially rehabilitated and qualified to receive pensions and other benefits. There were accusations by Jewish organizations that some of the rehabilitated were notorious Nazi collaborators. In this case the Lithuanian response was constructive, offering to review controversial cases in cooperation with Israel. By 1994 about 50,000 cases of rehabilitation had been reviewed by the Lithuanian High Court and in about 1,000 of these the rehabilitation was not confirmed.[49]

These were individuals who had already been convicted of crimes. A more pressing and difficult issue was the question of alleged Lithuanian war criminals who had escaped to the West during the Second World War and had never faced prosecution. The activities of the nazi-hunting Simon Wiesenthal Foundation, in co-operation with the American authorities, produced *prima facie* evidence of collaboration with the Nazis on the part of a handful of Lithuanian-Americans. These either returned to Lithuania voluntarily or were expelled from the United States for presenting false information in their residence applications decades earlier. The best known of these individuals was Aleksandras Lileikis, who during the Nazi occupation was in charge of the security department of the Lithuanian police in the Vilnius region. There is no space to discuss the individual cases. Suffice it to say that the Lileikis example offered an opportunity to test Lithuania's willingness to confront its past. Failure to bring charges was attributed by the Lithuanian prosecutor's office to the absence of hard evidence, and by the Simon Wiesenthal Foundation to foot-dragging by the Lithuanian authorities. The question was complicated by Lileikis's illness, and whether he was fit to stand trial. The issue of prosecution of war criminals became a *cause célèbre*, involving the Israeli Parliament, the Israeli government, the American Jewish Foundation, the United States Department of Justice, and the US House of Representatives. Lithuanian receptivity to the requests of foreign representatives was not assisted by the hyperbolic language adopted by Efraim Zuroff of the Simon Wiesenthal Foundation. It is undeniable that there was great anger in Israel at the failure to bring prosecutions against alleged war criminals living in Lithuania, despite Brazauskas's promise two years earlier to bring them to justice. Although the Lithuanian prosecutor's office faced genuine difficulties in bringing 'cast-iron cases' five decades after the alleged events, observers such as Emanuelis Zingeris claimed that it was not utilising all the available evidence. It appears that Brazauskas's apology of 1995 did not command widespread popular support in

Lithuania and was therefore ahead of its time. In the light of the extreme problems associated with the war criminals issue perhaps an increasing number of Lithuanians are now prepared to accept Zingeris's judgment that Lithuania would only seriously come to grips 'with this chapter from its war-time past' when Lileikis and other alleged war criminals were brought to trial and Lithuanians showed that they were taking the issue of war crimes seriously.[50]

For those seeking improved relations between Lithuania and the world Jewish community there are a number of initiatives which seek to demonstrate Lithuanian good will, and to show that the traditional harmonious relations between Lithuanians and Jews over many centuries can be recreated, despite the distortions created by the Nazi and Soviet occupations. The impressive official celebrations in Vilnius to mark the two hundredth anniversary of the death of the famous Jewish scholar, Gaon Elijahu, in 1997, though marred by controversy, were a good first step in this direction. Holocaust education is being introduced in pilot schemes into the school curriculum. The Ministry of Culture, in association with the National Library, is creating a new section to accommodate the Vilnius collection of thousands of Hebrew and Yiddish books. Finally an *ad hoc* committee set up in Vilnius to review Jewish issues reported in December 1998 that it would establish two sub-committees to coordinate and supervise the experts investigating crimes committing during the Holocaust and the Soviet occupation.[51]

In the light of Lithuania's ambitions to join the whole panoply of European and Euro-Atlantic organizations it is difficult to believe that the progress made in consolidating democratic values and in protecting human and minority rights will be reversed. Two parliamentary and two presidential elections since 1992 have shown a mature acceptance of popular verdicts at the polls. Political corruption, however, though diminishing, has proved hard to eradicate, and continues to threaten the reputation of the democratic system. President Adamkus has made honesty and conscientiousness in government a central feature of his presidency. For younger generations of Lithuanians he is likely to provide an important example of democratic practice and pluralistic values. Although governmental treatment of the Polish minority receives international approval, and there seems to be no strong groundswell of support for Polish Electoral Action, the main political organization of the Lithuanian Poles, the Polish question continues to have the capacity to inflame relations between the majority and minority communities. The controversy over Lithuanian participation in the Holocaust is far from being settled. There needs to be a continuing effort to investigate and publicize the Lithuanians' role during the Nazi occupation, and to continue the recent attempts to improve relations between the ethnic Lithuanian and the Lithuanian-Jewish communities. When the full implications of membership in the European Union become widely known, political opposition to the application for membership is likely to grow, and this will strengthen the nationalist elements in the body politic. The impact of global homogenization on Lithuanian culture and national identity will be a cause of anxiety among the older generations of Lithuanians at least, with the possibility that the disaffected will mobilize to defend core values. Furthermore, membership of the EU, with its inevitable transfer of an increasing proportion of national sovereignty to Brussels, is likely to alarm Lithuanians who have only recently recovered their powers of self-determination. The close association of Lithuanian culture, the countryside, and rural life will enhance the struggles of farmers to defend their livelihoods against threats from foreign imports of food, particularly from the EU. Overall, the election of Adamkus was very significant in consolidating progress towards democracy and in restraining the less tolerant elements of Lithuanian nationalism. Adamkus is likely to remain a critical influence in promoting the idea that Lithuanian citizens of whatever ethnicity are equal and equally respected members of the Lithuanian body politic.

1 Opinion polls at the time of writing show that Adamkus has gained the confidence and support of the overwhelming majority of the Lithuanian population

2 Advocates of the idea of the restitution of the Lithuanian state were embarrassed that the last Constitution before the Soviet occupation, that of 1938, provided an authoritarian model of government

3 Algimantas Jankauskas, 'The Re-emergence of Multi-Partism in Lithuania' in Wojciec Lukowski and Konstanty Adam Wojtaszczyk eds., *Reform and Transformation in Eastern Europe* (Warsaw: Elipsa Press, 1996), pp.94–5; Ole Norgaard et al., *The Baltic States after Independence* (Cheltenham: Edward Elgar, 1996), pp.3, 64–5, 72–5; *The Baltic Independent* (hereafter *TBI*), 20–26 March, 10–16 April, 29 May–6 June, 1992; *The Baltic Times* (hereafter *TBT*), 30 October–5 November, 1997

4 Two have been passed since then: municipalities were elected for three years instead of two; and in 1996 the sale of land to foreigners was permitted under pressure from the EU

5 Brazauskas asserted that, until there was a change, the president would continue to be a 'paper tiger': *TBT*, 8–14 August, 1996, 30 October–5 November, 1997, 12–18 February, 1998; BBC, *Summary of World Broadcasts* (hereafter *SWB*) SU/2681 E/3 3 August, 1996; Norgaard et al., p.75

6 Failure requires a run off between the top two candidates in the poll

7 Norgaard et al., pp.79–80; *TBT*, 4–10 July, 15–21 August, 1996; Terry D. Clark, 'The Lithuanian Political Party System: A Case Study of Democratic Consolidation', *East European Politics and Societies*, vol.9, no.1, Winter 1995, p.46

8 Tomas Venclova, 'A Fifth Year of Independence: Lithuania, 1922 and 1994', *East European Politics and Societies*, vol.9, no.2, Spring 1995, p.352; *TBI*, 16–22 July, 5–11 November, 1993; *SWB* SU/2659 E/2 9 July, 1996

9 *TBI*, 20–26 December, 1991, 7–13 May, 1993, 24 December–6 January, 1994; *SWB* SU/1877 E/1 21 December, 1993; Venclova, pp.352–5

10 Clark, pp.41–2, 61–2; *TBI*, 17–23 October, 1996; Norgaard et al., pp.117–8

11 'Treaty between the Republic of Lithuania and the Russian Soviet Federated Socialist Republic on the Basis for Relations between States', *Lithuanian Foreign Policy Review*, 98/1, p.121; Norgaard et al., pp.183–4; Graham Smith, Aadne Aasland and Richard Mole, 'Statehood, Ethnic Relations and Citizenship' in G. Smith ed., *The Baltic States: the National Self-Determination of Estonia, Latvia and Lithuania* (Basingstoke: Macmillan, 1994), pp.183, 189–90; *Izvestiia*, 1 November, 1991; *TBI*, 13–19 December, 1991

12 *TBT*, 22–28 July, 1999; *SWB* SU/2290 E/4 29 April, 1995; *Baltic News Service* (hereafter *BNS*), 25 April, 1997; *TBT*, 11–17 July, 1996

13 Exemptions were permitted for any KGB operatives who had worked only against 'real criminal elements' or who ceased employment with the organization no later than one day after the re-establishment of independence on 11 March, 1990

14 *TBT*, 11–17 June, 23–29 July, 1998, 11–17 March, 1999

15 *TBI*, 15–21 October, 1993, 24 December, 1993–6 January, 1994; 11–17 November, 1994, 17–23 February, 1995, 3–9 November, 17–23 November, 1995, 12–18 January, 1996; *TBT*, 4–10 March, 1999

16 *TBI*, 26 May–1 June, 1995; *Lithuania and the European Union: Towards Accession Negotiations* (Vilnius: Du Ka UAB, 1998), p.4; Smith, Aasland and Mole, pp.190–2; Norgaard et al., pp.185–7; Halina Kobeckaite, 'Ethnic Groups and National Minorities in Lithuania', *The East Express*, no.22, July, 1991

17 Venclova, p.361; *Izvestiia*, 17 February, 1993

18 *TBT*, 13–19 February, 24–30 April, 11–17 December, 1997

19 Terry D. Clark, 'Coalition Realignment in the Supreme Council of the Republic of Lithuania and the Fall of the Vagnorius Government', *Journal of Baltic Studies*, vol.XXIV, Spring, 1993, pp.60–2

20 Venclova, pp.353–4; *TBI*, 30 October–5 November, 1992; Clark, 'The Lithuanian Political Party System', pp.41, 59; Clark, 'Coalition Realignment', pp.61–2

21 *sajūdis*'s employment of a bishop in a TV broadcast to denounce a vote for the LDDP as a vote for the Devil was unlikely to have won many converts to the right wing cause: *TBI*, 20–26 November, 1992

22 *Baltic Observer*, 26 November–3 December, 1992; *The Economist*, 21 November, 1992; *TBI*, 30 October–5 November, 1992, 23–29 July, 1993; *TBT*, 10–16 October, 1996 (Article by Lars Johannsen); Jankauskas, pp.94–5; *TBI*, 25 September–1 October, 1992; Clark, 'The Lithuanian Political Party System', pp.55–8; Clark, 'Coalition Realignment', pp.60–1

23 *The Times*, 16 February, 1993

24 *SWB*, SU/2026 E/2 20 June, 1994; Terry D Clark, 'The 1996 Elections to the Lithuanian Seimas and their Aftermath', *Journal of Baltic Studies*, vol.XXIX, no.2, Summer, 1998, p.135; *The Economist*, 12 October, 1996; *The Times*, 29 October, 1996

25 *TBI*, 7–13 January, 1994; *TBT*, 12–18 September, 17–23 October, 1996 (the latter contains an excellent analysis by Lars Johannsen

26 *TBI*, 1–7 July, 1994, 27 October–2 November, 3–9, 17–23 November, 1995, 12–18 January, 26 January–1 February, 1996; *TBT*, 12–18 September, 1996; *SWB* SU/2087 E/1 30 August, 1994

27 *SWB* SU/2264 E/2 29 March, 1995; *TBI* 31 March–6 April, 1995; Norgaard et al., pp.95–6

28 *SWB* SU/2651 E/7 29 June, 1996; *TBT*, 11–17 July, 1996

29 *The Economist*, 12 October, 1996; *The Times*, 29 October, 1996; *SWB* SU 2758 E/1 1 November, 1996, SU/2754 E/2 28 October, 1996; Clark, 'The 1996 Elections', pp.137–44

30 Clark, 'The 1996 Elections', pp.144–5; Clark, 'The Lithuanian Political Party System', pp.56–9

31 Clark, 'The 1996 Elections', p.142; *SWB* SU/2479 E/1 22 October, 1996; *TBT*, 21–27 March, 1996, 12–18 September, 17–23 October, 28 November–4 December, 14–20 November, 12–18 December, 1996

32 *TBT*, 4–10, 11–17 March, 1999

33 *SWB* SU/2724 E/3 23 September, 1996; *BNS*, 2, 15 September, 1997; Bank of Finland, *Economic Report*, 12 January, 1998; *The Economist*, 10 January, 1998; *Financial Times*, 6 January, 1998; *Le Soir*, 6 January, 1998; *Le Figaro*, 22 December, 1997; *TBT*, 27 March–2 April, 13–19 November, 11–17 December, 1997, 8–14 January, 1998

34 Adamkus's 'presidential' qualities compared favourably with those of Landsbergis, whose 'exasperating intransigence' and 'aristocratic cold manners' and 'theoretical debates' had alienated Lithuanians, despite their recognition of his great role in the achievement of independence: *TBT*, 30 April–6 May, 1998; *Financial Times*, 6 January, 1998; *Le Figaro*, 22 December, 1997; *Associated Press*, 30 January, 1998

35 *TBT*, 6–12, 13–19, 20–26 May, 1999

36 For general background to minorities issues and to the legal framework for the protection of minority rights see Ministry of Foreign Affairs and Department of Nationalities of the Republic of Lithuania, *National Minorities in Lithuania* (Vilnius: Centre of National Researches of Lithuania, 1992). See also *Financial Times*, 10, 16 September, 1991; Kobeckaite, loc.cit; V. Stanley Vardys and Judith B. Sedaitis, *Lithuania: the Rebel Nation* (Boulder, CO Westview Press, 1997), pp.112–14; Clark, 'The Lithuanian Political System', pp.57–8; Vesna Popovski, 'Ethnic Minorities in Lithuania', paper delivered at the Lithuania Studies Day, School of Slavonic and East European Studies, University of London, 24 January, 1997

37 Norgaard et al., pp.185–6, Kobeckaite, loc.cit; *SWB* SU/2191 E/4 3 January, 1995, SU/2218 E/1 3 February, 1995; *TBT*, 29 January–4 February, 1998

38 About half of ethnic Poles in Lithuania described relations between them and ethnic Lithuanians as normal, 46 per cent of Poles said relations were good or very good, and nearly 5 per cent described them as confrontational and hostile. Only 1.4 per cent of ethnic Poles living in Lithuania regard themselves as discriminated against: Popovski, op.cit; Stephen R. Burant, 'Overcoming the Past: Polish-Lithuanian Relations, 1990–1995', *Journal of Baltic Studies*, vol.XXVII, no.4 Winter 1996, pp.316–7; Norgaard et al., pp.176–7, 184–5; *SWB* SU/2090 E/5 2 September 1994, SU/2251 E/4 14 March 1995

39 *SWB* SU/2191 E/4 3 January, 1995, SU/2218 E/1 3 February, 1995; *TBT*, 8–14, 22–28 January, 1998; Norgaard et al., pp.185–7

40 *TBI*, 29 May–4 June, 1992, 23–29 July, 1993; *SWB* SU/2651 E/5 29 June, 1996; *TBT*, 13–19 June, 1996

41 *TBI*, 13–19 November, 1992

42 *TBT*, 15–21 August, 1996

43 *SWB* SU/1983 E/2 28 April, 1994

44 *Index on Censorship*, October, 1992; *SWB* SU/2570 E/3 26 March, 1996. SU/2722 E/1 20 September, 1996; *TBI*, 3–9 November, 1995; Norgaard et al., pp.186–7

45 Popovski, op.cit; *TBT*, 4–10 July, 1996

46 *TBI*, 10–16 March, 17–23 November, 1995; *TBT*, 4–10 July, 1996; 18–24 September, 1997, 15–21 January, 29 October–4 November, 5–11 November, 1998; Venclova, pp.362, 364. Dov Levin has described numerous pogroms in Lithuanian towns before the Nazis arrived

47 *TBI*, 7–13 October, 11–17 November, 1994, 17–23 February, 1995, 12–18 November, 1998

48 *TBI*, 22–28 April, 30 September–6 October, 1994, 10–16 March, 1995, 17–23 November, 1995

49 *TBI*, 2–8 July, 2–8 September, 11–17 November, 1994; *SWB* SU/1939 E/4 7 March, 1994, SU/2351 E/2 10 July, 1995

50 *TBI*, 7–13 October, 1994, 17–23 February, 10–16 March, 1995; *TBT*, 4–10 July, 1996, 6–12 February, 6–12 March, 14–20 August, 27 November–3 December, 1997, 5–11 February, 21–27 May, 1998. The Lithuanian historian, Arvydas Anusauskas, has found no evidence that Lileikis was involved in the genocide of the Jews; indeed, he claims that some of Lileikis's subordinates were saving Jews. Of course, as head of the security police in the Vilnius area, Lileikis could not avoid collaborating with the Germans

51 *TBT*, 4–10, 18–24 September, 1997, 26 November–2 December, 1998

Chapter 5

THE LITHUANIAN ECONOMY AFTER INDEPENDENCE

The tasks facing the political leadership in the newly independent Lithuania were immense and unprecedented. To be sure, the former Soviet satellite states of Central and Eastern Europe faced similar challenges, but the position of the ex-Soviet republics such as Lithuania was even more complex since they had not possessed the modest degree of economic and political independence, nor the separate economic and financial institutions, enjoyed in Poland, Czechoslovakia and Hungary. Moreover, although the latter had set out on the path of economic transition a little earlier than the Baltic states, they had accumulated too little historical experience to offer a clear and uncontentious guide to the latecomers. Again, though there were many international agencies such as the World Bank and the International Monetary Fund (IMF) which were well qualified to offer advice on economic development and anti-inflation strategies, they too had little or no experience of plotting the transition from Soviet-type command economies to liberal capitalism. Essentially, Lithuanian leaders knew, or thought they knew, the desired outcome of their economic reforms, but reaching that destination was a formidable and problematic task.

What then were the economic objectives of the Lithuanian political leadership? Shortly after their independence was recognized the Baltic prime ministers, meeting on 30 October, 1991, declared their aim to be a successful transition to a market economy involving a fast and radical implementation of market reforms. This transition, they believed, could not be separated from overall democratic reforms nor from the international integration of the Baltic economies. The political imperative was to move rapidly away from the command economy which had demonstrably failed, to consolidate political independence by means of economic reforms, and to re-establish the trading and financial links with the West which Lithuania had enjoyed in the inter-war period. However, establishing a liberal market economy was also recognized as the essential condition for achieving high and sustainable economic growth and prosperity.[1]

The following discussion focuses on two stages of economic reforms in post-independence Lithuania. The first is concerned with the economics of transition—the process of disentangling the Lithuanian economy from the institutions and restraints of the Soviet economic system and creating the essential elements of a free market economy. This process involved three crucial developments: liberalization, stabilization and privatization. The first meant freeing economic decision making from the centralized planning and price controls of the Soviet system; the second involved establishing a new economic equilibrium after the demolition of the Soviet system, based on free market institutions and mechanisms; and the third meant transferring state ownership of property to private owners and encouraging the formation of new privately-owned enterprises. The second stage is the period through which Lithuania is currently progressing. During it, successive Lithuanian governments have had to choose policies for maximising economic growth. To what extent have they succeeded in generating prosperity, and what more needs to be done if Lithuania is gradually to narrow the gap in living standards between herself and the advanced economies of the Baltic region?

THE ECONOMICS OF TRANSITION

In approaching the question of transition the Lithuanian leadership was aware of the so-called 'shock-therapy' approach adopted in the neighbouring state of Poland under Finance Minister Balcerowicz. Moreover, an economic blueprint for Lithuania had been developed by *Sajūdis* economists as early as 1988, and ways of implementing it had been the subject of considerable discussion during the debates on economic autonomy in 1988 and 1989. The general approach of the International Monetary Fund (IMF) was also understood in Lithuanian government circles. The question at issue was how rapidly the first stage reforms should be carried out. There were people in *Sajūdis* and the government who were intent on a 'big bang' approach. But scattered through the bureaucracy and in the management structures of the large state enterprises, whose personnel were largely survivors of the previous Soviet regime, were influential individuals who either opposed reform or preferred a gradualist approach. Others had seen the devastating effects of shock therapy on production, employment and wages in other countries, and wished to moderate the process in the interests of employees' welfare and the maintenance of output levels. There is widespread agreement that the transition process in Lithuania was slower overall than in Estonia and slightly slower than in Latvia. However, the difference in pace should not be exaggerated since there was no more than a year of delay in achieving certain targets, what the IMF chose to call 'conditionality criteria'. The delay might have been longer had not the IMF vigorously involved itself in the discussions about policy and held out various inducements as well as sanctions before successive Lithuanian governments.

Lithuanian society was generally agreed on the objectives of the reform process but there were disagreements about tactics. Those who preferred rapid and radical transformation claimed that gradual and sequential reforms would not work, the reason being that all aspects of the economy were interrelated and change in one part depended on corresponding changes elsewhere if the process was to be effective. Moreover, gradualism would give time for half-hearted reformers to mobilize against the reform process, delay privatization and thereby strengthen government influence in the economy. As against these views, the gradualists argued that shock therapy was so radical in its economic and social consequences that it would create a reaction against all reforms among the population. A slower transition would alleviate the unnecessary hardships inflicted on the population, moderate inflation and limit the falls in production. But, as Grennes has argued, a faster pace of marketization elsewhere produced lower inflation and smaller falls in income. Lithuania, choosing a slower liberalization, suffered one of the greatest declines in output and standard of living. The most aggressive shock therapists, the Estonians, achieved the most successful programme of reform because they recognized that the reduction in inflation was a precondition of growth and increasing employment.[2]

A complex of factors explains the more gradualist pace adopted by the Lithuanians There was, to begin with, a larger heavy industrial sector in Lithuania, composed of all-Union enterprises which had traditionally been under the direct management of Moscow ministries. As Norgaard has convincingly argued, the local managers of these enterprises effectively opposed change through their personal and party connections in the ministries, with the aim of maintaining their traditional subsidies from the state budget. They also wanted to retain good relations with Russia and the other former Soviet republics so as to safeguard their traditional markets, and their supplies of raw materials and energy. But not all the responsibility for the slower pace of change can be placed on the shoulders of industrial managers and bureaucrats. By offering all residents the right of citizenship, and therefore the possibility of voting, the Lithuanian government increased the opposition to its reform process. The groups most adversely affected were the Russians, Poles and other minorities, who worked disproportionately in the heavy industries and agriculture and who, unlike most of their counterparts in Estonia and Latvia, had the vote and used it against the government, thus slowing down reform.[3]

The ethnic Lithuanians themselves, so it is argued, also were ambivalent about economic reform. Just as they had tried to oppose orders from Moscow, so now they resisted models of reform derived from western sources. This was possibly connected, so it is argued, to a xenophobic rural-based Catholicism which resisted outside ideas and had a powerful influence on the Lithuanian value system, helping to perpetuate egalitarian and collective values. Moreover, opinion polls showed that in 1991 only about a fifth of the population had liberal views on radical economic reform while a substantial majority supported some aspects of reform and opposed others. Three years later only 43 per cent of the population agreed that prices of commodities should reflect market forces; around 40 per cent of the respondents wanted prices to be set by the government. There was also a strong popular suspicion of entrepreneurs, who were equated with the black marketeers of the Soviet system. These surveys reveal a deep alienation from aspects of the reform process among the population as a whole. At the same time, as Palm has argued, bureaucrats and industrial managers were being asked to dismantle an economic system which they had operated and where they had made their careers, in order to replace it with a privatized system, without any knowledge of competition, marketing or incentives. Simultaneously they were required to create a reformed educational system offering instruction in market economics and business for a new generation of entrepreneurs and managers. It would not be surprising if their performance was characterized by uncertainty, indifference, or downright opposition.[4]

The transition process was further bedevilled by an extraordinarily unfavourable economic legacy from the Soviet Union. Actual or repressed inflation was a consequence of shortages of goods in the shops combined with high levels of savings, constituting what was commonly called a 'monetary overhang'. Enterprises were compelled to observe centrally imposed plans and targets, and the allocation of resources was determined by planners rather than by the market. Businesses were not rewarded for productivity gains nor were they penalised for failure. The latter could often be obscured by obfuscation or deception in the making of statistical returns. Almost all trade was conducted with other Soviet republics and

Lithuania was overwhelmingly dependent on them for its supplies of energy and raw materials, and for the markets for its foodstuffs and industrial products. Industrial plants had obsolete, inefficient and labour-intensive equipment and consumed around twice as much energy, wood and metals per unit of output as their Finnish and Swedish counterparts. Heating of living space required three times as much fuel in Lithuania as in Finland. Managers were expected to follow directives and not to experiment or to show initiative. Agricultural yields, both arable and livestock, were in general substantially lower than in neighbouring Scandinavian countries, and in the period 1981–87 Lithuanian labour productivity in agriculture was believed to be only about one quarter of the American figure. Services were mostly treated as a second-rate activity employing around one half the proportion of the workforce of Western developed economies. Overcoming this legacy would be a phenomenally difficult and complex task.[5]

LIBERALIZATION

The first steps in the transition from planned to market economy included the liberalization of prices, trade and banking. The government set out to achieve this by devolving economic decision making to entrepreneurs and business managers, ensuring that decisions about prices, products and investments reflected market forces rather than centralized planning decisions. In November 1991 the prices of foodstuffs and industrial products were freed, though the government still retained control of some food prices, rents, and the costs of transport, telecommunications and energy. By the autumn of 1992 85 per cent of prices were determined by market enterprises, including the price of credit. The excess consumer demand reflected in very high levels of savings ensured that prices would rise significantly. Queues and shortages quickly vanished, real wages fell despite the efforts of workers to link wages and prices, and inflation reached astronomical levels, receiving additional fuel from the imposition by Russia of world market prices for her exports of raw materials and energy supplies. The monopoly position of some Lithuanian producers, such as meat packers, also permitted exaggerated price increases. Currency depreciation in late 1992 and early 1993 further contributed to price inflation. The end result was that, although some commodities were still subsidised from the state budget at the end of 1992, consumer prices were estimated to have risen by a little over 1,000 per cent in 1992 compared with an increase of some 380 per cent in 1991. This was the high point of the inflationary tide and in 1993 the level dropped by three fifths, to around 400 per cent. Nevertheless, inflation fell less steeply than in Estonia and Latvia, largely as a result of Lithuania's less rigorous stabilization policies in 1992 and 1993.[6]

The Soviet market dominated Lithuanian trade until independence. Freed from Moscow's controls Lithuanian producers were now given the freedom to trade in western markets. This was indispensable since export markets in the former Soviet Union quickly contracted and imports became unpredictable and subject to political interference. The Lithuanian government moved quickly to create a favourable environment for international trade by concluding trade agreements with European countries and the United States, thus providing alternative export and import markets for Lithuanian companies. At the same time banks were freed from central Soviet controls and interest rates were determined by supply and demand. Like prices in general, the cost of borrowing rose significantly to reflect credit scarcity. The government, having created the Bank of Lithuania in 1990, endowed it with the normal responsibilities of a central bank operating in a market economy, namely the emission of money and securities, control of exchange rate policy, management of foreign exchange reserves and the regulation of commercial banks. The commercial operations of the Bank were removed from it as part of the process of separating commercial and regulatory functions.[7]

Although the government took significant steps to liberalize the economy to bring it into line with free market standards, it responded to political pressures by retaining control over some prices and continuing to offer subsidies to energy costs. In addition, commercial companies in financial difficulties were not subject to rigorous bankruptcy legislation. Unable to borrow at the prevailing high levels of interest, they obtained a form of subsidy by not paying their suppliers, and the result was a cycle of indebtedness in the economy which was very difficult to break. This imperfect liberalization was characteristic of the first two or three years after independence. By 1994–95, most of the limitations were removed and the economy could begin to grow on firm free market foundations.

STABLILIZATION

Liberalization and stabilization were not sequential processes. Attempts to stabilize the economy, particularly by squeezing inflation out of the system and reversing the serious fall in industrial and agricultural output, began contemporaneously with the freeing of prices, but persisted for some time after liberalization had been achieved. By most criteria the process of stabilization was completed by 1995 and it is worth examining how this was done. The government's use of monetary and fiscal policy, its adoption of a Lithuanian currency, the *litas*, the value of which was stabilized by means of a currency board system, and the establishment of a trading regime which offered substitutes for the loss of former Soviet markets, all made significant contributions to the stabilization process.

It was a condition of effective stabilization policies that Lithuania have its own currency. Continuing to use the rapidly depreciating Russian rouble was a guarantee of galloping inflation. At the end of 1991 the *Seimas*, the Lithuanian parliament, decided to establish a separate currency and appointed a currency reform committee to recommend how and when this should be done. On one side were arguments that the introduction of a national currency would, in itself, have a major impact on the stabilization process. On the other, that the new currency would only be effective in contributing to stabilization if it maintained a stable value. In a sense both these propositions were true; a new currency could hardly be more inflationary than the Russian rouble, but it would fail to eradicate inflation unless it was accompanied by rigorous monetary policies.

It became obvious in 1992 that Lithuania had to be isolated from rouble depreciation. Since the *litas* was not yet ready for introduction, coupons (*talonas*) were introduced on 1 May, 1992 to supplement a shortage of roubles circulating in Lithuania. By October the rouble ceased to be legal tender and was entirely replaced by coupons. However, this provisional currency rapidly lost value, falling from 250 to the US dollar to 550 in the period October 1992 to May 1993. This was clear evidence that a new currency in itself would not extinguish inflationary pressures. At this point the IMF made one of its first critical interventions in steering the Lithuanian government to adopt a more stringent anti-inflationary policy in preparation for the introduction of the *litas* in 1993. A policy of stabilizing the coupons was finally adopted in April 1993 and had immediate effect, inflation falling from 25 per cent in April to 6.2 per cent in June. And so, when the *litas* was finally introduced on 25 June, it had a favourable launch pad of relatively low inflation. According to government assurances, the country's foreign currency reserves guaranteed that the *litas* would not be devalued.[8]

In practice the new currency was allowed to float. Beginning at 4.5 to the US dollar, it had risen to 3.2 by September and fallen back to 3.9 by December. Exporters protested at this appreciation which was adversely affecting exports, and complained about the fluctuations in value which made foreign trade and stabilization more difficult. In December 1993 President Brazauskas advocated tying the *litas* to a western currency to prevent rapid fluctuations in value. His prime minister Adolfas Šleževičius recognised that pegging the currency would reduce the role of the central bank and prevent the government influencing currency markets. It would, however, remove uncertainty about prices, help to bring down interest rates, and remove or reduce political pressures by government and industry to influence the value of the currency. In the first three months of 1994 the government and Bank followed an anti-inflationary policy, preparing the way for tying the *litas* to a western currency with effect from 1 April, 1994. The mechanism chosen for this purpose was the currency board system which had a two fold stabilizing effect. The *litas* was fixed at a ratio of 4 to 1 to the US dollar. Internally the amount of currency in circulation was strictly related to the amount of dollars and gold in the Bank's reserves. The Bank was compelled by law to purchase all *litai* offered to it with dollars, thus ensuring that the national currency was fully backed by foreign exchange. In this way the government could keep tight control of the money supply and resist inflationary pressures. In a sense the mechanism adopted depoliticized monetary policy. On the other hand, the currency board system was the subject of mounting criticism over the next few years, despite its obvious success in curbing inflation, securing full convertibility, and lowering interest rates.[9]

The control of inflation correlates with two other indices of economic recovery. Dale has provided evidence that inflation stabilization in some transition economies leads to credit expansion, and that the more successful the stabilization, the greater the expansion. Fischer et al. have shown that once inflation is brought under control output begins to recover and growth becomes positive. Those countries with the highest rates of inflation are usually, as Grennes has argued, the ones to experience the greatest reductions in real income. This was true of Lithuania among the Baltic states. Although inflation was ultimately controlled, the introduction of an effective monetary policy and a separate currency was delayed in comparison with Estonia and Latvia, and consequently economic recovery took longer to achieve. Yet the IMF, though urging faster progress, was happy that Lithuania was moving in the right direction. Each year from 1993 there was a reduction in inflation and an increase in gross national product (GNP). Inflation was over 1,000 per cent in 1992, around 400 per cent in 1993, about 70 per cent in 1994, and the figures for 1995 and 1996 were 35 per cent and 13 per cent respectively. From 1997 to the present the volumes were in single figures. At the same time GNP, after falling catastrophically between 1989 and 1993 to only 43 per cent of the 1989 level, one of the deepest falls anywhere in post-Soviet transition experience, began to recover in 1994 when output grew by 1 per cent, followed by figures of 3 per cent, 4.2 per cent and 5.7 per cent for each of the next three years and a forecast of 7 per cent for 1998. Monetary policy was not, however, the only factor affecting the level of inflation. Fiscal policy and import prices were also important and require brief discussion.[10]

In the pre-independence period government spending on subsidies for housing, transport and foodstuffs created substantial inflationary pressures. Stabilization policy involved cutting subsidies and generally reducing government expenditure in the interests of achieving a balanced budget. Research suggests that an improvement in public finances correlates well with the reduction of inflation and GDP growth. High public deficits are associated with increased inflation and lower growth. Balanced budgets also encourage lower interest rates and increase confidence of external investors in the stability of the economy. In the immediate aftermath of independence the cutting of most subsidies, accompanied by prudent fiscal management, kept budgets in balance. The first deficit appeared in 1993. The parliamentary elections of 1992 and the presidential election of 1993 installed the Democratic Labour Party in power in response to the dissatisfaction of the electorate with the reform programme. There were considerable pressures on the new government to meet the increased social needs of the population, particularly the pensioners and the low paid, by subsidizing domestic heating costs and increasing social benefits. This increased spending had to be financed by means of loans, thus adding to inflationary pressures. As the demands grew for increased social expenditure the drastic fall in output cut tax revenues substantially. This fall was exacerbated by an ineffective and poorly administered tax collection system, which was riddled with exemptions. The customs system needed drastic reform as increasing evidence of corruption at the border posts came to light. The big state enterprises tried to avoid paying tax as a means of obtaining subsidies by other means. Private companies often dealt in cash to evade tax bills. The IMF frequently lectured the Lithuanian government on the necessity to stick to fiscal targets and to limit deficits to agreed amounts. In particular, the higher cost of imported energy had to be passed on to consumers in full, and other budget spending had to be restricted. IMF warnings and threats of withholding stabilization loans gradually had their effect. Value Added Tax was introduced in 1994 and tax collection gradually improved. The remaining budget subsidies were steadily eliminated, leaving only subsidies to agriculture in the second half of the 1990s. The recovery of economic growth produced higher tax yields. By 1995 the gap between income and expenditure in the budget as a percentage of GDP had been narrowed and the planned budget deficit for 1996 was within IMF limits. This did not mean that the tax system was satisfactory in all respects and there was ample scope for reform to stimulate business and the economy. But fiscal policy and more effective tax administration had, by 1995–96, made a significant contribution to stabilizing the economy.[11]

Finally, trade stabilization was of critical importance to economic recovery. The almost total integration of the Lithuanian and the USSR economies ensured that independence would pose major problems for Lithuanian industrial and agricultural producers. At the end of the 1980s Lithuania imported almost 90 per cent of its fuel and energy requirements from the Soviet Union. Over 80 per cent of Lithuanian exports went to the Soviet market and even more Lithuanian imports came from there. The interruption or loss of these markets would have incalculable consequences for the Lithuanian economy. Indeed, the worst fears of the Lithuanian government were realised. Despite the conclusion of trade and economic cooperation treaties between the Baltic republics and the other republics of the former Soviet Union in the autumn of 1991, Lithuania received only 40–50 per cent of the fuel promised by the end of the year and had to introduce a system of rationing. There was also a reduced supply of metals and other products through the traditional links as a result of a complex system of export quotas and licences in Russia. The shortfall in raw material deliveries from Russia threatened to close a number of industrial plants while others were on short-time working. These shortages, some Lithuanians alleged, were part of a campaign of political pressure against Lithuania, a continuation of the economic blockade by other means. On the other hand, the severe dislocations in production in Russia itself also contributed to the reductions in supply, and heavy Russian export taxes in 1992 on timber, newsprint and aluminium pipes and other products drove up Lithuanian production costs. The most important contribution to increased costs for Lithuanian industry was the decision by Russia to charge world market prices for energy exports in the second quarter of 1992.[12]

Difficulties in receiving supplies from former trading partners were matched by equal problems in maintaining exports to traditional markets. In the immediate post independence period Russia imposed high duties on imports of Lithuanian foodstuffs. Later, it doubled protective duties as a negotiating tactic in the discussions over transit to Kaliningrad. In 1992 Moscow had agreed in principle to a most favoured nation (MFN) trade treaty with Lithuania. A detailed agreement was concluded in 1993 but Russia refused to implement it until a satisfactory accord was reached with Lithuania over the transit of military equipment and personnel to and from Kaliningrad on Lithuanian railways. Accordingly the implementation of the MFN accord did not begin until January 1995. This agreement provided the foundation for a new trade regime with Russia, whose markets remained important to Lithuanian exporters.[13]

However, stabilization of external trade no longer depended entirely on regaining traditional export and import markets. The aftermath of independence had produced such damaging consequences for Lithuanian trade and industrial production that commercial enterprises had been forced to accelerate the re-orientation of their trade to non-Russian markets. This reorientation would have been necessary even if Russian trade had continued on former lines since modernization of Lithuanian industry depended on its restructuring and the development of new markets. Integration with world markets would help to align domestic prices, and competition from imports would send powerful signals to Lithuanian producers, particularly to monopolies, about quality and pricing policy. Exports to the west would pay for the necessary investment in new equipment, and more frequent contacts with western enterprises would increase the flow of expertise and strengthen awareness of new technology. More broadly, closer trade and investment relations with the states of western Europe and north America would offer Lithuania a valuable form of 'soft' security. Also, access to western markets through trade agreements would provide an important inducement to foreign investors. Finally, it was obvious that until alternative sources of energy supplies could be developed, Lithuanian exports to the West as well as earnings from the transit trade would be needed to purchase Russian oil and gas.[14]

The government's ambition to become less dependent on Russian markets and suppliers after independence was realised at an unexpected speed. Stabilization of production following the dramatic downturn in the early 1990s required that alternative markets and suppliers should become available. For all these reasons the government moved quickly to negotiate trade agreements with western countries. Lithuania's first such agreement with a western state was signed with Sweden in March 1992, when tariffs were removed on all industrial commodities. This was followed by a series of similar agreements with other European states. Lithuania obtained preferential status with the European Union (EU) and the United States at the beginning of 1992 and in the same year she entered into a bilateral free trade agreement on most industrial products with European Free Trade Area (EFTA) countries with the exception of Austria and Iceland. Two years later Lithuania and the EU signed a free trade agreement in industrial products which came into force on 1 January, 1995. An association agreement with the EU was concluded in June 1995. The result of this trade re-orientation was remarkable. Between 1990 and 1995 the share of Lithuanian exports to the former Soviet Union fell from almost 85 per cent of the total to 46.7 per cent and the import share decreased from well over 70 per cent to 50 per cent. In the same period the share of Lithuanian exports going to western markets rose from 15.3 per cent to around 53 per cent and imports from the west increased as a proportion from 27.5 per cent to almost 50 per cent. This increase in trade with the west was attributable to the free trade and MFN agreements, to the introduction and stabilization of the *litas*, and to the fall in domestic inflation in 1993 to 1995.[15]

PRIVATIZATION

Capitalism needs capitalists, writes Haavisto. Private ownership of firms in a market economy encourages, in King's words, responsibility, accountability and rationality among managements. All things being equal, it stimulates enterprise, efficiency, and a dynamic response to market opportunities. A study by the World Bank in 1992 offered solid evidence that privatization as compared with state ownership resulted in a net increase in wealth, higher investments, managerial innovation, better pricing of services, and labour efficiency. But privatization in itself is not a panacea; the transformation of publicly-owned monopolies into privately-owned ones is unlikely to develop desirable entrepreneurial qualities among managers. Newly privatized firms may be under-capitalized or have inadequate access to external investment. This handicaps their efforts at modernization and expansion. Additionally, successful privatization requires a favourable business environment, and therefore demands concomitant changes in the framework of law and institutional rules concerning, for example, property rights, taxation and the establishment of capital markets. Some economists also question the importance of ownership as opposed to management, arguing that effective management displaying the admired entrepreneurial virtues can co-exist with either state or private ownership. Pelikan disagrees on two counts. Firstly, he suggests that in the transition period privatization is indispensable to the restructuring and regeneration of ex-Soviet enterprises. Secondly, the quality of management strongly depends on the competence of the owners. If you put the question, which form of ownership of firms makes it more likely that firms will be owned by competent persons, the answer lies in private and tradeable, not state, ownership.[16]

The consensus of opinion in Lithuania was that privatization was indispensable for a successful transition to a liberal market economy.[17] Indeed, the pace of privatization in Lithuania was extremely rapid, and represented one of the most successful transitions from publicly-owned to private property, not only in the former Soviet Union, but in East Central Europe as well.[18] At the same time there were serious reservations about transferring to private control certain large state-owned enterprises in the energy, transport and communications sectors since it was felt that this could have an adverse impact on state security. Five methods of privatization were used in Lithuania: public subscription for shares; public auction; public tender; sale by direct negotiations in the case of only one bidder at auction or tender; lease with option to buy. One might add the method of restitution which applied particularly to agricultural property. Initially the methods adopted were public subscription and public auction, and these dominated in the first period of privatization from 1991 to 1995. By that time the shortcomings of these methods in regard to large publicly-owned enterprises were becoming clear and the second stage of privatization was embarked on in 1996, when the emphasis shifted to public tender and direct negotiations as more effective methods of inducing foreign investment in Lithuanian companies. The emphasis now was less on stabilization and the creation of the foundations of a market economy than on accelerating the pace of economic growth.[19]

Privatization was a central feature of the economic reform process in Lithuania from the beginning. Its centrepiece was the use of investment vouchers which gave an egalitarian cast to the process. Vouchers were credited to all residents on a sliding scale related to age, over 35s receiving vouchers to the value of 5,000 roubles, 18 year olds 1,000. Vouchers could be exchanged for shares in all state property except farmland. Alternatively, vouchers could be used in auctions. Citizens were also permitted to buy shares with cash but only up to the value of the vouchers they held. It was expected that vouchers would be used mainly in the purchase of housing, for which 80 per cent of the price could be covered by vouchers, with the remainder being paid in cash.[20]

Small scale services and industry were quickly privatized by means of public subscription of shares and auctions. By October 1992 most eligible housing and small enterprises had been sold. By the end of 1994 70 per cent of state property had been privatized. 24 per cent of the vouchers had been used for housing, 66 per cent for the purchase of enterprises. A law of 28 January, 1993 enabled workers to use vouchers to privatise up to 50 per cent of the shares of the enterprises they worked for before the firm was opened for auction. The government set a nominal price for these 'employee' shares. From early 1993 to November 1994 out of 950 large enterprises privatized, some 80 per cent had at least 50 per cent employee ownership. Investors often took advantage of this system to buy up shares from employees at favourable prices. A much criticised aspect of this process was the opportunity it offered to managers and other senior figures in the former Communist Party to purchase shares at a discount and to retain control of the firms which they had run in the Soviet period. This form of 'crony capitalism' became much more prevalent after the Democratic Labour Party came to power in 1992 and Landsbergis, as leader of the newly-formed opposition Conservative Party, attacked it as 'nomenklatura privatiza-tion'. The fairly widespread employee and management ownership distorted the statistics of privatization. The denominated non-state sector included privatized minority holdings in only partly privatized enterprises and private companies owned and run by managers and employees. The managerial approach and the investment potential of these firms would probably have changed very little. It has been suggested that many continued to operate in the old ways with the expectation that losses would be made good by government or central bank subsidies. However, as the economy emerged from the transition stage, these companies could no longer count on such traditional largesse and either had to come to terms with the new conditions or go under. The great success of the first stage of privatization lay in the transfer to private owners of small enterprises such as restaurants and cafes, small shops and services. Also, and sometimes overlooked, was the freedom to create new businesses of which thousands of Lithuanian entrepreneurs took advantage.[21]

A very important sector for privatization was agriculture, which employed around 25 per cent of the Lithuanian work force in massive collective and state farms. It was a priority for the *Sajūdis* government after independence to restore this collectivized land to private owners as quickly as possible. The principle adopted was to return the land to its former owners or their descendants, provided they were resident in Lithuania. The results were not economically beneficial in the short term. Many plots were now extremely small and inefficient; others were owned by city dwellers who lacked farming skills and capital and wanted to sell, some plots had been used for non-agricultural purposes such as factory buildings or roads. Provision was made for the purchase of land, the price to vary according to fertility. Unfortunately the cost was too high for many former employees of the collectives who were granted small and uneconomic plots of less than 4 ha., often with no machinery or livestock. The policy caused bitter resentment in the countryside and was a decisive factor in the electoral defeat of *Sajūdis* in the 1992 parliamentary election.[22]

There was another relatively unsuccessful aspect to the first stage of privatization, namely the policy of auctioning enterprises for hard currency which began in August 1992. Initially most of these enterprises were in the service sector, such as hotels, restaurants, and large stores. Very few of the listed companies received adequate bids, and only 9 out of 114 were sold by the end of the year. The government included large-scale industrial enterprises among the listed firms from January 1993. Among those purchased by foreign investors were Klaipėda Tobacco, bought by Philip Morris, Kaunas Confectionery by Kraft Jacobs Suchard, Audėjas Textiles and the Klaipėda Dairy. The largest hotel in Lithuania, the Lietuva in Vilnius, did not receive a high enough bid and was withdrawn from sale. These purchases draw attention to a vital and increasingly important aspect of the privatisation process which produced a radical change of policy by government in 1996–1997. Philip Morris and other foreign investors, in addition to purchasing enterprises for hard currency, also invested heavily in new technology, restructured the organization, changed management practices, and established excellent training for staff. In doing so they benefited from the first-class technical education of many of their employees and from the low wage and salary structure in comparison with western Europe.[23]

This and a few similar examples drew attention to the necessity not only of privatizing firms but of reshaping and refinancing them if they were to be competitive in international markets. The World Bank recommended in 1993 that the government should encourage foreign investment in Lithuanian enterprises by, to take one example, allowing them to purchase rather than lease land. The Chair of the Central Privatization Commission stressed the desirability of foreign investment which would attract modern marketing and management, lead to an influx of capital for restructuring and new technology, and open up export markets.[24] Three years later the World Bank was still critical of the way in which Lithuanian governments had deterred foreign investment up to the end of 1995 by giving preference to managers and employees in the privatization process. The voucher system, though admirably egalitarian, was unsuited to privatisation of large Soviet industrial plants and utilities.[25] The international agencies repeatedly advised that the emphasis in Lithuanian economic policy, once stabilization had been achieved, should be on accelerating the pace of economic growth in order to boost living standards. Privatization policy, it was urged, should reflect this emphasis on growth. Transfer of ownership, especially in the case of the remaining very large state-owned companies, should be accompanied by injections of capital, restructuring of management, and retraining of workers to ensure international standards of performance. Only in this way could Lithuanian exports compete in international markets.

By 1995 the first stage of independent Lithuania's economic development, the economics of transition, had been largely completed. Before leaving this subject it is worth recording the critically important role of international agencies in ensuring that the transition process was not derailed by inappropriate government policies. The role of the World Bank, the European Bank for Reconstruction and Development, the Nordic Investment Bank and a number of other credit organizations were indispensable in the reconstruction process, focusing their efforts on infrastructural investment, environmental clean-ups, the supply of technical assistance, agricultural investment and banking reform. It was the IMF, however, which had most influence in shaping government transition policy since its approval was a pre-requisite for the international credibility of the reform programme. In refusing to endorse aspects of the policy as being counter to the stabilization process the IMF had considerable influence over the direction taken by the government. For example, it advised caution in the introduction of the *litas*, warning that mechanisms had to be in place to defend the value of the currency before it was issued. In 1993 it prohibited the government from using loans to increase salaries and pensions by a substantial 40 per cent. In early 1994 it warned the government and Bank of Lithuania against loosening monetary policy, and later in the year it cautioned against allowing the budget deficit to rise to unacceptable levels. Perhaps its most categoric warnings were against the highly inflationary proposal to index link depreciated savings in 1994. These were some of the more vivid examples of IMF intervention, taking place when the consensus had broken down. Generally, however, the government and IMF were engaged in a continuous dialogue in a joint effort to establish the correct lines for stabilization policy during the transition stage.[26]

THE STRATEGY FOR GROWTH

In the second and current stage of economic development there emerged a changed emphasis in economic policy. It would be a mistake to draw too sharp a distinction between the two stages since there are obvious elements of continuity. Stabilization remains important in the second stage, just as considerable efforts were made to arrest the decline in economic growth in the first. But as a broad generalization we can say that the emphasis in the second stage has been on the creation of a favourable environment for economic growth. This involved the central recognition that Lithuania lived by international trade and that its entire economic policy should be geared to improved competitiveness in international markets. In turn, this depended on significant new investments in industry, agriculture and services. These could be realised by a variety of methods, notably the encouragement of foreign participation in ownership and management, the development of the transit trade to and from the former Soviet republics in the East as a significant earner of foreign exchange, the establishment of a framework of law and regulation to give confidence to business, the maintenance of sound monetary and fiscal policies, and the development of new sources of energy in order to lessen dependence on former Soviet suppliers. Though space precludes their detailed consideration, the discussion which follows will touch on each of these aspects of the growth strategy.

Essential for investors and hence for growth are legally protected property rights. In a market economy investors and entrepreneurs need to be confident of obtaining future profits from current investments. Investors require stability and predictability; they must be able to manage and trade their property. Without these conditions the economy will be regarded as high risk, which will push up interest rates to the point where necessary and profitable investments may not be made. Government, therefore, has to show a credible commitment to the protection of property rights, not only through legislation but by an effective civil service, law enforcement agencies, and judiciary. Society also has to recognise that highly complex laws and regulations, constantly changing legislation, and high tax rates increase both uncertainty and the opportunities for administrative corruption. It is vital that systems are put in place to provide high standards of legal and administrative training and adequate salaries for personnel. In the transition period there were frequent complaints by foreign investors in Lithuania about constantly changing and confusing regulations, corruption at customs posts, fluctuations in the value of the currency, and inadequate tax administration. Efforts were made to correct all these deficiencies and much time and money went into training administrators, police, customs officers, and tax inspectors. A new director of customs was appointed in January 1997 with immediate beneficial effects.[27] After the banking crash in 1996 the National Bank tightened up its regulations for commercial banks, and banks and other creditors were encouraged to use the bankruptcy legislation to gain access to at least part of the assets of defaulting companies. Article 47 of the Lithuanian Constitution prohibited foreigners from owning land, allowing them leasehold rights only. A campaign to amend the constitution to permit foreigners to own non-agricultural land was successful in 1996, though largely as a result of pressure from the European Union. These changes in law and practice were indispensable if foreign and domestic investors were to commit their capital to long-term economic projects.[28]

An interesting case study of the adverse effects on economic development of inadequate regulation is the banking crisis which began in December 1995. During the transition process in East Central Europe banks were often inadequately capitalised, their main source of funds being short-term demand deposits. Accordingly, bank lending tended to be characterized by short-term loans at high rates of interest. By contrast, long-term loans at reasonable rates, which could have provided a stimulus to economic development, were in short supply. Borrowers were encouraged by the system to invest in high risk, high return projects, usually in commerce, which by their very nature had a high chance of failure. This left the banks with a portfolio of bad loans which made them quite vulnerable. In 1995 investigations showed that over 20 per cent of loans extended by several large banks in Lithuania were either bad or uncertain. In addition, three-quarters of loans were for less than one year, with an average interest rate of 30 per cent. Disincentives to invest long term included high inflation rates in the period 1991–1995, government insurance of deposits in state-owned banks which encouraged increased risk taking by private banks, and the voucher system of privatization which impeded restructuring of firms. The problems were compounded by inexperienced managers and supervisors, and minimal capital requirements for banks. In 1995 nine of Lithuania's small banks closed and two larger ones needed temporary state help. In December 1995 the Bank of Lithuania suspended operations at two of the largest banks, Litimpex and Innovation. The policy followed by the Bank and the government in the aftermath of this crisis had adverse consequences for economic growth, at least in the short term, but had the advantage of tightening the regulations for the operation of Lithuanian commercial banks.[29]

The government could have chosen to allow the two banks to go bankrupt, placing the costs of failure on the banks' depositors. Instead it decided to offer the depositors as much protection as possible, and placed the costs on the taxpayers and new borrowers. It proposed to restructure the banks, finding the investment from the issue of long term domestic securities, and to pay a minimum compensation to the depositors in the failed banks. However, the effects of the crisis were negative for economic growth, at least in the short term. Accounts were frozen and the money supply contracted as people converted *litai* into dollars or other foreign currencies. Room had to be found in the budget to meet the costs of compensation and restructuring, which in practice meant spending less on other items and increasing the fiscal burden. At the least the economic recovery was interrupted though the longer term impact on growth was marginal.[30]

The exchange rate of the Lithuanian currency was critically important to Lithuanian competitiveness in international markets. The currency board system ensured stability in the value of the *litas* and effective control of domestic inflation, to the benefit of inward investment and foreign trade. Nevertheless opposition to the system emerged among opposition MPs in 1996 and there were calls for devaluation, amid claims that the currency had been effectively revalued at the expense of Lithuanian exporters. When the Conservative coalition came to power in 1996 their intention was to replace the currency board and restore to the central bank its conventional functions of managing the exchange rate and controlling the emission of currency. Although there were banking arguments in favour of this course, perhaps more important among MPs was the belief that the currency board was demeaning to Lithuania owing to its infringement of national sovereignty. Yet polls showed public opinion to be in favour of maintaining the board to avert the resumption of inflation and devaluation. This was complemented by international concern that replacing the board at that time could damage international confidence and hence economic recovery in Lithuania. Consequently a compromise was agreed with the IMF under which the central bank would gradually regain money supply functions though the currency board system would be kept in place for the immediate future. Meanwhile there would be consideration of whether the *litas* should be pegged to a European currency rather than the American dollar. Slay's comment that the currency board could not be maintained indefinitely since it effectively meant transferring control to the US Federal Reserve leaving little room for discretionary monetary policy, overlooked the monetary implications of Lithuania's ambition to join the EU, namely the transfer of monetary control and management to Frankfurt. The compromise arrangement referred to seemed in the best interests of the Lithuanian economic recovery in the immediate post-transition stage.[31]

Similarly the government's fiscal policy aimed to balance the budget by cutting expenditure on energy subsidies and grants to enterprises and by increasing income through more efficient tax collection. The aim was to reduce borrowing, making more funds available for investment in production, and to encourage fairness and stability in the tax system by reducing tax breaks and arbitrary exemptions. But importantly the strategy was also designed to encourage growth. A new Finance Minister, Gediminas Šemeta, appointed in 1997, made significant changes in the audit and tax collection procedures. Calculations showed that inadequate revenue collection had cost the Finance Ministry some 30 per cent of total tax yield. An increased yield would enable the government to stimulate business and investment by cutting VAT and income tax. The result of this activity was a steady reduction in the budget deficit with the aim of achieving balance by the year 2000.[32]

Significant contributions to the revenue were generated by the second stage of privatization which came into effect in September 1996. Although the vast majority of Lithuanian companies had been privatized by 1996, private capital accounted for only about 36 per cent of Lithuania's total capital. The aim of the new law was to attract Lithuanian and foreign investors to the minority of very large state-owned companies. These tended to be companies formerly directed from Moscow, a large proportion of which were in the crucial fields of energy, transport and communications. Privatization of these companies was the only way of attracting the large investments required to modernise and rationalise production and to increase output. Since domestic sources of capital were inadequate to meet both the purchase price and the re-financing demands of these enterprises, recourse to international corporations was essential. Privatization of these large companies would consolidate the formation of a new middle class with a stake in maintaining an effective system of property rights and progressive economic policies.[33]

Perhaps for this very reason, progress was disappointingly slow from the passage of the law until the election victory of the conservatives in autumn 1996. It was alleged that management in the large state-owned companies, and bureaucrats in the responsible ministries, were reluctant to cede control, and that workers in these companies feared redundancy and a reduction in social benefits. In the vast majority of companies for sale, only minority shares were made available, and this proved unattractive to potential investors. A change in the political climate occurred after the banking crash in late 1995–early 1996 which acted as a catalyst for accelerated reform. Even so, there was strong feeling in parliament that 'strategic' facilities in sectors such as energy and transport should not be privatised. An important objection to this view came from Eduardas Vilkas, the chair of the Privatization Commission, who declared in January 1997 that it was time to remove such items from the non-privatisation list. The conservative government took a more vigorous and energetic approach to privatization which was made possible in part by a supportive majority in parliament. It decided to proceed by means of public tenders and private negotiations as appropriate, and called in foreign consultants to advise on all aspects of the sales, including the valuation of the shares. In April 1997 the government announced the names of 14 of the largest state-owned enterprises for privatisation, including Lithuanian Telecom, the Lithuanian Shipping Company, Lithuanian Oil, Lithuanian Gas, Klaipėda Stevedoring Company, the Agriculture Bank and the Savings Bank, and perhaps the most sensitive of all owing to its close links with its Russian suppliers, the Mažeikiai oil refinery, along with the Biržai pipeline and the Būtingė oil terminal, then in process of construction. The government had in fact made a strategic choice regarding these 'strategic' enterprises: it decided that modernization by means of foreign investment and the resulting budget contributions from the proceeds of the sales were more important than maintaining state ownership, which could not provide the investment needed for restructuring. Moreover, ownership or substantial shareholdings by western companies offered a form of 'soft' security and, potentially, a lower dependence on Russia, particularly for energy supplies.[34]

The most successful of these privatisations was the sale of Lithuanian Telecom to Amber Teleholdings, a consortium of Telia from Sweden and Sonera of Finland. The consortium agreed to purchase 60 per cent of the shares for $510 million and to invest an additional $200 million or so to modernize the equipment. The most controversial was the sale of a one-third stake in the merged energy concern of the Mažeikiai oil refinery, Būtingė oil terminal and Biržai oil pipeline to the Williams company of the United States. Originally Banque Paribas was chosen by the Lithuanian government to act as consultant and adviser on this privatisation but it soon emerged that the Ministry of Economics had signed a letter of intent with Williams undercutting the agreed process. In an agreement of October 1998 Williams was recognised as a strategic investor and granted tax guarantees and other financial concessions in return for taking a 33 per cent stake in the concern, with 57 per cent of the shares being retained by the state prior to subsequent disposal, leaving the state with a 25 per cent interest. It was agreed that Williams would pay $150 million in two stages. There were three vital considerations in this sale: United States government support for the agreement; the injection of substantial additional funds by Williams for the modernization of Mažeikiai and the completion of Būtingė; and finally the possibility of obtaining crude oil for refining in Mažeikiai by way of the Būtingė terminal, reducing dependence on Russia for essential fuel supplies, which constituted a major part of Lithuania's imports from Russia. Williams' involvement was believed to offer 'soft' security guarantees, particularly in the light of active American support for the investment, and an encouragement to other foreign investors to purchase Lithuanian companies.[35]

Foreign direct investment (FDI) made a considerable contribution to the acceleration in growth rates in Lithuania in the post-transition period. During the transition itself, Lithuania had lagged behind the other Baltic states. As late as 1996 FDI in Lithuania was only 1.5 per cent of GDP as compared with Estonia 2.5 per cent, Latvia 4.6 per cent, Hungary 4.5 per cent and Poland 2.1 per cent. In terms of value Hungary had received an average of some $200 per person, Estonia about $75 and Lithuania a mere $50. In the period 1989–94 Estonia received $471 million in FDI, Latvia $249 million and Lithuania only $173 million. A number of factors have been adduced for this investment lag, namely bureaucratic and administrative barriers, popular suspicion of foreign owners, vague and inconsistent legislation, a relatively small number of consumers (but Estonia had even fewer), high inflation, and an undeveloped and under-regulated banking system. Restrictions on land ownership for foreigners and widespread criminality were other inhibiting factors. The fact that Lithuania's rate of inflation was the highest in the Baltics in 1995 and had been for the previous half decade was a major deterrent to investment, as was Lithuanian government and bureaucratic preferences for management and employee privatisation as opposed to privatisation for cash.[36]

Foreign investment began to grow quite rapidly in the second half of the 1990s. By 1998 FDI in Lithuania was the fastest growing in the region. This is illustrated by the figures for cumulative investment for the 1990s as follows: 1991 $8 million; 1992 $19 million; 1993 $149 million; 1994 $310 million; 1995 $352 million; 1996 $700 million; 1997 $1,041 million; 1998 $2,200 million. The sale of Lithuanian Telecom made a major contribution to the doubling of cumulative FDI in 1998; clearly the second stage of the privatization programme was much more friendly to foreign investors than the first had been. Part of the FDI growth was due to the effects of stabilization, notably in fiscal and monetary policy. But foreign investors became aware of other advantages accruing from investment in Lithuania such as excellent infrastructure, an ice free and rapidly modernizing port in Klaipėda, the establishment of three Free Enterprise Zones and the new law on foreign capital investment of June 1995 which offered a more favourable climate for foreign investment. A highly qualified but low cost labour force and cheap rents, services, and overheads added to the attractions of Lithuania for foreign business. Improved international credit ratings confirmed these impressions. A poll of investors in 1998 confirmed that 87 per cent would invest in Lithuania again, and this was borne out by the fact that more than 60 per cent of new FDI came from investors who were expanding their existing operations. There is no space to discuss at length the major commitments of foreign companies in Lithuania. Just to list some of the major investors shows the attractiveness of Lithuania to international companies. Their presence can only reassure other potential investors that the investment climate in Lithuania had become very favourable by the late 1990s. Among the companies with a stake in Lithuania are Philips, Philip Morris, Siemens, Scandinavian Enskilda Bank, Amber Teleholdings, Statoil, LUKoil, Williams International, Partek Insulation, Lancaster Steel, Motorola, Millicom, Shell and, inevitably, Coca-Cola.[37]

Lithuanian producers had long depended on foreign markets (we may include Soviet markets in this category); independence offered the opportunity, after half a century, to re-orient trade towards the west at a time when trade with Russia and the other former Soviet republics was actually contracting. By 1995 Lithuanian trade with the former Soviet Union had contracted to about 40 per cent of the total.[38] This major trade re-orientation provided Lithuanian exporters and importers with a very promising springboard for further growth in the West while not ignoring the continued importance of the former Soviet markets, particularly for exports of Lithuanian foodstuffs. The growth of trade was a major engine for economic growth in Lithuania in the period since 1995. The continued importance of the Russian market was shown by the serious downturn in exports and imports as a result of the Russian financial crisis in the summer of 1998. The success of Lithuanian exports to the west depended on quality (which was continually improving as a result of investment in modernization and training), lower prices based on cheap labour costs, and trade agreements with most western trading partners providing for free trade or low tariffs. Hard currency earnings were also increasing from the ever growing transit trade through the port of Klaipėda. Major exports in 1997 were agricultural products, minerals, chemicals, textiles and textile articles, machinery and electrical equipment and vehicles and transport. The largest categories of imports were prepared foods, mineral products, chemicals, plastics, textiles, base metals, machinery and vehicles.[39]

There are numerous examples of Lithuanian companies successfully adapting to the challenges of the new markets. A case in point was the furniture company, Šilutė, which, after the contraction of the furniture industry following the loss of Soviet markets, adapted to the new demands and as early as 1994 was selling 70 per cent of its products to the West, up from 14 per cent in 1992. The company's successful marketing strategy was to target medium and low income families who appreciated the quality of Šilutė's production.[40] The Alytus—based Snaigė, a manufacturer of refrigerators, exported 90 per cent of its output in 1995. Its success was based on high quality, low prices, and a wide range of different models. Vilnius Vingis became a subcontractor for Samsung of Berlin, producing components which competed successfully on quality, deadlines and price. Each of these companies, and many more like them, adapted remarkably quickly to new challenges. However, the major export industry of the inter-war period, the food processing industry, was unlikely to regain its former eminence owing to highly restrictive agricultural markets in the west, especially in Lithuania's major trading partner, the EU.

Establishing a Baltic free trade area and a common market has been the subject of much time and effort on the part of Baltic governments, even though the advantages of such arrangements are more likely to be political than economic. The markets are neither extensive nor complementary as reflected in Lithuanian trade statistics for 1997, which showed that Estonia and Latvia took about 11 per cent of Lithuanian exports and provided about 3 per cent of Lithuanian imports. These figures benefited from the establishment of a Baltic free trade area which came into force in April 1994 and from the introduction of free trade in agricultural products, with some limitations, at the beginning of 1997. Although the potential for trading growth was limited, the establishment of these agreements was viewed favourably by the EU which looked for evidence of regional cooperation among potential members.[41]

Trade with Lithuania's southern neighbour Poland was also quite small at 5 per cent of total trade turnover, but the potential here was very much greater owing to Poland's membership in the Central European Free Trade Area (CEFTA). The free trade agreement with Poland of 28 June, 1996 (though agricultural tariffs were reduced, not eliminated) prepared the way for similar agreements with the other CEFTA members. The Lithuanian Association Agreement with the EU and negotiations on membership of the World Trade Organization (WTO) augured well for ultimate membership in CEFTA which has a combined population in excess of 100 million. Yet trade policy suffered from the decisions by the Lithuanian government to impose price floors on imports of agricultural goods from the other Baltic states in February 1999, as well as export subsidies and certain non-tariff measures on imports, all of which were incompatible with WTO rules and delayed Lithuanian entry into the WTO with its accompanying trade advantages. Continued protection of agriculture by trading measures will lose the advantages in trade growth flowing from membership of the WTO, and diminish the rate of overall economic expansion.[42]

The trade deficit, which had been growing steadily in the 1990s, had reached worrying proportions by the end of the decade. Potentially it had serious implications for the rate of Lithuanian economic growth. The excess of imports over exports was partly attributable to the heavy import of investment goods such as machinery. A contributory factor, however, was the overvaluation of the *litas* in relation to European currencies. This handicapped exports and cut the price and increased the volume of imports. Similarly, part of the inflow of FDI financed the import of consumer goods which, unlike the imports of equipment, did not generate additional exports to pay for them. The government attempted to tackle the persistent and worsening trade deficits by discouraging short-term borrowing and encouraging the repatriation of Lithuanian savings held abroad. If these measures prove to be unsuccessful it might ultimately be necessary to tighten monetary policy to reduce domestic demand and channel a greater proportion of production into exports. This deflationary policy was an alternative to devaluation of the *litas* which would have helped to increase exports but at the cost of rising inflation at home and a reduction in foreign investment. In either case there was a serious risk of a reduction in overall growth rates.[43]

The establishment of a viable energy policy was a major condition of sustained economic growth. Lithuania had few domestic sources of energy and relied overwhelming on imports to meet domestic needs. This was a heavy burden on the external trading account. Moreover, supplies of oil and gas from the former Soviet Union were frequently interrupted during the transition period, with adverse effects on production. It was imperative to develop new sources of energy, to find alternative supplies without adverse environmental effects, and to decrease costs by more efficient use of fuel. The economics of a proposal to build oil and gas pipelines from Norway via Sweden and Finland to the Baltic states proved to be unviable at existing levels of consumption. Lithuania's own oil reserves, which had barely been exploited, were modest in comparison with annual consumption. There were three major options before the Lithuanian government: to make major economies in consumption and in this they were helped by substantial grants and loans from international agencies such as the World Bank; to keep the Ignalina nuclear power station in service for longer than the original projection; and, perhaps most important, to build an oil terminal at Būtingė, near the Latvian border, which could be used both for export and import purposes, and thus reduce dependence on Russian supplies. So long as Russia remained the major supplier of energy the Lithuanian economy would be subject to the dictates of Russian policy. Russia had used oil as a major political card in the transition period and was not averse from doing so in the later part of the decade, notably interrupting the supply of crude oil to the Mažeikiai refinery in April 1999. Press reports suggested that Moscow was prepared to cut supplies further in protest against Williams being permitted to take a 66 per cent stake in the refinery, oil terminal and Biržai pipeline. It was obviously tempting to the Lithuanian government to offer the Russian oil company LUKoil a share in Mažeikiai if that would guarantee continuity of supplies, but there was ample evidence from the past to suggest that the Russian government would find ways to interrupt supply if that proved to be politically advantageous to it. The entry of Williams into the equation offered Lithuania freedom from dependence on Russian oil. The Williams deal was a strategic step of the first importance, guaranteeing continuity of supplies, economic stability, higher growth rates, and increased security. However, it did not solve the long-term problem of power supplies. Here the major problem was the future of the Ignalina nuclear power station.[44]

The Chernobyl-type reactors at Ignalina were understandably regarded as dangerous in the first part of the decade. Substantial investments were injected into the plant to improve safety procedures, and considerable management and employee training was undertaken. As a condition of various grants and loans from the EBRD the first reactor was scheduled to close down in 2005 and the second in 2010. The EU insisted that this original schedule had to be adhered to if Lithuania were to make progress in its negotiations for entry. The Lithuanian government, by contrast, believed that significant improvements in safety procedures under international supervision had outdated the original closure schedule. Moreover, the cost of closure was extremely high and could not be met by Lithuania. Finally, since one reactor could supply up to 90 per cent of Lithuania's electricity consumption at the lowest cost in Europe, the closure would mean that new power supplies would have to be developed, again at high cost. Existing geo-thermal plants could, if operated at full capacity, supply most of Lithuania's needs but at an annual increase in costs of some $200 million. Closure would also lose export earnings from Belarus, and the potential to sell electricity to Poland and western European markets through an extended power grid. This issue highlighted Lithuania's dependence on external political factors in shaping its energy policy. It also raised the question of increased international assistance for decommissioning Ignalina and investing in new power supplies. Evidently failure to solve the energy problem would pose a hazard to future economic development.[45]

By contrast transport was a dynamic, fast-growing and and rapidly-modernizing sector which provided a major source of national revenue and tremendous growth potential. Lithuania is an important transit centre, standing at the intersection of major North–South and East–West arteries. As a result of its Soviet inheritance the most developed rail and road links run from west to east, whereas the north south links require considerable further investment. Lithuania figured largely in the decision of the EU in March 1994 to create nine priority corridors, so-called Trans-European networks, for road and rail traffic in East Central Europe, two of which traversed Lithuanian territory. The better-known of these was usually referred to as the Via Baltica, running north–south from Tallinn in Estonia, with a spur to St. Petersburg, through Riga and Kaunas to Warsaw. This offered an alternative access to Central and East European markets for Scandinavian and Russian traders. In 1994 up to 300 lorries per day were crossing the Polish-Lithuanian border; this figure was forecast to rise two or three-fold by the year 2000. Growth in traffic depended on significant road improvements, such as the construction of by-passes and dual carriageways, service stations, rescue facilities and improved border crossings, where customs and immigration services had often created bottlenecks. Development of the Via Baltica is overseen and promoted by a high level Via Baltica Monitoring Committee involving senior officials from all the countries involved including representatives from the EU.[46]

Intersecting the Via Baltica near Riga is the projected Via Hanseatica, another route in the Trans-European network, which would extend from Kiel to St. Petersburg via Hamburg, Gdansk and Kaliningrad and then through Lithuania to Riga. European transport planners gave priority to road as opposed to rail transport on north–south routes through the Baltic states, reflecting the low quality and light usage of existing rail routes. There was one exception to this, namely the plan to construct a standard gauge rail line from the Polish border to Kaunas, where transhipment of goods to the Russian gauge would take place. This proposal would strengthen Kaunas's position as a major transport hub for road, rail and air traffic, and as the site of a free enterprise zone.[47]

The north-south transit routes made a significant contribution to national wealth in the second half of the 1990s, with the prospect of dynamic growth in the future. Significant as they were, however, they were relatively unimportant in comparison with the extremely heavy transit traffic on east–west routes. Here the transport infrastructure was excellent, probably the best in the Baltic states, and was dominated by rail freight connections between Klaipėda or Kaliningrad and Belarus, Russia, Ukraine and the Central Asian republics. Klaipėda had been an important Soviet port for shipment of raw materials, oil, grain, and other heavy products such as steel. At the time of independence it required a heavy programme of investment to make its facilities more diverse and more suited to the requirements of a modern commercial seaport. It has subsequently adapted to fundamental changes in East–West trade and increased its provision for the transit of manufactured consumer goods. It has developed new container, passenger and oil terminals, new facilities for Ro-Ro traffic, and a new quay for bulk cargo such as fertilizers. Rail links in and to the port area have been reconstructed and the harbour has been deepened to take much larger ships.[48]

The continuing modernization consolidated Klaipėda's position as the fifth largest port on the Baltic sea, taking some 25 per cent of the volume of international trade passing through the ports of the eastern Baltic and Kaliningrad. Its capacity of 21 million tons in 1996 was forecast to rise to 30 million tons by the beginning of the next century. Around 90 per cent of goods entering and leaving the port were transported by rail. From the point of view of Lithuanian economic development the most important aspect of Klaipėda's role was that 78 per cent of exports and 84 per cent of imports in 1997 were transit goods going to and from Belarus, Ukraine and other parts of the former Soviet Union. The fastest growing, though in absolute terms still quite small, sectors in the port were Ro-Ro, container and passenger traffic. Klaipėda offered one of the cheapest routes from the Baltic to Belarus, Ukraine, and south–west Russia, and its ice-free status ensured uninterrupted access throughout the year. The income generated by the port, the railways, and pipelines made a considerable contribution to Lithuanian national and local revenue and to its continued economic development. This trend was likely to continue. Since 80 per cent of the transit trade was to Belarus and Ukraine it was less susceptible to a decline in the Russian market or to the imposition of Russian trade sanctions.[49]

By contrast, Lithuanian agriculture in the 1990s was unable to recover from the impact of the transition and required constant state support. There was little optimism that agriculture could play a leading role in economic development as it had in the inter-war period. The reasons for this were relatively clear. In the Soviet period Baltic farmers exported agricultural products to Russia and received in return substantial subsidies to purchase feed grain and energy. Independence meant that a substantial part of this market was lost and the subsidies ended. At the same time de-collectivization created a mass of very small farms which were ill-equipped to meet the challenge of free markets. As Lithuanian GDP fell the domestic market contracted and was not matched by comparable export growth. As a result rural incomes at the end of 1995 were a catastrophic 29 per cent of December 1990 levels. Although one fifth of the Lithuanian labour force worked in agriculture it accounted for only 7 per cent of GDP. The monthly per capita income of urban residents was 43 per cent higher than rural residents in 1995. Given the limited growth and small size of the domestic market, agriculture needed to look abroad to dispose of its surplus products. The Russian market was uncertain, and to be competitive in western markets the price and quality of Lithuanian products had to improve. Unfortunately this proved very difficult to achieve. The major obstacle was the low productivity of Lithuanian agriculture which could only be increased by mechanization, an updating of skills, and greater specialization of production in crops where Lithuania had a comparative advantage. The necessary changes required injections of short- and long-term capital, greater expertise on the part of banks, and an expansion of rural training programmes both to update agricultural skills and to provide vocational training for those wishing to leave agriculture. Farmers also had to confront the problems of monopoly among their suppliers and their food processing plants. EU policies of restricting agricultural imports and extremely stringent health and safety regulations dampened demand for Lithuanian agricultural products.[50]

Although the Soviet legacy was equally strong in Estonia agricultural productivity there was much higher than in Lithuania. Less than one quarter of the work force produced the same value of agricultural products. Another obvious difference lay in the average size of farms— 8.4 ha. in Lithuania compared with 23.1 ha. in Estonia. This productivity gap may be accounted for in part by the determination of the Estonian government not to protect their farmers against external competition. Successive Lithuanian governments have been more disposed to offer minimum prices, subsidies, and external protection to the farm sector which has resulted in a slower pace of adjustment to market forces. Certainly, strikes and threats of strikes, political obligations, and the power of the rural vote disposed governments to be wary of rejecting farmers' claims out of hand. The end result was a weak sector where the pace of change had not kept up with the revolution in other sectors, nor with the demands of the outside world. The best that could be hoped for was a steady evolution of the agricultural sector in the direction of greater productivity, lower costs and heightened specialisation. Meanwhile, agriculture would continue to be regarded as a lagging sector making few contributions to the increasing efficiency of the Lithuanian economy.[51]

ECONOMIC PROSPECTS

In the period leading up to the Russian financial crisis in the summer of 1998 the Lithuanian economy was performing well. There was, as we have seen, strong inward investment, the currency was stable, inflation was under control and moving downwards, there had been a major shift in foreign trade patterns, and the privatization process for the major utilities was on a sound basis. The liberalization and stabilization of the economy had been successfully accomplished. Although not in the first round of European Union enlargement in East Central Europe, Lithuania was making steady progress towards meeting EU targets for inclusion in the negotiating process for full membership. Foreign investment was on a substantial and growing scale, and was fuelling the technological modernization of the Lithuanian economy which was required if the country was to sustain its relatively rapid rate of economic growth. Efficient production methods and effective training for the workforce were conditions for the growth of foreign trade, on which Lithuanian prosperity depended.

The Russian financial crisis had a damaging short-term impact on Lithuanian economic growth. The proportion of Lithuanian trade with Russia, though seriously diminished from earlier levels, was still relatively high and Russian markets were valuable, particularly for Lithuanian agriculture. The instability of the Russian economy strengthened the existing motivation to become less dependent on Eastern markets, but it was recognised that the potential of the Russian economy was enormous and that Lithuanian economic growth would be more rapid if the options for trade in that direction remained open. The speedier recovery of the Russian economy than had been generally forecast provided a useful boost to the Lithuanian economy in 1999. The Ukraine, Belarus and the Central Asian economies offered quite significant opportunities for Lithuanian traders, though they too were affected by events in Russia. One of the major effects of the crisis was its adverse impact on the very profitable transit trade, perhaps the most dynamic sector of the Lithuanian economy, offering great potential for future growth.

Towards the end of 1999 the Lithuanian economy could be said to be at a crossroads. In the previous decade it had taken important steps towards the achievement of a liberal, and increasingly prosperous, market. Yet the gap in living standards measured in incomes per capita between Lithuania and the western developed world was very considerable and would take decades to bridge, even with sustained and rapid economic growth in Lithuania. Furthermore, though the IMF approved of the direction and pace of economic change, there were clouds on the horizon which darkened the economic prospects. The Williams contract to invest in Mažeikiai, though sound from a strategic perspective, was so expensive for the Lithuanian government finances that it brought about a political crisis, leading to the resignation of the Prime Minister in the Autumn of 1999. A second enormous demand on both government finances and on the supply of energy was anticipated in the first years of the new century when the first reactor at Ignalina was expected to be decommissioned. The loss of cheap electricity from this source will increase energy costs, impact on actual and potential energy exports, and have a damaging effect on manufacturers' and consumers' costs. It will, in addition, place an extra and substantial burden on public finances, with consequential effects on other government expenditure. Maintaining a balanced budget in these circumstances, necessary though it would be to maintain control of inflation, would have adverse effects on the government-financed social safety net. Cuts in this area might shift the balance of opinion against the reform process. Moreover, since the closure of Ignalina had been demanded by the EU, it is conceivable that anti-EU sentiments might grow as a consequence, to the benefit of the nationalist wing in politics. If, however, the EU and other international bodies were to assist Lithuania in meeting the costs of closure in the interests of nuclear safety in the Baltic Sea region, the economic and social consequences might be containable. Nevertheless, it is certain that energy costs will rise, to the detriment of Lithuanian foreign trade. The economic challenges facing Lithuania in the medium term are, therefore, formidable. But, in the light of Lithuania's historical resilience in the face of enormous vicissitudes, the international community should be confident that, once again, Lithuanians will rise to the economic challenge, as they did in the inter-war period.

1 Ole Norgaard et al., *The Baltic States after Independence* (Cheltenham: Edward Elgar, 1996), pp.122–3; *The Baltic Independent* (hereafter *TBI*), 8–14 November, 1991; *The Baltic Times* (hereafter *TBT*), 6–12 November, 1997

2 Norgaard et al., pp.5–9; Alice H. Amsden, Jacek Kochanowicz and Lance Taylor, *The Market Meets its Match: Restructuring the Economies of Eastern Europe* (Cambridge, Mass: Harvard University Press, 1994), pp.vi–ix; George J. Neimanis, 'Baltic Politics and Economics in Transition: Review Essay', *Journal of Baltic Studies* (hereafter *JBS*), vol.XXVIII, no.1, Spring 1997, p.99; Review by Thomas Palm of Raphael Shen, *Restructuring the Baltic Economies*, *JBS*, vol.XXVI, no.1, Spring 1995, p.67; Stanley Fischer, Ratna Sahay and Carlos A. Vegh, 'Stabilization and Growth in Transition Economies: the Early Experience', *Journal of Economic Perspectives*, vol.10, no.2, Spring 1996, pp.46–50

3 Norgaard et al., pp.146–50, 162

4 *TBT*, 27 June–3 July, 1996; Thomas Palm, 'Institutional Economics and Institution Building in the Baltic States', *JBS*, vol.XXVIII, no.1, Spring 1997, p.108; Neimanis, p.95; Thomas Palm, George J. Viksnins, Juris Niemanis and Valdas Samonis, 'The Economic Transformation of the Baltic States: A Panel Discussion', *JBS*, vol.XXII, no.3, Fall 1992; p.305; J. Thad Barnowe, Gundar J. King and Eli Berniker, 'Personal Values and Economic Transition in the Baltic States', *JBS*, vol.XXIII, no.2, Summer 1992, pp.179–89; Rasa Alisauskiene, Rita Bajaruniene and Birute Sersniova, 'Policy Mood and Socio-Political Attitudes in Lithuania', *JBS*, vol.XXIV, no.2, Summer 1993, pp.141–3; *TBI*, 12–18 March 1993, 7–13 January, 1994

5 Neimanis, pp.95–8; Niels Mygind, 'A Comparative Analysis of the Economic Transition in the Baltic Countries – Barriers, Strategies, Perspectives', in Tarmo Haavisto ed., *The Transition to a Market Economy: Transformation and Reform in the Baltic States* (Cheltenham and Brookfield: Edward Elgar, 1997), pp.18–22

6 Amsden et al., pp.6–8; Piritta Sorsa, 'Trade Issues in Transition', in Constantine Michalopoulos and David G. Tarr eds., *Trade in the New Independent States* (Washington, D.C.: The World Bank in association with the United Nations Development Programme, 1994), pp.157–9; *Izvestiia*, 11 and 18 November, 1991; *TBI*, 9 October–15 October, 1992, 15–21 January, 1993; Lithuanian Ministry of the Economy (in association with PricewaterhouseCoopers), *Lithuania: Country Profile*, August 1998, pp.9–10

7 Rasa Dale, 'Inflation Stabilization and Credit Availability in the Baltic States', *JBS*, vol.XXVIII, no.1, Spring 1997, pp.42, 47–53; Norgaard et al., pp.132, 159; *TBI*, 20–26 March, 1992; *TBI*, 2–8 October, 1992

8 *TBI*, 19–25 March, 25 June–1 July, 2–8 July, 10–16 December, 1993; BBC, *Summary of World Broadcasts* (hereafter *SWB*), SU/1851, E/4, 20 November, 1993; *TBI*, 16–22 July, 1993; *SWB* SU/ 2031 E/4 25 June, 1994; *TBI*, 8–14 November, 1991; Tauno Tiusanen, 'The Baltic States in

transition', *International Politics*, vol.33, March 1996, pp.85–7; Michael Bradshaw, Philip Hanson and Denis Shaw, 'Economic Restructuring' in Graham Smith ed., *The Baltic States: the National Self-Determination of Estonia, Latvia and Lithuania* (Basingstoke: Macmillan, 1994), p.175; Seija Lainela, 'Currency Reforms in the Baltic States', *Communist Economies and Economic Transformation*, vol. 5, no.4, 1993, pp.427–9, 440–2

9 Thomas Grennes, 'Inflation and Monetary Policy during two periods of Lithuanian Independence', *JBS*, vol.XXVII, no.2 Summer 1996, pp.137–8; Sorsa, p.158; Lainela, pp.441–2; *TBI*, 17–23, 24–30 September, 1993, 25–31 March, 1994; *TBT*, 20–26 November, 1997; Norgaard et al., p.128

10 Lithuanian Development Agency, *Advantage Lithuania*, 1998; Bank of Finland, *Institute for Economies in Transition*, 22 January, 1999; *TBI*, 12–18 January, 1996; Fischer et al., pp.46–51; Lithuanian Ministry of the Economy, p.11; Dale, pp.39–41, 56

11 Gundar J. King and Nicholas Wolfgang Balabkins, 'Fundamental Thought on Four Basic Economic Goals', *JBS*, XXVIII, no.1, Spring 1997, pp.4–5; Thomas Grennes, 'The Economic Transition in the Baltic Countries', idem, p.16; Norgaard et al., pp.129–31; *SWB*, SU/1995 E/6 12 May, 1994, SU/2127 E/2 15 October, 1994, SU/2226 E/1 13 February, 1995; *TBI*, 7–13 January, 6–12 May, 13–19 May, 20–26 May, 1994; Sorsa, p.159

12 Bradshaw et al., pp.170–9; Norgaard et al., pp.158–60; *TBI*, 27 September–3 October, 15–21 November, 1991, 3–9 April, 1992; Ramunas Vilpisauskas, 'Lithaunia's Membership in the European Union; Possible Effects on its External Trade Policy', in Foreign Policy Research Center, *Lithuanian Foreign Policy Review*, 98/1, 1998, pp.70–6

13 *SWB* SU/2047 E/1 14 July, 1994, SU/2131 E/1 21 October, 1994, SU/2239 B/7 28 February, 1995; *TBI*, 8–14 April, 15–21 April, 6–12 May, 3–9 June, 17–23 June, 15–21 July, 1994; 21–27 April, 1995

14 Vilpisauskas, pp.70–5

15 *TBI*, 20–6 March, 1992, 21–27 April, 1995; Norgaard et al., pp.158–60

16 Haavisto, 'Introduction' in Haavisto ed., p.11; Pavel Pelikan, 'State-Owned Enterprises after Socialism: Why and How to Privatize them Rapidly', in Haavisto ed., p.161; *The Economist*, 13 June, 1992; King and Balabkins, p.6

17 Public opinion polls taken in 1993, however, showed that Lithuanians were less likely to support privatization of large enterprises, less than 10 per cent believing they should be run by private companies and 85 per cent wanting them run by the state or by a state-private partnership: Alisauskiene et al., pp.141–3

18 The World Bank reported in July 1993 that only Lithuania had placed privatization at the very heart of the government's economic programme from the beginning. However, it also recommended that monopolies should be broken up before privatization to ensure competition in the market: *TBI*, 25 June–1 July 1993

19 *SWB*, SU/2191 E/3 3 January, 1995; Lithuanian Development Agency, 1998, np

20 Restrictions on the amount of cash and vouchers to be used as a means of payment were lifted in September 1992: *TBI*, 6–12 June, 1991, 2–8 October, 1992; *The Economist*, 11 May, 1991; Norgaard et al., pp.132–6

21 *TBI*, 2–8 October, 1992, 29 January–4 February, 19–25 February, 26 February–4 March, 8–14 October, 1993, 7–13 July, 1995; *TBT*, 27 June–3 July, 1996; Mygind, pp.39–40, 47–52; King and Balabkins, p.6

22 Norgaard et al., 136–8; *Ekonomika i Zhizn*, no.40, October, 1991; Lidija Sabajevaite, 'Political Parties and the Political Situation in Lithuania', in Wojciech Lukowski and Konstanty Adam Wojtaszczyk eds., *Reform and Transformation in Eastern Europe* (Warsaw: Elipsa Press, 1996)

23 *TBI*, 8–14 October, 1993, 13–19 May, 1994, 7–13 July, 1–7 September, 1995

24 *SWB*, SU/1851 E/3 20 November, 1993

25 The accountants Coopers and Lybrand declared that the main idea of voucher privatization was to limit the influence of foreigners in the privatization process: *TBI*, 20–26 May, 1994

26 *TBI*, 15–21 November, 1991, 1–7 May, 14–20 August, 30 October–5 November, 27 November–3 December, 1992; *SWB* SU/1797 E/2 18 September, 1993, SU/1915 E/2 7 February, SU/1995 E/6 12 May, 1994

27 Drastic measures were taken against smuggling, customs officers were frequently re-located, and in the first two months after the director's appointment an extra $25 million were transferred to the state budget: *TBT*, 3–9 April, 1997

28 Haavisto ed., pp.12–13; Gunnar Eliasson, 'Investment Incentives in the Formerly Planned Economies' in Haavisto ed., pp.184–97; Amsden et al., p.3; Palm, p.110; *SWB*, SU/2038 E/1 4 July, 1994, SU/2482 E/3 9 December, 1995, SU/2647 E/4 25 June, 1996; *TBI*, 29 October–4 November, 1993, 22 December–11 January, 1996; *TBT*, 24–30 October, 1996, 20–26 March, 3–9 April, 1997, 5–11 February, 1998

29 Dale, pp.41–2; *The Economist*, 20 January, 1996; *TBI*, 2–8 February, 16–22 February, 1–7 March 1996; *TBT*, 20–26 February, 1997; T. Ross, 'On the Lithuanian Banking Crisis', *The Baltic Review*, vol.10, Summer 1996, pp.38–9

30 *TBI*, 2–8 February, 16–22 February, 1–7 March, 25 April–1 May, 6–12 and 13–19 June, 1996

31 *TBT*, 26 September–2 October, 31 October–6 November, 5–11 December, 12–18 December, 1996, 20–26 November, 1997; Lithuanian Development Agency, np; Grennes, 'Inflation and Monetary Policy', pp.136–8; Ben Slay, 'External Transformation in the Post-Communist Economies: Overview and Progress' in Michael Kraus and Ronald D. Liebowitz eds., *Russia and Eastern Europe after Communism: the Search for New Political, Economic and Security Systems* (Boulder, CO: Westview press, 1996), pp.77; *SWB*, SU/1953 E/3 23 March, 1994; Grennes, 'The Economic Transition', pp.14–16

32 *TBT*, 16–22 January, 3–9 April, 30 October–5 November, 1997, 15–21 April, 1999, *Lithuanian Weekly*, 2–15 October, 1998

33 *TBI*, 22–28 September, 15–21 December, 1995; Bank of Finland, 9 February, 1999; *TBT*, 4–10, 18–24 July, 1–7 August, 1996, 3–9 April, 1997

34 *TBT*, 24–30 April, 5–11 June, 10–16 July, 21–27 August, 1997; *Lithuania and the European Union: Towards Accession Negotiations* (Vilnius: Du Ka UAB, 1998), p.7

35 *Lithuania and the European Union*, p.7; Lithuania Development Agency, np; *TBT*, 19–25 March, 26 March–1 April, 23–29 July, 6–12 August, 24–30 September, 1998, 25 February–3 March, 1999

36 *SWB*, SU/2235 E/3 23 February, SU/2246 E/3 8 March, 1995; *The Economist*, 1 November, 1997; Bank of Finland, 4 November, 1998; *TBI*, 24–30 July, 1992, 19–25 January, 1996; *TBT*, 11–17 April, 1996

37 An increased share of invested capital was contributed by the purchase of shares on the Lithuanian stock market; in 1996 foreigners bought shares worth about $430 million: Lithuanian Development Agency, 1998; *TBI*, 17–23 March, 7–13 April, 1995; Lithuanian Ministry of the Economy, p.12

38 Before the Russian financial crisis in the summer of 1998, Russia had only about 25 per cent of Lithuanian trade which decreased by a further 28 per cent in January and February 1999 compared with the same period in 1998. After the severe downturn in Russian foreign trade, 54 per cent of Lithuanian exports in January and February 1999 went to the EU and only 19 per cent to the former Soviet republics: *TBT*, 22–28 April, 1999

39 *TBI*, 26 August–1 September, 1994; Lithuanian Development Agency, 1998; Lithuanian Ministry of the Economy, pp.37–8

40 *TBI*, 3–9 June, 1994

41 Lithuanian Ministry of the Economy, pp.37–8; *TBT*, 30 May–5 June 1996, 9–15 January 1997; *SWB*, SU/1958 E/1 29 March, 1994

42 *SWB*, SU/2174 E/4 9 December, SU/2176 E/4 12 December, 1994, SU/2636 E/2 12 June, 1996, SU/2721, E/3 19 September, 1996; *TBT*, 6–12 June, 20–26 June, 4–10 July, 1996, 23–29 April, 1998; Vilpisauskas, pp.70–7; Stephen R. Burant, 'Overcoming the Past: Polish-Lithuania Relations, 1990–1995', *JBS*, vol.XXVII, no.4 Winter 1996, pp.322–4

43 Bank of Finland, 12 January, 4 November, 1998, 22 January, 1999; *TBT*, 2–8 April, 23–29 July, 1998, 21–27 January, 25 February–3 March, 1999

44 *TBI*, 19–25 March, 1993, 1–7 April, 1994, 2–8 June, 1995; *TBT*, 22 January–3 February, 18–24 March, 15–21 April, 22–28 April, 6–12 May, 1999

45 *TBI*, 25 February–3 March, 1994, 26 May–1 June, 18–24 August 1995; *TBT*, 9–15 July, 6–12 August, 26 November–2 December, 17–30 December, 1998, 4–10 March, 11–17 March, 18–24 March, 1999

46 *TBT*, 9–15 May, 11–17 July, 15–21 August, 1996; *TBI*, 4–10 November, 1994

47 *TBI*, 9–15 June, 1–7 September 1995

48 *TBI*, 9–15 July, 1993, 3–9 November, 1995; *TBT*, 9–15 May, 11–17 July, 3–9 October, 1996

49 *TBI*, 30 September–6 October, 1994; *TBT*, 9–15 May, 11–17 July, 3–9 October, 1996, 9–15 January, 6–12 February, 13–19 March, 10–16 April, 3–9 July, 4–10 September, 1997, 8–14 January, 5–11 March, 1998

50 Grennes, 'The Economic Transition', pp.21–2; *TBI*, 28 January–3 February, 29 July–4 August, 1994; *TBT*, 11–17, 18–24 April, 6–12 June, 17–23 October, 12–18 December, 1996, 28 August–3 September, 1997, 25 February–3 March, 1999

51 *TBT*, 12–18 December, 1996, 28 August–3 September, 1997; *SWB*, SU/2345 E/4 3 July, 1996

Chapter 6

LITHUANIA'S FOREIGN AND NATIONAL SECURITY POLICY

All the main political parties in Lithuania agree on the twin pillars of Lithuanian foreign and security policy, namely entry into the North Atlantic Treaty Organization (NATO) and the European Union (EU). This consensus was reached after a number of alternative policies were discarded for not fulfilling all the requirements of national security. In an unstable, competitive and even anarchic international system all states seek to protect their vital interests, to maintain their territorial integrity, and to secure the capacity to solve potentially destabilising problems, such as environmental pollution, crime, immigration, economic decline, and poverty. Security, therefore, is not exclusively associated with external defence capability (although that remains an indispensable element), but should be understood in a multidimensional way. In this chapter discussion will focus on the unremitting efforts of Lithuania since independence to enhance both its external and internal security, in particular its identification of NATO and EU membership as the surest means of achieving these objectives.[1]

The choice of NATO and the EU as guarantors of Baltic security arose out of the specific international context in which the newly-independent Lithuania found itself. Although she had chosen political independence, her economy, as we saw in Chapter Four, was heavily dependent on Russian markets, raw materials and energy. The diversification of her economic links was a precondition for the strengthening of her political independence. This occurred quite rapidly in the 1990s. Throughout this period, however, Lithuanian politicians remained sensitive to any evidence of Russia's attempting to strengthen its influence in the Lithuanian economy by investments and market power, particularly in the basic infrastructure of transport, telecommunications and energy. The agreement with the Williams oil company of the USA for a substantial investment in the Mažeikiai oil refinery showed the determination of the Lithuanian conservative government to avoid too great a dependence on Russia.

Yet, membership of international organizations, to which Lithuania aspires, also increases dependence and diminishes formal autonomy, while enhancing interdependence and partnership. Dependence *per se* was not, therefore, the issue. The question was whether one type of dependence, as opposed to another, increased Lithuania's security and strengthened its ability to defend its vital interests. The judgment of the vast majority of Lithuanians was that membership in western organizations, such as the EU, was in Lithuania's strategic interests. By contrast, Lithuania's past experience suggested that close links with Russia would result in subordination, not partnership.

However, it is worth remembering that, shortly before the disintegration of the Soviet Union, Russia had recognized the independence of Lithuania in a treaty signed on 29 July, 1991. This agreement, uniquely, referred to the 'annexation' of 1940, and invited the Soviet Union to withdraw its troops from the Baltic states. Recognizing Lithuania's right to independence and full sovereignty, including the right to determine its own security and defence arrangements, Russia promised to refrain from the use of force, and to respect Lithuania's territorial integrity. At that time many in the Lithuanian political leadership envisaged the maintenance of close economic links between the two states, and there was some support for the idea of neutrality rather than membership in western security organizations.[2] However, even though there was general agreement on the need to maintain good neighbourly relations with Russia, there was, from the start, a fixed determination not to become part of a new confederacy of the former Soviet republics. This decision was reinforced rather rapidly after the break up of the Soviet Union by the increasingly aggressive and inflexible tone of the Russian leadership on such pivotal issues as Russian minorities in the 'near abroad', the withdrawal of Russian troops from the Baltic states, the establishment of a transit agreement with Lithuania, and the legality of the annexation of Lithuania and the other Baltic states in 1940.[3] Andronik Migranjan, a Yeltsin adviser and a member of the Presidential Council, claimed that the speedy recognition by Russia of the independence of the Baltic states had been a mistake since more time had been needed to resolve various pressing problems, such as borders and the rights of Russian minorities. At the same time, radical Russian nationalists threatened that Baltic independence was an 'unpleasant aberration', which needed to be put right. These kinds of remarks reinforced the opinions of Lithuanians such as Vytautas Landsbergis that Russia was incorrigibly and fundamentally imperialist.[4]

Long before these expressions of Russian chauvinism Lithuania had made up its mind to reclaim its western heritage and to rejoin Europe, following its 'forcible transfer' to the East. Lithuania's western orientation was composed in equal parts of attraction and repulsion: attraction to Europe whose civilization it shared; repulsion from the Soviet Union and Russia, whose values and culture were alien, in part European, of course, but looked at in the round, Eurasian.[5] Lithuanians were also determined not to become politically or economically dependent on Russia since this would identify their country as part of Russia in western eyes, and make it more difficult to rejoin Europe.[6]

The fear of being pulled towards closer relations with Russia partly accounts for the virulence of the attacks by Homeland (the Lithuanian conservatives) and other right-of-centre parties on the new LDDP government of Šleževičius and Brazauskas after 1992. It was assumed that negotiations with Russia on troop withdrawals, military transit, Kaliningrad, or trade, by a government composed of, and supported by, former communists would inevitably lead to the betrayal of Lithuanian national interests. Landsbergis and his political allies seemed to believe that Lithuania and Russia were in a zero sum game, in which concessions made to Russia would inevitably damage Lithuanian interests. Any attempt by Brazauskas to improve bilateral relations was regarded as weakness, which would encourage Russia to make further demands without corresponding concessions. As a case in point, the Right quoted the negotiations over the trade agreements which were concluded in November 1993, giving Lithuania indispensable most favoured nation (MFN) status, and the establishment of joint commissions to discuss other vital issues between the two states, such as improving economic relations, the payment of debts to Lithuanian depositors in the former Soviet foreign trade bank, the return of former Lithuanian embassies in Paris and Rome, and co-operation in the supply of energy from Russia to Lithuania. When the Russian side refused to ratify the treaties until other issues were satisfactorily resolved, notably relations with Kaliningrad and the question of transit, Landsbergis and his allies accused the government of following a pro-Russian policy, which would lead to a Russian-controlled transport corridor across Lithuania and the control of major Lithuanian energy and transport enterprises by Russian organizations. Landsbergis saw evidence in these actions of Russian ambitions to tie Lithuania into a military union and to subjugate her economy to Russian interests. He summed up Lithuania's policy under Brazauskas as 'indulgence to another state'.[7]

The LDDP government would not have recognized this characterization of itself. The failure to obtain MFN status placed Lithuanian trade with Russia under an impossible handicap and profoundly damaged the interests of Lithuanian producers. It was indispensable to negotiate with the Russians to obtain concessions. Trade with the West was growing throughout the 1990s, but it was widely accepted that it would not completely displace Russian markets and supplies. Similarly, there had to be real negotiations with Russia over troop withdrawals and transit, on the understanding that international law and western governments would give their support to a moderate Lithuanian position. At the heart of the disagreements between Left and Right in Lithuania were different assessments of Russian attitudes and policy. Landsbergis and the Lithuanian nationalists were convinced that Russia would take every opportunity to draw the Baltic states back into its sphere of influence, and would manipulate the former communists in the Lithuanian government to achieve this aim. Brazauskas and his allies on the Left equally wanted Lithuania to be part of the West and a member of NATO and the EU, but knew that there were vital issues to be settled with Russia in a negotiating process which would inevitably necessitate some Lithuanian concessions. They also appreciated that there were political forces in Russia which supported economic and political reforms, and which wanted to achieve a lasting *détente* with the West and to draw on western help for the modernization process. It was believed that reform could best be consolidated in Russia if she had good relations with neighbouring states, particularly in the strategically sensitive area of the Baltic. The basically constructive course of Lithuanian–Russian relations in the latter part of the 1990s suggests that, in general, the LDDP strategy was well-judged.

Yet we should be careful of drawing too marked a distinction between the policies of the Left and Right in Lithuanian foreign policy. Both wanted a western orientation; both agreed on membership in western economic and security organizations; neither wanted to be drawn into a Russian sphere of influence. The question at issue was how to treat Russia; nationalists were basically Russophobic and doctrinaire, the Left was pragmatic and flexible. The difference in approach was partly to be understood in the light of contradictory signals coming out of Moscow, which made it very difficult to estimate accurately the direction of Russian policy. It was, therefore, no accident that the decision to apply for NATO membership on the part of the Baltic states followed a period of unpredictable actions and aggressive rhetoric on the part of the Russian leadership, which increased Baltic feelings of vulnerability. This was the context in which Lithuanian security policy was being formulated.

The most pressing and urgent problem in Lithuanian–Russian relations, the withdrawal of Russian military forces from their bases in Lithuania, demonstrated very well the uncertainties and frustrations of the Lithuanians in the face of prevarication, evasiveness, and unpredictability on the Russians' part. This is not to deny that the Russians had some genuine logistic difficulties and that the timing of withdrawal from the Baltic depended on the progress of the withdrawal from Germany and Poland, and on the supply of adequate housing in Russia for officers withdrawn from bases in Lithuania. But these problems do not entirely account for Russia's delay in carrying out its obligations under international law, or in responding to the call from the Conference on Security and Co-operation in Europe (CSCE) in July 1992 for early withdrawal. The charge of foot dragging and evasion cannot be dismissed. To give just one example, Landsbergis arrived in Moscow to sign an agreement on troop withdrawal in September 1992, and a more detailed schedule on the pull-out. Yeltsin, however, refused to sign the main agreement and only the schedule received a Russian signature.

Despite the withdrawal of Russian troops in the course of the next year, the absence of a signed agreement created uncertainty in the interim about whether Russia would try to maintain a presence in the country. Russia advanced objections to clauses in the agreement seeking compensation for damage caused by Soviet forces, and for the transfer of some military equipment to Lithuania, in return for Lithuanian arms and matériel confiscated by the Red Army in 1940. Twice Russia suspended withdrawal in an attempt to force Lithuania to drop these clauses. On 17 August, 1993, just two weeks before the scheduled completion of the withdrawal, Russia stopped the hand-over of its bases indefinitely, arguing that the troop withdrawal schedule was invalid unless the Lithuanians signed the general agreement with the claim for compensation withdrawn. Moreover, Russia was determined not to concede the principle, which was referred to in the agreement, that her withdrawal was in accordance with international law. This would have implied her acceptance, as the Soviet Union's successor state, of the illegality of the annexation of Lithuania in 1940, and of her obligation under international law to withdraw from an area where, under the concept of the 'near abroad', she claimed a special responsibility.[8]

It was this claim of special responsibility which particularly alarmed the Baltic states. It seemed to them to be the first step in a scheme to return former Soviet republics to the Russian sphere of influence, and to demand the right to act as policeman for this, as yet, ill-defined area. Hawkish statements from the unholy alliance of Russian communists and nationalists (the 'Red-Brown coalition'), and their growing strength at the polls, pushed Yeltsin and his formerly reformist and western-oriented foreign minister, Kozyrev, into more belligerent claims against adjacent territories in the so-called 'near abroad', and into the formulation of a special role for Russia in these regions. The withdrawal of Russia from areas in which she had had a centuries' old interest would, it was alleged, create a political vacuum which would be filled by Russia's enemies. Instability in the territory of the former Soviet Union would threaten Russia's vital interests and demand 'peacekeeping' intervention by Russian forces to restore stability. The key element in Russian rhetoric on this issue was the necessity for the Russian government to protect the interests of the many millions of Russian nationals still living outside Russia in former Soviet territory. Migranyan advanced this as a reason, along with geo-political considerations, for Russia's taking special responsibility in the Baltic states. The adoption of a more nationalist position was prompted, not only by the growing electoral threat of the 'Red-Brown Coalition', but also by Russian alarm at the West's discussions on the enlargement of NATO, Western criticism of the Russian role in the former Soviet republic of Moldova, and growing support in the West for the Bosnian Muslims.[9]

The conclusion to be drawn from the statements of Russia's political leaders about the 'near abroad' was that Russia had yet to come to terms with the loss of territory in the former Soviet Union. This explained her unwillingness to recognise the sovereignty of the newly independent republics, and her self-interested claims to intervene there to ensure that Russian interests were safeguarded. Stirring up trouble in these areas, and then offering herself as a peacemaker, provided Russia with the perfect opportunity for 'covert colonialism', to use Jonathan Eyal's term. The eminent Finnish commentator, Max Jacobson, concluded that Russia still possessed the imperial mentality, had not yet come to terms with the independence of the Baltic states, and remained insensitive to the aspirations of small nations. Throughout her history Russia had pursued territorial expansion as a means to security; the 'near abroad' doctrine was the application of a modified version of this territorial imperative. Inevitably the Lithuanians came to the same conclusion. Despite Landsbergis's charges of pro-Russian bias on the part of the LDDP government, Brazauskas spiritedly repudiated the

Russian military doctrine and Moscow's claim for special responsibility in the 'near abroad'. Russia's ambition to become a member of the Council of Europe and her existing membership in other international organizations, he believed, demanded 'high democratic standards' and the associated respect for the sovereignty of other independent states. In fact, Moscow's demands proved counter-productive, since they reinforced Lithuania's declared preference for collective security through membership in western organizations.[10]

This preference was further underlined by the important practical issue of military transit, which illustrated Lithuania's vulnerability to Russian economic and political pressure. In this instance Russia ruthlessly used her economic and diplomatic power to force the Lithuanian side to make concessions. Russian policy on transit was, in fact, not unreasonable, and this was recognised by the West which forced a compromise on the issue. The question of transit was inextricably linked to the maintenance of the Russian exclave of Kaliningrad, where there remained substantial Russian military forces. Supplying the military base with equipment, and transporting Russian troops to and from the region, required access to the railways across Lithuania. The question at issue was the conditions under which Russia could obtain access.

Lithuania's intention was to introduce rigorous regulations for the transport of military cargoes and personnel, since inadequate controls could have serious security implications. The LDDP government drew up regulations which it expected to be applied to every country transporting military matériel across Lithuania. These rules involved an advance application for approval 21 days before the proposed journey, limits on the number of unarmed servicemen to be transported on any one occasion, and the use of Lithuanian guards for each train. The regulations also prohibited the transportation of weapons of mass destruction. The Lithuanian State Council on Defence approved these regulations on 21 October, 1994, and expected that they would be incorporated in a treaty between Russia and Lithuania. This idea was vehemently opposed by the political opposition which believed that it would 'entangle Lithuania in nets of military co-operation with Russia' and facilitate Russian expansionism in the Baltic region.[11]

Russia's negative response to the proposed regulations included a point blank refusal to ratify the trade agreement concluded in November 1993, which awarded vital MFN status to Lithuania. Without this, Lithuanian exporters would continue to be deprived of some of their most important markets. This economic pressure was combined with a Russian appeal to the international community, to the effect that Lithuania had failed to take into account the peculiar military needs of Russia for more easy and rapid access to rail communications across Lithuania. This appeal drew a favourable response from the EU, which urged an agreement in the interests of the stability and security of the region, and of good neighbourly relations. The EU also gave a critically important guarantee to Lithuania that an agreement between Lithuania and Russia on this issue would not impair closer relations between Lithuania and the EU. Each of these factors, along with some intermediation from western sources, combined to persuade Lithuania to change its position and return to the temporary regulations which had been in place during the period when Russian troops were withdrawing from Germany and being conveyed across Lithuania by train. However, the new regulations would be applied to military shipments by any other state; Russia was, therefore, treated as an exceptional case. For Lithuania the great prize was the ratification of MFN status by the Russian side, which would release Lithuanian exporters from the strait jacket of the Russian import regime. Whatever the economic gains, the Lithuanian concessions could be portrayed as a climb down, though the LDDP government claimed a victory in the sense that Russia had been forced to abandon her original demands for an international transit corridor across the country and unrestricted military transport by road, rail and air. Prime Minister Šleževičius gave the credit for the modification in the Russian position to the proposed transit regulations.[12]

Lithuanian concessions, which paved the way for the implementation of the Russo-Lithuanian trade agreement, helped to normalize relations between the two states. From 1995, constructive negotiations took place on the delimitation of borders, and a treaty was signed during Brazauskas's visit to Moscow in October 1997, the first between Russia and a former Soviet republic. During the same visit a sheaf of other agreements on trade, energy, and taxation issues was presented for signature. Nonetheless, the claim that there were no fundamental disagreements between the states was probably exaggerated, given the significant differences on a number of important issues. As we shall see, there was a profound disagreement over the question of NATO enlargement in the Baltic states, to which Russia was implacably opposed. Similarly, the Russian offer of security guarantees in October 1997 was equally unwelcome to the Baltic states. Russia's refusal to admit that the

Soviet Union occupied the Baltic states in 1940 in violation of international law constituted another cause of tension and resentment. Zbigniew Brzezinski, the former US National Security Adviser, commented that Russia's failure to condemn the illegal occupation of Lithuania in 1940 bred insecurity and distrust on the Lithuanian side, and erected a barrier to the normalization of relations between the two states. At the same time, Russia was working vigorously to extend its influence over the Lithuanian economy, especially in the fields of energy and transport infrastructure, as part of a policy advocated by Yevgeni Primakov of bringing the former Soviet republics closer together by thorough economic integration, without at the same time violating Lithuanian sovereignty. These causes of tension were compounded by the problem of Russia's Kaliningrad exclave.[13]

In the Russia–Lithuania treaty of 1991 the borders of Kaliningrad were recognised, the rights of ethnic minorities respected, and co-operation in the fields of economics, trade and culture encouraged. But instability was injected into bilateral relations by the preferences expressed by some Lithuanian politicians for changing the future status of Kaliningrad, including free trade zones, reductions in the number of troops stationed there, and a reference of the demilitarization issue to international organizations. Lithuania could plausibly argue that the concentration of so much military power in Kaliningrad offered a threat to stability in the region. Yet the LDDP government was unprepared to widen the divisions over Kaliningrad by endorsing Rightist claims to the territory on historical, cultural, and geographical grounds. It preferred to work for increased trade and investment between Lithuania and the territory. The creation by the Russian government of a free trade zone in Kaliningrad in 1996 stimulated trade, and by 1998 Lithuania had risen to third place behind Germany and Poland in trade and investment with the region.[14] Practical co-operation to increase trade and to solve common problems such as drug trafficking, smuggling, illegal migration, and pollution should increase regional stability and contribute to confidence building between Lithuania, the Kaliningrad territory, and Moscow. Yet Russia's anxiety over Kaliningrad has been understandably increased by the prospect of NATO's enlargement to include the Baltic states, which would result in Kaliningrad being surrounded by the NATO member states of Poland and Lithuania. Although it is arguable that NATO's expansion into Lithuania would lead to an overall increase in stability in the Kaliningrad region, this is not a proposition which Moscow is ready to accept.[15]

One of the potentially most difficult and sensitive problems in Lithuania's foreign policy was the relationship with Poland. The largest state in East Central Europe, with a population of some 39 million, Poland offered the possibility of close political and economic ties based on a shared history and deep mutual understanding. Good neighbourly relations would offer Lithuania an advantageous counter-weight to Russia, and possible assistance in becoming a member of NATO and the EU, which Poland looked set to enter in the first round of enlargement. The initial difficulties in achieving good relations arose largely, though not entirely, on the Lithuanian side, which could not forget the frozen relations of the inter-war period following Zeligowski's seizure of Vilnius and the incorporation of south-east Lithuania in Poland. Nor could the Lithuanians forget their struggles to establish a nation state against Polish resistance, and their opposition to Polish cultural imperialism during the Lithuanian renaissance of the nineteenth century. The isolation of Lithuania from Poland imposed by the Soviet Union after 1945 ensured that Lithuanians could not fully appreciate the enormous changes which had taken place in Poland in the four post-war decades, even though many were familiar with, and admired, the activities of Solidarity and accompanying political changes at the end of the 1980s. Of course, there were noisy examples of nationalist Poles who regretted the loss of Wilno and clamoured for its return. There were also Polish visitors to Lithuania in the late 1980s and early 1990s who, Lithuanians alleged, behaved paternalistically, insisted on speaking Polish, and assumed that they would be welcomed as long-lost older brothers and sisters. It was obvious that nationalists on either side would not get along and would fuel each others' suspicions. On the other hand, members of the Polish intelligentsia and the leaders of Solidarity, such as Michnik, Geremek, and Skubiszewski, though referring to the common history of Poland and Lithuania, nonetheless emphasised the responsibility of Poles to respond to Lithuanian aspirations 'with particular sensitivity'. The principles underlying relations between Poles and Lithuanians should be 'the common values that apply to all of Europe ... freedom and democracy', to quote Michnik again, who, with Geremek and Wujec, wrote an open letter 'To Our Friends the Lithuanians' in 1989.[16]

Yet the opposition to Lithuanian independence of substantial numbers of Polish-Lithuanians, and their support for the continuation of the Soviet Union, which was discussed in a previous chapter, could not fail to arouse hostility among the ethnic Lithuanian majority. This antagonism was, to some extent, transferred to the Polish government when President Walesa and Foreign Minister Skubiszewski requested that the rights of the Polish minority be recognized, in particular the restoration of the suspended local councils in the Polish-inhabited areas.[17] These requests did not mean, however, that the separatist or autonomist ambitions of the Polish minority in Lithuania were supported by leading Polish politicians—indeed, there were many statements to the contrary.[18]

Furthermore, since 1989, the Polish government had declared on many occasions that it recognised the territorial integrity of its neighbouring states and renounced all claims to the Vilnius region. Many members of *Sajūdis* seemed unable or unwilling to believe this renunciation, despite Poland's signing agreements with all its other neighbours to this effect. This appears to lend credibility to Venclova's suggestion that Lithuanians, particularly Lithuanian intellectuals, displayed an 'inveterate complex' vis-a-vis the Poles and were unwilling to repudiate it. Lithuania's suspicion of Polish intentions, and Poland's worries about the treatment of the Polish minority in Lithuania, ensured that little or no progress was made in negotiations for a state treaty between the two countries until the LDDP government came to power in 1992 and Brazauskas replaced Landsbergis as President the following year. The victory of the LDDP was not in itself sufficient to change the course of Lithuanian policy, since the new government was very sensitive to the currents of public opinion on the Lithuanian-Polish relationship. Three factors swung Lithuanian opinion in favour of accepting Polish overtures and overcame hesitations on the Polish side. The first was the articulation of the draft Russian military doctrine, accompanied by Russian claims to police the 'near abroad'; the second was the acceptance by both sides that their ambitions to join western organizations such as NATO and the EU were dependent on the establishment of good neighbourly relations; and third was the restoration of local self-government to the ethnically Polish region of south-east Lithuania. In addition, there was a growing perception in Lithuania that the route to NATO and EU membership would be eased if Lithuania could harness Polish support for Lithuanian entry. This improved climate quickly led to a military agreement in June 1993 providing for Polish training of Lithuanian soldiers, co-ordinated air space control and the purchase of Polish armaments by the Lithuanian defence forces. However, history, or perhaps more accurately historical mythology, remained a major obstacle to the proposed treaty of friendship and co-operation which was intended to smooth relations between the two states.[19]

A major point at issue was the Lithuanians' insistence that the treaty should refer to the 'armed invasion of Poles in Vilnius' in 1920. The Polish Prime Minister, Hanna Suchocka, responded that the Polish government could never agree to incorporate 'such a one-sided assessment of complicated events', and anyway, history should be left to the historians. A half-way house of leaving historical references out of the treaty itself but including them in a separate declaration also proved abortive. Eventually, after protracted negotiations and heart searchings, a few neutral words were included in the treaty, indicating that both sides condemned the use of force in their previous relations, and regretted the conflicts which occurred after the First World War. It was agreed that historical issues would be left to the speeches of the two presidents during Wałęsa's visit to Vilnius in April 1994. The fact that this compromise took almost a year to reach, between the summer of 1993 and the Spring of 1994, showed the power of history to delay the achievement of major foreign policy and security goals. Brazauskas admitted that it took some time for the Lithuanians to accept that looking into the past would delay both the agreement with Poland and, consequently, the hoped-for integration into western organizations. It was feared that further delay could push Lithuania 'into the arms of Russia'. In a speech before parliament during Wałęsa's visit, Brazauskas claimed that the model of reconciliation between former antagonists which prevailed in western Europe after 1945 was being adopted in Lithuanian-Polish relations. Although he admitted that the two sides would probably never agree on history, they should not close the door on the past entirely. Meanwhile, they had created the foundations on which their relationship would rest: respect for each other's sovereignty; rejection of any territorial claims; and the adoption of European norms in formulating minorities policies. For the immediate future, he concluded, both sides should look forward to constructive dialogue and a series of agreements on trade, military co-operation and mutual security.[20]

Relations have in fact been extremely close and fruitful. There has been intensifying co-operation in foreign and security policy, defence, the economy, trade, transport, communications, culture, and education. So close and important has the relationship become that in 1997 it was announced that it had assumed the character of a 'strategic partnership'. Of great importance to Lithuania was the free trade agreement signed in June 1996 which went into effect in January of the following year. This gave much easier access to Polish markets for Lithuanian exporters, and was the first step towards Lithuanian membership in the Central European Free Trade Area (CEFTA). Also significant were a number of military agreements providing for joint training exercises and the formation of a joint peacekeeping unit LITPOLBAT to undertake UN and NATO peacekeeping and peace enforcement missions. Co-operation in customs and immigration controls on the problematic Polish-Lithuanian border was given high priority. These close relations were solidified by the creation in 1997 of joint institutions such as a parliamentary assembly, with 20 members from each parliament, a consultative committee of both presidencies, and a joint council of both governments set up by the prime ministers. Commentators could not help joking that the old Commonwealth was being resurrected. Certainly there was a good deal of mutual congratulation on the excellence of the relationship. Problems nevertheless remained. When parliamentarians from both sides met together in the Interparliamentary Assembly they spent a good deal of time criticising each other's treatment of ethnic minorities in the areas of language, spelling of names, and the production of school textbooks.[21] Nevertheless, the treaty of 1994 was an important building block in the creation of a Lithuanian security policy. Successive Lithuanian governments judged that membership of NATO and the EU were the central pillars of that policy, and that a close association with Poland and, through CEFTA, with the rest of East Central Europe, would facilitate membership. This represented a modification in Lithuania's tactical approach, placing greater emphasis on the central European route to the West and rather less on the Baltic/Nordic route. Landsbergis agreed that ties between Lithuania and Poland were 'even more intensive' than those between the three Baltic states. Nekrašas commented that Poland was much more important to Lithuania than Estonia and Latvia taken together, owing to its economic, demographic and military potential, and its influence in NATO, the EU and CEFTA. The advantages were not all one way in the sense that Poland's ambition to have an influential role in East Central Europe would be advanced by a close partnership with Lithuania.[22]

In its post-independence search for security Lithuania gave absolute priority to membership of NATO and the EU. Povilas Gylys, the former Lithuanian foreign minister, argued that only NATO membership would guarantee security and prosperity in the Baltic region as a whole, which in turn would enhance stability and economic growth in the rest of Europe. Adequate substitutes could not be found in other organizations such as the OSCE or the EU, or in bilateral security arrangements, or in some kind of Baltic region security arrangement, still less in a policy of neutrality.[23] Only membership of NATO could bring the benefits of security and stability to Lithuania. Only NATO, it was believed, was capable of resisting Russia's attempts to return the Baltic states to Moscow's sphere of influence.[24]

Accordingly, on 4 January, 1994, Lithuania formally applied for NATO membership and NATO expressed its preparedness to expand. Later in January Lithuania joined the Partnership for Peace (PfP), a NATO-inspired organization composed of members and aspirant members of the organization. PfP was designed to be an intermediary body, membership of which would prepare aspirants for NATO membership by helping to modernize their defence forces, would offer intensive training of their military personnel in an international environment, and assist them to achieve standardization and interoperability in military communications and equipment. At the Madrid summit of NATO states in July 1997 though, Lithuania was not admitted to the first round of NATO enlargement, though she and the other Baltic states were assured that the dialogue between them and NATO would continue, in part in the newly-established Euro-Atlantic Partnership Council of which they were members, partly in an enhanced Partnership for Peace, in which Lithuania would be expected to engage in as many planning, training and operational activities as feasible. Lithuania took full advantage of the opportunities for joint activities and co-operation with PfP partners, such as weapons standardization, joint military exercises, the establishment of a joint peacekeeping battalion (BALTBAT), an airspace monitoring system (BALTNET) and a joint naval squadron (BALTRON). Lithuania is demonstrating by her active involvement that she is, as the saying goes, producing security as well as consuming it, and has a positive contribution to make to the alliance. By these various initiatives NATO has shown that she is developing a 'pre-accession strategy' to meet the security needs of the Baltic states. Under these arrangements Lithuania has a form of associate membership of NATO. The aim of these joint activities is to ensure that integration in NATO military and political structures is as complete as possible, short of full membership. In practice, the gap between partner and member should become so thin as to ensure that the eventual transition from one to the other can be easily and smoothly accomplished.[25]

Two years after the Madrid summit, despite almost constant vocal opposition from Russia to the Baltic states' membership of NATO, influential American voices were calling for Lithuania's admission. A delegation of US Senators visiting Lithuania in December 1998 affirmed that Lithuania was a likely candidate for membership in the second round of NATO enlargement. The Congressional Research Service in Washington reported in March 1999 that Lithuania qualified in the top three among the applicants for membership. The following month 24 US Congressmen visiting Lithuania pledged support for Lithuania's membership in the light of her admirable participation in PfP activities and her increase in defence spending. At the Washington summit of NATO to mark the 50th anniversary of its foundation in May 1999, the Baltic states were gratified to find themselves listed by name as aspirant members. Former US National Security Adviser, Zbigniew Brzezinski, identified Lithuania as the best qualified of the Baltic states for entry: it was well on the way to meeting the objective criteria for entry; it did not have a problem with its Russian minority; and it had stabilised its relations with Poland. What it needed was a firm assurance that NATO membership would eventually be achieved. Both Brzezinski and Paul Goble thought Lithuania would be ready for NATO membership in the year 2002. In any case, it seems that NATO assistance to the Baltic states includes aid to achieve interoperability in command, communications and control. In this sense the difference between the new members and the aspirant states in interoperability may well be insignificant. As one American source noted, the Baltic states are moving millimetre by millimetre towards NATO.[26]

In this atmosphere of optimism it is worth remembering that membership of NATO on its own is insufficient to create lasting security in the eastern Baltic region. Confidence-building measures also include the development of a more effective partnership in military, political and economic affairs between Washington, NATO and Moscow. Despite, or because of, the set-back to Russian economic development arising from the financial collapse of the summer of 1998 and the numerous political problems which afflict the country, the West should, so far as possible, continue to strengthen its security partnership with Russia, assist her transition to democracy, and consolidate her economic reforms. The West's interest, and that includes Lithuania's interest, lies in the continuation of Russian reforms and the strengthening of her democracy; these will provide continuing assurance of good neighbourly relations and regional security.[27]

The second major pillar of Lithuanian security policy is membership of the EU, which is regarded as indispensable for economic progress, for good relations with other member states, and for ensuring that Lithuania's ambition to 'return to Europe' will result in tangible advantages for the Lithuanian people. The EU, from its beginnings as the European Economic Community, was not simply an economic organization; it was always perceived by its members as having broad security implications as well. Not least was the expectation that membership would, by stimulating economic growth, increase the welfare and prosperity of member states' populations. The increasingly close integration of member states' economies since the establishment of the single market and monetary union has enhanced economic and political interdependence, and produced, in Peter van Ham's words, 'a community of political solidarity', in which a threat to one member state would be seen to be directed against the whole community. Membership, therefore, offers to each partner a security guarantee, 'a form of non-military security', which might be as reliable as formal treaty commitments. In addition, the Western European Union (WEU), a security organization composed of the European member states of NATO, and recently the security arm of the EU, offered the Baltic states associate partnership; as a result they attend the Permanent Council and take part in other WEU working groups.[28]

Lithuania has made steady progress towards achieving full membership of the EU, though it has been excluded from the first list of applicants from East Central Europe to engage in negotiations for membership. Lithuania's first step in the direction of membership was a trade and economic co-operation agreement with the EU in 1992 which offered her MFN status. Accompanying this agreement was a political declaration avowing shared ideals and pledging foreign policy co-ordination. This first accord was supplanted by a free trade agreement, coming into effect in 1995. As its name implied, this set on foot a process of mutual tariff reductions, giving Lithuania a six year period to remove duties and quotas on imports of manufactured goods from the EU, while allowing her to retain tariffs on a limited range of goods, such as shoes, leather and electronics, after the six year period had elapsed. In return, the EU lowered its tariffs on Lithuanian goods, but at a more rapid pace. The next stage in the negotiating process was the achievement of an Association or Europe Agreement, which was signed in 1995 and came into effect in 1998. This was a far more comprehensive arrangement and constituted a major step for Lithuania towards full membership of the EU. The agreement provided for political dialogue, co-operation in establishing the free movement of people, goods and services, and the co-ordination of economic, financial and technical systems. Lithuania also began the important process of co-ordinating and harmonising its legislation with EU law and regulations as part of a pre-accession strategy, leading eventually to membership of the Union.[29]

Unfortunately, Lithuania failed to make sufficient progress in meeting the criteria for accelerated negotiations for her to be included in the list of front runners. These criteria were laid down by the EU Commission to ensure that there was adequate harmonization between the economies and legislation of the applicant countries and the EU. In political terms applicants were expected to have democratic institutions, to guarantee human rights, and to apply the rule of law. Economically, the aspirant states needed functioning market economies and the ability to show that they could accept all the obligations of membership, including the demands of EMU. In its document *Agenda 2000* of September 1997, the EU Commission identified the six states to be in the advance guard of negotiations. Lithuania's satisfactory progress in meeting the political criteria was noted, but the Commission expressed its doubts about the level of development of Lithuania's market economy, and whether she could sustain economic competition from the other member states if she became a member. Further steps towards the freeing of prices were required, accompanied by an accelerated privatization programme and improved bankruptcy legislation. The restructuring of agriculture, improved environmental protection, and a coherent energy strategy were also deemed to be necessary. The latter meant, as a first step, carrying out the closure of the two reactors at Ignalina nuclear power station.[30]

This rejection was extremely disappointing for Lithuania. Government spokesmen criticised the Commission for using dated statistics, and for not taking into account Lithuania's recent very rapid economic progress. Exclusion was allegedly based on biased political judgments, which were connected with maintaining trouble-free transit arrangements between Kaliningrad and Russia and with excluding cheap electricity exports from Lithuania. These improbable allegations arose mainly from frustration, which was also mirrored in a notable increase in anti-EU feeling in Lithuania in 1997. In an effort to cushion the blow, the Commission moved quickly to assure the excluded countries that they could join the negotiations as soon as they had made sufficient economic and political progress. Negotiations with Lithuania, the Commission affirmed, would not be long delayed if the existing pace of reform was maintained. Meanwhile, Lithuania would be able to take advantage of the so-called accession partnerships, which offered monetary and technical assistance to meet the Commission's criteria. The slogan 'differentiation without discrimination' was invented to summarise this strategy.[31]

As we noted at the beginning of this chapter, security is not confined to external defence capability. It also involves the capacity to solve potentially destabilising internal problems, or to deal collectively with issues of a broadly regional nature which can exacerbate relations between states and are not resolvable by individual state action. Examples of such issues are pollution and environmental hazard, organized crime, illegal migration, drug running, smuggling and terrorism. Similarly, some of the adverse consequences of economic development, such as increased poverty and insecurity for sections of the population, can often be tackled more effectively by co-ordinating policy and disseminating ideas of best practice. Most of these problems are common to the eastern Baltic region and have an impact on the Baltic Sea area as a whole. In the eight years since the fall of communism the states of the Baltic Sea region have co-operated closely in seeking solutions to these problems. Furthermore, sub-regional governments, city councils, chambers of commerce, business and arts organizations, and environmental protection agencies have also worked together to produce a co-ordinated approach to common problems. At the same time, the Baltic states themselves have attempted to co-ordinate their trading and defence policies, to achieve common practice in their customs and immigration procedures, and to agree on border delimitation.

Baltic states' and Baltic regional organizations have evolved realistically during this period. They do not seek to offer hard security guarantees; that is the task of NATO for those states which are members. Sweden prefers to maintain neutrality but, as a recent member of the EU, is happy to assist the Baltic countries to achieve membership. Finland is in a similar position. Denmark as a member of NATO has been particularly active among the Nordic states in offering defence co-operation. The Baltic states themselves do not see regional co-operation in security matters as a substitute for NATO or EU membership. Until 1993 Lithuania regarded participation in Nordic-Baltic regional organizations as useful preparation for membership in Euro-Atlantic bodies. Although she subsequently began to re-orient herself towards central Europe through her Polish connections, her destination, NATO and the EU, remained the same.

Baltic regional organizations and sub-regional organizations have accordingly developed a security agenda which complements, but does not compete with, that of the major western security and economic bodies. On the eastern Baltic shore, for example, during the early independence period, Lithuania, Latvia and Estonia restored the pre-war Baltic Co-operation Council. Within the ambit of this association their Presidents and Prime Ministers held regular meetings, and reached agreements in principle on questions such as investment priorities, a free trade area, visa free travel for their citizens, and ultimately a customs union. They tried also to co-ordinate their positions on relations with Russia, particularly on the withdrawal of Russian troops. They aspired to develop joint policies on currency, transport and energy. Critics said these meetings were very good at producing high-sounding declarations but ineffective in deciding joint policies. The same could be said of the Baltic Assembly, a forum of Baltic parliamentarians formed in 1991, which was strong on rhetoric and even useful as a consultative body, but seemed unable to exercise influence on the governments or parliaments of the member states. Some of its leading figures, such as the Lithuanian MPs Egidijus Bičkauskas and Romualdas Ozolas, advocated more power for the assembly, increased influence over the national parliaments, and closer liaison with the Council of Ministers. In their view common institutions should take a more active role in developing common policies, instead of acting merely as co-ordinating bodies. These proposals did not appeal strongly to the governments, particularly when Estonia was invited to enter the first round of negotiations for EU membership and was drawn increasingly close to Finland, at a time when Lithuania was pulling towards Poland.[32]

However, despite the frustration of some parliamentarians at the ineffectiveness of common Baltic institutions, in practice the Baltic states marked up a solid record of achievement in promoting security, tightening co-operation, and dealing effectively with many common problems. Serious progress began to be made with the formation in 1994 of the Baltic Council of Ministers, although a free trade agreement had been signed the previous year. This Baltic Council was a smaller and more effective body than its predecessors. The most pressing problem was the easing of customs regimes between the three states to facilitate the movement of goods and people. This, in turn, was conditional on the tightening of controls on the eastern borders with

Russia and Belarus. Council members agreed not only to co-ordinate the preparation of joint programmes in the fields of crime, social security, education, transport and energy, but to co-operate in their implementation. It was intended to develop a common mechanism for border checks in the fight against illegal immigration, terrorism, and drugs smuggling. Perhaps most successful of all was co-operation among defence ministers and commanders for improved training of service personnel, common participation in military and naval exercises, notably in the Baltic Battalion, and joint control of airspace.[33]

With the Nordic states there were an increasing number of bilateral ties supplemented by meetings of the Nordic/Baltic council, the so-called 5 + 3 group. This grouping was well suited to bringing about regional co-operation on a series of pressing issues affecting the stability of the Baltic Sea region, notably ecology, power supplies, transportation, illegal migration, and control of organized crime. An even broader Baltic regional grouping, the Council of Baltic Sea States (CBSS), embracing all the riparian states including Poland, Germany and Russia as well as the Nordic/Baltic group, was formed in 1992. Its objective was to 'initiate, support and develop multi-dimensional collaboration in the Baltic region' on such important regional issues as the environment, transport and energy. It also set out to foster the region's democratic and economic development, particularly in the former Soviet states, and to strengthen its ties with the EU. As a result of Russian insistence it also took under its remit the potentially destabilising question of human and minority rights, by appointing a Commissioner for Democratic Institutions and Human Rights in 1994 to monitor, and make recommendations about, the observance of these rights. The creation of a joint energy system for the Baltic region has been one of the priorities of the Council. It has also recommended improvements for the regional transport systems, including the transport of oil. The Council has promoted reform of the higher educational systems in the Baltic states by sponsoring courses in law, management, economics and administration. Environmental protection was one of the most important tasks of the Council. Within that brief the Council has undertaken studies and made recommendations on such issues as nuclear and radiation safety, the sharing of intelligence information, and the development of an environmental plan. Russian participation in these joint endeavours provides the opportunity to improve mutual relations and to identify common interests. As well as playing a full part in the Council's activities, Lithuania has also become a member of two euroregions fostered by the EU: the Niemen (1997) whose other members are Poland and Belarus with observer status for Kaliningrad, and the Baltic (1998) with a membership of Poland, Sweden and Russia. These subregions are intended to develop more intense forms of co-operation between the partner states.[34]

All this co-operative activity is designed to stabilise relations between states and sub-regions in the Baltic Sea region as a whole, to protect them from common dangers which threaten civic society, to narrow the gap between rich and poor by intensifying economic development in the former communist states, and to create political structures within which solutions to threatening problems can be devised and implemented. It is expected that this intense and positive collaboration will narrow the gap between countries and build up confidence and mutual trust. Baltic co-operation will certainly promote a more effective resolution of environmental problems and strengthen the general feeling of security. This is important since opinion polls suggest that a majority of Lithuanians regard organized crime, environmental pollution and illegal migration as more threatening to their security than the risk of an external military attack. However, the governments of Lithuania, whatever their political colour, have recognised that this situation may not last for ever, and their long-term security against potential external aggression is best guaranteed by membership in Euro-Atlantic defence structures.

For the foreseeable future Lithuania's foreign and security policy-making will be dominated by relations with Russia, by the drive for membership in NATO and the EU, and by considerations of Lithuania's place in the Baltic Sea region. Major uncertainties in dealing with Russia will remain owing to the unpredictability and instability in Russian politics. For this reason and for reasons of identity, Lithuania will continue to seek membership in western organizations offering both hard and soft security. Similarly, she will seek to ensure that Russian companies do not acquire a dominating interest in Lithuanian transport, energy and communications industries. Although there are increasing signs that the United States will not yield to Moscow in its determination that Lithuania and the other Baltic states will eventually become members of NATO, European members of the Alliance have not recently demonstrated the same interest and enthusiasm for Lithuanian membership, which will therefore remain somewhat problematic. By contrast, the prospects for EU membership are much brighter since they depend on Lithuania's achieving certain known economic criteria rather than on political calculations. Lithuania's progress in meeting these criteria has been encouraging, holding out hopes for EU entry in the medium term.

Lithuania's recent close association with Poland has encouraged the idea that her route to 'Europe' should be through Warsaw rather than through the Scandinavian or Norden states. This neglects the obvious fact that Lithuania, like Poland, is both a Baltic and a Central European state, and that her interests require her to keep both avenues open. Moreover, maintaining a dual-track approach would be entirely consistent with the development of regional identities in Europe, and with the establishment of trans-border regions. There is no reason why citizens of Lithuania should not have multiple identities, local and regional as well as national and supranational. In the context of a united Europe with many layers of identity, one can envisage minorities having close relations with fellow ethnics in trans-border regions of different states without any question of challenging existing borders.

The Baltic Sea Region as a whole offers enormous advantages to Lithuania in economic, environmental, educational, and cultural terms. While the Council of Baltic Sea States provides a conventional inter-state framework for joint consultation and policy formation, an even more promising structure is one based on regions or small states or even cities, since the disparities of size and economic power are not so marked. Ole Waever's vision of a new Baltic identity being developed as a result of the formation of close inter-regional links following the removal of the Cold War divisions in the area is a seductive one. A network of inter-regional links could transform the Baltic area into a trans-border region of considerable economic and cultural significance.[35]

This would also be a profitable way of resolving the tensions between, on the one hand, the political and economic homogenization resulting from membership in the EU and the need to maintain cultural identity. As the power of the territorial states diminishes vis-a-vis Brussels, ethnic and national identities remain and are not threatened by the EU. Put in Waever's terms there will be a dichotomy between the 'unit of identification'—the nation—and the unit of political organization—increasingly the EU. The nations will be able to come together in trans-border regions, giving a new layer to the existing sets of identities. Although it is impossible to predict precisely how Lithuania will develop in her relations with the EU and the Baltic region, it seems certain that much of her present sovereignty, so recently reclaimed from the now defunct Soviet Union, will be yielded up to the EU. At the same time, however, it should not be assumed by those who fear this development that the identity of Lithuania or of the various minorities within the country will be under threat from the same process. The nation states may crumble, but the nations will remain; and they are likely to have a far more positive relationship with each other in the new political context than the old.

1. Gintaras Tamulaitis, *National Security and Defence Policy of the Lithuanian State* (Geneva: United Nations Institute for Disarmament Research, 1994), Research Paper No.26, pp.11–12; Birthe Hansen, 'The Baltic States and Security Strategies Available', in Birthe Hansen and Bertel Heurlin eds., *The Baltic States in World Politics* (Richmond, Surrey: Curzon, 1998), p.91; Elzbieta Stadtmüller, 'Polish Policy towards the Baltic Region in the 1990s', paper presented to the Conference and First Convention of the CEEISA, Prague, 24–26 May, 1999, p.19.

2. This agreement may be interpreted as part of Russia's attempt to weaken the Soviet Union and strengthen her own position either within a modified Union or as an independent state: Vilenas Vadapalas, 'Relations between Lithuania and Russia: Some Legal and Political Questions', unpublished mss. in possession of author; 'Treaty between the Republic of Lithuania and the Russian Soviet Federated Socialist Republic on the Basis for Relations between States', in *Lithuanian Foreign Policy Review*, 98/1, 1998, pp.119–20.

3 Vitaly Churkin, a Deputy Foreign Minister, said in January 1994 that from a legal point of view 'the events of 1940 cannot be qualified as annexation or aggression or occupation': *The Baltic Independent* (hereafter *TBI*), 4–10 February, 1994

4 *TBI*, 4–10 February, 1994; Tomas Venclova, 'A Fifth Year of Independence: Lithuania, 1922 and 1994', *East European Politics and Societies*, vol.9, no.2, Spring 1995, p.357

5 Whether this Russophobia is justified is beside the point; it was a profound part of the Lithuanian mentality, especially on the Right, at the time of independence: Christopher Marsh, 'Realigning Lithuanian Foreign Relations', *Journal of Baltic Studies*, vol.XIX, no.2, Summer 1998, p.156; Evaldas Nekrasas, 'Is Lithuania a Northern or Central European Country?', *Lithuanian Foreign Policy Review*, 98/1, 1998, pp.19–23; *TBI*, 30 July–5 August, 1993

7 *Summary of World Broadcasts* (hereafter *SWB*) SU/1796 E/2 17 September, SU/1851 E/1 20 November, 1993; *TBI*, 30 July–5 August 1993; *SWB* SU/1877 E/1 21 December, 1993, SU 1967 E/1 9 April, 1994.

8 *Izvestiia*, 5 October, 1991; *The Times*, 4 December, 20 December, 1991, 24 August, 1993; *TBI*, 17–23 July, 1992, 20–26 August, 27 August–2 September, 3–9 September, 10–16 September, 1993; Darius K. Mereckis and Rimantas Morkvenas, 'The 1991 Treaty as a Basis for Lithuanian–Russian Relations', *Lithuanian Foreign Policy Review*, 98/1, 1998, p.11

9 One of the more moderate statements of the Russian position was enunciated by Yevgeny Primakov, who displaced Kozyrev as foreign minister. He offered the reassurance that Russia did not want to eradicate the sovereignty of the various republics, but should develop an element of integration in the economic and political spheres. This could, of course, be interpreted as a claim for a Russian sphere of influence: Astrid S. Tuminez, 'Russian Nationalism and the National Interest in Russian Foreign Policy', in Celeste A. Wallander, ed., *The Sources of Russian Foreign Policy after the Cold War* (Boulder, CO: Westview Press, 1996), pp.51–60; Bruce D. Porter, 'Russia and Europe after the Cold War: the Interaction of Domestic and Foreign Policies', in Wallander ed., pp.121, 131–5; *The Economist*, 10 December, 1994, 26 February, 1995; *SWB* SU/1856 B/11 26 November, 1993, SU/2492 B/6 21 December, 1995; *TBI*, 11–17 February, 1994

10 *SWB* SU/1861 E/1 2 December, 1993, SU/1976 E/1 20 April, 1994; *The Times*, 14 December, 1993 (article by Lawrence Freedman and report of an assessment by Charles Dick of the Russian military doctrine); International Institute of Strategic Studies, Strategic Survey 1993–94, reported in *The*

Times, 24 May, 1994; *The Economist*, 27 August, 1994; *TBI*, 27 August–2 September, 1993, 4–10 February, 1994; *Izvestiia*, 14 December, 1993

11 *SWB* SU/2082 E/5 24 August, SU/2095 E/1 8 September, SU/2115 E/1 1 October, SU/2121 E/4 8 October, SU/2131 E/1 21 October, SU/2134 E/4 24 October, SU/2140 E/3 31 October, SU/2158 E/3 21 November, 1994; *TBI*, 14–20 October, 1994.

12 *SWB* SU/2160 E/4 23 November, 1994, SU/2168 E/5 2 December, 1994; SU/2193 E/1 5 January, SU/2198 E/1 11 January, SU/2216 E/2 1 February, SU/2239 B/7 28 February, 1995; *TBI*, 13–19 January, 27 January–2 February 1995, 8–14 February, 1996; Mereckis and Morkvenas, p.12

13 Marsh, p.149; Mereckis and Morkvenas, pp.9–12; *SWB*, SU/2304 E/2 16 May, 1995; *Baltic News Service* (hereafter *BNS*), 5 May, 1997; *TBT*, 10–16 July, 4–10 September, 1997; 5–11 March, 18 June–2 July, 6–12 August, 26 November–2 December, 1998, 13–19 May, 1999.

14 Trade between Lithuania and Kaliningrad rose two and a half times between 1996–97: Mereckis and Morkvenas, pp.13–14.

15 Mereckis and Morkvenas, p.13; *SWB* SU/1830 E/2 27 October, 1993, SU/1995 E/5 12 May, 1994, SU/2001 E/3 1 June, 1994, SU/2156 B/9 18 November, 1994; *TBI*, 8–14 September, 1995; Address by Vygaudas Usackas, Political Director, Foreign Ministry of Lithuania, to Vilnius Conference 'Euro-Atlantic Integration as a Key Aspect of Stability', Vilnius, 3–4 September, 1998.

16 Stephen R. Burant, 'Polish-Lithuanian Relations: Past, Present, and Future', *Problems of Communism*, vol.XL, no.3, May–June, 1991, p.69

17 Stephen R. Burant, 'Overcoming the Past: Polish-Lithuanian Relations, 1990–1995', *Journal of Baltic Studies*, vol.XXVII, no.4 winter 1996, pp.314–16; *TBI*, 20–26 September, 1991; *Financial Times*, 10, 16 September, 1991

18 See, for example, the statement of Bronislaw Geremek after the Moscow coup attempt, August 1991, and the words of President Wałęsa to representatives of ethnic Poles in Lithuania: 'The Lithuanian state is your state. Its welfare is your welfare. Be worthy citizens. Care for your homeland': *TBI*, 16–22 August, 1991; Burant, 'Overcoming the Past', p.322; Stephen R. Burant and Voytek Zubek, 'Eastern Europe's Old Memories and New Realities: Resurrecting the Polish-Lithuanian Union', *East European Politics and Societies*, vol.7, no.2, Spring 1993, p.370

19 Venclova, p.360; *TBI*, 25 June–1 July, 1993; Burant, 'Overcoming the Past', pp.309–12

20 Tim Snyder, 'National Myths and International Relations: Poland and Lithuania 1989–1994', *East European Politics and Societies*, vol.9, no.2, Spring 1995, p.333; Stephen R. Burant, 'International Relations in a Regional Context: Poland and its Eastern Neighbours – Lithuania, Belarus, Ukraine', *Europe-Asia Studies*, vol.45, no.3, 1993, p.403–4; Burant, 'Overcoming the Past', pp.318–22; *SWB*, SU/1801 E/5 23 September, 1993, SU/1979 E/2 23 April, SU/1983 E/1 28 April, 1994; *TBI*, 30 July–5 August, 1993

21 Indicative of the general mood of good will and the determination to strengthen mutual relations was the establishment of the Adam Mickiewicz Foundation for the Development of Lithuanian-Polish Co-operation: *SWB*, SU/2555 E/1 8 March, SU/2636 E/2 12 June, SU/2723 E/1 21 September, 1996; *The Economist*, 12 October, 1996; *BNS*, 21 April, 18 June, 28 August, 15 September, 1997, 28 January, 1998; *Le Soir*, 5 September, 1997; *TBT*, 24–30 April, 1997, 25 February–3 March, 4–10 March, 22–28 April, 1999; Lithuanian Foreign Ministry web site, http://www.urm.lt/foreign/#HD NM 1, 29 January, 1999

22 *The Economist*, 12 October, 1996; *BNS*, 15 September 1997

23 A. Thomas Lane, 'The Baltic States, the Enlargement of NATO and Russia', *Journal of Baltic Studies*, vol.XXVIII, no.4, Winter 1997, p.297

24 Mare Haab, 'Potentials and Vulnerabilities of the Baltic States: Mutual Competition and Co-operation' in Hansen and Heurlin eds., p.5

25 Peter van Ham, 'The Baltic States and Europe: the Quest for Security', in Hansen and Heurlin, pp.34–5; Lane, p.305

26 *BNS*, 13 May, 26 June, 1997; *TBT*, 26 November–2 December, 1998; *TBT*, 29 April–5 May, 1999

27 Lane, p.306; Linas Linkevicius, 'NATO and Lithuania: What is Blocking Entry through an Open Door?', *Lithuanian Foreign Policy Review*, 98/1, 1998, pp.47–9; Lithuanian Foreign Ministry web site http://www.urm.lt/new/press/1998–11–24. htm, Press Release, 24 November, 1998

28 Van Ham, pp.25, 32

29 Lithuanian Foreign Ministry web site, http://www.urm.lt/foreign/#HD NM 1, 29 January, 1999; *Financial Times*, 9 September, 1991; *TBI*, 1–7 May, 15–21 May, 1992, 19–25 November, 1993, 22–28 July, 2–8 December, 1994; *SWB*, SU/2492 E/1 21 December, 1995; The European Commission, Background Report, *EU Relations with the Baltic States*, December 1995; *The Economist*, 21 March, 1998l

30 European Commission, Background Report; *The Economist*, 12 July, 1997; *BNS*, 16 September, 1997; *Lithuania and the European Union: Towards Accession Negotiations* (Vilnius: Du Ka UAB, 1998), pp.2–10; *TBT*, 24–30 July, 1997, 12–18, 19–25 November, 1998

31 *BNS*, 16 September, 1997; *Financial Times*, 10 September, 8 December, 1997; *TBT*, 5–11 June, 25 September–1 October, 1997, 2–8 April, 1998; Marsh, pp.158–9

32 Lithuanian Foreign Ministry, official website http://www.urm.lt/foreign/#HD NM 1, 29 January, 1999; *TBI*, 6–12 June, 27 September–3 October, 11–17 October, 8–14, 15–21 November, 1991, 18–24 September, 1992, 2–8 October, 1992, 17–23 February, 1995; *TBT*, 18–24 April, 1996; *Izvestiia*, 4 October, 1991; *SWB*, SU/1796 E/1 17 September, 1993

33 *SWB*, SU/2022 E/1 15 June, SU/1927 S1/4 21 February, SU/2101 E/1 15 September, SU/2162 E/1 25 November, 1994, SU/2259 S1/3 23 March, 1995, SU/2638 E/1 14 June, SU/2690 E/3 14 August, 1996; *BNS*, 17 October, 1997

34 Stadtmüller, pp.2–3, 19–21; *TBI*, 15–21 November, 1991, 13–19 March, 1992; *BNS*, 26 April 1997; *The Economist*, 18 April, 1998; *TBT*, 9–15 May, 6–12 June, 10–16 October, 1996, 12–18 June, 1997; Lithuanian Foreign Ministry, website, http://www.urm.lt/foreign/#HD NM1; *SWB*, SU/2008 E/1 28 May, 1994

35 Ole Waever, 'Europe since 1945: Crisis to Renewal' in Kevin Wilson and Jan van der Dussen eds., *The History of the Idea of Europe* (Milton Keynes, London and New York: The Open University and Routledge, rev.ed., 1995), pp.195–202

BIBLIOGRAPHY

DOCUMENTS

Bourne, Kenneth, D. Cameron Watt and Michael Partridge, General Editors, *British Documents on Foreign Affairs: Reports and Papers from the Foreign Office Confidential Print* Part II, From the First to the Second World War. Series F, *Europe, 1919–1939, Scandinavia and the Baltic States* edited by John Hiden and Patrick Salmon

Kaslas, Bronis J., ed., *The USSR-German Aggression against Lithuania* (New York: Robert Speller & Sons, Publishers, Inc., 1973)

Third Interim Report fo the Select Committee on Communist Aggression, House of Representatives Eighty-Third Congress, Second Session 1954, *Baltic States: A Study of their Origin and National Development; their Seizure and Incorporation into the USSR* third Reprint Edition, William S. Hein & Co. Inc., Buffalo, New York, 1972

République Polonaise, Ministère des Affaires Etrangères, *Documents Diplomatiques concernant les Relations Polono-Lithuaniennes* (Decembre 1918–Septembre 1920), Warsaw, 1920

British Foreign Office, files of the Northern Department, FO 371 series, 1938–1941. Also Foreign and Commonwealth Office, Background Brief 'The Baltic States: Forty Years under Soviet Rule', FO 973/111, September 1980

Lithuanian Government websites: www.president.lt; www.lrs.lt: www.urm.lt; www.finmin.lt

'Treaty between the Republic of Lithuania and the Russian Soviet Federated Socialist Republic on the Basis for Relations between States', in *Lithuanian Foreign Policy Review*, 98/1, 1998

'The Republic of Lithuania Law on Citizenship', 5 December 1991, in *The East Express*, no.4, February 1992

Lithuanian Ministry of Economy/PricewaterhouseCoopers, *Lithuania Country Profile*, August 1998

Lithuanian Development Agency, *Advantage Lithuania*, August/September 1998

Lithuanian Ministry of Foreign Affairs, *Public Opinion and Level of Awareness on Security Issues in the Baltic Countries*, 1998

Lithuanian State Property Fund, 'Privatisation in Lithuania' 1998

JOURNALS AND NEWSPAPERS

Atmoda
The Estonian Independent
The Baltic Independent
The Baltic Times
The Baltic Observer
The Baltic Review
Radio Free Europe/Radio Liberty *Research Reports*
Transition
BBC, *Summary of World Broadcasts*
International Herald Tribune
The Economist
The Times
The Guardian
Izvestiia
Pravda

BOOKS AND ARTICLES

Adirim, Itzchok, 'Realities of Economic Growth and Distribution in the Baltic States' in Loeber, Vardys and Kitching eds., *Regional Identity under Soviet Rule*

Alesandravicius, Egidijus, 'Political Goals of Lithuanians, 1863–1918', *Journal of Baltic Studies*, vol.XXIII, no.3, Fall 1992

Alisauskiene, Rasa, Rita Bajaruniene and Birute Sersniova 'Policy Mood and Socio-Political Attitudes in Lithuania', *Journal of Baltic Studies*, vol.XXIV, no.2, Summer 1993

Amsden, Alice H., Jocek Kochanowicz and Lance Taylor, *The Market Meets its Match: Restructuring the Economies of Eastern Europe* (Cambridge, Mass: Harvard University Press, 1994)

Anderson, Edgar, 'The Baltic Entente: Phantom or Reality?' in Vardys and Misiunas eds., *The Baltic States in Peace and War*

Applebaum, Anne, *Between East and West: Across the Borderlands of Europe* (London and Basingstoke: Papermac, 1995)

Arkadie, B. van and M. Karlssen, *Economic Survey of the Baltic States* (London: 1992)

Arumae, Heino, 'Moscow's Point of View of the 1939–1940 Events in the Baltic States', *Rahva Haal* (The People's Voice), 13 February, 1991

Barnowe, J. Thad., Gundar King and Eli Berniker, 'Personal Values and Economic Transition in the Baltic States', *Journal of Baltic Studies*, vol.XXIII, no.2, Summer 1992

Bilinsky, Yaroslav and Tonu Parming, 'Helsinki Watch Committees in the Soviet Republics: Implications for Soviet Nationality Policy', *Nationalities Papers*, vol.IX, Spring, Part 1, 1981

Binyon, Michael, *Life in Russia* (London: Panther, 1983)

Bourdeaux, Michael, *Land of Crosses: the Struggle for Religious Freedom in Lithuania 1939–1978* (Devon: Augustine Publishing Co., 1979)

Brada, Josef C., 'Privatization in Transition—Or Is It?', *Journal of Economic Perspectives*, vol.10, no.2, Spring 1996

Bradshaw, Michael, Philip Hanson and Denis Shaw, 'Economic Restructuring' in G. Smith ed., *The National Self-Determination of Estonia, Latvia and Lithuania*

Bremer, I. and R. Taras eds., *Nations and Politics in the Soviet Successor States* (Cambridge: Cambridge University Press, 1991)

Burant, Stephen R., 'Polish-Lithuanian Relations: Past, Present, and Future', *Problems of Communism*, May–June, 1991

Burant, Stephen R., 'Overcoming the Past: Polish-Lithuanian Relations, 1990–1995', *Journal of Baltic Studies*, vol.XXVII, no.4 (Winter 1996)

Burant, Stephen R. and Voytek Zubek, 'Eastern Europe's Old Memories and New Realities: Resurrecting the Polish-Lithuanian Union', *East European Politics and Societies*, Spring 1993

Burant, Stephen R., 'International Relations in a Regional Context: Poland and its Eastern Neighbours—Lithuania, Belarus, Ukraine', *Europe-Asia Studies*, no.3, 1993

Chronicle of the Catholic Church in Lithuania, no.60, 1 November, 1983, no.61, 6 January, 1984, no.68, 16 October, 1985, no.70, 23 April, 1986

Clark, Terry D., 'Coalition Realignment in the Supreme Council of the Republic of Lithuania and the Fall of the Vagnorius Government', *Journal of Baltic Studies*, vol.XXIV, no.1, Spring 1993

Clark, Terry D., 'The Lithuanian Political System: A Case Study of Democratic Consolidation', *East European Politics and Societies*, vol.9, no.1, Winter 1995

Clark, Terry D., 'The 1996 Elections to the Lithuanian *Seimas* and their Aftermath', *Journal of Baltic Studies*, vol XXIX, no.2, Summer 1998

Clemens, Walter C., Jr., *Baltic Independence and Russian Empire* (Basingstoke: Macmillan, 1991)

Crowe, David M., *The Baltic States and the Great Powers: Foreign Relations 1938–1940* (Boulder, CO: Westview Press, 1993)

Dale, Rasa, 'Inflation Stabilization and Credit Availability in the Baltic States', *Journal of Baltic Studies*, vol.XXVIII, no.1, Spring, 1997

Dallin, Alexander, 'The Baltic States between Nazi Germany and Soviet Russia', in Vardys and Misiunas, *The Baltic States in Peace and War*

Dassanowsky-Harris, Robert von, 'The Philosophy and Fate of Baltic Self-Determination', *East European Quarterly*, vol.XX, no.4, January 1987

Daviddi, Renzo, 'Macroeconomic Stabilization and Privatization in Central and Eastern Europe', European Institute of Public Administration, *EIPASCOPE*, no.3, 1994

Dellenbrant, Jan Ake, 'The Integration of the Baltic Republics into the Soviet Union', in Loeber, Vardys and Kitching eds., *Regional Identity under Soviet Rule*

Dunn, Dennis J., 'The Catholic Church and the Soviet Government in the Baltic States 1940–1941', in Vardys and Misiunas, *The Baltic States in Peace and War*

Dziewanowski, M.K., 'Piłsudski's Federal Policy, 1919–1921', part I, *Journal of Central European Affairs*, vol.10, no.2, July 1950; part II, vol.10, no.3, October 1950

Eidintas, Alfonsas, Vytautas Zalys, Alfred Erich Senn, ed. by Edvardas Tuskenis, *Lithuania in European Politics: the Years of the First Republic 1918–1940* (Basingstoke and London: Macmillan, 1997)

Eidintas, Alfonsas, 'The Meeting of the Lithuanian Cabinet, 15 June 1940' in John Hiden and Thomas Lane eds., *The Baltic and the Outbreak of the Second World War*

Fischer, S.R., R. Sahay and C.A. Vegh, 'Stabilization and Growth in Transition Economies: the Early Experience', *Journal of Economic Perspectives*, vol.10, no.2, Spring 1996

Foreign Office (UK), *Peace Handbooks*, vol.VIII, *Poland and Finland* (London: HMSO, 1918–1919)

Forgus, Silvia P., 'Manifestations of Nationalism in the Baltic Republics', *Nationalities Papers*, vol.VII, no.2, Fall 1979

Forwood, William, 'Nationalities in the Soviet Union', in George Schöpflin ed., *The Soviet Union and Eastern Europe: A Handbook* (London: Anthony Blond, 1970)

Gerner, Kristian and Stefan Hedlund, *The Baltic States and the End of the Soviet Empire* (London and New York: Routledge, 1993)

Gerner, Kristian, 'Between Sweden and Russia: the Baltic Rimland', *Journal of Baltic Studies*, vol.XVII, no.3, Fall 1986

Gerutis, Albertas, ed., *Lithuania 700 Years*, 7th edition (New York: Manyland Books, 1984)

Girnius, Kestutis K, 'The Opposition Movement in Postwar Lithuania', *Journal of Baltic Studies*, vol.XII, Part 1, 1982

Girnius, Saulius, 'The Lithuanian Communist Party versus Moscow', Radio Liberty, *Report on the USSR*, vol.2, no.1, 5 January, 1990

Girnius, Saulius, 'The Political Pendulum Swings Back in Lithuania', *Transition*, vol.3, no.2, 7 February, 1997

Goble, Paul, 'Moscow's Nationality Problems in 1989', Radio Liberty, *Report on the USSR*, vol.2, no.2, 12 January, 1990

Gregory, John D., *On the Edge of Diplomacy: Rambles and Reflections 1902–1928* (London: Hutchinson, 1929)

Grennes, Thomas, 'The Lithuanian Economy in Transition', *Lituanus*, vol.40, Summer 1994

Grennes, Thomas, 'Inflation and Monetary Policy during Two Periods of Lithuanian Independence', *Journal of Baltic Studies*, vol.XXVII, Summer 1996

Grennes, Thomas, 'The Economic Transition in the Baltic Countries', *Journal of Baltic Studies*, vol.XXVIII, no.1, Spring 1997

Grobel, O. and A.Lejins eds., *The Baltic Dimension of European Integration* (Riga: LIIA, IEWS, Royal Danish Embassy, 1996)

Haavisto, Tarmo ed., *The Transition to a Market Economy: Transformation and Reform in the Baltic States* (Cheltenham: Elgar, 1997)

Hansen, Birthe and Bertel Heurlin eds., *The Baltic States in World Poliics* (Richmond, Surrey: Curzon, 1998)

Hanson, Philip, *From Stagnation to Catastroika: Commentaries on the Soviet Economy 1983–1991* (New York: Praeger with the Center for Strategic and International Studies in cooperation with Radio Free Europe-Radio Liberty Inc., 1992)

Harrison, E.J., *Lithuania Past and Present* (London: T. Fisher Unwin Ltd., 1922)

Harrison, E.J., ed. and compil., *Lithuania 1928* (London: Hazell, Watson & Viney Ltd., 1928)

Harrison, E.J., *Lithuania's Fight for Freedom* (New York: Lithuanian American Information Center, 1952)

Hiden, John and Patrick Salmon, *The Baltic Nations and Europe* (London: Longman, 1991)

Hiden, John and Thomas Lane eds., *The Baltic and the Outbreak of the Second World War* (Cambridge: Cambridge University Press, 1992)

Hiden, John, 'Baltic Security Problems between the two World Wars' in John Hiden and Thomas Lane eds., *The Baltic and the Outbreak of the Second World War*

Hiden, John and Thomas Lane 'Facing both Ways: Baltic Economies after Independence', in Peter J. Buckley and Pervez N. Ghauri eds., *The Economics of Change in East and Central Europe: Its Impact on International Business* (London: Academic Press Ltd., 1994)

Hope, Nicholas, 'Interwar Statehood: Symbol and Reality' in G. Smith ed., *The Baltic States*

Idzelis, Augustine, 'Locational Aspects of the Chemical Industry in Lithuania: 1960–1970', *Lituanus*, vol.XIX, part 4, 1973

Idzelis, Augustine, 'Response of Soviet Lithuania to Environmenal Problems in the Coastal Zone', *Journal of Baltic Studies*, vol.X, no.4, 1979

Idzelis, Augustine, 'Institutional Response to Environmental Problems in Lithuania', *Journal of Baltic Studies*, vol.XIV, no.4, Winter 1983

Idzelis, Augustine, 'The Socioeconomic and Environmental Impact of the Ignalina Nuclear Power Station' *Journal of Baltic Studies*, vol.XIV, Fall, 1983

Ivinskis, Zenonas, 'Lithuania during the War: Resistance against the Soviet and the Nazi Occupants' in V. Stanley Vardys ed., *Lithuania under the Soviets*

Jankauskas, Algimantas, 'The Re-emergence of Multi-partism in Lithuania', in Lukowski and Wojtaszczyk eds., *Reform and Transformation in Eastern Europe*

Jervell, Sverre, Mare Kukk and Perrti Joenniemi eds., *The Baltic Sea Area: A Region in the Making* (Oslo: Europa-programmet, 1992)

Joenniemi, Pertti and Juris Prikulis eds., *The Foreign Policies of the Baltic Countries: Basic Issues* (Riga: Centre of Baltic-Nordic History and Political Studies, 1994)

Johnson, Hank, 'The Comparative Study of Nationalism: Six Pivotal Themes from the Baltic States', *Journal of Baltic Studies*, vol.XXIII, no.2, Summer 1992

Johnson, Hank, 'Religion and Nationalist Subcultures in the Baltics', *Journal of Baltic Studies*, vol.XXIII, no.2, Summer 1992

Kaminski, Bartolomiej, 'Trade Performance and Access to OECD Markets' in Constantine Michalopoulos and David G. Tarr eds., *Trade in the New Independent States: Studies of Economies in Transition* (Washington: The World Bank/UNDP, 1994)

Kaslas, Bronis J., 'The Lithuanian Strip in Soviet-German Secret Diplomacy 1939–1941', *Journal of Baltic Studies*, vol.IV, no.3, Fall 1973

Kaslas, Bronis J., *The Baltic Nations—the Quest for Regional Integration and Political Liberty: Estonia, Latvia, Lithuania, Finland, Poland* (Pittston, Pa.: Euramerica Press, 1976)

Kennan, George F., *Memoirs 1925–1950*, (Boston: Atlantic-Little, Brown, 1967)

Kibena-Zechanowitsch, Aimi, 'Neville Chamberlain and the Baltic States: A Retrospective Analysis', *Nationalities Papers*, vol.VIII, no.1, 1980

King, Gundar J. and David O. Porter, 'Moving to Markets: A Review Essay', *Journal of Baltic Studies*, vol.XXIV, no.4, 1993

King, Gundar J., and Nicholas W. Balabkins, 'Fundamental Thoughts on Four Basic Economic Goals', *Journal of Baltic Studies*, vol.XXVIII, no.1, Spring 1997

King, Lionel, 'The Absorption of the Baltic States into the USSR in 1940', *Contemporary Review*, vol.222, 1973, pp.239–242

Kirby, David, 'Incorporation: the Molotov-Ribbentrop Pact' in G. Smith ed., *The Baltic States*

Kirby, David, 'Morality or Expediency? The Baltic Question in British-Soviet Relations' in V. Stanley Vardys and Romuald Misiunas eds., *The Baltic States in Peace and War*

Kirch, Aksel, Marika Kirch and Tarmo Tuisk, 'Russians in the Baltic States: To Be or not To Be?', *Journal of Baltic Studies*, vol.XXIV, no.2, Summer 1993

Kolde, Endel-Jakob, 'Structural Integration of Baltic Economies into the Soviet System', *Journal of Baltic Studies*, vol.IX, no.2, Summer 1978

Kreindler, Isabelle T., 'Baltic Area Languages in the Soviet Union: A Sociolinguistic Perspective' in Loeber, Vardys and Kitching eds., *Regional Identity under Soviet Rule*

Krickus, Richard, 'Nationalism and Baltic Independence', *Problems of Communism*, vol.40, no.6, 1991

Krickus, Richard, *Showdown: the Lithuanian Rebellion and the Breakup of the Soviet Empire* (New York: Brasseys, 1997)

Kubilius, Vytautas, ed., *Lithuanian Literature* (Vilnius: Vaga Publishers for the Institute of Lithuanian Literature and Folklore, 1997)

Lane, A.Thomas, 'The Baltic States, the Enlargement of NATO and Russia', *Journal of Baltic Studies*, vol.XXVIII, no.4, Winter 1997

Lane, Thomas, 'NATO-Russian Relations after the Cold War', *Le Monde Atlantique*, no.65, December 1997

Lane, Thomas, 'The Baltic States, NATO and European Security', *Revue Baltique*, no.7, 1996

Lane, Thomas, 'Nationalism and National Identity in the Baltic States', *Journal of Area Studies*, vol.1, no.4, 1994

Lane, Thomas, 'Citizenship and Civil Rights in the Baltic States' in José Amodia ed., *The Resurgence of Nationalist Movements in Europe* (Bradford: Occasional Papers No. 12, 1993)

Lane, A. Thomas, 'The Winding Down of British Trade with the Baltic States, 1939–1940: the View from Whitehall' in Anders Johansson et al., eds., *Emancipation and Independence: The Baltic States as New Entities in the International Economy, 1918–1940* (Stockholm: Act Universitatis Stockholmiensis, 1994)

Lainela, Seija, 'Currency Reforms in the Baltic States', *Communist Economics and Economic Transformation*, vol.5, no.4, 1994

Lange, Fulke, 'The Baltic States and the CSCE', *Journal of Baltic Studies*, vol.XXV, no.3, Fall 1994

Laqueur, Walter, *The Dream that Failed: Reflections on the Soviet Union* (London: Oxford University Press, 1995)

Laurinavičius, Česlovas, 'The Baltic States between East and West: 1918–1940', *Lithuanian Foreign Policy Review*, 98/1, 1998

Levin, D., 'The Jews and the Election Campaigns in Lithuania 1940–41', *Soviet Jewish Studies*, 10, 1980, pp.39–51

Lieven, Anatol, *The Baltic Revolution: Estonia, Latvia, Lithuania and the Path to Independence* (New Haven and London: Yale University Press, 1993)

Linkevičius, Linas, 'NATO and Lithuania: What is Blocking Entry through an Open Door?', *Lithuanian Foreign Policy Review*, 98/1, 1998

Loeber, Dietrich, V.Stanley Vardys and Laurence P.A.Kitching, eds., *Regional Identity under Soviet Rule: the case of the Baltic States* (Hackettstown, NJ: Association for the Advancement of Baltic Studies, 1990)

Lord, Robert H., 'Lithuania and Poland', *Foreign Affairs*, vol.I, 1923

Lukowski, Wojciech and Konstanty Adam Wojtaszczyk eds., *Reform and Transformation in Eastern Europe* (Warsaw: Elipsa, 1996)

Maciuika, Benedict V., 'Contemporary Social Problems in the Collectivized Lithuanian Countryside', *Lituanus*, vol.XII, no.3, 1967

Maciuika, Benedict V., 'The Role of the Baltic Republics in the Economy of the USSR', *Journal of Baltic Studies*, vol.III, no.1, 1972

Maciuika, Benedict V., 'Acculturation and Socialization in the Soviet Baltic Republics', *Lituanus*, XVIII, no.4, 1972

Marsh, Christopher, 'Realigning Lithuanian Foreign Relations', *Journal of Baltic Studies*, vol.XXIX, no.2, Summer 1998

Meissner, Boris, 'The Baltic Question in World Politics', in Vardys and Misiunas, *The Baltic States in Peace and War, 1917–1945*

Melo, Martho de, Cevdet Denizer and Alan Gelb, 'Patterns of Transition from Plan to Market', *The World Bank Economic Review*, vol.10, no.3, 1996

Mereckis, Darius and Rimantas Morkvenas, 'The 1991 Treaty as a Basis for Lithuanian-Russian Relations', *Lithuanian Foreign Policy Review*, 98/1, 1998

Milosz, Czeslaw, *Beginning with My Strees: Baltic Reflections* (London and New York: I.B. Taurus, 1992)

Milosz, Czeslaw, *Native Realm: A Search for Self-Definition* (London: Penguin Books, 1988, first pub. 1968)

Minkus-McKenna, Dorothy, 'Lithuanian Attitudes towards Business Indicators', *Journal of Baltic Studies*, vol.XXVIII, no.1, Spring 1997

Misiunas, Romuald J. and Rein Taagepera, *The Baltic States: Years of Dependence, 1940–1990* (London: Hurst & Co., rev. edn, 1993, first pub. 1983)

Mostov, Julie, 'Endangered Citizenship' in Michael Kraus and Ronald D. Liebowitz eds., *Russia and Eastern Europe after Communism: The Search for New Political, Economic and Security Systems* (Boulder CO: Westview Press, 1996)

Murrell, Peter, 'Conservative Political Philosophy and the Strategy of Economic Transition', *European Politics and Society*, vol.6, no.1, 1992

Murrell, Peter, 'How Far has the Transition Progressed?', *Journal of Economic Perspectives*, vol.10, no.2, Spring 1996

Nahaylo, Bohdan and Victor Swoboda, *Soviet Disunion: A History of the Nationalities Problem in the USSR* (London: Hamish Hamilton, 1990)

Neimanis, George J., 'Baltic Politics and Economics in Transition: A Review Essay', *Journal of Baltic Studies*, XXVIII, no.1, 1997

Nekrasas, Evaldas, 'Is Lithuania a Northern or Central European Country?', *Lithuanian Foreign Policy Review*, 98/1, 1998

Newman, E.W. Polson, *Britain and the Baltic* (London: Methuen, 1929)

Norem, Owen J.C., *Timeless Lithuania* (Chicago, Ill.: Amerlith Press, 1943)

Norgaard, Ole, et al., *The Baltic States after Independence* (Cheltenham: Elgar, 1996)

Nurek, Mieczyslaw, 'Great Britain and the Baltic in the last months of peace' in John Hiden and Thomas Lane eds., *The Baltic and the Outbreak of the Second World War*

Olcott, Martha Brill, 'The Lithuanian Crisis', *Foreign Affairs*, Summer 1990

Page, Stanley W., *The Formation of the Baltic States: A Study of the Effects of Great Power Politics upon the Emergence of Lithuania, Latvia, and Estonia* (Cambridge, MA: Harvard University Press, 1959)

Palm, Thomas, George J. Viksnins, Juris Niemanis and Valdas Samonis, 'The Economic Transformation of the Baltic States: A Panel Discussion from the AABS Conference in Toronto', *Journal of Baltic Studies*, vol.XXII, no.3, Fall 1992

Palm, Thomas, 'Institutional Economics and Institution Building in the Baltic States—A Review Essay', *Journal of Baltic Studies*, vol.XXVIII, no.1, 1997

Parming, Tonu, 'Population Processes and the Nationality Issue in the Soviet Baltic', *Soviet Studies*, vol.XXXII, no.2, July 1980

Parming, Tonu, 'Roots of Nationality Differences' in Edward Allworth ed., *Nationality Group Survival in Multi Ethnic States* (New York: Praeger, 1977)

Parming, Tonu, '"Baltic Studies": The Emergence, Development and Problematics of an Area Studies Specialization' in Loeber, Vardys and Kitching eds., *Regional Identity under Soviet Rule*

Pick, F.W., *The Baltic Nations: Estonia, Latvia and Lithuania* (London: Boreas Publishing Co. Ltd., 1945)

Poland, Ministry of Foreign Affairs, *Documents diplomatiques concernant les relations polno-lituaniennes*, vol.1, (Warsaw, 1920)

Ratnieks, Henry, 'The Energy Crisis and the Baltics', *Journal of Baltic Studies*, vol.XII, no.3, Fall 1981

Rauch, Georg von, *The Baltic States: the Years of Independence. Estonia, Latvia, Lithuania 1917–1940* (London: C. Hurst & Co., 1974)

Rei, August, *The Drama of the Baltic Peoples* (Stockholm: Kirjastus Vaba Eesti, 1970)

Remeikis, Thomas, 'The Impact of Industrialization on the Ethnic Demography of the Baltic Countries', *Lituanus*, vol.XIII, part 1, 1967

Remeikis, Thomas, 'Modernization and National Identity in the Baltic Republics: Uneven and Multi-Directional Change in the Components of Modernization' in Ihor Kamenetsky ed., *Nationalism and Human Rights: Processes of Modernization in the USSR* (Littleton, CO:, 1977)

Remeikis, Thomas, *Opposition to Soviet Rule in Lithuania 1945–1980* (Chicago: Institute of Lithuanian Studies Press, 1980)

Rodgers, Hugh I., *Search for Security: A Study in Baltic Diplomacy, 1920–1934* (Hamden, CT.: Archon Books, 1975)

Ross, T., 'On the Lithuanian Banking Crisis: Influences of State Control', *The Baltic Review*, vol.10, Summer, 1996

Royal Institute of International Affairs, *The Baltic States: a Survey of the Political and Economic Structure and the Foreign Relations of Estonia, Latvia and Lithuania* (London: Oxford University Press, 1938)

Sabajevaite, Lidija, 'Political Parties and the Political Situation in Lithuania' in Lukowski and Wojtaszczyk eds., *Reform and Transformation in Eastern Europe*

Sabaliunas, Leonas, *Lithuania in Crisis: Nationalism to Communism 1939–1940* (Bloomington, Ind.: Indiana University Press, 1972)

Sabaliunas, Leonas, 'The Politics of the Lithuanian-Soviet Non-Aggression Treaty of 1926', *Lituanus*, vol.VII, no.4, pp.97–102

Sadunaite, Nijole, *A Radiance in the Gulag* (Manassas, Va.: 1987)

Sapiets, Jan, 'The Baltic Republics' in George Schöpflin ed., *The Soviet Union and Eastern Europe: A Handbook* (London: Anthony Blond, 1970)

Sapiets, Marite, 'Religion and Nationalism in Lithuania', *Religion in Communist Lands*, vol.7, no.2, 1979

Sapiets, Marite, 'The Baltic Churches and the National Revival', *Religion in Communist Lands*, vol.18, no.2, 1990

Savasis, J., *The War against God in Lithuania* (New York: Manyland Books, 1966)

Segal, Zvi, 'Jewish Minorities in the Baltic Republics in the Postwar Years', in Loeber, Vardys and Kitching eds., *Regional Identity under Soviet Rule*

Senn, Alfred E., *The Emergence of Modern Lithuania* (New York: Columbia University Press, 1959)

Senn, Alfred Erich, *The Great Powers, Lithuania and the Vilna Question 1920–1928* (Leiden: E.J. Brill, 1966)

Senn, Alfred Erich, *Lithuania Awakening* (Berkeley: University of California Press, 1990)

Senn, Alfred Erich, 'Toward Lithuanian Independence: Algirdas Brazauskas and the CPL', *Problems of Communism*, vol.39, no.2, 1990

Senn, Alfred Erich, 'Lithuania's Path to Independence', *Journal of Baltic Studies*, vol.XXII, no.3, 1991

Senn, Alfred Erich, 'The Political Culture of Independent Lithuania: A Review Essay', *Journal of Baltic Studies*, vol.XXIII, no.3, Fall 1992

Senn, Alfred Erich, 'Comparing the Circumstances of Lithuanian Independence 1918–1922 and 1988–1992', *Journal of Baltic Studies*, vol.XXV, no., year

Senn, Alfred Erich, 'Lithuania's First Two Years of Independence', *Journal of Baltic Studies*, vol.XXV, no.1, Spring 1994

Senn, Alfred Erich, *Gorbachev's Failure in Lithuania* (New York: St. Martin's Press, 1995)

Senn, Alfred Erich, 'Struggle and Impasse: Polish-Lithuanian Relations between the Wars', *Lituanus*, 2, (1983), pp.72–6

Shafir, Gershon, 'Relative Overdevelopment and Alternative Paths of Nationalism: a Comparative Study of Catalonia and the Baltic Republics', *Journal of Baltic Studies*, vol.XXIII, no.2, Summer 1992

Shen, Raphael, *Restructuring the Baltic Economies: Disengaging Fifty Years of Integration with the USSR* (New York: Praeger, 1994)

Shtromas, Aleksandras, 'The Baltic States', in R. Conquest ed., *The Last Empire: Nationalities and the Soviet Future* (Stanford CA: Hoover Institute Press, 1986)

Shtromas, Aleksandras, 'The Baltic States as Soviet Republics: Tensions and Contradictions', in Graham Smith ed., *The Baltic States*

Shtromas, Aleksandras, 'Prospects for Restoring the Baltic States' Independence: A View on the Prerequisites and Possibilities of their Realization', *Journal of Baltic Studies*, vol.XVII, no.3, Fall 1986

Simutis, Anicetas, *The Economic Reconstruction of Lithuania after 1918* (New York: Columbia University Press, 1942)

Slay, Ben, 'External Transformation in the Post-Communist Economies: Overview and Progress', in Michael Krause and Ronald D. Liebowitz eds., *Russia and Eastern Europe after Communism: the Search for New Political, Economic and Security Systems* (Boulder CO: Westview Press, 1996)

Smith, Graham ed., *The Nationalities Question in the Soviet Union* (London: Longman, 1990, rept. 1992)

Smith, Graham ed., *The Baltic States: the National Self-Determination of Estonia, Latvia and Lithuania* (Basingstoke: Macmillan, 1994)

Smith, Graham, 'The Resurgence of Nationalism' in Smith ed., *The Baltic States*

Smith, Graham, Aadne Aasland and Richard Mole, 'Statehood, Ethnic Relations and Citizenship' in G. Smith ed., *The Baltic States*

Smith, Graham, *The Post-Soviet States: Mapping the Politics of Transition* (London: Arnold, 1999)

Snyder, Tim, 'National Myths and International Relations: Poland and Lithuania, 1989–1994', *East European Politics and Societies*, Spring 1995

Sorsa, Piritta, 'Lithuania: Trade Issues in Transition' in Constantine Michalopoulos and David G. Tarr eds., *Trade in the New Independent States: Studies of Economies in Transformation* (Washington DC: The World Bank/UNDP, 1994)

Steen, Anton, 'Consolidation and Competence: Research in the Politics of Recruiting Political Elites in the Baltic States', *Journal of Baltic Studies*, vol.XXVII, no.2, Summer 1996

Suny, Ronald Grigor, *The Soviet Experiment: Russia, the USSR and the Successor States* (New York: Oxford University Press, 1998)

Suziedelis, Saulius, 'Alfonsas Eidintas, Vytautas Zalys, "Lithuania in European Politics: The Years of the First Republic, 1918–1940"', *Lithuanian Foreign Policy Review*, 98/1, 1998

Sveics, Vilnis V., 'How Stalin Got the Baltic States', publication of Jersey City State College, NJ, USA, 1991 (in possession of author)

Swettenham, John Alexander, *The Tragedy of the Baltic States: A Report Compiled from Official Documents and Eyewitnesses' Stories* (London: Hollis and Carter, 1952)

Taagepera, Rein, 'Who Assimilates Whom? The World and the Baltic Region', in Loeber, Vardys and Kitching eds., *Regional Identity under Soviet Rule*

Taagepera, Rein, 'National Differences within Soviet Demographic Trends', *Soviet Studies*, vol.20, no.4, 1968–69

Taagepera, Rein, 'Civic Culture and Authoritarianism in the Baltic States 1930–1940', *East European Quarterly*, vol.VII, no.4, 1973

Taagepera, Rein, 'Baltic Population Changes, 1950–1980', *Journal of Baltic Studies*, vol.XII, no.1, 1981

Taagepera, Rein, 'The Population Crisis and the Baltics', *Journal of Baltic Studies*, vol.XII, no.3, Fall 1981

Tamulaitis, Gintaras, *National Security and Defence Policy of the Lithuanian State*, Research Paper No. 26 (Geneva: UN Institute for Disarmament Research, 1994)

Tarulis, Albert, *Soviet Policy towards the Baltic States 1918–1940* (South Bend, IN: Notre Dame University Press, 1959)

Tarulis, Albert N., 'A Heavy Population Loss in Lithuania', *Journal of Central European Affairs*, no.4, January 1962

Tauras, K.V., *Guerrilla Warfare on the Amber Coast* (New York: Voyages Press, 1962)

Taylor, Lance, 'Market Met Its Match: Lessons for the Future from the Transition's Initial Years', *Journal of Comparative Economics*, vol.19, August 1994

Tiusanen, Tauno, 'The Baltic States in Transition', *International Politics*, vol.33, March, 1996

Trapans, Jan Arveds ed., *Toward Independence: The Baltic Popular Movements* (Boulder and Oxford: Westview Press, 1991)

Urdze, Andrejs, 'Nationalism and Internationalism: Ideological Background and Concrete Forms of Expression in the Latvian SSR', in Loeber, Vardys and Kitching eds., *Regional Identity under Soviet Rule*

Vardys, V. Stanley, *The Catholic Church, Dissent and Nationality in Soviet Lithuania* (Boulder & New York: East European Quarterly, distributed by Columbia University Press, 1978)

Vardys, V. Stanley and Romuald J. Misiunas eds., *The Baltic States in Peace and War 1917–1945* (University Park & London: Pennsylvania State University Press, 1978)

Vardys, V. Stanley, 'The Partisan Movement in Postwar Lithuania', *Slavic Review*, vol.XXII, no.3, September 1963

Vardys, V. Stanley, 'Recent Soviet Policy toward Lithuanian Nationalism', *Journal of Central European Affairs*, vol.XXII, 1963, pp.313–32

Vardys, V. Stanley, 'How the Baltic Republics Fare in the Soviet Union', *Foreign Affairs*, vol.44, no.3, April 1966

Vardys, V. Stanley, 'Democracy in the Baltic States, 1918–1934: the Stage and the Actors', *Journal of Baltic Studies*, vol.X, no.4, 1979

Vardys, V. Stanley, 'Polish Echoes in the Baltic', *Problems of Communism*, vol.32, no.4, 1983

Vardys, V. Stanley ed., *Lithuania under the Soviets: Portrait of a Nation, 1940–1965* (New York: Praeger, 1965)

Vardys, V. Stanley 'Lithuanian National Politics', *Problems of Communism*, July–August, 1989, pp.53–76

Vardys, V. Stanley, 'Lithuanians', in Graham Smith ed., *The Nationalities Question in the Soviet Union*

Vardys, V. Stanley, 'The Role of the Churches in the Maintenance of Regional and National Identity in the Baltic Republics' in Loeber, Vardys and Kitching eds., *Regional Identity under Soviet Rule*

Vardys, V. Stanley, 'Modernization and Baltic Nationalism', *Problems of Communism*, XXIV, 5, 1975

Vardys, V. Stanley, 'Human Rights Issues in Estonia, Latvia and Lithuania', *Journal of Baltic Studies*, vol.XII, no.3 (Fall 1980), pp.275–98

Vardys, V. Stanley and Judith B. Sedaitis, *Lithuania: the Rebel Nation* (Boulder, CO: Westview Press, 1997)

Vebra, Rimantas, 'Political Rebirth in Lithuania 1990–1991: Events and Problems', *Journal of Baltic Studies*, vol.XXV, no.2, 1994

Venclova, Tomas, 'The Years of Persistence', *Lituanus*, XXVII, no.2, 1981

Venclova, Tomas, 'A Fifth Year of Independence: Lithuania, 1922 and 1994', *East European Politics and Societies*, vol.9, no.2, Spring 1995

Viksnis, George J., Review Article on *The Emergence of Market Economies*, *Journal of Baltic Studies*, vol.XXIII, no.2, Summer 1992

Vilpisauskas, Ramunas, 'Lithuania's Membership in the European Union: Possible Effects on its External Trade Policy', *Lithuanian Foreign Policy Review*, 98/1, 1998

Vital, David, *The Survival of Small States: Studies in Small Power/Great Power Conflict* (London: Oxford University Press, 1971)

Vitas, Robert A., *The United States and Lithuania: the Stimson Doctrine of Nonrecognition* (Westport, CT: Greenwood Press, 1990

Waever, Ole, 'Europe since 1945: Crisis to Renewal' in Kevin Wilson and Jan van der Dussen eds., *The History of the Idea of Europe* (London: Routledge, 1993, rept. 1996)

Wallender, Celeste A., ed., *The Sources of Russian Foreign Policy after the Cold War* (Boulder CO: Westview Press, 1996)

Weeks, Theodore R., 'Lithuanians, Poles and the Russian Imperial Government at the Turn of the Century', *Journal of Baltic Studies*, vol.XXV, no.4, 1994

White, James D., 'Nationalism and Socialism in Historical Perspective' in Graham Smith ed., *The Baltic States: the National Self-Determination of Estonia, Latvia and Lithuania*

White, James D., 'National Movements in the Baltic Provinces', *International Politics*, vol.33, no.1, March 1996

Winiecki, Jan, 'Are Soviet-Type Economies entering an Era of Long-Term Decline?, *Soviet Studies*, vol.XXXVIII, 3, July 1986

Zamascikov, Sergei, 'Soviet Methods and Instrumentalities of Maintaining Control over the Balts', in Loeber, Vardys and Kitching eds., *Regional Identity under Soviet Rule*

Ziedonis, Arvids Jr., Rein Taagepera and Mardi Valgemae eds., *Problems of Mininations: Baltic Perspectives*, (San Jose, CA: Association for the Advancement of Baltic Studies, 1973)

Ziugzda, Robertas, 'Lithuania in International Relations in the 1920s', in John Hiden and Aleksander Loit eds., *The Baltic in International Relations between the Two World Wars* (Uppsala: Acta Universitatis Stockholmiensis, 1988)

Index

Adamkus, Valdas, 131, 137, 144,
    147–8–9, 152, 158, 159
Afanasyev, Yuri, 107
Agenda 2000, 215
Algirdas, (6)
Alksnis, Colonel Viktor, 117
Amber Teleholdings, 184
American Jewish Foundation, 157
Andropov, Iurii, 68, 96
Apžvalga xxxi
Assimilation
   Russian policy of, xxvii
   Soviet policy of, 1, 68, 70
Atlantic Charter, 691
Aušra, xxx, 75, 91

Baden, Prince Max von, 5
Bakatin, Vadim, 116
Balcerowicz, Leszek, 164
Balfour, Arthur, 31
BALTBAT, 212, 218
BALTNET, 212
BALTRON, 212
Baltic Assembly, 106, 217
Baltic Co-operation Council, 217
Baltic Council of Ministers, 217–18
Baltic Entente, 34–5
Baltic Free Trade Area, 187–8
Baltic identity, 220
Baltic League
   potential members of, 33
   difficulties in way of, 33
   objectives of, 33–4
   reasons for failure, 34
Baltic neutrality, 35–6
Banking crisis 146, 180–1
Bank of Lithuania, 168, 170, 180–1
Basanavičius, Jonas, xxx, xxxiii, 22

Bermondt-Avalov, Pavel
   Mikhailovich, 6, 8,
Bičkauskas, Egidijus, 217
Biržai oil pipeline, 184, 189
Bolsheviks,
   aggression against Lithuania, 6, 7
   attitudes to self-determination of
      subject nationalities, xvii, 2
   failure in Lithuania 1919, 7–8
Book smuggling, xxix
Borisevičius, Bishop Vincentas, 71
Boruta, Bishop Jonas, 156
Brazauskas, Algirdas, 102–3, 106,
    108, 110, 111, 114, 120, 132,
    132, 139, 142, 147, 153, 156,
    170, 201, 202, 210
Brest-Litovsk, Treaty of (1918), 5
Brezhnev, Leonid, 68, 70, 89
Bulduri conference, 33
Būtingė oil terminal, 184, 189
Butler, R.A., 36

Calendar demonstrations, 99, 100
Capital punishment, 137
Carr, E.H., 20, 31
Catholic Church, xxix, xxx, 21, 28,
    54, 71–2, 89–90
   identification of, with resistance to
      Russification, xxix
Catholic Committee for the Defence
    of Believers' Rights, 90
Central European Free Trade
    Association (CEFTA), 187, 211
Centre Union, 134, 141, 145
Černius, General Jonas, 28
Chernobyl, 81, 100, 189
Christian Democratic Party, 21–4, 28,
    134, 141, 144, 145

Chronicle of Current Events, 89
Chronic of the Catholic Church in Lithuania, 89–90
Churchill, W.S., speech at Fulton, 65
Citizenship Laws 135–6, 149
Čiurlionis, Mikalojus Konstantinas, 18
Collectivization, 49, 53, 63
Committee of National Salvation, 122
Communist Party, 22, 26, 56
Communist Party of the Soviet Union (CPSU), Nineteenth Congress, 102
Confederation of Industrialists, 146
Conference on Security and Co-operation in Europe (CSCE), 90, 92, 203
Congress of Peoplés Deputies 1989, 105–6
   elections to, 105
   and the Molotov-Ribbentrop Pact, 106–7
Conservationists, 93
Constituent Assembly. See Lithuania
Constitution, 1922. See Lithuania
Co-operatives in inter-war period, 14, 17, 26
Council of Baltic Sea States (CBSS), 218
Council of Europe, 42, 131, 137, 139, 156
Coup of 1926, 20–3
Coup attempt in Moscow 1991, 111–12, 119, 125
Courish lagoon, 80–1
Crèvecoeur, J. Hector St. John de, 74
'Crony capitalism', 176
Currency Board, 146, 170, 181–2
Curzon, Lord (George Nathaniel), 9

Dainos, xxiii
Danilowicz, Ignac, xxviii
Daukantas, Simonas, xxv, xxviii
Decollectivization, 141, 177, 192
Dekanozov, Vladimir, 40

Denmark, support of for Lithuania, 216
Deportations, 52, 56–7, 61–3
Desovization, 140–1
Dievas ir Tėvynė, 91
Dissidents, 87, 89–92
Dmowski, Roman, 30
Donelaitis, Kristjonas, xxv

Eastern Locarno, 34
East Prussia, 6, 30, 32
Economic cost accounting, 98, 106
Economic liberalization, 141
Elections, Soviet conduct of 1940, 40–1
Elections after 1991, 134–5, 141–2, 144–6, 148
Energy policy. See Lithuanian economy since 1991, energy policy
Environmental damage in Lithuania, 79, 80
Environmental protection groups, 79–80, significance of, 82, 92, 99–100
Ethnic minorities 17–19, 23, 25–6, 138, 149
   autonomy of, 19, 152
   Belarussian minority, 20, 138
   education 140, 150
   election of representatives to *Seimas*, 19
   emigration of Jews, 155
   Jewish minority, xxvii, 20, 26–7, 55–6, 138, 154
      participation of, in Communist security services alleged, 155–6
      protection of, by ethnic Lithuanians, 156
   Jewish minoritýs treatment by Soviet government, 51, 56, 155–6
   language rights of, 20, 151–2
   legislation on, 138, 139, 153

Polish minority, xxvii, 20, 27, 135, 138, 151
  declarations of autonomy, 152
  local councils of, dissolved, 152
  opposition of, to changed electoral boundaries, 153
  opposition of, to enlargement of Vilnius boundaries, 152
  opposition of, to national governments, 151–2, 208
  representativeness of leadership of, 153
  support of, for Gorbachev's referendum, 153
  support of, for Lithuanian independence, 153
  populations of, in Vilnius region between the wars, 30–1
  rights of, in Constitution of 1922, 2
  rights of, in Constitution of 1992, 131, 135, 138, 149; Administration of, 139
Russian minority, xxvii, 136, 138, 149–50
Ukrainian minority, 138
voting rights of, 136, 149, 153
Ethnic nationalism, xvii, xxv, xxix, xxxi–xxxii, xxxiv, 1, 21, 26, 31, 135, 151
Euro-Atlantic Partnership Council, 212
European Bank for Reconstruction and Development, 178
European Free Trade Association (EFTA), 174
European Institute for Dispersed Ethnic Minorities, 140
European Union (EU) relations of Lithuania with, xiv, 144, 159, 174, 187, 192, 193 199, 214–5, 219, 220–21
Europe agreement of, with Lithuania, 214

free trade agreement of, with Lithuania, 214
Euroregions (Niemen and Baltic), 219

Farmers' Union, 21
Finkelstein, Eitan, 91
Foreign investment in Lithuania, 178, 185–6
'Four man proposal', 98
Free Enterprise Zones, 185, 191

Gaon Elijahu, 158
Gediminas, (5)
Gedvilas, Mečys, 40, 60
Genocide. See Holocaust
Geremek, Bronislaw, 208
German aggression against Poland 1939, 36
Germanization, 2, 57
German policy in Lithuania in First World War, 2, 5
German-Polish Non-Aggression Pact 1934, 34
Geruckas, the Reverend Karolis, 91
Ginzburg, Aleksandr, 89
glasnost, 1, 60, 88, 95, 97, 99
Goethe, J.W. von, xxvi
Gorbachev, Mikhail, 1, 67, 69, 82, 87–8, 96, 101,
  acceptance by, of a weakenion after failed Lithuanian coup, 118
  and the meaning of perestroika, 102–3
  belief of, that republican independence a threat to perestroika, 109
  conciliation by, of his conservative critics, 116, 125
  decline in international reputation of, 123–4
  failure of his economic reforms, 113
  growth of conservative opposition to, 116–19, 121
  negotiations of, for a new union treaty, 116, 118–19

Gorbachev, Mikhail (*continued*)
  primary aim of, 112
  proposals for constitutional reform,
    103, 105–6, 109, 116, 118,
    121
  opposition of, to decentralization,
    104, 114
  receives Nobel Peace Prize, 124
  referendum of, on constitutional
    reforms, 1991, 123
  responsibility of, for attempted
    coup in Lithuania, 117, 124
  support of the West for, 124–5
  support of, for Shatalin plan, 116
  visit of, to Lithuania, January 1990,
    106, 108
  weakness of, after failed Moscow
    coup, 1991, 119
  western pressure on, 124–5
  and Lithuaniás voluntary
    membership in the USSR,
    107, 109
  attitude of, to Lithuanian
    declaration of independence,
    114, 121
Grinius, President Kazys, 23
Griskevičius, Petras, 100
Grodno (Gardinas), 6, 30
Gulf War, 122, 125

Helsinki conference, January, 1920, 33
Helsinki Conference on Security and
  Co-operation in Europe, 90
Helsinki process, 92
Herder, J.G.von, xxvi
Heritage movement, 100
Hitler, Adolf, 32, 34, 55
Holocaust education, 158
Holocaust in Lithuania, 51, 55, 58,
  154–6
  degree of Lithuanian involvement
    in, 155–7
  role of Lithuanian security police
    in, 155, 157
Holocaust studies, 139, 158–9

Holsti, Rudolf, 34
Homeland, or Lithuanian
  Conservatives, 134, 135,
  144–6
  attacks on LDDP government's
    foreign policy, 201
  economic achievements of, in
    government, 147
  programme of, 146–7
  privatization policy of, 183
Human rights, 136–8, 218
Hurd, Douglas, 125
Hussain, Saddam, 122
Hymans, Paul, 31

Identity, of Lithuanians, viii, xviii,
  xix, xxx, 1, 18, 29, 50, 60, 67,
  75, 92–3, 100–1, 220–1
Ignalina nuclear plant, 76, 81, 100,
  189–90, 194, 215
Initiative Group, 101
Innovation Bank, 181
International agencies, impact of,
  135–6
International Monetary Fund (IMF),
  163–5, 169, 172, 178–9, 182
Investment vouchers, 176

Jadwiga, xxi
Jews in Lithuania. See Ethnic
  minorities
Jogaila, xxi
Jonava fertiliser works, 81, 114
Juozaitis, Arvydas, 101
Jurbarkas, 81
Juršėnas, Česlovas, 143

Kalanta, Romas, 88,
Kaliningrad (Koenigsberg), xxv, xxvi,
  32, 173, 191, 201, 207, 215
Karaites, 138
Kaunas, 146, 191
Kaunas faction of *Sajūdis*, 104, 120
Keştutis, xxi
KGB, 137, 141

Khrushchev, Nikita, 68, 70, 72, 78
Klaipėda, 3, 9, 28, 29–30, 32, 34, 36, 81, 146, 185–6, 191
Knights of the Cross, xx
Knights of the Sword, xx
Koenigsberg. See Kaliningrad
Kovalev, Sergei, 89
Kovno Province, xxvii, 9
Kozyrev, Andrei, 204
Krėvė-Mickevičius, Vincas
Kruglov, General Sergei, 65
Kudirka, Vincas, xxxi
Kuzmickas, nius, 101
Kwasniewski, Alexander, 147

Labour Federation, 21
Laisvės Šauklys, 91
Landsbergis, Vytautas, xiii, 101, 104, 109, 120–1, 135, 137, 140
    election of, to Chair of Supreme Soviet, 110
    elimination of, from presidential election 1997, 147
    initial caution of, on Lithuanian declaration of independence, 110
    leadership of, during attempted coup in Lithuania, 122–3
    leadership of, during attempted Moscow coup, 122–3
    meeting of, with Yeltsin June 1990, 115
    opinion of Russia, 200–2
    opposition of, to foreign policy of the LDDP, 202
Law on Preventive Detention, 138
Laws on language, 151–2
League of Nations
    interventions of, in Polish-Lithuanian territorial disputes, 31
Legal continuity of the Lithuanian state, 132
Lelewel, Joachim, xxviii
Lenin, Vladimir Ilich, nationalities policy of, xvii, 68

Lessing, G.E., xxvi
Lileikis, Aleksandras, 157–8
Lingys, Vitas, 137
Lietūkis, 15
Lithuania
    achievement of independence, 119, 125
    anti-Semitism in, 29, 55
    army in, 23, 26
    attempted coup in, January 1991, 117–18, 122; reasons for failure, 117, 123
    character of government in, 1926–1939, 24–6
    civil rights in, 19, 26, 28, 69, 135, 151
    consequences for, of failed Moscow coup, 1991, 119, 125
    consequences for, of Soviet annexation 1940, 49
    Constituent Assembly of, 6, 8, 19
    Constitution of 1922, 19–21, 24, 30
    Constitution of 1928, 24
    Constitution of 1938, 28, 132
    Constitution of 1992, 131–3, 138
    conversion to Christianity, xxi–xxiii
    Declaration of Independence, 1918, 5
    Declaration of Independence, 1990, 110–11; reasons for, 110
    dependence on German help 1919, 6–7
    destruction in, by Russian army in World War I, 10
    divisions in government during drive for independence, 120–1
    education under Soviet occupation, 70
    educational reforms in inter-war period, 17–18
    effects of educational reforms, 27
    elections in, 21, 134–5
    electoral system in, 133–4, 152–3

Lithuania *(continued)*
  ethnic composition of, between the
    wars, 9
  ethnographic boundaries of, 6, 30
  Executive Council, 3
  expatriate communities of, xxviii,
    4, 93–4, 148
  fascism in, 24–5, 27–8
  foreign policy of, inter-war period,
    34–5
  foreign policy of, from 1991, 140,
    144, 199, 206–12
  German occupation of, during
    World War I, 2, 5
  German occupation of, during
    World War II, 55–8
  historic boundaries of, 6, 30
  illiteracy level in, 1923, 20
  influence of Solidarity movement
    in, 94
  institutions of government under
    Soviet rule, 51
  intermarriage in, 74
  land reforms in, 1920–25, 15–17
  linguistic boundaries of, 30
  medieval period of, xx–xxi
  membership of EU sought for, 212,
    214–5
  membership of NATO sought for,
    202, 211–13
  national renaissance of. See
    Lithuanian National
    Renaissance
  paganism in, xx, xxii–xxiii, xxx
  party system after 1992, 135–6
  political parties in inter-war period,
    20–22, 28
  population movements in, 73–4, 93
  promotion of the arts in, 18
  Provisional Government in, during
    World War II, 56
  purges in, after World War II, 60
  referendum on independence, 123
  relations with Germany, in
    inter-war period, 32–4
  relations with Poland in inter-war
    period, 30–1, 34
  relations with Poland after 1991,
    144, 153, 208–11
  relations with Soviet Union in
    inter-war period, 33, 35
  repression in, by Tsarist
    governments, xxvii
  resistance to rule by Moscow, 49, 50
  restraint of, in face of Soviet
    provocations, 119–22
  restoration of independence after
    World War I, 1–2, 4–5
  return to Europe of, xiii
  role of army in, 21
  role of Russians in, during Soviet
    period, 74
  Russian incorporation of, after
    Partitions, xxiv
  security policy, 199, 211–15
  social reforms in inter-war period,
    17, 19
  Supreme Committee of Liberation
    in, 59
  threats to language rights, 94
  western influences in, 93–4, 201
  western recognition of
    independence, 124–5
Lithuania Minor, xxv, 6, 30
Lithuania-Poland military agreement,
  1993, 209
Lithuania-Poland state treaty, 1994,
  210–11
Lithuania-Russia state treaty, 1991,
  136, 200
Lithuanian Activist Front, 50, 54–6
Lithuanian Assembly, (19); manifesto
  of, (19–20)
Lithuanian Central Bank, 146
Lithuanian Communist Party (CPL)),
  95, 100, 102, 105;
  decision to secede from the Soviet
    Communist Party (CPSU),
    106, 108, 142
  endorsing a reform programme, 106

impact of defeat in elections to the Congress of Peoplés Deputies, 105
reformers in, 100
and acceptance of a multi-party state, 108, 134
and loss of leading role, 108
Lithuanian culture, impact of globalization on, 159
Lithuanian Democratic Labour Party (LDDP), 120
apprehensive of a strong presidency, 132
character of, 134, 141, 143, 176
corruption in, alleged, 134, 143–4
changed position on the powers of the President, 133
defeat in 1995 local elections, 144
defeat in 1996 parliamentary election, 134, 144
defeat in 1997 presidential election, 144
policy towards Russia, 202
programme of, 141–2
record of, in government, 142–5
support of, for a new constitution, 132
support of, by minorities, 136, 141, 150, 153
victory in 1992 parliamentary election, 134, 141–2
victory in 1993 presidential election, 142
Lithuanian economy in inter-war period
agricultural exports of, 12
agricultural holdings, size of, 17
backwardness of, at independence, 10
composition of exports and imports, 13
composition of manufacturing industry, 13
devastation of, during World War I, 9–10

exports of, to Germany, 33
growth of, 11–12, 29
growth rate of agriculture, 29
growth rate of industry, 29
government investment in, 12
land distribution 1914, 10
litas valuation of, 11, 52
macro-economic policy of, 11–12
productivity in agriculture of, 14
tariff protection in, 12–13
trading partners of, 13
Lithuanian economy in the Soviet period,
agricultural output of, 64, 77
agricultural productivity in, 76–7
agricultural specialization in, 64
centralized control of, 78
collectivization in, 77–8
composition of industrial output in, 76
composition of labour force in, 77
economic growth of, 79, 94
economic policy of, 76
economic self-management in, 98, 106
education and training investment in, 76–7
energy policy of, 76
food processing in, 78
industrial growth in, 75
interdependence of, with the other Soviet republics, 78
investment in, 75–6
levels of personal savings in, 95
model for economic reform of, 98
productivity in, 64
proportion of industrial output in GNP of, 75–6
role of private plots in total agricultural output of, 77–8
share of agriculture in GNP of, 76
standard of living in, 79, 94–5

Lithuanian economy post 1991,
141–3
agricultural protection in, 193
agriculture, relative backwardness
of, 192
debate on reform strategy of, 165
economics of transition in, 163–4
energy policy in, 188–9, 195
fiscal policy of, 171–2, 180, 182
food processing in, 187
foreign direct investment (FDI) in,
185–6
foreign ownership of land
permitted in, 180
foreign trade of, 168, 172–3–4,
186, 194
free trade agreement of, with
Poland, 211
growth strategy of, 179–80
hyper-inflation in, 167–8
law on foreign capital investment
in, 185
legacy of the Soviet command
economy in, 166–7
liberalization of, 164–5, 167–8
introduction of the *litas in*, 169
monetary policy in, 169–70, 182
most favoured nation (MFN)
agreement of, with Russia,
173, 201, 205
objectives of reformers in, 163–5
opponents of reform in, 164–6
privatization. See Privatization
progress of reform in, 171
protection of property rights in,
180
stabilization in, 164, 168–74
trade agreements of, 174, 211
trade of, with Estonia and Latvia.
187
trade of, with Poland, 187
trade of, with Russia, 193–4
transit trade of, 190–2, 194
transport industry in, 190

Lithuanian folk culture, xix, xxii, xxv,
xxvi
Lithuanian Freedom League, 95, 99,
101–2
Lithuanian Helsinki Group, 90–1
Lithuanian Human Rights Centre,
139
Lithuanian Israeli relations, 156–7
Lithuanian-Jewish relations, 158
Lithuanian language, xxii, xxix, xxxii
Lithuanian national renaissance, xviii,
xxv, xxvii–xxix, xxxi, xxxiii,
1, 5, 7
politicization of, xxxi–xxxii
Lithuanian National Party, 134
Lithuanian Parliament, defence of,
122
Lithuanian Peasants' Party, 134
Lithuanian population
birth rate of, 73, 93
in Soviet Socialist Republic, 7, 73
inter-war distribution of, by
economic sector, 9
involvement in Holocaust of, 55–6
preservation of ethnic Lithuanian
character of, 67, 73, 76, 93
urban/rural distribution of, 9, 77
Lithuanian press, xxxiii, 18, 138
Lithuanian-Russian relations,
199–207
Lithuanian Scientific Society, xxxiii
Lithuanian Society of Fine Arts, xxxiii
Lithuanian Supreme Soviet, elections
to, 1990, 106, 110
Lithuanian Telecom, 184, 185
Lithuanian Territorial Defence Force,
57–8, 65
Lithuanian TV tower, attack on by
Soviet forces, 115, 122
Lithuanian Union of Peasants, 144
Lithuanian Union of Political
Prisoners and Deportees, 134
Litimpex, 181
LITPOLBAT, 211

Litvinov, Maxim, 36, 37
Lozoraitis, Stasys, Senior, 34,
Lozoraitis, Stasys, Junior, 142, 148
Lublin, Treaty of (1569), xix, xxiv, 7,
    30
Lukauskaitė-Poškienė, Ona, 91
LUKoil, 189
Luther, Martin, 10

Madrid Summit of NATO states, 212
Maistas, 15
Mažeikiai oil refinery, 81, 114, 184,
    189, 194, 199
Medvedev, Roy, 107
Memel. See Klaipeda
Memel Convention, 32
Merkys, Antanas, 39–40
Michnik, Adam, 208
Migranjan, Andronik, 200, 204
Military transit, 144, 200–1, 205–6
Minorities. See Ethnic minorities
Molotov, Vyacheslav, 35–7, 39–40
Molotov-Ribbentrop Pact 1939,
    36–7, 91, 95, 99, 102, 106, 110
Moratorium on Lithuaniás
    declaration of independence,
    121
Muscovy, threat of, xxxiii–xxxiv
Mutual Assistance Pacts, 37–9, 49

Nationalist Party, 22–5
Nature Protection Association, 82
Naujienos, xxxi
Near abroad, 200, 203–4, 209
Nemunas, pollution of the, 80–1
Niessel, General Henri, 8
NKVD, 60
'Nomenklatura privatization', 176
Nordic-Baltic Council, 218
Nordic Investment Bank, 178
North Atlantic Treaty Organization
    (NATO), 199, 202, 206,
    212–3
  enlargement of, 207, 219

Okinczyc, Czesław, 153
Onacewicz, Z. I., xxvii
Operation Barbarossa, 54–5
Opposition to Soviet rule in
    Lithuania, 88–9
Organization for Security and Co-
    operation in Europe (OSCE),
    131, 212
Ozolas, Romualdas, 101, 143, 217

Paderewski, Ignacy Jan, 30
Paksas, Rolandas, 149, resignation of,
    194
Paleckis, Jus, 40, 51, 60
Panevėžys chemical plant, 81
Partnership for Peace (PfP), 212
Party of Rural People. See Populists
Paulauskas, Artūras, 144, 147–8
Perestroika, 60, 79, 95–7, 112, 155
Perspektyvos, 91
Petkus, Viktoras, 91
Pienocentras, 15
Piłsudski, Marshal Jozef, 7, 23, 30
Plechavičius, General Povilas, 57
Pogroms in Lithuania, 51, 154–5
Poland,
  Partitions of, xiii, xvii, xxiv, 3, 32
  attitude of, to Lithuanian
      independence, xvii–xviii, 3,
      7, 30
  policy to Lithuania after 1991,
      208–9
Polish Congress, 153–4
Polish Electoral Action, 134, 151–3
Polish-Lithuanian Commonwealth,
    xiii, xvii–xviv, xxxiv, 3, 30
Polish-Lithuanian dynastic union,
    1385, xxi–xxii
Polish-Lithuanian revolts against
    Tsarism, xxvi–xxvii
Polish ultimatum to Lithuania, 1938,
    28, 32
Pollution of air, 81
Pollution of water, 80–81

Polonization, xix, xxiv, xxx,
     xxxiii–xxxiv, 23
Pope John Paul II, 90, 94
Populist Party (Party of Rural People),
     21–3, 28
Poška, Dionizas, xxviii
Primakov, Yevgeni, 207
Privatization, 140–41, 143, 164,
     174–78, 183, 185
  contribution of, to security, 184
  foreign investments in Lithuanian
     companies, 177–8, 183
  methods of, in Lithuania, 175
  progress in, 176–7, 183–4
Prunskienė, Kazimiera, 101, 111,
     120–1
Prussians, xx
Pugo, Boris, 116

Radviliskis, 8
Reformation, The, in Lithuania,
     xxv
Rei, August, 19, 28
Reinis, Archbishop Mecislovas
Resistance, The, 49, 57–9, 64–7
  impact of, on Lithuania, 66–7
Restitution of property, 140, 152,
     177
Romantic movement in Lithuanian
     lands, xxvi
Russia,
  attacks of, on Lithuania, after
     World War I, 6, 8
  attitude to Lithuanian
     independence of the Whites,
     2
  co-operation of, with Lithuania,
     under Yeltsin
  declaration of sovereignty of, 1990,
     114
  February Revolution in, 19
  failure of, to condemn occupation
     of Lithuania, 1940, 207
  foreign policy of, after 1991,
     202–6

security partnership of, with West,
     213
  serf emancipation in, 1861, xxviii, 10
Russian Duma, Lithuanian
     representation in, xxxiii
Russian financial crisis, 1998, effect
     on Lithuania, 193
Russian speakers in Lithuania, 132
Russification policy in Poland-
     Lithuania, xxvi–xxvii, xxix
Russo-Japanese War, political effects
     of, xxxii

Sajūdis, 59, 92, 95
  calls of, for Lithuanian
     independence, 107
  character of, 101
  defeat in parliamentary election of,
     1992, 134–5, 141, 177
  defeat of, in presidential election,
     1993, 135
  ethnic nationalism in, 135
  founding congress of 1988, 104
  fragmentation of, 134
  leaders of, 101, 120
  membership characteristics of,
     104
  objectives of, 101–2, 104, 177
  programme of, in first
     administration, 140
  radicalization of, 101–2, 104–5
  support for a new constitution,
     132–3
  support for a strong presidency,
     132, 135
  victory in elections to Congress of
     People's uties 1989, 105;
     impact of the victory, 105
  victory in election to Lithuanian
     Supreme Set, 1990,
Sakharov, Andrei, 89, 91
Šalčininkai, 151
Samizdat, 89, 91
Sapieha, Prince E, 19
Sauliu Sajunga, 26

Sblizhenie, 68–9
Security policy of Lithuania, 211–16
  many aspects of, 216–19
Seimas
  elections to, under Constitution of
    1992
  powers of, in Constitution of 1922,
    19
  replacing the Supreme Soviet, 140
Šemeta, Gediminas, 182
Sharansky, Natan, 97
Shevardnadze, Eduard, 116
Shock therapy, 164, 165
Šiauliai, 8, 146
Šilutė Furniture Company, 187
Simon Wiesenthal Foundation, 157
Škirpa, Colonel Kazys, 54
Skubiszewski, Krysztof, 208–9
Šleževičius, Adolfas, 144, 156, 170, 206
Sliianie, 68–9
Smetona, Antanas, 5, 12, 19, 22–8,
    38–40, 135
Snaigė Refrigerator Company, 187
Sniečkus, Antanas, 51, 60, 66
Social Democratic Party, xxxi, 22–3,
    28, 134, 141, 145
Songaila, Ringaudas, 100, 102
Soviet annexation of Lithuania 1940,
    24, 39–40, 42, 49–2, 107, 200,
    203
Soviet annexation of Lithuania 1944,
    50, 58–61
Soviet campaign against the Catholic
    Church, 54, 71–2
Soviet constitution 1977, 68, 107–8
  Gorbachev's proposals for reform
    of, 1988, 103
  Gorbachev's reform proposals
    1990, 109, 116–17
  right to secede under, 103, 109
Soviet economic crisis, 112–13
Soviet encouragement of population
    movements, 73–4
Soviet foreign policy in 1930s, 36
Soviet invasion of Poland, 1939, 29

Sovietization, 49–54, 59–60, 67–75,
    89
Soviet-Lithuanian Non-Aggression
    Treaty 1926, 23, 35, 38
Soviet-Lithuanian Treaty, 1920, 107
Soviet negotiations with Britain,
    France and Germany, 1939, 36
Soviet-Polish Non-Aggression Pact
    1934, 34
Soviet Sanctions against Lithuania
    1990, 113–14
Sovnarkhoz, 80
Soyuz, 117
Stalin, Joseph, 1–2, 37–8, 60, 68, 71,
    72
Stanevičius, Simonas, xxviii
Stankevičius, Ceslovas, 120
Steponavičius, Archbishop Julijonas,
    72, 89
Stolypin reforms, (20), 10
Suchocka, Hanna, 210
Suslov, Mikhail, 61
Suwalki, xxvii–xxviii, 6, 9, 30–1
Švenčionys, 151
Šviesa, xxx
Swedish neutrality, 216

Tamošaitis, Antanas, 39
Taryba, 4, 5
Tatars, xx, 138
Tautos Vadas, 25
Temporary Basic Law, 132
Teutonic Order, xx–xxi
Tėvynės Sargas, xxxi
Trakai, 151
Trans-European Networks, 190
Transit agreement, 200
Treaty of Moscow, 31, 38
Troop withdrawals, 144, 150, 200,
    203,
Truman Doctrine, 65
Tūbelis, Juozas, 12

Ukininkas, xxxi
Union of Poles, 152, 154

University of Vilnius, xxv,
xxvii–xxviii, 18
Urbšys, Juozas, 37, 38

Vagnorius, Gediminas, 135, 140–1, 149
Vairininkai, 25
Vaisvila, Zigmas, 101
Valančius, Bishop Motiejus, xxix
Varpas, xxxi, 91
Vatican, Concordat with Poland, 22
Venclova, Tomas, 56, 91, 139, 209
Via Baltica, 190
Vilkas, Eduardas, 101, 183
Vilniaus Žinios, xxxii
Vilnius (Vilna, Wilno), xxi, xxiii,
xxvii, 6, 30, 55, 152
Vilnius Assembly, 4
Vilnius region, 9, 22, 27, 29, 30–1,
33, 37, 151
Vilnius Vingis Electronics Company,
187
Voldemaras, Augustinas, 6, 23, 25, 34
Voroshilov, Kliment, 38
Vytautas the Great, xxi–xxii
Vytautas Magnus University, 18

Waffen-SS, 57
Wałesa, Lech, 91, 209–10
'War against God', 54, 71
war criminals, 155–7
failure to convict, alleged, 157
rehabilitation of, alleged, 157
Warsaw conference, 1922, 34
Washington summit of NATO states,
213
Weimar Republic, constitution of, 19
Welles, Sumner, 42
Western attitudes to Soviet Union,
118, 124–5
refusal to extend *de jure*
recognition of Soviet rule in
Lithuania, 124
Western European Union, 214
Williams International, 184, 189,
194, 199

Wilson, President Woodrow, 3
World Bank, 163, 174, 178
World Trade Organization, 187–8
Wujek, Henryk, 208

Yakovlev, Alexander, 102, 107, 116
Yanaev, Gennady, 116, 119
Yazov, Marshal Dmitry, 116
Yčas, Martynas, 6, 22
Yedinstvo, 96, 149–50
Yeltsin, Boris, 113
claims of, on the near abroad
demand of, for a voluntary
confederation of Soviet
republics, 118
demands of, for sovereignty for
Russia, 115
effect of, on Gorbachev's
discussions with Lithuania,
115
election as chair of Russian
Supreme Soviet, 1990, 114
election as President of the Russian
Federation 1991, 118
meeting with Landsbergis in June
1990, 115
opposition of, to centralization,
115, 116
opposition of, to attempted coup in
Lithuania, 117
resistance to attempted Moscow
coup, August 1991, 119
support of, for independence of the
Soviet republics, 113
western attitudes towards, 125
Yudenich, Nikolai, 8

Zaijdlerowa, Zoë, 62
Žalgiris (Grunwald, Tannenberg),
battle of, xxi
Zeligowski, General Lucjan, 29, 31,
208
Zhdanov, Andrei, 35
Zingeris, Emanuelis, 140, 157–8
Zuroff, Efraim, 157